Environmental Applications of Nanomaterials

Synthesis, Sorbents and Sensors

2nd Edition

Environmental Applications of Nanomaterials

Synthesis, Sorbents and Sensors

2nd Edition

editors

Glen E Fryxell
Pacific Northwest National Laboratory, USA

Guozhong Cao
University of Washington, USA

Imperial College Press

ICP

Published by

Imperial College Press
57 Shelton Street
Covent Garden
London WC2H 9HE

Distributed by

World Scientific Publishing Co. Pte. Ltd.
5 Toh Tuck Link, Singapore 596224
USA office: 27 Warren Street, Suite 401-402, Hackensack, NJ 07601
UK office: 57 Shelton Street, Covent Garden, London WC2H 9HE

British Library Cataloguing-in-Publication Data
A catalogue record for this book is available from the British Library.

ISBN 978-1-84816-803-9
ISBN 978-1-84816-804-6 (pbk)

Typeset by Stallion Press
Email: enquiries@stallionpress.com

Printed by FuIsland Offset Printing (S) Pte Ltd Singapore.

Preface to the First Edition

Nanotechnology has attracted a lot of attention recently, particularly in the research and industrial communities. It offers many unprecedented opportunities for advancing our ability to influence not only our day-to-day lives, but also the very environment in which we live. The ability to design, synthesize and manipulate specific nanostructured materials lies at the very heart of the future promise of nanotechnology. Nanomaterials may have unique physical and chemical properties not found in their bulk counterparts, such as unusually large surface-area-to-volume ratios or high interfacial reactivity. Such properties give hope for new chemical capabilities arising from exciting new classes of nanomaterials. Indeed, as this book summarizes, nanomaterials have been developed for specific applications that involve interfacial reactions and/or molecular transport processes.

The industrial revolution of the late 19th and early 20th century led to unprecedented economic growth in Europe and the United States. However, it also produced unprecedented environmental pollution. In those simpler, more naive times, contamination of the environment was largely ignored, a common sentiment being simply that "dilution is the solution to pollution". The economic benefits of increased industrial production outweighed the emerging environmental problems, and the vastness of the wilderness and ocean (along with the lower populations of the day) allowed this industrial contamination to dissipate to levels that made it relatively easy to ignore...for a while. Today, we see other countries (e.g. China) going through similar growing pains, and experiencing similar environmental damage.

The 20th century also brought an unprecedented arms race, which in turn brought its own unique set of environmental concerns and needs. Of particular importance to the environment are the legacy wastes arising from 40 years of nuclear weapons production, as well as the vast stockpiles of chemical weapons, throughout the globe. Our parents devised and built these devastating weapons in order to fight back against the ruthless tyrants that threatened their world. They did not have the luxury of planning ahead for the eventual disposal of these deadly materials; they needed to fight, and they needed to fight NOW. Their success ultimately led to improved standards of living throughout much of Europe and the United States (eventually spreading to other parts of the world, as well). However, the issues raised by the presence of these difficult waste materials are still unresolved.

This new-found quality of life was starting to be threatened by industrial pollution in the 1960s and early 1970s, and society quickly realized that we must take a more active stance in terms of pollution management and prevention. In the years since, governments and industry have learned to work together (albeit awkwardly at times), monitoring industrial effluents and limiting new releases of toxic materials into the environment. Remediation methods have been developed to repair some of the damage that previously took place. We are learning. We have a responsibility, both to future generations, and to our global neighbors, to share these insights — both regulatory and remediatory.

The last 10 to 15 years have seen a remarkable explosion of research in the design and synthesis of nanostructured materials — nanoparticles, nanotubes, nanorods, etc. Early work largely focused on making different shapes, or different sizes; then work started to focus on making a variety of compositions, and multicomponent materials. Tailoring the composition or interface of a nanomaterial is a key step in making it *functional*. This book is concerned with functional nanomaterials — materials containing specific, predictable nanostructure whose chemical composition or interfacial structure enable them to perform a specific job — destroy, sequester or detect some material that constitutes an environmental threat. Nanomaterials have a number of features that make them ideally suited for this job — high surface area, high reactivity, easy dispersability, rapid diffusion, etc. The purpose of this book is to showcase how these features can be tailored to address some of the environmental remediation and sensing/detection problems faced by mankind today. A number of leading researchers have contributed to this volume, painting a picture of diverse synthetic strategies, structures, materials and methods. The intent of this book is to showcase the current state of environmental nanomaterials in such a way as to be useful either as a research resource, or as a graduate-level textbook. We have organized this book into sections on nanoparticle-based remediation strategies, nanostructured inorganic materials (e.g. layered materials like the apatites), nanostructured organic/inorganic hybrid materials, and the use of nanomaterials to enhance the performance of sensors. The materials and methods described herein offer exciting new possibilities in the remediation and/or detection of chemical warfare agents, dense non-aqueous-phase liquids (DNAPLs), heavy metals, radionuclides, biological threats, CO_2, CO and more. The chemistries captured by these authors form a rich and colorful tapestry. We hope the final result is both valuable and enjoyable to the reader.

Glen Fryxell, Ph.D.
Pacific Northwest National Laboratory, USA

Guozhong Cao, Ph.D.
University of Washington, USA

March 2006

Preface to the Second Edition

Recent years have seen a number of exciting advancements in the field of functional nanomaterials and their use to address issues related to the environment. Generally speaking, these nanomaterial-based strategies can be grouped into three broad categories: nanomaterials designed to capture chemical species of concern from the environment (or prevent their release into the environment), nanomaterial-enhanced sensors for making more sensitive measurements of chemical species in the environment, and nanomaterial-enhanced biomedical technologies (i.e. technologies to deal with the results of environmental exposure). Specific examples of nanomaterials for capture would include: the use of crystalline silicotitanates to sequester cesium from nuclear waste, the use of hybrid nanomaterials to remove toxic anions from water, new classes of nanostructured polysilsesquioxanes that capture toxic metal ions, and hybrid nanoporous materials tailored for the selective and efficient capture of CO_2 (a major global warming concern). Recent examples of the nanomaterial-enhanced sensors would include work with carbon nanotube and graphene-based sensors tailored for detecting organophosphate pesticides, semiconductor oxide nanotubes for gas sensors for toxic gases (e.g. CO, NO_2, etc.), as well as electrochemical sensors based on TiO_2 nanotube arrays. Functional nanomaterials are also being evaluated for a variety of biomedical applications, examples of this would include: the use of functional nanomaterials as an orally administered alternative to chelation therapy, and electrochemical sensor systems designed to perform heavy metal determinations in biological fluids like urine and blood. We felt that the readers would benefit from discussion of these exciting results, so the second edition of this book has been expanded to include these topics and we believe that the readers will find this expanded second edition to be both a valuable research reference, as well as a useful teaching tool.

We thank the authors for their excellent contributions.

Glen E. Fryxell
Pacific Northwest National Laboratory

Guozhong Cao
University of Washington

April 2012

CONTENTS

Chapter 3.   Synthesis, Characterization, and Properties of Zero-Valent
Iron Nanoparticles   49

Donald R. Baer, Paul G. Tratnyek, You Qiang,
James E. Amonette, John Linehan, Vaishnavi Sarathy,
James T. Nurmi, Chongmin Wang and J. Antony

Nanostructured Inorganic Materials   **87**

Chapter 4.   Formation of Nanosized Apatite Crystals in Sediment for
Containment and Stabilization of Contaminants   89

Robert C. Moore, Jim Szecsody, Michael J. Truex,
Katheryn B. Helean, Ranko Bontchev and Calvin Ainsworth

Nanoporous Organic/Inorganic Hybrid Materials   **121**

Chapter 9.   Chemically Modified Mesoporous Silicas and Organosilicas
for Adsorption and Detection of Heavy Metal Ions   227

Oksana Olkhovyk and Mietek Jaroniec

Chapter 10.   Hierarchically Imprinted Adsorbents   261

Hyunjung Kim, Chengdu Liang and Sheng Dai

Chapter 11. Functionalization of Periodic Mesoporous Silica and its Application to the Adsorption of Toxic Anions   287

Hideaki Yoshitake

Chapter 12. Layered Semi-crystalline Polysilsesquioxane: A Mesostructured and Stoichiometric Organic-Inorganic Hybrid Solid for the Removal of Environmentally Hazardous Ions   327

Hideaki Yoshitake

Chapter 13. A Thiol-functionalized Nanoporous Silica Sorbent for Removal of Mercury from Actual Industrial Waste   359

Shas V. Mattigod, Glen E. Fryxell and Kent E. Parker

Chapter 14. Functionalized Nanoporous Silica for Oral Chelation Therapy of a Broad Range of Radionuclides   369

Wassana Yantasee, Wilaiwan Chouyyok, Robert J. Wiacek, Jeffrey A. Creim, R. Shane Addleman, Glen E. Fryxell and Charles Timchalk

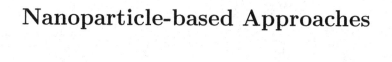

Nanoparticle-based Approaches

Chapter 1

Nanoparticle Metal Oxides for Chlorocarbon and Organophosphonate Remediation

Olga B. Koper*, Shyamala Rajagopalan*, Slawomir Winecki*
and Kenneth J. Klabunde*,†

*NanoScale Corporation, Inc., Manhattan, KS, USA
†Kansas State University, Manhattan, KS, USA

1.1 Introduction

The nanotechnology revolution has affected many areas of science, including chemistry and chemical engineering. Although nanotechnology advances in electronics or in the manufacturing of nanomachines are rather recent developments, nanoscale chemical structures are much older. Materials known and widely used over the past several decades, such as high-surface-area carbons, porous inorganic metal oxides, and highly dispersed supported catalysts, all fall into the category of nanostructures. Although a few decades-old publications describing these materials rarely used the word "nano", they deserve proper credit for providing the foundation for current advances in nanochemistry.

Since approximately the 1970s, enormous advances in the synthesis, characterization, and understanding of high-surface-area materials have taken place and this period can be justifiably described as the nanoscale revolution. The development and commercial use of methods like sol-gel synthesis, chemical vapor deposition, and laser-induced sputtering allowed for the manufacturing of countless new nanomaterials both at laboratory and commercial scales. Traditional chemical manufacturing methods, used in catalyst synthesis or in the production of porous sorbents, were gradually improved by systematic development of methodologies, as well as quality control measures. In addition, the characterization of nanostructures was revolutionized by the availability of new instruments. Various high-resolution microscopic tools were developed and became available commercially. For the first time, pictures of nano-objects became easily obtainable. Today, they appear not only in scientific literature but also in popular magazines and textbooks. Measurement of specific surface area, once a tedious and lengthy laboratory procedure, is done today routinely using automatic yet affordable instruments. Numerous companies develop, manufacture, and sell scientific equipment specifically for nano research, resulting in more instruments being available and at reduced costs. The understanding of nanostructures is growing at a phenomenal rate due to the efforts of countless research groups, academic institutions, and commercial entities. Since approximately 1990, nanoscience has been recognized as a distinct field of study with tremendous potential. As a result, nanotechnology research has proliferated, scientific journals dedicated to nanoscience were created, nanotechnology research centers were established, and universities have opened faculty positions related to nano research. These developments were, and continue to be, heavily supported by the federal and local governments in the form of research grants and other funding opportunities. Private organizations, including all major chemical manufacturers, started robust nanotechnology programs and investments.

This chapter is intended as a brief description of the technology and products developed in the laboratories of one of the authors (Kenneth J. Klabunde) at Kansas State University, Manhattan, KS; and by NanoScale® Corporation, Manhattan, KS. The description will start with a reminder of a few technical concepts related to all nanoscale materials intended for chemical uses.

Nanomaterials have large specific surface areas and a large fraction of atoms are available for chemical reaction. Figure 1.1 presents a high-resolution transmission electron microscopy (TEM) image of aerogel-prepared nanocrystalline MgO. The image demonstrates the nanomorphology of this material with rectangular crystals, 2–4 nm in size, a length-scale that is only an order of magnitude larger than distances between atoms of this oxide. It is apparent that significant fractions of atoms are located on crystallite surfaces as well as on edges and corners. These atoms/ions are partially unsaturated (coordination less than six) and can readily interact with chemical species that come into contact with the nanocrystallites. Such interaction can be a physical adsorption, if caused by van der Waals attraction, or a chemical reaction if chemical bonds are formed or modified. Simple estimations presented in Figure 1.2 demonstrate that a reduction in the size of nanocrystallites results in a sharp increase of specific surface areas, as well as fractions of atoms located on surfaces, edges, and corners. For instance, a material with 2 nm crystallites has more than half of the atoms located on their surface, approximately 10% of atoms/ions on edges, 1% of

Figure 1.1. High-resolution transmission electron microscopy (TEM) image of nanocrystalline MgO.

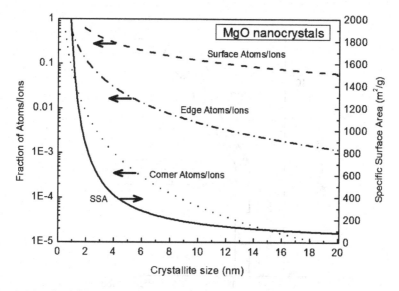

Figure 1.2. Specific surface area and fractions of atoms/ions located on surface, edges, and corners of nanocrystalline MgO.

atoms/ions placed on corners, and with specific surface area approaching $1,000 \, m^2/g$. It is important to recognize that the trends demonstrated in Figure 1.2 apply to all porous nanocrystalline and amorphous solids; and are the basis of the unique reactivity of nanocrystalline materials in chemical applications. Amorphous materials can be viewed as a limiting case of nanocrystalline materials with the crystallite size approaching inter-atomic distances.

One of the major important features of nanocrystalline materials is related to the size of pores between crystallites. It is apparent from Figure 1.1 that most pores between crystallites are similar in size to the crystallites themselves; therefore, nanocrystalline materials have a large fraction of pores below 10 nm, that are traditionally described in the literature as micropores. Fluids, particularly gases, behave differently inside micropores as compared to normal conditions. The first difference involves Knudsen diffusivity, conditions where collisions between fluid molecules are less frequent than collisions between fluid and pore walls; this is likely to dominate pores below the 10–100 nm size range, as shown in Figure 1.3. This mode of diffusivity is significantly slower than regular diffusivity for an unbound gas and will result in slower mass transfer within nanocrystalline materials. Gas diffusivities in micropores may be two to four orders

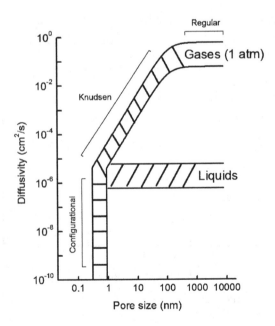

Figure 1.3.   Influence of the pore size on diffusivities of gases and liquids.

of magnitude smaller than diffusivities known for gases and may approach liquid diffusivities. The reduced diffusivities need to be taken into account in the engineering design. The second consequence of the micropores is the possibility of capillary condensation of vapors and gases and enhanced adsorption by nanocrystalline materials. The capillary condensation phenomenon is traditionally described by the Kelvin equation that relates reduction of the equilibrium vapor pressure inside pores to pore size and the physical properties of the chemical. The reduction of the equilibrium vapor pressures for several common substances, predicted by the Kelvin equation, is shown in Figure 1.4. It is apparent that the equilibrium pressure reduction and capillary condensation effects become pronounced for micropores. On the other hand, nanocrystalline materials, when aggregated into microstructures or pellets, also possess numerous mesopores. Thus, both micropores and mesopores need to be considered in describing the behavior of this new class of materials (Table 1.1).

Physical properties of select nanocrystalline metal oxides (NanoActive®️ materials) manufactured by NanoScale are shown in Table 1.1. The surface areas range from 20 to over $600\,m^2/g$ for the NanoActive Plus metal oxides. Figure 1.5 shows the powder X-ray diffraction (XRD) spectra of commercial MgO, NanoActive MgO, and NanoActive MgO Plus. The Plus material has the broadest peaks, indicating the smallest crystallites. These nanometer-sized small crystallites, due to the high surface reactivity, aggregate into

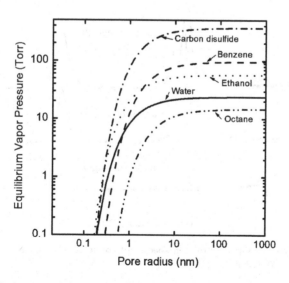

Figure 1.4. Reduction of the equilibrium vapor pressures inside cylindrical pores for several common substances, as predicted by the Kelvin equation.

Table 1.1.   Physical properties of nanocrystalline metal oxides.

NanoActive Material	Powder Appearance	Surface Area (m^2g^{-1})	Crystallite Size (nm)	Average Pore Diameter (Å)	Mean Particle Size (μm)
MgO Plus	White	≥ 600	≤ 4	30	12
MgO	White	≥ 230	≤ 8	50	3.3
CaO Plus	White	≥ 90	≤ 20	110	4
CaO	White	≥ 20	≤ 40	165	4
Al_2O_3 Plus	White	≥ 550	Amorphous	110	5
Al_2O_3	White	≥ 275	Amorphous	28	1.5
CuO	Black	≥ 65	≤ 8	85	6
ZnO	Off-white	≥ 70	≤ 10	170	4
TiO_2	White	≥ 500	Amorphous	32	5
CeO_2	Yellow	≥ 50	≤ 7	70	9.5

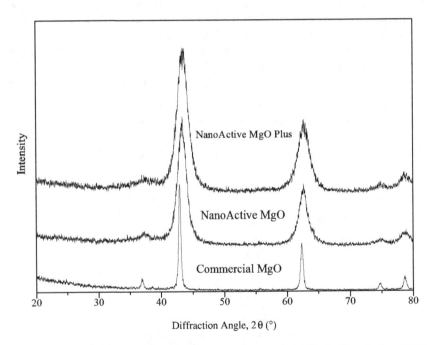

Figure 1.5.   Powder X-ray diffraction spectra of commercial MgO, NanoActive MgO, and NanoActive MgO Plus.

larger particles (micron size) that preserve the high chemical reactivity, yet are much easier to handle.

1.2 Environmental Applications of NanoActive Materials

There are many areas where nanocrystalline materials can be used, but the environmental applications of these materials are particularly significant. Air and water pollution continues to be a challenge in all parts of the world. In many instances, the pollution control technology is limited by the poor performance and/or high costs of existing sorbent materials; and considerable research activity is directed towards the development of new sorbents. The following list gives a few examples of such cases:

- Control of chemical spills and accidental releases of harmful chemicals in various locations and settings: Accidental chemical releases are common occurrences in laboratories, industrial locations, and all places where chemicals are used. Spills and accidental releases often result in environmental contaminations. NanoScale developed a specialized, safe sorbent formulation, FAST-ACT® (First Applied Sorbent Treatment — Against Chemical Threats), a powder capable of safe spill treatment and environmentally responsible disposal.
- Protection from chemical hazards: Nanocrystalline metal oxides can be incorporated into protective textiles or skin creams to provide additional protection from chemical and biological hazards. The toxic agents are destroyed on nanoparticles, eliminating the threat of off-gassing and secondary contamination.
- Indoor air quality control in buildings and vehicles: Currently, various air filtration technologies are available commercially; including particulate filters, electrostatic precipitators, and activated-carbon filters. There remains a large group of high-volatility chemicals that cannot be effectively controlled by these approaches. Attempts to develop more effective sorbents or catalysts for these pollutants are ongoing.
- Removal of elemental and oxidized mercury from combustion gases generated by electrical utilities using coal as a source of energy: Recent regulations enacted by the U.S. Environmental Protection Agency (EPA) spurred widespread development of new mercury sorbents, including activated carbons and metal oxides.
- Control of arsenic, perchlorate, and methyl-*t*-butyl ether (MTBE) in drinking water: The presence of these compounds in water causes serious health concerns; and has resulted in mandatory maximum contaminant levels (MCL) for arsenic and may trigger new EPA regulations.

Development of new sorbents for economical and efficient removal of these pollutants is an active area of research.

The above applications, as well as numerous others, can benefit from the use of nanocrystalline sorbents, owing to enhancement in reaction kinetics, increased removal capacities, and permanent destruction of harmful chemicals.

Nanocrystalline materials exhibit a wide array of unusual properties, and can be considered as new materials that bridge molecular and condensed matter.[1] One of the unusual features is enhanced surface chemical reactivity (normalized for surface area) toward incoming adsorbates.[2] For example, nanocrystalline MgO, CaO, TiO_2, and Al_2O_3 adsorb polar organics such as aldehydes and ketones in very high capacities, and substantially outperform the activated-carbon samples that are normally utilized for such purposes.[3] Many years of research at Kansas State University, and later at NanoScale, have clearly established the destructive adsorption capability of nanoparticles towards many hazardous substances, including chlorocarbons, acid gases, common air-pollutants, dimethyl methylphosphonate (DMMP), paraoxon, 2-chloroethylethyl sulfide (2-CEES), and even military agents such as GD, VX, and HD.[4-9] The enhanced chemical reactivity suggests a two-step decomposition mechanism of the adsorbates on nanoparticles (first step, adsorption of toxic agent on the surface by means of physisorption, followed by the second step, chemical decomposition). This two-step mechanism substantially enhances the detoxification abilities of nanoparticles because it makes the decomposition less dependent on the rate (speed) of chemical reaction. The rate of chemical reaction depends on the agent-nanoparticle combination; therefore, for some agents the rate may be quite low. In addition, the reaction rate strongly decreases at lower temperatures. For these reasons, any detoxification method that relies only on chemical reactivity would not work for many toxic agents and would not be effective at low temperatures. Reactive nanoparticles do not have this drawback because the surface adsorption sites remain active even at very low temperatures. In fact, the physisorption of the potential toxic agents is enhanced at low temperatures. In this way, the toxins are trapped and eventually undergo "destructive adsorption."

1.3 Destructive Adsorption of Hazardous Chemicals by Nanocrystalline Metal Oxides

A wide variety of chlorinated compounds such as cleaning solvents, plasticizers, lubricants, and refrigerants are used by society in many beneficial functions. While some of these chlorinated compounds are being replaced by less harmful chemicals, many continue to be used because of the lack of suitable replacements or as a result of economic considerations. Therefore, considerable interest exists in developing methods for the safe disposal of chlorinated and other problematic wastes. It has been shown that nanocrystalline metal oxides are particularly effective decontaminants for several classes of environmentally problematic compounds at elevated temperatures; enabling complete destruction of these compounds at considerably lower temperatures than that required for incineration.[10]

Koper *et al.*[11–15] have examined the reaction of aerogel-prepared CaO with carbon tetrachloride and other chlorinated hydrocarbons. The primary carbon-containing product was CO_2, with the CaO converted to $CaCl_2$. Conventionally prepared CaO, however, produced little CO_2. A three-step mechanism was proposed. Initially, CO_2 and $CaCl_2$ are produced from metathesis of CaO and CCl_4; CO_2 then combines with CaO in a second step to yield $CaCO_3$; finally, metathesis of $CaCO_3$ and CCl_4 generates $CaCl_2$ and CO_2. Phosgene, $COCl_2$, is an intermediate, which reacts with the remaining CaO to form calcium chloride and carbon dioxide. Low-temperature infrared studies of CCl_4 monolayers on CaO demonstrated that CCl_x intermediates begin to form at temperatures above 113 K; at 200 K, CCl_4 and CCl_x are no longer observed.[13] Although the reactions of chlorocarbons with metal oxides are often thermodynamically favorable, liquid-solid or gas-solid reactions are involved only on the surface of the metal oxide. Kinetic parameters involving ion (Cl^-/O^{2-}) migration inhibit complete reaction. Therefore, the use of ultrahigh-surface-area metal oxides allows reasonably high capacities for such chlorocarbon destructive adsorption processes[16]: Li *et al.*[17,18] have studied the reactions of aerogel-prepared CaO and MgO with chlorinated aromatics at 500–900°C. Upon destruction of chlorobenzene, a biphenyl was formed as the reaction by-product. Destruction of chlorinated aromatics occurred to a much greater extent, and at significantly lower temperatures, relative to destruction in the absence of the nanoscale metal oxides. Trace toxins such as dibenzo-*p*-dioxins were not observed under any conditions when aerogel-prepared MgO was used, nor when aerogel-prepared CaO was used in the presence of oxygen in the carrier gas. However, when low-surface-area CaO was used in the presence of oxygen, small amounts of dibenzo-*p*-dioxins were produced. Hooker and Klabunde[19] have studied the reaction of iron(III) oxide with carbon tetrachloride in a fixed-bed pulse reactor, from 400 to 620°C. The main carbon-containing product was CO_2; other carbon-containing products included C_2Cl_4 and graphite. It was found that the extent of reaction was greater than what could be accounted for by reaction only on the surfaces of Fe_2O_3; this suggested that Fe_2O_3 on the surface was regenerated. The apparent regeneration of Fe_2O_3 on the surface of the particles suggested that coating other metal oxide nanoparticles with iron(III) oxide might enable more complete utilization of the core metal oxide. Accordingly, aerogel-prepared MgO overlaid with Fe_2O_3 (designated as $[Fe_2O_3]MgO$) was prepared, and its reactions with carbon tetrachloride were examined by Klabunde et al.,[20] Khaleel and Klabunde,[21] and Kim et al.[22] It was found that when $[Fe_2O_3]MgO$ reacted with CCl_4, nearly all of the metal oxides (MgO) were consumed; in contrast, Fe_2O_3 treatment of conventionally prepared MgO did not give this enhancement. Since the iron oxide coating is regenerated by the core MgO, the Fe_2O_3 coating can be considered to be catalytic. The scheme below indicates the reaction of carbon tetrachloride with nanocrystalline MgO and iron-coated nanocrystalline MgO.

$$\left(\text{MgO}\right) + CCl_4 \xrightarrow[-CO_2]{400°C} \left(\text{MgO}\right)\!-\!MgCl_2$$

$$\left(\text{MgO}\right)\!\!\diagup^{Fe_2O_3} + CCl_4 \xrightarrow[-CO_2]{400°C} \left(\text{MgCl}_2\right)\!\!\diagup^{FeCl_xO_y}$$

Further work, using X-ray photoelectron spectroscopy (XPS), showed that the best catalysts were those where Fe_2O_3 was coated on the surface of the MgO,[23] and was highly dispersed (as shown by Mössbauer spectroscopy[24] and XRD). Simple mixtures of Fe_2O_3 with MgO were not as effective and solid-state mixtures of Fe_2O_3-MgO were also less effective. Indeed, a layered $[Fe_2O_3]MgO$ structure worked best. Figure 1.6 shows the destructive adsorption of carbon tetrachloride on two forms of nanocrystalline CaO coated with iron oxide and vanadium oxide. Again, the same trends were observed, where the sol-gel-based form of CaO exhibited higher chemical reactivity as compared to nanocrystalline CaO (nano), whose reactivity was much higher than the commercially available CaO. Furthermore, a coating of the transition metal oxide imparted additional reactivity, driving the reaction closer to its stoichiometric limit.

Other metal oxides were also found to be effective coatings for enhancing the utilization of aerogel-prepared MgO. Indeed, Fe_2O_3 enhanced the catalytic properties of both MgO and SrO.[25,26] In an attempt to gain more detailed information on the chemical state of the $Fe_2O_3/FeCl_x$ before, during, and after CCl_4 reaction, a series of $[Fe_2O_3]SrO$ nanocrystals and microcrystals were studied by Jiang et al., using extended X-ray absorption fine-structure (EXAFS) spectroscopy.[27] Strontium was chosen as the base oxide because of the available synchrotron energies. The results of these experiments were quite surprising. First of all, the data showed that SrO

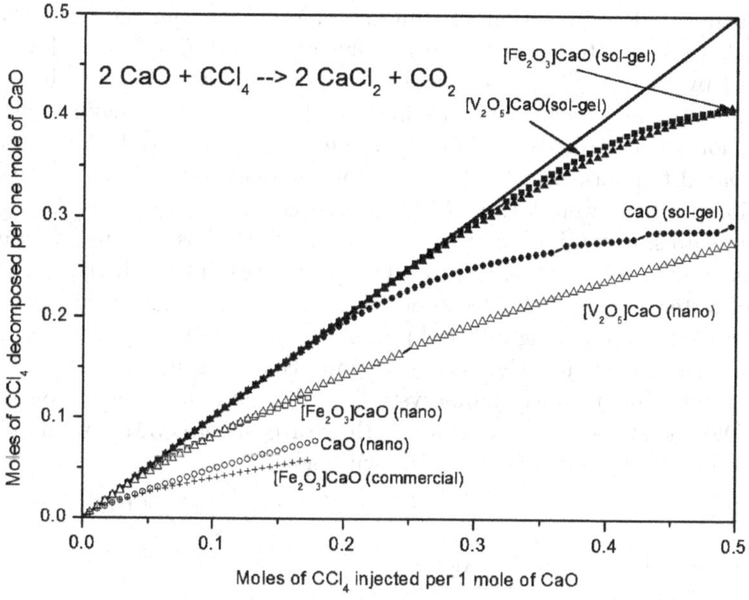

Figure 1.6. Decomposition ability of CaO and $[V_2O_5]$CaO towards CCl_4. In a stoichiometric reaction 1 mol of CaO can decompose 0.5 mol of CCl_4.

itself was more reactive than CaO, which is known to be more reactive than MgO (per unit surface area, although MgO can be prepared with the highest surface area). Also, it was found that the Cl^-/O^{2-} exchange was extremely facile, and even after near-stoichiometric exposure to CCl_4, Fe_2O_3 remained as Fe_2O_3. In other words, if any SrO was still available, it readily gave up its O^{2-} to $FeCl_x$ to reform Fe_2O_3.[27]

Besides CCl_4, other chlorocarbons have been examined. The decomposition of CCl_3F by several vanadium oxides and vanadium oxide-coated aerogel-prepared MgO was studied by Martyanov and Klabunde.[28] V_2O_3 was consumed in the reaction, with some vanadium-halogen species being produced. Capture of the volatile vanadium-halogen species by MgO was thought to be responsible for the catalytic effect of the vanadium compounds on MgO. Furthermore, Fe_2O_3 exchange catalysis was found to work well for 1,3-dichlorobenzene, but not for trichloroethylene. Obviously, the catalytic effect of transition metal oxides depends a great deal on the intimate mechanistic details that are, of course, different for each chlorocarbon under study. Interestingly, this type of catalytic action is not restricted to O^{2-}/Cl^- exchange, but also operates in O^{2-}/SO_x^{n-} and O^{2-}/PO_x^{n-} systems as well.[24]

Chemical decomposition of dimethyl methyl phosphonate (DMMP), trimethyl phosphate (TMP), and triethyl phosphate (TEP) on MgO was studied by Li et al.[29-31] The agents were allowed to adsorb on the MgO samples both in a vacuum environment and in a helium stream. Substantial amounts of strongly chemisorbed agents were observed at room and at elevated temperatures (500°C). At low temperatures, the main volatile reaction products were formic acid, water, alcohols, and alkenes. At higher temperatures, CO, CH_4, and water predominated. Phosphorous-containing products remained immobilized at all temperatures. Interestingly, the addition of water enhanced the decomposition abilities of nanocrystalline MgO.

Detailed studies using infrared photoacoustic spectroscopy and isotope labeling confirmed that the chemisorption occurs through the P=O bond destruction. For instance, nanocrystalline MgO is able to hydrogen bond with DMMP at room temperature; with hydrolysis of DMMP occurring to produce surface-bound species as shown below.

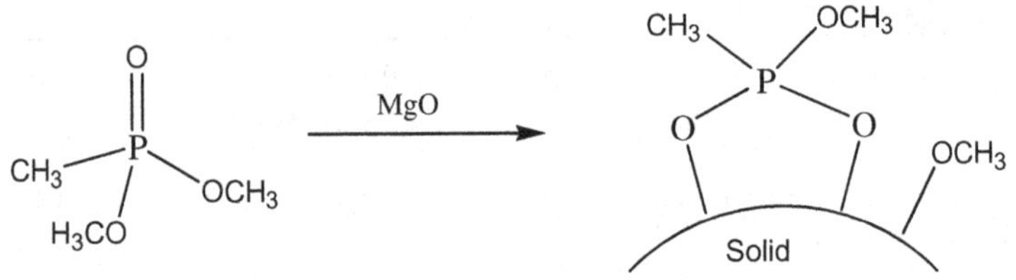

Further studies on the decomposition of phosphonate esters $(RO)_2IP=O$ identified two important reactions, nucleophilic substitution at P and nucleophilic substitution at the alkyl carbon of an alkoxy group.[32] Treatment of DMMP with MgO yielded formic acid (HCOOH) as the major volatile product; other volatile products included methanol (CH_3OH), dimethyl ether (CH_3OCH_3), and ethane (CH_3CH_3). Carbon dioxide, carbon monoxide, water, hydrogen, and phosphoric acid were not observed as products. All phosphorus-containing products were immobilized on the MgO surface; elemental analysis was consistent with O_{solid}-Mg-O-P(OMe)(Me)-O-Mg-O_{solid}, with an O_{solid} between the two Mg atoms. For this structure, the PO bond order of the O-P-O linkage is 1.5, i.e. the two P-O bonds share one double bond.[33] Proton abstraction of a β-hydrogen of an ethoxy group can lead to ethylene production, which is sometimes observed in reactions of ethyl esters.[34] The reactivity of nanocrystalline MgO towards a range of organophosphates $(RO)_3P=O$, organophosphites $(RO)_3P$, and organophosphines R_3P was studied by Lin and Klabunde.[34] Phosphorous compounds were allowed to adsorb on thermally activated MgO at room temperature and at elevated temperatures reaching 175°C. Most of the phosphorous compounds were adsorbed and chemically decomposed in large quantities, and in some cases the reactions were essentially stoichiometric. In the most favorable cases, approximately one phosphorus molecule was decomposed for every two MgO molecules present in the bulk. Infrared studies performed on spent sorbents indicated that phosphates adsorb very strongly through the P=O bond (the P=O bond is destroyed upon adsorption) accompanied by net electron loss from the RO groups. Phosphites adsorb through the phosphorous atom with a net electron density gain by the RO groups. Phosphines adsorb less strongly through the phosphorous atom. Decomposition of the phosphorous compounds yielded volatile hydrocarbons, ethers, and alcohols. Products containing phosphorous were strongly retained by the MgO surface which prevented product desorption and release into the gas phase:

$$3DMMP + 6MgO \rightarrow CH_3OH + H_2O + 2HCOOH + [OH]_{adsorbed}$$
$$+ 3[-P(OCH_3)CH_3-]_{adsorbed}.$$

As another example of the extremely high reactivity of nanocrystalline metal oxides, ambient-temperature destructive adsorption of paraoxon is illustrated. Paraoxon is an insecticide and is a suitable simulant for nerve warfare agents. Figure 1.7 illustrates the rate of removal of paraoxon from a pentane solution over time employing various adsorbents. The removal was monitored using UV-Vis spectroscopy and reduction of the paraoxon (265–270 nm) peak was observed over time. Nanocrystalline MgO nanoparticles achieved complete adsorption within two minutes of exposure to

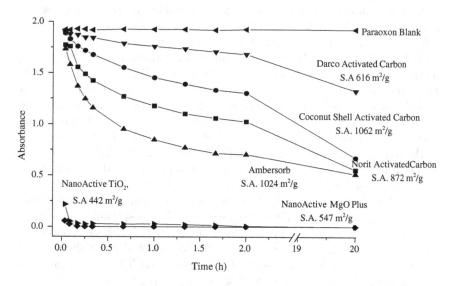

Figure 1.7.   Rate of adsorption of paraoxon by various samples.

paraoxon while all the activated carbon and ion-exchange resin (IER) samples were significantly less adsorptive and unable to adsorb paraoxon completely, even after 20 hours of exposure. In addition, with activated carbons the agent was merely adsorbed, whereas with nanoparticles it was first adsorbed onto the metal oxide surface and then decomposed into phosphate and p-nitrophenol:

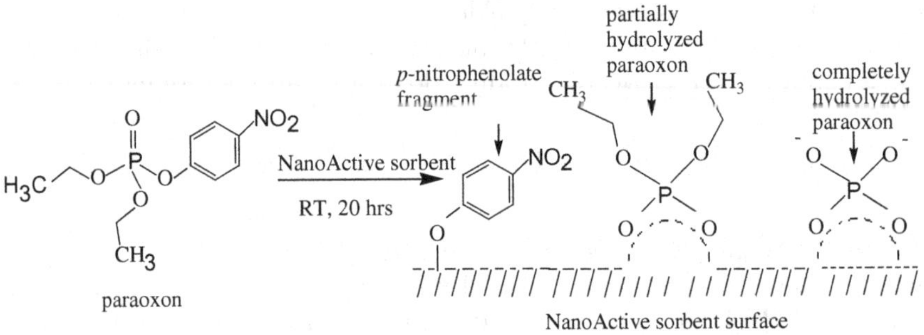

The adsorption and hydrolysis of paraoxon is readily visible on MgO nanoparticles due to the change in color of the powder from white to bright yellow. Note that paraoxon is a light yellow, oily substance and its color does not duplicate the bright yellow observed after contact with the nanoparticles. The anion $O_2NC_6H_4O^-$ is bright yellow, and the rapid color formation clearly shows that the anion is formed quickly on the surface of the nanoparticles. More definitive evidence for the destructive adsorptive capability of metal oxide nanoparticles comes from nuclear magnetic resonance (NMR) studies. ^{31}P NMR spectra of intimately mixed dry nanocrystalline metal oxide/paraoxon mixtures after 20 hours are displayed in Figure 1.8. Paraoxon in deuterochloroform solvent exhibits a signal around -6.5 ppm and the product derived via complete hydrolysis of paraoxon, namely, the phosphate ion (PO_4^{3-}), is expected to show a signal around 0 ppm. However, it should be noted that the exact chemical shift values are sensitive to the nature of the metal oxide employed. Results from NMR studies with MgO nanoparticles over time indicate that destructive adsorption starts immediately and continues over a long period of time.

The NMR spectrum of the MgO nanoparticles/paraoxon mixture contains at least four peaks. The outer peaks are due to spinning side bands. The major peak at $\delta = 0.8$ ppm is due to completely hydrolyzed paraoxon and the peak around $\delta = -8.2$ ppm is due to adsorbed but unhydrolyzed paraoxon. Upon close examination of the NanoActive TiO$_2$/paraoxon mixture spectrum, three signals were observed: a sharp peak around -8.4 ppm attributed to unhydrolyzed paraoxon, a broad peak around -5 ppm due to

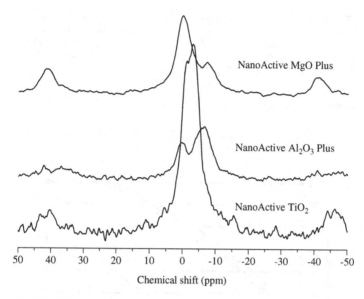

Figure 1.8. ^{31}P NMR spectra of nanocrystalline metal oxide/paraoxon mixtures after 20 hours.

partially hydrolyzed paraoxon, and a shoulder peak between 0 and -2 ppm was attributed to completely hydrolyzed paraoxon. The spectrum of the NanoActive Al_2O_3 Plus/paraoxon mixture displayed two major peaks. The upfield peak at $\delta = -6.7$ ppm is attributed to partially hydrolyzed paraoxon, and the peak at 0 ppm is due to completely hydrolyzed paraoxon. In short, adsorption of paraoxon by nanoparticles occurs instantly with the destruction of the agent as an ongoing active process.

Nanocrystalline materials can be utilized in powdered or compacted/granulated form. Compaction of the nanocrystals into pellets does not significantly degrade surface area or surface reactivity when moderate pressures are employed, ensuring that these nanostructured materials can be utilized as very fine powders or as porous, reactive pellets.[35] As shown in Table 1.2, pressure can be used to control pore structure. It should be noted that below about 5,000 psi the pore structure remains relatively unchanged, and these pellets behave in adsorption processes essentially identical to the loose powders. For example, both NanoScale MgO and Al_2O_3 (Figure 1.9) vigorously adsorb acetaldehyde with much higher rates and capacities than activated carbon. This is the case whether the samples are loose powders or compressed pellets. As noted in Table 1.2, it was only when very high compression pressure was used that the adsorption process was hindered.

Table 1.2.   Effects of compaction pressure on nanocrystalline MgO.

Load (psi)	Surface Area ($m^2 g^{-1}$)	Total Pore Volume ($cm^3 g^{-1}$)	Average Pore Diameter (nm)
0	443	0.76	6.9
5,000	434	0.57	5.3
10,000	376	0.40	4.2
20,000	249	0.17	2.7

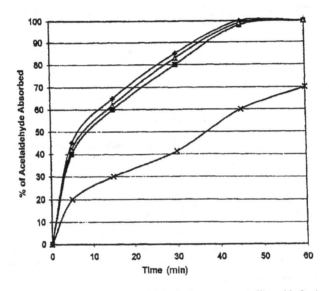

Figure 1.9.   Rate of adsorption of acetaldehyde by nanocrystalline Al_2O_3 in powder or pellet form. The following pressures were used to compact and form the pellets at room temperature: ♦ = powder, ■ = 5,000 psi, △ = 10,000 psi, X = 20,000 psi.

The application of nanocrystalline materials as destructive adsorbents for acid gases such as HCl, HBr, CO_2, H_2S, NO_X, and SO_X has been investigated by Klabunde et al.[36-38] These materials are much more efficient than commercially available oxides. This is seen with nanocrystalline ZnO, which reacts in a stoichiometric molar ratio of 1 mol of hydrogen sulfide to 2.4 mol of ZnO, whereas the commercial ZnO reacts in a molar ratio of 1 mol of hydrogen sulfide to 32 mol of ZnO. This indicates that nanoparticles are chemically more reactive at room temperature than their commercial counterparts. For example, nanocrystalline ZnO stays at near-stoichiometric ratio at room temperature, whereas the commercial ZnO does not.[39]

In addition to high-surface-area pure metal oxide nanoparticles, the development of intermingled metal oxide nanoparticles has been reported which yield special advantages.[40] The adsorption properties of intermingled metal oxide nanoparticles were found to be superior when comparing their reactivity to that of individual metal oxide nanoparticles and physical mixtures. For example, the nanoparticles of Al_2O_3/MgO mixed product have enhanced chemical reactivity over pure metal oxides of Al_2O_3 or MgO for the adsorption of SO_2. In comparison, the intermingled Al_2O_3/MgO showed an adsorption capacity of 6.8 molecules for SO_2 (of SO_2/nm^2 adsorbed), while the adsorption capacity of nanoparticles of Al_2O_3 and MgO was observed at 3.5 molecules and 6.0 molecules, respectively.[41]

1.4 Destructive Adsorption of Chemical Warfare Agents (CWAs) by Nanocrystalline Metal Oxides

Nanocrystalline metal oxides not only neutralize toxic industrial chemicals, but also destroy chemical warfare agents (V-, G-, and H-series) through hydrolysis and/or dehydrohalogenation. The G-series tend to be volatile and highly toxic by inhalation, while the V-agents are relatively non-volatile, persistent, and highly toxic by the percutaneous route. HD is an acronym for mustard gas and it belongs to the vesicant class of chemical warfare agents.

Wagner *et al.*[7-9] reported that NanoScale metal oxide nanoparticles are very effective in the destructive adsorption of VX, GD, and HD. These studies were done at room temperature using the pure agent on a column of dry metal oxide nanoparticles. It was found that the products formed in reactions with HD were the less toxic thioglycol (TG) and divinyl (DVHD) compounds while the nerve agents afforded the surface-bound hydrolyzed species. Destruction of nerve agents (GD and VZ) and the blistering agent (HD) was studied using solid-state NMR, as well as extraction, followed by GC-MS analysis. The destruction products of GD are pinacolyl methylphosphonic acid (GD-acid) that converts into surface-bound methylphosphonic acid (MPA), and HF, which is acidic and is neutralized by the metal oxide surface as well.

GD   **GD-acid**   **MPA**

The phosphonate destruction product of VX is ethyl methylphosphonic acid (EMPA) which converts into surface-bound MPA. It should be noted that during this reaction the toxic EA-2192 (S-(2-diisopropylamino)ethyl methylphosphonothioate) does not form, in contrast to basic VX hydrolysis in solution:

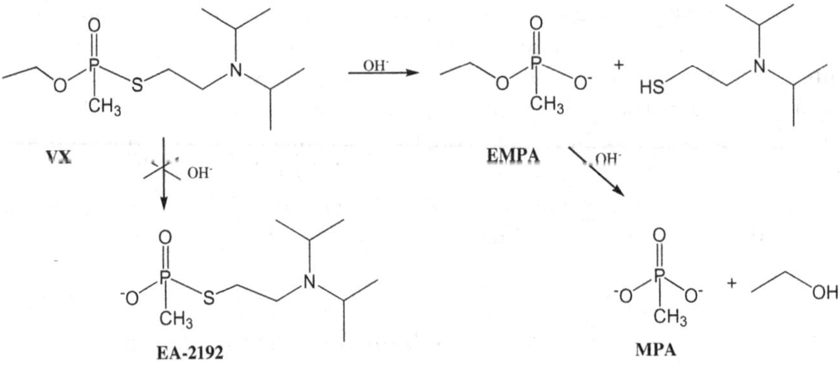

VX

EMPA

EA-2192

MPA

The decomposition product of HD is promoted by nucleophilic substitution on the β-carbon of a 2-chloroethyl group, resulting in a 2-hydroxyethyl group — thiodiglycol (TG) — whereas the decomposition products of HD via elimination are 2-chloroethyl-vinyl sulfide (CEVS) and divinyl sulfide (DVS). Both processes convert mustard into considerably less toxic compounds:

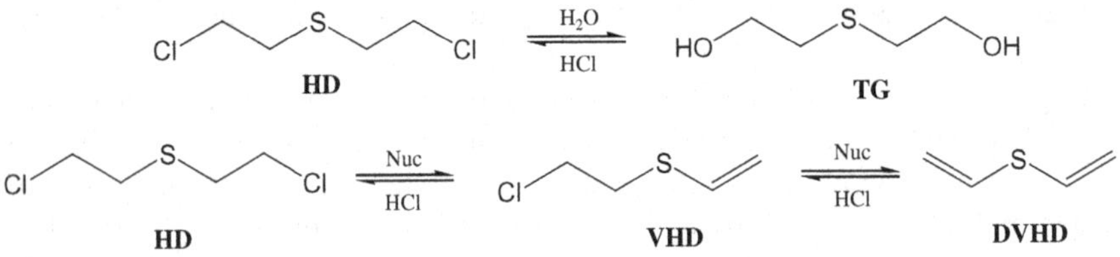

HD   **TG**

HD   **VHD**   **DVHD**

The off-gassing experiments for reactive nanoparticles after exposure to HD, GD, and VX indicated, through the absence of any major chromatographable peaks, that the reaction products are either low-molecular-weight (less than 60 amu), small gaseous materials that were eluted with the air peaks, or compounds that are tightly bound to the nanomaterials.

NanoScale has developed a nanocrystalline metal-oxide-based product, FAST-ACT, that is utilized as a chemical hazard containment and neutralization system. This product can be applied manually as a dry powder for treating liquid spills, or in aerosol form for treating toxic vapors, decontamination of vertical surfaces, or contaminated suits (Figure 1.10). Nanocrystalline metal oxides have low bulk density and remain suspended in the air for a prolonged period of time.

Figure 1.10. Decontamination of a first responder's suit utilizing FAST-ACT.

The effectiveness of FAST-ACT has been verified against chemical warfare agents GD (Soman), VX, and HD by two independent laboratories: Battelle Memorial Institute (Battelle), Columbus, OH; and the United States Army Soldier and Biological Chemical Command (SBCCOM), Edgewood, MD. All studies were conducted using controlled protocols at ambient room conditions (temperature, pressure, and humidity). The agent was placed on a glass surface and FAST-ACT was applied. Within 90 seconds FAST-ACT was validated to remove over 99.9% of HD and GD; and over 99.6% (detection limit) of VX from the surface (as determined by surface extraction followed by GC-MS analysis). Upon contact with FAST-ACT, the agent was quickly adsorbed and then destroyed. The destruction was confirmed by changes in the NMR spectra (SBCCOM) and by the inability to extract the agent from the powder (Battelle). In 10 minutes 99% of GD and over 99.9% of VX is destroyed, while in about 60 minutes 70–80% of HD is destroyed.

1.5 Safety of NanoActive Materials

Considering the high chemical reactivity of nanocrystalline metal oxides, a question to pose is how safe are humans or animals if exposed to these materials? NanoScale has conducted rigorous toxicity testings at independent laboratories studying the oral, dermal, pulmonary, and ocular effects of NanoActive metal oxides. The testings have revealed that there are no safety hazards associated with the nano nature of these materials. Dermal LD_{50} (rabbit) was $> 2\,g/kg$ and oral LD_{50} (rabbit) was $> 5\,g/kg$. The formulation has also been tested for inhalation toxicity and proven to be non-toxic to rats.

In addition, it was determined that in excess of 99.9% of the particles are captured by standard NIOSH particulate filters (flat media, Model 200 Series, N95 NIOSH, 30981J; and pleated filter with flat media, NIOSH Pro-Tech respirator, PN G100H404 OVIP-100). The non-toxic behavior of these particles in the inhalation testings, as well as the high removal efficiency, can be explained by the formation of weak aggregates of nanomaterials. The nanocrystalline metal oxides manufactured at NanoScale aggregate into micron-sized particles due to their high surface reactivity. Micron-sized particles do not penetrate into the alveoli, but are stopped in the bronchi. NanoScale's products utilized for environmental applications are made from inherently non-toxic materials (magnesium oxide, calcium oxide, titanium dioxide, aluminum oxide) and possess higher solubility than their non-nano counterparts. Therefore, even if introduced into a human or animal body, they will dissolve and be expelled. Overall, the nanocrystalline materials produced at NanoScale were found to be no more toxic than the respective commercially available metal oxides.[42]

1.6 Conclusions

Nanocrystalline metal oxides are highly effective adsorbents towards a broad range of environmental contaminants, ranging from acids, chlorinated hydrocarbons, organophosphorus and organosulfur compounds to chemical warfare agents. These materials do not merely adsorb, but actually destroy many chemical hazards by converting them into much safer byproducts under a broad range of temperatures. Metal oxides produced by NanoScale were proven to be no more toxic than their non-nano commercial counterparts and continue to be a great choice for abating environmental pollutants.

References

1. L. Interrante and M. Hampden-Smith (eds.), *Chemistry of Advanced Materials*, Chap. 7 (Wiley-VCH, New York, 1998), pp. 271–327.
2. K. J. Klabunde, J. V. Stark, O. Koper, C. Mohs, D. G. Park, S. Decker, Y. Jiang, I. Lagadic and D. Zhang, *J. Phys. Chem. B* **100**, 12142–12153 (1996).

3. A. Khaleel, P. N. Kapoor and K. J. Klabunde, *Nanostruct. Mater.* **11**, 459–468 (1999); E. Lucas and K. J. Klabunde, *Nanostruct. Mater.* **12**, 179–182 (1999); O. Koper and K. J. Klabunde, U.S. Patent 6,057,488; May 2 (2000).

4. O. Koper, K. J. Klabunde, L. S. Martin, K. B. Knappenberger, L. L. Hladky, S. P. Decker, U.S. Patent 6,653,519; November 25 (2003).

5. K. J. Klabunde, U.S. Patent 5,990,373; November 23 (1999).

6. S. Rajagopalan, O. Koper, S. Decker and K. J. Klabunde, *Chem. Eur. J.* **8**, 2602 (2002).

7. G. W. Wagner, P. W. Bartram, O. Koper and K. J. Klabunde, *J. Phys. Chem. B* **103**, 3225 (1999).

8. G. W. Wagner, O. B. Koper, E. Lucas, S. Decker and K. J. Klabunde, *J. Phys. Chem. B* **104**, 5118 (2000).

9. G. W. Wagner, L. R. Procell, R. J. O'Connor, S. Munavalli, C. L. Carnes, P. N. Kapoor and K. J. Klabunde, *J. Am. Chem. Soc.* **123**, 1636 (2001).

10. S. Decker, J. Klabunde, A. Khaleel and K. J. Klabunde, *Environ. Sci. Technol.* **36**, 762–768 (2002).

11. O. Koper, Y. X. Li and K. J. Klabunde, *Chem. Mater.* **5**, 500 (1993).

12. O. Koper and K. J. Klabunde, *Nanophase Materials*, eds. G. C. Hadjipanayis and R. W. Siegel (Kluwer Academic Publishers, Dordrecht, 1994), pp. 789–792.

13. O. B. Koper, E. A. Wovchko, J. A. Glass Jr., J. T. Yates Jr. and K. J. Klabunde, *Langmuir* **11**, 2054–2059 (1995).

14. O. Koper, I. Lagadic and K. J. Klabunde, *Chem. Mater.* **9**, 838–848 (1997).

15. O. Koper and K. J. Klabunde, *Chem. Mater.* **9**, 2481–2485 (1997).

16. R. J. Hedge and M. A. Barteau, *J. Catal.* **120**, 387–400 (1989).

17. Y. X. Li, H. Li and K. J. Klabunde, *Nanophase Materials*, eds. G. C. Hadjipanayis and R. W. Siegel (Kluwer Academic Publishers, Dordrecht, 1994), pp. 793–796.

18. Y. X. Li, H. Li and K. J. Klabunde, *Environ. Sci. Technol.* **28**, 1248–1253 (1994).

19. P. D. Hooker and K. J. Klabunde, *Environ. Sci. Technol.* **28**, 1243–1247 (1994).

20. K. J. Klabunde, A. Khaleel and D. Park, *High Temp. Mater. Sci.* **33**, 99–106 (1995).

21. A. Khaleel and K. J. Klabunde, *Nanophase Materials*, eds. G. C. Hadjipanayis and R. W. Siegel (Kluwer Academic Publishers, Dordrecht, 1994), pp. 785–788.

22. H. J. Kim, J. Kang, D. G. Park, H. J. Kweon and K. J. Klabunde, *Bull. Korean Chem. Soc.* **18**, 831–840 (1997).

23. K. J. Klabunde, A. Khaleel and D. Park, *High Temp. Mater. Sci.* **33**, 99–106 (1995).

24. S. Decker and K. J. Klabunde, *J. Am. Chem. Soc.* **118**, 12465–12466 (1996).

25. Y. Jiang, S. Decker, C. Mohs and K. J. Klabunde, *J. Catal.* **180**, 24–35 (1998).

26. S. Decker, I. Lagadic, K. J. Klabunde, J. Moscovici and A. Michalowicz, *Chem. Mater.* **10**, 674–678 (1998).

27. Y. Jiang, S. Decker, C. Mohs and K. J. Klabunde, *J. Catal.* **180**, 24–35 (1998).

28. I. N. Martyanov and K. J. Klabunde, *J. Catal.* **224**, 340–346 (2004).

29. Y. X. Li and K. J. Klabunde, *Langmuir* **7**, 1388–1393 (1991).

30. Y. X. Li, J. R. Schulp and K. J. Klabunde, *Langmuir* **7**, 1394–1399 (1991).

31. Y. X. Li, O. Koper, M. Atteya and K. J. Klabunde, *Chem. Mater.* **4**, 323–330 (1992).

32. M. K. Templeton and W. H. Weinberg, *J. Am. Chem. Soc.* **107**, 774–779 (1985).

33. Y. X. Li and K. J. Klabunde, *Langmuir* **7**, 1388–1393 (1991).

34. S. T. Lin and K. J. Klabunde, *Langmuir* **1**, 600–605 (1985).

35. E. Lucas, S. Decker, A. Khaleel, A. Seitz, S. Fultz, A. Ponce, W. Li, C. Carnes and K. J. Klabunde, *Chem. Eur. J.* **7**, 2505–2510 (2001).

36. K. J. Klabunde, J. V. Stark, O. B. Koper, C. Mohs, D. G. Park, S. Decker, Y. Jiang, I. Lagadic and D. Zhang, *J. Phys. Chem.* **100**, 12142 (1996).

37. J. V. Stark and K. J. Klabunde, *Chem. Mater.* **8**, 1913–1918 (1996).

38. C. L. Carnes, J. Stipp, K. J. Klabunde and J. Bonevich, *Langmuir* **18**, 1352–1359 (2002).

39. C. L. Carnes and K. J. Klabunde, *Chem. Mater.* **14**, 1806 (2002).

40. G. M. Medine, V. Zaikovskii and K. J. Klabunde, *J. Mater. Chem.* **14**, 757–763 (2004).

41. C. L. Carnes, P. N. Kapoor, K. J. Klabunde and J. Bonevich, *Chem. Mater.* **14**, 2922–2929 (2002).

42. USCHPPM Reports: 85-XC-01BN-03; 85-XC-5302-01; 85-XC-5302-03; 85-XC-03GG-11-05-01-02c; 85-XC-03GG-05; 85-XC-03GG-10-02-09-03. (Publication in preparation)

Chapter 2

Nanoscale Zero-Valent Iron (nZVI) for Site Remediation

Daniel W. Elliott*, Hsing-Lung Lien[†] and Wei-xian Zhang*

*Department of Civil and Environmental Engineering
Lehigh University, Bethlehem, PA 18015, USA

†Department of Civil and Environmental Engineering
National University of Kaohsiung
Kaohsiung, Taiwan, ROC

2.1 Introduction

Over the span of little more than a decade, the multi-disciplinary nanotechnology boom has inspired the creation and development of powerful new tools in the ongoing challenge of addressing the industrialized world's legacy of contaminated sites.[1] These might include improved analytical and remote-sensing methodologies, novel sorbents, and pollution-control devices, as well as superior soil- and groundwater-remediation technologies. The nanoscale zero-valent iron (nZVI) technology described in this chapter represents an important, early-stage achievement of the burgeoning environmental nanotechnology movement.

Since 1996, our research group at Lehigh University has been actively engaged in developing new nanometal materials, improving the synthetic schemes, performing both bench-scale and field-scale assessments, and extending the technology to increasing numbers of amenable contaminant classes. We have tested nZVI technology with more than 75 different environmental contaminants from a wide variety of chemical classes. These include chlorinated aliphatic hydrocarbons (CAHs), nitroaromatics, polychlorinated biphenyls (PCBs), chlorinated pesticides like lindane and DDT, hexavalent chromium, and perchlorate.

nZVI technology may prove to be useful for a wide array of environmental applications, including providing much-needed flexibility for both *in situ* and *ex situ* applications. Successful direct *in situ* injection of nZVI particles, whether under gravity-fed or pressurized conditions, to remediate chlorinated hydrocarbon-contaminated groundwater has already been demonstrated.[2,3] In addition, nZVI particles can be deployed in slurry reactors for the treatment of contaminated soils, sediments, and solid wastes; and can be anchored onto a solid matrix such as activated carbon and/or zeolite for enhanced treatment of water, wastewater, or gaseous process streams (Figure 2.1).

In this chapter, an overview of nZVI technology is provided, beginning with a description of the process fundamentals and applicable kinetic models. This is followed by a discussion of the synthetic schemes for the nZVI types developed at Lehigh University. Next, a summary of the major research findings is provided, highlighting the key characteristics and remediation-related advantages of nZVI technology versus granular/microscale ZVI technology. A discussion of fundamental issues related to the potential applications of nZVI technology and economic hurdles facing this technology is also included.

2.2 Overview and ZVI General Process Description

Zero-valent iron (Fe^0) has long been recognized as an excellent electron donor, regardless of its particle size. Even the non-scientifically inclined have observed the familiar corrosion (oxidation) of iron-based structures, implements, and art objects for millennia. In the rusting process, Fe^0 is oxidized to various ferrous iron (Fe^{+2}) and ferric iron (Fe^{3+}) salts (i.e. various oxidation products collectively known as rust) while atmospheric oxygen is reduced to water. Interestingly, while iron corrosion has been observed and known for millennia, the potential application of the iron corrosion process to environmental remediation remained largely unknown until the mid-1980s.[4]

2.2.1 *Applications of ZVI to environmental remediation*

Gillham and co-workers serendipitously observed that chlorinated hydrocarbon solvents in contaminated groundwater samples were unstable in the presence of certain steel and iron-based well casing materials.[4] This discovery triggered enormous interest in the potential applications of relatively inexpensive and essentially non-toxic iron materials in the remediation of contaminated groundwaters. Since that time, exhaustive research on various aspects of the topic has been undertaken by numerous research groups around the world.

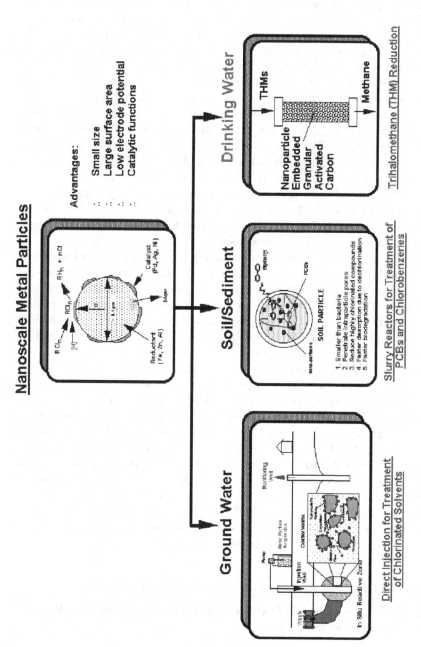

Figure 2.1.   Applications of nanoscale zero-valent iron technology for environmental remediation.

The majority of the ZVI research conducted thus far has focused on the bench-scale degradation of relatively simple one- and two-carbon chlorinated hydrocarbon contaminants in either batch aqueous systems or column studies. Including such notorious solvents and feedstock chemicals as perchloroethene (PCE), trichloroethene (TCE), carbon tetrachloride (CT), and vinyl chloride (VC), these contaminants pose a considerable and well-documented threat to groundwater quality in many of the world's industrialized nations. In batch-test studies, aqueous contaminant concentrations ranging from approximately 1 mg/L to near saturation levels were degraded by variable granular ZVI dosages of about 20–250 g/L in timescales on the order of hours.[4-7] Reductive dehalogenation of the contaminant to typically non-halogenated end-products (e.g. ethene, ethane, methane) was accompanied by profound changes in the observed water chemistry, particularly in solution pH and standard reduction potential. In many studies, the experimental data was modeled according to pseudo first-order kinetics although others used more complex approaches.[8,9]

Other investigators have demonstrated the efficacy of ZVI technology towards other notorious contaminants including hexavalent chromium, nitrate, nitroaromatics, energetic munitions compounds, azo dyes, and pesticides like DDT.[10-15] One of the common themes of this rich and extensive body of research is the essential role of the iron surface in mediating the contaminant degradation process[5,16] although other factors including sorption[17,18] also play an integral role.

In addition to the laboratory-scale research, applied investigations at the field pilot scale have also been under way since the early 1990s. More than 100 ZVI field deployments of granular ZVI in permeable reactive barriers (PRBs) have been completed at industrial, commercial, and governmental sites in North America, Europe, and in Asia.[19] In essence, PRBs consist of sub-surface walls of granular iron; often constructed in a funnel-and-gate design, perpendicular to the flow of groundwater. The funneling mechanism directs contaminated groundwater through the reactive gate(s). The gates function as a plug flow reactor with contaminant degradation occurring throughout the thickness of the ZVI zone. For potable aquifers, PRBs are often designed to theoretically treat groundwater contamination to acceptable US EPA maximum contaminant levels (MCLs).

2.2.2 *ZVI process description: Chemical fundamentals*

As previously mentioned, Fe^0 exhibits a strong tendency to donate electrons to suitable electron acceptors:

$$Fe^{2+} + 2e^- \rightarrow Fe^0. \tag{2.1}$$

Zero-valent iron and dissolved ferrous iron form a redox couple with a standard reduction potential (E^o) of -0.440 V.[5] Metallic ZVI can couple with several environmentally significant and redox-amenable electron acceptors, including hydrogen ions (i.e. protons), dissolved oxygen, nitrate, sulfate, and carbonate.[5] Under aerobic conditions typical of vadose-zone soils or shallow, oxygenated groundwaters, ZVI can react with dissolved oxygen (DO) as follows:

$$2Fe^0_{(s)} + 4H^+_{(aq)} + O_{2(aq)} \rightarrow 2Fe^{2+}_{(aq)} + 2H_2O_{(l)}, \qquad (2.2)$$

yielding ferrous iron and water. The overall E^o for this reaction (E^o_{rxn}) is $+1.71$ V at $25°C$, indicating a strongly favorable reaction from a thermodynamic perspective.[20] Implicit in the stoichiometry is the transfer of four electrons from the iron surface and associated increase of solution pH (based on the consumption of protons). Assuming that residual DO levels remain, the ferrous iron would be expected to undergo relatively facile oxidation to ferric iron, Fe^{3+}. The increasing pH favors the formation of one or more iron hydroxides, or carbonate-based precipitates, and can have the effect of passivating the reactivity of the metal surface.[5]

In anaerobic or low DO groundwater environments, which are more characteristic of deeper aquifers and contaminated plumes, ZVI also forms an effective redox couple with water, yielding ferrous iron, hydroxide, and hydrogen gas:

$$Fe^0_{(s)} + 2H_2O_{(l)} \rightarrow 2Fe^{2+}_{(aq)} + H_{2(g)} + 2OH^-_{(aq)}. \qquad (2.3)$$

Unlike the reduction of DO, the iron-mediated reduction of water is not thermodynamically favored as indicated by the E^o_{rxn} of -0.39 V at $25°C$.[20] Thermodynamic considerations notwithstanding, the kinetics of these and other reactions in natural waters tend to be rather sluggish and, consequently, chemical equilibrium is generally not attained.[21]

In addition to the common environmentally relevant electron acceptors, ZVI also readily reacts with a wide variety of redox-amenable contaminants. Using a generalized chlorinated hydrocarbon, RCl, as an example, the ZVI-mediated transformation of RCl to the corresponding hydrocarbon, RH, can be represented as:

$$RCl + H^+ + Fe^0 \rightarrow RH + Fe^{2+} + Cl^-. \qquad (2.4)$$

From a thermodynamic perspective, the large positive E^o_{rxn} values imply that a spontaneous reaction with ZVI should occur. In terms of the chlorinated hydrocarbons, the degree of favorability increases with the number of chlorine substituents. For many chlorinated hydrocarbons, the E^o_{rxn} is on the order of $+0.5$ to $+1.5$ V at $25°C$.[5,22]

2.2.3 *ZVI process description: Mechanistic aspects*

An examination of Equations (2.2) through (2.4) reveals that multiple species exist in the systems under investigation which are capable of serving as reducing agents. The electron-donating potential of ZVI has already been discussed. However, ferrous iron can serve as a reductant by donating an electron to a suitable electron acceptor yielding ferric iron, Fe^{3+}. Hydrogen gas, too, is a well-known reducing agent, particularly in the realm of microbiology where it is often characterized as the "universal electron donor". The roles of these three electron donors in the reduction of chlorinated hydrocarbons were studied. For example, Matheson and Tratnyek[5] proposed three possible mechanisms: (1) direct reduction at the metal surface, (2) reduction by ferrous iron, and (3) reduction by hydrogen with catalysis. They studied the potential roles of these reductants and found that ferrous iron, in concert with certain ligands, can slowly reduce the chlorinated hydrocarbons; and that dissolved hydrogen gas, in the absence of a suitable catalytic surface, failed to reduced the chlorinated hydrocarbon.

In the presence of granular ZVI, Matheson and Tratnyek[5] and numerous other research groups observed the rapid transformation of various contaminants, confirming the validity of the direct surface reduction model. Weber[16] elegantly confirmed these findings in a study using 4-aminoazobenzene (4-AAB), an aromatic azo dye which readily undergoes ZVI-mediated reduction. In this work, 4-AAB which was immobilized by electrophilic derivatization to a solid support was not reduced by ZVI because it could not associate with the iron surface while the control, non-derivatized 4-AAB, was rapidly transformed.[16] Thus, these studies clearly demonstrated the fact that the degradation of contaminants by ZVI is a surface-mediated process via one or more heterogeneous reactions. The nature of these reactions, whether occurring sequentially or in concert, depends upon the particular degradation pathways involved; which in turn are a function of the specific contaminant(s).

2.2.4 *Kinetic models of the ZVI process*

Although many investigations have invoked more-complex models recognizing the heterogeneous nature of the ZVI degradation process,[9] first-order kinetics was the most common model used to explain the experimental datasets from batch degradation studies. As has been previously described, these datasets covered a wide variety of chemical classes (e.g. chlorinated solvents, chlorinated aromatics, pesticides, PCBs, nitroaromatics, and metals such as hexavalent chromium). To the extent feasible, first-order kinetics were presumed to be applicable to a wide variety of reactions of environmental significance primarily because of the relative simplicity of the mathematical treatment. Thus, it is often the first model utilized to fit the experimental data for a new environmental process. As a result, our discussion mainly focused on the first-order kinetic model.

In a system in which first-order kinetics is determined to prevail, the rate of contaminant loss, or degradation, is proportional to its concentration in solution, as expressed in Equation (2.5):

$$\frac{dC}{dt} = -kC, \tag{2.5}$$

where C is the contaminant concentration at time t, and k is the constant of proportionality known as the first-order rate constant. In order for the units to balance, k is reported in units of reciprocal time, $1/t$.

It stands to reason that ZVI-mediated transformations should not only depend upon the contaminant concentration but on the iron concentration as well. After all, the reaction involves two chemical entities: the reductant and the reductate. In this case, the first-order kinetic expression must be modified as follows:

$$\frac{dC}{dt} = -k[C][Fe^0], \tag{2.6}$$

where the rate of contaminant transformation now depends on both the contaminant and iron concentrations in solution. Thus, Equation (2.6) is applicable to reactions characterized by second-order kinetics. However, in the vast majority of cases involving the ZVI-mediated degradation of contaminant(s), the concentration of iron is appreciably larger than that of the aqueous contaminant. That is, $[Fe^0] \gg [C]$. For the vast majority of the peer-reviewed ZVI literature regarding granular, microscale, or nanoscale iron, this simplifying assumption can be made. If $[Fe^0]$ is large enough such that it does not change meaningfully over the course of the observed changes in $[C]$, it can be said to remain constant and Equation (2.7) can now be represented as:

$$\frac{dC}{dt} = -k_{obs}[C]. \tag{2.7}$$

The observed first-order rate constant, k_{obs}, is related to k from Equation (2.6) in that $k_{obs} = k[Fe^0]$. Equation (2.7) is referred to as a pseudo first-order expression and, as has just been described, results from applying a valid simplifying assumption to a system formally characterized by second-order kinetics.

Although Equation (2.7) is indeed applicable to the ZVI-mediated degradation process, investigators such as Johnson *et al.*[8] noted that k_{obs} data from the literature for specific contaminants in batch degradation studies under differing experimental conditions exhibited variability of up to three orders of magnitude. In an effort to account for the majority of the experimental factors accounting for this variability, they expanded Equation (2.8) as follows:

$$\frac{dC}{dt} = -k_{SA}\rho_a C, \qquad (2.8)$$

where k_{SA} is defined as the surface-area-normalized rate constant and is reported in units of liters per hour per meter squared (L/hr-m^2) and ρ_a is the surface area concentration of ZVI in square meters per liter (m^2/L) of solution.[8] The iron surface area concentration is related to the specific surface area of ZVI, a_s, and mass concentration of iron, ρ_m, by Equation (2.9):

$$\rho_a = a_s\rho_m, \qquad (2.9)$$

where the units of a_s are square meters per gram (m^2/g) and those of ρ_m are grams per liter of solution (g/L). The specific surface area concentration is also related to k_{obs} by Equation (2.10) as follows:

$$k_{obs} = k_{SA}\rho_m a_s. \qquad (2.10)$$

In using Equations (2.8) through (2.10) which describe the kinetics of the ZVI-mediated transformation of various chlorinated hydrocarbon contaminants, Johnson et al.[8] found that most of the variability could be eliminated. After replotting the literature-derived dataset using these expressions, variability of one order of magnitude was still observed, probably associated with uncertainty in the methodology and/or measurement of ZVI surface area.

2.3 Overview of Major Methodologies for Synthesizing nZVI

At least three distinct methods have been used to prepare nanoscale ZVI (nZVI). All involved the reduction and precipitation of zero-valent iron from aqueous iron salts using sodium borohydride as the reductant. These included one approach in which the iron salt was iron(III) chloride (referred to as the chloride method), and two schemes where iron(II) sulfate was the principal ZVI precursor salt (termed the sulfate method). Each of these methodologies is discussed in detail in the following sub-sections. It is important to note that the laboratory-scale synthetic protocols were not optimized and that potential significant batch-to-batch variability in terms of characteristics, behavior, and performance is to be expected.

2.3.1 Type I nZVI using the chloride method

The chloride method for synthesis represents the original means of producing nanoscale ZVI at Lehigh University.[24-26] Consequently, the chloride-method iron, also referred to as Type I nZVI, was the earliest generation of nanoscale iron used in experimental and in field-scale work.

In this synthesis, 0.25 molar (M) sodium borohydride was slowly added to 0.045 M ferric chloride hexahydrate in aqueous solution under vigorously mixed conditions such that the volumes of both the borohydride and ferric salt solutions were approximately equal (i.e. a 1:1 volumetric ratio). The mixing time utilized was approximately one hour. This reaction is shown in Equation (2.11) as follows[25]:

$$4Fe^{3+}_{(aq)} + 3BH^-_{4\ (aq)} + 9H_2O_{(l)} \rightarrow 4Fe^0_{(s)} + 3H_2BO^-_{3\ (aq)} + 12H^+_{(aq)} + 6H_{2(g)}.$$

$$(2.11)$$

The ratio between the borohydride and ferric salt exceeded the stoichiometric requirement by a factor of approximately 7.4.[25] This excess is thought to help ensure the rapid and uniform growth of the nZVI crystals. The harvested nano-iron particles were then washed successively with a large excess of distilled water, typically $> 100\,mL/g$. The solid nanoparticle mass was recovered by vacuum filtration and washed with ethanol. The residual water content of the nZVI mass was typically on the order of 40–60%. A transmission electron microscopy (TEM) micrograph of Type I nZVI is shown in Figure 2.2.

If bimetallic particles were desired, the ethanol-wet nZVI mass was soaked in an ethanol solution containing approximately 1% palladium acetate as indicated in Equation (2.12):

$$Pd^{2+} + Fe^0 \rightarrow Pd^0 + Fe^{2+}.$$

$$(2.12)$$

The chloride method proved to be readily adapted to any standard chemical laboratory assuming that fume-hoods and adequate mixing were provided to dissipate and vent the significant quantities of hydrogen gas produced. The synthetic scheme was expensive, due to the high cost of sodium borohydride and lack of production scale.

2.3.2 Types II and III nZVI using the sulfate method

The development of the sulfate method for producing nZVI arose from two fundamental concerns associated with the chloride method: (1) potential health and safety concerns associated with handling the highly acidic and very hygroscopic ferric chloride salt; and (2) the potential deleterious

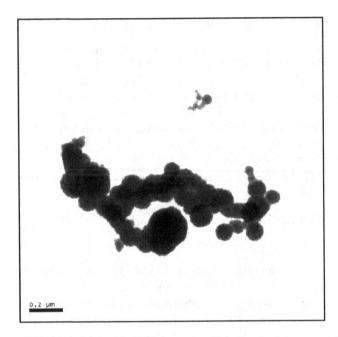

Figure 2.2.   TEM image of Type I nZVI aggregates. Note the size bar represents 200 nm.

effects of excessive chloride levels from the nZVI matrix in batch degradation tests where chlorinated hydrocarbons are the contaminant of concern. In addition, the reduction of the iron feedstock from Fe(II) requires less borohydride than the chloride method, in which Fe(III) is the starting material which may favorably enhance the overall process economics. Because this method represented the second generation of iron nanoparticles developed at Lehigh University, the iron is referred to as Type II nZVI.

Sulfate-method nZVI was prepared by metering equal volumes of 0.50 M sodium borohydride at 0.15 L/min into 0.28 M ferrous sulfate according to the following stoichiometry:

$$2Fe^{2+}_{(aq)} + BH_4^-{}_{(aq)} + 3H_2O_{(l)} \rightarrow 2Fe^0_{(s)} + H_2BO_3^-{}_{(aq)} + 4H^+_{(aq)} + 2H_{2(g)}.$$

$$(2.13)$$

The stoichiometric excess of borohydride used in Type II nZVI synthesis was about 3.6, considerably less than that with Type I nZVI. Owing to this process change, the rate of borohydride addition was extended to approximately two hours to help control particle size. Thus, the reduction in production costs was partially offset by the longer synthesis time. Process improvements were achieved, particularly during the concerted production of 10 kg, dry weight basis, of Type II nZVI for a field demonstration in the Research Triangle Park, NC, in 2002.[3] Specifically, these included enhancements to the mixing, nanoparticle recovery, and storage elements of the synthesis.

As was the case with the chloride method, the synthesis was carried out in a fume-hood in open five-gallon (18.9 L) polyethylene containers fitted with variable-speed, explosion-resistant mixers. No attempt was made to exclude air from the reaction mixture. The freshly prepared nZVI particles were allowed to settle for approximately one hour and were then harvested by vacuum filtration. The harvested nanoparticles were washed with copious amounts of distilled water (> 100 mL/g), then by ethanol, purged with nitrogen, and refrigerated in a sealed polyethylene container under ethanol until needed. The residual water content of Type II nZVI was typically on the order of 45–55%, very similar to that observed for Type I iron. A representative TEM image of Type II nZVI is shown in Figure 2.3.

Type III nZVI was also synthesized using the sulfate method. As procedure for synthesizing Type III iron was very similar to that for Type II nZVI, the details are not repeated here. It represented the latest generation

200 nm

Figure 2.3.   TEM image of Type II nZVI aggregates. Note the size bar represents 200 nm.

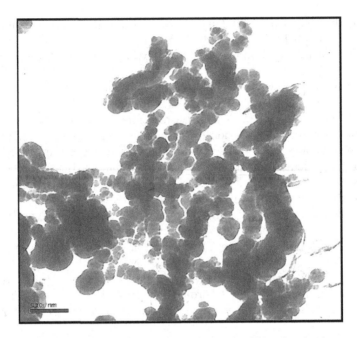

Figure 2.4. TEM image of Type III nZVI aggregates. Note the size bar represents 100 nm.

of nZVI and exhibited an average particle size of approximately 50–70 nm, very similar to that observed for Type II nZVI. However, the moisture content was appreciably lower than observed for the previous nZVI types: 20–30% versus 40–60% for Types I and II. The basis for this difference is not known. A TEM image of Type III nZVI is shown in Figure 2.4.

2.4 Characterization of nZVI

The average particle size of the chloride-method nZVI was in the order of 50–200 nm and the specific surface area was measured by mercury porosimetry to be approximately $33.5\,m^2/g$.[24] X-ray diffraction of Type I palladized nZVI (nZVI/Pd) surface composition indicated that the major surface species of freshly prepared nanoscale Pd/Fe is Fe^0 (44.7°), with lesser quantities of iron(III) oxide, Fe_2O_3 (35.8°), and Pd^0 (40.1°) (Figure 2.5). The presence of iron(III) oxide, which results from air exposure of nZVI, visually appeared as a surficial rust patina on the surface of the iron and typically did not extend into the bulk iron mass. Not surprisingly, the XRD diffractogram for nZVI/Pd aged for 48 hours shows

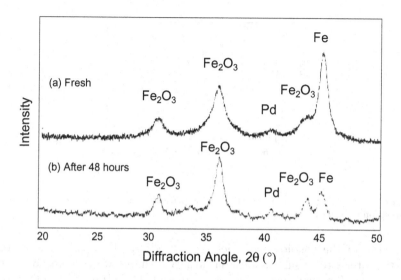

Figure 2.5.   X-ray diffractogram of nZVI/Pd particles.

relatively larger peaks for the iron oxide but still an appreciable zero-valent iron peak (Figure 2.5).

The average particle size of the sulfate-method nZVI was on the order of 50–70 nm. Analysis of over 150 individual particles and clusters yielded a mean diameter of 66.6 ± 12.6 nm with more than 80% being smaller than 100 nm and fully 30% being smaller than 50 nm.[26] Moreover, the BET specific surface area of the iron was 35 ± 2.7 m^2/g, approximately equivalent to that of Type I nZVI.[26]

Using an electroacoustic spectrometer (Dispersion Technology DT-1200), the zeta potential of a 0.85% (by weight) slurry of Type III nZVI in water was -27.55 mV at a pH of 8.77 (Figure 2.6). According to the Colloidal Science Laboratory, Inc. (Westampton, NJ), colloidal particles with zeta potential values more positive than $+30$ mV or more negative than -30 mV are considered stable, with maximum instability (i.e. aggregation) occurring at a zeta potential of zero. Thus, using this benchmark, Type III nZVI would be considered metastable.

Theoretically, the zeta potential refers to the potential drop across the mobile or diffuse portion of the classical electric double layer surrounding a particle in solution.[27] It provides an indication of the stability of the colloidal suspension and is a function of many variables, including the nature of the particle surface, ionic strength, pH, and presence of other substances that can interact with the surface (e.g. surface-active polymers). It has been well established that solution pH strongly influences the zeta

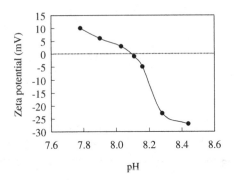

Figure 2.6.   Titration of zeta potential versus pH using 2.0 N sulfuric acid for a 10 g/L suspension of nZVI in water.

potential of the colloidal particles. Specifically, as the pH increases, the particles tend to acquire additional negative charge which translates into a decreased (i.e. more negative) zeta potential. The pH corresponding to a zeta potential of zero, known as the isoelectric point, represents the area of minimum particle stability. Figure 2.6 depicts a zeta potential versus pH titration for a chloride-method nZVI of 50–70 nm average particle size in water. In the neutral pH range, the relatively low zeta potential of nZVI in solution supports the observation of particle aggregation which could adversely affect sub-surface mobility.

2.5   Summary of nZVI Research and Applications

As demonstrated in the research of Zhang *et al.*, nZVI technology exhibits enhanced reactivity and superior field deployment capabilities as compared with microscale and granular iron as well as other *in situ* approaches. The enhanced reactivity stems from the appreciably greater specific surface area of the iron. The colloidal size of the iron nanoparticles, their amenability to direct sub-surface injection via gravity feed conditions (or under pressure, if desired), and need for substantially less infrastructure all contribute to the technology's portability and relative ease of use.

2.5.1   *Laboratory batch studies*

Since 1996, Zhang *et al.* have investigated the ability of nZVI to degrade a wide variety of environmental contaminants including PCBs, chlorinated aliphatic and aromatic hydrocarbons, hexavalent chromium, chlorinated pesticides, and perchlorate.

2.5.1.1 *Chlorinated hydrocarbons*

Degradation of chlorinated hydrocarbons using nZVI, mZVI (microscale ZVI), and nZVI/Pd has been extensively studied. A wide array of chlorinated hydrocarbons including aliphatic compounds (e.g. carbon tetrachloride, trichloroethylene, and hexachloroethane), alicyclic compounds (e.g. lindane), and aromatic compounds (e.g. PCB and hexachlorobenzene) have been tested. In general, nZVI/Pd showed the best overall performance followed by nZVI and then mZVI. Degradation of carbon tetrachloride by different types of ZVI represents a typical example. The surface-area-normalized rate constant (k_{SA}) data derived from the experimental datasets followed the order nZVI/Pd > nZVI > mZVI. The surface-area-normalized rate constant of nZVI/Pd was two orders of magnitude higher than that of mZVI. Furthermore, the same reaction products including chloroform, dichloromethane, and methane were observed in the use of mZVI, nZVI, and nZVI/Pd; however, product distributions were significantly different. The highest yield of methane (55%) and lowest production of dichloromethane (23%) were found in the use of nZVI/Pd. In comparison, the accumulation of dichloromethane, a more toxic compound than the parent, accounted for more than 65% of the initial carbon tetrachloride and a yield of methane of less than 25% (Figure 2.7).

The superiority of nZVI/Pd can be attributed to two key factors. First and foremost, a significantly increased surface area of nZVI. As compared to mZVI, the enhanced reactivity of nZVI can be attributed to the smaller average particle size which translates into a much larger specific surface

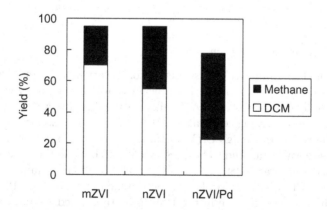

Figure 2.7. Comparison of yields of ethane and dichloromethane for the dechlorination of carbon tetrachloride with nZVI/Pd, nZVI, and mZVI. The doses of nZVI/Pd, nZVI, and mZVI were 5 g/L, 5 g/L, and 400 g/L, respectively.

area, $33.5\,m^2/g$ versus $<0.9\,m^2/g$, for the irons studied.[24] Second, the presence of palladium resulting in the catalytic reducing power of nZVI/Pd. It should be pointed out that palladium not only enhances the reactivity but also alters the product distribution, implying that different reaction mechanisms are likely involved.[28,29]

In comparison, chlorinated alicyclic and aromatic compounds have received less attention from investigators thus far. Very few systematic studies have been reported in the peer-reviewed literature, partially due to complicated transformation processes as they possess relatively large molecular structures. The fact that numerous degradation-related intermediates are often produced is another factor. Nevertheless, Xu and Zhang[30] demonstrated that nZVI and nZVI/Ag (iron-silver nanoparticles) were quite effective in transforming chlorinated aromatics through a sequential dechlorination pathway. For example, nZVI/Ag, with a measured specific surface area of $35.0\,m^2/g$, transformed hexachlorobenzene (HCB) to a series of lesser chlorinated benzenes including the following principal products: 1,2,4,5-tetrachlorobenzene, 1,2,4-trichlorobenzene, and 1,4-dichlorobenzene. No chlorobenzene or benzene was observed as reaction products. HCB concentrations were reduced below the detection limit $(<1\,\mu g/L)$ after four days.

However, conventional mZVI powder at an iron-to-solution ratio of $25\,g/100\,mL$ produced little reaction with HCB under similar experimental conditions. After 400 hours of elapsed time, the total conversion of HCB to products was approximately 12% with primarily 1,2,4,5-tetrachlorobenzene and 1,2,4-trichlorobenzene detected as intermediates.[30] Clearly, nZVI/Ag is more well suited than mZVI in degrading polychlorinated aromatics like HCB.

2.5.1.2 Hexavalent chromium

Hexavalent chromium, Cr(VI), is a highly toxic, very mobile, and quite common groundwater contaminant. The efficacy of nZVI technology was evaluated in batch aqueous systems containing soils and groundwater affected by chromium ore processing residuals (COPR) at a former manufacturing site in New Jersey. The average Cr(VI) concentration in groundwater samples from the site was measured to be $42.83\pm0.52\,mg/L$ while the concentration in air-dried soils was $3,280\pm90\,mg/kg$.[31] The total chromium concentration, that is Cr(III) plus Cr(VI), was determined to be $7,730\pm120\,mg/kg$. Due to the presence of lime in the COPR-contaminated media, the pH of groundwater typically exceeded 10–11. The basis for this reaction involves the very favorable reduction of Cr(VI) to Cr(III), a relatively non-toxic, highly immobile species which precipitates (i.e. Cr(III) oxyhydroxides) from solution at alkaline pH.

$$\frac{3}{2}Fe^0_{(s)} + CrO_4^{2-}{}_{(aq)} + 5H^+_{(aq)} \rightarrow \frac{3}{2}Fe^{2+}_{(aq)} + Cr(OH)_{3(s)} + H_2O_{(l)}. \quad (2.14)$$

Batch solutions containing 10 g of COPR-contaminated soils and 40 mL of groundwater were exposed to 5–50 g/L (89.5–895 mM) Type I nZVI under well-mixed conditions (Figure 2.8). Not surprisingly, once the COPR-contaminated soils were placed in the reactors, the Cr(VI) concentration increased from 42 mg/L to approximately 220 mg/L, demonstrating that substantial desorption and dissolution of hexavalent chromium occured during the course of the experiment. Within a timeframe of up to six days, Cr(VI) concentrations in solution containing Type I nZVI were generally less than the detection limit, $< 10\,\mu g/L$ (Figure 2.8). In contrast, the Cr(VI) concentration was observed to increase substantially in the reactors containing microscale iron (Fisher, $10\,\mu m$) during the course of the reaction.

The reductive capacity of Type I nZVI was found to be on the order of 84–109 mg Cr(VI) per gram of iron, approximately two orders of magnitude greater than the reaction with mZVI. Given the highly heterogeneous nature of the COPR materials and the highly alkaline pH, it is likely that Cr(VI) desorbing from the matrix is reduced and rapidly precipitated as chromium(III) hydroxide, $Cr(OH)_3$. This precipitate enmeshes the COPR soils, forming a shell that tends to encapsulate any Cr(VI) remaining in the potentially still reactive core region.

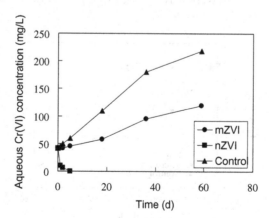

Figure 2.8.   Results of batch Cr(VI) reduction by mZVI and nZVI. The doses of mZVI and nZVI were 150 g/L and 5 g/L, respectively.

2.5.1.3 Perchlorate

Perchlorate (ClO_4^-) emerged as a high-profile environmental contaminant in the late 1990s when improved analytical techniques revealed widespread and previously undetected contamination in water supplies, particularly in the Western United States.[32] Perchlorate concentrations in excess of $100\,\mu g/L$ have been detected in Nevada's Lake Mead, well beyond the U.S. EPA guidance level of $1\,\mu g/L$.[33]

While the iron-mediated reduction of perchlorate, shown in Equation (2.15), is a strongly thermodynamically favored process (having a large negative value for the standard Gibbs free energy of the reaction, ΔG_{rxn}, of $-1{,}387.5\,kJ/mol$), one recent study reported that it is generally not reactive with ZVI.[34]

$$ClO_4^- + 4Fe^0 + 8H^+ \rightarrow Cl^- + 4Fe^{2+} + 4H_2O. \tag{2.15}$$

In general, ZVI transformations require either direct contact with the reactive iron surface or through suitable bridging groups. The structure of perchlorate makes this difficult because its reactive chlorine central atom, Cl(VII), is shielded by a tetrahedral array of bulky oxygen substituents, which also fully delocalize the oxyanion's negative charge.

Representative results of the reaction at $25°C$ are shown in Figure 2.9. Batch reactors containing nitrogen-purged deionized water spiked with $1–200\,mg/L$ sodium perchlorate and $1–20\,g/L$ Type II sulfate-method nZVI. After 28 days of elapsed time, negligible reaction with $20\,g/L$ microscale iron (Aldrich, $4.95\,\mu m$) was observed while significant removal of perchlorate was measured in the reactor containing $20\,g/L$ Type II nZVI. As depicted in Figure 2.10, progressively better removal of perchlorate was observed as

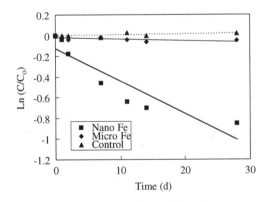

Figure 2.9.   Degradation of $200\,mg/L$ aqueous perchlorate by nZVI ($20\,g/L$) at $25°C$.

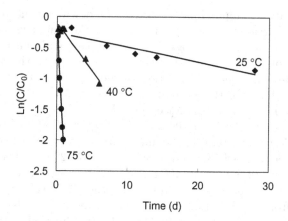

Figure 2.10.   A plot of Ln (C/C$_o$) versus time for the reduction of perchlorate by nZVI (20 g/L) at various temperature conditions.

the ambient temperature was increased to 40°C and finally to 75°C. The most interesting feature of these plots is the dramatic reduction in the transformation timescale achieved with increasing the temperature from 25°C to 75°C. The half-life for perchlorate declined from 18–20 days at 25°C to about 80–100 hours at 40°C and finally to approximately eight hours at 75°C. Thus, temperature was clearly an important factor in catalyzing the reduction of perchlorate in aqueous solution.

The appearance of chloride as the major end-product demonstrated that perchlorate transformation was occurring and not due to mere sorption or other surface-associated sequestration. Besides chloride, the only intermediate identified was chlorate, ClO_3^-, and only at trace levels.[32] Values of 0.013, 0.10, 0.64, and 1.52 mg perchlorate per gram of nZVI per hour ($mg\text{-}g^{-1}\text{-}hr^{-1}$) were calculated at temperatures of 25°C, 40°C, 60°C, and 75°C, respectively. The experiments conducted at different temperatures enabled calculation of the activation energy for the nZVI-mediated degradation process. The activation energy was determined to be 79.02 ± 7.75 kJ/mol, a relatively formidable value for aqueous reactions.

2.5.2 *Field testing demonstration*

Several *in situ* field demonstrations of nZVI technology have been conducted at sites with contaminated groundwater since 2000, including the first field pilot test in Trenton, New Jersey.[2] A subsequent field demonstration in North Carolina involved the injection of approximately 12 kg of chloride-method nZVI/Pd into a fractured bedrock aquifer. The test area was situated approximately 38 meters downgradient of a former waste disposal area, which had groundwater contaminated with chlorinated solvents.[26] The site hydrogeology is quite complex, with much of the contaminated groundwater traversing through fractures in the sedimentary sandstones and siltstones. In addition to the injection well, the test area infrastructure included three monitoring wells at distances of 6.6, 13, and 19 meters downgradient of the injection well.

Approximately 6,056 L (1,600 gal) of a 1.9 g/L nZVI/Pd slurry in tap water was injected at a rate of 2.3 L/min (0.6 gal/min) over a nearly two-day period from September 13 through September 15, 2002.[26] The slurry was prepared on-site in a 1,500 L (400 gal) heavy-duty polyethylene tank and mixed during the course of the injection process. Within approximately seven days from the injection, a reduction of more than 90% of the pre-injection TCE concentration of 14,000 μg/L was observed at the injection well and nearest monitoring well. These results are comparable to those from the first nZVI field demonstration conducted in New Jersey two years earlier.[2] Groundwater quality standards for TCE, PCE, and cis-DCE were generally achieved within six weeks of the injection at these two monitoring wells without increases in VC being observed. Although pre-injection groundwater ORP levels were in the range of +50 to −100 mV, indicative of iron-reducing conditions, post-injection ORP readings plunged to −700 mV in the injection well and −500 mV in nearby monitoring wells. The radius of influence measured at the injection well was approximately 6–10 meters.[26]

This demonstration successfully showed that nZVI can travel distances of more than 20 meters in groundwater, that the longevity of reactivity can reach periods of greater than 4–8 weeks in the field, and that very high degrees of contaminant removal can be achieved. However, due to the significant costs of fieldwork of this nature, the test was not conducted long enough to observe significant recovery of the contaminant levels within the test area.

2.6 Challenges Ahead

As nZVI technology moves rapidly from laboratory research to real-world implementation, a number of fundamental issues relating to the potential applications of iron nanoparticles for environmental remediation remain. These include (1) mobility, (2) environmental impact, and (3) cost-effectiveness. An important attribute of nZVI is their potential mobility in porous environmental media, especially in the sub-surface environment. Due partly to their very small surface charge, empirical evidence suggests that iron nanoparticles tend to undergo strong aggregation and thereby form much larger chains, rings, or assemblages of nZVI particles. Stable dispersions of iron particles can be achieved by modifying the iron surface or using microemulsion mixtures to protect nanoparticles. The surface-modified (i.e. charged) iron nanoparticles can potentially remain in suspension for extended periods.[35] More recently, He and Zhao[36] reported the use of a new starch-based stabilizer which facilitated the suspension of nZVI/Pd in water for days without precipitation or agglomeration. Only a stable dispersion of iron particles offers the possibility of facile injection, mixing, and rapid transport within the slow-flowing groundwater.

Thus far, no studies on the eco-toxicity of low-level iron in soil and water have been reported in the peer-reviewed literature. However, it is the authors' opinion that systematic research on the environmental transport, fate, and eco-toxicity of nanomaterials is needed to help overcome increasing concerns in the environmental use of nanomaterials; so as to minimize any undesirable consequences. The use of iron nanoparticles in such studies could actually provide a valuable opportunity to demonstrate their expected positive overall effect on environmental quality. Iron is the fifth most used element; only hydrogen, carbon, oxygen, and calcium are consumed in greater quantities during our daily activities. Iron is also typically found at the active center of many biomolecules and plays a key role in oxygen transfer and other life-essential biochemical processes. Iron in the body is mostly present as iron porphyrin or heme proteins, which include hemoglobin in the blood, myoglobin, and the heme enzymes.[37] The heme enzymes permit the reversible combination with molecular oxygen. By this mechanism, red blood cells carry oxygen from one part of the body to another. Nature has evolved highly organized systems for iron uptake (e.g. the siderophore-mediated iron uptake), transport (via transferrins), and storage of iron (e.g. in the form of ferritin). The intracellular concentration of free iron (i.e. iron not bound to organic ligands) is tightly controlled in animals, plants, and microbes. This is partly a result of the poor solubility of iron under aerobic conditions in aqueous solution: the solubility production of $Fe(OH)_3$ is 4×10^{-38}. Within the cell, free iron generally precipitates as polymeric hydroxides. The challenge for us is to determine the transport, aquatic, and intraorganism biochemical fate of manufactured iron nanoparticles.

In the early phase of our work (prior to 2000), no commercial vendors or manufacturing techniques existed for the production of iron nanoparticles. Small quantities (< 10 kg) were prepared on a batch-wise basis in laboratory for bench- and pilot-scale tests. Not surprisingly, the production cost was rather high ($> \$200$/kg). Over the past few years, several vendors have emerged and are in the process of developing and scaling up different nZVI manufacturing techniques. These include both bottom-up chemical synthetic schemes similar to the borohydride method used at Lehigh as well as more advanced top-down particle-size reduction approaches. Consequently, the price is coming down rapidly (e.g. $\ll \$50/\text{kg}$ as of mid-2005). Even with the increased supply and reduced price, the general perception is that the nanoparticles are still too expensive for real-world applications. A careful analysis suggests the contrary, that iron nanoparticles potentially offer superior economics based on their large specific surface area. The specific surface area (SSA) can be calculated according to Equation (2.16) as follows:

$$SSA = (Surface\ Area/Mass) = \frac{\pi d^2}{\rho \frac{\pi}{6} d^3} = \frac{6}{\rho d}, \qquad (2.16)$$

where ρ is the density of the particle, $7.8\,\text{kg/m}^3$. With diameters on the order of $0.5\,\text{mm}$, the theoretical SSA of granular iron used in typical PRBs can be calculated to be approximately $1.5\,\text{m}^2/\text{g}$. In contrast, $50\,\text{nm}$ nZVI particles would exhibit SSA values on the order of $15,000\,\text{m}^2/\text{kg}$. Although granular iron is considerably less expensive that nZVI, approximately $\$0.50/\text{kg}$ versus $\$100/\text{kg}$, the cost-effectiveness expressed in terms of surface area per dollar is dramatic: $3\,\text{m}^2/\text{dollar}$ for granular iron versus $150\,\text{m}^2/\text{dollar}$ for nZVI.

Research has demonstrated that iron nanoparticles have much higher surface activity per unit of surface area. For example, the surface-area-normalized reaction rates for TCE and DCE dechlorination for iron nanoparticles are one to two orders of magnitude higher than bulk ($\sim 10\,\mu\text{m}$) iron powders.[38] Accordingly, nZVI particles have much lower cost per unit surface area and consequently should offer much higher returns in terms of performance per unit iron-mass. The key hurdle for the application of nZVI particles is perhaps not the cost, but demonstrating the field efficacy of the technology at field scale.

References

1. T. Masciangioli and W.-X. Zhang, *Environ. Sci. Technol.* **37**, 102A–108A (2003).
2. D. W. Elliot and W.-X. Zhang, *Environ. Sci. Technol.* **35**, 4922–4926 (2001).
3. R. Glazier, R. Venkatakrishnan, F. Gheorghiu and W.-X. Zhang, *Civil Engineering* **73**, 64–69 (2003).
4. R. W. Gillham and S. F. O'Hannesin, *Ground Water* **32**, 958–967 (1994).
5. L. J. Matheson and P. G. Tratnyek, *Environ. Sci. Technol.* **28**, 2045–2053 (1994).
6. B. Deng, D. R. Burris and T. J. Campbell, *Environ. Sci. Technol.* **33**, 2651–2656 (1999).

7. J. Farrell, M. Kason, N. Melitas and T. Li, *Environ. Sci. Technol.* **34**, 514–521 (2000).

8. T. L. Johnson, M. M. Scherer and P. G. Tratnyek, *Environ. Sci. Technol.* **30**, 2634–2640 (1996).

9. W. F. Wüst, O. Schlicker and A. Dahmke, *Environ. Sci. Technol.* **33**, 4304–4309 (1999).

10. R. M. Powell, R. W. Puls, S. K. Hightower and D. A. Sabatini, *Environ. Sci. Technol.* **29**, 1913–1922 (1995).

11. D. P. Siantar, C. G. Schreier, C.-S. Chou and M. Reinhard, *Water Res.* **30**, 2315–2322 (1996).

12. J. F. Devlin, J. Klausen and R. P. Schwarzenbach, *Environ. Sci. Technol.* **32**, 1941–1947 (1998).

13. J. Cao, L. Wei, Q. Huang, S. Han and L. Wang, *Chemosphere* **38**, 565–571 (1999).

14. J. Singh, S. D. Comfort and P. J. Shea, *Environ. Sci. Technol.* **33**, 1488–1494 (1999).

15. G. D. Sayles, G. You, M. Wang and M. J. Kupferle, *Environ. Sci. Technol.* **31**, 3448–3454 (1997).

16. E. J. Weber, *Environ. Sci. Technol.* **30**, 716–719 (1996).

17. D. R. Burris, T. J. Campbell and V. S. Manoranjan, *Environ. Sci. Technol.* **29**, 2850–2855 (1995).

18. R. M. Allen-King, R. M. Halket and D. R. Burris, *Environ. Sci. Technol.* **16**, 424–429 (1997).

19. EnviroMetal Technologies, Inc., http://www.eti.ca (2005).

20. V. L. Snoeyink and D. Jenkins, *Water Chemistry* (John Wiley & Sons, New York, NY, 1980).

21. D. Langmuir, *Aqueous Environmental Geochemistry* (Prentice Hall, Upper Saddle River, NJ, 1997).

22. T. M. Vogel, C. S. Criddle and P. L. McCarty, *Environ. Sci. Technol.* **21**, 722–736 (1987).

23. G. N. Glavee, K. J. Klabunde, C. M. Sorensen and G. C. Hadjipanayis, *Inorg. Chem.* **34**, 28–35 (1995).

24. C.-B. Wang and W.-X. Zhang, *Environ. Sci. Technol.* **31**, 2154–2156 (1997).

25. H.-L. Lien and W.-X. Zhang, *J. Environ. Eng.* **125**, 1042–1047 (1999).

26. W.-X. Zhang, *J. Nanoparticle Res.* **5**, 323–332 (2003).

27. W. Stumm and J. J. Morgan, *Aqueous Chemistry*, 3rd edn. (John Wiley & Sons, Inc., New York, 1996).

28. J. H. Brewster, *J. Am. Chem. Soc.* **76**, 6361–6363 (1954).

29. T. Li and J. Farrell, *Environ. Sci. Technol.* **34**, 173–179 (2000).

30. Y. Xu and W.-X. Zhang, *Ind. Eng. Chem. Res.* **39**, 2238–2244 (2000).

31. J. Cao and W.-X. Zhang, *J. Hazard. Mater.* **B132**, 213–219 (2006).

32. J. Cao, D. W. Elliott and W.-X. Zhang, *J. Nanoparticle Res.* **7**, 499–506 (2005).

33. E. T. Urbansky, *Perchlorate in the Environment* (Kluwer Academic/Plenum Publishers, New York, 2000).

34. B. E. Logan, *Environ. Sci. Technol.* **35**, 482A–487A (2001).

35. B. Schrick, B. W. Hydutsky, J. L. Blough and T. E. Mallouk, *Chem. Mater.* **16**, 2187–2193 (2004).
36. F. He and D. Zhao, *Environ. Sci. Technol.* **39**, 3314–3320 (2005).
37. D. P. Ballou (ed.), *Essays in Biochemistry: Metalloproteins* (Princeton University Press, Princeton, NJ, 1999).
38. H.-L. Lien and W.-X. Zhang, *Colloids Surf. A: Physicochem. Eng. Asp.* **191**, 97–106 (2001).
39. J. Cao and W.-X. Zhang, *J. Mater. Res.* **20**, 3238–3243 (2005).
40. H.-L. Lien and W.-X. Zhang, *J. Environ. Eng.* **131**, 4–10 (2005).

Chapter 3

Synthesis, Characterization, and Properties of Zero-Valent Iron Nanoparticles

Donald R. Baer*, Paul G. Tratnyek[†], You Qiang[‡],
James E. Amonette*, John Linehan*, Vaishnavi Sarathy[†],
James T. Nurmi[†], Chongmin Wang* and J. Antony[‡]

*Pacific Northwest National Laboratory, Richland, WA, USA

[†]Department of Environmental Science and Engineering
Oregon Health Science University, Beaverton, OR, USA

[‡]Department of Physics, University of Idaho, Moscow, ID, USA

3.1 Introduction

The chemical reactivity of nanometer-sized materials can be quite different from that of either bulk forms of a material or the individual atoms and molecules that comprise it. Advances in our ability to synthesize, visualize, characterize, and model these materials have created new opportunities to control the rates and products of chemical reactions in ways not previously possible. One application of this sort of tuning of materials for optimum properties is in remediation of environmental contamination. In fact, it has been suggested that nanotechnology applications for energy production and storage, and water treatment and remediation, will prove to be the number one and number three most important applications of nanotechnology for developing countries.[1]

Nanoparticles of various types (metal, oxide, semiconductor, or polymer) can be applied in a variety of ways to deal with environment-related issues. They may be used for their sorption properties,[2,3] or for their ability to facilitate chemical reactions, including those involving catalytic or photocatalytic behaviors.[4] Particles may be distributed by transport in solution[5] or immobilized in a polymer composite[2] or some type of membrane.[6]

Nanoparticles that rely on surface sorption sites have a high but finite storage capacity because of the large surface area of nanoparticles. For particles with catalytic properties, the desirable behaviors may continue indefinitely, assuming that some type of poisoning process does not occur. A third type of chemistry is where the particle is a reactant that reacts with contaminants, transforming them into more desirable products. The main example of this is zero-valent iron which reduces many contaminants, resulting in oxidation of the particle. In general, reaction occurs until all of the accessible metal is oxidized, and this determines the useful lifetime of such materials. For nanoparticles, all of the metal contained in the particles can be oxidized, so its capacity to reduce contaminants is limited by the total quantity of metal.[7] In other cases, especially for larger particles, they may become passivated by a shell of unreactive material which prevents the metal core from reacting further. In this case, the capacity of iron to reduce contaminants will be limited by the rate and degree of passivation.

Zero-valent iron (ZVI), including non-nanoparticle forms for iron, is one of the most promising remediation technologies for the removal of mobile chlorinated hydrocarbons and reducible inorganic anions for groundwater.[8–10] Suggestions that iron nanoparticles have enhanced reactivity and may be relatively easily delivered to deep contamination zones[5] have great appeal to the engineering community and summaries of field demonstrations have already been completed and described in the literature.[11,12] In addition to enhanced reactivity and particle mobility in the sub-surface, iron nanoparticles potentially offer the ability to select the reaction products formed. The prospect for better remediation technologies using nanoparticles of iron, iron oxides, and iron with catalytic metals (i.e. bimetallics) has potentially transformative implications for the environmental management of contaminated sites around the world. Of particular interest is the potential to avoid undesirable products from the degradation of organic contaminants by taking advantage of the potential selectivity of nanoparticles to produce environmentally benign products.

Although ZVI nanoparticles may have great potential to assist environmental remediation, there are significant scientific and technological questions that remain to be answered. Work in our research groups has confirmed that different types of nanoparticulate iron do lead to different reaction pathways.[13] However, the factors controlling the pathway are not yet understood and therefore not readily controlled. We need to know more about the characteristics of the nanoparticles and what properties most influence the reaction pathways. Because of the high reactivity of iron in many environments, iron nanoparticles are either precoated with some type of protecting layer or such a layer forms upon exposure to air or water. This outer layer is sometimes called the shell, leading to a core-shell structure with the metal being in the core. Many of the properties of these core-shell nanoparticles will be controlled or at least moderated by the transport and chemical nature of the shell. Thus core-shell nanoparticles containing ZVI are inherently more complex than the "pure" single-material non-reactive nanoparticles that are often studied. Among the complications are the time dependence and environmental dependence of the particle composition, structure, and properties.

Understanding reactive metal core-shell nanoparticles requires the use of particles that are as well characterized and understood as possible. We have, therefore, undertaken a series of studies that include synthesis of particles, analytical characterization of the particles, and finally measurements of their chemical properties. Current efforts include understanding how the particles evolve in time during storage and when reacting with contaminants in aqueous solution. The remainder of this chapter describes the approaches we (and others) have used and are extending to determine and ultimately control properties of reactive metal nanoparticles for environmental remediation.

3.2 Synthesis

The terms "top-down" and "bottom-up"[14] have become relatively common descriptions for an engineering approach of making small objects from larger ones (top-down) or the building up of larger objects from smaller ones (bottom-up). Although most nanoparticles are made by self-assembly processes, it is possible to use grinding and ball-milling processes to produce nanostructured particles, including metallic iron nanoparticles.[15-18] Because they are more common, we focus here on a limited set of self-assembly methods.

Many different self-assembly approaches can be used to produce metallic iron nanoparticles, including wet chemical and gaseous or vacuum routes. The different approaches often produce particles with variations in physical shape and coating structure. Vacuum-based methods can allow a high level of control of the particle composition and some ability to alter the particle coatings. The coatings on vacuum- or gaseous-formed particles are often relatively well-defined oxide structures. Such particles are well-suited for fundamental mechanistic studies. However, they are difficult to produce in large quantities that would be required for actual remediation work. In all cases, a variety of materials and shapes may appear and impurities may be introduced. A close coupling of synthesis and characterization is required to determine the true nature of the particles actually produced.

Solution synthesis and the hydrogen reduction of oxide nanoparticles have been used to produce enough nanoparticles for in-ground remediation tests. These particles may be less well-defined in several ways, including shape and size uniformity, and many have coatings that contain a residue from or organics deliberately added during the synthesis process. The oxides formed on solution produce particles that often appear to be less crystalline and may contain elements in addition to iron and oxygen. We have used different types of synthesis processes to allow studies that enable us to address fundamental questions as well as understanding the important properties of the types of particles used in field studies.

3.2.1 *Vapor-phase nanoparticle synthesis*

Nanoparticle synthesis is an important component of many rapidly growing research efforts in nanoscale science and engineering. They are the starting point for many bottom-up approaches for preparing nanostructured materials and devices. Nanoparticles of a wide range of materials can be prepared by a variety of methods. This section describes methods for preparing nanoparticles in the vapor phase. There are significant variations in the nature and quantities that can be prepared by each method. The production method must be matched to the need. Mechanistic studies often require very well-defined and uniform materials while applications need an adequate supply of material that functions in the appropriate manner. Detailed information can be found in the book edited by Granqvist *et al.*[19]

In vapor-phase synthesis of nanoparticles, conditions are created where the vapor-phase mixture is thermodynamically unstable relative to the formation of the solid material to be prepared in nanoparticulate form. This usually involves generation of a supersaturated vapor. If the degree of supersaturation is sufficient and the reaction condensation kinetics permits, particles will nucleate homogeneously. Once nucleation occurs, the remaining supersaturation can be relieved by condensation or reaction of the vapor-phase molecules on the resulting particles, and particle growth will occur rather than further nucleation. Therefore, to prepare small particles, one wants to create a high degree of supersaturation, thereby inducing a high nucleation density, and then immediately quench the system, either by removing the source of supersaturation or slowing the kinetics, so that the particles do not grow. In most cases, this happens rapidly taking only milliseconds in a relatively uncontrolled fashion, and lends itself to continuous or quasi-continuous operation.

3.2.1.1 *Flame synthesis*

The most commercially successful approach to nanoparticle synthesis — producing millions of tons per year of carbon black and metal oxides — is flame synthesis. One carries out the particle synthesis within a flame, so that the heat needed is produced *in situ* by the combustion reactions. However, the coupling of the particle production to the flame chemistry makes this a complex process that is rather difficult to control. It is primarily useful for making oxides since the flame environment is quite oxidizing. Recent advances are expanding flame synthesis to a wider variety of materials and providing greater control over particle morphology. Janzen and Roth[20] recently presented a detailed study of flame synthesis of Fe_2O_3 nanoparticles, including a comparison of their results to a theoretical model.

3.2.1.2 *Chemical vapor synthesis*

In analogy to the chemical vapor deposition (CVD) processes used to deposit thin solid films on surfaces, vapor-phase precursors during chemical vapor synthesis are brought into a hot-wall reactor under conditions that favor nucleation of particles in the vapor phase rather than deposition of a film. This method has tremendous flexibility in producing a wide range of materials and can take advantage of the huge database of precursor chemistries that have been developed for CVD processes. The precursors can be solid, liquid, or gas at ambient conditions, but are delivered to the reactor as a vapor (from a bubbler or sublimation source, as necessary). There are many good examples of the application of this method in the recent literature.[21,22]

3.2.1.3 *Physical vapor synthesis*

Four different physical vapor deposition processes are described: inert gas condensation, pulsed laser ablation, spark discharge generation, and sputtering gas-aggregation.

3.2.1.3.1 Inert gas condensation

This straightforward method is well suited for the production of metal nanoparticles, since many metals evaporate at reasonable rates at attainable temperatures. By heating a solid to evaporate it into a background gas, one can mix the vapor with a cold gas to reduce the temperature, and produce nanoparticles. By including a reactive gas, such as oxygen, in the cold gas stream, oxides or other compounds of the evaporated material can be prepared. Wegner *et al.*[23] presented a systematic experimental and modeling study of this method, as applied to the preparation of bismuth nanoparticles, including both visualization and computational fluid dynamics simulation of the flow fields in their reactor. They clearly showed that they could control the particle size distribution by controlling the flow field and the mixing of the cold gas with the hot gas carrying the evaporated metal.

3.2.1.3.2 Pulsed laser ablation

Rather than simply evaporating a material to produce supersaturated vapor, one can use a pulsed laser to vaporize a plume of material that is tightly confined, both spatially and temporally. This method can generally only produce small amounts of nanoparticles. However, laser ablation can vaporize materials that cannot readily be evaporated.

3.2.1.3.3 Spark discharge generation

Another means of vaporizing metals is to charge electrodes made of the metal to be vaporized in the presence of an inert background gas until the breakdown voltage is reached. The arc (spark) formed across the electrodes then vaporizes a small amount of metal. This produces very small amounts of nanoparticles, but does so relatively reproducibly. Weber et al.[24] recently used this method to prepare well-characterized nickel nanoparticles for studies of their catalytic activity in the absence of any support material. By preparing the nanoparticles as a dilute aerosol they were able to carry out reactions on the freshly prepared particles while they were still suspended. Metal oxides or other compounds can be prepared by using oxygen or another reactive background gas.

3.2.1.3.4 Sputtering gas-aggregation

A schematic view of a newly developed sputter-gas-aggregation (SGA) nanocluster deposition apparatus[25-27] that has been used to produce Fe nanoparticles important to our research is shown in Figure 3.1.[29] We label materials produced by this process as Fe^{SP}. The system uses a combination of magnetron-sputtering and gas-aggregation techniques. The cluster beam deposition apparatus is mainly composed of three parts: a cluster source, an e-beam evaporation chamber, and a deposition chamber. The sputtered Fe atoms from a high-pressure magnetron-sputtering gun are decelerated by collisions with Ar gas (the flow rate: 100–500 sccm) injected continuously into the cluster growth gas-aggregation chamber, which is cooled by chilled water. The clusters formed in this chamber are ejected from a small nozzle by differential pumping and a part of the cluster beam is intercepted

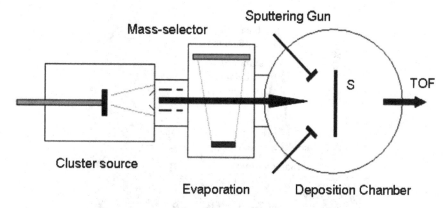

Figure 3.1. Schematic drawing of a three-chamber nanocluster deposition system. This is a third generation particle source with increased particle production rates. Particles are produced in the cluster source and mass selector areas. They can be coated or reacted with a gas in the evaporation chamber and can be exposed to gas and collected on a substrate in the deposition chamber. After Ref. 28.

by a skimmer, and then deposited onto a sample holder in the deposition chamber. The mean size of clusters, from 1 nm to 100 nm, is easily varied by adjusting the aggregation distance, the sputter power, the pressure in the aggregation tube, and the ratio of He to Ar gas flow rate. The aggregation distance and the ratio of He to Ar gas flow rate are important parameters for getting a very high-intensity (> 5 mg/h), monodisperse nanocluster beam. A major advantage of this type of system is that the clusters have a much smaller size dispersion than grains obtained in a typical vapor deposition system. Studies with transmission electron microscopy (TEM) and time-of-flight (TOF) mass spectrometry have shown that the observed lognormal size distribution has a standard deviation of below 10%. When oxygen gas is introduced into the deposition chamber during processing, uniform iron oxide shells covering the Fe clusters are formed. For a constant flow of Ar gas, the gas pressure in the deposition chamber can be adjusted by changing the flow rate of oxygen gas (O_2). In this circumstance, the iron clusters have an oxide shell before they are collected on a substrate. The clusters land softly on the surface of substrates at room temperature and retain their original shape. This process ensures that all Fe clusters are uniformly oxidized before the cluster films are formed.

TEM images of the core-shell Fe cluster assemblies deposited on a TEM grid are shown in Figure 3.2. In this example the mean cluster diameter (D) is 6.65 nm and the standard deviation is less than 7%. High-resolution TEM (HRTEM) images of four Fe nanoparticles with sizes ranging from

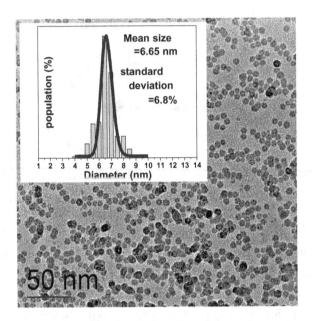

Figure 3.2.   TEM image and size distribution of iron/iron oxide nanoclusters deposited on a TEM grid by the SGA nanocluster source. The inset shows the size distribution of the particles as determined from TEM measurements. After Ref. 29.

3 nm to 85 nm deposited on a carbon film of microgrids are shown in Figure 3.3. The smaller nanoparticles appear almost spherical while a crystal shape begins to appear for larger particles. The smallest particle is fully oxidized as are all of the air-exposed particles with a diameter of 7 nm or less. The dark center is an iron core region while the light-gray shell is oxide. To the extent we can measure by TEM, the oxide shells on these particles do not significantly grow when exposed to air.

3.2.2 Solution-phase synthesis

Incipient wetness is one of the oldest and most time-honored methods of producing nanoparticles on surfaces, especially for catalytic purposes.[30] In a typical procedure, a solid support is "wetted" with a metal salt solution, the solvent is removed, and the metal ions are treated to produce the desired materials (metals, metal oxides, metal sulfides). The morphology of the particles produced depends upon the metal salts, solvents, solution ionic strength, co-precipitation agents (if any), temperature, and supports. The typical particles can be up to micron-sized and polydisperse, but for

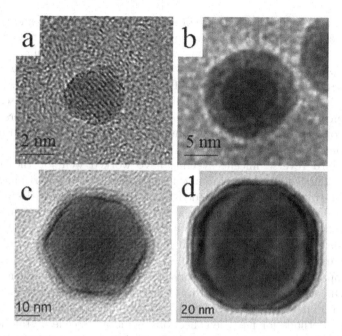

Figure 3.3.   HRTEM micrographs of the Fe clusters with diameters from around 3 nm to 85 nm prepared on carbon microgrids by the SGA cluster source. The smaller particles are round while crystallographic faces begin to appear at larger sizes. Particles below 8 nm are fully oxidized. All particles have an oxide shell. After Ref. 28.

precious metals with low loadings the particles can be nanometer-sized and surprisingly monodisperse.

3.2.2.1 *Micellar techniques*

Micellar techniques, which include normal oil-in-water microemulsions and reverse water-in-oil systems, are intuitively attractive methods of controlling particle size by controlling the size of the microemulsion droplets in which the particles are formed.[31,32] These nanoreaction vessels are tailored for core sizes up to 50 nm. In micellar and reverse micellar systems, a surfactant (or amphiphilic compound) is dissolved in a continuous phase and the second phase is mixed with the dissolved surfactant to form a thermodynamically stable, optically transparent, homogeneous solution. This solution actually consists of monodisperse droplets of the second phase in the nanometer to hundred-nanometer size range. The size of the droplets and the shape of the droplets are controlled by the ratio of the second phase to the surfactant.

Reverse micelles allow the use of aqueous soluble reagents and aqueous reaction chemistries to produce nanoparticulates. The water-to-surfactant ratio actually determines the shape and the size of the micellar droplets. Ternary phase diagrams, such as those found in Darab et al.,[33] are maps of the phase stability and can be maps of the shape and size of the micelle core. Working in the non-spherical region of the phase diagram allows researchers to produce mesoporous templates as described in other chapters. Here, we are concerned primarily with spherical aqueous droplets in an organic continuous phase.

Standard reverse micelle systems have worked well for the production of milligram amounts of quantum crystallites.[34,35] Unfortunately the high-ionic-strength solutions needed to produce gram quantities of particles destabilize the reverse micelles. To counter this negative effect, co-surfactants have been successfully utilized to stabilize reverse micelles at high ionic strengths.[33] This method has allowed the production of multi-gram quantities of nanoparticles.[36]

Nanoparticles produced using micellar methods are usually monodisperse. One of the problems with the micellar method is particle collection as the particles produced want to stay suspended in the micellar solution. Methods for particle collection, including micelle destabilization, usually lead to particle agglomeration. The isolated particles from micellar methods are almost always coated to some degree with the surfactants. Almost any type of aqueous reaction can be adapted to the reverse micellar system to produce nanoparticles. The types of particles which can be produced include metals, metal oxides, metal sulfides, core-shell particles, mixed metal-containing materials, and organic coated particles. Supercritical CO_2 reverse micellar systems have also been used to produce nanoparticles.[37]

A variant on the micellar particle synthesis method is to produce nanoparticles in the presence of a coating material which can again be a surfactant, a polymer, or another reagent which attaches itself to the surface of the particle. This technique has been successfully exploited using thermal or sonochemistry to produce metallic particles from metallic carbonyl or strained organometallic precursors.[38,39] In these syntheses, milligram to gram quantities of coated nanoparticles can be produced. The coatings can be beneficial in that they can stabilize the small particle size and they stabilize the thermodynamically unstable species, i.e. nanometallics in the presence of oxygen. In some cases it is not necessary to add a surfactant as the particles produced are self-protecting. This is the case for the iron particles produced by Zhang and co-workers using an aqueous/ethanol borohydride reduction of ferrous salts.[40,41] In this case, a boron-rich layer is formed on the iron metal core which passivates the particles to growth and oxidation.

We have examined some of this material produced by the Zhang group and material we have synthesized at the Pacific Northwest National Laboratory (PNNL). This material is labeled Fe^{BH} in our studies.

3.2.2.2 *Hydrothermal synthesis*

Hydrothermal synthesis either in batch or in continuous mode can be useful in producing multigram quantities and nanocrystalline materials. As the name implies, these reactions are run with metal salts dissolved in water at high temperatures in pressure vessels. These reactions produce metal oxides by hydrolysis of the metal ions. Batch reactions can yield highly crystalline materials but the long heat-up and cool-down times of large pressure vessels preclude the production of nanoparticulates. The ionic strength of the solution does determine to some extent the size of the crystals and particulates produced.

An alternative to the batch hydrothermal method is the rapid thermal decomposition of solutes (RTDS) method in which the metal ions in solution under pressure are subjected to hot reaction zones for very short times (seconds) and then immediately cooled as the superheated liquid is passed through a nozzle.[42] The RTDS method allows one to control the particle size of metal oxide products through temperature, reaction time, pressure, and by adding co-reactants to the reactant stream. This is an especially good method to produce mixed metal or doped species. For example, one can add 1% of a nickel salt to a ferric salt solution and obtain a 1% uniform nickel oxide doping of the iron oxide produced.

3.2.2.3 *Post-synthesis processing*

In all of the above methods (solution- and vapor-phase) one can produce particles which may or may not be reduced. Further processing, either through gas-phase reduction with hydrogen or another solution reduction step can lead to reduced particles, even to the point of producing metallic particles. During the subsequent steps, care must be taken to ensure minimal particle growth which usually entails mild reduction conditions. Residues from earlier synthesis or processing steps can have a significant influence on product shape and surface chemistry, increasing the need for particle characterization. In unique cases, the reduction can quickly follow the oxide particle formation. A commercial product produced by Toda Kogyo Corp. (Schaumburg, IL) known as RNIP is produced by the reduction of goethite and hematite particles with H_2 (200–600°C).[43] We have conducted some experiments on this type of material which is indicated by the label Fe^{H2} in our work.

3.3 Particle Handling and Characterization

In this section and the following one, we first show by example the application of some methods for *ex situ* characterization and then provide a summary of methods used to characterize reactions in solution. Before looking at specific techniques or examples, it is important to consider some of the general issues and challenges related to characterizing and measuring the properties of reactive metal nanoparticles. Readers unfamiliar with some of the basics of the methods described are referred to the *Encyclopedia of Materials Characterization* by Brundle, Evans and Wilson.[44]

3.3.1 *Challenges for nanoparticle handling and characterization*

3.3.1.1 *General issues*

Sample handling, environmental stability, and contamination issues that are sometimes problems for micron-scale or bulk materials are of significantly greater importance for nanoparticles. It is possible, for example, for nanopores to retain solvents that would evaporate from surfaces.[45] This solvent retention, which could occur during a standard solvent cleaning procedure before analysis, could alter sample analysis in several ways and lead to confusion in interpreting results. Surface contamination may be a minor issue for a bulk material but is of significant concern to nanoparticles. Even relatively stable (in comparison to metallic iron) nanoparticles may change in different environments. For example, the structure of ZnS nanoparticles has been shown by Zhang *et al.*[46] to change due to the presence or absence of water in the surrounding environment. This environmental effect is not an accident, but a characteristic of small systems: "the thermodynamics of small systems is highly dependent on the environment".[47] The size, structures, and even composition of nanoparticles will significantly depend on the history and processing of the particles. For nanoparticles, any set of physical and chemical measurements is incomplete without a relatively complete history of the sample handling and processing.

In addition to environmental effects and specimen history, there is a more general instability or dynamic behavior of nanoparticles. The existence of many local minima for nanostructures, and the shape and physical structure of nano-sized objects may be readily altered by any incident radiation, including incident electron[48] or X-ray flux.[49] TEM images of Au nanoparticles show that the structure is not constant but varies with time.[48] This dynamic behavior of nanoparticles can be important and it has been shown, as in one example, that the changes in the structure of catalyst nanoparticles as a function of time (at somewhat higher than ambient temperatures) are an important property of catalyst particles during the growth of carbon nanotubes (CNTs).[50] Particles large enough to minimize the shape changes are not useful catalysts for the CNTs.

It is also useful to recognize that some electronic and magnetic characteristics of individual nanoparticles are altered by close proximity to other nanoparticles. Proximity effects might be expected for magnetic interactions. As one example, inter-particle interactions are observed to significantly change the superparamagnetic relaxation behavior of NiO nanoparticles.[51] However, proximity effects also influence electronic properties of nanoparticles. For instance, Au nanoparticles in dilute solution have a size-dependent surface plasmon resonance that can be observed in UV-Vis absorption measurements. When Au particles reversibly aggregate in solution, an additional extended plasmon band is observed in addition to the size-dependent surface plasmon peak.[52] Packing or dispersing nanoparticles for measurement purposes may in some circumstances alter the properties being measured.

3.3.1.2 *Special issues for reactive metals*

Many of the issues noted above are present for most nanoparticles. However, because of their chemical reactivity, iron and other reactive metal nanoparticles give rise to additional complexity as they react to solutions (or other environments) to which they are exposed, including the chlorinated hydrocarbons of particular interest. Conceptually it is sometimes useful to recognize that passive corrosion films may react or evolve with time. This might be called an intra-particle effect since it does not necessarily involve extensive interaction with the surrounding environment (e.g. the so-called autoreduction between core and shell material[53]). In addition, the particles will interact with the environment, including both major components of the environment (e.g. water and water components) and contaminants (usually relatively minor species in the water). These might be called extra-particle interactions. These different types of interactions may occur at different rates and with different sensitivities to environmental conditions, so their distinctions can be useful, even if they are all — ultimately — interrelated.

To a significant degree, time is an additional variable not independent of environmental effects. Since metallic iron is thermodynamically unstable in many environments, it is likely that with long-term storage there will be additional oxidation of the iron or other changes in the material. Therefore, properties of the nanoparticles will likely vary somewhat as a function of time. Although a small amount of corrosion may not

significantly alter the bulk behavior of a piece of iron, the same amount
of corrosion may significantly alter the nature of nanoparticles. Although
we generally store and handle specimens in relatively inert environment
(anoxic, dry, and/or vacuum), we often see time-dependent changes in the
particles.

Even though the oxides that formed on the sputtering-produced iron
nanoparticles, as mentioned earlier, can be described as stable, "stable" is
only a relative term. It is nearly impossible to produce a metallic particle
with no coating and the particles have a tendency to continue to oxidize
over time. We have examined solution-formed particles (with their oxide
coatings) using surface-sensitive techniques with and without exposure to
air; and found an additional amount of oxidation for samples exposed to
air for even a few minutes. Questions about environmental stability frequently require careful sample preparation and handling so as to minimize
the effects of air exposure or time on the analysis results. Control specimens
and other test procedures may be needed to provide some bounds on the
environmental effects on the results.

3.3.1.3 *Sample preparation and handling*

It is safe to say that any characterization tool that has been applied to any
type of material is currently (or soon will be) applied to nanostructured
materials. Many of the tools useful for the most detailed characterization
of materials, surfaces, and nanoparticles require removing the material from
the environment of interest to an atmosphere useful for the analysis. For
example, electron microscopy and spectroscopy methods usually require
moving the material into vacuum. Although there should always be concern about the impacts of the change in environment, much critical useful
information is obtained from these *ex situ* methods. A few methods, such
as optical spectroscopy, can more easily be applied in the environment of
interest (or something similar). These *in situ* methods are very useful, but
often provide only a subset of the information needed. Ideally it would be
possible to collect all of the information we would like to obtain about a
nanoparticle in the environment of interest as a function of time. Since
this is not possible we must combine the use of *in situ* and *ex situ* methods to piece together information about the particles. Since each technique
has limitations and no one method provides all the desired information,
it is helpful to know the strengths and limitations of each experimental
tool, including the possible effects of sample handling and probe-induced
damage. It is hard to overemphasize the importance of applying multiple
and complementary methods for obtaining an accurate understanding of
nanoparticles.

The reactivity of metallic iron makes sample preparation and handling an important challenge. In general, we presume that these materials will be most stable when stored dry and under an inert (anoxic) atmosphere. However, some samples are received (or synthesized) in water or in air. The wet samples must be dried for analysis by some characterization methods (e.g. TEM) and the dry samples must be immersed to allow for other characterization methods (e.g. electrochemistry). All of these manipulations will affect the results of characterization and not necessarily in a consistent or (at this time) predictable way. To manage this issue, we have adopted a protocol for removing nanoparticles from solutions that we call flash drying, and applied this to most of the wet samples we have received or synthesized.[54] Then, when necessary, the flash-dried powder can be weighed and re-immersed in solution — along with samples that were received or synthesized dry. This process made the samples easier to handle and process and enhanced the reproducibility of the measurements without altering the properties (physical or chemical) significantly.[13] Additional comments about specific specimen handling methods associated with specific methods are discussed below.

3.3.2 *Ex situ examples of iron nanoparticle characterization*

The techniques to be used and the order in which they are used depend upon the questions being asked. When characterizing nanoparticles, natural questions include size, size distribution, and surface area; surface and bulk composition and chemical state; electronic and physical structure, including the presence of defects. General characteristics of these materials and other samples used in much of our work are included in Table 3.1. In Table 3.2 we list a set of the techniques applied for the characterization of iron nanoparticles and a rough indication of the type of information they can be used to obtain. The lighter symbols indicate that the information is not a primary result of the method. The "X" indicates that we routinely apply the technique in our work and that examples of the use will be presented in the text.

To demonstrate some of the similarities and differences between different types of nanoparticles and how these differences can be identified, several examples are provided from three types of iron nanoparticles that we have examined most extensively.[13] Although the methods have not been described in great detail, each of these materials was mentioned in the synthesis section. For ease of identification, they are labeled in the following way: Materials produced by the cluster sputter process are identified as Fe^{SP}; nanoparticles produced by the borohydride-reduction process are identified as Fe^{BH}; and particles produced by the H_2 reduction of iron oxides are labeled Fe^{H2}.

Table 3.1.   Nanoparticles and characteristics.

Name	Source	Method	Particle Size (dia.)	Surface Area	Major Phase	Minor Phase
Fe^0/ Disc		High purity polished Fe^0			Fe^0	
Fe^{H2}	Toda America, Inc.	High temp. reduction of oxides with H_2	70 nm	$29\,m^2/g$	α-Fe^0	Magnetite
Fe^{BH}	W.-X. Zhang, Lehigh Univ.	Precip. w/$NaBH_4$	10–100 nm	$33.5\,m^2/g$	Fe^0	Goethite, wüstite
Fe^{EL}	Fisher Scientific	Electrolytic	150 μm	0.2–$1\,m^2/g$	99% Fe^0	
Fe^{SP}	Y. Qiang, Univ. Idaho	Sputter deposition	2–100 nm		Fe^0	Magnetite, maghemite
Fe_3O_4	PNNL	Precip. from $FeSO_4$ w/KOH	30–100 nm	4–$24\,m^2/g$	Fe_3O_4	
Fe_2O_3	Nanotek, Corp.	Physical vapor synthesis (PVS)	23 nm	$50\,m^2/g$	γ-Fe_2O_3	

Adapted from Ref. 13.

3.3.2.1   *Transmission electron microscopy (TEM)*

Transmission electron microscopy is an essential tool for characterization of the natural or synthesized nanoparticles.[55] When the particles are deposited at relatively low density on a TEM grid, it is possible to gain information about the size, shape, and structure of the particles. TEM images of different sized sputter-grown particles were shown earlier. With careful work, it is possible to examine the structure of the particles and particle coatings. Figure 3.4 shows a high-resolution TEM image of a relatively large Fe^{SP} nanoparticle where the crystal structure has an effect on the particle shape. At these larger sizes, the nanoparticles are transforming from round shapes (characteristic of smaller particles) to a more crystallographic structure. It is relevant to remember that the TEM images are a projection of the particle shape on one plane. Unless tomography measurements are taken or the diffraction patterns are modeled, the relationship between the two-dimensional projection and the three-dimensional shape is not necessarily certain. In this case, the diffraction image has been modeled and the data is

Table 3.2. Partial listing of techniques and information that can be obtained about nanoparticles or collections of nanoparticles.

Techniques	Information Obtained								
	Size & Shape	Surface Area	Grain Size	Physical Structure	Composition	Surface Chemistry	Electronic Structure	Molecular Structure	Chemical Reactivity
• Scanning Electron Microscopy + EDS	✓				✓				
• Transmission Electron Microscopy + EDS + EELS	✗	✗	✓	✓	✓		✓		
• X-ray Diffraction			✓	✓					
• Small Angle X-ray Scattering	✓								
• Gas Sorption (BET)		✗							
• X-ray Photoelectron Spectroscopy					✗	✗	✓		
• Auger Electron Spectroscopy					✓	✓			
• X-ray Adsorption Spectroscopy and Microscopy					✓	✓	✓		
• Scanning Probe Microscopy	✗								

(*Continued*)

Table 3.2. (*Continued*)

Techniques	Information Obtained								
	Size & Shape	Surface Area	Grain Size	Physical Structure	Composition	Surface Chemistry	Electronic Structure	Molecular Structure	Chemical Reactivity
• Secondary Ion Mass Spectrometry						✓		✓	
• Raman Spectroscopy								✓	
• Fourier Transform Reflection Infrared Spectroscopy								✓	
• Light Scattering	✓								
• Mossbauer Spectroscopy				✓					
• Electron Paramagnetic Resonance									✗
• Electrochemical Methods									✗
• Batch Reaction Studies									✗

The symbols (✓ and ✗) indicate the type of information that can be obtained by the techniques listed. Lighter symbols indicate information that is not a primary result of the method. The ✗s indicate techniques that we routinely apply in our work and for which examples of use of the technique are presented in this chapter.

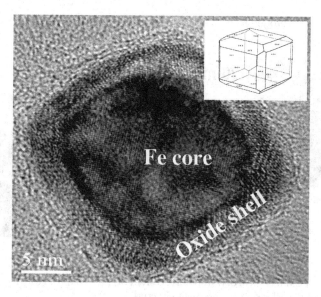

Figure 3.4. HRTEM image of a core-shell structured iron oxide nanoparticle. The inset shows the particle morphology following the Wulff construction. The particle is defined by six {100} planes and truncated slightly at the edges by the {110}-type planes.

consistent with a slightly truncated cube defined by {100} planes. The particle is also covered by an iron oxide shell. In other work, we have identified vacancies in the particles[56] that result from their oxidation.

TEM images of Fe^{H2} and Fe^{BH} materials removed from their as-received storage conditions (dry inert gas for this particular Fe^{H2} and in an alcohol solution for the Fe^{BH}) and quickly[57] inserted into the TEM to minimize sample changes are shown in Figure 3.5. The shapes of the Fe^{H2} particles, one example of which is shown in Figure 3.5(a), vary significantly and the particles are covered by a crystalline oxide. The Fe^{BH} particles appear (Figure 3.5b) almost as bubbles, usually round but of varying size. The coating on these particles does not appear to be crystalline.

The TEM images in Figure 3.5 are from material before exposure to water or water containing a chlorinated hydrocarbon. The TEM image of Fe^{H2} specimen exposed to water for 24 hours is shown in Figure 3.6. The oxide shell around the metal core appears to have become somewhat thicker and less uniform. There is also a growth of fibrous oxide — likely goethite — around the particles.

When an electron energy loss spectrometer is incorporated in an instrument, it is possible to obtain compositional and chemical information from the particles. Energy dispersive X-ray spectroscopy (EDS or EDX) can also

Hydrogen Reduced – Iron Nanoparticles – Dry Stored
As received

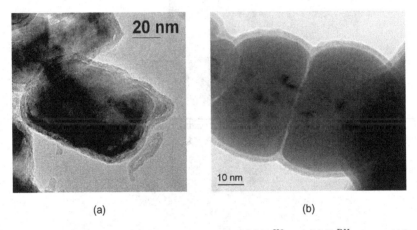

(a) (b)

Figure 3.5.   TEM images of dry, stored-as-received (a) Fe^{H2} and (b) Fe^{BH} nanoparticles. Both types of particles show a core-shell structure.

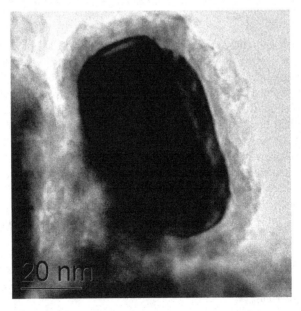

Figure 3.6.   TEM image of Fe^{H2} nanoparticle after 24 hours of exposure to DI water. The core-shell structure remains, but the shell appears thicker and less uniform. Fibrous iron oxide is observed around the metal particles.

be used to gather elemental information about the particles. When many images are collected and analyzed, it is possible to obtain information about particle size distribution. As discussed below, TEM measurements of particle size can be correlated to surface area determined by other methods.

There are several important challenges or limitations to the TEM measurements. First, the current densities used can damage specimens. We have observed that some particles shells can be damaged and evaporated by the electron beam while in other circumstances the shells can grow. There can also be a sample analysis bias or selectivity in TEM because TEM measurements can only be made with a specimen that is transparent to the electrons. Consequently, it can be useful to compare TEM results with those obtained by other methods to verify consistency. Inconsistent results may not be an indication of error, but provide useful information related to particle size distribution.

3.3.2.2 X-ray diffraction (XRD)

X-ray diffraction is also an important tool for analysis of the crystal structure and grain size of the nanoparticles. Specimens of Fe^{H2} paste or Fe^{BH} slurry were prepared for XRD analysis by spreading $\sim$25 mg of material on a zero-background slide and allowing it to dry for 2–3 days in an anoxic glove box that was continuously purged with N_2. In an attempt to protect them from oxidation in air during transport and the diffraction measurements, the dried specimens were coated with glycerol by applying a few drops of 10% glycerol in 95% ethanol, and allowed to dry for 1–2 additional days. The coated specimens were analyzed in ambient air using Cu-K_α radiation in a Philips X'Pert MPD diffractometer (PW3040/00) operated at 40 KVP and 50 mA. Continuous scans from 2–75 $°2\theta$ were collected at a scan rate of about 2.4 $°2\theta$ min^{-1}.

The XRD patterns for as-received Fe^{BH} and Fe^{H2} specimens are shown in Figures 3.7(a) and (b) respectively.[58] Both samples have a major peak at a 2-theta value of approximately 44°, characteristic of metallic iron. The difference in peak width is readily apparent. Based on the Scherrer equation, after correction for instrumental broadening, the relatively wide main diffraction peak from the Fe^{BH} samples is consistent with Fe^0 crystallites with an average grain size of 1.5 nm. There is also a broad diffraction peak that may be from goethite. The narrower peaks from the Fe^{H2} samples indicate Fe^0 with a 30 nm average grain size and magnetite with a 50 nm grain size.

To test for sample stability, successive scans were taken at 30 minutes, 60 minutes and 24 hours of exposure to ambient air. No change in the structural properties of the glycerol-coated specimens was observed, verifying

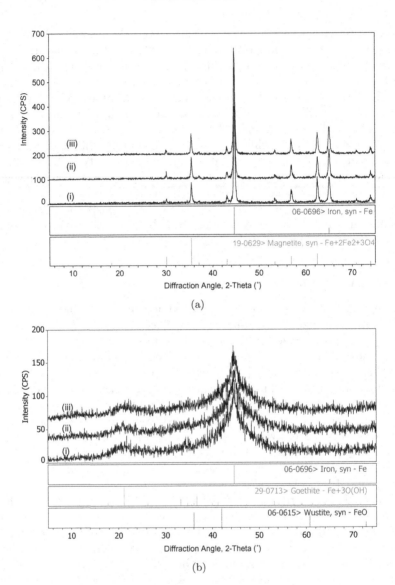

Figure 3.7. (a) X-ray diffraction pattern from Fe^{H2} particles received in a slurry. The patterns show the presence of iron metal with an average grain size of approximately 30 nm and magnetite with grain sizes of approximately 50 nm. (b) X-ray diffraction pattern from Fe^{BH} particles received in a solution. The patterns show the presence of iron metal with an average grain size of ~1.5 nm and possibly some goethite (after Ref. 13, Figure S1). The different scans (i)–(iii) in the figure were collected (i) 30 minutes, (ii) 60 minutes, and (iii) 24 hours after removal from the glove box as described in the text.

either that the protective effect of the glycerol film with respect to oxidation or that the drying process itself was sufficient to slow the reactivity of the nanoparticles over the time frame of the XRD measurement. Mean crystallite dimension was estimated using the Scherrer equation, after correction for instrumental broadening.

Useful information can be obtained by combining information about the two types of material from TEM and XRD data. The Fe^{H2} sample shown in Figure 3.5(a) has a particle size observed by TEM quite similar to the grain sizes estimated from the XRD peak width. This implies that the Fe^{H2} nanoparticles are highly crystalline with each particle made up of one or a few metal grains. In contrast, the Fe^{BH} particles as shown in Figure 3.5(b) are approximately the same size as observed in TEM, but they are made up of 1.5 nm metallic grains.

It is also possible to estimate the ratio of metal to oxide in a collection of materials. We have made XRD and TEM measurements on a series of Fe^{H2} samples (in collaboration with Greg Lowry at Carnegie Mellon University) that have been exposed to water for different periods of time. The fraction of Fe in the magnetite phase is plotted for both TEM and XRD data in Figure 3.8. Several approximate assumptions are needed in both TEM and

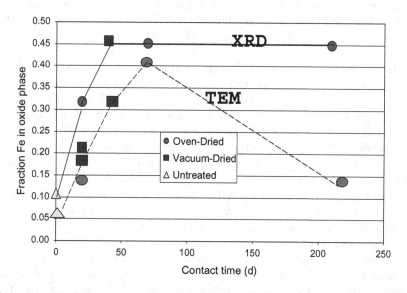

Figure 3.8. Fraction of measured Fe in the magnetite phase as a function of time exposed to water. Up to about 70 days, the TEM and XRD data are relatively consistent. The differences at longer times are likely due to selective sampling of the TEM. In particular, larger oxide particles are not transparent to the electron beam.

XRD to determine the fraction of Fe in magnetite. However, up to about 70 days, the trends in both types of data are very consistent. The longer time difference between XRD and TEM data is likely due to growth in size of the oxide grains to the point that they become opaque to the electron beam from the TEM.

The major limitations of XRD include sensitivity (the amount of material needed) and limitation to detection of crystalline phases. Micro-XRD can help minimize the amount of material needed for analysis. Small-angle X-ray scattering (SAXS) can provide information about particle size, shape, and pore size.

3.3.2.3 *X-ray photoelectron spectroscopy (XPS)*

X-ray photoelectron spectroscopy is particularly useful for examining shell composition and the presence of impurities in or on nanoparticles. Because of the surface sensitivity of XPS, extra care is required to minimize changes in analysis results due to contamination or air exposure. A glove bag has been connected to the spectrometer to allow transfer from a glove box to the glove bag from which the sample is inserted into the spectrometer. The Fe 2p photoelectron peaks from a Fe^{BH} sample transferred without air exposure and after air exposure are shown in Figure 3.9. The as-received material had a strong Fe^0 peak with an indication of Fe^{+2}. After air exposure the Fe^0 was still present, but significantly decreased in size. The surface iron was oxidized and a significant amount of Fe^{+3} was present.

In addition to looking for changes in surface chemical state, the XPS measurements provide an overall composition. The initial composition (in atomic %) of the as-received Fe^{BH} sample was 20% Fe, 49% O, 16% B, 15% Na, 0.5% S; while the initial composition of the Fe^{H2} was 51% Fe, 44% O, 3.0% Na, 1.9% S. The significant differences in the Fe are due to the presence of B in the coating of the Fe^{BH} material. The B is mostly oxidized, indicating that the outer shell of this material has a significant BO_3 component.

The combination of TEM, XRD, and XPS data indicates that even though both the Fe^{H2} and Fe^{BH} are made up of mostly Fe^0 particle tens of nanometers in size, their grain size, shell structures, and shell compositions are totally different.

Significant limitations of XPS include the need to place the samples in vacuum and to assemble a collection of nanomaterials to facilitate analysis. Usually, quantitative information extracted from XPS assumes that the sample is ideally flat. It is possible to combine TEM or other shape and size data to extract more elemental distribution information from XPS data through the use of particle and signal intensity models.[49,59]

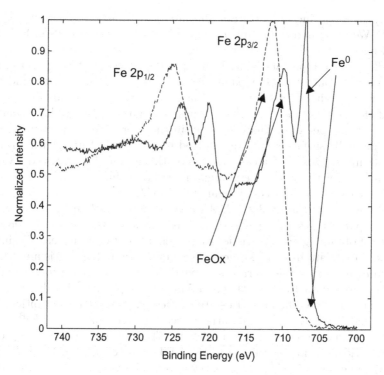

Figure 3.9. XPS 2p photoelectron peaks from Fe^{BH} material after anaerobic transfer (solid line) from the shipping container and the same sample after exposure to air for five minutes (dashed line).

3.3.2.4 *Surface area determination*

Particle size (and distribution) and the surface area associated with a collection of particles (or other nanostructured materials) are highly important and useful parameters regarding nanomaterials. The standard method for determining surface area is referred to as BET (Brunauer–Emmett–Teller); and involves N_2 (or Kr) physisorption and desorption as a function of heating a known weight of material. Although a common method, the procedures have sensitivity, pore volume, and other complications that can limit accuracy. Because heating is involved, it is possible for nanoparticles to sinter or grow during heating. For many types of materials, the final temperature of the BET analysis needs to be lower than that used for routine analysis. The most widely used alternative method of determining specific surface area is calculation — usually assuming spherical geometry — of surface areas from particle size distributions determined from TEM images. This approach also has advantages and disadvantages and rarely gives specific surface areas that are identical to those determined by BET. We compared the results obtained by these two methods in Ref. 13.

Consistent with the earlier discussion about particle stability, sample handling is an important issue. Damp powder placed on filter paper in air can start to burn the filter paper, clearly destroying the powder and the ability to obtain a surface area. We have found that surface area measurements for Fe^{BH} and Fe^{H2} stored as a paste or slurry appear to have smaller surface areas per gram than reported in the literature. In particular, the nominal Fe^{H2} surface areas are reported to be about $29\,m^2/g$ with a range of 4–$60\,m^2/g$. Our BET measurements on this material determined a value of $3\,m^2/g$. For the Fe^{BH}, the nominal literature value is $33\,m^2/g$ while our BET measured value was approximately $5\,m^2/g$.

As shown in Figure 3.2, TEM measurements are frequently taken to obtain particle size and particle size distribution. Because of the importance of obtaining good statistics to get reliable data, accurately obtaining surface area data from TEM measurements is very tedious. It is important to note that using the average size particle is not adequate to determine surface area data from TEM measurements, as particle size distribution can make a large difference. In spite of these significant uncertainties, our BET and TEM estimates of surface areas between one set of Fe^{BH} and Fe^{H2} material were relatively consistent. TEM surface area was approximately $7.5\,m^2/g$ in comparison to the BET value of $5\,m^2/g$. For Fe^{H2} the TEM value was $3.5\,m^2/g$ in comparison to a value of $3\,m^2/g$ from the BET measurement.

It is also possible to get particle size information from light scattering measurements, although there is uncertainty and some disagreements regarding the accuracy at lower particle sizes. We found that our iron particles tended to aggregate in the solution used for the light scattering measurements; and we primarily learned about the size of the aggregates. Small-angle X-ray scattering can be used to get average particle size, particle shape, and pore size information. Because of differences in the measurement process as well as sample handling, it is strongly recommended that multiple methods be used to obtain particle size, aggregation, size distribution, and surface area information.

3.3.2.5 *Other methods*

Several other techniques are potentially important and have been used for iron nanoparticle characterization. We note particularly the use of Mössbauer spectroscopy to characterize the core-shell structures of iron nanoparticles by Kuhn *et al.*,[60] who were able to show that the oxide shell structure changed depending on the oxidation conditions.

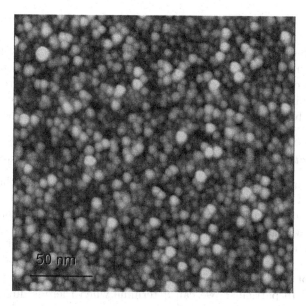

Figure 3.10. Atomic force microscopy image of Fe^{SP} particles for the same material as shown in Figure 3.2.

Synchrotron-based X-ray absorption spectroscopy has also proven very useful for characterizing the particle size dependence of the shell structure for iron core-shell particles.[61]

Scanning probe microscopy, particularly atomic force microscopy (AFM) and scanning tunneling microscopy (STM) have been major tools for characterizing nanostructures. AFM is a useful but not primary tool for our characterization of nanoparticles or films produced by collecting nanoparticles on a substrate. An AFM image of a particle film — for the same conditions as the TEM image in Figure 3.2 — is shown in Figure 3.10.

Although we have not applied them frequently, laser Raman spectroscopy,[62] Fourier transform infrared spectroscopy (FTIR), and time-of-flight secondary ion mass spectrometry (TOF-SIMS) are each sensitive to molecular species and can sometimes be very useful for characterizing molecular species on particle surfaces.

3.4 Methods for Characterizing Reactivity in Solution

Reactions between nanoparticles and an aqueous medium can be viewed — and studied — from the perspective of the particles or that of the reactants in solution. *In situ* spectroscopies, such as electron paramagnetic resonance

(EPR), Raman, and Fourier transform infrared (FTIR), as well as electrochemical methods can be used to monitor reactivity in near-real time and their results are used to infer changes in the properties of the particles. When the focus is on changes in solution composition such as the destruction of contaminants, periodic sampling and analysis of the solution from well-mixed batch reactors containing slurries of the particles or from flow-through reactors containing supported nanoparticles can be used.

3.4.1 *Electron paramagnetic resonance (EPR) spectroscopy*

Chemical and physical properties of the shells surrounding the metallic iron cores of the nanoparticles are likely to play an important role in determining the chemical reaction pathway, as noted in the introduction. The availability of electrons to transfer from the nanoparticle to a chlorinated hydrocarbon and the presence of hydrogen are two important factors that may influence the pathway. We have used EPR to monitor the changes in the mode of electron transfer by iron nanoparticles as they age in aqueous solutions.[63] In particular, we can determine when there is direct electron transfer from the particle to a spin-trapping agent and when there is atom transfer via H atoms generated by corrosion of the iron. The electron transfer mode is associated with the production of relatively harmless reduction products whereas the hydrogen transfer mode is associated with the greater production of undesirable products such as chloroform.

The approach involves the use of spin-trapping agents that react with reductants to form adducts having distinctive EPR signatures. Initial experiments demonstrating this technique were conducted using 5,5-dimethyl-1-pyrroline-N-oxide (DMPO) as a nitrogen-centered spin trap. This agent yields no EPR signal until it is reduced, and if reduced by a radical species, an adduct is formed that yields a distinctive EPR signal. In these experiments, portions of two solutions — (1) a suspension of initially dry Fe^{H2} nanoparticles in buffered (pH 7 bis-tris propane) deoxygenated deionized (DODI) H_2O, and (2) a solution of DMPO in DODI water — were mixed and injected into a flow-through flat cell located in the sample cavity of the EPR. The intensity and nature of the EPR signal from the spin-trapping agent is then monitored over time periods of up to 24 hours, before additional solution is inserted into the cell. The initially dry Fe nanoparticles evolve with time (Figures 3.6 and 3.8) while suspended in the buffered DODI H_2O. Thus, the experiment can be used as an *in situ* monitor of the successive changes in the reactivity of the Fe.

The results of this experiment demonstrate that Fe^{H2} nanoparticles reduce DMPO in two ways (see Figure 3.11). First, direct electron transfer to the DMPO occurs. At a later time atomic H is generated and

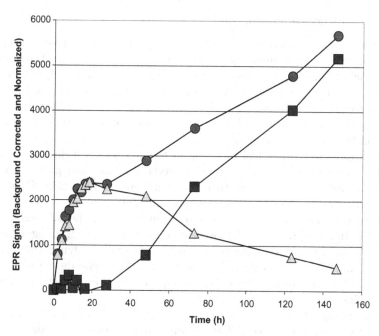

Figure 3.11.   Changes in the relative proportion of direct electron-transfer and hydrogen-transfer products with aging time in H_2O. The triangles (△) are for direct electron transfer, the squares (■) are for atom (hydrogen) transfer, and the circles (●) are the total spin transfer.

H-associated electron transfer takes place. These produce distinct EPR signals that can be used to reveal the change in the reductive process (between the particles and DMPO) as a function of the time the particles have aged in solution. During the first 24 hours essentially all reduction occurs by direct electron transfer. After this period, H production begins and reduction by H transfer surpasses direct electron transfer after about 60 hours.

The particular nanoparticles used in this experiment consist of two phases, a magnetite phase associated with an oxide-coated zero-valent Fe phase. Our interpretation of the results suggests that the oxide shell initially present on the zero-valent Fe phase shields the zero-valent Fe core for about 24 hours and all electron transfer occurs at the surface of the magnetite phase. Thereafter, breaches in the oxide shell occur to allow direct reaction of solution with zero-valent Fe and production of atomic H (see Figure 3.6). Over the next several days, total reductive capacity doubles and the H transfer assumes dominance as oxide coatings form on the magnetite surface and decrease the contribution from the direct electron transfer mechanism.

3.4.2 *Electrochemical methods*

Electrochemical methods offer a unique perspective on the properties and reactivity of nanostructured materials. For the present purpose, key advantages include (1) applicability to small quantities of material, such as is obtained by the gas-phase vacuum deposition methods described above, (2) control of mass transport so its contribution to reaction kinetics can be quantified, and (3) control over the formation of the oxide shell(s) that form on the particle's metal core.

To apply electrochemical methods to the characterization of nanoparticles, it is necessary to make working electrodes from the particles with minimal alteration to the particle structure and composition. One way to do this involves molding the nanoparticles into a disk electrode using binders and other additives. For example, by adapting the soft-embedded rotating disk electrode (RDE) described by Cha et al.[64], Long et al.[65] made a "sticky carbon" electrode specifically for studying the corrosion of nano-sized catalysts. Subsequently, this electrode design was used to study the aqueous corrosion of nanoparticulate Fe metal and Fe-Ni bimetallics.[66,67]

In addition to the soft-embedded electrodes, there is a range of alternative electrode designs. For example, it is possible to make powder electrodes of nanoparticulate metals and metal oxides by compaction (without binders) in cavities on microelectrode supports.[64,68] Although most applications of powder electrodes have been to noble metals and semiconductors, one early study reported success with powder electrodes of iron metal.[69] An advantage of these methods for electrode fabrication is that they are used to support nanoparticles that are prepared separately, by whatever means and for whatever purpose. In other cases, it is desirable to form nanoparticles directly on electrodes, and this can be done by chemical vapor deposition,[70] electrochemical deposition,[70] or *in situ* precipitation.[71] The latter was used in making thin-film wall-jet electrodes of microcrystalline iron oxides for studies of contaminant reduction by Fe(II) surface sites.[72,73] The former — chemical vapor deposition — we have tested using the method described earlier.

We have applied the methods mentioned in order to make electrodes out of a variety of nano-sized iron powders (Table 3.1), so that electrochemical methods can be applied to the characterization of these materials. Selected results obtained by linear sweep voltammetry are shown in Figure 3.12. The data show that electrodes made by gas-phase deposition of iron nanoparticles directly onto the electrode (Fe^{SP}) give anodic polarization curves that are similar to what is obtained with electrodes packed with nano-sized iron

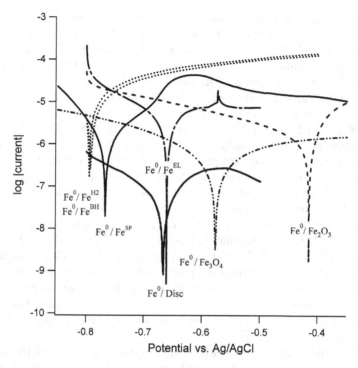

Figure 3.12. Log absolute current versus potential plot from anodic polarization curves obtained with electrodes made from nano-sized iron and iron oxides. All data are for powder disk electrode (PDE) except for Fe^{SP}, where the particles were applied directly by vapor deposition and conventional polished Fe^0 disk. All data at scan rate = 0.1 mV/s in anoxic aqueous borate buffer (pH 8.4).

(Fe^{H2} and Fe^{BH}), and this is indicative of the higher reactivity that is expected for nano-sized particles.[13,41,73–76] Ultimately, we hope to model the polarization curves obtained using these electrodes in the presence and absence of dissolved contaminants, and thereby extract kinetic data for contaminant reduction on iron powders, as we have done previously with polished iron disk electrodes.[77,78]

3.4.3 Batch and column methods

Continuous and well-mixed batch reactors are the most common way to study the reaction of solutes with (nano)particle suspensions. This experimental configuration offers a number of potential advantages, including long contact times, high particle loads and therefore high total surface areas, controlled mixing to reduce external mass transfer effects, and sampling for external analysis. The latter is essential for characterization of environmentally relevant concentrations of contaminants and their degradation products (by, for example, gas chromatography).

While the implementation of batch experiments to study the reactivity of nanoparticles with solutes is straightforward, rigorous consideration of the design and interpretation of such experiments has shown that they are fraught with many ambiguities and complications. A major issue of this sort is whether to control pH and risk secondary buffer effects[79,80] or avoid buffers and risk confounding the results with the effects of drifting pH.[13,81] This dilemma applies to batch studies of all particle sizes, but has an added dimension with nanoparticles because of the likely effect of pH on nanoparticle aggregation. Another example of the challenges in designing good-quality batch experiments is whether to perform analysis of organic residues on headspace[79,81] or the aqueous phase.[13,80] While the former is more amenable to gas chromatography, it can give misleading results when the rate of reaction with the particles in solution is fast relative to the rate of analyte exchange between the solution and headspace. This effect is more likely with nanoparticles, in so far as they produce faster reaction rates than coarser particles.

As part of an extensive investigation of iron nanoparticle reactivity with aqueous contaminants, we have performed a series of carefully designed batch experiments using nanoparticles of zero-valent iron to reduce carbon tetrachloride (CT). These experiments were done without buffer, but with 24 hours of equilibration between deionized water and the particles which produces a stable pH during contact with CT; without headspace and with aggressive mixing to minimize mass transport effects; with frequent sampling and global fitting of a mechanistic kinetic model for CT degradation and product formation. The results include surface-area-normalized first-order rate constants for degradation of CT (k_{SA}) and yields of the major product chloroform (Y_{cf}).

Figure 3.13 provides a summary of our data for CT degradation by three types of zero-valent iron, two nanoscale materials (Fe^{BH} and Fe^{H2}) and one micron-sized granular iron (Fe^{EL}) for comparison. Clearly, the material produced by reduction of nano-sized iron oxides with H_2 (Fe^{H2}) has the lowest Y_{cf}, which is advantageous because CF is a persistent and toxic by-product. Contrary to the popular perception that nanoparticles are intrinsically more reactive than bulk particles, surface-area-normalized rate constants are higher for Fe^{EL} than for the two types of nano-iron studied here. We think this result is definitive for CT, but we do not yet know how it applies to other contaminants because our preliminary analysis of

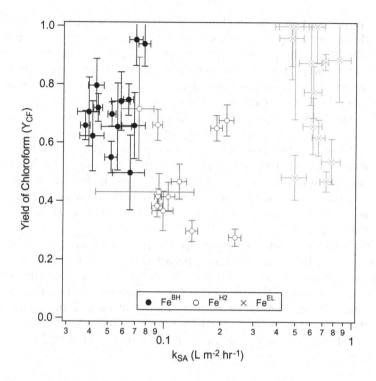

Figure 3.13.   Yield of CF (toxic by-product of CT degradation by iron) versus surface-area-normalized pseudo first-order rate constant for CT reduction by zero-valent iron. Fe^{H2} and Fe^{BH} are nano-sized iron synthesized by H_2 and $NaBH_4$ reduction of iron oxides respectively, and Fe^{EL} is a micron-sized iron obtained from Fisher Scientific. For more details on these materials, see Table 3.1. Conditions for these experiments were similar to those described in Nurmi *et al.*[13]

k_{SA} data for trichloroethylene suggest more differences between nano- and micron-sized Fe^0 particles.[82]

3.5 Summary

The field represented by the subject of this chapter is largely new and therefore its standard practices and fundamental principles are only just beginning to take shape. Nevertheless, we have tried to provide a comprehensive snapshot of the current state of theory and experimental practice in the area of nanoparticle applications to the decontamination of environmental materials.

In principle, there are fundamental differences between nanoparticles and other types of materials, and these differences make the materials scientifically and technologically interesting. However, the most interesting fundamental properties of nanomaterials tend to be unsustainable under environmental conditions and this makes it difficult to characterize or utilize those properties in environmental contexts.

The examples of the synthesis and characterization provided in this chapter show the variety of ways that metal iron nanoparticles can be produced. Because of the variety of synthesis methods used, the likely presence of contamination and coatings, it is important to apply tools (surface spectroscopies and optical methods) that can sense molecular or elemental surface impurities. It is clear that adequate understanding of the behavior of these particles requires a more detailed characterization than often has been reported.

Two major elements in proper characterization of these materials are proper sample handling and the application of multiple characterization methods. Although many techniques are routinely applied to the characterization of nanomaterials, it is important to understand the strengths and limitations of each method. It is often important to combine information from the different methods to produce a "smart" analysis. Modeling of the experimental data is important for many techniques, including TEM, XRD, and XPS.

Acknowledgments

In addition to samples synthesized as part of the project, samples of nano-iron were provided by Greg Lowry (Carnegie Mellon University), K. Okinaka (Toda Kogyo Corp.), and W.-X. Zhang (Lehigh University). We thank EMSL staff Mark Engelhard, David McCready, and Cathy Chin for assistance with the XPS, XRD, and BET analysis, respectively; and Prof. Paul Davis of Pacific Lutheran University for discussions on Au particle aggregation. This work was supported by the U.S. Department of Energy (DOE) Office of Science, Offices of Basic Energy Sciences and Biological and Environmental Research. Parts of the work were conducted in the William R. Wiley Environmental Molecular Sciences Laboratory (EMSL), a DOE User Facility operated by Battelle for the DOE Office of Biological and Environmental Research.

References

1. F. Salamanca-Buentello, D. L. Persad, E. B. Court, D. K. Martin, A. S. Daar and P. A. Singer, *PLoS Med.* **2**, 300–303 (2005).

2. L. Cumbal and A. K. Sengupta, *Environ. Sci. Technol.* **39**, 6508–6515 (2005).
3. W. Tungittiplakorn, L. W. Lion, C. Cohen and J.-Y. Kim, *Environ. Sci. Technol.* **38**, 1605–1610 (2004).
4. G. Vissokov and T. Tzvetkoff, *Eurasian Chem. Technol. J.* **5**, 185–191 (2003).
5. N. Saleh, T. Phenrat, K. Sirk, B. Dufour, J. Ok, T. Sarbu, K. Matyjaszewski, R. D. Tilton and G. V. Lowry, *Nano Lett.* **5**(12), 2489–2494 (2005).
6. J. Xu and D. Bhattacharyya, *Environ. Prog.* **24**, 358–366 (2005).
7. Y. Q. Liu, S. A. Majetich, R. D. Tilton, D. S. Sholl and G. V. Lowry, *Environ. Sci. Technol.* **39**(5), 1338–1345 (2005).
8. R. M. Powell, R. W. Puls, D. W. Blowes, J. L. Vogan, R. W. Gillham, P. D. Powell, D. Schultz, R. Landis and T. Sivavic, *Permeable Reactive Barrier Technologies for Contaminant Remediation*, U.S. Environmental Protection Agency, EPA/600/R-98/125 (1998).
9. P. G. Tratnyek, M. M. Scherer, T. J. Johnson and L. J. Matheson, in *Chemical Degradation Methods for Wastes and Pollutants: Environmental and Industrial Applications*, ed. M. A. Tarr (Marcel Dekker, New York, 2003), pp. 371–421.
10. D. W. Blowes, C. J. Ptacek, S. G. Benner, C. W. T. McRae, T. A. Bennett and R. W. Puls, *J. Contam. Hydrol.* **45**, 123–137 (2000).
11. J. Quinn, C. Geiger, C. Clausen, K. Brooks, C. Coon, S. O'Hara, T. Krug, D. Major, W. S. Yoon, A. Gavaskar and T. Holdsworth, *Environ. Sci. Technol.* **39**(5), 1309–1318 (2005).
12. D. W. Elliott and W. X. Zhang, Field assessment of nanoscale bimetallic particles for groundwater treatment, *Environ. Sci. Technol.* **35**(24), 4922–4926 (2000).
13. J. T. Nurmi, P. G. Tratnyek, V. Sarathy, D. R. Baer, J. E. Amonette, K. H. Pecher, C. Wang, J. C. Linehan, D. W. Matson, R. L. Penn and M. D. Driessen, *Environ. Sci. Technol.* **39**(5), 1221–1230 (2005).
14. X. Zhang, C. Sun and N. Fang, *J. Nanopart. Res.* **6**, 125–130 (2004).
15. J. Z. Jiang, Y. X. Zhou, S. Moeup and C. B. Koch, *Nanostruct. Mater.* **7**(4), 401–410 (1996).
16. H. Hermann, W. Gruner, N. Mattern, H. D. Bauer, F. Fugaciu and T. Schubert, in *Mechanically Alloyed, Metastable and Nanocrystalline Materials: Part 1*, Material Science Forum, Vols. 269–272 (Trans Tech Publications, Switzerland 1998), pp. 193–198.
17. D. B. Vance, *Pollution Engineering* **37**(7), 16–18 (2005).
18. S. S. Suthersan, D. Vance, P. Palmer, U.S. Patent Application 890,066 (2005).
19. C. Granqvist, L. Kish and W. Marlow (eds.), *Gas Phase Nanoparticle Synthesis* (Springer, New York, 2004).
20. C. Janzen and P. Roth, *Combust. Flame* **125**, 1150 (2001).
21. H. Hahn, *Nanostruct. Mater* **9**, 3–12 (1997).
22. M. L. Ostraat, J. W. De Blauwe, M. L. Green, L. D. Bell, H. A. Atwater and R. C. Flagan, *J. Electrochem. Soc.* **148**, G265–G270 (2001).
23. K. Wegner, B. Walker, S. Tsantilis and S. E. Pratsinis, *Chem. Eng. Sci.* **57**, 1753 (2002).
24. A. P. Weber, M. Seipenbusch and G. Kasper, *J. Phys. Chem. A* **105**, 8958 (2001).

25. Y. Qiang, Y. Thurner, Th. Reiners, O. Rattunde and H. Haberland, *Surf. Coat. Technol.* **101**(1–3), 27–32 (1998).
26. Y. Qiang, R. F. Sabiryanov, S. S. Jaswal, Y. Liu, H. Haberland and D. J. Sellmyer, *Phys. Rev. B* **66**, 064404 (2002).
27. Y. Xu, Z. Sun, Y. Qiang and D. J. Sellmyer, *J. Appl. Phys.* **93**, 8289 (2003).
28. Y. Qiang, J. Antony, A. Sharma, S. Pendyala, J. Nutting, D. Sikes and D. Meyer, *J. Nanoparticle Res.* **8**, 489 (2006).
29. J. Antony, Y. Qiang, D. R. Baer and C. Wang, *J. Nanosci. Nanotechnol.* **6**, 568–572 (2006).
30. J. R. Anderson and M. Boudart, *Catalysis, Science & Technology* (Springer-Verlag, Berlin, 1996).
31. T. Dwars, E. Paetzold and G. Oehme, *Angew. Chem. Int. Ed.* **44**, 7174–7199 (2005).
32. V. Uskokovic and M. Drofenik, *Surf. Rev. Lett.* **12**, 239–277 (2005).
33. J. G. Darab, J. L. Fulton, J. C. Linehan, M. Capel and Y. Ma, *Langmuir* **10**, 135–141 (1994).
34. A. R. Kortan, R. Hull, R. L. Opila, M. G. Bawendi, M. L. Steigerwald, P. J. Carroll and L. E. Brus, *J. Am. Chem. Soc.* **112**, 1327–1332 (2005).
35. M. L. Steigerwald, A. P. Alivistos, J. M. Gibson, T. D. Harris, R. Kortan, A. J. Muller, A. M. Thayer, T. M. Duncan, D. C. Douglass and L. E. Brus, *J. Am. Chem. Soc.* **110**, 3046–3050 (1988).
36. D. W. Matson, J. C. Linehan, J. G. Darab and M. F. Buehler, *Energ. Fuel.* **8**, 10–18 (1994).
37. J. L. Fulton, M. Ji, C. M. Chen and J. Wai, *J. Am. Chem. Soc.* **121**, 2631 (1999).
38. N. A. D. Burke, H. D. H. Stover and F. P. Dawson, *Chem. Mater.* **14**, 4752 (2002).
39. K. S. Suslick and G. J. Price, *Annu. Rev. Mater. Sci.* **29**, 295–326 (1999).
40. C.-B. Wang and W.-X. Zhang, *Environ. Sci. Technol.* **31**, 2154–2156 (1997).
41. H.-L. Lien and W.-X. Zhang, *Colloid. Surface. A. Physicochem. Eng. Aspect.* **191**(1–2), 97–105 (2001).
42. D. W. Matson, J. C. Linehan, J. G. Darab, M. R. Buehler, M. R. Phelps and G. G. Neuenschwander, in *Advanced Techniques in Catalyst Preparation*, ed. W. R. Moser (Academic Press, New York, 1996), pp. 259–283.
43. M. Uegami, J. Kawano, T. Okita, Y. Fujii, K. Okinaka, K. Kakuya and S. Yatagai, Iron particles for purifying contaminated soil or groundwater. Process for producing the iron particles, purifying agent comprising the iron particles, process for producing the purifying agent and method of purifying contaminate soil or groundwater. Toda Kogyo Corp. U.S. Patent Application (2003).
44. C. R. Brundle, C. A. Evans and S. Wilson (eds.), *Encyclopedia of Materials Characterization* (Butterworth-Heinemann, Stoneham, MA, 1992).
45. D. J. Gaspar, M. H. Engelhard, M. C. Henry and D. R. Baer, *Surf. Interface Anal.* **37**, 417–423 (2005).
46. H. Zhang, B. Gilbert, F. Huang and J. F. Banfield, *Nature* **424**, 1025 (2003).
47. T. L. Hill, *Nano Lett.* **1**, 273–275 (2001).

48. M. J. Yacaman, J. A. Ascencio, H. B. Liu and J. Gardea-Torresdey, *J. Vac. Sci. Technol. B* **19**, 1091 (2001).

49. D. R. Baer, M. H. Engelhard, D. J. Gaspar, D. W. Matson, K. H. Pecher, J. R. Williams and C. M. Wang, *J. Surf. Anal.* **12**, 101–108 (2005).

50. P. M. Ajayan, *Nature* **427**, 402–403 (2004).

51. F. Bodker, M. F. Hansen, C. B. Koch and S. Morup, *J. Magn. Magn. Mater.* **221**(1–2), 32–36 (2000).

52. T. J. Norman, C. D. Grant, D. Magana, J. Z. Zhang, J. Liu, D. Cao, F. Bridges and A. V. Buuren, *J. Phys. Chem. B* **106**, 7005–7012 (2002).

53. K. Ritter, M. S. Odziemkowski and R. W. Gillham, *J. Contam. Hydrol.* **55**, 87–111 (2002).

54. R. Miehr, P. G. Tratnyek, J. Z. Bandstra, M. M. Scherer, M. Alowitz and E. J. Bylaska, *Environ. Sci. Technol.* **38**, 139–147 (2004).

55. D. J. Burleson, M. D. Driessen and R. L. Penn, *J. Environ. Sci. Health A. Tox. Hazard. Subst. Environ. Eng.* **39**(10), 2707–2753 (2004).

56. C. M. Wang, D. R. Baer, L. E. Thomas, J. E. Amonette, J. Anthony, Y. Qiang and G. Duscher, *J. Appl. Phys.* **98**, 094308 (2005).

57. Although the "quick" sample transfer from glove box to microscope appears to provide reproducible and useful results, we are currently implementing a controlled atmosphere transfer system to further minimize any changes that might occur during transfer.

58. These XRD patterns are from the supplementary material Figure S1 for Ref. 13.

59. Multiquant is an example of one program that includes sample shape information in the analysis. www.chemres.hu/aki/XMQpages/XMQhome.htm.

60. L. T. Kuhn, A. Bojesen, L. Timmermann, M. M. Nielsen and S. Morup, *J. Phys. Condens. Matter* **14**, 13551–13567 (2002).

61. L. Signorini, L. Pasquini, L. Savini, R. Carboni, F. Boscherini, E. Bonetti, A. Giglia, M. Pedio, N. Mahne and S. Nannarone, *Phys. Rev. B* **68**, 195423 (2003).

62. L. Guo, Q. Huang, X.-y. Li and S. Yang, *Phys. Chem. Chem. Phys.* **3**(9), 1661–1665 (2001).

63. J. E. Amonette, V. Sarathy, J. C. Linehan, D. W. Matson, C. Wang, J. T. Nurmi, K. Pecher, R. L. Penn, P. G. Tratnyek and D. R. Baer, *Geochim. Cosmochim. Acta* **69**, A263, Supplement S. Presented at 15th Annual Goldschmidt Conference, Moscow, ID, May 21–25, (2005).

64. C. S. Cha, C. M. Li, H. X. Yang and P. F. Liu, *J. Electroanal. Chem.* **368** (1–2), 47–54 (1994).

65. J. W. Long, K. E. Swider, C. I. Merzbacher and D. R. Rolison, *Langmuir* **15**(3), 780–785 (1999).

66. S. M. Ponder, J. G. Darab, J. Bucher, D. Caulder, I. Craig, L. Davis, N. Edelstein, W. Lukens, H. Nitsche, L. Rao, D. K. Shuh and T. E. Mallouk, *Chem. Mat.* **13**(2), 479–486 (2001).

67. B. Schrick, S. M. Ponder and T. E. Mallouk, *Extended Abstracts, Division of Environmental Chemistry*, Vol. 42, No. 2 (American Chemical Society, Washington, D.C., 2000), pp. 639–640.

68. V. Vivier, C. Cachet-Vivier, B. L. Wu, C. S. Cha, J. Y. Nedelec and L. T. Yu, *Electrochem. Solid-State Lett.* **2**(8), 385–387 (1999).
69. G. V. Zhutaeva, N. D. Merkulova, M. R. Tarasevich and V. V. Surikov, *J. Appl. Chem. USSR* **61**(1), 56–59 (1988).
70. H. O. Finklea, *Studies in Physical and Theoretical Chemistry Semiconductor Electrodes* (Elsevier, Amsterdam, 1988).
71. J. S. Gao, T. Arunagiri, J. J. Chen, P. Goodwill, O. Chyan, J. Perez and D. Golden, *Chem. Mat.* **12**(11), 3495–3500 (2000).
72. B. A. Logue, Ph.D. thesis, Oregon State University, Corvallis, OR (2000).
73. B. A. Logue and J. C. Westall, *Environ. Sci. Technol.* **37**, 2356–2362 (2003).
74. S. Choe, Y. Y. Chang, K. Y. Hwang and J. Khim, *Chemosphere* **41**(8), 1307–1311 (2000).
75. H.-L. Lien and W.-X. Zhang, *J. Environ. Eng.* **125**(11), 1042–1047 (1999).
76. B. Schrick, J. L. Blough, A. D. Jones and T. E. Mallouk, *Chem. Mat.* **14**(12), 5140–5147 (2002).
77. M. M. Scherer, J. C. Westall, M. Ziomek-Moroz and P. G. Tratnyek, *Environ. Sci. Technol.* **31**(8), 2385–2391 (1997).
78. M. M. Scherer, K. Johnson, J. C. Westall and P. G. Tratnyek, *Environ. Sci. Technol.* **35**(13), 2804–2811 (2001).
79. S. H. Joo, A. J. Feitz, D. L. Sedlak and T. D. Waite, *Environ. Sci. Technol.* **39**(5), 1263–1268 (2005).
80. M. M. Scherer, J. C. Westall, M. Ziomek-Moroz and P. G. Tratnyek, *Environ. Sci. Technol.* **31**, 2385–2391 (1997).
81. T. E. Meyer, C. T. Przysiecki, J. A. Watkins, A. Bhattacharyya, R. P. Simondsen, M. A. Cusanovich and G. Tollin, *Proc. Natl. Acad. Sci. USA* **80**, 6740–6744 (1983).
82. P. G. Tratnyek, V. Sarathy and B. Bae, *Extended Abstracts, Division of Environmental Chemistry*, American Chemical Society, 230th National Meeting, Washington, D.C., August 28–September 1, Vol. 45, No. 2, (2005), p. 4.

Nanostructured Inorganic Materials

Chapter 4

Formation of Nanosized Apatite Crystals in Sediment for Containment and Stabilization of Contaminants

Robert C. Moore*, Jim Szecsody[†], Michael J. Truex[†],
Katheryn B. Helean*, Ranko Bontchev* and Calvin Ainsworth[†]

*Sandia National Laboratory, Albuquerque, NM, USA
[†]Pacific Northwest National Laboratory
Richland, WA, USA

4.1 Introduction

Apatite is a calcium phosphate mineral that has a strong affinity for sorption of many radionuclides and heavy metals. A large number of laboratory studies and field applications have focused on the use of apatite for remediation of contaminated sites.[1] Recently, numerous studies have appeared in the literature focused on the development of apatite with a nanoporous structure. The majority of these studies are focused on biomedical applications. However, the development of a high-surface-area, nanoporous apatite sorbent with an enhanced capacity for contaminant uptake is highly significant for environmental applications. In this chapter, we describe a method for forming nanosized apatite crystals in sediment using aqueous solutions containing calcium, phosphate and the citrate ligand (Figure 4.1). The method has applications in the construction of *in situ* reactive barriers for groundwater remediation and stabilization of contaminated sites.

Apatite, or hydroxyapatite, $Ca_{10}(PO_4)_6(OH)_2$, in its purest form, has been the focus of many laboratory studies and has been used as a radionuclide sorbent in reactive barriers and for groundwater remediation.[1-3] Apatite has been demonstrated to be a strong sorbent for uranium,[4-6] neptunium,[7] plutonium,[8] strontium[9-11] and lead.[12-14] Apatite can be synthesized through precipitation reactions from supersaturated solutions of calcium phosphate,[15] high-temperature solid-state reactions[16] and high-temperature treatment of animal bone.[17] An excellent source of information

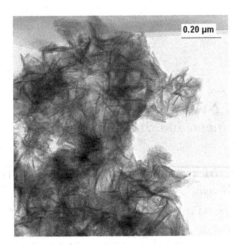

Figure 4.1. Nanosized apatite crystals formed in sediment through precipitation of aqueous solution. Transmission electron microscopy image taken in bright field mode.

on calcium phosphates including apatite is given by LeGeros.[18] Mineral apatite, phosphate rock, is mined throughout the world, but typically is associated with significant amounts of impurities, including radionuclides and heavy metals.

Permeation grouting, high-pressure jet grouting and excavation followed by backfilling with sorbent are methods typically used for placing solids, including apatite, in sediment for the construction of reactive barriers and contaminant immobilization. For certain sites, it may not be practical or even possible (i.e. deep placement below 50 feet) to use these conventional methods. For example, the presence of obstacles such as buried large waste tanks or geological formations can make solids placement difficult. Additionally, if the sorbent cannot be homogeneously dispersed, the formation of a continuous barrier cannot be ensured and may result in preferential flow around the reactive material.

An alternative to using a solid sorbent for contaminant immobilization is to inject a reagent in aqueous solution into the sediment that converts the contaminant to a less toxic form or reacts with the contaminant to form an insoluble precipitate. The use of phosphoric acid for lead stabilization in sediment through the formation of sparingly soluble lead phosphates is an example. A novel approach for forming apatite in sediment using an aqueous solution of calcium chloride, sodium phosphate, sodium citrate and nutrients for microbial activity has been reported by Moore and others,[7,19] allowing sufficient time to inject the solution into an aquifer (0.5–4 days) before citrate is degraded and apatite precipitates.

Calcium phosphates are sparingly soluble in water and precipitation of calcium phosphates from a supersaturated solution is rapid. However, citrate forms strong complexes with calcium and can be used to prevent precipitation or slow the precipitation kinetics of calcium phosphates. When the aqueous solution containing calcium citrate and phosphate is then mixed with sediment, the indigenous soil microorganisms biodegrade the citrate, freeing up the calcium, so amorphous calcium phosphate solids precipitate. Over time (1–3 weeks), these solids change into the thermodynamically favored crystalline apatite structure. The method has been used to form 100 μm apatite crystals in sediment and studied for the immobilization of ^{90}Sr.[7,19]

4.2 Calcium Phosphate Precipitation

The formation of a particular calcium phosphate solid through precipitation from aqueous solution is strongly dependent on solution conditions and the rate of solution mixing. Pure hydroxyapatite cannot be synthesized at ambient temperatures through precipitation. However, apatite or apatitic calcium phosphates with varying degrees of crystallinity and purity can be easily prepared by precipitation. Certain species are known to promote or inhibit the formation of calcium phosphate crystalline phases. Known promoters are fluoride and chloride ions whereas citrate, aluminum, tin and carbonate are known to inhibit apatite formation. Additionally, various substitutions can be made into the apatite structure. These include Na^+ and Sr^{2+} for Ca^{2+}, CO_3^{2-}, and HPO_4^{2-} for PO_4^{3-}, and F^- for OH^-. The extent of substitution of a specific ion into apatite is dependent on the concentration of that ion in solution.[18] Substitutions into the apatite structure can significantly affect physical and chemical properties. Carbonate substitution, for example, increases apatite aqueous solubility whereas fluoride decreases apatite solubility.[20,21]

In sediment and groundwater, significant amounts of ionic species are available for substitution into apatite forming through precipitation *in situ*. The composition of natural sediment/water systems can widely vary. Typical arid region sediments tested are Ca/Mg-carbonate dominated, with groundwater containing Ca (53 mg/L), Mg (13 mg/L), Na (25 mg/L) and K (8.2 mg/L) that are held on oxide and clay surfaces (in somewhat higher concentrations than in solution) primarily by ion exchange. Anion composition is carbonate dominated (166 mg/L) with some sulfate (67 mg/L) and chloride (24 mg/L) present. Even in a high Ca-carbonate system, the aqueous concentration of Ca (1.3 mM) and surface Ca concentration (1.8 mM/g) is an insufficient amount to complex with the injected phosphate to form apatite; therefore Ca must be injected in a stoichiometric ratio with phosphate to form apatite.

4.3 Apatite Formation in Sediment

Citrate is a tetraprotic organic ligand with three carboxylate groups and forms strong complexes with calcium and is easily biodegraded by soil microorganisms.[23] The products of aerobic degradation of citrate by soil microorganisms are reported to be acetate, lactate, propionate and formate.[24] Equilibrium conditions for a metal, M, bonding with an organic ligand, L, is expressed by the complexation constant given by:

$$\beta = [M - L]/[M][L]. \tag{4.1}$$

The stability constant for citrate is two orders of magnitude higher than that for the ligands with a single carboxylate group. For example, the logarithm of the complexation constants for citrate and acetate are 3.5 and 0.35, respectively.[25] Van der Houwen and Valsami-Jones[26] have reported that citrate significantly increases the degree of supersaturation of calcium phosphate solutions whereas acetate has little effect. The method for forming apatite in sediment takes advantage of this chemistry. By chelating calcium with citrate, calcium phosphate precipitates do not form in solutions containing phosphate.

The formation of apatite in sediment using solutions of sodium citrate, calcium chloride and sodium phosphate have been shown to be stable for weeks without any observed formation of precipitates.[7] At pH 7.0 to 10, a stable solution requires a ratio of 2.5 citrate molecules for each calcium ion to keep all of the calcium complexed (i.e. some free citrate present). A typical formulation for apatite formation in sediment is given in Table 4.1. Higher concentrations of reagents, $> 100 \, \text{mM}$ citrate, result in the formation of precipitates almost immediately. The ratio of reagents in the formulation

Table 4.1. Composition of apatite-forming solution. Solution pH is adjusted to 7.2 with HCl and NaOH.

Component	Concentration (mM)
Trisodium citrate	50
Calcium chloride	25
Disodium phosphate	18
Ammonium nitrate	5
Hoagland's solution	1 mL added to 1 L of solution

is significant. The ratio of sodium citrate to ammonium nitrate or another nitrate source is set at approximately $10:1$ to support microbial growth. Hoagland's solution, as modified by Johnson *et al.*,[27] contains the necessary micronutrients for microorganisms. The ratio of phosphate to calcium is set at approximately $3:5$, the same as apatite. A critical parameter is solution pH which must be adjusted to between 7 and 8. At pH lower than 7.0 and higher than 9.5, precipitates other than amorphous calcium phosphate and apatite can form.

The solutions can be injected into sediment or allowed to infiltrate from the surface. Over time, approximately one to four weeks depending on the initial reagent concentrations, temperature and sediment, the soil microorganisms convert the citrate to acetate, lactate and formate under anaerobic conditions or carbon dioxide under oxic conditions. Oxic aquifers are driven to reducing conditions quickly even with oxygen saturated ($8.4\,mg/L$) due to low oxygen solubility in water. The calcium, chelated with the single carboxylated ligands, reacts with phosphate and precipitation of calcium phosphate occurs. Initially, amorphous calcium phosphate forms. Over time, the amorphous material changes into apatite. An application utilizing the process is illustrated in Figure 4.2 for the formulation of an *in situ* reactive barrier. The formulation is injected into the sediment in the migration path of the contaminant. Apatite forms in the sediment and as the metal-contaminated groundwater passes through the barrier the contaminant is initially sorbed by the apatite, then over weeks incorporated into the apatite structure.

Phosphate in the formulation not only serves as a reagent in the reaction to form apatite, but also buffers the pH at a value where conditions are reported to be very favorable for apatite formation. The use of a constant composition method for the precipitation of apatite is described by Nancollas and Mohan.[28] For our system, the pH of the solution is buffered at 7.2 as given by the following reactions:

$$H_2PO_4^- + OH^- \rightarrow HPO_4^{2-}$$
$$HPO_4^{2-} + H^+ \rightarrow H_2PO_4^-. \tag{4.2}$$

Other than preventing the premature formation of calcium phosphate precipitates before the solution can be injected into sediment, citrate may play an additional role in apatite formation. Citrate is reported to strongly interact with apatite.[29,30]

In a recent study, Van der Houwen *et al.*[30] indicated that citrate affects apatite crystal size and can increase the amount of impurities in the apatite structure. This effect may be significant when precipitation of apatite occurs in the presence of sediment.

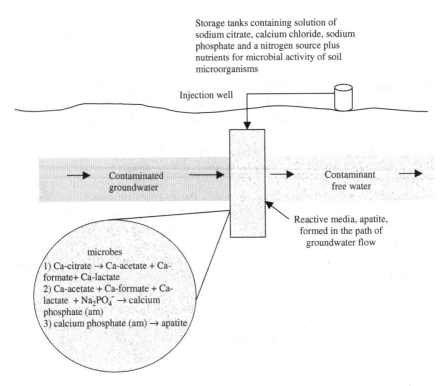

Figure 4.2. Formation of apatite in sediment to create a reactive barrier for removing contaminants in groundwater. Apatite is formed through the biodegradation of citrate resulting in the precipitation of amorphous calcium phosphate that changes to a poorly crystallized apatite over time.

4.4 Biodegradation of Citrate

The mechanism of apatite formation in sediment using the described method is driven by the biodegradation of citrate by the indigenous soil microorganisms. Citrate can be microbially metabolized under both aerobic and anaerobic conditions. Typically, citrate is added well in excess of the amount that can be degraded by any dissolved oxygen present in an aquifer, but not in the vadose zone. Additionally, aerobic degradation of the citrate portion of a calcium citrate complex results in the formation of calcium carbonate (Figure 4.3). Aerobic biodegradation of citrate shows 108.5 mM citrate generating a maximum of 70 mM carbonate, so we have accounted for only 11% of the carbon mass from the citrate. Anaerobic degradation processes for citrate are most relevant to *in situ* apatite formation in aquifers where oxygen supply is limited. Under

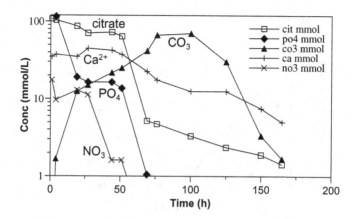

Figure 4.3. Citrate biodegradation in an aerobic sediment/water system forming predominantly carbonate.

anaerobic conditions, fermentation of citrate is a primary means of citrate degradation.

4.4.1 *Citrate fermentation*

Citrate ferments to form acetate and formate, as has been observed in unamended sediments from the Hanford site and in these same sediments amended with a known citrate-fermenting microorganism (Figures 4.4(a) and (b)). A lag time of about three days prior to fermentation observed in the unamended sediments was likely due to low initial populations of microorganisms. While acetate and formate are the primary fermentation products, the amended sediments showed small amounts of lactate and propionate produced in addition to the acetate and formate. Citrate concentration remained constant in the sterile control over the duration of the incubation period. In these experiments (Figures 4.4(a) and (b)), citrate fermentation followed the reaction stoichiometry given by:

$$C_6H_8O_7 + H_2O + OH^- \rightarrow 2C_2H_4O_2 + CH_2O_2 + HCO_3^-. \qquad (4.3)$$

Biomass in the pre-test unamended sediment sample was 1.51×10^5 cells/ g-soil and an average of 2.25×10^8 cells/g-soil after the 16-day incubation period. Using a generic cell formula of $C_5H_7O_2N$, the molecular weight of cells is 113 mg/mmol. These anaerobic citrate biodegradation experiments show that 73% of the carbon from citrate can be accounted for in formate and acetate, and it is likely that the other 25% is carbon dioxide, as shown in Equation (4.3). With this cell molecular weight, the

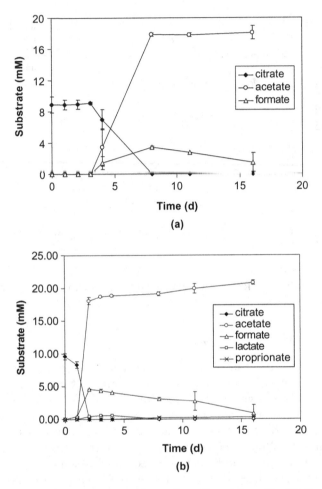

Figure 4.4. Citrate fermentation in a composite sediment from the 100 North Area of the Hanford Site: (a) without inoculation (unamended); and (b) inoculated with approximately 10^7 cells/mL of *Klebsiella pneumoniae* (amended). Error bars are for results from three replicate treatments. Initial solution composition for each experiment was 10 mM calcium citrate, 1 mM ammonium chloride and 5 mM phosphate buffer adjusted to pH 7.3. The soil to solution ratio was 10:3.

estimated yield from citrate is 0.06 mmol-cells/mmol-citrate (0.04 mg-cells/ mg-citrate). The biomass yield is important to consider for *in situ* citrate degradation as part of determining whether the added citrate will cause enough biomass growth to significantly decrease the permeability of the aquifer.

4.4.2 *Modeling citrate biodegradation and temperature effect*

Citric acid is utilized by many organic systems as part of the TCA (Krebs) photosynthetic process, where the citrate (a C6 organic acid) is converted to C6, C5 and C4 organic acids producing CO_2 and H^+, then cycled from oxaloacetic acid (C4) to citric acid.[31] Citrate can also be further degraded to acetic acid (C2), formaldehyde, formic acid (C1) and CO_2. For the purpose of this study, citrate is used to complex Ca, so only the decrease in citrate concentration (by biodegradation) is of significance, as the lower-molecular-weight organic acids only form weak complexes with Ca. Two different modeling approaches were considered to quantify citrate biodegradation: a first-order model and a Monod model. A first-order model is an empirical approach that describes citrate removal with a single reaction rate coefficient. A Monod model is also an empirical approach that describes citrate removal externally to microbial organisms with a similar mathematical form of enzyme degradation of a compound (Michaelis–Menten kinetics). Monod kinetics are utilized when the observed data clearly show a considerable slowing of reaction rate at low concentrations that cannot be accounted for using the simpler first-order kinetic model.

Citrate biodegradation experiments showed a slower rate at colder temperatures and a slower rate at higher citrate concentrations (Figure 4.5). At 10 mM citrate concentration, citrate was completely gone by 200 h (21°C) to 300 h (10°C, Figure 4.5(a)). At 50 mM citrate concentration, citrate was gone by 250 h (21°C) to 450 h (10°C, Figure 4.5(b)). At 100 mM citrate concentration, a small amount of citrate remained at 300 h (21°C) to 600 h (10°C, Figure 4.5(c)). At each concentration, duplicate experiments showed similar results. A first-order model (lines, Figure 4.5) showed good fits, and indicated that in some cases, citrate biodegradation may be somewhat more rapid at lower concentrations than a first-order approximation. For example, the 100 mM citrate data at 10°C (Figure 4.5(c)) showed a good first-order fit to 500 h, but then citrate more rapidly degraded. This effect is observed for all citrate concentrations at 10°C, but not at 21°C. A Monod kinetic model would describe the data equally as well with small half-saturation constants, but would describe the data more poorly with higher concentration half-saturation constants, which would slow citrate biodegradation at low concentration, the opposite effect of that observed. Therefore, a pseudo first-order model was used to quantify the rate data (Table 4.2). The citrate biodegradation rate was 3.0 times slower (10°C data) to 3.3 times slower (21°C data) as the citrate concentration increased from 10 mM to 100 mM. The citrate biodegradation rate averaged 3.3 times slower as the temperature decreased from 21°C to 10°C. The activation energy estimated from

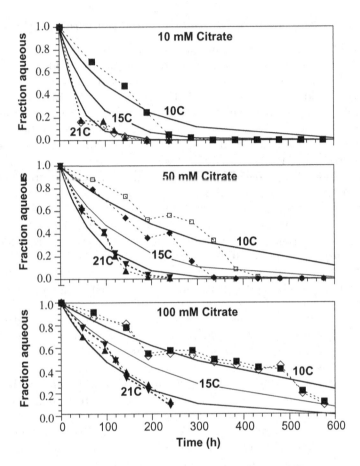

Figure 4.5.   Citrate biodegradation by Hanford 100N sediment at different temperatures for citrate concentrations of (a) 10 mM, (b) 50 mM and (c) 100 mM.

Table 4.2.   Citrate biodegradation rates (1/h) as a function of temperature and initial citrate concentration.

Concentration (mM)	Rate Constant (10°C)	Rate Constant (15°C)	Rate Constant (21°C)
10	0.0071	0.013	0.025
50	0.0074	0.0101	0.013
100	0.0024	0.0042	0.0075

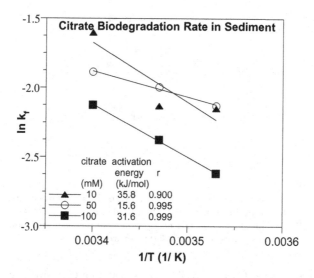

Figure 4.6. Arrhenius plot of citrate biodegradation rate change with absolute temperature with calculated activation energy.

the reaction rate change with temperature is 35 kJ/mol for the 10 mM citrate data, 16 kJ/mol for the 50 mM citrate and 32 kJ/mol for the 100 mM citrate (Figure 4.6). These activation energies indicate the rate is controlled by the biochemical reaction and not diffusion, which is expected.

4.5 Analysis of Apatite Precipitates Formed in Sediment

The mineral apatite, $Ca_5(PO_4)_3(F, Cl, OH)$, is the most abundant, ubiquitous phosphorous-bearing mineral on Earth and is used broadly by Earth scientists in the study of igneous, metamorphic and sedimentary petrogenetic processes.[32] In addition to its geologic utility, apatite is also used in industry for such diverse applications as a source for fertilizer, as a component of fluorescent lamps and lasers, as nuclear waste forms and in biomedical applications.[33-35] In fact, hydroxyapatite is the primary mineral component in bone and teeth.[35] The apatite structure and its chemical facility provide the basis for its broad and varied application.

Apatite can refer to three specific end-member minerals: fluorapatite, chloroapatite and hydroxyapatite. All can be viewed as slight modifications of the $P6_3/m, Z = 2$ structure. For a complete description and structural details see Hughes *et al.*[36] Several features of the structure are noteworthy. PO_4 tetrahedra are found in a hexagonal arrangement within the $(00l)$ planes defining columns parallel to the z-axis (Figure 4.7). The PO_4 oxygens

Figure 4.7.   The crystal structure of apatite ($P6_3/m, Z = 2$) projected on the (001) plane. PO_4 tetrahedra lie parallel to the plane and are coordinated to two different calcium sites. The anions F, Cl and OH occupy the columns that run along the z-axis. (Structure rendered using CrystalMaker® 7.0 software.)

are coordinated with Ca in two different sites. The Ca1 site is intercalated between the (00l) planes and coordinated to nine oxygens. The Ca2 site is coordinated to six oxygens and the column anions (typically F, Cl or OH). The Ca2 site shows the greatest degree of structural distortion upon chemical substitution. The apatite structure exhibits extensive solid solution with respect to both cations and anions. Metal cations including actinides, K, Na, Mn, Ni, Cu, Co, Zn, Sr, Ba, Pb, Cd and Fe substitute for Ca; and oxyanions (e.g. AsO_4^{3-}, SO_4^{2-}, CO_3^{2-}, SiO_4^{4-}, CrO_4^{2-}) replace PO_4^{3-} through a series of coupled substitutions to preserve electroneutrality.[32]

Multiple characterization techniques were employed to assess the crystal chemistry of the apatite formed by the microbial digestion of Ca-citrate in sediment from the US DOE Hanford reservation. The apatite crystals are shown in Figure 4.1. High-resolution transmission electron microscopy (HRTEM) and powder X-ray diffraction (XRD) were used to assess apatite crystallinity and to document the transformation from an amorphous calcium phosphate to nanocrystalline apatite. Energy dispersive spectroscopy (EDS) and Fourier transform infrared (FTIR) spectroscopy were used to analyze the chemical constituents. The apatite was formed in sediment collected from the 100 North Site on the US DOE Hanford Reservation by treatment with a solution of 50 mM sodium citrate, 25 mM calcium chloride, 20 mM sodium phosphate and nutrients for microbial activity at pH 7.4. The blade-like crystals are in an amorphous matrix and are approximately 0.1 μm in size. Figure 4.8(a) is the XRD pattern for the apatite crystals shown in Figure 4.1. The observed broad overlapping peaks in the pattern at 2θ of approximately 32° are characteristic of poorly crystallized apatite. The remaining peaks in the XRD pattern correspond to

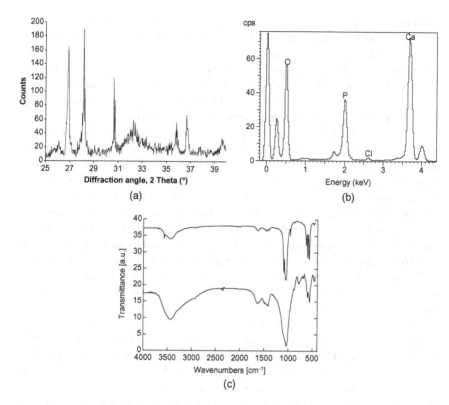

Figure 4.8. Chemical characterization of apatite crystals by (a) XRD, (b) EDS and (c) FTIR of nanocrystalline apatite formed in Hanford sediment by microbially mitigated Ca-citrate degradation in the presence of aqueous phosphate.

components of the sediment. The FTIR spectra is given in Figure 4.8(c) along with the spectra for a pure hydroxyapatite produced by precipitation and heat treatment at 700°C. The lower resolution of the PO_4^- bands confirms the lower crystallinity of the sample, as observed by both HRTEM and XRD. The bands at $1,455\,\text{cm}^{-1}$ and $879\,\text{cm}^{-1}$ indicate the presence of carbonate in the apatite structure. The TEM-EDS (Figure 4.8(c)), spectrum identifies calcium and phosphate as the major components with a stoichiometric apatite ratio of approximately 5:3.

4.6 Sorption of ^{90}Sr by Apatite Formed in Sediment

Moore *et al.*[19] studied the sorption of ^{90}Sr by apatite precipitates in sediments. The study consisted of equilibrating apatite-treated sediment, collected from the Albuquerque desert, with solutions containing a range of ^{90}Sr concentrations. Untreated soil was also tested. The untreated soil sorbed 6.4–38% of the ^{90}Sr from the solutions whereas the apatite-treated soil sorbed 76–99.2% of the ^{90}Sr. Possibly even more significantly, the untreated soil released 20.4–41% of the sorbed ^{90}Sr back into solution when mixed with fresh water. The treated soil released only 1.7–11.2% of the ^{90}Sr, illustrating the strong sorptive properties of the apatite.

The sorption of Sr sorbed to apatite formed in sediment has also been studied by Szecsody.[37] A sub-surface sediment composite used in experiments was collected from the ^{90}Sr-contaminated 100N site on the Hanford Reservation in eastern Washington. At the 100N site, sub-surface Sr contamination is the result of once-through cooling water disposed from the Hanford 100N Reactor into the two disposal trenches. Sediments collected in a bore-hole drilled in 2004 showed that the ^{90}Sr concentration was greatest in the deep vadose zone (12′ to 15′ depth) and shallow groundwater (15′ to 22′ depth). Two sediments used in this study were a composite of sediments from all depths and the 13′ depth, which had the greatest ^{90}Sr concentration (280 pCi/g). The ratio of Sr/^{90}Sr averages 1.4×10^6 in the contaminated sediment.

The ^{90}Sr adsorption rate was measured for Hanford sediments under both natural groundwater conditions and in a solution of 10 mM citrate, 4 mM Ca and 2.4 mM PO$_4$ mixed with the Hanford groundwater. Because ^{90}Sr is retained by Hanford 100N sediments primarily by ion exchange,[38] it was expected that the amount of ^{90}Sr adsorption by sediments in the higher ionic strength solution would be less. These batch experiments were conducted using 116 g of sediment and 70 mL of water to achieve a soil/water ratio near that in saturated porous media. The aqueous solution in both experiments contained 0.2 mg/L Sr (natural groundwater), and the experiment was initiated by the addition of ^{90}Sr to achieve 45,000 pCi/L. At specified time intervals, 0.5 mL of water was removed, filtered with a 0.45 micron PTFE filter (13 mm diameter), and placed in a scintillation vial with scintillation fluid. These samples were counted after 30 days after ^{90}Sr/^{90}Yt secular equilibrium was achieved and the ^{90}Sr activity was half of the total activity.

Sequential extraction experiments were conducted to determine if ^{90}Sr retention by sediments changed with the addition of apatite solution containing 100 mM citrate, 40 mM Ca and 24 mM PO$_4$. The three different systems used for these experiments were (1) untreated 100N sediment treated with an ^{90}Sr spike (described above), (2) apatite-laden sediment treated with a ^{90}Sr spike (described above) and (3) ^{90}Sr-laden sediment from the 13′ depth (80 pCi/g) that was then treated with an apatite solution for 30 days. The first system is the control, and should show no change in ^{90}Sr ion exchange over time. The second and third systems are hypothesized to show a change from ^{90}Sr retention by ion exchange to slow incorporation into apatite. Because the ^{90}Sr in the third system has been in contact with sediment for decades, there may be some resistance to desorption from sediment onto the apatite. Therefore differences between the second and third systems may reflect the influence of ^{90}Sr/sediment aging.

Sequential extractions consisted of selected chemical extraction to remove ion-exchangeable ^{90}Sr, organic-bound ^{90}Sr, carbonate-bound ^{90}Sr and remaining (residual) [90]Sr. Both Sr and ^{90}Sr were analyzed in extractions to determine if the Sr was retained differently from the [90]Sr. It was expected that Sr was geologically incorporated into many different sediment minerals, so should be more difficult to remove compared with ^{90}Sr, which was recently added to the systems. Sequential extractions were conducted on 5 g sediment per water samples removed from the 120 g per 70 mL systems at specified time intervals from a few hours to 4,000 h. The ion-exchangeable extraction consisted of the addition of 0.5 M KNO$_3$ to the sediment sample for 16 hours.[39] The organic-bound extraction conducted after the ion-exchangeable extraction consisted of 0.5 M NaOH for 16 hours.[40] The carbonate-bound extraction conducted after the organic-bound extraction consisted of the addition of 0.05 M Na$_3$EDTA for six hours.[41,42] The residual extraction conducted after the carbonate-bound extraction consisted of the addition of 4 M HNO$_3$ at 80°C for 16 hours.[41] Apatite dissolution rates are highest at low pH,[43] so this extraction is expected to remove ^{90}Sr that is incorporated into the apatite. For every extraction, the mobilized Sr and ^{90}Sr was measured. The amount of carryover from one extraction to another was accounted for in calculations.

[90]Sr added to aqueous solutions in contact with Hanford sediments was rapidly adsorbed (Figure 4.9). The sediment with no additional treatment (i.e. no apatite) showed that adsorption equilibrium (Figure 4.9, square

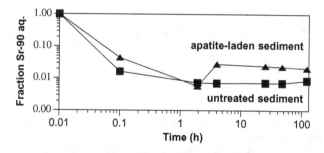

Figure 4.9.   [90]Sr adsorption rate to untreated and apatite solution-laden Hanford 100N sediment. The resulting retention by sediments, described by a distribution coefficient (K_d), was 24.7 cm^3/g for the untreated sediment and 7.6 cm^3/g for the apatite-laden sediment.

symbols) was reached witin two hours, giving a Sr K_d value of 14.8 cm^3/g. For the sediment that contained precipitated apatite, there was initial rapid uptake of ^{90}Sr (more than one hour) that was likely adsorption. In this case, the solution ^{90}Sr continued to decrease even after 120 hours, likely due to some uptake by the apatite. The apparent K_d at 120 hours was 7.6 cm^3/g, which was smaller than the untreated sediment due to the higher solution ionic strength (apatite solution ionic strength 97 mM versus groundwater ionic strength 11 mM). In a study of the influence of major ions on Sr retention by Hanford sediments, Routson and others[44] reported a Sr K_d of 49 cm^3/g (0.001 M NaNO$_3$) and a K_d of 16 cm^3/g (0.1 M NaNO$_3$), so there was a four times decrease in the K_d value by increasing the ionic strength with Na$^+$, which is the predominant cation in the spent apatite solution.

Solid-phase extractions showed that the addition of apatite to sediment resulted in stronger retention of ^{90}Sr, which is presumed to be caused by some incorporation into the apatite structure. For the untreated sediment, solid-phase extractions showed that 83.6±3.3% of the ^{90}Sr was bound by ion exchange (0.5 M KNO$_3$, Figure 4.10(a)), 0.0% organic-bound; 5.2 ± 0.7%

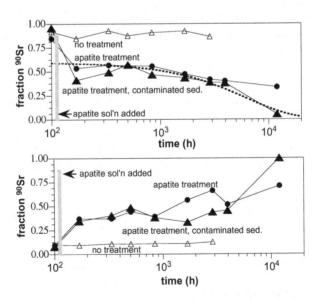

Figure 4.10.   Solid-phase ^{90}Sr extractions on sediment without and with apatite treatment showing changes in ^{90}Sr mobility: (a) mobile fraction — combined aqueous and ion-exchangeable extractions, and (b) immobile fraction — combined carbonate (EDTA) and mineral dissolution (HNO$_3$) extractions. Extractions are shown for the Hanford 100N sediment with no treatment (open triangle), apatite solution addition to sediment with ^{90}Sr added (circles), and apatite solution addition to ^{90}Sr field-contaminated sediment (solid triangles). The apatite solution was in the experimental system for the entire time.

by a carbonate extraction (0.05 M EDTA); and $5.8 \pm 0.6\%$ by a mineral dissolution extraction (4 M HNO_3 at 80°C). The remaining 5.6% ^{90}Sr was aqueous. Therefore, ^{90}Sr was retained on untreated Hanford sediments predominantly by ion exchange, as expected.

The addition of the apatite solution to sediment and subsequent apatite precipitation resulted in ^{90}Sr being incorporated into the apatite structure by one or more mechanisms: (i) during initial precipitation of apatite, some ^{90}Sr is incorporated into the apatite structure substituted for Ca; (ii) solid-phase Ca-hydroxyapatite can incorporate $Sr/^{90}Sr$ by solid-phase-dissolution/reprecipitation; (iii) solid-phase Ca-hydroxyapatite can incorporate $Sr/^{90}Sr$ by crystal lattice rearrangement of adsorbed Sr. The rate of incorporation of ^{90}Sr during the initial precipitation phase (mechanism (i)) should be fairly rapid (the precipitation rate, tens to hundreds of hours), whereas the solid-phase incorporation rate is slow (thousands of hours).

A significant amount of ^{90}Sr (40–50%) was no longer in the ion-exchangeable fraction by 200 h (Figure 4.10(a), both circles and solid triangles), likely indicating ^{90}Sr uptake during the initial apatite precipitation. There were corresponding increases in EDTA- and HNO_3-extractable amounts (Figure 4.10(b)). This could not be due to sorption onto the apatite, which would show up in the ion-exchangeable fraction), so is likely due to ^{90}Sr incorporation into apatite during the initial precipitation phase. The organic-bound extraction (0.5M NaOH) extracted no ^{90}Sr in all cases, as there was no natural organic matter in these sub-surface sediments.

Over the course of thousands of hours (months) the ^{90}Sr became less mobile as shown by the mobile fraction decreasing (Figure 4.10(a)), and the immobile fraction increasing (Figure 4.10(b)). The rate of ^{90}Sr immobilization (defined by the extraction change) was well approximated by a simple first-order reaction (black dashed line in Figure 4.10(a)) with a half-life of 180.5 days (4,332 hours). It is hypothesized that this ^{90}Sr uptake is caused by the slow dissolution of Ca-apatite and precipitation of ^{90}Sr-apatite. The actual rates that would be observed in the field would be dependent on the aquifer temperature, the mass of apatite in the sediment, and the presence of the high-ionic-strength apatite solution. For example, an aquifer temperature of 15°C is likely to have a two times slower rate than observed in these 22°C laboratory studies. A greater mass of apatite loading in the sediment (by higher-concentration or multiple apatite injections) will uptake ^{90}Sr more rapidly. Solid-phase extractions shown (Figure 4.10) were conducted with the high-ionic-strength solution remaining in contact with the sediment for the entire experiment, whereas other experiments in which the solution was removed showed more rapid ^{90}Sr uptake. In the field, the spent apatite solution will migrate downgradient, which would result in higher adsorption of the ^{90}Sr to the apatite, and as a consequence more rapid uptake in the apatite structure.

4.7 Conclusions

A method has been described for the formation of nanosized apatite crystals in sediment that has applications for the formation of reactive barriers and contaminated site stabilization. The method is based on complexing calcium in calcium phosphate solutions to prevent precipitation of calcium phosphate solids before the solution can be injected into the sediment to be treated. As the indigenous soil microorganisms degrade the citrate, the calcium is made available to form calcium phosphate precipitates. These precipitates change to apatite over time. In batch sorption experiments, the apatite demonstrated strong sorption and retention of [90]Sr. Based on sorption data from the literature, other radionuclides and heavy metals including U, Np, Pu and Pb should exhibit the same behavior. The method offers an alternative approach to conventional methods such as jet grouting, excavation and backfilling for the construction of reactive barriers and contaminated site stabilization. It also may be applied for the remediation of certain problematic sites where conventional methods are not feasible.

Acknowledgments

This work was supported by Sandia National Laboratories under contract with the US Department of Energy under Contract DE-AC04-94AL85000, and the Pacific Northwest National Laboratory operated by Battelle for the US Department of Energy under Contract DE-AC06-76RL01830.

References

1. R. D. Spence and C. Shi (eds.), *Solidification and Stabilization of Hazardous, Radioactive and Mixed Wastes* (CRC Press, Boca Raton, FL, 2005).
2. P. H. Ribbe (ed.), *Reviews in Mineralogy & Geochemistry, Vol. 48: Phosphates, Geochemical, Geobiological and Materials Importance* (Mineralogical Society of America, Washington D.C., 2002).
3. United States Department of Energy, *Passive Reactive Barrier: Subsurface Contaminant Focus Area*, Innovative Technology Summary Report, http://www.em.doe.gov/ost (2002).
4. J. S. Arey, J. C. Seaman and P. M. Bertsch, *Environ. Sci. Technol.* **33**(2), 337–342 (1999).

5. C. C. Fuller, J. R. Bargar, J. A. Davis and M. J. Piana, *Environ. Sci. Technol.* **36**(2), 158–165 (2002).
6. J. Jeanjean, J. C. Rouchaud, L. Tran and M. Fedoroff, *J. Radioanal. Nucl. Chem. Lett.* **201**(6), 529–539 (1995).
7. R. C. Moore, C. Sanchez, K. Holt, P. Zhang, H. Xu and G. Choppin, *Radiochim. Acta* **92**(9–11), 719–723 (2004).
8. R. C. Moore, M. Gasser, N. Awwad, K. Holt, F. Salas, A. Hasan, M. Hasan, H. Zhao and C. Sanchez, *J. Radioanal. Nucl. Chem.* **263**(1), 97–101 (2005).
9. S. Lazic and Z. Vukovic, *J. Radioanal. Nucl. Chem.* **149**(1), 161–168 (1991).
10. R. Z. LeGeros, G. Quirolgico and J. P. LeGeros, *J. Dent. Res.* **58**, 169 (1979).
11. Z. Vukovic, S. Lazic, I. Tutunovic and S. Raicevic, *J. Serb. Chem. Soc.* **63**(5), 387–393 (1998).
12. S. Bailliez, A. Nzihou, E. Beche and G. Flamant, *Process Saf. Environ. Prot.* **82**(2), 175–180 (2004).
13. E. Mavropoulos, A. M. Rossi, A. M. Costa, C. A. C. Perez, J. C. Moreira and M. Saldanha, *Environ. Sci. Technol.* **36**(7), 1625–1629 (2002).
14. I. Smiciklas, A. Onjia, J. Markovic and S. Raicevic, *Mater. Sci. Forum* **494**, 405–410 (2005).
15. E. Andronescu, E. Stefan, E. Dinu and C. Ghitulica, *Key Eng. Mat.* **206**(2), 1595–1598 (2002).
16. A. D. Papargyris, A. I. Botis and S. A. Papargyri, *Key Eng. Mat.* **206**(2), 83–86 (2002).
17. A. J. Tofe, U.S. Patent 5,711,015 (1998).
18. R. Z. LeGeros, *Calcium Phosphates in Oral Biology and Medicine* (S. Karger Basel, Switzerland, 1991).
19. R. C. Moore and R. Bontchev, *Apatite Formation in Sediment for In Situ Reactive Barrier Formation,* SAND 2005–6869. Sandia National Laboratories (2005).
20. R. P. Shellis, A. R. Lee and R. M. Wilson, *J. Colloid Interface Sci.* **218**(2), 351–358 (1999).
21. A. B. Barry, H. H. Zhuang, A. A. Baig and W. A. Higuchi, *Cal. Tissue Int.* **72**(3), 236–242 (2003).
22. P. A. W. van Hees, D. L. Jones and D. L. Godbold, *Soil. Biol. Biochem.* **34**(9), 1261–1272 (2002).
23. L. Brynhildsen and T. Rosswall, *Water Air Soil Poll.* **94**(1–2), 45–57 (1997).
24. M. J. Truex, Pacific Northwest National Laboratory, unpublished data (2005).
25. A. E. Martell and R. M. Smith, *Critical Stability Constants, Vol. 3: Other Organic Ligands* (Plenum Press, New York, NY, 1979).
26. J. A. M. van der Houwen and E. Valsami-Jones, *Proceedings of the 11th Annual V.M. Goldschmidt Conference* (2001).
27. C. M. Johnson, P. R. Stout, T. C. Boyer and A. B. Carlton, *Plant Soil* **8**, 337–353 (1957).
28. G. H Nancollas and M. S. Mohan, *Arch. Oral. Biol.* **15**, 731–745 (1970).
29. D. N. Misra, *J. Dent. Res.* **75**(6), 1418–1425 (1996).

30. J. A. M. G. van der Houwen, Cressy, B. A. Cressey and E. Valsami-Jones, *J. Cryst. Growth* **249**, 572–583 (2003).

31. J. E. Bailey and D. F. Ollis, *Biochemical Engineering Fundamentals*, 2nd edn. (McGraw-Hill, Inc., New York, 1986), pp. 246–250.

32. J. M. Hughes and J. Rakovan, in *Phosphates: Geochemical, Geobiological and Materials Importance, Reviews in Mineralogy and Geochemistry*, Vol. 48 (Mineralogical Society of America, Washington, D.C., 2002), pp. 1–12.

33. G. Waychunas, *Rev. Mineral* **18**, 638–698 (1989).

34. R. C. Ewing, *Can. Mineral.* **39**, 697–715 (2001).

35. J. C. Elliot, P. E. Mackie and R. A. Young, *Science* **180**, 1055–1057 (1973).

36. J. M. Hughes, M. Cameron and K. D. Crowley, *Amer. Mineral.* **74**, 870–876 (1989).

37. J. Szecsody, C. Burns, R. Moore, J. Fruchter, V. Vermeul, D. Girvin, J. Phillips and M. Williams, *Hanford 100N Area Apatite Emplacement: Laboratory Results of Ca-Citrate-PO4 Solution Injection and Sr-90 Immobilization in 100N Sediments*, PNNL-16891. Pacific Northwest National Laboratories Report (2005).

38. R. J. Serne and V. L. LeGore, *Strontium-90 Adsorption-Desorption Properties and Sediment Characterization at the 100N-Area.* PNL-10899/UC-702. Pacific Northwest National Laboratories (1996).

39. C. Amrhein and D. L. Suarez, *Soil Sci. Soc. Am. J.* **54**, 999–1007 (1990).

40. G. Sposito, L. Lund and A. Chang, *Soil Sci. Soc. Am. J.* **46**, 260–264 (1982).

41. G. Sposito, C. LeVesque, J. LeClaire and A. Chang, *Soil Sci. Soc. Am. J.* **47**, 898–902 (1983).

42. C. Steefel, in *Water-Rock Interaction*, eds. E. Wanty and P. Seal (Taylor and Francis Group, London, 2004), pp. 999–1002.

43. C. Chairate, E. Oelkers, S. Kohler, N. Harouiya and J. Lartigue, in *Water-Rock Interaction*, eds. E. Wanty and P. Seal (Taylor and Francis Group, London, 2004), pp. 671–674.

44. R. C. Routson, G. Barney, R. Smith, C. Delegard and L. Jensen, *Fission Product Sorption Parameters for Hanford 200 Area Sediment Types*, RHO-ST-35. Rockwell Hanford Operations, Richland, Washington (1981).

45. C. C. Fuller, J. R. Bargar and J. A. Davis, *Environ. Sci. Technol.* **37**(20) 4642–4649 (2003).

46. R. C. Moore, K. C. Holt and K. B. Helean, *Temperature Dependence of Citrate Degradation by Soil Microorganisms*, SAND 2005–6870. Sandia National Laboratories (2005).

47. R. C. Moore, K. C. Holt, H. Zhao, A. Hasan, N. Awwad, M. Gasser and C. Sanchez, *Radiochim. Acta* **91**, 721–727 (2003).

48. G. H. Nancollas and B. B. Tomazic, *J. Phys. Chem.* **79**, 2218–2225 (1974)

49. K. Onuma, A. Oyane, K. Tsutsui, K. Tanaka, G. Treboux, N. Kanzaki and A. Ito, *J. Phys. Chem. B* **104**(45), 10563–10568 (2000).

50. R. P. Singh, Y. D. Yeboah, E. R. Pambid and P. Debayle, *J. Chem. Eng. Data* **36**, 52–54 (1991).

51. D. L. Sparks (ed.), *Methods of Soil Analysis, Part 3: Chemical Methods* (Soil Science Society of America Book, Madison, WI, 1996).

Chapter 5

Functionalized Nanoporous Sorbents for Adsorption of Radioiodine from Groundwater and Waste Glass Leachates

Shas V. Mattigod, Glen E. Fryxell and Kent E. Parker

Pacific Northwest National Laboratory
Richland, WA, USA

5.1 Introduction

Radioactive iodine is one of the fission products from the nuclear industry that ends up encased in waste forms for ultimate disposal. Iodine in the environment is highly mobile because it does not either form solids of limited solubility or irreversibly adsorb onto soil mineral surfaces.[1] Radioiodine, in particular [129]I, is of significant environmental concern due to its very long half-life (1.7 × 10[7] years). Therefore, considerable effort has been focused on finding suitable adsorptive materials that can immobilize or delay the transport of radioiodine that would be released from physically and chemically degrading waste packages. For instance, metallic Cu and its oxides,[2] sulfide minerals,[3-11] cementitious forms,[12] various types of clay and oxide minerals,[6,12-18] modified zeolites,[19] and organophilic clays[20-25] have been tested and evaluated for their adsorptive properties for radioiodine.

When exposed to groundwater, vitrified or cement waste forms used for immobilizing radioactive waste will leach high concentrations of dissolved constituents containing moderate to high alkalinity. Under such conditions, several complex geochemical reactions are known to occur in waste forms, neighboring engineered structures, and the surrounding sediments, which include the following: dissolution of several carbonate and silicate minerals, precipitation of secondary and tertiary mineral phases, radionuclide adsorption onto minerals (primary, secondary, and tertiary), and sequestration of radionuclides into secondary and tertiary mineral phases.

Under Eh-pH conditions typically encountered in ground and surface waters, iodine exists mainly as iodide. In highly oxidizing and alkaline conditions, however, iodine may exist as iodate. Therefore, a significant fraction of iodine in highly oxidizing alkaline leachates from glass and cement waste forms may also exist in the form of iodate. A moderate pH value (<8.0) and the absence of any strong oxidants in the glass leachate used in our experiments indicated that iodine in solution would exist principally as iodide.

According to the criteria enumerated by Vlacil,[26] successful radionuclide sorbent materials should (1) have at least moderate adsorption/immobilization properties; (2) be stable for a long time in the post-closure environment; (3) not adversely affect water chemistry; and (4) not be prohibitively expensive. The objective of this study was to evaluate novel adsorbent materials we have developed as high-affinity sorbents for radioiodine and to compare their adsorptive performance with selected natural minerals. In this study, we investigated the radioiodine sorptive performance of functionalized nanoporous materials.

5.2 Experimental

The iodide-specific sorbents consisted of a series of materials developed from nanoporous ceramics. These nanoporous ceramic substrates have very high surface areas ($\sim$1,000 m^2/g) and specially tailored pore sizes that range from 20 to 100 Å. These high-surface-area substrates are functionalized with monolayers of well-ordered functional groups that have a high affinity and specificity for specific types of free or complex cations or anions. These self-assembled monolayers on mesoporous silica (SAMMS) materials with high adsorption properties have been tested successfully on a series of heavy metal cations (Hg, Cu, Cd, and Pb) and oxyanions (As, Cr, Mo, and Se). Detailed descriptions of the synthesis, fabrication, and adsorptive properties of these novel materials have been published.[27–34] The nanoporous sorbents selected for these tests consisted of thiol-functionalized SAMMS materials. One of these samples was prepared by saturating the sorption sites with mercury and the other was generated by site saturation with silver ions.

Certain sulfide minerals are known to adsorb radioactive iodide. Therefore, six sulfide minerals were selected for tests of their adsorptive performance with synthetic novel sorbents (Table 5.1). All mineral samples except argentite were obtained from Ward's Geology, Rochester, New York. The

Table 5.1. Experimental matrix used in adsorption experiments.

^{125}I Spike (Bq/mL)	Solution Matrix	Sorbents
~36,500	Groundwater	Hg-thiol SAMMS, Ag-thiol SAMMS
~32,500	Glass leachate	Hg-thiol SAMMS, Ag-thiol SAMMS, argentite (Ag_2S), chalcocite (Cu_2S), chalcopyrite ($CuFeS_2$), cinnabar (HgS), galena (PbS), stibnite (Sb_2S_3)

argentite was synthesized by adding 1 g of reagent-grade silver nitrate to 1.4 g of reagent-grade sodium sulfide monohydrate dissolved in 200 mL of DI water. The precipitate was separated and washed three times with deionized distilled water and dried overnight at 105°C. The mineral samples were prepared for adsorption experiments by crushing with an agate mortar and pestle and then sieving to separate particles between <0.150 and >0.075 mm.

The groundwater used in these experiments was collected from an uncontaminated well located in the Hanford Site. The groundwater was filtered through a 0.45-micron filter and then analyzed using standard techniques. Inductively coupled plasma-atomic emission spectroscopy and ion chromatography were used to determine dissolved cation and anion concentrations, respectively. The composition of the groundwater indicated that calcium, magnesium, sodium, and potassium are the principal cationic constituents with chloride, sulfate, and bicarbonate the dominant anions (Table 5.2).

Glass leachate was prepared by equilibrating, for more than seven days, a quantity of crushed simulated waste glass with groundwater from the Hanford Site. Following equilibration, the glass leachate was filtered through a 0.45-micron filter (Table 5.2).

Adsorption experiments were conducted by equilibrating each sorbent with aliquots of groundwater or simulated glass leachate spiked with ^{125}I (Table 5.1). Solution-to-solid ratios of ~100, ~500, ~1,000, ~5,000, and ~10,000 mL/g were used to evaluate the degree of radionuclide loading. A positive control containing the ^{125}I-spiked groundwater and sorbent was used to evaluate ^{125}I sorption to labware and filters. The mixture was gently agitated for ~20 hours at $25 \pm 3°C$ and portions of equilibrated solutions were removed, filtered, and counted for residual ^{125}I activity. Analysis of ^{125}I in liquid samples was conducted by gamma-ray spectrometry, using a calibrated Wallac® 1480 Wizard™ 3-in NaI detector with built-in software.

Table 5.2.   Chemical composition of groundwater samples and waste glass leachate.

Constituent	Groundwater (mg/L)	Glass Leachate (mg/L)
pH (SU)	8.1	7.8
Cond. (mS/cm)	0.40	0.56
Alk (as CO_3^{2-})	54.1	67.5
Cl	7.8	22.0
Br	0.10	<0.01
F	0.17	<0.01
I	<0.005	<0.005
NO_2^-	0.68	<0.01
NO_3^-	27.2	1.7
PO_4^{-3}	<0.01	0.3
SO_4^{-2}	82.5	108.0
Al	0.01	0.02
B	<0.05	0.11
Ba	<0.03	0.03
Be	<0.01	0.01
Ca	49.5	61.4
Fe	0.07	<0.05
K	1.7	8.3
Mg	14.6	16.1
Mn	0.17	0.03
Na	13.2	46.0
Si	16.5	16.6

5.3 Results and Discussion

The adsorption data indicated that the nanoporous sorbents (Hg-thiol SAMMS and Ag-thiol SAMMS) had adsorbed as much as ~3 × 10^8 Bq/g of radioiodine (Table 5.3). Among the sulfide minerals, the highest observed adsorption for cinnabar, argentite, chalcocite, and chalcopyrite were ~7 × 10^7 Bq/g, ~1.8 × 10^8 Bq/g, ~3.5 × 10^6 Bq/g, and ~6.9 × 10^5 Bq/g, respectively (Table 5.3). A positive control experiment indicated that the labware and filters did not adsorb any radioiodine. The other two sulfide minerals that were tested, galena and stibnite, did not adsorb iodide to any measurable extent. These data showed that the novel sorbents had the highest iodide adsorption capacities that were about two to four times more than that of cinnabar and argentite and about two to three orders of magnitude higher than that of chalcocite and chalcopyrite (Table 5.3).

The selectivity (affinity) of a sorbent for a contaminant is typically expressed as a distribution coefficient (K_d) which defines the partitioning

Table 5.3.   Iodide-125 adsorption by novel sorbents and minerals.

Soln./Sol. Ratio (mL/g)	Eq. Act. (Bq/mL)	Ads. (Bq/g)	Ads. (%)	Soln./Sol. Ratio (mL/g)	Eq. Act. (Bq/mL)	Ads. (Bq/g)	Ads./ (%)
			Groundwater				
	Hg-thiol SAMMS				Ag-thiol SAMMS		
99	1,093	3.51×10^6	97	98	108	3.58×10^6	100
489	611	1.76×10^7	98	495	409	1.79×10^7	99
1,007	942	3.59×10^7	97	1,021	848	3.64×10^7	98
5,255	1,911	1.82×10^8	95	5,068	2,299	1.73×10^8	94
9,618	2,801	3.24×10^8	92	10,432	2,961	3.50×10^8	92
			Glass Leachate				
	Hg-thiol SAMMS				Ag-thiol SAMMS		
480	163	1.55×10^7	100	96	7	3.13×10^6	100
968	165	3.13×10^7	100	485	64	1.57×10^7	100
4,860	427	1.56×10^8	99	973	124	3.15×10^7	100
9,745	647	3.10×10^8	98	4,825	2,188	1.46×10^8	94
				9,741	2,274	2.94×10^8	93
	Cinnabar				Argentite		
97	1,078	3.03×10^6	97	96	67	3.12×10^6	100
469	2,402	1.41×10^7	93	488	781	1.55×10^7	98
953	7,471	2.38×10^7	77	832	2,119	6.77×10^7	93
4,253	21,060	4.86×10^7	35	4,057	7,461	1.02×10^8	77
10,950	26,020	7.08×10^7	14	8,026	9,528	1.84×10^8	71
	Chalcocite				Chalcopyrite		
98	18,614	1.35×10^6	43	96	29,562	2.82×10^5	9
444	28,206	1.90×10^6	13	461	31,587	4.35×10^5	3
941	29,380	2.92×10^6	10	953	31,766	6.86×10^5	2
8,920	32,096	3.48×10^6	1	4,046	33,067	0	0
				8,109	32,559	0	0

of the contaminant between sorbent and solution phase at equilibrium. The distribution coefficient is the measure of an exchange substrate's selectivity or specificity for adsorbing a specific contaminant or a group of contaminants from matrix solutions, such as waste streams. The distribution coefficient (sometimes referred to as the partition coefficient at equilibrium) is defined as a ratio of the adsorption density to the final contaminant concentration in solution at equilibrium. This measure of selectivity is defined as

$$K_d = \frac{(x/m)_{eq}}{c_{eq}}, \qquad (5.1)$$

Table 5.4.   Iodine-125 adsorption affinity parameters K_d (mL/g) for novel sorbent materials and minerals.

Adsorbent	Groundwater[a]	Surface Water[b]	Glass Leachate[a]	Concrete Leachate[b]
Hg-thiol SAMMS	3×10^4 to 1×10^5	2×10^3 to 2×10^6	8×10^4 to 5×10^5	8×10^3 to 6×10^5
Ag-thiol SAMMS	3×10^4 to 1×10^5	2×10^4 to 3×10^5	6×10^4 to 5×10^5	9×10^3 to 6×10^5
Cinnabar (HgS)	—	7×10^2 to 3×10^3	2×10^3 to 7×10^3	7×10^1 to 1×10^2
Argentite (Ag$_2$S)	—	3×10^2 to 8×10^4	1×10^4 to 5×10^4	2×10^4 to 7×10^5

[a]This study.
[b]Ref. 11

where K_d is the distribution coefficient (mL/g), $(x/m)_{eq}$ is the equilibrium adsorption density (Bq of iodide per gram of adsorbing substrate), and c_{eq} is the iodide concentration (Bq/mL) in contacting solution at equilibrium.

The novel sorbents exhibited very high K_d for ^{125}I in both Hanford groundwater and the simulated glass leachate (Table 5.4). The iodide K_d value for Hg-thiol SAMMS substrate in groundwater (3×10^4 to 1×10^5 mL/g) was noticeably lower than the K_d value in glass leachate (8×10^4 to 5×10^5 mL/g). This difference in the I adsorption affinity of Hg-SAMMS can be attributed to the probable surface sites that exist in these two solution matrices. Based on Hg hydrolysis and speciation data of Baes and Mesmer,[35] we estimated that the iodide binding sites on Hg-SAMMS equilibrated in the groundwater matrix were likely to be R-Hg(OH)$_2$, whereas, due to lower pH and higher concentration of chloride, the binding sites on Hg-SAMMS reacting in glass leachate medium were probably in the form of R-HgOHCl. If this were the case, the I adsorption in these two cases can be represented by the following ligand exchange reactions:

$$R\text{-}Hg(OH)_2^0 + 2I^-\,(aq) = 2OH^-\,(aq) + R\text{-}HgI_2^0 \tag{5.2}$$

$$R\text{-}HgOHCl^0 + 2I^-\,(aq) = Cl^-\,(aq) + OH^+\,(aq) + R\text{-}HgI_2^0. \tag{5.3}$$

Although association constants for surface species (R-Hg(OH)$_2^0$, R-HgOHCl^0, R-HgI$_2^0$) are unknown, using association constants for aqueous species,[36] the magnitude of I adsorption constants (ligand exchange) for these reactions were estimated to be $\sim 7 \times 10^2$ and $\sim 4 \times 10^6$ respectively. Using these constants, the free energies of iodide adsorption were calculated to be ~ -16 kJ/mol and ~ -38 kJ/mol respectively. These calculated adsorption free energy values indicated that although iodide adsorption onto both the sites (R-Hg(OH)$_2^0$ and R-HgOHCl^0) are energetically favored, adsorption through ligand exchange onto R-HgOHCl^0 is likely to occur with significantly higher free energy release than the ligand exchange reaction with R-Hg(OH)$_2^0$. These qualitative differences in the energetics of ligand exchange reaction appeared to be reflected in the higher distribution coefficient (K_d) observed in glass leachate medium.

The iodide K_d values for Ag-thiol SAMMS in both the groundwater and glass leachate matrices are 3×10^4 to 1×10^5 mL/g and 6×10^4 to 5×10^5 mL/g respectively. The magnitude of these values is similar to that of Hg-thiol SAMMS. Therefore, the adsorption mechanisms of iodide onto Ag-thiol SAMMS sites is expected to be similar to the mechanism suggested for Hg-thiol SAMMS sites.

As compared to novel sorbents, the iodide K_d values for cinnabar and argentite were 2×10^3 to 7×10^3 mL/g and 1×10^4 to 5×10^4 mL/g, respectively. These data indicate that the novel sorbents performed significantly better in that they adsorbed two to four times more ^{125}I and exhibited significantly higher adsorption affinity (distribution coefficients one to two orders of magnitude higher) than the sulfide minerals, cinnabar and argentite. Such improved performance can be attributed to higher numbers of accessible metal sites on novel sorbents as compared to the accessible adsorption sites on metal sulfides. Previous measurements indicated that Hg-thiol and Ag-thiol SAMMS have about 2 and 2.8 mmol of adsorption sites per gram of substrate. By comparing the adsorption performance of novel sorbents and the minerals, we estimated that cinnabar has a maximum of ~0.45 mmol of iodide adsorption sites per gram of material, whereas argentite has a maximum of ~1.47 mmol of iodide adsorption sites per gram of material. Therefore the observed maximum ^{125}I loading in these experiments (~7×10^{-8} mmol/g for novel getters, and 2×10^{-8} to 4×10^{-8} mmol/g for mineral sorbents) indicate occupancy of a negligible fraction of potential I adsorption sites for both the novel and mineral sorbents.

Although the energetics of iodide adsorption through ligand exchange reactions are similar for Hg-SAMMS and cinnabar, the higher maximum adsorption and distribution coefficient exhibited by the novel sorbents can be attributed to its two times higher number of available sites per unit mass (~2 mmol/g). Similarly, the higher maximum adsorption and distribution coefficient exhibited by Ag-SAMMS in comparison to argentite can also be ascribed to its significantly higher number of potential I adsorption sites.

For comparison, iodide (^{129}I) data for these same sorbents obtained in a previous study[11] in surface water and cement leachate matrices are also

Table 5.5.   Iodide distribution coefficient data for natural and synthetic sorbent materials.

Sorbent	K_d (mL/g)	Reference
Hg-thiol SAMMS	3×10^4 to 5×10^5	This study
	2×10^3 to 2×10^6	11
Ag-thiol SAMMS	3×10^4 to 5×10^5	This study
	2×10^4 to 5×10^5	11
Cinnabar	2×10^3 to 7×10^3	This study
	1×10^1 to 1×10^2	4
	0 to 5×10^1	7
	7×10^1 to 3×10^3	8, 9
	7×10^2 to 3×10^3	11
	4×10^3 to 2×10^4	15
Argentite	1×10^4 to 5×10^4	This study
	3×10^2 to 8×10^4	11
Other sulfides: bornite, chalcocite, chalcopyrite, galena, pyrite, enargite	1×10^1 to 3×10^3	This study, 3, 4, 5, 6, 8, 9, 11, 15
Oxides: Al-, Cu-, Fe-, Pb-, Ti-	0 to 1×10^2	2, 4, 5, 6, 16
Clays: allophane, attapulgite, bentonite, clinoptilolite, kaolinite, illite, sepiolite, vermiculite	0 to 3×10^1	4, 6, 15, 17
Modified clays	2×10^0 to 5×10^3	21, 22, 24, 25

included in Table 5.4. These data confirmed that the novel sorbents are capable of adsorbing radioiodine with high specificity. For instance, these experiments were conducted in different media with pH and solution conductivity ranging from a low of 5.65 SU and 0.021 mS/cm for surface water to 11.8 SU and 4.9 mS/cm for cement leachate respectively. These data suggest that both Hg-thiol SAMMS and Ag-thiol SAMMS can adsorb radioiodine from diverse media with differing pH, chemical composition, and ionic strength.

To gain some understanding of the iodide adsorption behavior of the two novel sorbents (Hg-thiol and Ag-thiol SAMMS), the data obtained in our tests were compared with the published iodide performance data for natural and modified mineral sorbents (Table 5.5). These data show that most of the natural minerals, except sulfides, have K_d values of $<10^3$ mL/g, whereas sulfide minerals such as cinnabar and argentite have K_d values that typically range from 10^2 to 10^4 mL/g. Modified clays were reported to exhibit enhanced affinity (2×10^0 to 5×10^3 mL/g) to adsorb iodide. However, these data also showed that the performance of these modified minerals was no better than the iodide adsorption performance of natural sulfide minerals such as cinnabar and argentite. Additionally, most of these sorbent tests were conducted in solution matrices in which the competing halide ions (Cl, Br, and F) were either absent or their concentrations were low.

The novel sorbents Hg-thiol and Ag-thiol SAMMS showed very high affinities for adsorption of radioiodine ($K_d \sim 10^4$ to 10^5 mL/g). The performance of these novel sorbents was remarkable because these sorbents were tested in matrices of groundwater, surface water, waste glass, and concrete leachates, which are typical environments in which these sorbents are expected to perform. Also, these novel sorbent tests were conducted in matrices of natural environmental solutions that contain other halide ions such as Cl, Br, and F in concentrations that far exceeded the concentrations of radioiodine. Calculated molar concentration ratios showed that Cl, Br, and F concentrations in these natural solution matrices exceeded radioiodine concentrations by eight to ten orders of magnitude. Therefore, the magnitude of the measured K_d values indicated that the novel sorbents were adsorbing iodide with very high selectivity.

5.4 Conclusions

In summary, performance tests were conducted using novel sorbent materials. These tests revealed that these sorbent materials can immobilize or delay the transport of radioiodine that would be released during physical and chemical degradation of solidified low-level waste packages. The results showed that metal-capped novel sorbents such as Hg-thiol and Ag-thiol SAMMS, designed specifically to adsorb soft anions such as I^-, had very high affinities for adsorption of radioiodine ($K_d \sim 1 \times 10^4$ to 4×10^5 mL/g). The iodide adsorption performance of these novel sorbents was one to two orders of magnitude better than that of many natural mineral and modified mineral sorbents. These data indicate that the novel nanoporous sorbent materials are capable of significantly retarding the mobility of radioiodine leaching from physically and chemically weathered low-level waste packages during various physical and chemical weathering reactions expected during long-term disposal.

Acknowledgments

The Pacific Northwest National Laboratory is a multiprogram national laboratory operated for the U.S. Department of Energy by Battelle Memorial Institute under Contract DE-AC06-76RLO 1830.

References

1. J. C. Wren, J. M. Ball and G. A. Glowa, *Nucl. Technol.* **129**, 297–325 (2000).
2. Z. Haq, G. M. Bancroft, W. S. Fyfe, G. Bird and V. J. Lopata, *Environ. Sci. Technol.* **14**, 1106–1110 (1980).
3. R. Strickert, A. M. Friedman and S. Fried, *Nucl. Technol.* **49**, 253–266 (1980).
4. G. Allard, B. Torstenfelt, K. Andersson and J. Rydberg, Possible retention of iodine in the ground, in *Material Research Society Symposium Proceedings*, Vol. 176 (Material Research Society, Pittsburgh, Pennsylvania, 1981), pp. 673–680.
5. Z. Huie, Z. Jishu and Z. Lanying, *Radiochim. Acta* **44/45**, 143–145 (1988).
6. D. Rancon, *Radiochim. Acta* **88**, 87–193 (1988).
7. Y. Ikeda, M. Sazarashi, M. Tsuji and R. Seki, *Radiochim. Acta* **65**, 195–198 (1994).
8. S. D. Balsley, P. V. Brady, J. M. Krumhansl and H. L. Anderson, *Environ. Sci. Technol.* **30**, 3025–3027 (1997).
9. S. D. Balsley, P. V. Brady, J. M. Krumhansl and H. L. Anderson, [129]*I and* [99]*TcO₄ Scavengers for Low Level Radioactive Waste Backfills*, SAND 95-2978. Sandia National Laboratories, Albuquerque, New Mexico (1997).
10. S. D. Balsley, P. V. Brady, J. M. Krumhansl and H. L. Anderson, *J. Soil Contam.* **7**, 125–141 (1998).
11. D. Kaplan, S. V. Mattigod, K. E. Parker and G. Iversen, *Experimental Work in Support of the* [129]*I-Disposal Special Analysis*, WSRC-TR-2000-00283, Rev 0. Westinghouse Savannah River Company, Savannah River Site, Aiken, South Carolina (2000).
12. M. Atkins and F. P. Glasser, Encapsulation of radioiodine in cementitious waste forms, in *Material Research Society Symposium Proceedings*, Vol. 176 (Material Research Society, Pittsburgh, Pennsylvania, 1990), pp. 15–22.
13. R. A. Couture and M. G. Seitz, *Nucl. Chem. Waste Man.* **4**, 301–306 (1983).
14. P. Taylor, *A Review of Methods for Immobilizing Iodine-129 Arising from a Nuclear Fuel Cycle Plant, with Emphasis on Waste-Form Chemistry*, AECL-10163. Atomic Energy Canada Ltd., Whiteshell Nuclear Research Establishment, Pinawa, Canada (1990).
15. M. Sazarashi, Y. Ikeda, R. Seki and H. Yoshikawa, *J. Nucl. Sci. Tech.* **31**, 620–622 (1994).
16. N. Hakem, B. Fourest, R. Guillaumont and N. Marmier, *Radiochim. Acta* **74**, 225–230 (1996).
17. D. Kaplan, R. J. Serne, K. E. Parker and I. V. Kutnyakov, *Environ. Sci. Technol.* **34**, 399–405 (2000).
18. M. J. Kang, K. S. Chun, S. W. Rhee and Y. Do, *Radiochim. Acta* **85**, 57–63 (1999).
19. G. D. Gradev, *J. Radioanal. Nucl. Chem.* **116**, 341–346 (1987).
20. J. Bors, *Radiochim. Acta* **51**, 139–143 (1990).
21. J. Bors, *Radiochim. Acta* **58/59**, 235–238 (1992).
22. J. Bors, A. Gorny and S. Dultz, *Radiochim. Acta* **66/67**, 309–313 (1994).
23. J. Bors, A. Gorny and S. Dultz, *Radiochim. Acta* **74**, 231–234 (1996).

24. J. Bors, S. Dultz and A. Gorny, *Radiochim. Acta* **74**, 231–234 (1998).
25. M. Sazarashi, Y. Ikeda, R. Seki, H. Yoshikawa and Y. Takashima, Adsorption behavior of I ions on minerals for geological disposal of ^{129}I wastes, in *Material Research Society Symposium Proceedings*, Vol. 353 (Material Research Society, Pittsburgh, Pennsylvania, 1995), pp. 1037–1043.
26. B. Viani, *Assessing Materials ("Getters") to Immobilize or Retard the Transport of Technetium Through the Engineered Barrier System at the Potential Yucca Mountain Nuclear Waste Repository*, UCRL-ID-133596. Lawrence Livermore National Laboratory, Livermore, California (1999).
27. G. E. Fryxell, J. Liu, A. A. Hauser, Z. Nie, K. F. Ferris, S. V. Mattigod, M. Gong and R. T. Hallen, *Chem. Mater.* **11**, 2148–2154 (1999).
28. G. E. Fryxell, J. Liu and S. V. Mattigod, *Mat. Tech. and Adv. Perf. Mat.* **14**, 188–191 (1999).
29. G. E. Fryxell, J. Liu, S. V. Mattigod, L. Q. Wang, M. Gong, T. A. Hauser, Y. Lin, K. F. Ferris and X. Feng, Environmental applications of interfacially-modified mesoporous ceramics, in *Proceedings of the 101st National Meetings of the American Ceramic Society* (1999).
30. K. M. Kemner, X. Feng, J. Liu, G. E. Fryxell, L. Q. Wang, A. Y. Kim, M. Gong and S. V. Mattigod, *J. Synchrotron Radiat.* **6**, 633–635 (1999).
31. J. Liu, G. E. Fryxell, S. V. Mattigod, M. Gong, Z. Nie, X. Feng and K. N. Raymond, *Self-Assembled Monolayers on Mesoporous Support (SAMMS) Technology for Contaminant Removal and Stabilization*, PNL-12006. Pacific Northwest National Laboratory, Richland, Washington.
32. J. Liu, G. E. Fryxell, S. V. Mattigod, T. S. Zemanian, Y. Shin and L. Q. Wang, *Stud. Surf. Sci. Catal.* **129**, 729–738 (2000).
33. S. V. Mattigod, X. Feng, G. E. Fryxell, J. Liu, M. Gong, C. Ghormley, S. Baskran, Z. Nie and K. T. Klasson, *Mercury Separation from Concentrated Potassium Iodide/Iodine Leachate Using Self-Assembled Mercaptan on Mesoporous Silica Technology*, PNNL-11714. Pacific Northwest National Laboratory, Richland, Washington (1997).
34. S. V. Mattigod, J. Liu, G. E. Fryxell, S. Baskaran, M. Gong, Z. Nie, X. Feng and K. T. Klasson, *Fabrication and Testing of Engineered Forms of Self-Assembled Monolayers on Mesoporous Silica (SAMMS) Material*, PNNL-12007. Pacific Northwest National Laboratory, Richland, Washington (1998).
35. C. F. Baes and R. E. Mesmer, *The Hydrolysis of Cations* (John Wiley and Sons, New York, 1976).
36. R. M. Smith and A. E. Martell, *Critical Stability Constants* (Plenum Press, New York, 1976).

Nanoporous Organic/Inorganic Hybrid Materials

Chapter 6

Nature's Nanoparticles: Group IV Phosphonates

Abraham Clearfield

Texas A&M University, College Station, TX, USA

6.1 Introduction

6.1.1 *Natural nanoparticle formation*

From the time nanoscience and technology became fashionable, practitioners have devised a very large number of methods by which nano-sized particles or systems can be produced. In my own work it is nature that has been the producer of nanoparticles. We have just been the recipient. This has been true for zirconium phosphates, several types of zirconium phosphonates, aluminum phosphonates and our latest discovery, tin phosphonates. These compounds have been utilized to develop layer-by-layer films, as ion exchangers, additives for proton-conducting fuel cells, as catalysts and catalyst carriers and in many diverse uses as nanoparticles. We shall begin our odyssey with a description of the synthesis, structure and properties of α-zirconium phosphate, $Zr(O_3POH)_2 \bullet H_2O$.

6.1.2 *α-zirconium phosphate and nanoparticles*

6.1.2.1 *History and structure*

The emergence of zirconium phosphate as a compound of interest arose from work done at Oak Ridge National Laboratory in the 1950s.[1] There was a need for ion exchangers to remove radioactive species from reactor cooling water. The organic ion-exchange resins of that time were degraded by the radioactive species in hot reactor water. Therefore, a search was under way worldwide for inorganic ion exchangers that would not be so affected. Hydrous oxides were considered and it was observed that hydrous zirconia sorbed large amounts of phosphate, becoming a cation exchanger.[1] Subsequently, zirconium phosphate was prepared as a gelatinous amorphous precipitate on addition of phosphoric acid to a soluble zirconium salt. The dried gel possessed interesting cation exchange properties. However, in hot water a significant loss of phosphate ion resulted due to hydrolysis.

At that time, I was employed by a branch of NL Industries that manufactured zirconium chemicals. I suggested a study of a family of zirconium-based ion exchangers but met with only lukewarm enthusiasm. However, I was teaching in the evening school at Niagara University and proposed the project to a M.S. candidate, James Stynes. The result was that we were able to convert the zirconium phosphate gel to crystals and establish the composition as $Zr(HPO_4)_2 \cdot H_2O$ and the layered nature of the compound.[2] The crystal structure was determined first by film methods[3a] and later by automated diffractometry.[3b] The unit cell dimensions are $a = 9.060(2)$, $b = 5.297(1)$, $c = 15.414(3)$ Å, $\beta = 101.71(2)$, monoclinic space group $P2_1/n$. A schematic drawing of the structure is shown in Figure 6.1. The Zr atoms are slightly above and below the mean plane of the layer and are six-coordinate to oxygens from six phosphate groups. Each monohydrogen phosphate bonds to three Zr atoms arranged at the apices of a near equilateral triangle. The P-OH group points into the interlayer space and hydrogen bonds to the water molecule. The water in turn hydrogen bonds to a framework oxygen in the same layer. There are no interlayer hydrogen bonds but only van der Waals forces holding the layers together.

6.1.2.2 Crystal growth and ion-exchange behavior

In order to grow single crystals for the first X-ray study[3a] the gel was held in a sealed quartz tube with $12\,M\ H_3PO_4$ at 170°C for four weeks. Subsequently, crystals could be grown in a day or two in an HF solution.[4] These differences in the rate of crystal growth were of interest because we had observed a rather pronounced difference in the ion-exchange behavior of batches of crystals grown under different conditions. Examination of the dried gel particles showed that the particles had no crystalline-type shape and were on the order of 10–40 nm in size. Even when refluxed in $0.35\,M$ H_3PO_4 for several days the particles did not grow significantly. Figure 6.2 illustrates the condition of the crystals grown in increasing concentrations of phosphoric acid. The slow crystal growth is the result of the low solubility of zirconium phosphate in the solutions of low phosphoric acid concentration. The estimated solubility of crystalline zirconium phosphate in $1\,M\ H_3PO_4$ at 25°C is estimated to be on the order of 10^{-6} g/L. A graph of solubility in phosphoric acid concentration of 7 M and greater at several temperatures

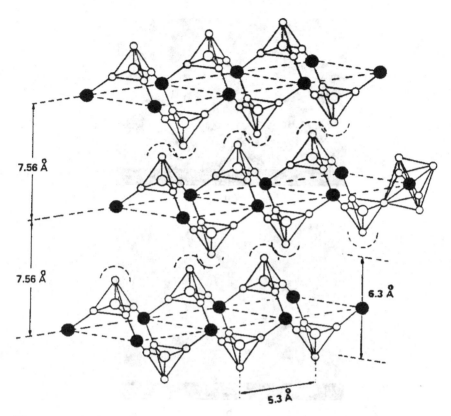

Figure 6.1. Schematic representation of α-zirconium phosphate layers as viewed down the *b*-axis direction: Zr, ●; P, ○; O, °.

is given in Figure 6.3.[5] The crystallization mechanism is that of Ostwald ripening in accord with the solubility shown in Figure 6.3. This fact is illustrated by the X-ray patterns in Figure 6.4.[6]

The picture that arises from our studies is that of crystal perfection accompanying particle growth. The behavior of the particles is strongly tied to their crystallinity. We shall illustrate by their ion-exchange behavior and intercalation and exfoliation properties. Because there are several additional phases of zirconium phosphate, we shall refer to the present one as the alpha phase or α-ZrP. The variation of the ion-exchange curves as a function of crystallinity is shown in Figures 6.5(a) and 6.5(b). To understand why the titration curves change in shape as the crystallinity increases we first consider the behavior of a strong acid polystyrene sulfonic acid ion-exchange resin. These resins are influenced by the degree of cross-linking by divinyl benzene. For an 8% cross-linked resin, the cross-links

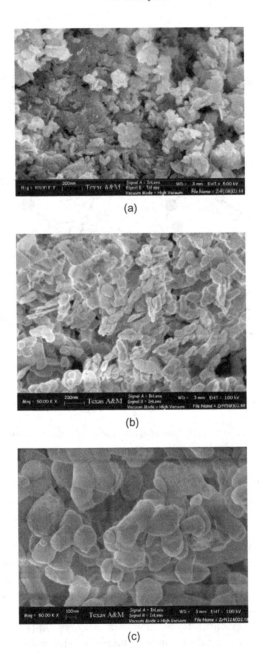

Figure 6.2.   Scanning electron micrographs of α-ZrP as grown by refluxing the gel in increasing concentrations of H_3PO_4. Conditions: $10\,g$ $ZrOCl_2 \cdot 8H_2O$ in $100\,ml$ H_3PO_4 of molarity (a) 3, (b) 9 and (c) 12.

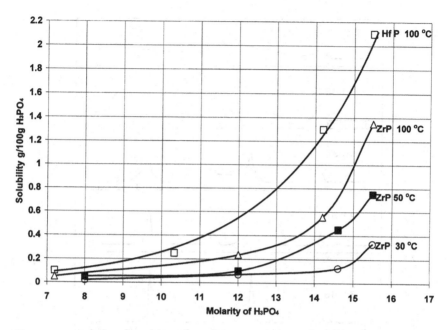

Figure 6.3. Solubility of α-ZrP and α-HfP in grams per 100 g of aqueous phosphoric acid.

between the linear polystyrene chains create cavities that are about 50 Å in diameter. The cavities are filled with water, causing the resin beads to swell. Consider the exchange of Na^+ for protons of the sulfonic acid groups, which are present in the cavities as hydronium ions. As the protons are displaced to the outer solution, the Na^+ spreads uniformly throughout the resin bead so that only one solid phase is present. Using the terminology of the phase rule, the system has three components. One choice of components is the hydrogen ion displaced or the pH, the total sodium ion added and ion-exchange capacity which gives us the amount of H^+ left in the solid phase. At constant temperature and pressure the phase rule equation $f = c - p + 2$ becomes $f = c - p + 0$, where f = degrees of freedom, c = number of components, p = number of phases. There are two phases present, the solid exchanger and the solution phase, so that the system has a degree of freedom. Thus, for each addition of NaOH, the pH increases slightly until the capacity is reached, whereupon the pH increases sharply. The titration curve is analogous to that of a strong acid-strong base titration.

For the ZrP gel and also for 0.5:48 (Figure 6.4) where the crystallinity is poorly developed, the added Na^+ spreads throughout the entire ZrP

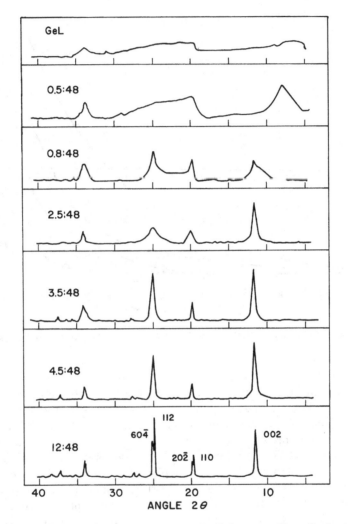

Figure 6.4.   X-ray diffraction powder patterns of α-ZrP prepared by refluxing the gel in different concentrations of H_3PO_4 for 48 hours. The concentration of H_3PO_4 in molarity is the number preceding 48.

nanoparticle. Thus, the pH increases with each addition because of the one degree of freedom. The pH rises more steeply than for the sulfonic acid resin because the P-OH groups are weak acid groups and some sodium ion hydrolysis occurs, leaving some NaOH in the solution phase. An interesting feature of the exchange is the fact that as the 0.5:48 sample becomes infused with water, the interlayer spacing originally at ~8 Å

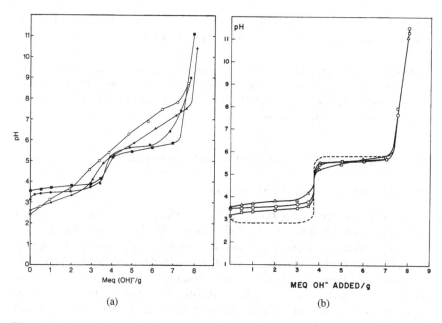

Figure 6.5. Potentiometric titration curves of α-ZrP for (a) samples of low crystallinity and (b) intermediate crystallinity. Dashed line represents the theoretical curve for a fully crystalline sample. Conditions: (a) amorphous gel refluxed in H_3PO_4 for 48 hours in 0.8 M (o), 2.5 M (▲), 4.5 M (•), 12 M (■); (b) amorphous gel in 12 M H_3PO_4 for 48 hours (Δ), 96 hours (o), 190 hours (□); tritrant 0.1 M (NaCl + NaOH).

increases to 11.2 Å. The sodium ion is then able to diffuse equally throughout the particle. In the case of the fully crystalline phase, which requires about three weeks of refluxing in 12 M H_3PO_4 to achieve this level of crystallinity,[7] a second phase of composition $Zr(NaPO_4)(HPO_4) \cdot 5H_2O$ with an 11.8 Å interlayer spacing forms at the first uptake of Na^+. This phase spreads inward from the edges until the crystallite is completely converted to the half-exchanged phase. Because two solid phases are always present, there are no degrees of freedom and the system is invariant until the endpoint is reached at 3.53 mequiv/g of $Zr(HPO_4)_2$. A second exchange process converts the half-exchanged phase to $Zr(NaPO_4)_2 \cdot 3H_2O$. The ideal titration curve would then resemble the dashed line in Figure 6.5(b). In between the gel-like phases and the fully crystalline phase there are two phases formed at low sodium ion uptake, and one of them is a solid solution. Eventually the phase with a low level of Na^+ is converted to the solid solution phase which proceeds to completion of the sodium uptake. As the crystallinity of the exchanger increases, the solid solution ranges become narrower but the number of changes of phase increase. The situation is quite complex.[8] A good summary of these and other aspects of the ion-exchange processes has been presented by Alberti.[9]

The individual particles in Figure 6.2(a) are as small as 20–50 nm parallel to the layers, with some particles as large as 200 nm but of poor crystal shape. The thickness of the particles is much less and may range as low as 2–5 nm or three to seven layers.

6.2 Nanoparticle Formation and its Consequences

6.2.1 Intercalation and exfoliation

By intercalation we mean the process by which molecules are inserted between the layers topotactically. This process should be reversible. In contrast, exfoliation is a process by which the layers are separated into individual colloidal particles. The thickness of the colloidal particles are generally on the order of 1 nm, i.e. one layer thick. Many layered compounds undergo intercalation reactions and may also be exfoliated. An excellent introduction to the subject is provided in the books by Whittingham and Jacobson[10] and Müller-Warmuth and Schollhorn.[11] α-ZrP is a solid acid and as such readily intercalates amines. The same is true for the other isostructural group IV phosphates and arsenates. Extensive chapters on amine intercalation by the group IV phosphates are provided in the aforementioned books[10,11] as well as in Comprehensive Supramolecular Chemistry.[12]

The maximum uptake of n-alkylamines by α-ZrP is two moles per formula weight or one amine molecule per P-OH group. For most amines, transfer of the phosphate proton to the amine group accompanies the intercalation reaction with formation of a bilayer. If the amine is added as a titrant, several phases are found to form before the saturation point.[13] Titration to the half-endpoint with propylamine results in swelling of the layers and addition of water with sonication results in exfoliation of the layers. Exfoliation may also be induced by intercalation of ammonium salts such as tetramethyl ammonium hydroxide or the tetrabutylammonium cation. Exfoliation takes place because the attraction of the layers for each other becomes very weak as the layers are spread apart by the intercalation reaction. There is also a very low van der Waals attraction for the propylamine chains for each other at the half-exchange point because the amines are 10.6 Å apart, twice the lateral distance of P-OH groups on the layers. However, we have found some very interesting features in the process.

One of the interesting aspects of the exfoliation of α-ZrP is that the sheets lose phosphate to the solution.[14] The phosphate is replaced by hydroxyl groups and the process proceeds from the edges inward. By lowering the temperature to near zero degrees, the rate of hydrolysis is reduced. Hydrolysis has been observed in other ways. Addition of even the highly crystalline α-ZrP to water results in slow hydrolysis with accompanying lowering of pH. In carrying out measurements of heats of exchange[15] or titrations, the amount of hydrolysis needs to be considered.[16] Hydrolysis can be largely prevented by addition of phosphate to the solution. Furthermore, the lower the degree of crystallinity the less exfoliant is required for complete exfoliation;[17] see below.

6.2.2 *Polymer-zirconium phosphate nanocomposites*

One of our interests in exfoliation was as a means to prepare polymer-layered inorganic nanocomposites. In the conventional usage, smectite clays, generally montmorillonites, are incorporated into the polymers. To effect a good distribution of the clay in the polymer, amines are first intercalated into the clay so as to move the layers apart and increase the hydrophobicity of the clay-amine intercalate. The composites exhibit large changes in their physical and mechanical properties.[18] For example, improvements in modules, yield strength and elongation at break as well as resistance to oxygen diffusion have been observed. Generally three outcomes of the clay incorporation into the polymer have been noted (Figure 6.6): conventional, where stacks of layers are inserted into the polymer; intercalated, in which the layers are spread apart by an intercalant; and exfoliated, where a true nanocomposite is formed with the exfoliated layers.

The use of clays has several drawbacks. It is extremely difficult to achieve 100% purity of the clay or narrow particle size distribution and therefore a controlled aspect ratio of the clay nanofiller. In most cases, the clay is

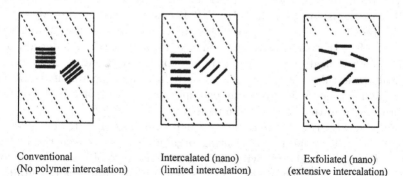

| Conventional | Intercalated (nano) | Exfoliated (nano) |
| (No polymer intercalation) | (limited intercalation) | (extensive intercalation) |

Figure 6.6. Schematic illustrations of the three possible types of polymer-clay composites (from Ref. 18a with permission).

not completely exfoliated and is rarely uniformly dispersed throughout the polymer.[19,20] It is therefore difficult to achieve reproducibility of the physical properties and to determine the effect of changing the aspect ratio of the layered filler on the properties of the nanocomposites. Our idea was to use a synthetic layered material such as α-zirconium phosphate to prepare the composites. To make the layers compatible with an epoxy resin, we choose to intercalate polyether amines into α-ZrP. Jeffamines are commercially available polyether amines containing both ethylene and isopropylene groups. They are designated as, for example, M300 where M stands for monoamine and 300 is the approximate formula weight. It was found that the ease of amine intercalation and subsequent exfoliation depended upon the crystallinity of the α-ZrP sample.

Samples of α-ZrP were prepared by refluxing 10 g of zirconyl chloride, $ZrOCl_2 \cdot 8H_2O$, in 100 ml of 3, 6, 9, 12 M H_3PO_4 for 24 hours.[17] Preparations were also carried out hydrothermally and refluxed with addition of HF. Hydrofluoric acid solubilizes the α-ZrP and speeds up the crystal growth process, as does the hydrothermal procedure. M715 was chosen as the intercalant and added to the α-ZrP samples. The results for the refluxed samples are shown in Figure 6.7 and not surprisingly the lower the crystallinity, the less M715 is required for complete layer intercalation. As little as 0.13 mol of M715 per mol of Zr was required for full expansion of the layers, whereas 1.5 mol was required for the 12 M preparation.

The reaction was performed in methyl ethyl ketone (MEK) and the ratio of M715 to α-ZrP was 2:1.[21] A transparent yellowish gel was obtained that gave an X-ray pattern with nine orders of 00ℓ reflections with $d_{001} = 73$ Å. The epoxy monomer was also dissolved in MEK to which 1.9% by volume of M715, α-ZrP gel was added. The solvent was removed and polymerization effected at 130°C. A transmission electron microscopy (TEM) micrograph of the resultant composite is shown in Figure 6.8. It shows a relatively uniform dispersion of exfoliated α-ZrP layers throughout the polymer.[22] The composite exhibited improved the mechanical properties but suffered from the fact that the glass transition temperature (Tg) was reduced from 227°C for the neat epoxy to 90°C in the composite. Control experiments indicated the culprit was the Jeffamine. Reduction of the amount of M715 to 0.5 mol resulted in an epoxy nanocomposite with a Tg of 160°C. Further improvement in the Tg is being pursued.

As a general rule, we may find systems of nanocomposites for a range of polymers by properly substituting organic groups between the α-ZrP layers that are compatible with the organic structure of the polymer; for example, ethylphenyl group for polystyrene. Organic groups can be grafted onto the exfoliated zirconium phosphate[23] layers as a means of gaining compatibility with the polymer. Proton-conducting properties of α-ZrP

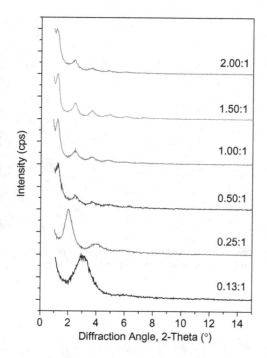

Figure 6.7. XRD powder patterns of α-ZrP prepared in three molar H_3PO_4 intercalated by Jeffamine M715 in the ratios indicated.

and their use in fuel cell membranes will be described in the section on sulfophosphonates.

6.2.3 *Nanoparticle α-zirconium phenylphosphonate*

Zirconium phenylphosphonate (ZrPP) was first prepared by Alberti *et al.*[24] Although its composition is stoichiometric, $Zr(O_3PC_6H_5)_2$, its X-ray powder pattern contained only a handful of broad peaks. Refluxing in the presence of HF improves the crystallinity but not sufficiently to obtain single crystals. In fact, it was not until 1993 that the crystal structure was solved utilizing crystals that had been grown by Prof. Alberti hydrothermally in HF at 200°C for 30 days (Figure 6.9). The layers are like those of α-ZrP but with phenyl groups replacing the -OH pendant groups in the phosphate. The reason for the 30° tilt of the phenyl rings arises from the fact that the Zr atoms are in the plane; whereas in α-ZrP, the metal atoms are slightly above and below the mean plane. The particle size as determined

Figure 6.8. Transmission electron micrograph of α-ZrP/epoxy polymer at high magnification showing the uniform dispersion of exfoliated α-ZrP layers.

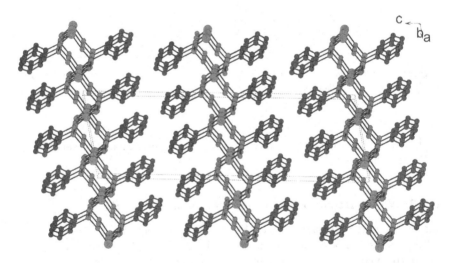

Figure 6.9. Structure of zirconium phenylphosphonate as viewed down the *b*-axis direction.

by scanning electron microscopy (SEM) was in the range of 20–60 nm when prepared hydrothermally at 140°C for a three-day hold time. These particles may be as thin as 7 nm or about nine layers thick (Figure 6.10). Smaller particles are produced by milder treatment.

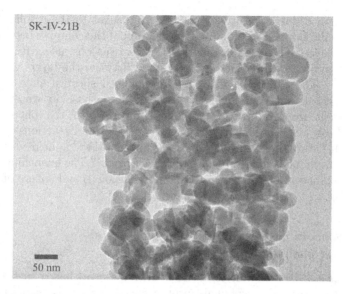

Figure 6.10. Scanning electron micrograph of zirconium phenylphosphonate particles grown hydrothermally at 140°C, three days, HF/Zr = 3.

6.2.4 *Sulfonated zirconium phenylphosphonate*

ZrPP can be sulfonated in fuming sulfuric acid to give $Zr(O_3PC_6H_4SO_3H)_{2-x}(O_3PC_6H_5)_x$.[26] However, the distance between phenyl groups along the layers is 5.3 Å. Because of the ring thickness (3.2 Å) there is barely enough room to insert an -SO$_3$H group in the meta-position relative to the P-C bond. The sulfonation reaction is more facile if a mixed derivative $Zr(O_3POH)(O_3PC_6H_5)$ is used. As a sulfonate the ZrPP becomes a polyelectrolyte with complete ionization of the sulfonic acid groups. This property results in the complete exfoliation of the sulfonate in aqueous media.[26] We carried out light scattering examination of one of these colloids with the result that about 95% of the particles could be assigned a layered shape with the in-plane dimensions of 4.5 nm × 4.5 nm and 1.5 nm thick.[27] This size is considerably smaller in length and breadth than the particles pictured in Figure 6.10 which may result from the aggressive sulfonation treatment. Further investigation is in progress. Several applications of the sulfonates are under examination. Their strong acid character imparts ion-exchange capabilities to these compounds.[28] The ion-exchange selectivity was found to increase as the cation size and charge increases. Free energies, $\Delta G°$ in kJ/mol, for the reactions $H^+ - M^+$ (M = Na^+, Cs^+) and M^{2+} (M = Mg^{2+}, Ba^{2+}) were determined (28b) as 0.27 (Na^+), -9.7 (Cs^+), -3.2 (Mg^{2+}) and -11.2 (Ba^{2+}).

Addition of large, positively charged complexes to the exfoliated sulfonate results in precipitation by sequestration of the complex between the sulfonate layers. For example, charge-transfer reactions, using $Ru(bipy)_3^{2+}$ (where bipy = 2,2'-bipyridine) encapsulated between the layers, were carried out.[29] The $Ru(bipy)_3$ exhibited pronounced spectral shifts in both its absorption and emission spectra relative to those in water. These shifts result from interactions of the complex with the phenyl rings of the host and the bipyridine. The excited state reactions between $Ru(bipy)_3^{2+}$ and the quencher, methylviologen (MV^{2+}), held within the zirconium sulfophenylphosphonate were measured.[30] The quenching occurs via a combination of diffusional (dynamic/collisional) and sphere of action quenching.

6.2.5 *Sensors based on zirconium phosphate and phenylphosphonate*

Alberti *et al.*[31,32] were able to devise hydrogen sensors based on zirconium phosphate and sulfophosphonate proton conductors. A layer of titanium hydride (TiH_x) was formed on the surface of a small disk of titanium metal by heating at 600°C in the presence of H_2. The hydride was then covered by a thin layer of exfoliated α-ZrP or $Zr(O_3PC_6H_4SO_3H)_x(HPO_4)_{2-x}$. A second electrode was added by deposition of a thin layer of platinum over the protonic conductor. The electromotive force (E) of this cell is given by:

$$E = -0.029 \log(P_{H_2}) + C, \qquad (6.1)$$

where P_{H_2} is the pressure exerted by H_2 and C is a constant.

In the presence of oxygen, a mixed potential:

$$E = (RT/aF)\ln(P_{O_2}{}^{1/4})(P_{H_2})^{-1/2} + C \qquad (6.2)$$

is measured (P_{O_2} is the pressure of oxygen and F is Faraday's constant). If the pressure of H_2 is held constant by use of a zirconium hydride electrode, the cell may be used as an oxygen sensor. Sensors for NO_x and SO_x are also under development.

6.3 Layer-by-Layer Constructs

6.3.1 *Thin films*

Thin films have great practical value in many applications such as optical coatings and microelectronics. One technique for growth of thin films that allows for exquisite control of the process is to grow the film layer by layer. A general summary of the techniques and results has been provided in a new book on nanochemistry.[33] We will limit ourselves to the use of phosphonates to construct multilayer thin films by a Langmuir–Blodgett (LB) films-type procedure. Mallouk and co-workers[34,35] pointed out the similarity between LB films and those formed from zirconium alkylphosphonates and devised a scheme to prepare thin films based upon this similarity. A substrate such as a gold film or silica surface is primed by binding one end of a long-chain molecule to the surface (Figure 6.11). The other end contains a phosphonic acid group. This surface is then dipped into a Zr(IV) solution, which then binds to the free phosphonic acid end of the monolayer. That the metal is actually bonded is evidenced by its not being removed on washing and subsequent adsorption of an α, ω-bis(phosphonic acid). The process is repeated as many times as desired. Characterization by ellipsometry and angle-resolved X-ray photoelectron spectroscopy (XPS) demonstrated that the thickness of each deposited layer was close to that of the layer spacing in the analogous microcrystalline solid.

Talham and co-workers[36] prepared zirconium phosphonate films by a LB technique. The surface of a silicon wafer was coated with a monolayer of octadecyltrichlorosilane (OTS). Then a single layer of octadecylphosphonic acid was transferred from an LB trough to the surface of the OTS-silicon wafer held in the trough. This procedure produced a monolayer of the phosphonic acid in a tail-to-tail arrangement with the OTS on the silicon wafer. The wafer was then removed from the trough and dipped into a beaker containing a 5 mM solution of Zr^{4+} to self-assemble the zirconium at the organic template. To complete the layer, the wafer was rinsed in water and then replaced into the LB trough where a new octadecylphosphonic acid film was compressed and transferred to the substrate. The film was built up by repetition of this three-step process. The position of the asymmetric methylene (v_a CH_2) band at $2{,}918\,cm^{-1}$ with a full width at half-maximum (fwhm) of $20\,cm^{-1}$ indicated that in this case an all-trans, close-packed template formed. The shape and position of this band remained unchanged as the multilayer film was built up. Ellipsometry and X-ray diffraction were in agreement as to the thickness of the bilayer (51–52 Å).

The films described in the foregoing paragraphs are centrosymmetric. Putvinski *et al.*[37] devised a method for depositing layers with polar order. 11-Hydroxyundecanethiol was deposited as a monolayer on a gold substrate. The hydroxyl groups were then phosphorylated with an acetonitrile solution of $POCl_3$ containing triethylamine to yield the corresponding phosphonic acid. The acid was in turn treated with Zr^{4+} that in turn was treated with 11-hydroxyundecyl phosphonic acid. The multilayers so built up are polar in the sense that in one layer the Zr is bonded to a PO_4 group and to a PO_3 group in the next layer.

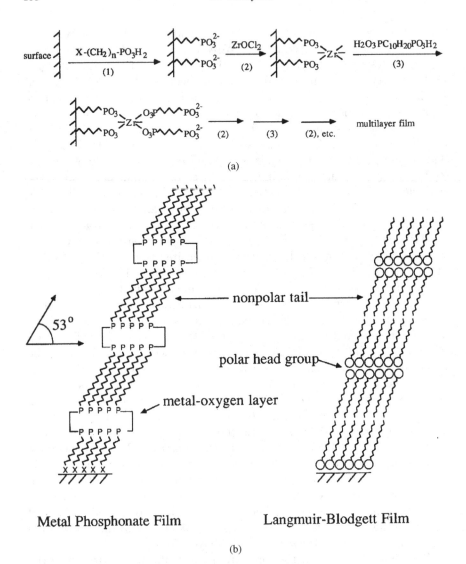

Figure 6.11. (a) Procedure for step-by-step assembly of zirconium diphosphonate thin films and (b) structural analogy between Langmuir–Blodgett films and the layer-by-layer films (from Ref. 35 with permission).

A similar procedure was utilized to fix azo dyes into repeating non-centrosymmetric layers.[38,39] These films exhibited non-linear optical behavior comparable in magnitude to that of $LiNbO_3$ and were stable to 150°C. Subsequently, films were prepared from a variety of non-linear chromophores.[39,40] The intensities of the second harmonic generated waves were proportional to the square of the number of layers.

The zirconium-phosphonate bonding in the layer-by-layer thin films is thought to yield the same arrangement as in the layers of α-zirconium phosphate. The area available to the phosphonate is 24 Å^2.[3] This requirement is met for most alkyl and linear aryl chains. However, the AT&T group has prepared films from phosphonated porphyrins and other ligands[41] whose cross-sectional areas exceeded 24 Å^2. The XPS analysis of such films shows that the amount of zirconium in the film is larger than expected for an α-ZrP-type structure. In final section, we shall show that in the preparation of bulk zirconium phosphates with ligands exceeding the required 24 Å^2, new structure types form.

Recent work on thin-film growth has involved the adsorption of oppositely charged polyelectrolytes.[42] Mallouk and co-workers[43] extended this technique to layering of structurally well-defined, two-dimensional colloidal polyanions and polymeric cations. The α-ZrP layers were completely exfoliated by intercalating a small amine [$MeNH_2$, $CH_3CH_2CH_2NH_2$, $(Me)_4NH^+$, etc.] between the layers, diluting and sonicating the suspended solid.[44] When such a suspension is placed in contact with an amine-modified gold or silicon surface, the surface-NH_3^+ groups displace the loosely held amine associated with one side of the ZrP layer. This procedure electrostatically attaches the α-ZrP layer to the gold or silicon surface. A number of polymeric or oligomeric cations were then attached to the exposed side of the α-ZrP layer by an ion-exchange reaction. Among the ions utilized were the aluminum Keggin ion $[Al_{13}O_4(OH)_{24}(H_2O)_{12}]_7^+$, polyallylamine hydrochloride (PAH) and cytochrome c.

As an example, we consider the PAH film in which the PAH was tagged with dye molecules, fluorescein and rhodamine B. Sequential adsorption of these polymers and α-ZrP was carried out on glass slides to form the films.[43] Energy transfer between the two dyes proceeds by a Förster mechanism and provides a quantitative probe of intermolecular distances within the film. By varying the number of α-ZrP/PAH spacer layers between the two dyes it was possible to measure the energy transfer efficiencies from steady-state emission spectra of the thin films. Energy transfer efficiencies near unity were achieved with properly designed systems. Based on these results, creation of more complex systems which combine both energy and electron transfer were contemplated. A valuable review has been presented by Mallouk *et al.*[45] in which many methodologies of preparing thin films and their potential applications are described. We present only a few in what follows.

6.3.2 Applications of thin films

6.3.2.1 Non-linear optical (NLO) thin films

We have already mentioned the zirconium phosphonate films containing azo dyes as exhibiting NLO behavior. The three-step adsorption used to prepare these films ensures the polar orientation of these molecules in the film. Because the bonding is covalent they are stable to 150°C, a relatively high temperature for NLO media.[40]

6.3.2.2 Sensors

Mallouk et al.[46] prepared a sensor for ammonia using a copper phosphonate layer-by-layer thin film anchored to a gold substrate. The external layer was terminated by a Cu^{2+} layer. The sensing device was a quartz crystal microbalance (QCM) to which an oscillating current was applied, causing a shear oscillation of the crystal. Mass changes on either electrode of the balance resulted in a dampening of the QCM frequency. Ammonia was sorbed onto the Cu atoms and the change in frequency recorded, which was calibrated to the mass uptake.

A similar technique was used to design a carbon dioxide sensor.[47] Three molecules that react with CO_2 to form carbonates and car-bamates, 3-aminopropanol (3-aminopropyl) methyldihydroxysilane and p-xylylenediamine were individually incorporated into the $Cu_2[O_3P(CH_2)_8 PO_3]$ thin films and found to respond to the presence of CO_2.

We shall conclude this section by discussing a recent paper where many of these thin-film techniques have been utilized to address the question of light-to-chemical energy conversion.[48] The stumbling block as with all photochemical systems is to prevent the back electron transfer of the pro-moted electron. "It follows that successful photosystems must maintain sensitizers, electron relays, and catalysts in the proper spacial arrange-ment and chemical environment to prevent the recombination reactions."[48] The Mallouk group developed onion-type composites rather than linearly stacked thin films.[49,50] Fumed silica particles of ~50 nm average diame-ter were derivatized with (3-aminopropyl) trimethoxysilane. The particles were then coated with exfoliated sheets of α-ZrP about $15 \times 15 \times 0.8$ nm in size. The next layer consisted of methylviologen-functionalized polystyrene followed sequentially by another layer of α-ZrP sheets and a polystyrene layer functionalized with a Ru(II) poly(pyridyl) photosensitizer p-[Ru] (p-bpy)(bpy)$_2$]^2, (p-Ru^{2+}) where bpy is 2,2'-bipyridine and p-bpy is the 2,2'-bipyridine ligand attached to the polymeric backbone.

Visible light absorption by p-Ru^{2+} produced a metal-to-ligand charge-transfer (MLCT) excited state, p-[Ru^{III}(p-bpy)(bpy)(bpy•$^-$)]$^{2+}$, that is a potent reductant capable of transferring an electron to p-MV^{2+}. However, by adding methoxy-N,N-bis (ethylsulfonate), MDESA^{2-}, to the system a more satisfactory charge-separated state was achieved with a quantum yield of ~30%. However, it is necessary to have good contact between the p-Ru layer and the MDESA^{2-}.

Another interesting layer-by-layer system was developed by Thompson *et al.*[51] They used thin films prepared from viologen derivatives and porphyrins. The arrangement was dictated by the redox potentials and optical energy gaps of the two types of layers. These films were photoactive when exposed to visible light producing photocurrent quantum yields of ~4% and a fill factor of 50%.

6.4 Pillared Porous Zirconium Diphosphonates

6.4.1 *Nanoparticle synthesis*

Dines *et al.*[52] conceived the idea of producing porous materials by cross-linking diphosphonic acids. The stoichiometric compound, $Zr(O_3PC_{12}H_8PO_3)$, was depicted ideally as in Figure 6.12. Because the pillars are 5.3 Å apart (as in α-ZrP) such materials should have no internal pores. However, by introducing spacer molecules such as the phosphorous ion O_3PH^{2-}, the Dines group was able to obtain porous "house of cards"-type products with high surface areas. Unfortunately, they did not measure the complete N_2 sorption isotherms to obtain information about the pores. One of their interesting results was the fact that the fully pillared compound, in which no spacer group was added, also exhibited a very high surface area. Subsequent work by us has shown that if the reactions are carried out in aqueous media the pore size distribution is very broad and mainly of the mesoporous type,[53] indicative of external interparticle pores.

Alberti *et al.*[54] prepared a number of zirconium diphosphonates in aqueous media that exhibited low surface areas (<10 m^2/g). However, with the addition of excess H_3PO_3 and DMSO as solvent, high surface areas were obtained (300–500 m^2/g) with the bulk of the pores in the 20–40 Å range. Such large pores must result from stacking of very small particles to produce interparticle mesoporosity. Subsequent work by us showed that considerable microporosity can develop in the particles if prepared in DMSO

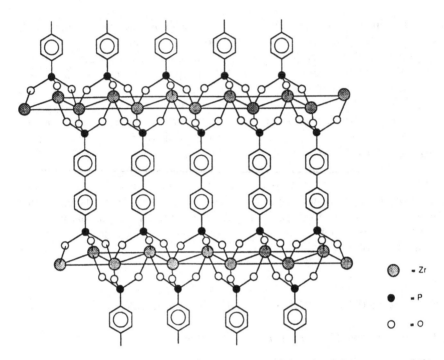

Figure 6.12.   Idealized depiction of $Zr(O_3PC_{12}H_8PO_3)$ based on the structure of the α-ZrP layer. The biphenyl groups may be tilted if the layer is similar to that in zirconium phenylphosphonate.

without the addition of H_3PO_3.[55] This microporosity was accompanied by a considerable percentage of mesoporosity. We also found, as will be detailed later, that the micron-sized particles were aggregates of much smaller nano-sized particles.

Our group discovered that the use of excess zirconium in reactions with 4,4′-biphenyl and 4,4′-terphenyldiphosphonic acid in DMSO or DMSO-ethanol mixtures with added HF produces layered microporous hybrids.[56,57] Surface areas of 350–$420\,m^2/g$ and pore sizes in the 10–$20\,\text{Å}$ diameter range are routinely obtained. Figure 6.13 shows the TEM micrographs of the particles as synthesized and as ground into a powder. We note that the individual particles contain about seven to nine layers and are 60–$80\,nm$ in length. The X-ray powder patterns provide an interlayer spacing of $13.65\,\text{Å}$; and this is in accord with the sum of the thickness of the layer, $6.6\,\text{Å}$, and the length of the biphenyl pillar, $\sim7\,\text{Å}$. Only two or three additional broadened reflections are observed at higher 2θ values. The very few number of layers and a somewhat out-of-phase positioning of the stacks of small

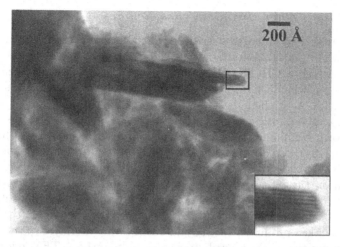

Figure 6.13. (a) Scanning electron micrograph of zirconium biphenyldisphosphonate prepared solvothermally in DMSO at 80°C for three days and (b) a transmission electron micrograph of the same sample as a fine powder. The inset shows a portion of an individual particle containing eight layers.

particles may be the reason for the few diffraction peaks observed. However, a mechanism by which internal pores of the observed size may form has been presented.[56,57]

6.4.2 *Sulfonated zirconium biphenyldiphosphonate*

The aromatic rings may be readily sulfonated in fuming sulfuric acid. The washed, dried solid has been found to be a strong Brønsted acid with acid strengths greater than the zeolite ZSM-5.[57] Catalytic studies in terms of alkylation and hydrocarbon isomerization are in progress. In the sulfonic acid form, the cavities fill with water and behave as strong acid ion exchangers. The difference between such exchangers and the commercially available sulfonic acid resins is that the phosphonates have much smaller pores: 15 Å diameter versus about 50 Å for 8% cross-linked sulfonic acid resins. Furthermore they are non-swelling resins, both characteristics conducive to higher selectivities. The possibility that the non-sulfonated materials can be used as molecular sieves with control of the hydrophobic-hydrophilic character of the pores by judicious functionalization of the aromatic rings is also suggested. The formation of nanoparticles in these and other metal systems appears to be a common occurrence for a range of ligands as detailed later.

6.4.3 Crown ether zirconium phosphonates

Previously we had prepared a series of aza-crown ethers essentially attached to zirconium phosphate-type layers. The compounds were of two types: those to which a $CH_2-PO_3H_2$ group was attached to a single aza group and those to which the attachment was to a 1,4-diazacrown. These phosphonic acids were then reacted with Zr(IV) in the presence of varying amounts of phosphoric acid.[58] A schematic of the results for the cross-linked diaza-crown compounds is presented in Figure 6.14. The X-ray patterns are similar to those obtained for the zirconium biphenyldiphosphonic acid materials. For the mono-aza-compound with 1:1 phosphate to phosphonate composition the interlayer spacing is 20 Å, indicative of a bilayer formation. The compound prepared with excess phosphate has a 13 Å interlayer spacing which arises from interdigitation of the crown ethers. In the case of the cross-linked products, 186211 and 186214, the interlayer spacing is the same irrespective of the phosphate-phosphonate ratio. A TEM micrograph

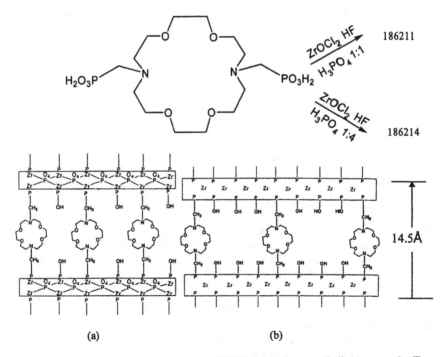

Figure 6.14. Zirconium complexes of 1,10-bis(phosphonomethyl)-18-crown-6. Top: schematic of preparation procedure; (a) at low HPO_4^{2-} content (186211) with PO_4 groups within the layers and (b) at high HPO_4^{2-} content with an α-ZrP-type layer. Oxygen atoms within the layers have been omitted.

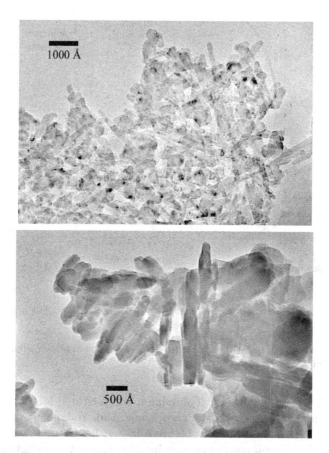

Figure 6.15.   Transmission electron micrographs of samples 186211 (top) and 186214 (bottom) showing the crystal morphologies of the cross-linked zirconium azacrown phosphonates.

of 186214 is shown in Figure 6.15. The particles are about 50–200 nm long and 20–40 nm wide. However, they may be only 10–20 nm or 8–16 layers thick. It should be mentioned that these compounds were prepared at 60–70°C and the biphenyl derivatives at 80–90°C in sealed vessels. Higher temperatures result in mixtures.

6.4.4   *Non-porous diphosphonates*

It was also discovered that if the zirconium biphenyl-bis(phosphonic acid) is prepared in ethanol or ethanol-water mixtures with no HF added, a high-surface-area product is also obtained but the bulk of this surface area is

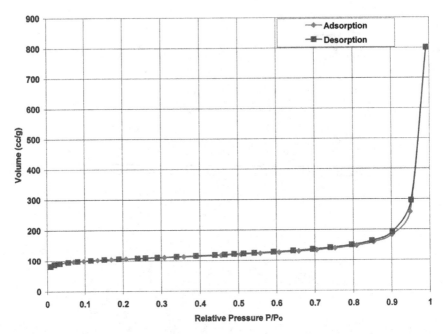

Figure 6.16. An N_2 sorption-desorption isotherm of aluminum biphenyldiphosphonate. S.A. $= 417\,\mathrm{m}^2/\mathrm{g}$ that is mainly external.

external. The same is the case if aluminum is substituted for zirconium but in an 8:3 mol ratio of metal to phosphonic acid. A typical N_2 sorption-desorption isotherm is shown in Figure 6.16 and has all the characteristics of a type IV isotherm with almost no internal porosity. An electron micrograph of an aluminum biphenylene diphosphonate is shown in Figure 6.17 demonstrating that the particles contain about five to eight layers. Surface areas of $150–400\,\mathrm{m}^2/\mathrm{g}$ are obtained based on the preparative method. These particles must have a very high level of phosphonic acid groups on the surface. If the groups are capped by metal atoms, then it should be possible to remove the metal by acid treatment, leaving phosphonic acid groups on the surface free to react with cations, charged complexes, bases, etc.

6.5 Proton-conducting Fuel Cell Membranes

6.5.1 Background

There is currently a great effort directed towards the development of workable fuel cells because of their energy efficiency. Efficiencies as high as

Figure 6.17. A transmission electron micrograph of an aluminum biphenyldiphosphonate showing particles with eight or less cross-linked layers.

70–80% are possible for fixed site units and 40–50% for transportation applications versus the current 20–35% with internal combustion engines.[59] For transportation purposes, proton-exchange membrane fuel cells (PEMFC) using H_2 as the fuel are preferred. A schematic drawing of such a fuel cell is given in Figure 6.18. The key ingredient in the PEMFC is the solid polymer membrane which conducts protons from the anode to the cathode. The most preferred membranes are fluorocarbon in nature to which are affixed sulfonic acid groups. These membranes are excellent electronic insulators with high proton conductivities on the order of 10^{-2} S cm^{-1} (siemens per cm or (ohm cm)$^{-1}$). The membranes require water as the main proton conduction medium and so work best at temperatures below 100°C. The problem arises from the fact that the H_2 must be generated by an on-site fuel processor from methane or methanol. In the process, CO is also generated which poisons the anode catalyst, platinum.[60] Operation of the fuel cell at 120–130°C would eliminate this problem but the membrane requires a considerable external pressure to avoid water loss.[59,60] At temperatures of 140–160°C direct methanol fuel cells (DMFC), where methanol is used directly as the fuel, are possible. Suitable membranes are not yet available for use at these temperatures. In the most recent work, polymer composites containing zirconium phosphates or sulfonated zirconium phenylphosphonate show considerable promise.

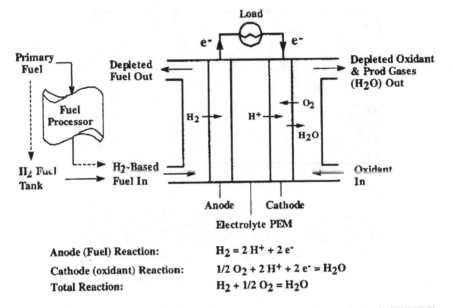

Figure 6.18.   A schematic diagram of a proton-exchange membrane fuel cell (PEMFC) using an on-board fuel cell processor (from Ref. 59 with permission).

With the advent of nanochemistry, research has developed with the aim of creating polymer composites in which inorganic nanoparticles are dispersed within a polymer.[61] We have already described polymer-clay composites and our work with exfoliated α-ZrP and an epoxy resin. Similar work has been done with Nafion and other proton-conducting polymers. We briefly review the proton conduction behavior of α-ZrP and the several sulfonated phenylphosphonates as a preliminary to describing the membranes.

6.5.2 Proton conductivity of α-zirconium phosphate and sulfophosphonates

The proton conductivity of α-ZrP depends upon its crystallinity and the relative humidity. Table 6.1 summarizes the conductivity of α-ZrP of different crystallinities. The specific conductance decreases as the crystallinity increases. The explanation is that the surface area decreases in the same order[63] and proton conduction is higher on the surface because of higher surface water content and greater number of surface protons. The effect of humidity on conductivity was determined for a highly crystalline α-ZrP sample that was exfoliated with propylamine, deintercalated to condense

Table 6.1. Specific conductance of α-ZrP samples obtained by different methods of preparation and ordered according to their degree of crystallinity (T = 25°C).

Sample	Preparation Method	Specific Conductance $(\Omega^{-1} \text{ cm}^{-1})$
1	Precipitation at room temperature, amorphous	8.4×10^{-3}
2	Precipitation at room temperature, amorphous	3.5×10^{-3}
3	Refluxing method (7:48)[a], semicrystalline	6.6×10^{-4}
4	Refluxing method (10:100), crystalline	9.4×10^{-5}
5	Refluxing method (12:500), crystalline	3.7×10^{-5}
6	Slow precipitation from HF solutions, crystalline	3.0×10^{-5}

[a]Numbers in parentheses indicate the concentration of H_3PO_4 in molarity and the number of hours refluxed, respectively.

the sheets and then compressed into discs. The water content of the discs averaged 0.7 mol for a relative humidity (RH) of 5% to 1.3 mol for 90% RH. The dc conductives at 20°C were found to lie in the range of $3–0.9 \times 10^{-4} \text{ S cm}^{-1}$ when the layers were oriented parallel to the electric field and seven times lower when the layers were perpendicular to the electric field.[64]

We have already described the synthesis of $Zr(O_3C_6H_4SO_3H)$ (O_3POH).[26] Alberti *et al.*[65] prepared a similar compound of composition $Zr(O_3C_6H_4SO_3H)_{0.73}(O_3PCH_2OH)_{1.27}$. This compound was shown to be a pure proton conductor. Arrhenius plots at several RH are shown in Figure 6.19. The highest conductivity at 295 K was $1.65 \times 10^{-2} \text{ S cm}^{-1}$. Somewhat higher conductivities were obtained at 278 K for more highly sulfonated Zr and Ti compounds as shown in Table 6.2. The zirconium and titanium phenylphosphonates were sulfonated in fuming sulfuric acid[66] and recovered by addition of methanol to the diluted acid followed by centrifugation.

The titanium compound experiences a certain amount of cleavage of the P-C bond as the formula derived from elemental analysis and thermogravimetric analysis was $Ti(O_3POH)_{0.25}(O_3PC_6H_5)_{0.12}$-$(O_3PC_6H_4SO_3H)_{1.63}\bullet3.64H_2O$ as opposed to the fully sulfonated zirconium phosphonate, $Zr(O_3PC_6H_4SO_3)_2\bullet3.6H_2O$. These sulfonated derivatives are among the best-known proton conductors.

6.5.3 *Membrane preparation*

Nafion membranes are among the most used as proton conductors. However, they suffer from a decrease in proton conductivity due to water loss as the temperature of the operation increases.[67] This limits the use of Nafion

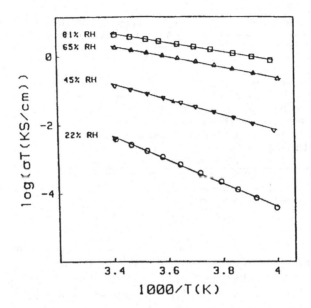

Figure 6.19. Arrhenius plots of log(σT) as a function of 1,000/T for Zr(O$_3$PC$_6$H$_4$SO$_3$H)$_{0.73}$(O$_3$PCH$_2$OH)$_{1.27}$ at different relative humidities (reprinted from Ref. 65 with permission).

Table 6.2. Conductivity in reciprocal Ω cm at 5°C as a function of relative humidity for zirconium and titanium sulfophenylphosphonates.

Sample	Relative Humidity (%)				
	20(3)[a]	30(3)	50(4)	65(4)	85(5)
EWS-3-89 (Zr)	5×10^{-6}	2×10^{-4}	1.1×10^{-3}	7.8×10^{-3}	2.1×10^{-2}
EWS-4-1 (Ti)	4×10^{-5}	3×10^{-3}	1.2×10^{-2}	7.2×10^{-2}	1.3×10^{-1}

[a]Numbers in parentheses are the estimated errors in the humidity measurement.

membranes to about 80°C[68] in fuel cells. Thus, either new membrane types need to be developed or the Nafion membranes must be improved to operate above 80°C, particularly above 120°C. This has been done by dispersing insoluble inorganic solids such as metal oxides, phosphates and phosphonates within the Nafion. The use of nanoparticles such as 50 nm silica spheres or α-ZrP lamellae or sulfonated zirconium and titanium phenylphosphonates ensures good distribution and strong interactions between the particles and the membrane. Much improved Nafion membranes have been produced in terms of water retention and conductivity by forming such composites especially with the phenylsulfonates.

A similar technique to form composites with other polymers has been similarly successful.[69] Still, membranes that operate in actual fuel cells above 120°C are not yet commercial, but continued research using lamellar nanoparticles is being actively pursued.

6.6 Tin(IV) Phosphonates

6.6.1 *Synthesis and nanoparticle formation*

A recent paper that caught our attention described the synthesis of tin(IV) phenylphosphonate prepared in the presence of sodium decylsulfonate (SDS) as a template.[70] It was found that this compound had a composition $Sn(O_3PC_6H_5)_2$, a surface area of $255\,m^2/g$ and an average pore size of $3.3\,nm$. What was strange was that preparations without templates also had high surface areas (197 and $135\,m^2/g$) but the pore distribution range was 20–$30\,nm$. We repeated this preparation and found a quite regular micropore distribution ($\sim10\,\text{Å}$ diameter) and a surface area of $\sim300\,m^2/g$ with a near type I isotherm.[71] The question arises as to how such a structure can form from a presumably layered compound similar to zirconium phenylphosphonate (based upon X-ray diffraction patterns). Scanning electron micrographs at high magnification revealed that the solid consists of nanoparticles that aggregate into spheres that are micron-sized (Figure 6.20). The pores arise from the "house of cards" face-to-edge arrangement of the particles. The particles are very small and must have a narrow size distribution to yield such regular pores. Small-angle neutron scattering confirmed the nano-size of the particles. The phenomenon appears to be general as we have now made about 20 derivatives and all of them show similar characteristics as shown in Table 6.3. We note that different average pore sizes result from different ligands. The only compound that was different was the methylphosphonate. It crystallized as platelets aggregating into rosettes. By longer reaction times and higher temperatures, the tin methylphosphonate yielded an X-ray powder pattern amenable to solution. The compound is eight-coordinate with bonding similar to that in rare earth phenyl phosphonates.[72] Continued hydrothermal treatment at 180°C for eight days produced larger platelets of average dimensions $50\,nm$ long and 10–$15\,nm$ thick or 6–10 layers thick. Thus, we conclude that the nano-sized feature is due to their very low solubility in aqueous media, even in the presence of HF. It should also be noted from Table 6.3 that the pore size varies with the type of ligand used to prepare the tin phosphonate. It also should vary with preparative conditions and this has been observed by us. Details will be published subsequently.

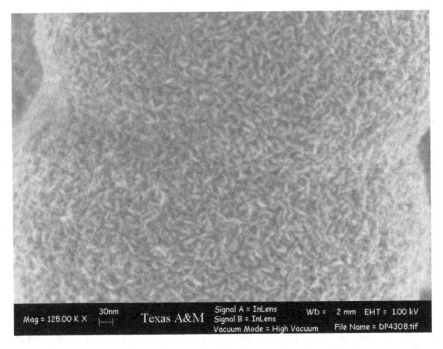

Figure 6.20. Transmission electron micrograph of $Sn(O_3PC_6H_5)_2$ at high magnification showing the aggregates of small particles and the porous nature of the spherical aggregate.

Table 6.3.   Ideal formulas and properties of selected tin(IV) phosphonates.

Ideal Formula	Surface Area (m^2/g)			Max in Pore Size	TGA wt. Loss (%)
	d_{001} (Å)	Total	Micro	Distribution Curve (Å)	
$Sn(O_3PC_6H_5)_{1.95}F_{0.05} \bullet H_2O$	15.35	367	350	9	34.09
$Sn(O_3PC_6H_5)_{0.95}(O_3PCH_3)_{0.95}$ $F_{0.10} \bullet 0.4H_2O$	15.22	514	493	11	24.05
$Sn(O_3PCH_3) \bullet 0.4H_2O$	9.74	68	0		6.8
$Sn(O_3PC_{12}H_8PO_3) \bullet 0.8H_2O$	13.8	416	388	17	36.0
$Sn(O_3PC_{10}H_{21})_2$	28.07	264	220	9	47.6
$Sn(O_3PCH_2CH_2COOH)_2 \bullet 0.8H_2O$	11.07	224	102	27	33.09

6.7 Conclusions

We have only mentioned γ-zirconium phosphate in a peripheral way for reasons given in what follows. It was discovered shortly after the alpha phase[73] and its structure determined much later.[74] The reason is that no single crystals were grown and only when X-ray powder methods were developed was a solution of the structure obtained. The formula for γ-ZrP is $Zr(PO_4)(H_2PO_4) \cdot 2H_2O$. It is layered with a 12.2 Å interlayer spacing. The orthophosphate group resides within the layers bonding to four Zr atoms forming linear chains. The chains are bridged into layers utilizing the two unprotonated oxygens of the dihydrogen phosphate groups. The -OH groups project away from the layers into the interlayer space, forming hydrogen bonds with the two water molecules. It is almost impossible to form phosphonate compounds of γ-ZrP by direct synthesis procedures. Instead, they are obtained by ester interchange reactions of the following type:

$$Zr(PO_4)(O_2P(OH)_2) \cdot 2H_2O + RPO_3H_2 \rightleftharpoons Zr(PO_4)\left(O_2P{\overset{R}{\underset{OH}{<}}}\right) + H_3PO_4.$$

$$(6.3)$$

As a result, the particle size is generally the particle size of the starting γ-ZrP sheets. A good introduction to the chemistry of γ-ZrP is given in a review article by Alberti.[75] Undoubtedly, a rich nanochemistry exists in this system.

The recognition that the group IV phosphates and phosphonates form nanoparticles because of their great insolubility in aqueous media is key in understanding and controlling their chemistries. For example, the ease of exfoliation of the layers depends upon the total force between layers and thus on the size of the layers. By controlling the size, we are able to exfoliate α-ZrP even in non-aqueous systems for use in dispersion in polymers. The size of the particles may also play a significant role in proton-conducting polymer composites for fuel cells. It will surely play a role in catalysis where control of surface area, pore size and active groups per unit of surface area are important variables. We are just beginning to explore the Sn(IV) phosphonates and many surprises may still be in store if a similar silicon chemistry is possible. It could hardly be imagined that the discovery of the crystalline zirconium phosphates in the early 1960s and the subsequent synthesis of the first zirconium phosphonates in 1978 that such a vast new field of chemistry would result. But it is not finished, many new developments will certainly arise from continued research in this field.

Acknowledgments

We gratefully acknowledge the financial support of the National Science Foundation, Grant No. DMR-0332453, and the Welch Foundation, Grant No. A-0673.

References

1. (a) K. A. Kraus and H. O. Phillips, *J. Amer. Chem. Soc.* **78**, 694 (1956); (b) K. A. Kraus, H. O. Phillips, T. A. Carlson and J. S. Johnson, *Proc. 2nd Int. Conf. Peaceful Uses of Atomic Energy*, Geneva, 1958, paper No. 15, p. 1832.
2. A. Clearfield and J. A. Stynes, *J. Inorg. Nucl. Chem.* **26**, 117–129 (1964).
3. (a) A. Clearfield and G. D. Smith, *Inorg. Chem.* **8**, 431–436 (1969); (b) J. M. Troup, A. Clearfield, *Inorg. Chem.* **16**, 3311–3314 (1977).
4. G. Alberti and J. Torracca, *J. Inorg. Nucl. Chem.* **30**, 317–318 (1968).
5. A. Clearfield and J. R. Thomas, *Inorg. Nucl. Chem. Lett.* **5**, 775–779 (1969).
6. A. Clearfield, *Annu. Rev. Mater. Sci.* **14**, 205–229 (1984).
7. A. Clearfield, L. Kullberg and A. Oskarsson, *J. Phys. Chem.* **78**, 1150–1153 (1974).
8. A. Clearfield, A. Oskarsson and C. Oskarsson, *Ion Exch. Membr.* **1**, 91–107 (1972) (journal out of print).
9. G. Alberti, *Acc. Chem. Res.* **9**, 163–170 (1976).
10. M. S. Whittingham and A. J. Jacobson (eds.), *Intercalation Chemistry* (Academic Press, New York, NY, 1982).
11. W. Müller-Warmuth and R. Schollhorn (eds.), *Progress in Intercalation Research* (Kluwer Academic Publ., Dordrect, 1994).
12. A. Clearfield and U. Costantino, in *Comprehensive Supramolecular Chemistry*, eds. G. Alberti and T. Bein, Vol. 7 (Pergamon Press, New York, NY, 1996), p. 107.
13. R. M. Tindwa, D. K. Ellis, G.-Z. Peng and A. Clearfield, *J. Chem. Soc. Faraday Trans.* **81**, 545–552 (1985).
14. D. M. Kaschak, S. A. Johnson, R. E. Hooks, H. N. Kim, M. D. Ward and T. E. Mallouk, *J. Am. Chem. Soc.* **120**, 10887–10894 (1998).
15. A. Clearfield and L. H. Kullberg, *J. Phys. Chem.* **78**, 1812–1817 (1974).
16. A. Clearfield and A. Oskarsson, *Ion Exch. Membr.* **1**, 205–213 (1974).
17. L. Sun, W. J. Boo, R. L. Browning, H.-J. Sue and A. Clearfield, *Chem. Mater.* **17**, 5606–5609 (2005).
18. (a) T. Lan, P. D. Kavirnatna and T. J. Pinnavaia, *Chem. Mater.* **7**, 2144–2150 (1995); (b) P. B. Messersmith and E. P. Giannelis, *Chem. Mater.* **6**, 1719–1725 (1994).
19. T. Lan and T. J. Pinnavaia, *Chem. Mater.* **6**, 2216–2219 (1994).
20. Y. Kurokawa, H. Yasuda, M. Kashiwagi and A. Oyo, *J. Mater. Sci. Lett.* **16**, 1670–1672 (1997).

21. N. Bestaoui, N. A. Spurr and A. Clearfield, *J. Mater. Chem.*, DOI:10.1039/ b511351b.

22. H.-J. Sue, K. T. Gam, N. Bestaoui, N. A. Spurr and A. Clearfield, *Chem. Mater.* **16**, 242–249 (2004).

23. S. Yamanaka, *Inorg. Chem.* **15**, 2811–2817 (1976).

24. G. Alberti, U. Costantino, S. Allulli and N. Tomassini, *J. Inorg. Nucl. Chem.* **40**, 1113–1117 (1978).

25. D. M. Poojary, H.-L. Hu, F. L. Campbell, III and A. Clearfield, *Acta Crystallogr.* **B49**, 996–1001 (1993).

26. C.-Y. Yang and A. Clearfield, *React Polym.* **5**, 13–21 (1987).

27. A. Clearfield (unpublished data).

28. (a) L. H. Kullberg and A. Clearfield, *Solv. Extr. Ion Exch.* **7**, 527–540 (1989); (b) L. H. Kullberg and A. Clearfield, *Solv. Extr. Ion Exch.* **8**(1), 187–197 (1990).

29. J. L. Colon, C.-Y. Yang, A. Clearfield and C. R. Martin, *J. Phys. Chem.* **92**, 5777–5781 (1988).

30. J. L. Colon, D. S. Thakur, C.-Y. Yang, A. Clearfield and C. R. Martin, *J. Catal.* **124**, 148–159 (1990).

31. G. Alberti and R. Polambari, *Solid State Ionics* **35**, 153–156 (1989).

32. G. Alberti, M. Casciola and R. Polambari, *Solid State Ionics* **52**, 291–295 (1992).

33. G. A. Ozin and A. C. Arsenault (eds.), *Nanochemistry: A Chemical Approach to Nanomaterials* (RSC Publishing, Cambridge, 2005), Ch. 3, pp. 95–125.

34. H. Lee, L. J. Kepley, T. E. Mallouk, H. G. Hong and S. Akhter, *J. Phys. Chem.* **92**, 2597–2601 (1988).

35. H. Lee, L. J. Kepley, H. G. Hong and T. E. Mallouk, *J. Am. Chem. Soc.* **110**, 618–620 (1988).

36. H. Byrd, J. K. Pike and D. R. Talham, *Chem. Mater.* **5**, 709–715 (1993).

37. T. M. Putvinski, M. L. Shilling, H. E. Katz, C. E. D. Chidsey, A. M. Mujsce and A. B. Emerson, *Langmuir* **6**, 1567–1571 (1990).

38. H. E. Katz, G. Scheller, T. M. Putvinski, M. L. Schilling, W. L. Wilson and C. E. D. Chidsey, *Science* **254**, 1485–1487 (1991).

39. H. E. Katz, M. L. Schilling, S. B. Ungashe, T. M. Putvinski and C. E. D. Chidsey, in *Supramolecular Chemistry*, ed. T. Bein (American Chemical Society Symposium Series 499, Washington, D.C., 1992), p. 24.

40. H. E. Katz, W. L. Wilson and G. Scheller, *J. Am. Chem. Soc.* **116**, 6636–6640 (1994).

41. S. B. Ungashe, W. L. Wilson, H. E. Katz, G. R. Scheller and T. M. Putvinski, *J. Am. Chem. Soc.* **114**, 8717–8719 (1992).

42. S. W. Keller, H.-N. Kim and T. E. Mallouk, *J. Am. Chem. Soc.* **116**, 8817–8818 (1994).

43. D. M. Kaschak and T. E. Mallouk, *J. Am. Chem. Soc.* **118**, 4222–4223 (1996).

44. M. Fang, D. M. Kaschak, A. C. Sutorik and T. E. Mallouk, *J. Am. Chem. Soc.* **119**, 12184–12191 (1997).

45. T. E. Mallouk, H.-N. Kim, P. J. Olliver and S. W. Keller, in *Comprehensive Supramolecular Chemistry*, ed. G. Alberti, Vol. 7 (Pergamon Press, New York, NY, 1996), pp. 189–217.

46. L. C. Brousseau and T. E. Mallouk, *Anal. Chem.* **69**, 679–687 (1997).

47. L. C. Brousseau, D. J. Aurentz, A. J. Benesi and T. E. Mallouk, *Anal. Chem.* **69**, 688–694 (1997).

48. P. G. Hoertz and T. E. Mallouk, *Inorg. Chem.* **44**, 6828–6840 (2005).

49. S. E. Keller, S. A. Johnson, E. S. Brigham, E. H. Yonemato and T. E. Mallouk, *J. Am. Chem. Soc.* **117**, 12879–12880 (1995).

50. D. M. Kaschak, S. A. Johnson, C. C. Waraksa, J. Pogue and T. E. Mallouk, *Coord. Chem. Rev.* **186**, 403–416 (1999).

51. F. B. Abddrazzaq, R. C. Kwong and M. E. Thompson, *J. Am. Chem. Soc.* **124**, 4796–4803 (2002).

52. M. B. Dines, P. M. Di Giacomo, K. P. Callahan, P. C. Griffith, R. H. Lane and R. E. Cooksey, in *Chemically Modified Surfaces in Catalysis and Electrocatalysis*, ed. J. S. Miller (Amer. Chem. Soc. Symposium Series 192, Wash. D. C., Amer. Chem. Soc., 1982), pp. 223–240.

53. A. Clearfield, in *Design of New Materials*, eds. A. Clearfield and D. A. Cocke (Plenum Press, New York, NY, 1986), pp. 121–134.

54. G. Alberti, U. Costantino, R. Vivani and P. Zappelli, in *Synthesis/Characterization and Novel Applications of Molecular Sieve Materials*, eds. R. L. Bedard *et al.*, Vol. 233 (Materials Research Society Symp. Proc., 1991), pp. 101–106.

55. A. Clearfield, *Chem. Mater.* **10**, 2801–2810 (1998).

56. A. Clearfield and Z. Wang, *J. Chem. Soc. Dalton Trans.* 2937–2947 (2002).

57. Z. Wang, J. M. Heising and A. Clearfield, *J. Am. Chem. Soc.* **125**, 10375–10383 (2003).

58. B. Zhang and A. Clearfield, *J. Am. Chem. Soc.* **119**, 2751–2752 (1997).

59. C. Song, *Catal. Today* **77**, 17–49 (2002).

60. G. Alberti, M. Casciola, L. Massinelli and B. Bauer, *J. Membrane Sci.* **185**, 73–81 (2001).

61. R. Dagni, *Chem. Eng. News* **77**, 25–37 (1999).

62. G. Alberti, M. Casciola, U. Costantino, G. Levi and G. Ricciardi, *J. Inorg. Nucl. Chem.* **40**, 533–537 (1978).

63. A. Clearfield and J. Berman, *J. Inorg. Nucl. Chem.* **43**, 2141–2142 (1981).

64. M. Casciola and U. Costantino, *Solid State Ionics* **20**, 69–73 (1986).

65. G. Alberti, M. Casciola, U. Costantino, E. Peraio and E. Montonero, *Solid State Ionics* **50**, 315–322 (1992).

66. E. W. Stein, A. Clearfield and M. A. Subramanian, *Solid State Ionics* **83**, 113–124 (1996).

67. Y. Sone, P. Ekdunge and D. Simonsson, *J. Electrochem. Soc.* **143**, 1254–1259 (1996).

68. G. Alberti and M. Casciola, *Annu. Rev. Mater. Res.* **33**, 129–154 (2003).

69. B. Bonnet, D. J. Jones, J. Roziére, L. Tchicaya and G. Alberti, *J. New Mater. Electrochem. Syst.* **3**, 87–92 (2000).

70. N. K. Mal, M. Fujiwara, Y. Yamada and M. Masahiko, *Chem. Lett.* **32**, 292–293 (2003).

71. A. Subbiah, D. Pyle, A. Rowland, J. Huang, R. A. Narayanan, P. Thiyagarajan, J. Zon and A. Clearfield, *J. Am. Chem. Soc.* **127**, 10826–10827 (2005).

72. R. C. Wang, Y.-P. Zhang, H. Hu, R. R. Frausto and A. Clearfield, *Chem. Mater.* **4**, 864–871 (1992).

73. A. Clearfield, R. H. Blessing and J. A. Stynes, *J. Inorg. Nucl. Chem.* **30**, 2249–2258 (1968).

74. (a) A. N. Christensen, E. K. Anderson, I. G. Anderson, G. Alberti, M. Nielsen and M. S. Lehmann, *Acta*; (b) D. M. Poojary, B. G. Shpeizer and A. Clearfield, *J. Chem. Soc. Dalton Trans.* 111–113 (1995).

75. G. Alberti, *Comprehensive Supramolecular Chemistry*, eds. G. Alberti and T. Bein (Pergamon, Elsevier, Tarrytown, NY, 1996), pp. 151–188.

Chapter 7

Twenty-five Years of Nuclear Waste Remediation Studies

Abraham Clearfield

Texas A&M University, College Station, TX, USA

7.1 Introduction

It is a known fact that the size of the economy of a nation can be tracked by the amount of energy consumed by that nation. At present the United States is the world's largest energy consumer. About 20% of our electrical power is derived from nuclear reactors and 2–3% from renewable energy. The remainder is derived from fossil fuels, coal, oil, and natural gas. The world's population is increasing and several countries such as India and China are rapidly expanding their economies. Thus the demand for energy is expected to increase by 50% by 2050. In order to maintain a 4% annual increase in the economy we will also need to double our energy consumption. It is insanity to expect that this high increase could or should be supplied by fossil fuels alone. Petroleum will continue to increase in price as demand increases and fossil fuels are highly polluting. Coal contains toxic metals such as lead, arsenic, mercury, cadmium, and more. While the quantities of impurities are small, when you burn 700 million tons of coal, thousands of tons of these toxic elements are released into the environment. Coal is inexpensive only because there is no charge for the damage done to the environment, and the thousands of respiratory illnesses and cancers that result from coal power plant emissions. Petroleum contains sulfur, nitrogen and trace amounts of other pollutants responsible for acid rain, ground-level ozone, and respiratory problems. Use of fossil fuels will also greatly accelerate global warming. Prudence would dictate that a large portion of future energy needs be derived from renewables such as nuclear, solar, wind, geothermal, tidal, and any other type that can be harnessed. Our country should seize this opportunity to expand these new industries so as to create new markets and achieve energy independence.

Nuclear energy should be one of the growth areas to supply electrical energy. There are currently 104 power-generating nuclear reactors spread throughout the country. They supply roughly 20% of our needed electrical power and do not contribute to global warming. However there are a number of environmental problems related to nuclear power generation and weapons production. This chapter will recount our preparation of ion exchange materials that can be used to remediate the waste that currently exists and deal with the spent rods that are stored under water. Such considerations are timely in relation to the destruction of most of the nuclear power stations in northern Japan.

The most critical waste is that accumulated as a result of our nuclear weapons program and is termed high-level waste (HLW) because it is highly radioactive. This waste arose as by-products of the processes utilized for the separation of uranium and plutonium and is stored in steel tanks underground.[1] To prevent corrosion of the tanks by the highly acid waste, excess sodium hydroxide was added to the tanks. This procedure precipitated insoluble hydroxides but held in solution quantities of alumina and silica. The precipitate exists as sludge at the bottom of the holding tanks. The fluid portion in the tank contains ^{137}Cs and smaller amounts of ^{90}Sr and actinides as well as aluminum and silicon species and is roughly 6 M in sodium nitrate and nitrite. Over the years of storage, the fluid was evaporated to make room for additional waste, forming a salt cake that exists above the fluid layer. There are 177 such tanks at the Hanford Site in Washington with $\sim$208,000 m^3 of waste and about 60% as much at the Savannah River Site in South Carolina.[1,2] However, the waste at the Savannah River Site contains much less aluminum and silica species.

Current thinking in terms of tank waste treatment is to remove the cesium, strontium, and actinides (Pu, Np) from the fluid portion with possible dissolution of the salt cake portion for similar treatment.[3,4] These solids would then be encased in a special boron glass held in sealed steel vessels and stored underground at the Yucca Mountain repository.[5] The effluent from this waste treatment would then be made into a cement or grout and stored above ground. The huge amount of sludge that remains at the bottom of the tank also needs to be treated. The composition of this sludge is largely unknown, because in the intervening residence time it has formed mixed oxides, hydroxides, and oxohydroxides. At the Savannah River Site the sludge is being dried and mixed with solid glass nodules, heated to 1,100°C in steel tanks and sealed. At present these tanks are stored on-site to be shipped to Yucca Flat for storage. But the state of Nevada is contesting the use of this repository so its use is on hold.

A second environmental problem is the spent rods from power generation. These rods are highly radioactive and are stored under 25 feet of water at the reactor sites. One of the great concerns in Japan is that the water supply would be shut off and the spent rods would heat up to the point of explosion. If all our spent rods were encased in steel and buried at Yucca Flat, they would fill the entire repository. Alternatively the rods could be treated to recover the usable fuels which represent 95% of the original fuel value with only 5% waste. This would seem to be the correct course of action but no decision on this matter has been forthcoming.

7.2 Ion Exchange Removal of Cesium and Strontium from Nuclear Waste Solutions

7.2.1 *Sodium titanium silicate*

In 1991 I was invited to participate in a workshop at the Hanford works to discuss possible new separations to apply to the nuclear waste problem. Shortly thereafter I was awarded a grant to examine methods of removing Cs^+ and Sr^{2+} or $Sr(OH)^+$, because of the high pH of the waste solutions, from tank wastes. In our first effort of any consequence we focused on a sodium titanium silicate first synthesized at the Sandia National Laboratories.[6-10] These reports indicated a high selectivity for Cs^+ but only fragmentary structural data. Therefore, we synthesized this compound and determined its structure and that of the Cs^+ phase using X-ray powder data.[11] The ideal composition is $Na_2Ti_2O_3SiO_4 \cdot 2H_2O$ but often has part of the sodium ion replaced by H^+ due to hydrolysis. The crystals are tetragonal with $a = 7.8082(2)$ Å, $c = 11.9735(4)$ Å, space group $P4_2/mcm$ with four molecules in the unit cell. Since the structure of the sodium titanosilicate, popularly termed crystalline silicotitanate (CST), is key to understanding the ion exchange properties, we shall give a brief description of its structure.

The titanium atoms occur in clusters of four grouped about a 4_2 axis, two up and two down rotated by 90°. Each titanium atom is octahedrally coordinated, sharing edges in such a way that an inner core of four oxygen atoms and four Ti atoms form a distorted cubane-like structure (Figure 7.1).[11] These cubane-type structures are bridged to each other through silicate groups along the a- and b-axis directions. The titanium-oxygen clusters are 7.81 Å apart in both the a- and b-axis directions with the Si atoms at $Z = 1/4, 3/4 \, c$. In the c-axis direction, the Ti atoms are bridged by oxo-groups. The c-axis is approximately 12 Å long, which is twice the distance from the center of one cubane-like cluster to its neighbor in the c-axis direction. These two views of the framework are shown in Figures 7.2 and 7.3.

A. Clearfield

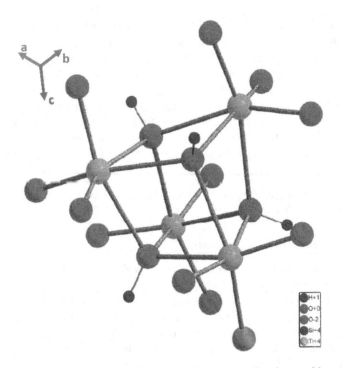

Figure 7.1. A portion of the titanosilicate structure showing the cluster of four titanium-oxygen octahedra sharing edges to form the cubane-like Ti_4O_4 group. The oxygens within the cubane group are each bonded to a proton, making them hydroxo groups. (Reproduced from Ref. 54, Figure 1, with permission from Elsevier.)

The net result of this framework arrangement is that tunnels are formed that are one-dimensional, running along the c-axis direction. Perpendicular to the tunnels are vacancies in the faces of four sides of the tunnels. These cavities are just the right size to enclose sodium ions. Four silicate oxygens bond to the sodium ion at a distance of 2.414(5) Å (Figure 7.3). The sodium ion coordination is completed by bonding to two water molecules in two adjacent tunnels at a bond distance of Na-O of 2.765(1) Å. Half the Na^+ ions are thus accounted for in the framework sites as there are two sodiums in each face over one c-axis cell length for a total of four out of the eight required per unit cell. The remaining sodiums reside within the tunnels along with the water molecules.

The Na-O bond distances within the tunnels are longer than the sum of the ionic radii (2.42 Å)[12] at 2.76(1) Å. This bond distance measurement was made with only 64% of the sodium ion sites occupied. The deficiency of sodium arises from hydrolysis during washing so that the actual formula

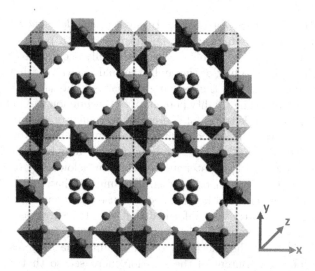

Figure 7.2. Top view (down the c-axis) of sodium titanium silicate showing the clusters of four Ti-O_6 octahedra bridged by silicate groups with oxygens. The tunnels are filled with Na$^+$ and water molecules. The Na$^+$ on top of the tetrahedra symbolizes the Na$^+$ ions sandwiched between silicate groups within the framework. (Reproduced from Ref. 54, Figure 2, with permission from Elsevier.)

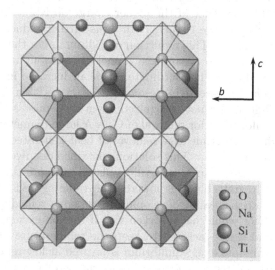

Figure 7.3. Polyhedral representation of Na-TS structure as viewed down the a-axis, showing the hexagonal-shaped framework cavities in which half of the Na$^+$ reside. The water molecules reside within the 8MR tunnels. (Reproduced with kind permission from Springer Science + Business Media, as adapted from Ref. 24, Figure 49:11.)

was $Na_{1.64}H_{0.36}Ti_2O_3(SiO_4)\cdot1.8H_2O$. Because of the deficiency of Na^+ the sodium ion positions were found to be disordered with partial occupancy by water molecules. It is possible to obtain the fully occupied sodium phase by not washing the product of the hydrothermal reaction with water.

The location of Cs^+ in the exchanged phase of $Na_{1.49}Cs_{0.2}H_{0.31}Ti_2O_3$ $(SiO_4)\cdot H_2O$ is interesting. Eighty percent of the cesium ion was in the tunnel at $1/4, 3/4\,c$, and bonded to eight framework oxygen atoms, four above and four below the cation. The second site was also in the tunnel but at about 0.14 and $0.64\,c$. The Cs^+ in this site is bonded to four framework oxygen atoms from above and two water molecules from below. Any one unit cell cannot have cesium ions in both sites. This means that only 25% of the total exchange capacity can be attained because of the large size of Cs^+.

Interestingly we later were informed that the structure was known as a mineral, with several impurities in the lattice, including partial substitution of Nb for Ti.[13] Also, the selectivity of the sodium phase decreased as the pH and sodium ion content of the solution increased so that it was ineffective in Cs^+ sequestration in the actual basic waste solutions. However, the Sandia group had also prepared a phase in which 25% of the Ti^{4+} was replaced in the lattice by Nb^{5+}. This phase was selective enough to be usable at high levels of basic media.[7] The reason for this salutary effect became apparent when the Cs^+ ion structure of the niobium-containing phase was determined.[14] The presence of the higher charge on Nb^{5+} reduces in half the amount of Na^+ in the tunnel. Hydrolysis further reduces the sodium ion to a very small level. This reduction of sodium ion in the tunnel allows a higher level of water to occupy the tunnel so that four water molecules bond to Cs^+, making it 12-coordinate. Such an increase in coordination must produce a large decrease in free energy for the Cs^+ ion exchange reaction. This was generally not recognized when trial pilot scale runs were made at the Hanford Site for removal of Cs^+. A large-diameter column was utilized that did not permit easy removal of heat generated by accumulation of ^{137}Cs in the column. This solution contained dissolved alumina and silica, so as the temperature rose zeolites plated out onto the CST pellets, clogging them. Regrettably, this lowered the value of CST in the eyes of the engineers responsible for developing a workable remediation process. I believe that CST would have worked well if the ion exchange tubes had been narrow and water jacketed to keep the temperature low. This also would benefit the uptake as equilibrium is favored at low temperature.

Some additional remarks on the ion exchange of Cs^+ by CST are in order. The cesium ions can only fit into the tunnel sites. They are too large for the framework sites. Furthermore, while a maximum of four monovalent ions can reside in the tunnel of one unit cell, only two Cs^+ ions can fit.

Thus, only 25% of the exchange capacity can be utilized for Cs^+ uptake as shown by titration.[15] However, sodium and lithium ions are small enough to also occupy the framework sites and therefore have a much higher uptake.[16] The high selectivity for Cs^+ with K_d values in the 10^5–10^6 range results from the 12-coordination and from the fact that the Cs-O bond distances are equal to the sum of their respective ionic radii. However, in the high sodium ion waste media the K_d values may be as low as 500 ml/g. The very high level of sodium ions in these waste solutions ensures that the framework sites are occupied but also that some of the tunnel sites are occupied, denying one of the four water molecules access to bond to Cs^+.

7.2.2 Mechanism of cesium ion exchange process

One of the drawbacks of CST is that Cs^+ could not be recovered by treatment with strong acid solutions. Furthermore, there are two exchange sites for Cs^+. The minor site is a six-coordinate site. Why would Cs^+ prefer a six-coordinate situation when eight- and twelve-coordination is energetically more likely? To answer these questions we turned to a procedure of *in situ* time-resolved X-ray studies as the exchange process was taking place. This study was done at the National Synchrotron Light Source, Brookhaven National Laboratory. It was a collaborative effort with John Parise, Professor of Geology and Chemistry, Stony Brook University. The CST sample was held in a quartz capillary tube open at both ends. A high-intensity X-ray beam recorded the structure changes as a 0.01 M solution of CsCl flowed through the CST sample. It should be recognized at this juncture that we chose to use the protonic phase of CST so that there would be no sodium ion present to influence the reaction.

A neutron diffraction study carried out earlier revealed that a structural change takes place upon protonation.[17] The diagonal of the sodium ion phase becomes the a- and b-axis of the proton phase (Figure 7.4).[18] In the process the circular or octagonal shape of the tunnel becomes elliptical. The driving force for this transformation is that it allows the cubane-bonded protons to hydrogen bond to the water molecules populating the tunnel (Figure 7.5). The Cs^+ ion is too large to overcome the barriers provided by the elliptical shape. Thus, in the first minutes of exchange the Cs accumulated at the six-coordinate site.[19,20,21] It could not penetrate beyond the first group of silicate oxygens in the narrowed channel. However, as this site became filled, the crystals reverted back to the original 7.8 Å axis. Then the major 12-coordinate site at 1/4, 3/4 c began to fill. The diffraction patterns contained both phases to the end of the experiment. The final product composition was $Cs_{0.36}H_{1.64}Ti_2O_3SiO_4$ or 72% of

A. Clearfield

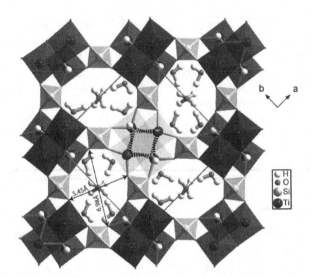

Figure 7.4.   Crystal structure of H-CST ($H_2Ti_2SiO_7 \cdot 1.5H_2O$). H_2O found in the 8MR parallel to [001]. The Ti_4-O_4 cubane-like unit is illustrated in the center of the diagram where bold segmented bonds outline the cubane geometry. (Reproduced from Ref. 19, Figure 2, with permission from Elsevier.)

the possible maximum cesium ion uptake. This is a remarkable uptake in view of the fact that the CsCl solution concentration was 0.01 M.

The next step in uncovering the mechanism was undertaken by Aaron Celestian as part of his Ph.D. dissertation. He carried out an *in situ* neutron diffraction study of a partially deuterated CST at the Rutherford Appleton Laboratory, UK.[22] This study allowed for observing the movement of protons, both those bonded to the framework and those belonging to water molecules. The entrance of Cs^+ into the six-coordinate site compels the water molecules, which are hydrogen-bonded to the cubane protons through oxygen atoms, to flip so that the negatively charged water oxygen points towards the positively charged cesium ion. In flipping, the water protons encounter the bonded framework protons and repel them, resulting in a rotation of 5.8° to return to the more symmetrical, more circular shape of the tunnel. Two water molecules bond to the cesium. The cesium ion is then able to diffuse into site Cs1 from Cs2 (Figure 7.6). The Cs^+ ion is then able to diffuse through the tunnels with both sites being occupied at an 80-20 ratio.

One would expect that the Cs^+ ion would have exchanged for a hydronium ion. But this would have required the cubane protons to transfer to water molecules, forming a positively charged species in competition with the Cs^+ ion. Instead we theorize that hydronium ions only form at the

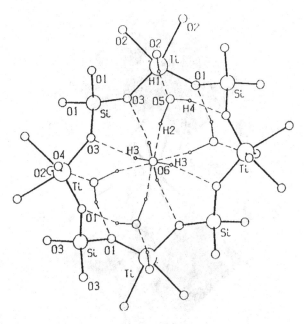

Figure 7.5. Plot of the atoms outlining the channel of $Ti_2(OH)_2OSiO_4 \cdot 1.5H_2O$ as viewed down the c-axis showing the arrangement of water molecules in the tunnel. Hydrogen bonds are shown by dashed lines. (Reproduced from Ref. 17, Figure 3, with permission from the American Chemical Society.)

surface of the CST particles and then enter the surrounding solution. This mechanism requires the cubane protons to hop along framework sites until reaching the surface, and may explain why the Cs^+ cannot be removed from the crystals with acid treatment. Protons would then begin to migrate into the cubane sites causing the Cs^+ to migrate from the Cs1 site to Cs2. The reverse flipping can then occur to return to the elliptical-shaped tunnel.[22,23]

Although the mechanism described above has shed light on the behavior of CST, it is also of importance to consider exchange with the sodium phase in the presence of sodium ion. In this case a 1 mM solution was used. As the Cs^+ ion exchanged with the Na^+ there was no change in the geometry of the crystal. However, there was a change in the unit cell dimensions.[24] The unit cell parameters $a = b$ showed a continuous increase from 7.8060(1) to 7.8435(1) Å and the c-axis decreased from 11.9599(2) to 11.9054(4) Å. This represents an increase in volume of 0.5%. The first step, at 245 minutes (1.5 minutes per step), involved occupancy of Cs^+ in the Cs2 sites (six-coordinate) to an occupancy level of 0.116(5). Simultaneous with the increased occupancy of this site, water occupancy in the tunnel began to decrease (Figure 7.7).[24] Obviously the exchange was between Na^+ and Cs^+

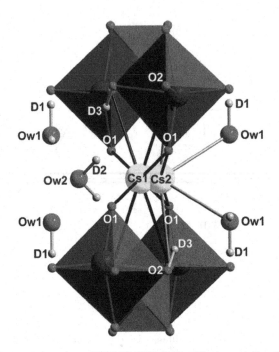

Figure 7.6.   Cross-section view of (Cs, D)-CST 8MR with channel direction in the plane from left to right. Silica tetrahedra in-plane with Cs sites, and one Cs2 overlapped with Ow2, were removed for clarity. Sites D3 and D1 are positionally disordered with H3 and H1. (Reproduced from Ref. 22, Figure 9, with permission from the American Chemical Society.)

but this water loss was essential for the next step to occur. This second step occurred from 245 to 375 minutes and involved the movement of Cs^+ into the Cs1 site. The final composition was $Na_{1.06}Cs_{0.26}H_{0.68}Ti_2O_3(SiO_4) \cdot H_2O$. The high proton content was the result of hydrolysis. The remaining Na^+ content is for the sodium ion in the small framework hexagonal pore. This sodium remains intact as Cs^+ is too large to fit this site and the sodium ion in this site is too strongly bonded to be hydrolyzed.

The final part of our *in situ* studies involved Cs^+ exchange with the Nb-CST. The Ti/Nb fractional occupancies refined to 77/23, close enough as a representative sample for the 25% Nb phase. Interestingly, when the Na^+ was exchanged for H^+ no phase change occurred.[24] The unit cell dimensions initially increased to minute 70 then showed a continuous volume decrease but less than 1%. Again exchange occurred in two steps. The first from 0 to 197 minutes involved movement of Cs^+ into site 1, with fractional occupancy to 0.115. Then 0.5 mole of water left in 1 minute. This was followed by uptake of Cs^+ from 197 to 700 minutes.

Fractional occupancy

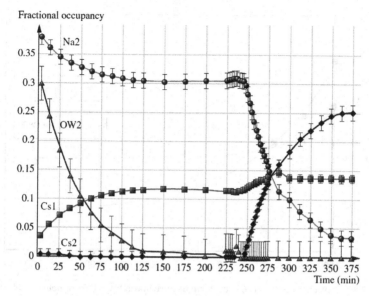

Figure 7.7. Results of fractional occupancy refinement for Na2, OW2, Cs2, and Cs1 during the Cs$^+$ ion exchange into Na-TS. (Reproduced with kind permission from Springer Science + Business Media, as adapted from Ref. 24, Figure 49:15.)

The final formula was $H_{1.33}Cs_{0.17}Nb_{0.5}Ti_{1.5}O_3(SiO_4)\cdot1.5H_2O$. The Nb/Ti was idealized to 25% Nb. It is ironic that as we approached the more realistic part of *in situ* studies, which would be Cs$^+$ exchange in simulated nuclear waste, our funding was not renewed.

7.2.3 *Structure of strontium phases of CST*

The uptake of Sr^{2+} in CST and Nb-CST as a function of Na$^+$ concentration is shown in Figure 7.8. Strontium uptake shows two distinct regions.[14,25] Below 1.0 M sodium ion the uptake of Sr^{2+} in Nb-CST is higher than that of CST. However, as the concentration of sodium ion increases, the decrease in uptake falls more rapidly for Nb-CST. This opens the possibility that CST may be effective for the removal of strontium from nuclear waste solutions. To seek answers as to why this is the case we resorted to structural studies.

A series of Nb-CSTs were prepared with 25%, 16%, and 5% Nb in mole percent. The 25% phase (400 mg) was added to 100 ml of a 0.025 M solution of SrCl$_2$ and equilibrated for 12 hours. The final composition was $Sr_{0.19}Na_{0.61}H_{0.5}Nb_{0.5}Ti_{1.5}O_3(SiO_4)$. Structure solution based on XRPD data gave a framework as previously detailed for Na-CST. The Sr^{2+} ions were not in the center of the tunnel but at 0.4,0.4,0. They are

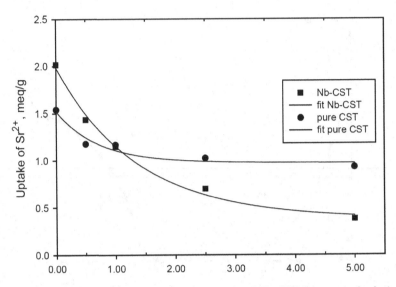

Figure 7.8.   Uptake of Sr^{2+} in sodium form of pure and Nb-CST from neutral solutions.

seven-coordinate, being bonded to five framework oxygen atoms and two water molecules.[25] The remaining members of the series were converted to protonic form by treatment with 0.1 M HCl. The H-CST was exchanged to the 25 and 55% Sr^{2+} levels and the 16% Nb-CST to the 50% Sr^{2+} exchange level. The sorting out of the correct unit cells was a problem. The 25% CST was orthorhombic with $a = 11.04$, $b = 11.16$, $c = 11.87$ Å. These unit cell dimensions are close to those for H-CST but $a \neq b$, making the cell orthorhombic. The 55% loaded Sr-CST consisted of 82% tetragonal and 18% orthorhombic phases. This means that the 82% phase contains $\sim$61% Sr^{2+}. The strontium ions are located in two sites, one at coordination number (CN) 10 and the other between 9 and 10. The 50% level 16% Nb-CST had a CN of 9 while the 25% Nb-CST had a CN of 7 (Figure 7.9(b)). In general, then, we may conclude that the non-Nb-CST is more selective than the 25% Nb-CST because of its higher strontium coordination number. More data on ion exchange and kinetics of exchange are presented in subsequent sections.

7.2.4 Kinetics of the exchange reactions

Given the very large quantities of waste solutions that need to be dealt with, the rate at which the exchange reaction takes place becomes an important consideration. We have found that it is possible to speed up the reaction by using less-crystalline, fine-grained CST particles to carry out the uptake of

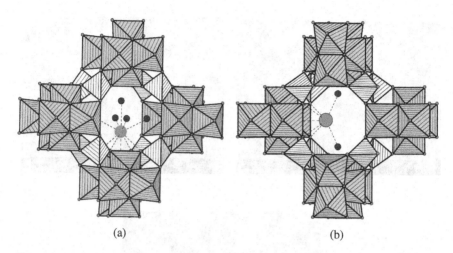

(a)   (b)

Figure 7.9.   A perspective view of the polyhedral representation in 25% Sr-CST (a) and Nb-CST (b). The eight-ring tunnel along the c-axis depicts a ten-coordinate Sr complex. Sr1 is coordinated by five framework oxygen atoms and five water molecules. The polyhedral representation of the 25% Nb-CST shows a seven-coordinate Sr^{2+} species. (Adapted from Ref. 25, Figures 3 and 5, with permission of Elsevier.)

Cs^+ and Sr^{2+} or $Sr(OH)^+$. The contrast in particle size of the poorly crystalline titanosilicate (P-CST) and the crystalline variety (C-CST) is shown in Figure 7.10 and the X-ray patterns in Figure 7.11. Rates of exchange were carried out by the radiotracer method in either neutral solutions or in simulants that were 4–5 M in Na^+, 1 M in NaOH and 0.43 M in $Al(NO_3)_3$. This composition approximates the actual waste at the Savannah River repository.

Figure 7.12 shows the contrast between the rates of exchange of Cs^+ in the crystalline and poorly crystalline varieties and the same for Sr in Figure 7.13. Tables 7.1 and 7.2 list the rates of uptake of Sr^{2+} and Cs^+ respectively in simulated waste located at the Savannah River Site.[26]

A surprising fact is that the P-CST with no niobium substitution was still effective for Cs^+ removal: almost 97% of the Cs^+ was taken up in 15 minutes. The uptake by the Nb-CST was somewhat slower but in the end the total uptake was greater: 99.3% versus 96.5%. Another surprise is the results for strontium. In this highly basic solution strontium is present as $Sr(OH)^+$ and therefore the exchange capacity for this ion is double that for Sr^{2+}. This is especially true for P-CST as shown in Figure 7.14. We know from X-ray diffraction studies that in Nb-CST with 25% Nb the CN of Sr^{2+} is 7 but in the non-niobium CST the CN is 10.[14,25] This change in coordination arises from a complex series of phase changes as explained in

A. Clearfield

(a)   (b)

(c)

Figure 7.10.   SEM images of P-CST (a, b) and C-CST (c) phases of the sodium titanium silicate.

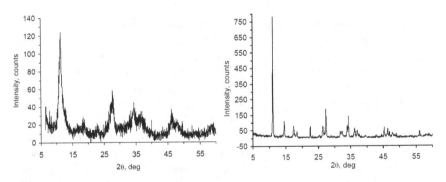

Figure 7.11.   X-ray powder pattern of sodium form of poorly crystalline P-CST (left) and highly crystalline C-CST (right).

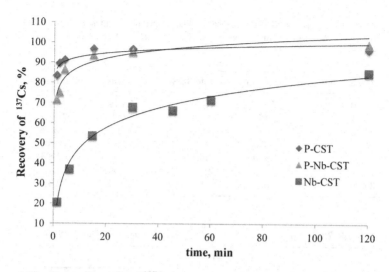

Figure 7.12. Rates of uptake of ^{137}Cs$^+$ from solution by poorly crystalline P-CST and P-Nb-CST, and crystalline Nb-CST.

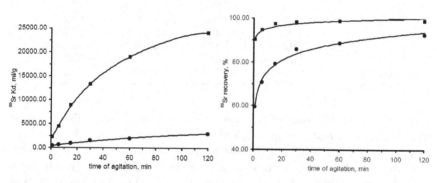

Figure 7.13. K_d values (left) and recovery of Sr-89 (right) from groundwater simulant as a function of time of agitation with C-CST (•) and P-CST (■) phase.

the cited references. Thus, in general the CST is more selective to Sr than is Nb-CST.

Our results show that the use of poorly crystallized forms of CST and Nb-CST greatly accelerates the rates of exchange of Cs$^+$ and Sr(OH)$^+$. Not enough consideration has been given to this point in developing schemes for removal of Cs$^+$ and (SrOH)$^+$ from basic nuclear waste systems. The higher rate for the poorly crystalline CST must be due to the much higher surface area and smaller particle size. Diffusion through the channels of the crystals

Table 7.1.   Uptake of Cs^+ from a nuclear waste simulant solution 7.17 M in Na^+, 5 M in OH^- in a mixture of 10^{-3} M Cs^+ and 10^{-3} Sr^{2+}.

	P-CST		P-Nb-CST	
Time/minutes	CTS/min	Sample 3-28-1	CTS/min	Sample 1-40a
0	15,800	0%	17,748	0%
1	2,600	83.5%	5,035	71.6%
2	1,650	89.6%	4,401	75.2%
4	1,370	91.3%	2,381	86.6%
15	500	96.8%	1,104	93.8%
30	523	96.7%	809	95.4%
3,600	550	96.5%	132	99.3%

Table 7.2.   Uptake of Sr^{2+} from the solution described in Table 7.1.

	P-CST		P – Nb-CST	
Time/minutes	CTS/min	Sample 3–28-1	CTS/min	Sample 1–40a
0	20,953	0%	17,478	0%
1	4,991	77.5%	4,916	71.9%
2	4,670	79.1%	4,496	74.3%
5	2,784	88.2%	2,312	86.8%
15	2,019	91.9%	2,004	88.5%
30	1,998	92.0%	1,975	88.7%
3,600	2,076	91.7%	1,948	88.8%

is slow but with the smaller particle sizes many more unit cells are on or near the surface and readily available for exchange. Thus, I believe that this single mix of P-CST and P-Nb-CST may well be used to remove both Cs^+ and $Sr(OH)^+$ from current waste solutions simultaneously. The best procedure would be to add the exchangers to a stirred tank of waste solution for a given time and recover the exchanged solid by centrifugation. This procedure would obviate the need to produce pellets or to use a column technique.

7.3   Sodium Nonatitanate (SNT) for Selective Removal of Strontium from Basic Solutions

A number of sodium titanates of unknown structure exhibited high uptakes of Sr^{2+} from ion mixtures.[27,28] A major interest in these materials was for the removal of ^{90}Sr from nuclear waste solutions. Subsequently, one of the principal student workers in this area, Jukka Lehto, came to work

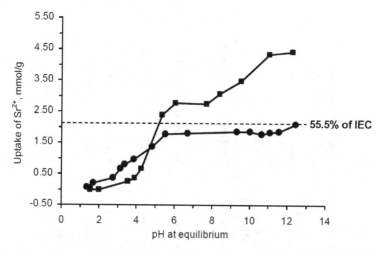

Figure 7.14. Potentiometric titration curves of highly crystalline (●C-CST) and poorly crystalline (■P-CST) phases of $H_2Ti_2O_3SiO_4 \cdot 1.5H_2O$. Titrant: 0.05 M $(SrCl_2\text{-}Sr(OH)_2)$. To achieve values of pH below 3, a 0.1 M HCl solution was used.

in my laboratory. We established the conditions of preparation and the composition of this strontium sorber[29] as shown in Table 7.3. Although there was some variation in composition, the most crystalline preparations were best designated as $Na_4Ti_9O_{20} \cdot nH_2O$. The compound is layered with an interlayer spacing of 6.9 Å in the anhydrous state, 8.6 Å in a semihydrous form, and 10.0 Å fully hydrated. Because many of the preparations listed in Table 7.3 were carried out at high temperatures we explored the preparation under milder hydrothermal conditions. In general it was found that lower temperatures and shorter heating times produced products with poorer crystallinity. The hydrothermal conditions of preparation are shown in Table 7.4 and the X-ray patterns for a few of the products are shown in Figure 7.15. One of the main factors determining crystallinity is the mole ratio of NaOH : Ti. In general the higher the ratio the more crystalline the product (13B and 90A in Figure 7.15).[30] The less crystalline products were at the low end of this ratio (67A, 67B, and 71A) or had short reaction time (64B). The fibrous nature of the more crystalline samples is shown in Figure 7.16,[31] but less evident with amine addition.

Titration of the proton phase, $H_4Ti_9O_{20}$, by the batch method with $SrCl_2 + Sr(OH)_2$ yielded an ion exchange capacity (IEC) of 5.32 ± 0.23 meq/g.[32] However, the Na^+ recovered was 4.54 ± 0.58 meq/g as compared to a theoretical IEC of 4.74 meq/g. It was concluded that the exchange reaction was

$$Na_4Ti_9O_{20} + 2\,Sr^{2+} \rightleftharpoons Sr_2Ti_9O_{20} + 4\,Na^+. \tag{7.1}$$

Table 7.3. Products obtained from titanium isopropoxide and NaOH by the sol-gel method followed by hydrothermal treatment of the gel.[a]

T(°C)	Na/Ti Ratio in the Starting Materials (mole)						
	2/9	4/9	3/4	1/1	4/3	2/1	3/1
320	Anatase $\gg Na_x TiO_2$	Brookite $+Na_x TiO_2$	$Na_x TiO_2$	$Na_x TiO_2$	$Na_x TiO_2$ $>Na_4 Ti_9 O_{20}$	$Na_x TiO_2$	?
280	Anatase $\gg$ Brookite	Brookite $\gg Na_4 Ti_9 O_{20}$ $\gg Na_x TiO_2$	$Na_x TiO_2$ $>Na_4 Ti_9 O_{20}$	$Na_4 Ti_9 O_{20}$ $+Na_x TiO_2$	$Na_4 Ti_9 O_{20}$	$Na_4 Ti_9 O_{20}$	$Na_4 Ti_9 O_{20}$
240	Anatase $\gg$ Brookite	Brookite $>Na_4 Ti_9 O_{20}$	$Na_4 Ti_9 O_{20}$	$Na_4 Ti_9 O_{20}$	$Na_4 Ti_9 O_{20}$	$Na_4 Ti_9 O_{20}$	$Na_4 Ti_9 O_{20}$
200	Anatase $\gg$ Brookite	Brookite $>Na_4 Ti_9 O_{20}$	$Na_4 Ti_9 O_{20}$	$Na_4 Ti_9 O_{20}$	$Na_4 Ti_9 O_{20}$	$Na_4 Ti_9 O_{20}$	$Na_4 Ti_9 O_{20}$
160	Anatase	$Na_4 Ti_9 O_{20}$	$Na_4 Ti_9 O_{20}$	$Na_4 Ti_9 O_{20}$	$Na_4 Ti_9 O_{20}$	$Na_4 Ti_9 O_{20}$	$Na_4 Ti_9 O_{20}$

Note: +, Products formed in the same amount range; >, latter product is minor product; $\gg$, latter product is found only in very small amounts.

[a] The $Na_4 Ti_9 O_{20}$ listed in the table was always obtained in a hydrated condition with $d_{001} \approx 8.$ Å.

Table 7.4. Reaction conditions for the synthesis of the sodium nonatitanates.

Sample	Conc. of Ti Soln. (M)	Conc. of NaOH Soln. (M)	Ti: Na Mole Ratio	Reflux Time (hours)	Hydrothermal Temp. (Time)
RC-3-67B	0.75	5.1	1:6.8	1	200°C (4 days)
RC-3-67A	0.43	0.87	1:2	3.5	145°C (4 days)
RC-3-71A	0.69	1.4	1:2.1	3	200°C (7 days)
RC-2-62A	1.6	2.5	1:1.56	3	190°C (20 hours)
RC-4-11B	1.0	2.0	1:2	3	200°C (20 hours)
RC-4-11A	1.0	4.0	1:4	3	200°C (20 hours)
RC-4-48A	0.63	3.1	1:5	2	170°C (22.5 hours)
RC-4-48B	0.70	3.3	1:4.7	0	170°C (23 hours)
RC-4-64A	0.65	4.5	1:6.9	3.2	193°C (19.7 hours)
RC-4-64B	0.58	5.1	1:8.9	3.2	193°C (19.7 hours)
RC-4-13B	0.90	9.0	1:10	3.2	190°C (21 hours)
RC-4-23B	0.92	9.7	1:10.5	0	190°C (22 hours)
RC-3-90A	0.44	8.35	1:19.4	16.5	200°C (1 day)

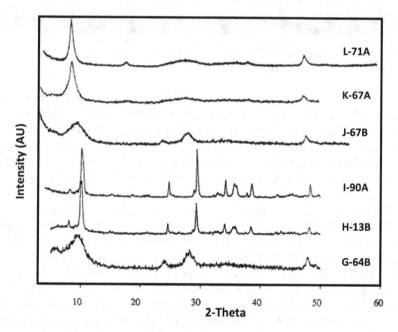

Figure 7.15. XRD powder patterns of a variety of sodium nonatitanates synthesized in an aqueous media with a reflux step prior to hydrothermal treatment.

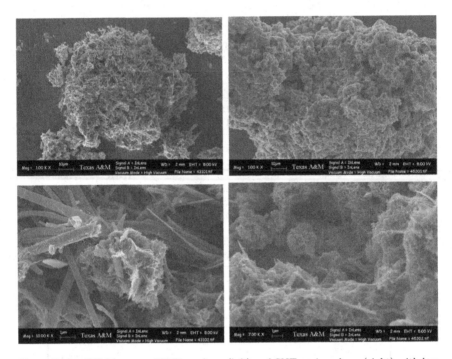

Figure 7.16.   SEM images of SNT-w phase (left) and SNT-amine phase (right) with low (top) and high (bottom) resolutions.

The high strontium uptake value was attributed to additional surface sorption of $SrCl_2$. Heating the exchanged SNT to 900°C yielded $SrTiO_3 + TiO_2$. More important is the behavior of SNT in strongly basic solutions but remember that the level of strontium in the waste solutions may be less than 10^{-4} M. At pH above about 12.8 all the strontium is present as $Sr(OH)^+$ or at least this appears to be the case from SNT ion exchange uptake. This is a much higher pH than predicted from formation constants.[33]

Extensive study was devoted to optimizing the strontium selectivity by SNT. A range of crystallinities was examined. Refluxing titanium isopropoxide with NaOH at ~120° for up to 24 hours yielded poorly crystalline, almost amorphous SNT of the type shown by sample 64B in Figure 7.15. Often in the preparation of SNT samples Na_2CO_3 is found to be present. This apparently results from exposure to CO_2 by the strong NaOH solution. IR spectra were used to show the presence of the sodium carbonate. This had a profound effect upon the K_d values. For example, samples 11A and 11B had K_d values of 27,3000 ml/g and 7,310 ml/g, respectively, in a simulant 5 M in $NaNO_3$, 1 M NaOH and 91–95 ppm Sr^{2+}. The samples were then washed in 0.2 M NaOH and further washed with a 60–40 ethanol-water

Table 7.5. Sr^{2+} K_d values using the simulated Hanford waste solution ($5\,M$ $NaNO_3$, $1\,M$ NaOH, and 91–95 ppm Sr^{2+}) on SNT samples that have been washed once (w) or twice (w_2) to remove carbonate.

Sample	$Sr^{2+}K_d(ml/g)$
RC-3-67BW	$\geq$133,000
RC-4-11AW	$\geq$132,000
RC-4-23BW	$\geq$132,000
RC-4-48BW	11,400
RC-4-64AW	$\geq$132,000
RC-4-64BW	$\geq$132,000
RC-3-92AW2	9,100
RC-3-67BW2	$\geq$133,000
RC-4-11AW2	$\geq$133,000
RC-4-23BW2	$\geq$134,000
RC-4-48BW2	20,400
RC-4-64AW2	$\geq$132,000
RC-4-64BW2	$\geq$132,000

mixture. This process was repeated a second time and K_d values determined in the same simulant. The results are collected in Table 7.5.[30] The increase in K_d values is remarkable.

One of the conditions of our grant from the PNNL Hanford works was to have an industrial partner. This was arranged with AlliedSignal and my counterpart there was Dr Stephen Yates. The results shown in Table 7.5 were imparted to AlliedSignal, now a part of Honeywell, and under Yates' guidance they produced a number of 10 kilogram samples. We also received the composition of the DSSF-7 simulant from Garrett Brown of the PNNL.[34] This simulant represents the composition of Tank 101-AW at Hanford diluted to $7\,M$ sodium ion (Table 7.6). Strontium was added to the simulant at a level reported in other similar Hanford wastes ($10^7\,M$). The DSSF-7 was diluted with deionized water to concentrations of $3.5\,M$ Na^+ and $1\,M$ Na^+ which were designated DSSF-3.5 and DSSF-1, respectively.

These solutions were spiked with ^{89}Sr by adding 0.3 ml of diluted ^{89}Sr tracer per 400 ml of solution. Unfortunately, due to the presence of greater than expected concentrations of inactive strontium carrier, this resulted in each of the DSSF solutions having a total strontium concentration of approximately 0.27 ppm ($\sim 3 \times 10^{-6}\,M$), which is significantly higher than the initial $1 \times 10^{-7}\,M$ strontium (0.009 ppm). Consequently, the strontium concentration in all of the simulations can be considered to be $\sim 3 \times 10^{-6}\,M$

Table 7.6. Initial composition
of the DSSF-7 simulant.

Component	Molarity
$NaNO_3$	1.162
KNO_3	0.196
KOH	0.749
$CsCl$	7×10^{-5}
Na_2SO_4	0.008
Na_2HPO_4	0.014
$NaOH$	3.885
$Al(NO_3)_3 \cdot 9H_2O$	0.721
Na_2CO_3	0.147
$NaCl$	0.102
$NaNO_2$	1.512
$Sr(NO_3)_2$	1×10^7

Table 7.7. Strontium K_d values for various titanate exchangers from DSSF solutions.

Sample ID	DSSF-1 K_d (ml/g)	DSSF-3.5 K_d (ml/g)	DSSF-7 K_d (ml/g)
8212-166-1	25,100	43,600	13,700
8212-166-2	110,500	27,500	8,700
8212-166-3	>670,000	148,000	75,500
8212-166-4	>670,000	254,300	127,400
8212-166-5	>670,000	307,700	144,900
8212-166-6	86,600	51,600	24,700
GMG-I-1	181,000	139,300	57,400

and is considerably higher than would be expected in actual tank wastes. A quantity of 0.05 g of exchanger was gently shaken for approximately 20 hours with 10 ml of solution spiked with [89]Sr (volume-to-mass ratio 200:1). The solution was then filtered through a Whatman No. 42 filter paper and the aqueous phase analyzed for ^{89}Sr using a liquid scintillation counter. The K_d values for the various titanate samples in the different DSSF feeds is given in Table 7.7. Also included in Table 7.7 is the data for the A&M titanate, GMG-I-1, which has proved to be one of our best titanates to date. All K_d values quoted are the average of at least two separate determinations.

It can clearly be seen from Table 7.7 that the best titanates are the AlliedSignal samples 8212-166-3, -4, and -5 which exhibit good K_d values in all of the solutions evaluated and exceed the performance of the benchmark titanate GMG-I-1 from our study. A K_d of 670,000 ml/g equates to reduction of the ^{89}Sr from an initial value of 100,000 cpm to 30 cpm.[35] A paper describing the process involved in preparing the large (10 kg) batches of

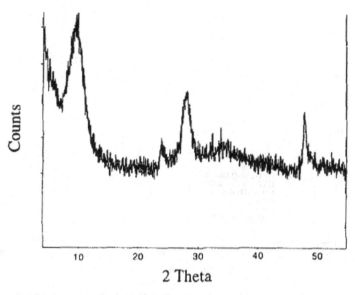

Figure 7.17. X-ray diffraction pattern of sodium nonatitanate (prepared by Allied-Signal). The d-spacing of the 001 reflection is 9.4 Å.

sodium nonatitanate was published.[35] The authors claim that they have made an even more selective titanate as shown by the XRPD pattern (Figure 7.17) and the very large broadness of the peaks indicative of small particles and a highly amorphous structure. They also confirmed the fibrous nature of the different batches and that the nonatitanate was not overly affected by addition of complexants such as EDTA, oxalic acid, or citric acid. Coupled with the fact that the nonatitanate was stable to radiation and almost totally unselective for alkali metal cations makes it an ideal candidate for removal of strontium from nuclear waste compositions. Furthermore, titration data show that the strontium is fully recoverable by treatment of the exchanged SNT with acid.[32,35] An AlliedSignal team was able to bind the particles into pellets for use in columns.[36,37] Samples were sent to various nuclear facilities for testing for strontium removal in actual stored nuclear waste. The results at the Savannah River Site were found to be very encouraging.[38] It was also found that SNT is effective for uptake of Pu and Np, as shown in Figures 7.18 and 7.19. MST or monosodium titanate will be described in the next section. The QAB samples are commercially prepared and marketed as SrTreat.[39] The JD and DM samples were prepared in our laboratory.

These positive results motivated AlliedSignal to file for patents.[36,37] Samples of SNT were sent to a variety of nuclear facilities for trial and found to be quite effective, as shown in a later section of this chapter.[38]

A. Clearfield

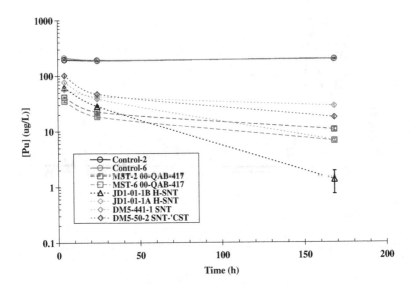

Figure 7.18. Plot of plutonium concentration versus time upon contact of TAMU-prepared SNT, MST, and SNT-CST materials with simulated SRS waste solution.

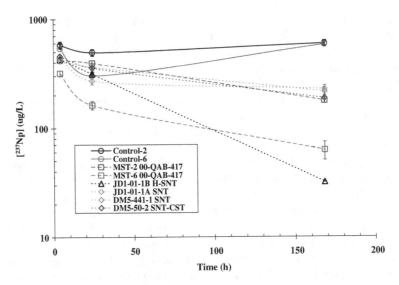

Figure 7.19. Plot of ^{237}Np concentration versus time upon contact of TAMU-prepared SNT and SNT-CST materials with simulated SRS waste solution.

Sodium nonatitanate is also effective for uptake of actinides, as shown in Figures 7.18 and 7.19. The protonated form had the largest long-term uptake of Pu and Np.

7.4 Monosodium Titanate, NaTiO$_5$H

In the late 1970s a new sodium titanate was prepared and reported to be especially effective for use in separation of radioactive nuclides from nuclear waste solutions.[40] The preparation method consisted of mixing sodium hydroxide in a methanol solution with titanium isopropoxide in isopropyl alcohol and refluxing for 1–2 hours. MST was precipitated by adding water. Lynch et al.[41] proposed the formula NaTiO$_5$H for the compound. This compound exhibited high selectivities as an ion exchanger for strontium ion as well as some of the actinides.[40] The structure and composition of MST were not characterized with a high degree of certainty as it is amorphous.

A baseline synthesis of MST as prepared at the Savannah River Westinghouse Research Center was prepared as well as several modifications.[42] Additional samples were prepared on a larger scale by Optima Chemical Company, Inc.[42] Some results obtained with the Savannah River simulant with added quantities of strontium, plutonium, neptunium and uranium are given in Table 7.8.

Because of the amorphous quality of MST and its surface with hair-like fringes (Figure 7.20) it was surmised that sorption occurs on the particle surface. Therefore, attempts to create high-surface-area particles by templating the preparations with polyethylene glycols and amines were carried out.[31]

Surface areas of the dried solids, as measured by the N$_2$ sorption BET method, and the Sr K_d values of these samples are provided in Table 7.9. Listed at the bottom in Table 7.9 are the results for a commercially prepared sample of MST, OO-QAB-417. A selection of the SEM images and corresponding N$_2$ absorption-desorption isotherms are provided in Figure 7.21. It is notable that no two isotherms are similar nor is there

Table 7.8. Typical strontium and actinide removal results upon contact of simulated waste solutions with $0.4\,gL^{-1}$ MST.

Sorbate	Initial Conc. (μM)	DF	% Removed	Loading on MST (μmole g^{-1})
Sr	10	100	99	25
Pu	1	15	93	2.3
Np	2	5	80	4.0
U	40	2.5	60	60

A. Clearfield

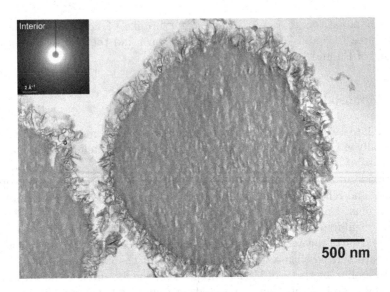

Figure 7.20. Low-magnification bright-field TEM image showing the typical morphology of MST particles with corresponding selected area electron diffraction (SAED) pattern for inner amorphous material.

Table 7.9. Summary of MST samples synthesized for strontium removal.

Prep. ID	Phase ID	Template	Slurry pH	BET SA (m^2/g)	Microporous SA (m^2/g)	Sr K_d (mL/g), Time to Eq.
DM4-46-1	MST-amine	Hexylamine	9.226	232	204.2	917,000 at 23.2 hours
DM5-261-1	MST-TEG-1.0	TEG*	9.532	288	143	301,000 at 24 hours
DM5-27-1	MST	No template	9.26	458	178	336,000 at 24 hours
DM5-271-1	MST-TEG-0.3	TEG	10.085	219	97	308,000 at 21.5 hours
DM5-28-1	MST-TEG-0.29	TEG	9.924	292	163	294,000 at 24 hours
DM5-281-1	MST-TEG-0.58	TEG	11.596	16	0	332,000 at 21.5 hours
00-QAB-417	Comm. MST	As received	11.87	141	0	259,598 at 22 hours

*Tetraethylene glycol.

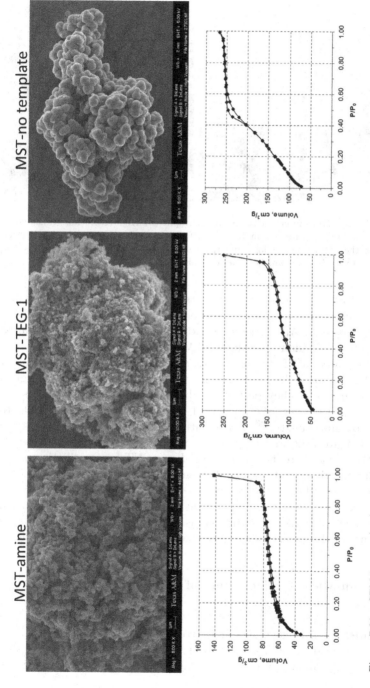

Figure 7.21. SEM images (top) and nitrogen adsorption isotherms (bottom) of MST samples. See Table 7.9 for surface area data and experimental section for synthetic procedures.

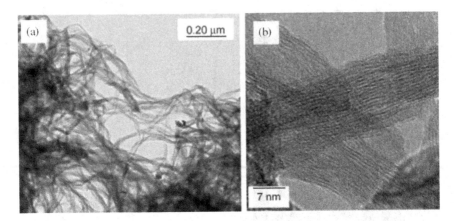

Figure 7.22. (a) Transmission electron micrograph. (b) High-resolution transmission electron micrograph of peroxo-titanate sorbent from the basic aqueous-peroxide synthesis, showing the fibrous morphology and the 7−8 Å layers. (Reproduced from Ref. 43, Figure 2, with permission from the American Chemical Society.)

a trend as to the observed surface areas. It is interesting that the commercially prepared sample has no micropore structure although the BET surface area is high. All the samples exhibited extremely high K_d values even though the simulant contained 7.5 M Na$^+$.

Non-templated MST samples were prepared by Dr May Nyman at the Sandia National Laboratories in several variants of the original Sandia preparations. They performed about as well for Sr uptake as the SNT samples. However, she conceived of the idea of adding H_2O_2 to the monosodium titanates.[43] This treatment changed the morphology of the MST (Figure 7.22) and increased the uptake of plutonium and neptunium to the point where it became favored over SNT for remediation of the Savannah River fluid waste. Some of these results are shown in Figure 7.23. A patent was issued to cover the use of peroxide-treated sorbents.[44]

One caveat for using the MST is that it has been found to exchange a broad spectrum of transition metal ions as well as lanthanides. In some cases the presence of these ions may reduce the selectivity for the targeted ions.

7.5 Sodium Micas for Treatment of Groundwater and Tank Leaks

As indicated, tank leakage should be treated as a serious problem as it is prone to spreading radioisotopes over large areas and local water supplies. Our original idea was to use clays as barriers to the spread of

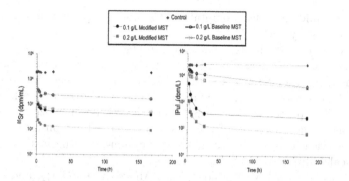

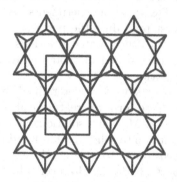

Figure 7.23.   Plot of concentrations of Sr and Pu in Savannah River Site simulant with respect to time, in contact with MST slurry and peroxide-treated MST slurry, at two different sorbent concentrations. The concentration is measured in disintegrations per minute (dpm) per unit volume (per milliliter for Sr and per liter for Pu). (Reproduced from Ref. 43, Figure 7, with permission from the American Chemical Society.)

Figure 7.24.   Hexagonal cavities formed by the sharing of corners of clay tetrahedra. The exchange ion fits into the cavity formed by six oxygen atoms from the layer above and an equal number from the layer below.

the radioisotopes. Smectitic clays have low ion exchange capacities and do not trap or hold the isotopes.[45] However, micas are thermally stable aluminosilicates.[46] They consist of negatively charged 2:1 layers that are compensated and bonded together by large unipositive ions. Most of the known natural micas exist in potassium form. This affinity is connected to the mica structure in that the outer surfaces of micas consist of SiO_4^{4-} in hexagonal arrangements of oxygen atoms that create pores that correspond ideally to the size of the K^+ ion (Figure 7.24). The K^+ ion has been retained over geologic time periods. Preliminary modeling calculations indicated that the ideal mica structure would also be selective for Cs^+. However, in order to achieve rapid and selective uptake of Cs^+ ion, the potassium would have to be replaced by Na^+.

Prior to our study K^+ had been replaced by Na^+ using sodium tetraphenylborate,[47] $NaB(C_6H_5)_4$. A slurry of finely divided mica and tetraphenylborate (TBP) is refluxed to effect the exchange. The resultant $KB(C_6H_5)_4$ is insoluble as is the mica, so the TBP is extracted with organic solvents. This procedure is expensive and the release of benzene is potentially hazardous, especially if the reaction is carried out on a large scale. Thus, we wished to develop an inexpensive, non-polluting method that is easily scaled up.

We chose to work with phlogopite, $KMg_3Si_3AlO_{10}(OH)_2$, and biotite, $K(Mg_2Fe^{2+})Si_3AlO_{10}(OH)_2$, which were purchased from Ward's Natural Science Establishment, Inc. Natural micas are abundant and inexpensive.[46] The interlayer distance in these micas is ~ 10 Å which leaves only a narrow slit opening that the large Cs^+ cannot access. Sodium ion is only loosely held by micas because this ion is too small to fill the hexagonal-shaped cavities of the mica surface. Two types of experiments were carried out to obtain sodium micas. In the first procedure the powdered micas were treated at elevated temperature with concentrated solutions of sodium chloride to replace K^+ with Na^+. The second method was to synthesize the micas from pre-prepared gel mixtures.

The conditions of conversion were determined by subjecting 1 g samples of K-biotite in powdered form to treatment at elevated temperatures, 140–240°C, in strong (0.5–2 M) NaCl solutions or mixtures of NaCl with NaOH.[48] The ratio of solution to solid was 250:1 and the time of hydrothermal treatment was varied from 2 to 24 hours. Biotite is converted to $\sim 70\%$ sodium phase in 24 hours at 190°C. However, phlogopite behaves differently. Almost no exchange (2–5%) occurs in the first treatment and four treatments are required for an 80% displacement of K^+ for Na^+. The process was followed by X-ray diffraction as shown in Figure 7.25 where the K^+ phase is represented by a peak at 10.4 Å and the Na^+ phase by an interlayer spacing of 12.6 Å. The progression of K^+ replacement for biotite is shown graphically in Figure 7.26. The uptake of Cs^+ in terms of its K_d value by the two micas in Na^+ form is shown in Table 7.10. The solutions were 10^{-3} M in cesium ion content with no interfering ions except Na^+. The results were extraordinarily good even for 0.1 M NaNO$_3$ but decreased rapidly with NaOH additions. However, these materials are intended to form barriers in the soil to prevent Cs^+ from migrating away from the storage site.

A collaboration with AlliedSignal Inc., sponsored by the Pacific Northwest National Laboratory, to screen products from our laboratory, was used to scale up the synthesis of the sodium micas. This effort at AlliedSignal

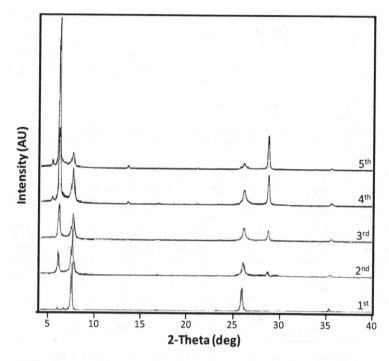

Figure 7.25. X-ray powder patterns indicating sodium replacement of potassium in phlogopite by successive treatments with 1M NaCl at 190°C.

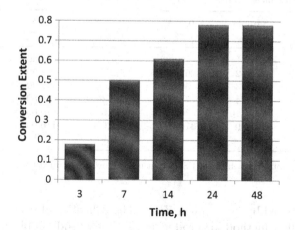

Figure 7.26. Kinetics of natural biotite conversion into the sodium form. 1M NaCl solution, T = 190°C.

Table 7.10.   Distribution coefficient values in ml/g for cesium exchange from model solutions containing 0.0001 M Cs$^+$, on mica samples.

Mica	H$_2$O	0.1 M NaNO$_3$	1 M NaNO$_3$	5 M NaNO$_3$	1 M NaOH	5 M NaNO$_3^+$ 1 M NaOH
Biotite-K	15	<5	<5	<5	<5	<5
Biotite-Na	>400,000	>100,000	14,300	750	—	450
Phlogopite-K	10	<5	<5	<5	<5	<5
Phlogopite-Na	>400,000	>100,000	6,250	500	500	120

Table 7.11.   The K_d values in ml/g for cesium uptake from aqueous solutions containing NaNO$_3$ by the mica samples. Initial Cs$^+$ content is $\sim 1 \cdot 10^{-3}$ M.

ID* and % Na$^+$	Sample No.	0.01 M NaNO$_3$	0.05 M NaNO$_3$	0.1 M NaNO$_3$	0.5 M NaNO$_3$	1 M NaNO$_3$
P1 20	M287-1	97,000	42,500	29,300	6,400	3,700
P2 50	M289-2	64,000	73,600	41,000	17,600	12,700
P3 75	M294-3	114,000	56,700	41,000	15,500	10,500
P4 100	M279-4	49,000	68,000	51,000	17,500	12,500
P5 100	M285-10	187,000	227,000	129,000	54,000	36,900
B1 100	M286-9	23,000	91,000	74,000	62,000	56,000
FS 100	JFM	11,000	7,300	4,800	690	260

*P = phlogopite, B = biotite, FS = synthetic fluorosilicic mica (JFM).

Table 7.12.   The K_d values for cesium uptake from aqueous solutions containing CaCl$_2$ by the mica samples. Initial Cs$^+$ content is $\sim 1 \cdot 10^{-3}$ M.

ID* and % Na	Sample	0.01 M CaCl$_2$	0.05 M CaCl$_2$	0.1 M CaCl$_2$	1 M CaCl$_2$
P1 20	M287-1	29,600	12,700	10,500	1,720
P2 50	M289-2	67,000	32,300	24,500	5,700
P3 75	M294-3	56,000	28,000	21,200	4,800
P4 100	M279-4	71,000	32,300	26,800	6,200
P5 100	M285-10	330,000	115,000	75,000	15,900
P6 100	M286-9	370,000	152,000	110,000	25,300
FS7 100	JFM	2,370	1,040	550	65

*P = phlogopite, B = biotite, FS = synthetic fluorosilicic mica (JFM).

was also directed by Dr Stephen Yates. The AlliedSignal team developed a continuous flow method for producing the desired sodium phase by removing the K$^+$ generated by a flow of fresh quantities of a 1 M sodium ion solution.[49] The K_d values for Cs$^+$ uptake by micas with different amounts of conversion to the sodium ion phase are presented in Tables 7.11 and 7.12.

These K_d measurements show that sodium ion micas are highly selective for Cs^+ ion in dilute $NaNO_3$ or dilute $CaCl_2$ solutions. Thus they should be useful for remediation of groundwater and some process waters. The K_d values are high even for the sample that was only 20% converted to the Na phase. This shows that the selectivity does not depend upon the extent of replacement of K^+. The two best samples were the phlogopite from Ward's at 100% K^+ replacement and the Ward's biotite. The K_d values were still respectable in 1 M $NaNO_3$ or 1 M $CaCl_2$.

To determine whether these sodium micas would be effective in removing Cs^+ ion from groundwater, a simulant for Hanford N-Springs ground water was prepared. The composition in molarity was Mg, 0.021; Na, 0.026; Sr, 1.48×10^{-4}; OH, 1.66×10^3; NO_3, 1.9×10^{-4}; SO_4^{2-}, 2.2×10^{-4}. This solution was spiked with ^{137}Cs to a level of 88,370 cts/min. The volume to mass was 200 and equilibration time was 20 hours. The exchangers consisted of Na-biotite, Na-phlogopite and a potassium hexacyanoferrate for comparison with the micas. The results in final cts/min were biotite, 115; phlogopite, 20; and KCoHex, 1,503. In terms of K_d values in ml/g they were 1.53×10^5, 5.87×10^5, and 1.2×10^4, respectively. From these results and the fact that the Cs^+ uptake is close to 1 meq/g indicates that the sodium micas would serve very well for the removal of cesium from groundwaters. A patent was issued on the synthesis of sodium micas.[49]

7.5.1 Amine-pillared micas

To increase the interlayer spacing between the layers of the sodium micas, amines were intercalated.[48] It was possible to expand the layers with the hydrochlorides of dimethyl, diethyl, propyl, and butyl amines. These ammonium ions expand the layers to allow easy access of Cs^+ to the exchange sites. Dimethylamine was not effective but the others exhibited K_d values of more than 10^5 in water as shown in Table 7.13. However, if Na^+ is present the K_d values decrease rapidly as a function of increased sodium ion content.

7.5.2 Direct synthesis of micas

A second approach to the use of micas is to prepare them synthetically. This has been done in a number of cases but they generally have not been selective for the ions of interest. Our reaction involved a mix of compounds containing metals or anions in concentrations equal to their occurrence in phlogopite but with 1 M NaOH and no potassium present. This mix was heated in the temperature range of 200–450°C for 3–10 hours. An X-ray pattern shown in Figure 7.27 matches that of natural phlogopite.

Table 7.13. Distribution coefficent values for Cs^+ sorption in ml/g on amine-pillared phlogopite and biotite samples. Initial Cs^+ concentration is 0.001 M.

Mica	Amine	H_2O	5 M $NaNO_3$	5 M $NaNO_3$+ 1 M NaOH
Phlogopite	$(CH_3)_2NH$	1,000	180	120
Phlogopite	$(C_2H_5)_2NH$	>100,000	240	80
Phlogopite	$C_3H_7NH_2$	>100,000	250	110
Phlogopite	$C_4H_9NH_2$	>100,000	460	60
Biotite	$C_3H_7NH_2$	>100,000	700	590

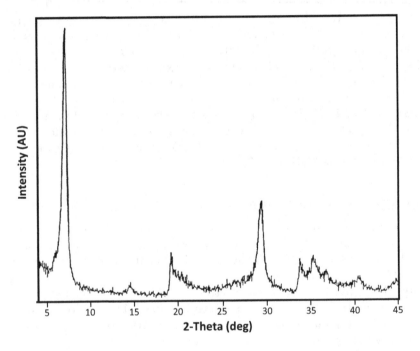

Figure 7.27. Synthetic sodium phlogopite prepared at 350°C in 6 hours.

This procedure may obviate the necessity to use successive sodium ion treatments of K-micas but rather to obtain phlogopite in a direct manner.

7.6 Pharmacosiderites

Our interest in pharmacosiderites is that they have tunnel structures somewhat resembling CST. Pharmacosiderite is a mineral of composition

K(FeOH)$_4$(AsO$_4$)$_3$.[50] However, a titanium version was prepared by Chapman and Roe[51] with a composition close to K$_3$H(TiO)$_4$(SiO$_4$)$_3$. We prepared a crystalline version and determined its structure from X-ray powder data.[52] In fact we were able to convert the potassium phase to the proton phase H$_4$(TiO)$_4$(SiO$_4$)$_3$·8H$_2$O and exchange this compound with Cs$^+$ to yield Cs$_3$H(TiO)$_4$(SiO$_4$)$_3$·4H$_2$O. All of these compounds are cubic with a = 7.8212(2) Å, 7.7644(3) Å, and 7.8214(6) Å for the Cs, K, and H phases, respectively. Indeed they have a similarity to CST as the framework is formed by a Ti$_4$O$_4$ cubane arrangement cross-linked by the silicate groups. The difference lies in the cubic nature of these compounds which requires silicate bridging along all three directions as shown in Figure 7.28. The unit cell dimension in the a and b directions of CST is 7.8082(2) Å, showing the closeness of the cubane structure in the two types of compounds. The difference is that the tunnel runs in all three directions and allows much more Cs$^+$ to be exchanged. An interesting sidelight is that

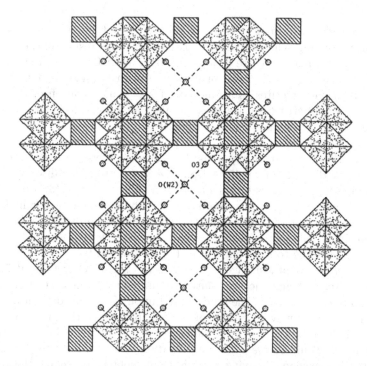

Figure 7.28.   Schematic drawing of the structure of H$_4$(TiO)$_4$(SiO$_4$)$_3$·8H$_2$O showing the network of hydrogen-bonded water molecules. The striped polyhedra represent the silicate groups which are also above and below the titanium-oxygen octahedra connecting these layers in the ± c-axis direction.

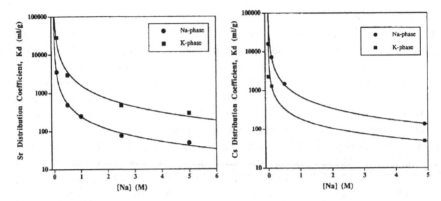

Figure 7.29.   (a) Strontium (left graph) and cesium (right graph) distribution coefficients K_d (ml/g) as a function of increasing sodium nitrate concentrations. The solid circles and squares depict K_d values for the sodium and potassium phases, respectively.

there are eight water molecules in the proton phase filling the tunnels. The hydrogen bonding scheme indicates that the protons reside on the cubane oxygen atoms, making these groups $(TiOH)_4$ as in CST.[53,54] Figure 7.29 shows distribution coefficients for exchange of Sr^{2+} and Cs^+ by the Na- and K-pharmacosiderites as a function of Na^+ concentration of the solution. Additional structural features, detailed as follows, led us to improve on the selectivity of these pharmacosiderites.

In the K^+ pharma the K^+ ion is precisely in the face centers of the unit cell. However, the siting of the Cs^+ ions is different in that they are located at a distance of $\pm 0.459\,\text{Å}$ from the face centers in a disorded manner. This siting produces four short $(3.14\,\text{Å})$ and four long $(3.41\,\text{Å})$ bonds to the silicate oxygen atoms. Similarly, the bonds to four water molecule oxygens are two short $(2.82\,\text{Å})$ and two long $(3.62\,\text{Å})$, providing a distorted 12-coordinate polyhedron. The expected Cs-O bond length for coordination 12, $3.24\,\text{Å}$, is close to the midpoint of the sum of the two observed values. Harrison et al.[55] synthesized single crystals of $Cs_3HTi_4O_4(SiO_4)_3$ at 750°C and 30,000 psi. Their structure solution exhibited the same type of disorder, confirming the validity of the study based upon X-ray powder data.

It is now clear why only three of the four protons of the protonic phase were exchanged by the Cs^+ or K^+ ions. Placing a fourth ion in the center of the unit cell would increase the cation-cation repulsive force to the extent that the ion exchange driving force is insufficient to overcome the repulsive forces. The smaller sodium and silver ions are able to produce the fully exchanged states.[56,57] However, the structures distort to rhombohedral phases to reduce the cation-cation repulsive force.

Ions situated in the center of the faces form eight strong bonds to the framework silicon oxygen atoms. This contrasts with cesium ion that forms four strong and four weak bonds. In addition the centered ions also form four strong bonds to water molecules. Eisenman has shown that in considering selectivity, two factors are most important[58]: the attractive force between the cation and the fixed negative groups of the exchanger and the hydration energy of the exchanging ions. The alkali metal selectivity series for $K_3H(TiO)_4(SiO_4)_3 \cdot 4H_2O$[59] is $Cs^+ > K^+Na^+ > Li^+$. This series is in decreasing order of hydration energy. Thus, if we could increase the strength of bonding of Cs^+, an increase in selectivity is to be expected. For this to occur it was felt that a slight expansion of the unit cell would allow for siting the cesium ion in the face centers. To accomplish this task we decided to replace the Si with the larger Ge. This was done,[59] and as shown in Table 7.14, the increase in K_d values for Cs^+, Sr^{2+}, Ba^{2+} uptake is remarkable. The data plotted in Figure 7.30 show that the poorest selectivity occurs when both the Si and Ti are replaced by Ge. Ti^{4+} is larger than Ge^{4+}, 0.605 versus 0.54 Å, so there is a net decrease in the unit cell size. However, the diagram in Figure 7.30 is a simplification of the actual situation. Both of the highly selective phases CsTiGe and CsTiSiGe are tetragonal with unit cell dimensions $a = b \cong 11.2$ Å, $c = 7.97$ Å. This is somewhat similar to the transformation of CST upon formation of the protonic phase where the new a- and b-axes are the diagonal of the sodium unit cell. For additional treatment of these phase changes the reader is referred to Refs. 54, 59, and 60. Another salutary effect is that the selectivity for Sr^{2+} is significantly increased. Table 7.14 presents the K_d values for both alkali and alkaline earth cations.

Another approach we tried was to partially replace the Ti^{4+} in the potassium pharmacosiderites with Nb^{5+}. A compound of composition

Table 7.14.   Distribution coefficients, K_d, in ml/g, and equilibrium pHs (in parentheses) for the alkali and alkaline earth metals using $HK_3(TiO)_{3.5}(GeO)_{0.5}(GeO_4)_{2.5}(SiO_4)_{0.5} \cdot 4H_2O$(KTiSiGe), $HCs_3(TiO)_{3.5}(GeO)_{0.5}(GeO_4)_{2.5}(SiO_4)_{0.5} \cdot 4H_2O$ (CsTiSiGe), $HK_3(TiO)_4(SiO_4)_3 \cdot 4H_2O$ (KTiSi), and $HK_3(TiO)_4(GeO_4)_3 \cdot 4H_2O$ (KTiGe).

Sample	Li	Na	Cs	Mg	Ca	Sr	Ba
KTiSiGe	120	280	32,000	4,900	5,480	7,900	72,700
	(8.4)	(9.3)	(9.3)	(8.1)	(8.2)	(7.8)	(7.9)
CsTiSiGe	10	90	Na	100	100	680	280
	(8.9)	(7.9)		(8.3)	(7.7)	(7.6)	(7.3)
KTiGe	200	100	46,200	6,100	>6,990	≫100,000	>100,000
	(8.5)	(8.2)	(8.5)	(8.2)	(8.3)	(7.2)	(7.3)
KTiSi	750	220	5,800	31,000	12,000	>52,000	3,100
	(10.1)	(10.7)	(9.7)	(10.6)	(10.7)	(9.6)	(9.4)

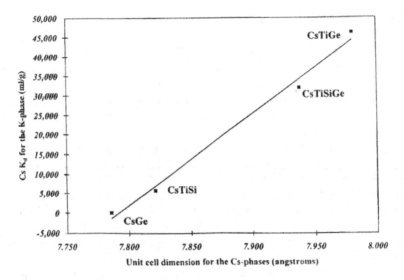

Figure 7.30. A plot of $Cs^+ K_d$ values as a function of the cubic unit cell dimension of germanium-substituted $Cs_3H(Ti_4O_4)(SiO_4)_3 \cdot 4H_2O$. Abbreviation as in Table 7.14. (Reproduced from Ref. 14, Figure 7, with permission from Elsevier.)

$HK_{2.2}(Ti_{3.2}Nb_{0.8}O_4)(SiO_4)_3 \cdot 4H_2O$ was prepared.[60] The difference in uptake of Cs^+ by this niobium-substituted phase and $HK_3Ti_4O_4(SiO_4)_3 \cdot 4H_2O$ is shown in Figure 7.30. The solution used was 0.05 M in NaOH and 0.05 M in $NaNO_3$. Thus, the results indicate the suitability of the Nb-substituted exchanger for use in groundwater systems.

7.7 Ion Exchangers for the Nuclear Fuel Cycle

We have indicated in the introduction that all the spent fuel rods accumulated from the very beginning of our nuclear program are stored under water at the reactor sites. They are termed "spent" because the amount of fissionable ^{235}U has been reduced to a low level with reduced generation of power. However, 95% of the fuel is still present because plutonium is generated in the reactor and may be recycled. Prudence would dictate that this fuel be recovered and the waste, amounting to about 5% of the total, be treated as for the weapons waste. This waste could be buried in the Yucca Mountain repository along with the weapons waste. To deal with this problem the Department of Energy's Office of Nuclear Energy is developing technologies to expand repository capacity, improve proliferation resistance, and recover energy that would otherwise be discarded.[61]

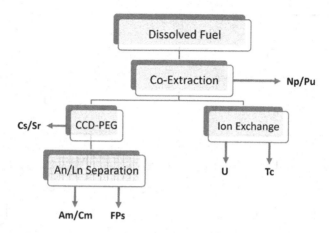

Figure 7.31. Schematic of the UREX+2 process.

The process to be used for fuel recovery of the usable fuel from the spent rods is termed UREX-Z. A schematic of the process is shown in Figure 7.31. All of the processes are accomplished by solvent extraction except separation of uranium and technetium.[62,63] Solvent extraction is carried out with tributyl phosphate dissolved in n-dodecane. However, there are some difficulties to be addressed. The lanthanides are neutron sorbers and need to be removed and separated from americium and curium. Furthermore the most desirable fuel is a combination of Pu, Np, and Am. These separations are difficult, especially the americium-curium separation. Curium is decidedly not wanted in the fuel.

In 1995, we published a paper describing a porous hybrid of zirconium phosphonate utilizing monophenyl diphosphonic acid to form a porous cross-linked material.[64] It is well known that zirconium phenylphosphonate, $Zr(O_3PC_6H_5)_2$, is layered with the phenyl rings forming a bilayer between the inorganic ZrO_6 layers.[65] By using a 1,4-diphosphonic acid, the ligand cross-links the inorganic layers with the phenyl group acting as a pillar. The resultant compound is porous but amorphous. Adding phosphoric acid to the synthesis incorporates — O_3POH groups between the phenyl pillars, increasing the porosity but also imparting ion exchange capabilities. A schematic portrayal of this inorganic-organic hybrid is shown in Figure 7.32. The composition of these materials may be altered over a broad solid solution range as indicated by the solid solution formula $Zr(O_3PC_6H_4PO_{3(1-x/2)}(O_3POH)_x \cdot nH_2O$.[66] The interesting property that concerns us is the fact that ZrPPP, as we shall designate the zirconium phenylphosphonate phosphate, has a high selectivity for 3+ lanthanides but is not selective for mono- and divalent cations.[64]

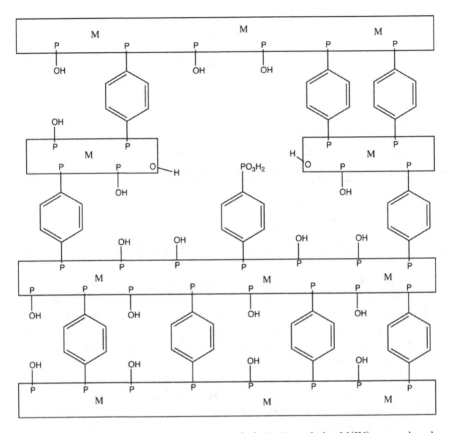

Figure 7.32. A schematic of how the mixed derivative of the M(IV) monophenyl-bisphosphonate phosphate may be formed.

We have been able to use this curious selectivity to separate lanthanides from actinides. In this connection it should be pointed out that Sn(IV) behaves similarly but with more regular pores.[67] The surface areas of both the tin and zirconium compounds with compositions close to $x = 1$ are given in Table 7.15. Note that the tin compounds have an essentially microporous structure whereas the Zr compounds have higher surface areas but a broader range of pore sizes.[66] Pores larger than 20 Å in diameter are considered to be mesopores. The porosity was determined by the N_2 sorption BET method with the tin compounds characterized as having Type I isotherms with small H4-type hysteresis loops. In contrast the Zr compounds have modified Type II isotherms with broader hysteresis loops, indicative of the broader pore size range.

Table 7.15. N_2 sorption surface area data determined from BET and percent microporosity by t-plot.

Synthesis Temperature	Zr Inorganic-Organic Hybrid Materials		Sn Inorganic-Organic Hybrid Materials	
	Total Surface Area (cc/g)	Microporous (%)	Total Surface Area (cc/g)	Microporous (%)
80°C	547.8	55.77	—	—
105°C	453.3	56.25	—	—
120°C	452.3	71.48	270.6	98.26
145°C	443.8	44.95	268.6	95.20
160°C	342.3	62.78	290.4	96.49
180°C	473.3	42.76	339.2	95.34
200°C	—	—	342.2	94.36

Table 7.16. K_d values reported in mL/g for selected lanthanides at pH $= 3.0 \pm 0.1$.

Sample	Nd^{3+}	Sm^{3+}	Ho^{3+}	Yb^{3+}
H-Zr Hybrid	29,030	80,463	112,020	89,769
Na-Zr Hybrid	1,942,619	1,315,828	454,729	452,270
H-Sn Hybrid	337,525	317,477	322,933	159,301
Na-Sn Hybrid	480,067	296,704	216,026	216,383

Distribution coefficients for Nd with these compounds prepared at different temperatures from 80 to 180°C resulted in values of 75,000–90,000 ml/g for ZrPPP and 30,000–60,000 ml/g for SnPPP. The compounds prepared at lower temperatures gave somewhat higher K_d values but as we shall see these differences are minor compared to two other factors.

As part of our synthetic procedures we utilized basic phosphates such as Na_3PO_4 in place of phosphoric acid.[68] The use of phosphoric acid and phosphonic acid ligands creates a highly acidic medium in which the solids precipitate. It turned out that the use of the trisodium phosphate brought about a remarkable increase in K_d values, as shown in Table 7.16. In addition the uptake of cations is very sensitive to pH as seen for Nd in Figure 7.33. The H-Zr and H-Sn hybrids were prepared with H_3PO_4. The highest uptake for all the lanthanides in Table 7.16 is at pH 3. Furthermore, there is a significant decrease in K_d for the heavier lanthanides Ho^{3+}, Yb^{3+} versus the lighter Nd and Sm. This may be the basis for a further separation of lanthanides as well as for separations from mono- and divalent cations.

Our study with actinides is in its infancy. However, we find that four valent actinides are readily taken up by our hybrids in addition to trivalent

A. Clearfield

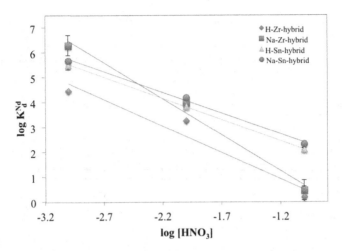

Figure 7.33. Extraction of Nd^{3+} as a function of $[HNO_3]$, the initial $[Nd^{3+}]$ was $\sim 10^{-4}$ M. No uptake was observed for $[HNO_3] > 0.1$.

ones. However, they are taken up at pH values of 2 and lower. At this low pH the uptake of lanthanides is much reduced. Furthermore, the oxidation state of actinides can be changed to produce -yl type compounds such as $[NpO_2]^+$ or plus-two charges. Preliminary studies with this ion have yielded K_d values at pH 2 of 80–478 ml/g and for $[PuO_2]^{2+}$ $\sim 1,000$–6,000 ml/g. These very preliminary data may be a route to group separations of lanthanides from actinides. Furthermore, by manipulating oxidation states we may be able to separate individual lanthanides or actinides from each other, especially Cm from Am.

Our work in this direction is continuing.

7.8 Summary, Conclusions, and Chernobyl

As a brief summary we have seen that a vast array of inorganic ion exchangers has been applied to nuclear systems by our group and others. We have applied these compounds to remediate a large number of groundwaters and HLW systems.[69–85] These articles are presented with titles so the reader can pick and choose those of interest. This listing may be of interest as the Japanese will have many years of clean-up of liberated radioisotopes. I base this conclusion on my experience at Chernobyl. You may remember the terrible explosion of nuclear reactor number 4 at the Chernobyl Power Complex. This disaster was not caused by nature but by human error. The site director wished to carry out an experiment, so all the safety devices were shut off. The experiment went wrong and when they finally realized it, it was too late.

I visited Chernobyl in the summer of 1998, 12 years after the explosion. When the Soviet Union disbanded, I obtained a grant to collaborate with a group at the Institute for Sorption and Problems of Endoecology in Kiev headed by the institute director Vladimir Strelko. The Soviets had developed types of porous carbon that could be fed to humans who had ingested radioisotopes. That worked fairly well in removing these ions. They also had robots inside the reactor building measuring levels of radioactivity and temperatures within the sarcophagus enclosing the reactor. There were about 200 scientists of all types dealing with the situation. One of the major problems was leakage of radioactivity. The molten core had fallen down into the basement and the rain penetrated through cracks in the sarcophagus. The molten reactor core on cooling also cracked. Water penetrated into the core and soaked up radioisotopes and leaked out into the surrounding fields. I saw workmen shoveling soil into trucks to dispose of in an unspecified area. Also some radioisotopes had already found their way into the Dnieper River. I asked Vladimir why they did not build a barrier to stop the flow of radioactive material. He just shrugged his shoulders. He said they had thought to plant potato-sized spheres of their special carbon with a potato-planting machine in the surrounding field. They could then monitor the flow of radioactive material across the field. Each month they could dig up a few carbon balls and determine the radioactive content. I suggested that some of our sorbents be added to the carbon spheres to pick up individual isotopes along with general sorption materials to pick up a range of isotopes. We wrote a proposal but were not funded.

During my meeting with the director of the Chernobyl Institute he introduced me to a biologist who was going to monitor the effect of radiation on the people living in the area. She indicated that several hundred people have returned, claiming they would rather die in their own home than in some strange place far away where the Soviet Government had shipped them. In fact, one of them had hitched a ride with us when we stopped at the entrance checkpoint to show our credentials.

In a recent article describing Chernobyl today, 25 years after the disaster,[86] there are mixed opinions on the effect. The Chernobyl Exclusion Zone covers more than 1,600 sq miles with the reactor site in the middle. The red pine forest that had died directly after the explosion has grown back; wolves, lynx, and elk roam the area and many species of birds live there. "It looks like a remarkable recovery in the ecology of the area."[86] However, the radioactivity ranges from 100 times to 1,000 times the normal level. Some scientists who studied the birds claim they show signs of genetic and brain alteration. Others, usually Ukrainian, say the opposite. It would be of immense value if the Ukrainian studies were made public, especially those on humans. Japan could use this information to good effect and we also would be better able to understand how to deal with a nuclear accident.

In planning a nuclear future, safety and protection of the environment should be paramount. Cutting costs at the expense of safety, as was done in Japan, should not be permitted. As we have seen in the BP fiasco in the Gulf of Mexico, why would anyone endanger the environment to save a few million dollars when the stakes are so high? In the end, not only does the environment suffer disastrously but the costs in monetary terms and human suffering are enormous. We need to develop nuclear power as part of our overall effort to supply energy for the future, but it should be done with the utmost of care.

Acknowledgment

The author gratefully acknowledges financial support from the Department of Energy Basic Energy Science for Grant DE-FG02-03ER 15420, Savannah River Nuclear Solutions — AC 70059-0 and earlier grants from both organizations.

References

1. T. L. Stewart, J. A. Frey, D. W. Geiser and K. L. Manke, in *Science and Technology for Disposal of Radioactive Tank Wastes*, eds. W. W. Schulz and N. J. Lombardo (Plenum Press, New York, 1998), Chapter 1, p. 3.

2. J. F. Keller and T. L. Stewart, *Proceedings of the First Hanford Separation Science workshop*, PNNL Richland, WA., 1993, PNL-SA-21775, I34.

3. G. M. Boyd, R. H. Gilbertson and A. Gritzke, in *Nuclear Site Remediation*, ACS Symposium Series 778, eds. P. G. Eller and W. R. Heineman (American Chemical Society, Washington, D.C., 2000), p. 2.

4. T. M. J. Brouns, A. Frey, T. L. Stewart, R. W. Allen and K. L. Manke, in *Science and Technology for Disposal of Radioactive Tank Wastes*, eds. W. W. Schulz and N. J. Lombardo (Plenum Press, New York, 1998), p. 25.

5. W. L. Lukens, R. C. Moore and K. H. Holt, *Workshop on Development of Radionuclide Getters for the Yucca Mountain Waste Repository: Proceedings*, Sandia National Laboratories, Albuquerque, N.M., 2006.

6. R. G. Anthony, C. V. Philip and R. G. Dosch, *Waste Manage.* **13**, 503–512 (1993).

7. R. G. Anthony, R. G. Dosch, D. Gu and C. V. Philip, *Ind. Eng. Chem. Res.* **33**, 2702–2705 (1994).

8. R. G. Anthony, R. G. Dosch and C. V. Philip, U.S. Patent 6,110,378 Aug. 29, 2000; U.S. Patent 6,479,427 Nov. 12, 2002.

9. R. G. Dosch, N. E. Brown, H. P. Stephens and R. G. Anthony, *Technology and Programs for Radioactive Waste Management and Environmental Restoration* **2**, 1751–1754 (1993).

10. Z. Zheng, C. V. Philip, R. G. Anthony, J. L. Krumhansl, E. D. Trudell and J. E. Miller, *Ind. Eng. Chem. Res.* **35**, 4246–4256 (1996).

11. D. M. Poojary, R. A. Cahill and A. Clearfield, *Chem. Mater.* **6**, 2364–2369 (1994).

12. C. T. Prewitt and R. D. Shannon, *Trans. Am. Cryst. Assoc.* **5**, 51–60 (1969).

13. E. V. Sokolova, R. K. Ratsvetaerva, V. I. Andrianov, Yu. K. Egorov-Tismenko and Yu. P. Men'shikov, *Dokl. Akad. Nauk SSSR* **307**, 114–117 (1989).

14. A. Tripathi, D. G. Medvedev, M. A. Nyman and A. Clearfield, *J. Solid State Chem.* **175**, 72–83 (2003).

15. A. I. Bortun, L. N. Bortun and A. Clearfield, *Solv. Extr. Ion Exch.* **15**, 909–929 (1997).

16. A. Clearfield, L. N. Bortun and A. I. Bortun, *React. Funct. Polym.* **43**, 85–95 (2000).

17. P. Pertierra, M. A. Salvado, S. Garcia-Granda, A. I. Bortun and A. Clearfield, *Inorg. Chem.* **38**, 2563–2566 (1999).

18. D. M. Poojary, A. I. Bortun and L. N. Bortun, *Inorg. Chem.* **35**, 6131–6139 (1996).

19. A. J. Celestian, D. G. Medvedev, A. Tripathi, J. B. Parise and A. Clearfield, *Nucl. Instrum. Methods Phys. Res. B* **238**, 61–69 (2005).

20. A. Clearfield, A. Tripathi, D. G. Medvedev, A. J. Celestian and J. B. Parise, *J. Mater. Sci.* **41**, 1325–1333 (2006).

21. A. J. Celestian and A. Clearfield, *J. Mater. Chem.* **17**, 4839–4842 (2007).

22. A. J. Celestian, J. B. Parise, R. I. Smith, B. H. Toby and A. Clearfield, *Inorg. Chem.* **46**, 1081–1089 (2007).

23. A. J. Celestian, D. J. Kubicki, J. Hanson, A. Clearfield and J. B. Parise, *J. Am. Chem. Soc.* **130**, 11689–11694 (2008).

24. A. J. Celestian, J. B. Parise and A. Clearfield, in *Handbook of Crystal Growth* (Springer-Verlag, Berlin, 2010), p. 1637.

25. A. Tripathi, D. G. Medvedev and A. Clearfield, *J. Solid State Chem.* **178**, 253–261 (2005).

26. A. Clearfield, D. G. Medvedev, S. Kerlegon, T. Bosser, J. D. Burns and M. Jackson, *Solv. Extr. Ion Exch.*, submitted.

27. O. J. Heinonen, J. Lehto and J. K. Miettinen, *Radiochim. Acta* **28**, 93–96 (1981).

28. J. Lehto, O. J. Heinonen and J. K. Miettinen, *Radiochem. Radioanal. Lett.* **46**, 381–288 (1981).

29. A. Clearfield and J. Lehto, *J. Solid State Chem.* **73**, 98–106 (1988).

30. R. Cahill, Ph.D. Dissertation, Texas A&M University, May 1996.

31. D. G. Medvedev, Ph.D. Dissertation, Texas A&M University, August 2004.

32. J. Lehto and A. Clearfield, *J. Radioanal. Nucl. Chem. Lett.* **118**, 1–13 (1987).

33. L. G. Sillen and A. E. Martell, *Stability Constants of Metal-Ion Complexes* (The Chemical Society, London, 1984), p. 42.

34. G. Brown, Pacific Northwest National Laboratory, private communication (1997).

35. I. DeFilippi, S. Yates, R. Sedath, M. Straszewski, M. Anderson and R. Gaita, *Separ. Sci. Tech.* **32**, 93–113 (1997).

36. R. Cahill, A. Clearfield, C. Andren, I. De Filippi, R. Sedath, G. Seminara, M. Straszewski, L. Wang and S. Yates, Patent Application Serial 08/546,041, Oct. 20, 1995.

37. I. DeFilippi, S. Yates, J. Shen, R. Gaita, R. Sedith, G. Seminara, M. Straszewski and D. Anderson, Patent Application Serial 08/546,448, Oct. 20, 1995.

38. D. J. McCake and B. W. Walker, *Examination of Sodium Titanate Applicability in the In-Tank Precipitation Process*, WSRC-TR-0015, Savannah River Technology Center, Aiken, SC, Apr. 25, 1997.

39. J. Lehto, L. Brodkin, R. Harjula and E. Tusa, *Nucl. Technol.* **127**, 81–87 (1999).

40. R. G. Dosch, *Use of Titanates in Decontamination of Defense Waste*, Report RS-8232-2/50 318, Sandia National Laboratories, Albuquerque, NM, 1978.

41. R. Lynch, R. G. Dosch, B. Kenna, J. Johnstone and E. Nowak, in *IAEA Symposium on the Management of Radioactive Waste* (International Atomic Energy Agency, Vienna, Austria, 1976), p. 360.

42. D. T. Hobbs, M. J. Barnes, R. L. Pulmano, K. M. Marshall, T. B. Edwards, M. G. Bronikowski and S. D. Fink, *Separ. Sci. Tech.* **40**, 3093–3111 (2005).

43. M. D. Nyman and D. T. Hobbs, *Chem. Mater.* **18**, 6425–6435 (2006).

44. M. D. Nyman and D. T. Hobbs, U.S. Patent 7,494,640 B1, Feb. 24, 2009.

45. G. Brown, in *Clay Minerals: Their Structure, Behavior and Use*, eds. L. Fowden, R. M. Barrer and P. B. Tinker (The Royal Society, London, 1984), pp. 1–20.

46. S. W. Bailey (ed.), *Micas: Reviews in Mineralogy* (Mineralogical Society of America, Washington, D.C., 1987), Vol. 13.

47. A. D. Scott, R. R. Hunziker and J. J. Hanway, *Soil Sci. Soc. Amer. Proc.* **24**, 191–194 (1960).

48. L. N. Bortun, A. I. Bortun and A. Clearfield, in *Ion Exchange Developments and Applications: Proc. of IEX '96*, Royal Society of Chemistry, London, 1996, p. 313.

49. S. S. Yates, I. DeFilippi, R. Gaita, A. Clearfield, L. N. Bortun and A. I. Bortun, U.S. Patent 6,114,269, Sep. 5, 2000.

50. M. J. Buerger, W. A. Dollase and I. Garaycochea-Wittke, *Kristallogr.* 92–108 (1967).

51. D. M. Chapman and A. L. Roe, *Zeolites* **10**, 730–737 (1990).

52. E. A. Behrens, D. M. Poojary and A. Clearfield, *Chem. Mater.* **8**, 1236–1244 (1996).

53. E. A. Behrens and A. Clearfield, *Micropor. Mater.* **11**, 65–75 (1997).

54. A. Clearfield, *Solid State Sciences* **3**, 103–112 (2001).

55. W. T. A. Harrison, T. E. Grier and G. D. Stuckey, *Zeolites* **15**, 408–412 (1995).

56. M. S. Dadachov and W. T. A. Harrison, *J. Solid State Chem.* **134**, 409–415 (1997).

57. M. A. Roberts and A. N. Fitch, *J. Phys. Chem. Solids* **52**, 1209–1218 (1991).

58. G. Eisenman, *Biophys. J. Suppl.* **2**, 259–323 (1962).

59. E. A. Behrens, D. M. Poojary and A. Clearfield, *Chem. Mater.* **10**, 959–967 (1998).

60. A. Tripathi, D. G. Medvedev, J. Delgado and A. Clearfield, *J. Solid State Chem.* **177**, 2903–2915 (2004).

61. J. J. Laidler and J. C. Bresee, in *Proceedings of WM-04*, Tucson, A.Z., February 27 to March 4, 2004.

62. S. Aase, S., Bakel, A, *et al.* Demonstration of the UREX+ Process on Irradiated Nuclear Fuel, 13th Annual Separations Science and Technology Conference, Gatlinburg, TN, October 2003.

63. G. F. Vandegrift, M. C. Regalbuto, S. Aase, A. Bakel, T. J. Battisti, D. Bowers, J. P. Byrnes, M. A. Clark, J. W. Emery, J. R. Falkenberg, A. V. Gelis, C. O. Pereira, L. Hafenrichter, Y. Tsai, K. J. Quigley and M. H. vander Pol, in *Proceedings of Atalante 2004*, Nimes, France, June 21–24, 2004.

64. R. Cahill, B. Shpeizer, G. Z. Peng, L. Bortun and A. Clearfield, in *Separations of f Elements*, eds. K. L. Nash and G. R. Choppin (Plenum, New York, 1995), p. 165.

65. M. D. Poojary, H.-L. Hu, F. L. Campbell III and A. Clearfield, *Acta Crystallogr. B* **B49**, 996–1001 (1993).

66. A. Clearfield, *Dalton Trans.* 6089–6102 (2008).

67. A. Subbiah, D. Pyle, A. Rowland, J. Huang, R. A. Narayanam, P. Thiyagarajan, J. Zon and A. Clearfield, *J. Am. Chem. Soc.* **127**, 10826–10827 (2005).

68. J. D. Burns and A. Clearfield, *Radioanal. Chem.*, submitted.

69. A. Clearfield, Inorganic ion exchange materials for nuclear waste effluent treatment, in *Industrial Environmental Chemistry*, eds. D. T. Sawyer and A. E. Martell (Plenum, New York, 1992), p. 289.

70. A. Clearfield, A. I. Bortun, L. N. Bortun and R. A. Cahill, Synthesis and characterization of a novel layered sodium titanium silicate $Na_2TiSi_2O_7.2H_2O$, *Solv. Extr. Ion Exch.* **15**, 285–304 (1997).

71. E. A. Behrens, P. Sylvester and A. Clearfield, Assessment of a sodium nonatitanate and pharmacosiderite-type ion exchanger for strontium and cesium removal from DOE waste simulants, *Environ. Sci. Technol.* **32**, 101–107 (1998).

72. A. I. Bortun, L. N. Bortun, S. A. Khainakov and A. Clearfield, Ion exchange properties of the sodium phlogopite and biotite, *Solv. Extr. Ion Exch.* **16**, 1067–1090 (1998).

73. P. Sylvester, A. Clearfield, The extraction of ^{137}Cs and ^{89}Sr from waste simulants using pillared montmorillonite, *Separ. Sci. Technol.* **33**, 1605–1615 (1998).

74. A. Clearfield, A. I. Bortun, S. A. Khainakov, L. N. Bortun, V. V. Strelko and V. N. Khryaschevskii, Spherically granulated titanium phosphate as exchanger for toxic heavy metals, *Waste Manage.* **18**, 203–210 (1998).

75. P. Sylvester and A. Clearfield, The removal of strontium and cesium from simulated Hanford groundwater using inorganic ion exchange materials, *Solv. Extr. Ion Exch.* **16**, 1527–1539 (1998).

76. A. I. Bortun, L. N. Bortun, S. A. Khainakov and A. Clearfield, Ion exchange properties of the sulfur-modified biotite, *Solv. Extr. Ion Exch.* **16**, 1541–1558 (1998).

77. E. A. Behrens, P. Sylvester, G. Graziano and A. Clearfield, Evaluation of a sodium nonatitanate, sodium titanosilicate, and pharmacosiderite-type ion exchangers for strontium removal from DOE waste and Hanford N-springs groundwater simulants, in *Science and Technology for Disposal of Radioactive Tank Wastes*, eds. W. W. Schulz and N. J. Lombardo (Plenum Press, New York, 1998), pp. 287–299.

78. P. Sylvester, A. Clearfield and R. J. Diaz, Pillared montmorillonites: cesium-selective ion-exchange materials, *Separ. Sci. Technol.* **34**, 2293–2306 (1999).

79. P. Sylvester and A. Clearfield, The removal of strontium from simulated Hanford tank wastes containing complexants, *Separ. Sci. Technol.* **34**, 2539–2551 (1999).

80. P. Sylvester, E. A. Behrens, G. M. Graziano and A. Clearfield, An assessment of inorganic ion-exchange materials for the removal of strontium from simulated Hanford tank wastes, *Separ. Sci. Technol.* **34**, 1981–1992 (1999).

81. A. Clearfield, Inorganic ion exchangers, past, present, and future, *Solv. Extr. Ion Exch.* **18**, 655–678 (2000).

82. P. Sylvester and A. Clearfield, The use of synthetic inorganic ion exchangers in the removal of cesium and strontium ions from nuclear waste solutions, in *ACS Symposium Series*, Vol. 778, eds. P. G. Eller and W. R. Heineman (American Chemical Society, Washington, D.C., 2001), pp. 133–145.

83. S. Solbra, N. Allison, S. Waite, S. V. Mikhalovsky, A. I. Bortun, L. N. Bortun and A. Clearfield, Cesium and strontium ion exchange on the framework titanium silicate M2Ti2O3SiO4.nH2O (M = H, Na), *Environ. Sci. Technol.* **35**, 626–629 (2001).

84. A. Clearfield, D. M. Poojary, E. A. Behrens, R. A. Cahill, A. I. Bortun and L. N. Bortun, Structural basis of selectivity in tunnel-type inorganic ion exchangers, *ACS Symposium Series*, Vol. 716, eds. A. H. Bond, M. L. Dietz and R. D. Rogers (American Chemical Society, Washington, D.C., 1999), pp. 168–182.

85. A. I. Bortun, L. Bortun, A. Clearfield, E. Jaimez, M. A. Villa-Garcia, J. R. Garcia and J. Rodriguez, Synthesis and characterization of the inorganic ion exchanger based on titanium 2-carboxyethylphosphonate, *J. Mater. Res.* **12**, 1122–1130 (1997).

86. A. Higginbotham, Chernobyl 25 years later, *Wired*, May, p. 160 (2011).

Chapter 8

Synthesis of Nanostructured Hybrid Sorbent Materials Using Organosilane Self-Assembly on Mesoporous Ceramic Oxides

Glen E. Fryxell

Pacific Northwest National Laboratory
Richland, WA, USA

8.1 Introduction

The single most important factor in determining quality of life in human society is the availability of pure, clean drinking water. Wars have been fought, and will continue to be fought, over access to and control of clean water. Drinking water has two major classes of contamination: biological contamination and chemical contamination. Bacterial contamination can be dealt with by a number of well-established technologies (e.g. chlorination, ozone, UV), but chemical contamination is a somewhat more challenging target. Common organic contaminants, such as pesticides, agricultural chemicals, industrial solvents and fuels, can be removed by treatment with UV/ozone, activated-carbon or plasma technologies. Toxic heavy metals like mercury, lead and cadmium can be partially addressed by using traditional sorbent materials like alumina, but these materials bind metal ions non-specifically and can easily be saturated with harmless, ubiquitous species like calcium, magnesium and zinc (which are actually nutrients, and do not need to be removed). Another weakness of these traditional sorbent materials is that metal ion sorption to a ceramic oxide surface is a reversible process, meaning they can easily desorb back into the drinking water supply.

A chemically specific sorbent material, capable of permanently sequestering these toxic metal ions from groundwater, or sequestering them from industrial waste, thereby preventing their entry into the groundwater, is needed. Ideally, the kinetics of heavy metal sorption should be fast, allowing for high throughput in the process stream, and a high binding capacity for the target heavy metal is clearly desirable. In addition, as acceptable drinking water contamination limits become lower and lower, more sensitive analytical methods are needed in order to detect such contamination, so if these sorbent materials could enhance the sensing and detection of these toxic analytes, that would be an added benefit.

The surfactant templated synthesis of mesoporous ceramics offers the synthetic chemist the ability to create highly uniform pore structures where the pores are only a few nanometers in diameter. This versatile and powerful methodology creates morphologies that condense a huge amount of surface area into a very small volume. For example, a typical procedure for making MCM-41 (an ExxonMobil catalyst support) provides a highly ordered hexagonal honeycomb of parallel 20–25 Å pores, with wall thicknesses of about 12–14 Å, and a surface area of about $1,000\,m^2/g$.[1,2] Thus, a 10 g sample of this material (about two rounded tablespoons) has more surface area than an entire football field, endzones included.

While a high surface area provides the promise of high binding capacity for a sorbent material, this backbone is still a simple ceramic oxide, and while ceramic oxides commonly physisorb heavy metals, they tend to do so non-selectively and reversibly, making them inefficient for waste remediation or water purification. What is needed is a way of enhancing the chemical selectivity of the interface, as well as making the binding chemistry more robust. This is done by chemically modifying these mesoporous materials with self-assembled monolayers that are terminated with chemically selective ligands (Figure 8.1). Molecular self-assembly allows the promise of high loading capacity to be realized by coating the mesoporous interface

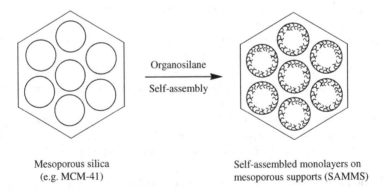

Mesoporous silica   Self-assembled monolayers on
(e.g. MCM-41)   mesoporous supports (SAMMS)

Figure 8.1.   Construction of a terminally functionalized self-assembled monolayer inside of a mesoporous ceramic creates a powerful and versatile foundation upon which to build chemically selective sorbent materials. By varying the nature of the interfacial ligand field, the chemical selectivity of the sorbent can be tuned for specific targets.

with a dense, covalently anchored functional coating. With this functional coating in place, the tethered ligand field provides an avenue for covalent chelation, delivering a significant driving force for sequestering a specific metal ion, and the close proximity of these ligands to one another allows for multiple metal-ligand interactions, further enhancing the ligand's affinity for a specific metal ion. By tailoring the nature of this interfacial ligand field, it is possible to selectively remove different classes of metal ions. The rigid, open-pore structure provided by the mesoporous ceramic backbone prevents solvent swelling (which can close off pores in polymeric resins). Since all of the binding sites are available all of the time, sorption kinetics are rapid. Although the mesoporous supports are more expensive than typical sorbent materials on a per-unit-weight basis (e.g. activated carbon, polymer-based ion-exchange resins), the very high functional density of the self-assembled monolayers on mesoporous supports (SAMMS) makes them cheaper to use in the final analysis as a result of their superior performance and much reduced waste disposal costs (which result from the fact that far less sorbent material is needed to effect the separation). Since the chemical selectivity of the SAMMS materials is dictated by the interfacial ligand field, the following discussion will be organized by ligand type.

8.2 Sulfur-based Ligands

When designing ligands to selectively bind specific metal species it is important to capture the appropriate stereochemical, electronic and steric attributes in the ligand field. A key concept here is that of the hard-soft acid-base theory, introduced by Pearson in the 1960s.[3,4] In short, this theory focuses on the polarizability of the acid-base pair, and the impact that it has on their affinity for one another. Highly polarizable Lewis acids (i.e. "soft" — large-atom, electron-rich species in which the electron cloud is readily distorted by an external influence) prefer to react with polarizable (i.e. soft) Lewis bases. For example, Hg species have a very high affinity for thiol (a.k.a. "mercaptan" for its ability to capture mercury) ligands. Non-polarizable Lewis acids (i.e. hard — a highly electropositive species in which the electron cloud is either non-existent or not easily distorted) prefer to react with non-polarizable (i.e. hard) Lewis bases. An example of a hard acid-base pair reaction would be that of a bare proton reacting with fluoride ion. Thus, when designing a sorbent material for a soft target ion (e.g. Hg), it is important to build it around a soft ligand field. When aiming towards a hard target ion, one should start with a hard ligand field. We will start this discussion on the early work done with Hg, using a ligand field of thiols within the mesoporous architecture.

The typical recipe that we have used to make thiol-SAMMS[5,6] involves pre-hydrating the mesoporous silica (65 Å pores and $900\,m^2/g$) interface by stirring a toluene suspension of the silica with two monolayers worth of water (based on available surface area, assuming a monolayer to be composed of 5×10^{18} molecules of water per square meter). Water plays a very important part in organosilane self-assembly.[7] While chloro- and alkoxysilanes can indeed react with surface silanols, Maciel determined many years ago that this is a slow process,[8] and that the relevant process generally involved hydrolysis of the silane by interfacially bound water, followed by condensation (acid catalyzed in the case of chlorosilanes) of the hydroxysilane with a surface silanol (or siloxane). The intermediate tris(hydroxy)silane is commonly overlooked in discussions of organosilane self-assembly, but to do so is to ignore the very species that makes the self-assembly process possible. After hydrolysis, the tris(hydroxy)silane is hydrogen-bound to the silica surface. Through a process of making and breaking H bonds, this species can "crab-walk" across the silica surface until it encounters another such molecule. The van der Waals attractive forces (as well as any other attractive forces, such as H bonds, dipole-dipole inter-actions) between these molecules drive the self-assembly process, forming a critical nucleus in the aggregation process. As this agglomerated assembly grows, it becomes less and less mobile. With the extended contact time and close proximity[9] of these tris(hydroxy)silanes and the silica surface, these molecules begin to undergo condensation processes, both with the surface and between themselves, ultimately forming a covalently anchored and cross-linked monolayer structure. For chlorosilanes, this process is catalyzed by the HCl formed by hydrolysis of the silane, while the alkoxysilanes generally require added heat for these condensation processes to take place. Without interfacial water, none of this would be possible, as the only reaction pathway available to the silane is simple capping of whatever silanols may be available.

This process results in a defective monolayer structure. There are a number of dangling hydroxyls left in a monolayer made by this route, as well as pinhole defects (vacancies). These defects can be preserved by any excess water, or by the alcohol resulting from silane hydrolysis. For example, a typical case of hydrated MCM-41 being treated with an excess of 3-mercaptopropyltrimethoxysilane (MPTMS) results in a mono-layer population density of about 4 silanes/nm^2.[10] If the reaction mix-ture is subjected to distillation (to remove the methanol by-product and water/toluene azeotrope) after the four- to six-hour reflux period, however, it is possible to increase the population density to about 5 silanes/nm^2 (this equates to about $3.3\,mmol/g$), as well as to significantly increase the degree of cross-linking.[10] By removing the water from the reaction mixture it is possible to drive the condensation equilibria further towards the ideal cross-linked structure, and by removing the H-bonding molecules (i.e. water and methanol), the pinhole defects can be filled in. The post-deposition distillation drives both of these processes.

The materials made by this route offer a high surface area that is densely covered by the thiol ligand field, thereby creating a very high population of Hg binding sites, while simultaneously keeping all of these binding sites at the sorbent interface and therefore accessible. The rigid, open-pore structure of the ceramic backbone prevents any solvent swelling phenomena, which can limit access to binding sites by closing off pores. This helps to maintain free access to all of the ligands and allows the sorption kinetics to be rapid. When thiol-SAMMS are used to remove Hg from a 10,000 ppb $Hg(NO_3)_2$ solution, over 98% of the Hg is sequestered in less than five minutes.[11] When a similar experiment was carried out using a thiolated resin (Rohm–Haas GT-73), it took eight hours to remove only 90% of the Hg. Both materials are using the same functionality to bind the Hg (a thiol), but the rigid, open-pore architecture of the thiol-SAMMS provides for a 500-fold rate difference by keeping all of the binding sites available all the time. Thiol-SAMMS was found to be able to sequester over 600 mg of Hg per gram of sorbent.[11]

The nanoporous honeycombed morphology of SAMMS not only provides high binding capacity and rapid sorption kinetics, it also blocks microbial access to the vast majority of the sorbent's active surface. This is a concern because certain bacteria can actually metabolize bound mercury, methylating it, making it not only more mobile, but also more toxic. By binding the Hg inside a nanoscale pore, a bacterium cannot gain access to it, thereby preventing these deadly processes.

Competition experiments revealed that thiol-SAMMS do not bind any of the ubiquitous cations commonly found in groundwater, like Na, K, Mg, Ca, Fe, Ni or Zn.[12] Thiol-SAMMS do bind other soft cations, like Cd, Ag and Au. Extended X-ray absorbance fine structure (EXAFS) studies of the Hg-thiol-SAMMS adduct revealed a 1:1 Hg/S stoichiometry, and that each Hg was interacting with 2 S atoms, with a third O ligand.[13] This is explained by a divalent Hg cation binding to one thiolate and one oxygenated ligand (i.e. either a nitrate anion or a water molecule), and coordinating to the S atom of a neighboring thiol (see Figure 8.2).

Pinnavaia and co-workers have developed a similar, but decidedly unique synthetic strategy for making some similar heavy metal sorbent materials.[14] In the Pinnavaia approach, neutral amines are used as the surfactants to form the micelles. Instead of calcining the as-synthesized green body to remove the surfactant, the template is removed via Soxhlet extraction (typically using ethanol). This is an expeditious method, and has the advantage

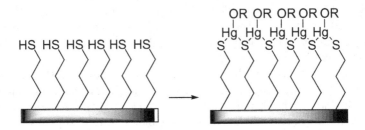

Figure 8.2.   Hg bonding in thiol-SAMMS.[13]

of leaving the pore walls fully hydroxylated (and hydrated), and therefore receptive to silanation. However, by avoiding calcination, this method also leaves the pre-ceramic pore walls incompletely cross-linked, and therefore of uncertain long-term stability. When these disordered assemblies of worm-like channels are subsequently decorated with a 3-mercaptopropylsilane coating, they found that these materials had a surface area of approximately $800 \, m^2/g$ and pore diameters on the order of 30 Å.[14] This material was found to contain approximately 1.5 mmol of thiol per gram of sorbent and had a Hg(II) capacity of 310 mg/g (these numbers are roughly half those of the PNNL material described earlier). This procedure was found to be both fast and easy, but led to incomplete functionalization.

A follow-up study compared the performance of these materials with other related sorbent materials.[15] In this study, the MCM-41-derived materials were prepared after being vacuum dried. In the absence of water there can be no possibility of a fully dense monolayer being formed. This procedure only allows the remaining surface silanols to be capped, ultimately producing a product with only $0.28 \, silanes/nm^2$ (or 0.57 mmol/g). The 1:1 binding stoichiometry of Hg/S was confirmed once again. No competition was observed from a series of representative transition metal cations.[16]

Pinnavaia subsequently streamlined this synthetic strategy even further by making it a "one-pot" co-condensation procedure using octylamine or dodecylamine as template.[17] The neutral amine template was once again removed via a separate Soxhlet extraction step (not calcination). Pores of the mercaptan derivatives were found to be in the range of 1.48 to 1.75 nm. Up to 8 mole-percent of the functional organosilane was successfully installed in these materials. This methodology was further optimized by moving to longer-chain alkyl amines and performing the assembly process at moderately elevated temperatures (e.g. 65°C).[18] Very high degrees of functional loading (i.e. 50 mole-percent organosilane) and framework cross-linking were obtained using this optimized procedure. Excellent hydrothermal stability was reported (material composed of 50 mole-percent organosilane was reported to be stable in boiling water for at least 10 hours).[18]

Materials made using these methods revealed that Hg can be bound in multiple ways.[19,20] One X-ray absorbance spectroscopy (XAS) study[19] found that the Hg-S coordination number never exceeded 1, and that at low Hg loading, Hg was bound as a monodentate S complex, with an oxygenated ligand. No evidence was found for any bidentate S-Hg-S complexes in this study. At higher Hg loadings, they found that condensation to form Hg-O-Hg-OH bridges became a significant pathway. A follow-up study,[20] using atomic pair distribution function (PDF) analysis of synchrotron X-ray powder diffraction data, found no evidence of any Hg-O bonds. At low Hg loadings the authors suggest that the Hg is tetrahedrally bound to four different S atoms. This interesting bonding scheme suggests either a great deal of flexibility of the mesoporous ceramic support, or that opposing faces of the functional pores structure are very close to one another. At higher Hg loadings, they see a 1:1 Hg/S stoichiometry, with the Hg bound to two bridging thiolate ligands,[20] somewhat similar to previous measurements.[13]

Metal-ligand binding is an equilibrium process. Solid-phase sorbent binding chemistry reflects not only the equilibrium between the metal and ligand, but is also complicated by the impact that the binding of the first few cations has on the binding chemistry of the final cations (i.e. the nature of the ligand field evolves as the binding process proceeds). After Pearson reported his concepts of hard-soft acid-base theory,[3,4] a number of efforts were made to quantify this model. None of these attempts were completely successful (for a discussion of these efforts, see Ref. 21). In 1967, Misono presented a model summarizing an extensive set of binding studies in which he fit the binding constants to a two-parameter model.[22] One of these parameters was called the metal cation's softness parameter. It was found that a useful correlation was obtained when the saturation binding capacity for a given cation was plotted against the Misono softness parameter.[23] One of the things that makes this correlation particularly intriguing is that some of the metal cations involved have notably different binding geometries from one another (square planar, octahedral, etc.), meaning that these cations interact with the thiolated interface in very different ways, and with different numbers of thiols. It is noteworthy that the correlation is found with the binding capacity expressed in terms of milliequivalents per gram rather than millimoles per gram, indicating that charge accumulation at the interface is a significant consideration in determining the position of equilibrium. It seems that the accumulation of Lewis acids on the monolayer interface decreases the effective basicity of the remaining thiol ligand field by shifting the equilibrium to alleviate the repulsive electropositive interactions between the cations. The impact of this effect is felt sooner for those Lewis acids with weaker binding affinities, and therefore saturation occurs at a lower loading density. While the lower end of the binding curve is dominated by equilibrium effects, the upper end of the curve is undoubtedly limited by stoichiometry, as it is unrealistic to suggest that the ligand field will have any significant effect after the ligand field is saturated via inner-sphere coordination. For divalent cations, this would suggest an asymptotic limit of approximately 6.6 meq/g (the population of thiols in these monolayers is approximately 3.3 mmol/g).

Similar thiol-functionalized mesoporous sorbents have also been used to bind Pt and Pd.[24,25] These materials clearly have promise in catalyst recovery.

Supercritical fluids have been found to be a particularly powerful and green reaction medium for organosilane self-assembly.[10,26,27] Supercritical carbon dioxide (SCCO$_2$) effectively solvates the siloxane monomers, but it also doesn't inhibit the attractive van der Waals forces between the hydrocarbon chains since CO$_2$ is a small linear molecule, thus allowing self-assembly to proceed smoothly. In addition, since self-assembly is an associative process, it can be accelerated by carrying it out under conditions of high pressure. This significantly speeds up the production of SAMMS, and at the same time virtually eliminates the waste stream resulting from production (a critical component of green manufacturing). In addition, the monolayers so formed have a lower defect density as a result of some novel annealing mechanisms that come into play under these reaction conditions. The low viscosity of SCCO$_2$ also facilitates mass transport of silane throughout the nanoporous ceramic matrix. When the monolayer deposition is complete, the pores of the SAMMS are clear and dry, and not filled with residual solvent that must be removed before the materials can be used.

Jaroniec and co-workers have also done some elegant work incorporating novel S-based ligands into mesoporous scaffolds, and the reader is referred to Prof. Jaroniec and Dr Olkhovyk's chapter in this book for further details.

Periodic mesostructured organosilicas (PMOs)[28] are a related class of functional mesoporous materials that has received a great deal of attention in recent years.[29] A remarkable example of Hg affinity was reported for a functional PMO containing a tetrasulfide linkage.[30] This material was made using P123 as the template and a disilane containing a tetrasulfide linkage was included up to 15 mole-percent. These materials were found to have surface areas in the range of 242–654 m^2/g, and surface area was found to decrease with increased tetrasulfide loading. The Hg loading capacity was found to range from 627 mg/g (for the 2 mole-percent material), up to an incredible 2,710 mg/g for the 15 mole-percent material. The Hg-laden materials were found to have Hg/S ratios of 2–3, suggesting that some sort of unusual binding scheme is taking place in these materials. It was possible to strip the Hg using concentrated acid. Subsequent Hg binding capacity was found to be approximately 40% of the original activity, suggesting that perhaps some ligand decomposition took place during regeneration.

8.3 Amines

Aminoalkylsiloxanes have been used as coupling agents for composite materials for several decades now. Historically, there have been some questions (and controversy) as to how these molecules interact with a silica surface, whether through acid/base chemistry between the amine and a surface silanol, through hydrogen bonding of the hydrolyzed siloxane to the silica surface, or through condensation chemistry between the hydrolyzed siloxane and surface silanol.[31] Extensive investigation, by many groups around the world, seems to have led to the consensus that each of these interactions is important, but each at a different stage of the overall process. The first thing that happens, as the aminoalkylsiloxane is sorbed to the surface, is a proton transfer between the surface silanol and the amine to form the ammonium silanolate salt. The surface silanol is more acidic[32] than is the ammonium ion, and this reaction is exothermic. If there is water adsorbed to the silica surface, then the siloxane anchor can undergo hydrolysis on the surface, generating the corresponding tris(hydroxy)silane that is so key to the self-assembly process. The mobility of the tris(hydroxy)silane may very well be limited due to the tethered ammonium silanolate (in a fully hydroxylated surface, this species should be able to move around easily as a result of proton hopping, but in more sparsely hydroxylated surfaces such mobility is likely to be more limited). As more and more of these tris(hydroxy)silanes accumulate on the surface they will begin to aggregate as a result of H-bonding and van der Waals interactions. It is important to recognize that many of these molecules will still be "bent over", with both the tris(hydroxy)silane and the ammonium ion head group interacting with the silica surface. As a result, when the condensation processes start, giving rise to the anchored, cross-linked monolayer structure, there will be a significant number of defects, both in terms of pinhole defects (vacancies) and in terms of cross-linking defects (dangling hydroxyls), both resulting from the blocking action of the "bent over" molecules. As a result, these monolayers tend to be somewhat more fragile than other systems (and as will be discussed later, these can be thermally cured post-complexation to enhance their stability).

Mesoporous silica has been functionalized with organosilanes terminated with a simple primary amine. For example, SBA-15 has been coated with 3-aminopropyltriethoxysilane and characterized.[33] Similar materials were prepared and evaluated as base catalysts for condensation chemistry.[34] Site-isolated behavior has been demonstrated for amine-functionalized meso-porous silicas.[35] These simple amine-terminated materials are versatile building blocks for creating more complex molecular architectures, but are not generally used directly for environmental applications.

Similar amine-modified mesoporous materials have also been used to graft various metal species to surfaces. For example, 2,2'-bipyridine and 1,10-phenanthroline complexes of Cu and Mn have been anchored to mesoporous silicas through an aminopropylsilane tether, and these complexes were studied spectroscopically and electrochemically.[36–41] The general focus of this work seems to be primarily catalyst oriented, but there is also potential for environmental application.

3-(2-Aminoethyl)-aminopropyl trimethoxysilane is commercially available and provides ready access to the chelating ethylenediamine (EDA) ligand. Use of this organosilane to construct EDA-SAMMS for the reversible capture of CO_2[42] is discussed in detail in the chapter by Dr Zheng and co-workers, and the reader is referred to that chapter for details.

An interesting aminated material was made in which macrocyclic cyclam ligands were incorporated into a PMO.[43] In this case the ligand loading density was approximately 10 mole-percent of the total siloxane content. The surface area of these macrocycle-containing PMOs ranged from 420 to $703 \, m^2/g$ and pore size varied from 51 to 87 Å. These materials were shown to be a versatile foundation upon which to build more complex ligand structures (macrocycles, phosphonates, etc.). There is clearly great promise in this area for the design and construction of sophisticated molecular architectures, perhaps with discriminating metal-binding properties.

8.4 Transition Metal Complexes

While the amine-terminated organosilanes tend to give rise to monolayers that have a fair number of defects in them, these amines can still be coordinated to a metal center, especially so for the chelating diamines as they form strong complexes with many transition metals. Once these amines are tied up with a metal center, then these defects are simply dangling hydroxyl groups and pinhole vacancies (i.e. the amine is no longer serving as a blocking group, preventing access to these defects). Therefore, it is possible to remove some of these defects by thermally curing the monolayer after complexation, promoting the condensation of some of the dangling hydroxyls, resulting in a more stable monolayer structure. This is conveniently done by refluxing the sample in toluene and removing the water using a Dean–Stark trap.

Since these chelating diamine ligands are neutral, they form cationic transition metal complexes. These are of interest since they can serve as anion-exchange materials. For example, EDA-SAMMS have been used to form complexes with a variety of transition metal cations. The Cu(II) complex is particularly interesting.[44] Being a d^9 species, an octahedral coordination environment would result in orbital degeneracy. Thus the complex "wants" to undergo some sort of Jahn–Teller distortion to alleviate this degeneracy. The bond angles and distances imposed by the chelated EDA ligand make this difficult. As a result, portions of the ligand field are more weakly held than the rest. Detailed EXAFS studies have shown that when a tetrahedral oxometallate anion comes in and associates with the Cu(II)-EDA complex, the primary amines from two adjacent EDA ligands are lost and the anion forms a direct, monodentate bond to the metal center[45] (see Figure 8.3). This results in a trigonal bipyramidal coordination geometry and alleviates the orbital degeneracy. This bond formation is potentially reversible, and offers a level of chemical selectivity for these sorbent materials. For tetrahedral anions that form soluble Cu(II) salts (e.g. sulfate), this process is readily reversed and the sulfate anion easily displaced from the complex; for tetrahedral anions that form insoluble Cu(II) salts (e.g. arsenate), this adduct does not readily dissociate and the anion is firmly held. This strategy has been successfully employed to sequester arsenate and chromate,[44] as well as pertechnetate.[46] This work has been built upon by the elegant research of Yoshitake and co-workers,[47–49] and the reader is referred to Chapter 11 of this book for details.

This approach can also be used as a functionalization strategy for subsequent elaboration of these metal-complex SAMMS. For example, treatment of Cu(EDA) SAMMS with sodium ferrocyanide results in the lavender-colored ferrocyanide adduct (Figure 8.4).[50] This bimetallic complex is of

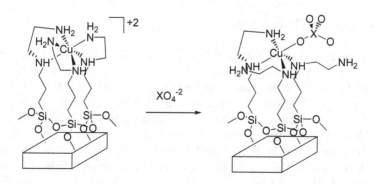

Figure 8.3.   Tetrahedral anion binding by the Cu(EDA)$_3$ complex.

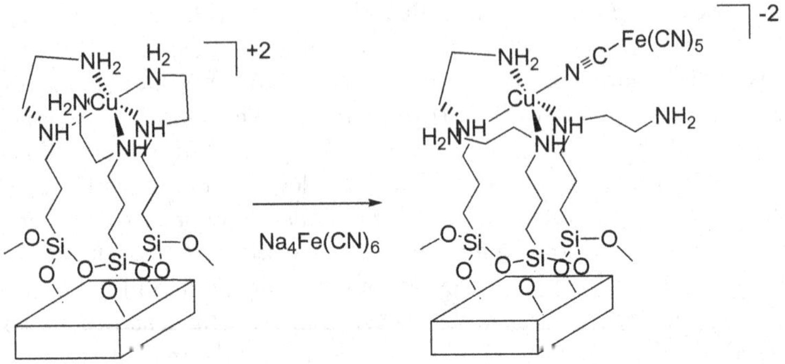

Figure 8.4.   Preparation of ferrocyanide-SAMMS.

interest because it has been long recognized that transition metal ferro-cyanide complexes tend to form insoluble cesium salts. Cs-137 has a half-life of about 30 years and is one of the primary radioactive isotopes of concern in nuclear waste and radioactive fallout.[51] Removing Cs in the presence of large excesses of Na (or K) has been a challenge in the past, although some success has been had with hydrophilic crown ether-based approaches[52] and the crystalline silicotitanates.[53] Ferrocyanide-SAMMS has been shown to have very high affinity for Cs, even in the presence of huge excesses of Na or K, and sorption kinetics are reasonably fast, with equilibrium being reached in less than an hour.[50]

Other metal complexes have also been used to remove other classes of anions. For example, the Ag and Hg adducts of thiol-SAMMS, being soft Lewis acids, have been demonstrated to be effective sorbents for removing iodide. The reader is referred to Chapter 13 of this book for those details.

8.5 Chelating Complexants

When constructing more complex ligand structures, amidation chemistry is a convenient and versatile strategy for joining the pieces. This can be done either before or after the silane has been anchored to the surface. However, it is worth pointing out that performing this linkage after the silane is anchored to the surface is an inherently limiting process, especially if the ligand has a large footprint, as it can then block adjacent amidation sites.[54] Thus, as a general rule, it is probably preferable to couple the fragments first, then perform the deposition onto the silica surface. This also allows the favorable H-bond interaction between the N-H and an adjacent C=O to help drive the self-assembly process.[55]

As a general rule, the rare earth cations tend to be fairly hard Lewis acids as a result of their high oxidation states. Thus, when designing a ligand field to selectively bind the lanthanide and actinide cations, it behooves the chemist to incorporate hard Lewis bases into the ligand structure. In addition, there is an interesting phenomenon that comes into play with these cations, in which the binding of one class of ligand to the cation enhances the binding affinity of another class of ligand. For the rare earth cations, the most important types of synergistic ligands are the amide carbonyl group and those ligands containing the phosphoryl group (P=O double bond, commonly loosely grouped under the term "phosphine oxide"). Given the versatility of the amide coupling process described in the previous paragraph, and the availability of the requisite precursor organosilanes, it is no surprise that amide linkage figures strongly in the ligand design strategy for the selective sequestration of rare earth cations.

Perhaps the most commonly encountered chelating complexants are those composed of one or more glycine sub-unit (e.g. EDTA), and indeed this strategy has been employed in these functional mesoporous sorbent materials. For example, treating the commercially available 3-isocyanatopropyltrimethoxysilane with glycine (in the presence of triethylamine as buffer) gives rise very cleanly to the corresponding glycinyl-urea silane, as shown in Figure 8.5.[56] Deposition of this silane in an MCM-41 matrix resulted in a monolayer population density of over 5 silanes/nm^2. This high population density seems to be due to the additional driving forces for self-assembly in this molecule (two NHs to interact with an adjacent C=O, dipole-dipole interactions, etc.). This ligand field was found to be an effective sorbent phase for both lanthanide and actinide cations, but only at pHs above about 4.5 (for the lanthanides, which is presumably the approximate pKa of the carboxylic acid)[56] and above about 2 for the actinides (the actinides are stronger Lewis acids than are the lanthanides, and actinide coordination enhances the dissociation, thereby increasing the acidity of the carboxylate).[57] Sorption kinetics were found to be rapid, with equilibrium being achieved in less than a minute.[56] There was no significant competition observed in the presence of excess transition metal cations, and it was found that this ligand field could easily be stripped and regenerated by a simple acid wash.[56]

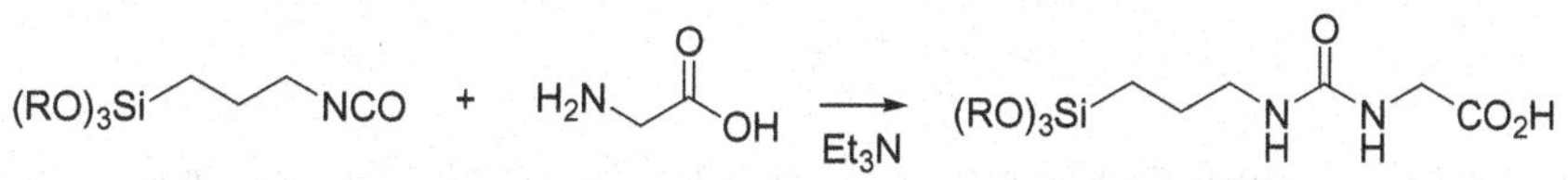

Figure 8.5. Preparation of the glycinyl-urea (Gly-UR) silane.

Given that the rare earth cations are hard Lewis acids, it comes as no surprise that they form strong complexes with hard Lewis bases like alkoxides. However, typical alcohols are not very acidic, and given that most lanthanide and actinide manipulations are done under acidic conditions (for solubility reasons), these rather basic ligands would be quickly protonated and cleaved at the pHs typically used to handle these materials. Thus it was necessary to design a chelating alkoxide that was sufficiently acidic to form stable complexes on the acid side. This was attempted by using a phenol instead of an alcohol, and by placing a carbonyl group in the ortho position, where it could both chelate the metal ion and enhance the phenol group's acidity (see Figure 8.6). Salicylamide- (Sa-) SAMMS was found to be an effective sorbent at near-neutral conditions (e.g. pH = 6.5), where it was particularly effective. Under more acidic conditions, its affinity for both lanthanide and actinide cations was quite limited.[56,57]

Detailed EXAFS analysis of the Eu(III) adduct of Sal-SAMMS revealed an average Eu-O coordination number of 8 and a Eu-O radial distance of 2.40 ± 0.015 Å.[56] These results are consistent with the Eu(III) cation being eight-coordinate (typical for the lanthanides), in either a cubic or distorted square antiprism geometry (see Figure 8.7). These EXAFS results support the conclusion that the close proximity of the ligands in the monolayer interface allow for multiple ligands to interact with a single metal cation. In this case we observe a 4 to 1 ligand : metal interaction, which clearly

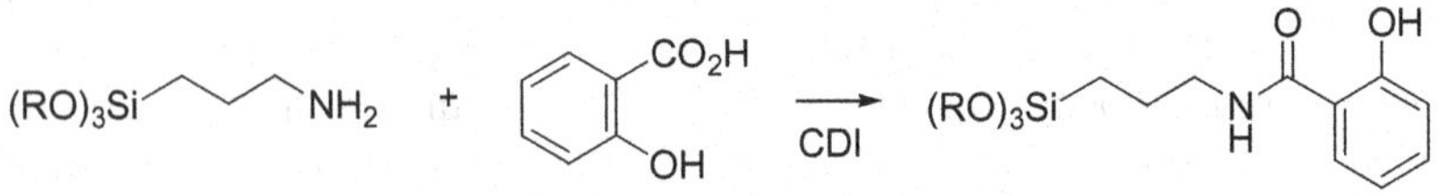

Figure 8.6.   Preparation of salicylamide (Sal) silane.

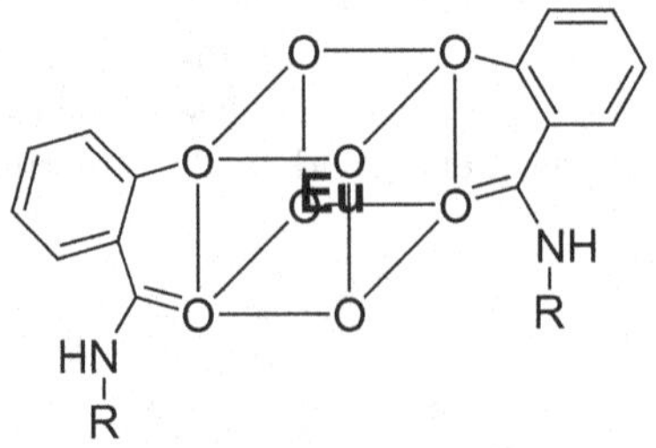

Figure 8.7.   Cartoon showing one possible bonding geometry of the 4:1 salicylamide/Eu(III) complex (only two ligands are shown for the purpose of clarity). The square anti-prism is also a possible geometry.

contributes to enhancing the binding affinity between a given ligand and metal cation.

Phosphates and phosphonates are known to have high affinity for rare earth cations, making these functionalities logical targets for incorporation into the SAMMS motif. Once again, the amide linkage was chosen for its desirable synergistic properties. The commercially available diethylphosphonoacetic acid was activated with carbonyldiimidazole (CDI), then subsequently treated with 3-aminopropyltrimethoxysilane (APS) to afford the acetamide phosphonate (Ac-Phos) silane cleanly and in high yield (see Figure 8.8).[58] Both the ester and the acid forms of the Ac-Phos-SAMMS were found to be effective sorbent materials for both the lanthanide and actinide species. In the case of the ester, it appears that the central enolizable methylene is sufficiently acidic to lose a proton upon complexation, even under fairly acidic conditions (e.g. pH = 2, or even lower in some cases). This is undoubtedly driven by the strength of the complex formed. The acid form of the Ac-Phos ligand has a very good affinity for lanthanide and actinide cations over a wide range of pHs, with no competition from common transition metal and alkaline earth metal cations.[56,57]

In an effort to probe the effect of chelate cavity size on the efficacy of these SAMMS materials, the homolog of Ac-Phos-SAMMS was also prepared (see Figure 8.9). Michael addition of diethylphophonous acid to trifluoroethylacrylate proceeded smoothly and in high yield.[58] Displacement of trifluoroethanol by APS was found to be clean and to afford the corresponding propionamide phosphonate (Prop-Phos) in virtually quantitative yield. Deposition of the Prop-Phos silane in MCM-41 resulted in a monolayer with a population density of 2.1 silanes/nm^2.[56] In contrast to the Ac-Phos ester SAMMS described above, the Prop-Phos ester SAMMS had no affinity at

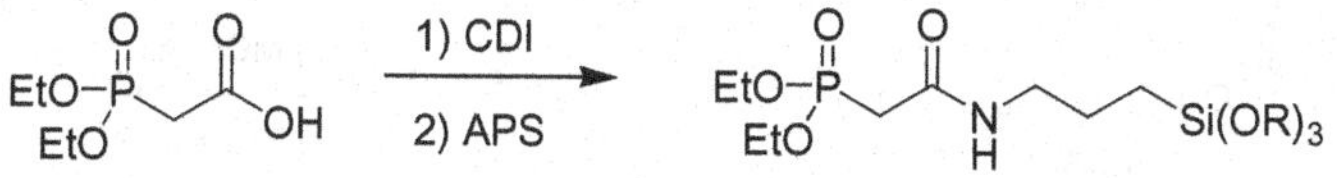

Figure 8.8.   Preparation of the acetamide phosphonate (Ac-Phos) silane.

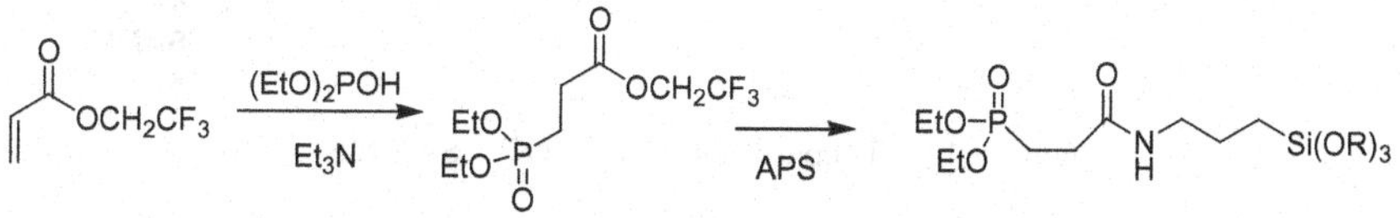

Figure 8.9.   Preparation of the propionamide phosphonate (Prop-Phos) silane.

all for the lanthanide cations.[56] In general, the Prop-Phos ester SAMMS had little affinity for the actinide cations, the exception being Pu (IV), for which this material demonstrated an excellent affinity. Like the Ac-Phos acid SAMMS, the Prop-Phos acid SAMMS were found to have very rapid sorption kinetics and good affinity for the actinides in general (i.e. U, Pu, Am, etc.). The larger chelation cavity of the Prop-Phos ligand was found to provide no significant advantage over the Ac-Phos ligand. Curiously, neither the Ac-Phos nor Prop-Phos acid SAMMS were effective sorbents for Np(V).

The Raymond group at Berkeley has spent over 20 years designing actinide-specific ligands for the selective chelation and *in vivo* extraction of these species.[59] One of the most powerful classes of ligands to result from these studies are the hydroxypyridinone (HOPO) class of ligands.[60] Due to the hydroxyl groups on these ligands, they must be installed in protected form to prevent competing reaction with the siloxane anchor. The benzyl-protected HOPO acids were activated with CDI, and subsequently coupled with APS, to afford the benzyl-protected silane.[61,62] This was deposited in the usual manner in MCM-41, then the benzyl-protecting group was removed using HBr in glacial acetic acid. Solid-state NMR studies suggested a population density of about 0.5 to 1 silanes/nm^2 for these systems (presumably due to the steric bulk of the benzyl-protecting groups during deposition).[62] Of the various HOPO-SAMMS tested, the 3,2-HOPO ligand (shown in Figure 8.10) was clearly superior, offering excellent binding affinity for all of the actinides studied, including Np(V).

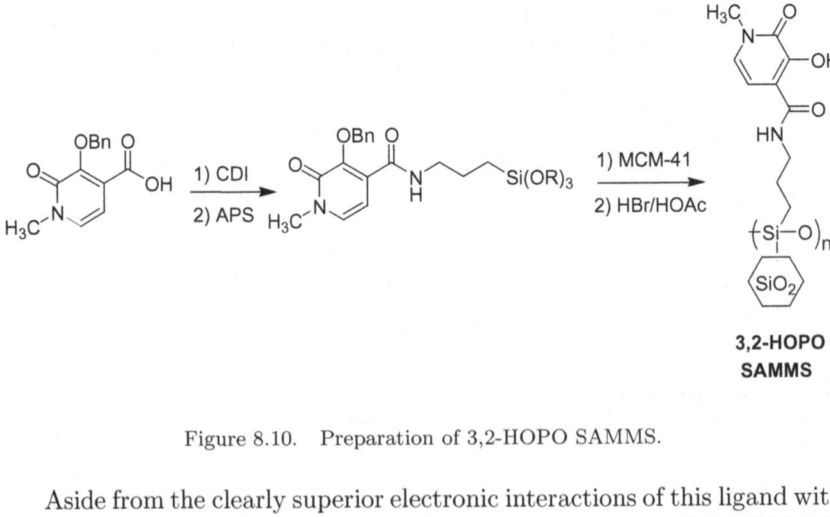

Figure 8.10.   Preparation of 3,2-HOPO SAMMS.

Aside from the clearly superior electronic interactions of this ligand with the actinide species, one possible contributing factor to 3,2-HOPO-SAMMS

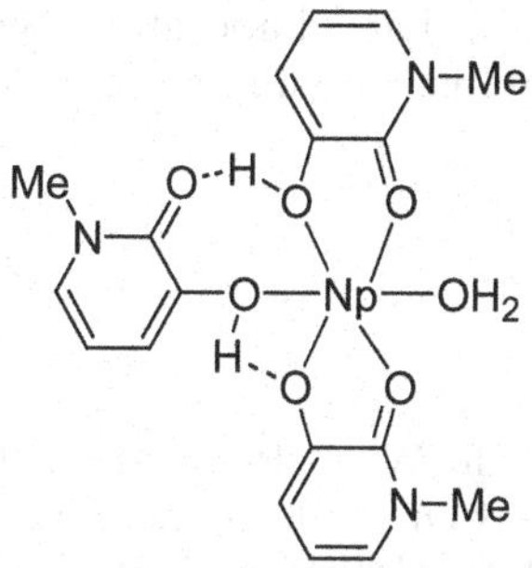

Figure 8.11. Chelation geometry of the 3,2-HOPO ligand binding the neptonyl (NpO_2^+) cation (looking down the O-Np-O axis, with the three HOPO ligands and a water molecule in the equatorial plane; the hydrogen bonds between the HOPO ligands contribute to the driving force for complexation).

affinity for Np(V) is the little bit of extra wiggle room left in the monolayer as a result of doing the deposition with the benzyl-protected form. Once this protecting group is removed there is some additional conformational freedom that allows these ligands to adopt a horizontal attitude, allowing three of them to chelate the Np(V) around its equatorial plane, in a hexagonal bipyramidal fashion, as shown in Figure 8.11.[62]

8.6 Conclusions

These new materials have been brought forth through the marriage of templated nanoporous materials and organic synthesis. New discoveries in this field are limited only by the creative application of ligand design and the construction of more complex organic architectures within the mesoporous framework. It is the self-assembly and molecular recognition provided by the organic ligand field that provide these materials with much of their remarkable performance. As new concepts and molecular architectures are dreamed up and reduced to practice we will see many exciting new discoveries in this area, for years to come.

Acknowledgments

I would like to thank my co-workers for all their dedicated efforts and creative thinking while contributing to the various projects that gave rise to many of these nanomaterials. Research support from the U.S. Department of Energy is gratefully acknowledged. Portions of this work were performed at the Pacific Northwest National Laboratory, which is operated for the DOE by Battelle Memorial Institute under contract DE-AC06-76RLO 1830.

References

1. J. S. Beck, J. C. Vartuli, W. J. Roth, M. E. Leonowicz, C. T. Kresge, K. D. Schmitt, C. T.-W. Chu, D. H. Olson, E. W. Sheppard, S. B. McCullen, J. B. Higgins and J. L. Schlenker, *J. Am. Chem. Soc.* **114**, 10834–10842 (1992).
2. C. T. Kresge, M. E. Leonowicz, W. J. Roth, J. C. Vartuli and J. S. Beck, *Nature* **359**, 710–712 (1992).
3. R. G. Pearson, *J. Am. Chem. Soc.* **85**, 3533–3539 (1963).
4. R. G. Pearson, *Science* **151**, 172–177 (1966).
5. X. D. Feng, G. E. Fryxell, L. Q. Wang, A. Y. Kim, J. Liu and K. Kemner, *Science* **276**, 923–926 (1997).
6. J. Liu, X. Feng, G. E. Fryxell, L. Q. Wang, A. Y. Kim and M. Gong, *Adv. Mater.* **10**, 161–165 (1998).
7. S. R. Wasserman, G. M. Whitesides, I. M. Tidswell, M. Ocko, P. S. Pershan and J. D. Axe, *J. Am. Chem. Soc.* **111**, 5852–5861 (1989).
8. D. W. Sindorf and G. E. Maciel, *J. Am. Chem. Soc.* **105**, 3767–3776 (1983).
9. F. M. Menger and U. V. Venkataram, *J. Am. Chem. Soc.* **107**, 4706–4709 (1985); F. M. Menger, *Acc. Chem. Res.* **18**, 128–134 (1985).
10. T. S. Zemanian, G. E. Fryxell, J. Liu, J. A. Franz and Z. Nie, *Langmuir* **17**, 8172–8177 (2001).
11. S. V. Mattigod, X. Feng, G. E. Fryxell, J. Liu and M. Gong, *Sep. Sci. Technol.* **34**, 2329–2345 (1999).
12. J. Liu, G. E. Fryxell, S. V. Mattigod, M. Gong, Z. Nie, X. Feng and K. Raymond, *Self-Assembled Mercaptan on Mesoporous Supports (SAMMS) Technology for Contaminant Removal and Stabilization*, PNNL-12006; September (1998).
13. K. Kemner, X. Feng, J. Liu, G. E. Fryxell, L.-Q. Wang, A. Y. Kim, M. Gong and S. V. Mattigod, *J. Synchrotron Rad.* **6**, 633–635 (1999).
14. L. Mercier and T. J. Pinnavaia, *Adv. Mater.* **9**, 500–503 (1997).
15. L. Mercier and T. J. Pinnavaia, *Environ. Sci. Technol.* **32**, 2749–2754 (1998).
16. J. Brown, L. Mercier and T. J. Pinnavaia, *Chem. Commun.* 69–70 (1999).
17. L. Mercier and T. J. Pinnavaia, *Chem. Mater.* **12**, 188–196 (2000).
18. Y. Mori and T. J. Pinnavaia, *Chem. Mater.* **13**, 2173–2178 (2001).
19. C. C. Chen, E. J. McKimmy, T. J. Pinnavaia and K. F. Hayes, *Environ. Sci. Technol.* **38**, 4758–4762 (2004).
20. S. J. L. Billinge, E. J. McKimmey, M. Shatnawi, H. J. Kim, V. Petkov, D. Wermeillle and T. J. Pinnavaia, *J. Am. Chem. Soc.* **127**, 8492–8498 (2005).
21. R. G. Parr and R. G. Pearson, *J. Am. Chem. Soc.* **105**, 7512–7516 (1983).

22. M. Misono, E. Ochiai, Y. Saito and Y. Yoneda, *J. Inorg. Nucl. Chem.* **29**, 2685–2691 (1967).
23. S. V. Mattigod, K. Parker and G. E. Fryxell, *Inorg. Chem. Commun.* **9**, 96–98 (2006).
24. T. Kang, Y. Park and J. Yi, *Ind. Eng. Chem. Res.* **43**, 1478–1484 (2004).
25. T. Kang, Y. Park and J. C. Park, *Stud. Surf. Sci. Catal.* **146**, 527–530 (2003).
26. Y. Shin, T. S. Zemanian, G. E. Fryxell, L. Q. Wang and J. Liu, *Micropor. Mesopor. Mater.* **37**, 49–56 (2000).
27. T. S. Zemanian, G. E. Fryxell, O. Ustyugov, J. C. Birnbaum and Y. Lin, Synthesis of nanostructured sorbent materials using supercritical fluids, in *Supercritical Fluids and Nanomaterials*, eds. A. S. Gopalan, C. M. Wai and H. K. Jacobs, Chap. 24 (American Chemical Society, Washington D.C., 2003), pp. 370–386.
28. T. Asefa, M. J. Maclachlan, N. Coombs and G. A. Ozin, *Nature* **402**, 867–871 (1999).
29. B. Hatton, K. Landskron, W. Whitnall, D. Perovic and G. A. Ozin, *Acc. Chem. Res.* **38**, 305–312 (2005).
30. L. Zhang, W. Zhang, J. Shi, Z. Hua, Y. Li and J. Yan, *Chem. Commun.* 210–211 (2003).
31. G. S. Caravajal, D. E. Leyden, G. R. Quinting and G. E. Maciel, *Anal. Chem.* **60**, 1776–1786 (1988).
32. R. K. Iler, *Chemistry of Silica* (John Wiley and Sons, New York, 1979).
33. A. S. M. Chong and X. S. Zhao, *J. Phys. Chem. B* **107**, 12650–12657 (2003).
34. C. Yang, X. P. Jia, Y. D. Cao and N. Y. He, *Stud. Surf. Sci. Catal.* **146**, 485–488 (2003).
35. M. W. McKittrick and C. W. Jones, *Chem. Mater.* **15**, 1132–1139 (2003).
36. S. Zheng, L. Gao and J. K. Guo, *J. Inorg. Mater.* **15**, 844–848 (2000).
37. S. Zheng, L. Gao and J. K. Guo, *J. Inorg. Mater.* **15**, 1015–1020 (2000).
38. S. Zheng, L. Gao and J. K. Guo, *J. Solid State Chem.* **152**, 447–452 (2000).
39. L. Gao and S. Zheng, *Mater. Trans.* **42**, 1688–1690 (2001).
40. S. Zheng, L. Gao and J. K. Guo, *Mater. Chem. Phys.* **71**, 174–178 (2001).
41. S. Zheng, L. Gao and J. K. Guo, *J. Inorg. Mater.* **16**, 459–464 (2001).
42. F. Zheng, D. N. Tran, B. Busche, G. E. Fryxell, R. S. Addleman, T. S. Zemanian and C. L. Aardahl, *Ind. Eng. Chem. Res.* **44**, 3099–3105 (2005).
43. R. J. P. Corriu, A. Mehdi, C. Reyé and C. Thieuleux, *Chem. Commun.* 1382–1383 (2002).
44. G. E. Fryxell, J. Liu, M. Gong, T. A. Hauser, Z. Nie, R. T. Hallen, M. Qian and K. F. Ferris, *Chem. Mater.* **11**, 2148–2154 (1999).
45. S. Kelly, K. Kemner, G. E. Fryxell, J. Liu, S. V. Mattigod and K. F. Ferris, *J. Phys. Chem. B* **105**, 6337–6346 (2001).
46. S. V. Mattigod, G. E. Fryxell, K. Alford, T. Gilmore, K. Parker and J. Serne, *Environ. Sci. Technol.* **39**, 7306–7310 (2005).
47. H. Yoshitake, T. Yokoi and T. Tatsumi, *Chem. Mater.* **14**, 4603–4610 (2002).
48. H. Yoshitake, T. Yokoi and T. Tatsumi, *Chem. Mater.* **15**, 1713–1721 (2003).

226 G. E. Fryxell

49. H. Yoshitake, E. Koiso, T. Tatsumi, H. Horie and H. Yoshimura, *Chem. Lett.* **33**, 872–873 (2004).
50. Y. Lin, G. E. Fryxell, H. Wu and M. Englehard, *Environ. Sci. Technol.* **35**, 3962–3966 (2001).
51. American Chemical Society, in *Cleaning Our Environment: A Chemical Perspective*, 2nd edn. (American Chemical Society, Washington D.C., 1978), pp. 378–452.
52. F. Chitry, S. Pellet-Rostaing, L. Nicod, J.-L. Gass, J. Foos, A. Guy and M. Lemaire, *J. Phys. Chem. A.* **104**, 4121–4128 (2000).
53. D. M. Poojary, R. A. Cahill and A. Clearfield, *Chem. Mater.* **6**, 2364–2368 (1994).
54. G. E. Fryxell, P. C. Rieke, L. L. Woods, M. H. Engelhard, R. E. Williford, G. L. Graff, A. A. Campbell, R. J. Wiacek, L. Lee and A. Halverson, *Langmuir* **12**, 5064–5075 (1996).
55. S. W. Tam-Chang, H. A. Biebuyck, G. M. Whitesides, N. Jeon and R. G. Nuzzo, *Langmuir* **11**, 4371–4382 (1995).
56. G. E. Fryxell, H. Wu, Y. Lin, W. J. Shaw, J. C. Birnbaum, J. C. Linehan, Z. Nie, K. Kemner and S. J. Kelly, *Mater. Chem.* **14**, 3356–3363 (2004).
57. G. E. Fryxell, Y. Lin, S. Fiskum, J. C. Birnbaum, H. Wu, K. Kemner and S. Kelly, *Environ. Sci. Tech.* **39**, 1324–1331 (2005).
58. J. C. Birnbaum, B. Busche, W. Shaw and G. E. Fryxell, *Chem. Commun.* 1374–1375 (2002).
59. A. E. V. Gorden, J. Xu, K. N. Raymond and P. W. Durbin, *Chem. Rev.* **103**, 4207–4282 (2003); J. Xu, P. W. Durbin, B. Kullgren, S. N. Ebbe, L. C. Uhlir and K. N. Raymond, *J. Med. Chem.* **45**, 3963–3971 (2002); J. Xu, B. Kullgren, P. W. Durbin and K. N. Raymond, *J. Med. Chem.* **38**, 2606 (1995).
60. J. Xu, D. W. Whisenhunt, Jr., A. C. Veeck, L. C. Uhlir and K. N. Raymond, *Inorg. Chem.* **42**, 2665–2674 (2003); J. Xu, E. Radkov, M. Ziegler and K. N. Raymond, *Inorg. Chem.* **39**, 4156–4164 (2000); P. W. Durbin, B. Kullgren, J. Xu, K. N. Raymond, M. H. Hengè-Napoli, T. Bailly and R. Burgada, *Rad. Prot. Dosim.* **105**, 503–508 (2003); D. L. White, P. W. Durbin, N. Jeung and K. N. Raymond, *J. Med. Chem.* **31**, 11 (1988).
61. W. Yantasee, G. E. Fryxell, Y. Lin, H. Wu, K. N. Raymond and J. Xu, *J. Nanosci. Nanotechnol.* **5**, 1537–1540 (2005).
62. Y. Lin, S. K. Fiskum, W. Yantasee, H. Wu, S. V. Mattigod, G. E. Fryxell, K. N. Raymond and J. Xu, *Environ. Sci. Tech.* **39**, 1332–1337 (2005).

Chapter 9

Chemically Modified Mesoporous Silicas and Organosilicas for Adsorption and Detection of Heavy Metal Ions

Oksana Olkhovyk and Mietek Jaroniec

Department of Chemistry, Kent State University
Kent, OH, USA

9.1 Introduction

The aim of this brief review is to give the current standpoint and trends in the development of nanomaterials for environmental applications, specifically for removal and detection of heavy metal ions from water. Recent advances in nanoscience and nanotechnology have been possible due to a tremendous development in the design of nanomaterials, i.e. materials consisting of nano-sized chemical domains and/or materials having nano-sized morphology and/or pores. Numerous research activities in this area are focused on the design and synthesis of nanomaterials with tunable structure and chemistry to achieve superior properties in comparison to conventional porous materials. An eminent example of this approach is the discovery of self-assembly (soft templating) synthesis, which afforded a new family of materials with ordered structures of nanopores[1] and allowed their post-synthesis decoration (functionalization) with various surface groups (see Figure 9.1). A growing interest in utilizing the unique properties of these materials for environmental applications has a stimulating effect on the development of novel nanomaterials, especially silica-based nanomaterials. Numerous interesting materials have been synthesized on the basis of the silica framework by introducing structural order, uniform nanopores and desired surface chemistry (Figure 9.1). With the discovery of ordered siliceous nanostructures,[1] the well-established modification methods for chemical immobilization of reactive ligands onto the silica surface got a second life. Nanomaterials designed by the aforementioned

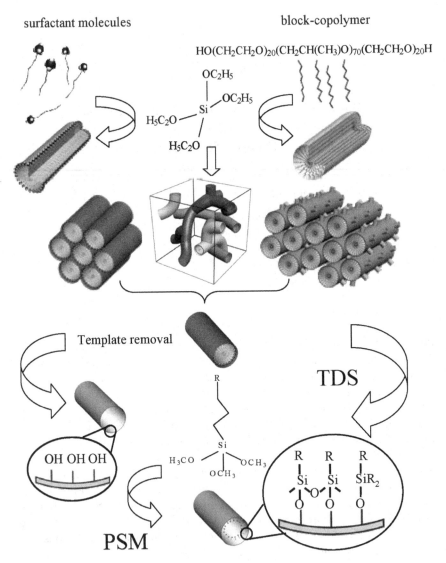

Figure 9.1.   Schematic illustration of the synthesis of ordered mesoporous silicas (OMS) via surfactant (left) or polymeric (right) templating routes. After the formation of a siliceous framework (showed as representative hexagonal or cubic arrangement of pores), post-synthesis modification (PSM) of the template-free OMS or template displacement synthesis (TDS) can be carried out to decorate pore walls with desired surface ligands (R).

chemical modification possess interactive surface groups on the walls of ordered mesopores. These novel nanomaterials, in addition to the stable and rigid silica framework, feature high and accessible specific surface area due to the presence of ordered and uniform nanopores of tailorable size, shape and surface chemistry, which make them much more attractive for many environmental applications in comparison to the conventional materials.

The first part of this review deals with the synthesis, characterization and comparison of properties and performance of mesoporous silicas with various organic ligands attached as well as their application for adsorption of heavy metal ions. There are some disadvantages of the post-synthesis surface modification of nanoporous silicas related to the reduction of accessible porosity for adsorption and limited selection of ligands to be introduced by this method, which narrow the range of possible applications of these materials. The first important disadvantage of the aforementioned modification is its multistep nature, which involves the synthesis of nanoporous silicas and their subsequent derivatization with desired ligands. Since among the existing silanols not all of them are available and suitable for chemical reaction, their complete replacement with other functional groups is a challenging task, which often requires a multistep process. Additionally, the geometrical factors related to the porous structure as well as size and shape of pores prevent organic ligands from fully replacing silanols. Therefore, one-step co-condensation synthesis and template-displacement synthesis with properly selected hydrolyzable silanes seem to be the methods of choice to reduce the cost and time of synthesis of novel mesoporous materials and simultaneously to retain the possibility of controlling their structural and surface properties.

The framework modification approach was considered as one that would permit overcoming the aforementioned disadvantages of the post-synthesis modification of porous silicas. The framework-modified porous silicas represent a successful combination of organic and inorganic building components and retain the template-predefined structural properties, while the organic fragments incorporated into the framework alter its chemical and mechanical properties (see Figure 9.2). The availability of organosilane precursors bearing desired functional groups and capable of self-assembly in the presence of templating agents is the only limitation in this rapidly growing area of research. The second part of this review is focused on the design and synthesis of framework-modified mesostructures, known as periodic mesoporous organosilicas (PMO), for sequestering metal ions. The PMO-related materials are extremely attractive for environmental applications because of the almost unlimited opportunities to design their chemical composition as well as interfacial and structural properties.

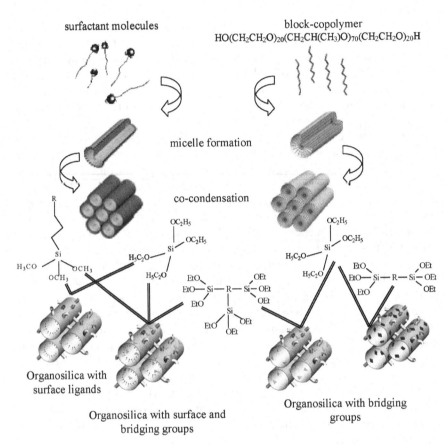

Figure 9.2. Schematic illustration of the co-condensation synthesis of periodic meso-porous organosilicas (PMO) via surfactant (left) or polymeric (right) templating routes. Co-condensation of silica precursors bearing organic functionality in the presence of organic template affords mesoporous organosilica with incorporated surface, bridging, or surface and bridging groups. The PMO structures are often formed via a cooperative mechanism, which does not involve prior formation of micelles by template species.

9.2 Environmental Significance and Main Concepts of Designing Adsorbents Based on Ordered Mesoporous Materials

The scientific and social significance of novel adsorbents for heavy metal ions stems from the extremely high toxicity of heavy metals. These toxic persistent pollutants are considered by the U.S. Environmental Protection Agency as major hazardous contaminants of aqueous ecosystems. A part per billion is the maximum allowable concentration of such major pollutants as mercury, cadmium and lead in water. The toxic effect of heavy metal ions on living cells occurs through their selective binding to important biomolecules such as glutathione, cysteine and homocysteine, which consequently affect the protein structure,[2] disturb the enzyme activity and intoxicate cell metabolic functions. The knowledge of the affinity and binding mechanism of particular heavy metal ions to functional groups, which are involved in the intoxication process, is helpful in designing proper functional materials that mimic the naturally occurring chelating mechanism and utilize it for the detection and specific adsorption of heavy metal ions. Such metal-sequestering ligands usually contain oxygen (phenolic, carbonyl, carboxylic, hydroxyl, ether, phosphoryl), nitrogen (amine, nitro, azo, diazo, nitrile), sulfur (thiol, thioether, thiocarbamate) and other functional groups that exhibit high affinity to heavy metal ions.[3] Various types of materials were employed as supports for the immobilization of functional groups capable of heavy metal ion adsorption. Polymers,[4–6] resins,[7–9] polystyrene,[10] fibers,[11] clays,[12] activated carbons,[13,14] natural zeolites such as clinoptilolite, mordenite, chabazite, erionite and phillipsite,[15] chitosan and other low-cost materials[13] along with currently available commercial materials such as TMT (trimercaptotriazine), STC (potassium/sodium thiocarbonate) and HMP-2000 sodium dimethylthiocarbonate (SDTC)[16] compose an enormous library of nanoporous adsorbents used for sequestering heavy metal ions. Among the aforementioned supports, various silicas (aerosils, aerosilogels, silica gels, porous glasses, etc.)[17–20] have been extensively used for the immobilization of various groups with high selectivity for and affinity towards heavy metal ions.[21,22]

Impregnation, covalent grafting, sol-gel synthesis and other methods for obtaining hybrid silica-based materials for selective adsorption, extraction and pre-concentration of heavy metal ions have been extensively studied (see Ref. 23 and references therein) because these materials possess high chemical specificity and selectivity due to the attached ligands as well as stable and accessible porosity due to the silica support. Thus, the key issue in the development of silica-based adsorbents is in the selection of proper surface ligands as well as silica supports of high surface area, accessible and well-developed porosity and high stability. Therefore, novel siliceous nanostructures are of high demand in the design of highly effective adsorbents for heavy metal ions that meet the EPA requirements and perform better than the existing materials.

9.3 Heavy Metal Ion Adsorbents Based on Ordered Mesoporous Silicas

A significant enhancement of the adsorption properties of porous solids for environmental applications was accomplished by the discovery of a new family of ordered nanoporous materials[1] synthesized via the bottom-up self-assembly approach. According to the IUPAC classification of porous systems,[24] the achievable pore sizes in these novel materials lie in the range of mesopores (widths from 2 to 50 nm), thus they are known as ordered mesoporous materials (OMM). Among them, materials having purely siliceous framework are known as ordered mesoporous silicas (OMS). Since the pioneering Mobil publications in 1992,[1,25] a novel family of materials have grown significantly with respect to the type of porous structure (network of pores of definite size and shape and its crystallographic symmetry) and chemical composition formed during the self-assembly of properly selected organic templates and inorganic and/or hybrid organic-inorganic precursors, e.g. tetraethyl orthosilicate (TEOS) and organosilanes. Surfactant- and polymer-templated OMS with hexagonal arrangement of mesopores are MCM-41,[25–27] SBA-2,[28] SBA-3,[29,30] HMS,[31] SBA-15[32,33] and FSM-16.[34] SBA-1 and MCM-48 possess cubic bicontinuous arrangement of interconnected and uniform channels.[28,35,36] There are also popular cage-like mesostructures (e.g. FDU-1 and SBA-16) prepared by using block copolymers as templates.[37] The availability of various precursors that are able to self-assemble in the presence of surfactants and block copolymers creates almost unlimited opportunities for the synthesis of OMM with tailorable properties.[38–40] A proper selection of synthesis conditions and the structure-directing agents allows one to control the pore size, ranging from $\sim$2 to 8 nm when surfactants are employed as templates and from $\sim$8 to 30 nm when block copolymers are used. In particular, the use of block copolymers was beneficial for the development of inexpensive and reproducible large-pore materials with channel-like and cage-like structures.[41–48]

    Ordered mesoporous organosilicas, due to the tunable surface chemistry and nanoporous structure, represent excellent substitutes for OMS in numerous applications.[49–51] The main advantage of these materials is their well-defined surface chemistry achieved by various surface modifications. Particular advantages arising from modification of OMS instead of conventional silicas are related to the structural organization of the former, high surface area (*ca.* 1,000 m^2/g), greatly reactive surface,[52] and large mesopores, which could be tuned by judicious choice of templates. The accessibility of binding sites,[53,54] effects of confined geometry on the ligand layer formation, and limited ligand bonding density[55] have been extensively studied to advance the design of OMM and improve their applicability for commercial uses[56] as well as to improve our understanding about the formation of these materials.[57–59]

9.3.1 *Post-synthesis modification of template-free OMS*

9.3.1.1 *OMS with monofunctional groups*

An attractive approach to obtaining high-capacity adsorbents on the basis of OMS considers their functionalization with organic groups of specific properties for desired environmental applications. The resulting materials should possess functional groups over the entire surface capable of interacting selectively with heavy metal ions. A high selectivity could be achieved by proper choice of functional group containing a metal-chelating center with high affinity for a particular heavy metal ion. Pearson's classification of soft acids and bases[60] suggests preferential binding of Lewis soft acids (e.g. mercury, cadmium, lead) to soft bases (sulfur), which is not affected significantly by the presence of hard acids such as magnesium, calcium and so on. The silane-type coupling agents possessing thiol groups were initially employed to obtain the OMS-based adsorbents for mercury ions; see the pioneering work from PNNL,[61] which initiated extensive research on the design and synthesis of functionalized OMS for adsorption of heavy metal ions and other pollutants.

A conventional approach to decorate the pore walls of OMS with various functional groups, such as metal-chelating ligands, involves the chemical reaction of the template-free OMS with organosilanes analogous to that used widely for the preparation of silica-based chemically bonded phases for chromatographic separations. Figure 9.1 shows a simplified chart of the OMS synthesis followed by the template removal (usually calcination or extraction) and its post-synthesis modification involving the reaction of the template-free OMS with proper organosilanes. Namely, in this modification silanols present on the pore walls of OMS are replaced by desired surface ligands via single- or multistep ligand attachment. Often a solvent (e.g. toluene) and/or catalyst (e.g. pyridine) are used to carry out the reaction of organosilanes with silanol groups. Subsequently, the resulting material is extensively washed with polar and non-polar solvents to eliminate the excess silane and reaction by-products. The main issue in this modification is to prevent the formation of silane multilayers on the mesopore walls instead of the desired dense monolayer. The self-assembled monolayers of thiol functionality on the mesopore walls of OMS afforded adsorbents with high specific surface area, in which *ca.* 75% of available silanols was replaced by thiol groups. An indicator of the high adsorption affinity of the thiol-modified OMS towards mercury ions is a large value of the distribution coefficient, $K_d = 3.4 \times 10^5$. This coefficient is defined as the amount of the adsorbed metal in micrograms per gram of adsorbent divided by the concentration of the metal ion (in micrograms per milliliter) remaining in the eluent after the adsorption experiment. The well-defined and open mesoporosity ensured a fast diffusion and mass transfer of metal ions. Up to 0.6 g of Hg^{2+} per gram of thiol-functionalized OMS was reported.[62] The self-assembled monolayers on mesoporous supports (SAMMS) consisting of thiol functionality were shown to be efficient adsorbents for the removal of mercury ions from water and reduction of their concentration to that allowed by the National Pollution Discharge Elimination System/($\sim$12 ppt). Also, these materials passed the Toxicity Characterization Leaching Procedure tests of the U.S. Environmental Protection Agency, which manifest their potential usage in environmental clean-up.[63]

The MCM-41 and MCM-48 silicas exhibiting hexagonal or cubic arrangement of mesopores as well as SBA-15 with hexagonally arranged large mesopores interconnected by small micropores were extensively used to prepare highly effective adsorbents for heavy metal ions.[64,65] The study of mercaptopropyl-modified adsorbents prepared by using silica supports of different structures suggested that three-dimensional networks give better adsorbents for heavy metal ions (Hg^{2+} and Pb^{2+}).[66] It was possible to reduce Hg^{2+} content to the ppb level (with initial ppm Hg^{2+} concentration) when passing the model water through a column packed with thiol-grafted mesoporous molecular sieves.[67] The factors affecting the desired characteristics of ideal heavy metal ion adsorbents, such as silanol density, channel dimensions, charge density and particle morphology, were studied to find the optimal properties of siliceous supports that would ensure their high performance in remediation processes.[68,69] At present, it is clear that inorganic-organic hybrid materials with functional monolayers have a great potential for environmental clean-up.[70,71] For instance, thiolated mesoporous solids showed a great selectivity for Hg^{2+} ions in the presence of other metal ions (e.g. Cd^{2+}, Pb^{2+}, Zn^{2+}, Co^{3+}, Fe^{3+}, Cu^{2+} and Ni^{2+}) in various aqueous environments[72,73] and were reported to adsorb tetrahedral oxyanions of As, Cr and other relevant toxic metals; various radionuclides[74,75]; electroplating metals (e.g. Cu, Ni, Ag)[76]; actinides[77]; and noble metal ions (Pt^{2+}, Pd^{2+}).[78] Also, thiolated HMS materials showed prospective applicability for the mining industry to recover rhodium ions from typical mine effluents containing also Cu, Ni and Zn elements; their high selectivity for rhodium ions in the presence of some reducing agents was observed, which permitted the recovery of these ions at ultra-low concentrations.[79] Also, incorporation of dithiocarbamate into the mesopores of MCM-41 afforded material with high adsorption capacity towards mercury ions too.[80] However, aminopropyl-modified mesoporous adsorbents were found to be capable of binding Cu^{2+}, Zn^{2+}, Cr^{3+} and Ni^{2+} ions,[65] while imidazole-functionalized nanostructures possessed great capability to preconcentrate noble metal ions.[81] The self-assembled mono-layers of carbamoylphosphonic acids, such as acetamide phosphonic acid and propionamide phosphonic acid, grafted onto the MCM-41-type of materials, were reported for adsorption of heavy and transition metal ions (Cd^{2+}, Co^{2+}, Cu^{2+}, Cr^{3+}, Pb^{2+}, Ni^{2+}, Zn^{2+} and Mn^{2+}); adsorption equilibration in these systems was achieved within a very short period of time.[82]

9.3.1.2 *OMS with multifunctional groups*

Despite the fact that an unprecedented ligand coverage[61] allowed the maximum 1:1 heavy metal ion:ligand ratio[68] and high adsorption capacity to be achieved, a main limitation of adsorbents with monofunctional groups is the ability to attract only one metal ion per ligand. In addition, the surface concentration of reactive silanols in ordered mesoporous silicas is usually smaller (often ~30%) than that in conventional silicas.[55] Thus, the concentration of ligands in the monolayer formed on the mesopore walls of OMS is not high due to the smaller concentration of silanols, pore curvature and related geometrical limitations. Therefore, a fully accessible porosity (three-dimensional structure) is a key factor in the design of adsorbents with enhanced transport properties. An interesting approach, the so-called pore expansion method,[83] utilizing expanders such as *N,N*-dimethylalkylamine afforded the MCM-41 analogues with controlled pore sizes from 3.5 up to 25 nm and pore volumes from 0.8 up to 3.6 cm^3/g suitable for adsorption of Co^{2+}, Ni^{2+} and Cu^{2+} ions and organic pollutants such as 4-chloroguaiacol or 2,6-dinitrophenol from water.[84] Also, their regeneration with alkylamines was possible.[84]

One of the possible alternatives to increase the adsorption capacity of the OMS-based adsorbents is to attach multifunctional ligands because they are able to bind more than one metal ion per ligand. Several multifunctional ligands such as 1-allyl-3-propylthiourea,[85] 1-benzoyl-3-propylthiourea,[86–88] and 2,5-dimercapto-1,3,4-thiadiazole ligands[89] were grafted onto the mesopore walls of MCM-41,[85,86] MCM-48[88] and SBA-15.[89] It was shown that those ligands possess the ability to coordinate more than one metal ion per ligand; consequently, the adsorption capacity of OMS with those ligands may exceed one gram of mercury(II) per gram of the adsorbent or 5.0 mmol Hg^{2+}/g. While previously reported adsorbents obeyed the Langmuir-type of adsorption,[90] the adsorption isotherm for mercury(II) ions on benzoyl-thiourea-modified OMS showed a more complex adsorption isotherm with two Langmuir-type constants of 1.41×10^5 L/mol and 1.08×10^8 L/mol, indicating a two-step adsorption process. In other words, the aforementioned adsorbent showed two groups of sites of different adsorption energies. Over 70% of the initial adsorption capacity was retained after mild regeneration of this adsorbent.[86] A comparative study of two-dimensional (MCM-41) and three-dimensional (MCM-48) mesostructures with bonded benzoyl-thiourea ligands showed that the performance of three-dimensional-type adsorbents is better than those having two-dimensional structures.[88]

The attachment of more complex ligands to the pore walls of silica, especially those that are not commercially available in the form of coupling agents, can be performed via two or more reactions. Initially, the silica surface is modified by bonding organosilane that contains a group inert towards silica but capable of attaching some specific molecular segments. This approach, which involves successive reactions, is especially convenient to build up large groups on the pore walls. In the case of large pores and good pore connectivity, prior chemical modification of the silica surface may not be necessary because the entire surface ligand can be attached in a single step by using a proper organosilane coupling agent, which is known as the homogeneous route of surface modification.[90] An example of this modification was the attachment of 2-mercaptopyridine onto the mesopore walls of MCM-41 and SBA-15, which afforded adsorbents for mercury ions that showed higher capacity than analogous materials prepared by multistep modification (heterogeneous route).[92]

The multistep surface modification provides much more flexibility for the design of surface functionality and avoids some undesired geometrical constraints. Initially, in this procedure some simple reactive groups such as chloroalkyl or aminoalkyl are attached to the mesopore walls via conventional chemical reaction of OMS with proper organosilanes, which is carried out in solvent and/or pyridine at elevated temperatures under reflux conditions.[93] This initial modification of OMS with reactive groups is followed by another modification step, which involves reaction of these groups with the proper chemical compounds in order to achieve the desired surface functionality. If the latter reaction proceeds to completion, the resulting material contains one type of surface groups only. In the case of an incomplete conversion of reactive groups, the resulting material contains some residual reactive groups in addition to the created multifunctional ligands. For instance, reaction of aminopropyl-OMS with benzoyl isothiocyanate resulted in ~70% conversion of aminopropyl groups into 1-benzoyl-3-propylthiourea ligands, giving bifunctional OMS with benzoylpropylthiourea ligands and residual aminopropyl groups.[86]

9.3.2 *Surface modification of OMS via co-condensation synthesis*

Incorporation of surface groups into OMS via direct co-condensation synthesis possesses numerous advantages over post-synthesis surface modification because it permits control of the hydrolysis and subsequent condensation of silica precursors such as TEOS or TMOS[94,95] and the achievement of a high loading of surface groups. Either the surfactant-templated co-condensation synthesis[91,96–102] or block copolymer-directed structure formation[103–105] were used to obtain high-capacity adsorbents for heavy metal ions.[59,106–110] The co-condensation synthesis is much simpler in comparison to the post-synthesis grafting because it reduces the number of synthesis steps. It permits the introduction of various reactive groups without pore blocking, which can be further derivatized to achieve the desired surface functionality. The co-condensation of primary (TEOS) and secondary (silane bearing reactive group) silica sources in the presence of structure-directing agents is also advantageous because it affords uniform distribution of these groups and simultaneously permits control of the structure and porosity of the resulting adsorbents.[111–114] For instance, direct incorporation of chloropropyl or aminopropyl reactive groups via co-condensation synthesis affords OMS with those groups, a further modification of which is feasible. This approach was used to create cyclam-type functionality and manganese(III) Schiff-base complexes in OMS.[115,116] Some recent reports on the incorporation of acidic groups (phosphonic, humic) into mesoporous materials afforded effective adsorbents for chromium(III)[117] and cadmium(II) ions[118] from aqueous solutions. In the latter case the hybrid adsorbent was prepared by using a microwave-assisted synthesis, which is much faster than the conventional procedure and especially suitable for high-throughput screening of a wide range of precursors and experimental conditions for OMS-based adsorbents.[119–122]

To increase the adsorption capacity of modified OMS, two or more surface groups could be introduced. Bifunctional OMS with aminopropyl and thiol groups synthesized via co-condensation were shown to have high adsorption capacity for several heavy metal ions.[123] Co-condensation synthesis afforded ordered mesostructures up to 60% of organic in the synthesis gel and worm-like mesostructures were obtained in the concentration range of 60% to 100% organosilane.[124] Since the aforementioned synthesis provided adsorbents with high ligand loading and tailorable porosity, their adsorption performance was shown to be superior, as evidenced by high values of the distribution coefficient (*ca.* 10^7) for several heavy metal ions.

In conclusion, co-condensation synthesis is a very simple and feasible approach to form a dense organic layer of functional groups and simultaneously create an ordered siliceous mesostructure of tailorable and accessible porosity. These mesostructures were found to be attractive not only in such traditional areas as adsorption, catalysis, separations and environmental clean-up but also in nanotechnology and biotechnology (see Refs. 125–128 and references therein).

9.3.3 Functionalization of template-containing OMS

Simultaneous template displacement and attachment of functional groups onto the pore walls of OMS is possible via so-called template displacement synthesis (TDS).[104,129–131] This method involves the removal (displacement) of organic templates from mesopores of the template-containing OMS accompanied by simultaneous attachment of desired surface ligands onto the pore walls. This one-step process involving template removal and surface modification utilizes the high affinity of the template-containing OMS towards reactive organosilanes, which results in the replacement of self-assembled template-silica interface by a covalently bonded layer of organic groups on the silica surface. In contrast to the post-synthesis grafting of the template-free OMS, the TDS process has been found to be attractive for direct incorporation of simple ligands into OMS that results in high-affinity adsorbents for mercury ions[132] because it does not require prior template removal by extraction and/or calcination (see Figure 9.1).

The combination of co-condensation synthesis with the template displacement process allows one to form more complex ligands inside mesopores of OMS. This approach was used to prepare high-capacity adsorbents for mercury; and yielded adsorbents that contain multifunctional 2,5-dimercapto-1,3,4-thiadiazole ligands, which are capable of interacting with mercury ions via at least three active sites.[89] This work shows that a two-step post-synthesis modification (PSM) of the template-free OMS afforded materials with low surface area and small ligand coverage. In contrast, the one-pot (co-condensation) synthesis (OPS) of chloropropyl-modified OMS followed by either TDS or PSM afforded materials with large surface area, open porosity and high concentration of multifunctional ligands, which were able to adsorb up to 8.5 mmol of mercury(II) per gram of the adsorbent having a ligand concentration of ~2.7 mmol/g. This high adsorption capacity suggests the presence of at least three binding sites for mercury(II) ions per attached ligand. The aforementioned adsorption capacity for the 2,5-dimercapto-1,3,4-thiadiazole-modified OMS studied is much higher than that for a conventional silica with analogous surface ligands.[133]

9.4 Periodic Mesoporous Organosilicas (PMO) as Adsorbents for Heavy Metal Ions

The first successful attempts to incorporate some heteroatoms into the silica framework were made in the early 1990s.[134,135] This area of research is still growing because metal-containing silica mesostructures are of great importance for heterogeneous catalysis. However, a tremendous potential of framework-modified silicas has been demonstrated by the discovery of the so-called periodic mesoporous organosilicas in 1999.[106,108,109,137] This perfect marriage of organic and inorganic chemistry had greatly accelerated the development of self-assembled mesostructured materials. In these materials not only a hybrid interface was created, but also a hybrid pore framework was achieved by co-condensation of organosilane precursors having the following general formula: $[(R'O)_3Si]_nR$, where R' refers to methyl or ethyl in a hydrolyzable alkoxy group and R is a functional group that constitutes an organic bridging unit within the inorganic framework of OMS. Thus, this organic group is homogeneously dispersed within the material framework. The co-condensation of $[(R'O)_3Si]_nR$ in the presence of surfactant or polymeric templates affords materials with a high concentration of bridging groups, accessible porosity and tailored surface and structural properties. The chemical and structural properties of $[(R'O)_3Si]_nR$ precursor, especially its bridging group, predefine the location of this group within the material's framework. Numerous $[(R'O)_3Si]_nR$ precursors, mainly those with relatively simple bridging groups, are able to self-assemble into ordered mesostructures in the presence of proper surfactant or polymeric templates. However, the self-assembly of more complex precursors, especially those with large bridging groups, requires the addition of easily hydrolyzable TEOS or TMOS in order to obtain organosilicas with ordered mesoporosity. Figure 9.2 shows a general synthesis route to obtain periodic mesoporous organosilicas, including those with surface, surface/framework or framework functionalities. Depending on the nature of silica precursors, PMO materials with an ultra-low dielectric constant,[138] unique electron acceptor ability, photo- and thermochromic responsiveness,[139] chiral activity,[140] electrogenerated chemiluminescence,[141] photoactivity and many other unique properties[142] can be obtained. The easiness of their synthesis was a major driving force in the development of novel heavy metal ions adsorbents containing ligands that could not be introduced via conventional post-synthesis modification due to their bulky structure. For instance, cyclam units and dibenzo-18-crown-6 ether moieties that are capable of forming complexes with Cu(II) and alkali cations (Na^+ and/or K^+), respectively, were incorporated into mesoporous silicas.[143–145] Organosilanes with trialkoxysilyl groups containing amine functionality connected via linear, square or tetrahedral bridges of coordination compounds of nickel, cadmium and zinc were used to synthesize hybrid mesoporous materials of various symmetries.[146] A novel method for the synthesis of lamellar silicas with high loading of amine groups was prepared by taking advantage of the reversible covalent binding of CO_2 to amines, where CO_2 gas acted as a structure-directing agent. The reported aminosilicas possess high adsorption capacity towards transition metals and lanthanide ions such as Cu(II), Eu(III) and Gd(III) ions.[147] Another study showed that mono-, di- and triamino-functionalized MCM-41 and SBA-1 silicas adsorb oxyanions (chromate and arsenate) depending on the ligand density and type of OMS structure.[148] Hybrid materials that have the capability to effectively bind heavy metal, inorganic, organic, charged and neutral mercury compounds from not only aqueous solutions but also from other media such as oil or gas were reported too.[149]

A challenging aspect of introducing multifunctional organics into a small volume of mesopores was accomplished by co-condensation synthesis of PMO. Multifunctional PMO with various concentrations of organic bridging groups, morphology of the mesoporous network, and structural symmetry of the pores were reported in the literature.[150–153] The co-condensation synthesis of organosilane mixtures[154,155] or co-condensation of precursors bearing multiple functionality[156] was investigated to develop PMO having multifunctional bridges within the silica framework. The main concern, arising from the bulky nature of bridging groups employed in the synthesis, was possible structure deterioration.[52,157–159] Recent efforts to design PMO with large bridging groups afforded nanomaterials with high surface area and ordered large pores.[160,161] Some large bridging groups such as 1,4,8,11-tetraazacyclotetradecane,[162] capable of adsorbing heavy metal ions, were reported to provide materials with a small percentage of these groups and relatively poor structural ordering.[163] Nevertheless, the attempt to graft a metal-N-triethoxysilylpropylcyclam complex inside the mesopores with subsequent incorporation of another metal salt into the framework afforded bifunctional material and showed the possibility for designing the surface and framework properties of mesostructures.[164]

The design of the framework composition of nanomaterials is an interesting and challenging area of research. Recently a PMO, obtained via one-pot synthesis of tris[3-(trimethoxysilyl)-propyl] isocyanurate as a secondary silica precursor in the presence of TEOS and poly(ethylene oxide)-poly(propylene oxide)-poly(ethylene oxide) block copolymer, was reported.[165] A successful incorporation of this heterocyclic bridging group afforded highly ordered mesoporous materials with hexagonal symmetry, large surface area of $\sim$600 m^2/g and large pores of $\sim$8–10 nm. The maximum adsorption of mercury(II) on these materials was $\sim$1.8 g of Hg^{2+}. The corresponding distribution coefficients were about 10^8, which indicates the high affinity of this bridging group towards mercury ions. Thioether-functionalized organic-inorganic mesoporous materials,[166] synthesized by co-condensation of (1,4)-bis(triethoxysilyl)propane tetrasulfide and tetraethoxysilane in the presence of a block copolymer, poly(ethylene oxide)-poly(propylene oxide)-poly(ethylene oxide), showed also high selectivity towards mercury ions in the presence of copper, cadmium, lead and zinc ions; their capacity was about 2.7 g of Hg^{2+} per gram of adsorbent. The latter result shows a superior performance of PMO-based adsorbents; for instance, the amount of mercury ions adsorbed by polythioether chelating resin would be one order of magnitude lower.[167]

A great potential of multifunctional versus monofunctional adsorbents was demonstrated recently for bifunctional OMS with anchored ethylenediamine and thiol groups synthesized by combining direct and post-synthesis chemical modification methods, where supercritical fluid was chosen as a reactive medium for chemical grafting. Introduction of thiol functionality into ethylenediamine-modified MCM-41 materials generated a high-capacity adsorbent for Cd^{2+} ions, which could be reused several times.[168]

Another study[169,170] shows the possibility of tailoring the interfacial and framework chemistry of PMO adsorbents by incorporation of two types of mercury-specific ligands. Introduction of thiol and isocyanurate ligands into PMO via co-condensation synthesis afforded a high-capacity adsorbent for mercury ions from aqueous solutions.[169] While the adsorption capacity of 1.8 g Hg^{2+}/g was obtained for isocyanurate-containing organosilica with a disordered mesoporous structure having a high concentration of bridging groups, the bifunctional adsorbent with relatively low concentrations of thiol and isocyanurate groups was structurally ordered and adsorbed 1.13 g of Hg^{2+}/g.[171] However, the introduction of an additional surface ligand into the PMO structure induced a structural change. Namely, the hexagonal ordering of channel-like mesopores in isocyanurate-containing PMO[170] changed to the cubic $I4_132$ symmetry group when additional mercaptopropyl surface groups were introduced during co-condensation.

Figure 9.3 illustrates the concept of the binding mechanism between heavy metal ions and multifunctional surface and bridging groups in PMO adsorbents. In this scheme, triangles and rectangles represent organic bridging groups in PMO that possess triple or quadruple binding sites for heavy metal ions. The adsorption capacity of PMO adsorbents can be further enlarged by the introduction of additional surface groups that are able to bind one or more heavy metal ions. Table 9.1 provides the list of metal-chelating ligands used to modify the siliceous mesopores, the methods used for their modification as well as the adsorption capacities of many OMS-based adsorbents for heavy metal ions. The chemical structures of these ligands are shown in Figures 9.4 and 9.5. Figure 9.4 presents the structures of the most common surface ligands, whereas Figure 9.5 shows the structures of the bridging groups used to design PMO adsorbents. The latter

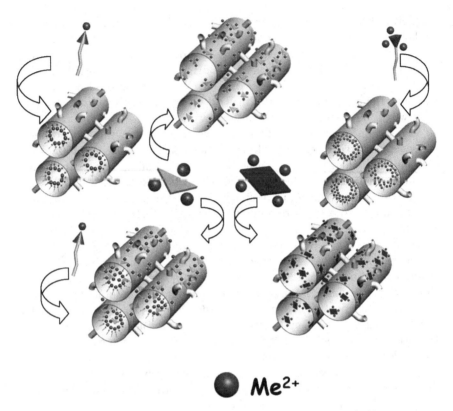

$\bullet$ **Me²⁺**

Figure 9.3.   Hypothetical models for binding of heavy metal ions (Me²⁺) by various types of functional groups present in mesoporous organosilicas.

belong to a large group of hybrid organic-inorganic ordered nanostructures, which continue to gain popularity because of their potential applications in many fields of science and technology (see Refs. 142, 172–175 and references therein).

9.5   A Brief Overview of Characterization of Mesostructured Adsorbents

Several techniques are used for characterization of the adsorption, surface and structural properties of mesoporous organosilicas. One of the key issues in this characterization is the identification of functional groups present on the silica surface and/or in its framework. It is important to know the chemical structure of these groups as well as the type of bonding involved,

Table 9.1.   List of functional ligands incorporated into ordered mesoporous materials via various synthesis methods and the corresponding metal ion adsorption capacities reported in the literature.

Ligand	Modification Method	Reference	Metal Ion	Adsorption (mmol/g)
(3-Mercaptopropyl)-trimethoxysilane	PSM	61	Hg^{2+}	1.04
		62	Hg^{2+}	2.99
		63	Hg^{2+}	3.04
		65	Hg^{2+}	1.46
		67	Hg^{2+}	1.54
		70	Hg^{2+}	2.99
		68	Hg^{2+}	1.5
Dithiocarbamate	PSM	80	Hg^{2+}	0.2
N-(3-triethoxysilylpropyl)-4,5-dihydroimidazole	PSM	81	Pt^{2+}	0.09
			Pd^{2+}	0.09
Acetamide phosphonic acid	PSM	82	Cd^{2+}	0.32
Propionamide phosphonic acid	PSM	82	Cd^{2+}	0.32
N,N-dimethyldecylamine	PSM	84	Co^{2+}	1.05
			Ni^{2+}	0.93
			Cu^{2+}	1.67
1-Allyl-3-propylthiourea	PSM	85	Hg^{2+}	1.49
1-Benzoyl-3-propylthiourea	PSM	86	Hg^{2+}	4.98
		88	Hg^{2+}	6.7
3-Trimethoxypropyl-thioethylamine	PSM	90	Co^{2+}	1.08(0.72)
			Ni^{2+}	1.20(1.74)
			Cu^{2+}	1.70(1.91)
			Pb^{2+}	1.34(2.19)
			Hg^{2+}	4.02(2.89)
				homogeneous (heterogeneous synthesis route)
2-Mercaptopyridine	PSM	92	Hg^{2+}	0.16
1,4,8,11-Tetraazacyclo-tetradecane	PSM	115	Cu^{2+}	0.77
			Co^{2+}	0.80
(3-Mercaptopropyl)-trimethoxysilane	OPS	91	Hg^{2+}	1.26
		94	Hg^{2+}	2.1
		97	Hg^{2+}	2.3
		98	Hg^{2+}	1.49
		99	Cd^{2+}	0.20
		100	Hg^{2+}	1.37
		112	Hg^{2+}	0.59

(*Continued*)

Table 9.1.　(*Continued*)

Ligand	Modification Method	Reference	Metal Ion	Adsorption (mmol/g)
Aminopropyl, [amino-ethylamino]propyl, [(2-aminoethylamino)-ethylamino]propyl	OPS	99	Cu^{2+} Ni^{2+} Co^{2+} Cd^{2+}	0.55 0.15 0.10 0.01
3-(2-Aminoethylamino)-propyltrimethoxysilane	OPS	114	Cd^{2+} Zn^{2+}	0.51 0.57
3-Aminopropyltrimethoxysilane	OPS	148	Chromate Arsenate	1.19 0.25
[1-(2-Aminoethyl)-3-aminopropyl]trimethoxysilane	OPS	148	Chromate Arsenate	0.27 0.28
(Trimethoxysilyl)-propyldiethylenetriamine	OPS	148	Chromate Arsenate	0.54 0.36
1-Allyl-3-propylthiourea	TDS	132	Hg^{2+}	0.80
2,5-Dimercapto-1,3,4-thiadiazole	TDS	89	Hg^{2+}	8.47
Trimethoxysilylpropyl diethylphosphonate	PMO	117	Cr^{2+}	1.58
Humic acid	PMO	118	Cd^{2+}	0.002
3-Aminopropyltriethoxysilane and 3-mercaptopropyltri-methoxysilane	PMO	123	Hg^{2+}	1.51
1,4,8,11-Tetrakis[3-(triethoxysilyl)-propyl]-1,4,8,11-tetraaza-cyclotetradecane	PMO	144	Cu^{2+}	1.31
N-(2-aminoethyl)-3-aminopropyltrimethoxysilane and *N*-(6-aminohexyl)-3-aminopropyl-trimethoxysilane	PMO	147	Cu^{2+} Eu^{3+}	6.5 4.8
(1,4)-Bis(triethoxysilyl)propane tetrasufide	PMO	166	Hg^{2+}	13.5
Tris[3-(trimethoxysilyl)propyl] isocyanurate	PMO	170	Hg^{2+}	8.9
Tris[3-(trimethoxysilyl)propyl] isocyanurate and (3-mercaptopropyl)-trimethoxysilane	PMO	169 171	Hg^{2+}	5.6

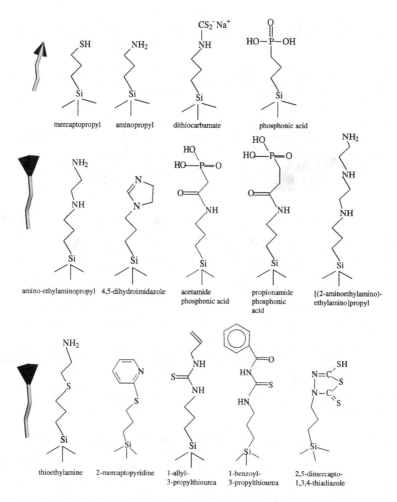

Figure 9.4. Chemical structures of the metal-chelating ligands used for surface modification of OMS silicas and co-condensation synthesis of PMO.

their location (pore walls, external surface, framework) and concentration. Depending on the type of adsorbent, its surface functionality is usually characterized by a combination of techniques such as elemental analysis, Fourier transform infrared (FTIR) spectroscopy, solid-state nuclear magnetic resonance (NMR), high-resolution thermogravimetry (TGA) and adsorption. Finally, the physicochemical properties of surface groups present and their accessibility determine the adsorbent's selectivity, adsorption capacity and its ability for regeneration and reuse.

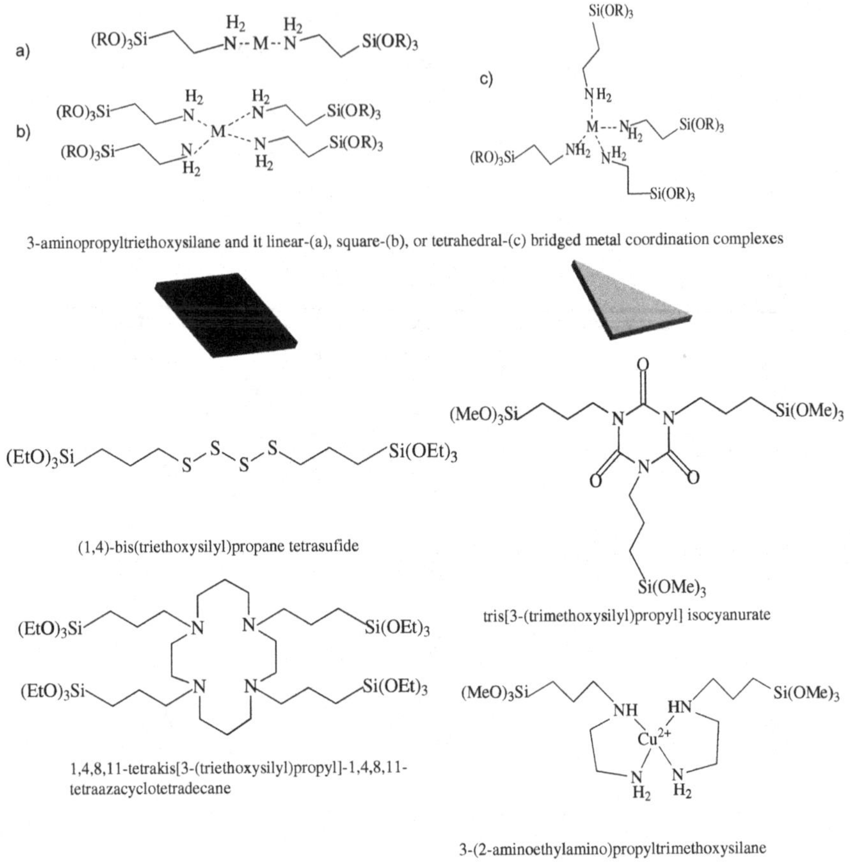

3-aminopropyltriethoxysilane and it linear-(a), square-(b), or tetrahedral-(c) bridged metal coordination complexes

(1,4)-bis(triethoxysilyl)propane tetrasufide

tris[3-(trimethoxysilyl)propyl] isocyanurate

1,4,8,11-tetrakis[3-(triethoxysilyl)propyl]-1,4,8,11-
tetraazacyclotetradecane

3-(2-aminoethylamino)propyltrimethoxysilane

Figure 9.5.   Chemical structures of the bridging groups used in the synthesis of PMO-based adsorbents.

The structural ordering of pores (hexagonal, cubic, lamellar), pore size and shape (cylindrical, cage-like, slit-like) as well as the material's morphology are usually assessed by transmission electron microscopy (TEM), scanning electron microscopy (SEM), atomic force microscopy (AFM), X-ray diffraction (XRD), small angle X-ray scattering (SAXS) and related techniques. Although all the aforementioned techniques provide extremely valuable information about the structural ordering of porous networks and surfaces, they are less convenient for assessment of the overall quality of nanoporous adsorbents. For instance, the alteration of porosity upon chemical modification, which is manifested by changes in the pore size, pore accessibility, pore size distribution as well as pore chemistry, can be effectively monitored by adsorption. This technique was shown to be an indispensable tool for the evaluation of the specific surface area, total pore volume, volume of micropores, pore size distribution, pore connectivity, surface heterogeneity and adsorption energy. Ordered mesoporous solids represent a family of nanoporous materials that exhibit characteristic types of nitrogen adsorption isotherms and hysteresis loops depending on their structural and surface properties. The relative pressure, at which the monolayer of an adsorbate is formed, is commonly used for evaluation of the specific surface area.[176] The standard procedure, the so-called BET method,[177] employs nitrogen adsorption data at 77 K at the relative pressure, ranging from ~0.05 to ~0.3, to evaluate the monolayer capacity according to the BET equation, which after multiplication by the molecular area of nitrogen is converted to the specific surface area of the adsorbent studied. The total pore volume is calculated by the conversion of the amount adsorbed at the relative pressure of about 0.99 to the volume of the liquid adsorbate. The presence and shape of hysteresis loops provides valuable information about the adsorbent's porosity including the size and shape of pores as well as pore connectivity.

Argon and nitrogen adsorption is a technique of choice to monitor the changes in the porous structure upon surface modification. For instance, a reduction in pore size upon silanization indicates a successful attachment of ligands onto the mesopore walls. A chemically bonded layer inside mesopores reduces their size, which is manifested by a shift of the capillary condensation step in the direction of lower relative pressures; also, its height is reduced, indicating a decrease in the volume of primary mesopores. Thus, the pore width can be easily tuned by selecting the proper length of the attached ligand. The steepness of the capillary condensation step indicates the pore size uniformity; a steep capillary condensation step reflects a narrow pore size distribution. This distribution can be evaluated from adsorption isotherms by using different methods, for example the KJS method,[178] which is based on the BJH algorithm[179] and was developed for hexagonally ordered mesostructures (MCM-41). This method is especially suited for the evaluation of pore size distribution in the range of small mesopores.

Figures 9.6 to 9.8 show typical adsorption isotherms and the corresponding pore size distributions for various types of mesoporous organosilicas. Nitrogen adsorption isotherms for various types of template-free (calcined) OMS are shown in Figure 9.6. The corresponding adsorption isotherms for those silicas subjected to post-synthesis modification with various organosilanes are presented in Figure 9.7. Namely, this figure shows isotherms for mesoporous organosilicas with 3-mercaptopropyl (SH), 1-allyl-3-propylthiourea (ATU) and 1-benzoyl-3-propylthiourea (BTU) ligands. For 2,5-dimercapto-1,3,4-thiadiazole (DMT)-modified material,

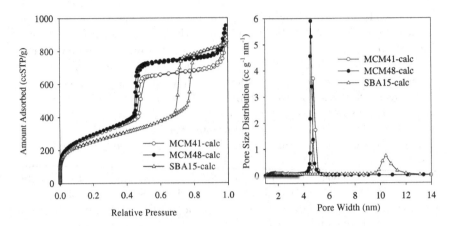

Figure 9.6.   Nitrogen adsorption isotherms and the corresponding pore size distributions for calcined ordered mesoporous silicas of different structures before surface functionalization. Adsorption data from Refs. 86 and 88.

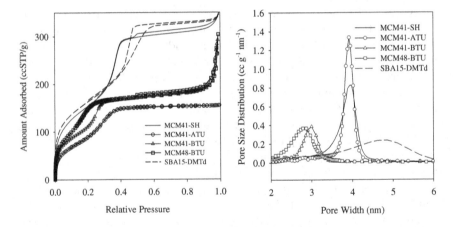

Figure 9.7.   Nitrogen adsorption isotherms and the corresponding pore size distributions for ordered mesoporous silicas after surface functionalization with multifunctional ligands: 3-mercaptopropyl (SH), 1-allyl-3-propylthiourea (ATU), 1-benzoyl-3-propylthiourea (BTU) and 2,5-dimercapto-1,3,4-thiadiazole (DMT). Adsorption data from Refs. 85, 86, 88 and 89.

template displacement synthesis was used as illustrated in Figure 9.2. The second set of adsorption isotherms corresponds to the materials obtained via co-condensation synthesis (Figure 9.8), which feature either one type of framework group, tris[3-(trimethoxysilyl)propyl] isocyanurate

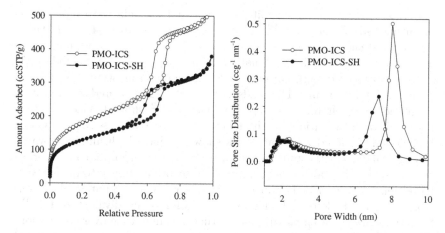

Figure 9.8. Nitrogen adsorption isotherms and the corresponding pore size distributions for periodic mesoporous organosilicas with multifunctional bridging and surface groups: 3-mercaptopropyl (SH) and tris[3-(trimethoxysilyl)propyl] isocyanurate (ICS). Adsorption data from Refs. 165 and 169.

(ICS), or two types of groups, 3-mercaptopropyl surface ligands and tris[3-(trimethoxysilyl)propyl] isocyanurate framework groups. As can be seen from Figures 9.6 and 9.7, post-synthesis surface modification of OMS can be easily monitored by the shift of the capillary condensation step in the direction of lower relative pressures, which indicates a successful immobilization of surface ligands inside mesopores. In contrast to post-synthesis modification, for the materials obtained by co-condensation synthesis the pore size is mainly determined by the type of template, and its dependence on the size and concentration of incorporated groups is much less pronounced. Therefore, organosilicas obtained by co-condensation synthesis feature high adsorption capacity and large pore size despite the high content of bridging groups; e.g. see adsorption isotherms in Figure 9.7 for modified silicas. However, further modification of this sample by the introduction of another segment to the chloropropyl ligand reduces the pore size and pore volume of the sample. A more detailed information about adsorption characterization of mesoporous materials can be found elsewhere.[86−89,118,165,169−171]

9.6 Metal-Containing Nanostructures as Heavy Metal Ion Adsorbents

While silica-based materials such as mesoporous organosilicas are regarded as the most promising adsorbents for heavy metal ions, other nanoporous materials such as surface-modified titania and zirconia were reported as promising adsorbents too.[180] Another example is MCM-41 with Fe^{3+}-coordinated chelating ligands, which possesses a high adsorption capacity for toxic oxyanions such as arsenate, chromate, selenate and molybdate.[181]

Nanostructured metals or metal-containing materials were also reported as potential heavy metal ion adsorbents. Poly(1-vinylimidazole)-grafted magnetic (maghemite, γ-Fe_2O_3) nanoparticles[182] showed selective binding of divalent metal ions in the following order: $Cu^{2+} \gg Ni^{2+} > Co^{2+}$. Hexagonally ordered zirconia[183] was shown to have high affinity towards Co(II), Ni(II), Cu(II) and Zn(II) ions. Titania-incorporated mesoporous composites[184,185] featured a remarkable adsorption affinity towards heavy metal ions such as Pb(II) and mercury, with particular potential to control a vapor of elemental mercury[186,187] and other divalent metals.[188] Some advantages of employing mesostructured materials over conventionally used La(III)-impregnated alumina and silica gels for arsenate removal were shown for SBA-15 material with various amounts of lanthanide oxide incorporated, for which an adsorption capacity about 10–14 times larger was obtained.[189] Also, porous mixed oxides with tailored properties derived from substituted layered double hydroxides were tested for adsorption of this pollutant.[190]

9.7 Mesostructures for Detection of Heavy Metal Ions

Since co-condensation synthesis of ordered mesoporous organosilicas (OMO) permits the introduction of a high amount of functional groups without blockage of the porous structure and significant reduction of the surface area and porosity in the resulting materials, their use in sensing devices allow one to lower detection limits and enhance response to the targeted substances. Therefore, there is great interest in the development of OMS-based materials for detection of heavy metal ions. For instance, high-surface-area and large-pore chemically modified SBA-15 materials with 5,10,15,20-tetraphenylporphinetetrasulfonic acid, anchored by N-trimethoxysilylpropyl-N,N,N-trimethylammonium chloride,[191] were capable of detecting Hg^{2+} at concentrations as low as 2.5×10^{-8} M. Also, a high detection limit (3.3×10^{-7} mol L^{-1} in water) and great selectivity for mercury ions was achieved in the presence of interfering cations such as Na^+, K^+, Ca^{2+}, Cu^{2+}, Cd^{2+} and Pb^{2+} on chemically modified SBA-15 material, in which calixarene bearing two dansyl fluorophores was an optically sensing moiety.[192] Mesoporous materials were also found to be very efficient in the determination of heavy metal ions via electrochemical methods. For example, a MCM-41-modified graphite paste electrode gave high selectivity for hydroxylated mercury(II) species.[193] A good voltammetric response for Hg(II), Pb(II), Ag(I) and Cu(II) cations was observed for mesoporous silica possessing thick walls.[194] The thiol monolayers grafted in silica mesopores, known for their exceptional affinity and selectivity for heavy metal ions, were used to prepare the electrode-sensing layer for detection of lead(II) in aqueous solutions. These electrodes exhibited fast response in the parts per billion (ppb) range of concentration when square-wave adsorptive stripping voltammetry technique was employed.[195] However, simultaneous detection of lead(II) and mercury(II) in aqueous solutions was possible with detection limits of 0.5 ppb for Pb^{2+} and 3 ppb for Hg^{2+} when a self-assembled monolayer of thiol groups on mesoporous silica combined with carbon paste was used as the electrode.[196] Glycinylurea and carbamoylphosphonic self-assembled monolayers on mesoporous silicas anchored to the carbon paste electrodes were shown to have detection limits of 1 ppb and a fast detection time of 2–3 minutes for lead, mercury, copper and cadmium ions using adsorptive stripping voltammetry.[197,198] The aforementioned studies distinctly show a great benefit of using functionalized ordered mesoporous materials for the development of sensing devices to detect heavy metal ions in water streams. Their high surface area, non-swelling and hydrothermally stable matrix, accessible mesopores with highly specific ligands ensure high selectivity and capacity towards heavy metal ions; moreover the tunable width of mesopores prevents adsorption of bulk organics from waste waters.

9.8 Mesoporous Materials for Capturing Toxic Pollutants Other Than Heavy Metals

The preceding sections described the great potential of properly designed ordered mesoporous organosilicas for adsorption and detection of heavy metal ions. It is noteworthy that these materials are also attractive for adsorption of other pollutants such as dioxins,[198] sulfides[200] and various volatile organic compounds (VOC) including carbon tetrachloride, *n*-hexane, benzene, toluene, *o*-xylene,[201,202] 4-*n*-heptylaniline, 4-nonylphenol, endocrine disrupters,[203,204] alkylphenols, alkylanilines,[205] chlorinated hydrocarbons,[206] carcinogenic nitrosamines[207,208] and other VOC.[209]

9.9 Conclusions

This review shows that both surface modification of ordered silica mesostructures as well as co-condensation synthesis of organosilanes in the presence of surfactant and polymeric templates can be used to design highly selective and high-capacity adsorbents for heavy metal ions. In particular, the latter method is feasible for the design and synthesis of adsorbents with high loading of surface and bridging groups without sacrificing their surface area, pore size and pore volume. Also, co-condensation synthesis in the presence of properly selected templates is the method of choice to generate mesostructures with highly accessible porosity.

Novel sources of non-toxic and biodegradable precursors and low-cost synthesis would play a key role in the development of methods for large-scale production of future materials, including those for environmental applications. Since nanocomposite and hybrid materials are mainly built up via the bottom-up approach, there is a great deal of attention to ensure that the resulting nanomaterials do not present a health risk and they are environmentally friendly, i.e. their synthesis and use fulfill requirements of green chemistry principles and safer technological processes. Since a major feature of nanomaterials is the possibility to introduce a variety of unique properties into confined nanospace, this research area is expected to grow rapidly in the coming years, especially in relation to advanced applications of nanomaterials.

Acknowledgment

We would like to acknowledge the support of the NSF under grant CTS-0086512.

References

1. J. S. Beck, J. C. Vartuli, W. J. Roth, M. E. Leonowicz, C. T. Kresge, K. D. Schmitt, C. T.-W. Chu, D. H. Olson, E. W. Sheppard, S. B. McCullen, J. B. Higgins and J. L. Schlenker, *J. Am. Chem. Soc.* **114**, 10834–10843 (1992).
2. B. Hultberg, A. Andersson and A. Isaksson, *Toxicology* **126**, 203–212 (1998).
3. E. S. Raper, *Coord. Chem. Rev.* **61**, 115–184 (1985).
4. R. V. Davies, J. Kennedy, E. S. Lane and J. L. Willans, *J. Appl. Chem.* **9**, 368–371 (1959).
5. C. Kantipuly, *Talanta* **37**, 491–517 (1990).
6. S. D. Alexandratos and D. L. Wilson, *Macromolecules* **19**, 280–287 (1986).
7. A. R. Tuker and A. Tunceli, *Fresen J. Anal. Chem.* **345**, 755–758 (1993).
8. A. Deratani and B. Sebille, *Anal. Chem.* **53**, 1742–1746 (1981).
9. R. J. Philips and J. S. Fritz, *Anal. Chem.* **50**, 1504–1508 (1978).
10. M. Griesbach and K. H. Lieser, *Angew. Makromol. Chem.* **90**, 143–153 (1980).

11. R. Liu, H. Tang and B. Zhang, *Chemosphere* **38**, 3169–3179 (1999).
12. R. Celis, M. Carmen Hermosin and J. Cornejo, *Environ. Sci. Technol.* **34**, 4593–4599 (2000).
13. S. Babel and T. A. Kurniawan, *J. Hazard. Mater. B* **97**, 219–243 (2003).
14. D. Mohan, V. K. Gupta, S. K. Srivastava and S. Chander, *Colloid Surface. A* **177**, 169–181 (2001).
15. M. J. Zamzow, B. R. Eichbaum, K. R. Sandgren and D. E. Shanks, *Separ. Sci. Technol.* **25**, 1555–1569 (1990).
16. M. M. Matlock, K. R. Henke and D. A. Atwood, *J. Hazard. Mater.* **92**, 129–142 (2002).
17. T. I. Tikhomirova, V. I. Fadeeva, G. V. Kudryavtsev, P. N. Nesterenko, V. M. Ivanov, A. T. Savitchev and N. S. Smirnova, *Talanta* **38**, 267–274 (1991).
18. H. Watanabe, K. Goto, S. Taguchi, J. W. McLaren, S. S. Berman and D. S. Russell, *Anal. Chem.* **53**, 738–739 (1981).
19. A. G. S. Prado, L. N. H. Arakaki and C. Airoldi, *Dalton Trans.* 2206–2209 (2001).
20. G.-Z. Fang, J. Tan and X.-P. Yan, *Anal. Chem.* **77**, 1734–1739 (2005).
21. E. F. Vansant, P. Van Der Voort and K. C. Vrancken (eds.), *Characterization and Chemical Modification of the Silica Surface*, Studies in Surface Science and Catalysis, Vol. 93 (Elsevier, Amsterdam, 1995), p. 572.
22. G. V. Lisichkin, G. V. Kudryavtsev, A. A. Serdan, S. M. Staroverov and A. Ya. Yuffa, *Modified Silicas in Sorption, Catalysis, and Chromatography* (Khimiya, Moscow, 1986), p. 248.
23. P. K. Jal, S. Patel and B. K. Mishra, *Talanta* **62**, 1005–1028 (2004).
24. J. Rouquerol, D. Avnir, C. W. Fairbridge, D. H. Everett, J. H. Haynes, N. Pernicone, J. D. F. Ramsay, K. S. W. Sing and K. K. Unger, *Pure Appl. Chem.* **66**, 1739–1758 (1994).
25. C. T. Kresge, M. E. Leonowicz, W. J. Roth, J. C. Vartuli and J. Beck, *Nature* **359**, 710–712 (1992).
26. J. S. Beck, J. C. Vartuli, W. J. Roth, M. E. Leonowicz, C. T. Kresge, K. D. Schmitt, C. T. W. Chu, D. H. Olson, E. W. Sheppard *et al.*, *J. Am. Chem. Soc.* **114**, 10834–10843 (1992).
27. C. Chen Yan, H. X. Li and M. E. Davis, *Microporous Mater.* **2**, 17–26 (1993).
28. Q. Huo, R. Leon, P. M. Petroff and G. D. Stucky, *Science* **268**, 1324–1327 (1995).
29. Q. Huo, D. I. Margolese, U. Ciesla, D. G. Demuth, P. Feng, T. E. Gier, P. Sieger, A. Firouzi, B. F. Chmelka, F. Schuth and G. D. Stucky, *Chem. Mater.* **6**, 1176–1191 (1994).
30. Q. Huo, D. I. Margolese, U. Ciesla, P. Feng, T. E. Gier, P. Sieger, R. Leon, P. M. Petroff, F. Schueth and G. D. Stucky, *Nature* **368**, 317–321 (1994).
31. P. T. Tanev and T. J. Pinnavaia, *Science* **267**, 865–867 (1995).
32. D. Zhao, J. Feng, Q. Huo, N. Melosh, G. H. Frederickson, B. F. Chmelka and G. D. Stucky, *Science* **79**, 548–552 (1998).

33. D. Zhao, Q. Huo, J. Feng, B. F. Chmelka and G. D. Stucky, *J. Am. Chem. Soc.* **120**, 6024–6036 (1998).

34. S. Inagaki, Y. Fukushima and K. Kuroda, *Chem. Commun.* 680–682 (1993).

35. A. Monnier, F. Schuth, Q. Huo, D. Kumar, D. Margolese, R. S. Maxwell, G. D. Stucky, M. Krishnamurty, P. Petroff, A. Firouzi, M. Janicke and B. F. Chmelka, *Science* **261**, 1299–303 (1993).

36. K. Schumacher, M. Grun and K. K. Unger, *Micropor. Mesopor. Mater.* **27**, 201–206 (1999).

37. C. Yu, Y. Yu and D. Zhao, *Chem. Commun.* 575–576 (2000).

38. Q. Huo, D. I. Margolese and D. Stucky Galen, *Chem. Mater.* **8**, 1147–60 (1996).

39. J. S. Beck and J. C. Vartuli, *Cur. Opin. Solid State Mater. Sci.* **1**, 76–87 (1996).

40. C. G. Goltner and M. Antonietti, *Adv. Mater.* **9**, 431–436 (1997).

41. C. Yu, Y. Yu, L. Miao and D. Zhao, *Micropor. Mesopor. Mater.* **44/45**, 65–72 (2001).

42. D. Y. Zhao, P. D. Dong, N. Melosh, F. Y. Lin, B. F. Chmelka and G. Stucky, *Adv. Mater.* **10**, 1380–1385 (1998).

43. P. Yang, D. Zhao, B. F. Chmelka and G. D. Stucky, *Chem. Mater.* **10**, 2033–2036 (1998).

44. C. Yu, B. Tian, J. Fan, G. D. Stucky and D. Zhao, *J. Am. Chem. Soc.* **124**, 4556–4557 (2002).

45. J. R. Matos, L. P. Mercuri, M. Kruk and M. Jaroniec, *Langmuir* **18**, 884–890 (2002).

46. P. Van Der Voort, M. Benjelloun and E. F. Vansant, *J. Phys. Chem. B* **106**, 9027–9032 (2002).

47. C. E. Tattershall, S. J. Aslam and P. M. Budd, *J. Mater. Chem.* **12**, 2286–2291 (2002).

48. J. R. Matos, M. Kruk, L. P. Mercuri, M. Jaroniec, L. Zhao, T. Kamiyama, O. Terasaki, T. J. Pinnavaia and Y. Liu, *J. Am. Chem. Soc.* **125**, 821–829 (2003).

49. G. S. Attard, J. C. Glyde and C. G. Goltner, *Nature* **378**, 366–368 (1995).

50. T. Linssen, K. Cassiers, P. Cool and E. F. Vansant, *Adv. Colloid Interface Sci.* **103**, 121–147 (2003).

51. A. Sayari, *Chem. Mater.* **8**, 1840–1852 (1996).

52. A. Vinu, K. Z. Hossain, and K. Ariga, *J. Nanosci. Nanotechnol.* **5**, 347–371 (2005).

53. A. Walcarius, M. Etienne and J. Bessiere, *Chem. Mater.* **14**, 2757–2766 (2002).

54. A. Walcarius, M. Etienne and B. Lebeau, *Chem. Mater.* **15**, 2161–2173 (2003).

55. B. P. Feuston and J. B. Higgins, *J. Phys. Chem.* **98**, 4459–4462 (1994).

56. S. Polarz and B. Smarsly, *J. Nanosci. Nanotechnol.* **2**, 581–612 (2002).

57. U. Ciesla and F. Schuth, *Micropor. Mesopor. Mater.* **27**, 131–149 (1999).

58. J. Y. Ying, C. P. Mehnert and M. S. Wong, *Angew. Chem. Int. Ed.* **38**, 56–77 (1999).

59. A. Sayari and S. Hamoudi, *Chem. Mater.* **13**, 3151–3168 (2001).
60. R. G. Pearson, *J. Am. Chem. Soc.* **85**, 3533–3539 (1963).
61. X. Feng, G. E. Fryxell, L.-Q. Wang, A. Y. Kim, J. Liu and K. M. Kemner, *Science* **276**, 923–926 (1997).
62. J. Liu, X. D. Feng, G. E. Fryxell, L. Q. Wang, A. Y. Kim and M. L. Gong, *Adv. Mater.* **10**, 161–165 (1998).
63. X. Chen, X. Feng, J. Liu, G. E. Fryxell and M. Gong, *Separ. Sci. Technol.* **34**, 1121–1132 (1999).
64. K. Moller and T. Bein, *Chem. Mater.* **10**, 2950–2963 (1998).
65. A. M. Liu, K. Hidajat, S. Kawi and D. Y. Zhao, *Chem. Commun.* 1145–1146 (2000).
66. Y. Kim, B. Lee and J. Yi, *Separ. Sci. Technol.* **39**, 1427–1442 (2004).
67. L. Mercier and T. J. Pinnavaia, *Adv. Mater.* **9**, 500–503 (1997).
68. L. Mercier and T. J. Pinnavaia, *Environ. Sci. Technol.* **32**, 2749–2754 (1998).
69. K. M. Kemner, X. Feng, J. Liu, G. E. Fryxell, L.-Q. Wang, A. Y. Kim, M. Gong and S. Mattigod, *J. Synchrotron Radiat.* **6**, 633–635 (1999).
70. J. Liu, X. Feng, G. E. Fryxell, L. Q. Wang, A. Y. Kim and M. Gong, *Chem. Eng. Technol.* **21**, 97–100 (1998).
71. Z. Luan, J. A. Fournier, J. B. Wooten, D. E. Miser and M. J. Chang, *Stud. Surf. Sci. Catal.* **156**, 897–906 (2005).
72. S. V. Mattigod, X. Feng, G. E. Fryxell, J. Liu and M. Gong, *Separ. Sci. Technol.* **34**, 2329–2345 (1999).
73. C. Lesaint, F. Frébault, C. Delacôte, B. Lebeau, C. Marichal, A. Walcarius and J. Patarin, *Stud. Surf. Sci. Catal.* **156**, 925–932 (2005).
74. G. E. Fryxell, J. Liu, T. A. Hauser, Z. Nie, K. F. Ferris, S. Mattigod, M. Gong and R. T. Hallen, *Chem. Mater.* **11**, 2148–2154 (1999).
75. G. E. Fryxell, J. Liu and S. Mattigod, *Mater. Technol.* **14**, 188–191 (1999).
76. G. E. Fryxell, J. Liu, S. V. Mattigod, L. Q. Wang, M. Gong, T. A. Hauser, Y. Lin, K. F. Ferris and X. Feng, in *Environmental Issues and Waste Management Technologies in the Ceramic and Nuclear Industries*, Ceramic Transaction, Vol. 107, ed. G. T. Chandler and X. Feng (American Ceramic Society, Westerville, OH, 2000), pp. 29–37.
77. G. Fryxell, Y. S. Fiscum, Y. C. Birnbaum and H. Wu, *Environ. Sci. Technol.* **39**, 1324–1331 (2005).
78. T. Kang, Y. Park and J. Yi, *Ind. Eng. Chem. Res.* **43**, 1478–1484 (2004).
79. A. Abughusa, L. Amaratunga and L. Mercier, *Stud. Surf. Sci. Catal.* **156**, 957–962 (2005).
80. K. A. Venkatesan, T. G. Srinivasan and P. R. Vasudeva Rao, *J. Radioanal. Nucl. Chem.* **256**, 213–218 (2003).
81. T. Kang, Y. Park, K. Choi, J. S. Lee and J. Yi, *J. Mater. Chem.* **14**, 1043–1049 (2004).
82. W. Yantasee, Y. Lin, G. E. Fryxell, B. J. Busche and J. C. Birnbaum, *Separ. Sci. Technol.* **38**, 3809–3825 (2003).
83. A. Sayari, *Angew. Chem. Int. Ed.* **39**, 2920–2922 (2000).
84. A. Sayari, S. Hamoudi and Y. Yang, *Chem. Mater.* **17**, 212–216 (2005).
85. V. Antochshuk and M. Jaroniec, *Chem. Commun.* 258–259 (2002).

86. V. Antochshuk, O. Olkhovyk, M. Jaroniec, I.-S. Park and R. Ryoo, *Langmuir* **19**, 3031–3034 (2003).

87. O. Olkhovyk, V. Antochshuk and M. Jaroniec, *Analyst* **130**, 104–108 (2005).

88. O. Olkhovyk, V. Antochshuk and M. Jaroniec, *Colloid. Surface. A* **236**, 69–72 (2004).

89. O. Olkhovyk and M. Jaroniec, *Adsorption* **11**, 205–214 (2005).

90. A. G. S. Prado, L. N. H. Arakaki and C. Airodi, *Green Chem.* **4**, 42–46 (2002).

91. R. I. Nooney, M. Kalyanaraman, G. Kennedy and E. J. Maginn, *Langmuir* **17**, 528–533 (2001).

92. D. Pérez-Quintanilla, I. del Hierro, M. Fajardo and I. Sierra, *Micropor. Mesopor. Mater.* **89**, 58–68 (2006).

93. C. P. Jaroniec, M. Kruk, M. Jaroniec and A. Sayari, *J. Phys. Chem. B* **102**, 5503–5510 (1998).

94. M. H. Lim, C. F. Blanford and A. Steinm, *Chem. Mater.* **10**, 467–470 (1998).

95. W. M. Van Rhijn, D. E. De Vos, B. F. Sels, W. D. Bossaert and P. A. Jacobs, *Chem. Commun.* **1998**, 317–318.

96. Y. Mori and T. J. Pinnavaia, *Chem. Mater.* **13**, 2173–2178 (2001).

97. J. Brown, R. Richer and L. Mercier, *Micropor. Mesopor. Mater.* **37**, 41–48 (2000).

98. A. Bibby and L. Mercier, *Chem. Mater.* **14**, 1591–1597 (2002).

99. L. Bois, A. Bonhomme, A. Ribes, B. Pais, G. Raffin and F. Tessier, *Colloid. Surface. A* **221**, 221–230 (2003).

100. H.-J. Im, C. E. Barnes, S. Dai and Z. Xue, *Micropor. Mesopor. Mater.* **70**, 57–62 (2004).

101. S. Huh, J. W. Wiench, J.-C. Yoo, M. Pruski and V. S.-Y. Lin, *Chem. Mater.* **15**, 4247–4256 (2003).

102. K. Kosuge, T. Murakami, N. Kikukawa and M. Takemori, *Chem. Mater.* **15**, 3184–3189 (2003).

103. D. Margolese, J. A. Melero, S. C. Christiansen, B. F. Chmelka and G. D. Stucky, *Chem. Mater.* **12**, 2448–2459 (2000).

104. H.-P. Lin, L.-Y. Yang, C.-Y. Mou, S.-B. Liu and H.-K. Lee, *New J. Chem.* **24**, 253–255 (2000).

105. Y.-H. Liu, H.-P. Lin and C.-Y. Mou, *Langmuir* **20**, 3231–3239 (2004).

106. S. Inagaki, S. Guan, Y. Fukushima, T. Ohsuna and O. Terasaki, *J. Am. Chem. Soc.* **121**, 9611–9614 (1999).

107. S. Guan, S. Inagaki, T. Ohsuna and O. Terasaki, *J. Am. Chem. Soc.* **122**, 5660–5661 (2000).

108. T. Asefa, M. J. MacLachlan, N. Coombs and G. A. Ozin, *Nature* **402**, 867–871 (1999).

109. B. J. Melde, B. T. Holland, C. F. Blanford and A. Stein, *Chem. Mater.* **11**, 3302–3308 (1999).

110. S. Inagaki, S. Guan, T. Ohsuna and O. Terasaki, *Nature* **416**, 304–307 (2002).

111. S. Dai, M. C. Burleigh, Y. H. Ju, H. J. Gao, J. S. Lin, S. J. Pennycook, C. E. Barnes and Z. L. Xue, *J. Am. Chem. Soc.* **122**, 992–993 (2000).

112. J. Brown, L. Mercier and T. J. Pinnavaia, *Chem. Commun.* 69–70 (1999).

113. D. J. Macquarrie, *Chem. Commun.* 1961–1962 (1996).

114. Y.-K. Lu and X.-P. Yan, *Anal. Chem.* **76**, 453–457 (2004).

115. R. J. P. Corriu, A. Mehdi, C. Reye and C. Thieuleux, *Chem. Mater.* **16**, 159–166 (2004).

116. P. Sutra and D. Brunel, *Chem. Commun.* 2485–2486 (1996).

117. K. H. Nam and L. L. Tavlarides, *Chem. Mater.* **17**, 1597–1604 (2005).

118. L. C. Cides da Silva, G. Abate, N. Andrea, M. C. A. Fantini, J. C. Masini, L. P. Mercuri, O. Olkhovyk, M. Jaroniec and J. R. Matos, *Stud. Surf. Sci. Catal.* **155**, 941–950 (2005).

119. B. L. Newalkar, S. Komarneni and H. Katsuki, *Chem. Commun.* 2389–2390 (2005).

120. B. L. Newalkar, J. Olanrewaju and S. Komarneni, *Chem. Mater.* **13**, 552–557 (2001).

121. S. E. Park, J. S. Chang, Y. K. Hwang, D. S. Kim, S. H. Jhung and J. S. Hwang, *Catal. Surv. Asia* **8**, 91–110 (2004).

122. H. Katsuki and S. Komarneni, *J. Am. Ceram. Soc.* **84**, 2313–2317 (2001).

123. B. Lee, Y. Kim, H. Lee and J. Yi, *Micropor. Mesopor. Mater.* **50**, 77–90 (2001).

124. A. Walcarius and C. Delacote, *Chem. Mater.* **15**, 4181–4192 (2003).

125. S. Mann, S. L. Burkett, S. A. Davis, C. E. Fowler, N. H. Mendelson, S. D. Sims, D. Walsh and T. W. Nicola, *Chem. Mater.* **9**, 2300–2310 (1997).

126. M. H. Valkenberg and W. F. Holderich, *Catal. Rev. Sci. Eng.* **44**, 321–374 (2002).

127. A. Stein, B. J. Melde and R. C. Schroden, *Adv. Mater.* **12**, 1403–1419 (2000).

128. H. Yoshitake, *New J. Chem.* **29**, 1107–1117 (2005).

129. V. Antochshuk and M. Jaroniec, *Chem. Commun.* 2373–2374 (1999).

130. S. Dai, Y. Shin, Y. Ju, M. C. Burleigh, J.-S. Lin, C. E. Barnes and Z. Xue, *Adv. Mater.* **11**, 1226–1230 (1999).

131. V. Antochshuk, A. S. Araujo and M. Jaroniec, *J. Phys. Chem. B* **104**, 9713–9719 (2000).

132. V. Antochshuk, M. Jaroniec, S. H. Joo and R. Ryoo, *Stud. Surf. Sci. Catal.* **141**, 607–614 (2002).

133. P. M. de Padilha, L. A. De Melo Gomes, C. C. F. Padilha, J. C. Moreira and N. L. Dias Filho, *Anal. Lett.* **32**, 1807–1820 (1999).

134. P. T. Tanev, M. Chibwe and T. J. Pinnavaia, *Nature* **368**, 321–323 (1994).

135. K. M. Reddy, I. Moudrakovski and A. Sayari, *Chem. Commun.* 1059–1060 (1994).

136. D. Antonelli and J. Y. Ying, *Angew. Chem. Int. Ed.* **35**, 426–430 (1996).

137. C. Yoshina-Ishii, T. Asefa, N. Coombs, M. J. MacLachlan and G. Ozin, *Chem. Commun.* 2539–2540 (1999).

138. S. Yang, P. A. Mirau, C.-S. Pai, O. Nalamasu, E. Reichmanis, J. C. Pai, Y. S. Obeng, J. Seputro, E. K. Lin, H.-J. Lee, J. Sun and D. W. Gidley, *Chem. Mater.* **14**, 369–374 (2002).

139. M. Alvaro, B. Ferrer, V. Fornes and H. Garcia, *Chem. Commun.* 2546–2547 (2001).
140. C. Baleizao, B. Gigante, D. Das, M. Alvaro, H. Garcia and A. Corma, *Chem. Commun.* 1860–1861 (2003).
141. J.-K. Lee, S.-H. Lee, M. Kim, H. Kim and W.-Y. Lee, *Chem. Commun.* 1602–1603 (2003).
142. G. Wirnsberger and G. D. Stucky, *Chem. Phys. Chem.* **1**, 90–92 (2000).
143. G. Dubois, R. J. P. Corriu, C. Reye, S. Brandes, F. Denat and R. Guilard, *Chem. Commun.* 2283–2284 (1999).
144. G. Dubois, C. Reye, R. J. P. Corriu, S. Brandes, F. Denat and R. Guilard, *Angew. Chem. Int. Ed.* **40**, 1087–1090 (2001).
145. C. Chuit, R. J. P. Corriu, G. Dubois and C. Reye, *Chem. Commun.* 723–724 (1999).
146. Z. Zhang and S. Dai, *J. Am. Chem. Soc.* **123**, 9204–9205 (2001).
147. J. Alauzun, A. Mehdi, C. Reye and R. J. P. Corriu, *J. Am. Chem. Soc.* **127**, 11204–11205 (2005).
148. H. Yoshitake, T. Yokoi and T. Tatsumi, *Chem. Mater.* **14**, 4603–4610 (2002).
149. C. Liu and J. Economy, *PMSE Preprints* **91**, 1037–1038 (2004).
150. S. R. Hall, C. E. Fowler, B. Lebeau and S. Mann, *Chem. Commun.* 201–202 (1999).
151. T. Asefa, M. Kruk, M. J. MacLachlan, N. Coombs, H. Grondey, M. Jaroniec and G. Ozin, *J. Am. Chem. Soc.* **123**, 8520–8530 (2001).
152. M. C. Burleigh, M. A. Markowitz, M. S. Spector and B. P. Gaber, *J. Phys. Chem. B.* **105**, 9935–9942 (2001).
153. Q. Yang, M. P. Kapoor and S. Inagaki, *J. Am. Chem. Soc.* **124**, 9694–9695 (2002).
154. W. J. Hunks and G. A. Ozin, *J. Mater. Chem.* **15**, 764–771 (2005).
155. S. Huh, J. W. Wiench, B. G. Trewyn, S. Song, M. Pruski and V. S.-Y. Lin, *Chem. Commun.* 2364–2365 (2003).
156. W. J. Hunks and G. A. Ozin, *Adv. Funct. Mater.* 259–266 (2005).
157. S. Huh, H.-T. Chen, J. W. Wiench, M. Pruski and V. S.-Y. Lin, *J. Am. Chem. Soc.* **126**, 1010–1011 (2004).
158. M. A. Wahab, I. Imae, Y. Kawakami and C.-S. Ha, *Chem. Mater.* **17**, 2165–2174 (2005).
159. R. P. Hodgkins, A. E. Garcia-Bennett and P. A. Wright, *Micropor. Mesopor. Mater.* **79**, 241 (2005).
160. M. Kuroki, T. Asefa, W. Whitnal, M. Kruk, C. Yoshina-Ishii, M. Jaroniec and G. A. Ozin, *J. Am. Chem. Soc.* **124**, 13886–13895 (2002).
161. K. Landskron, B. D. Hatton, D. D. Perovic and G. A. Ozin, *Science* **302**, 266–269 (2003).
162. R. J. P. Corriu, A. Mehdi, C. Reye and C. Thieuleux, *Chem. Commun.* 1382–1383 (2002).
163. R. J. P. Corriu, A. Mehdi, C. Reye and C. Thieuleux, *New J. Chem.* **27**, 905–908 (2003).
164. R. Corriu, A. Mehdi and C. Reye, *J. Organomet. Chem.* **689**, 4437–4450 (2004).

165. O. Olkhovyk and M. Jaroniec, *J. Am. Chem. Soc.* **127**, 60–61 (2005).
166. L. Zhang, W. Zhang, J. Shi, Z. Hua, Y. Li and J. Yan, *Chem. Commun.* 210–211 (2003).
167. Yu. A. Zolotov, O. M. Petrukhin, G. I. Malofeeva, E. V. Marcheva, O. A. Shiryaeva, V. A. Shestakov, V. G. Miskar'yants, V. I. Nefedov, Yu. I. Murinov and Yu. E. Nikitin, *Anal. Chim. Acta* **148**, 135–157 (1983).
168. W.-H. Zhang, X.-B. Lu, J.-H. Xiu, Z.-L. Hua, L.-X. Zhang, M. Robertson, J.-L. Shi, D.-S. Yan and J. D. Holmes, *Adv. Funct. Mater.* **14**, 544–552 (2004).
169. O. Olkhovyk, S. Pikus and M. Jaroniec, *J. Mater. Chem.* **15**, 1517–1519 (2005).
170. O. Olkhovyk and M. Jaroniec, *Stud. Surf. Sci. Catal.* **155**, 197–204 (2005).
171. O. Olkhovyk and M. Jaroniec, *SPIE Proceedings* **5929**, 176–183 (2005).
172. T. Asefa, C. Yoshina-Ishii, M. J. MacLachlan and G. A. Ozin, *J. Mater. Chem.* **10**, 1751–1755 (2000).
173. M. J. MacLachlan, T. Asefa and G. A. Ozin, *Chem. Eur. J.* **6**, 2507–2511 (2000).
174. W. Hunks and G. A. Ozin, *J. Mater. Chem.* **15**, 3716–3724 (2005).
175. B. Hatton, Landskron, K. W. Whitnall, D. Perovic and G. A. Ozin, *Acc. Chem. Res.* **38**, 305–312 (2005).
176. S. J. Gregg and K. S. W. Sing, *Adsorption, Surface Area and Porosity* (Academic Press, London, 1982).
177. S. Brunauer, P. H. Emmett and E. Teller, *J. Am. Chem. Soc.* **60**, 309–319 (1938).
178. M. Kruk, M. Jaroniec and A. Sayari, *Langmuir* **13**, 6267–6273 (1997).
179. E. P. Barrett, L. G. Joyner and P. H. Halenda, *J. Am. Chem. Soc.* **73**, 373–380 (1951).
180. R. C. Schroden, M. Al-Daous, S. Sokolov, B. J. Melde, J. C. Lytle, A. Stein, M. C. Carbajo, Fernandez, J. Torralvo and E. E. Rodriguez, *J. Mater. Chem.* **12**, 3261–3267 (2002).
181. T. Yokoi, T. Tatsumi and H. Yoshitake, *J. Colloid. Interface Sci.* **274**, 451–457 (2004).
182. M. Takafuji, S. Ide, H. Ihara and Z. Xu, *Chem. Mater.* **16**, 1977–1983 (2004).
183. R. F. de Farias, A. A. S. do Nascimento and C. W. B. Bezerra, *J. Colloid. Interface Sci.* **277**, 19–22 (2004).
184. Y.-M. Xu, R.-S. Wang and F. Wu, *J. Colloid. Interface Sci.* **209**, 380–385 (1999).
185. G. X. S. Zhao, J. L. Lee and P. A. Chia, *Langmuir* **19**, 1977–1979 (2003).
186. E. Pitoniak, C.-Y. Wu, D. Londeree, D. Mazyck, J.-C. Bonzongo, K. Powers and W. Sigmund, *J. Nanopart. Res.* **5**, 281–292 (2003).
187. E. Pitoniak, C.-Y. Wu, D. W. Mazyck, K. W. Powers and W. Sigmund, *Environ. Sci. Technol.* **39**, 1269–1274 (2005).
188. L. Lv, F. Su and G. X. S. Zhao, *Stud. Surf. Sci. Catal.* **156**, 933–940 (2005).
189. M. Jang, J. K. Park and E. W. Shin, *Micropor. Mesopor. Mater.* **75**, 159–168 (2004).

190. G. Carja, R. Nakamura and H. Niiyama, *Micropor. Mesopor. Mater.* **83**, 94–100 (2005).

191. T. Balaji, M. Sasidharan and H. Matsunaga, *Analyst* **130**, 1162–1167 (2005).

192. R. Metivier, I. Leray, B. Lebeau and B. Valeur, *J. Mater. Chem.* **15**, 2965–2973 (2005).

193. A. Walcarius and J. Bessiere, *Chem. Mater.* **11**, 3009–3011 (1999).

194. A. M. Bond, W. Miao, T. D. Smith and J. Jamis, *Anal. Chim. Acta* **396**, 203–213 (1999).

195. W. Yantasee, Y. Lin, X. Li, G. E. Fryxell, T. S. Zemanian and V. V. Viswanathan, *Analyst* **128**, 899–904 (2003).

196. W. Yantasee, Y. Lin, T. S. Zemanian and G. E. Fryxell, *Analyst* **128**, 467–472 (2003).

197. W. Yantasee, G. E. Fryxell, M. M. Conner and Y. Lin, *J. Nanosci. Nanotechnol.* **5**, 1537–1540 (2005).

198. W. Yantasee, Y. Lin, G. E. Fryxell and B. J. Busche, *Anal. Chim. Acta* **502**, 207–212 (2004).

199. R. T. Yang, R. Q. Long, J. Padin, A. Takahashi and T. Takahashi, *Ind. Eng. Chem. Res.* **38**, 2726–2731 (1999).

200. W. Z. Shen, J. T. Zheng and Q. J. Guo, *Stud. Surf. Sci. Catal.* **156**, 951–956 (2005).

201. X. S. Zhao, Q. Ma and G. Q. M. Lu, *Energ. Fuel.* **12**, 1051–1054 (1998).

202. Y. Ueno, T. Horiuchi, M. Tomita, O. Niwa, H-S. Zhou, T. Yamada and I. Honma, *Anal. Chem.* **74**, 5257–5262 (2002).

203. K. Inumaru, Y. Inoue, S. Kakii, T. Nakano and S. Yamanaka, *Chem. Lett.* **32**, 1110–1111 (2003).

204. K. Inumaru, J. Kiyoto and S. Yamanaka, *Chem. Commun.* 903–904 (2000).

205. K. Inumaru, Y. Inoue, S. Kakii, T. Nakano and S. Yamanaka, *Phys. Chem. Chem. Phys.* **6**, 3133–3139 (2004).

206. J. Lee, Y. Park, P. Kim, H. Kim and J. Yi, *J. Mater. Chem.* **14**, 1050–1056 (2004).

207. C. F. Zhou, Y. M. Wang, J. H. Xu, T. T. Zhuang, Y. Wang, Z. Y. Wu and J. H. Zhu, *Stud. Surf. Sci. Catal.* **156**, 907–916 (2005).

208. A. Ji, L. Y. Shi, Y. Cao and Y. Wang, *Stud. Surf. Sci. Catal.* **156**, 917–924 (2005).

209. D. P. Serrano, G. Calleja, J. A. Botas and F. J. Gutierrez, *Ind. Eng. Chem. Res.* **43**, 7010–7018 (2004).

Chapter 10

Hierarchically Imprinted Adsorbents

Hyunjung Kim, Chengdu Liang and Sheng Dai

Oak Ridge National Laboratory
Oak Ridge, TN, USA

10.1 Introduction

The production of selective adsorbents to detect and separate various entities in environmental waste is a growing field with broad applications and of critical importance. One of the important entities in environmental waste is metal ions. Industry produces a vast amount of metals and metal wastes. Nuclear energy and past weapons production facilities have also created challenges in the area of metal ion separation. Another significant component of environmental waste is biological molecules (e.g. proteins and viruses from wastewater). Recognition of these entities may be also integrated into sensors to detect bio-warfare agents or various environmental contaminants. This chapter describes the synthesis of these selective adsorbents to detect these various compounds in environmental applications using the hierarchical imprinting approach.

10.1.1 *Molecular imprinting*

The concept of molecular imprinting is based on an early lock-and-key concept, used to explain enzyme function and antibody formation.[1-3] Although this idea has since been proven incorrect, it was used by Frank Dickey in the 1940s to develop artificial counterparts to enzymes and antibodies originally in silica gels to develop selective silica adsorbents for different types of dye molecules.[4] By acidifying a silicate solution in the presence of methylorange dye, drying the resultant gel, and then washing the dye from the gel, Dickey obtained silica gels capable of specific adsorption of methyl-orange 1.4 times greater than of ethyl-orange.

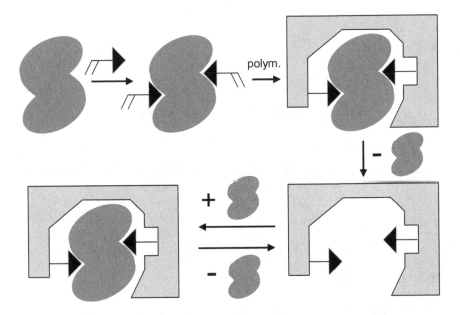

Figure 10.1.   Outline of the molecular imprinting process in organic polymers.

Building upon Dickey's findings, Wulff and Sarhan, in the 1970s, developed the current strategy to synthesize selective adsorbents by molecular imprinting in organic polymers.[5] This current molecular imprinting method to synthesize selective adsorbents in organic polymers is shown schematically in Figure 10.1. This molecular imprinting process utilizes solution interactions between a target molecule (template) and appropriate functional monomers, by either covalent[6] or non-covalent interactions[7,8] such as ionic, coordinative, hydrophobic, or hydrogen-bonding interactions. These pre-organized solution complexes are then immobilized into a polymer matrix by co-polymerization with a cross-linking monomer. Removal of the template leaves a cavity in the polymer matrix complementary to the shape, size, and functionalities of the template. This method of molecular imprinting in organic polymers has been used to imprint various templates from small metal ions to large biological molecules for applications in the areas of separation, catalysis, chemosensors, biomimic enzymes, drug delivery, and so on.[9]

10.1.2 Hierarchical imprinting

One major drawback associated with the current bulk molecular imprinting technique is that the pre-organized solution complexes are immobilized into heterogeneous organic polymer matrices without any independent control over structural parameters, such as pore size and surface area. In the current bulk molecular imprinting technique, the pore size and surface area are controlled by a porogenic solvent, where these structural parameters in polymers are determined by the rate and the extent of phase separation between growing polymer chains and the porogenic solvent. However, the reported "porogenic effect"[10–13] on these bulk imprinted polymers provides evidence that the imprinting effects on the polymers are affected by the different degrees of swelling of the polymers in the porogenic solvent, in addition to the desired effects from the immobilized solution complexes in polymer matrices. Also, the process of optimizing porogenic solvents to produce desired pore sizes and surface areas is countered by the strong interactions required between a template and functional monomers.[14] This lack of independent control on the structural parameters of bulk imprinted polymers has caused difficulties in optimizing imprinted materials for easy accessibility and fast mass transfer of substrates on the imprinted binding sites without affecting their binding properties.[15–19]

The hierarchical imprinting approach provides a solution to control the structural parameters independent of binding parameters, which are controlled by the immobilized solution complexes in solid matrices. The original concept of the hierarchical imprinting approach involves the imprinting synthesis using multiple templates over several discrete dimension scales for metal ions.[20] This concept has been extended from its original applications for metal ions to various target compounds.[21,22] Hierarchically imprinted structures can be assembled by two methodologies: (1) co-assembly and (2) stepwise assembly. The following sections address synthesis, characterization, and performance of the hierarchically imprinted adsorbents generated by these two methods.

10.2 Co-assembly Approach

The co-assembly approach to generate hierarchically imprinted adsorbents is based on the self-assembly of various structural components via simultaneous use of multiple templates. Surfactants have been used to control the structural parameters of the hierarchically imprinted silica gels and polymers.

10.2.1 *Mesoporous silica adsorbents*

Two templates (metal ions and surfactant micelles) on different length scales were used to generate hierarchically Cu^{2+}-imprinted silica adsorbents as shown in Figure 10.2.

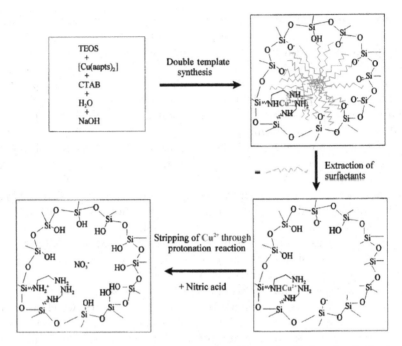

Figure 10.2.   Schematic diagram of the synthesis of hierarchically imprinted adsorbents using Cu^{2+} and CTAB simultaneously as templates.[20]

On the microporous level, the removal of the metal ion from the resultant silica adsorbent generates recognition sites for the metal ion. On the mesoporous level, removal of the surfactant templates results in the formation of relatively large cylindrical pores (diameters of 25–60 Å) that give the gel an overall porosity.

The synthetic process to produce these hierarchically imprinted adsorbents using Cu^{2+} and cetyltrimethylammonium bromide (CTAB) is as follows. The bifunctional ligand — 3-(2-amino-ethylamino) propyltrimethoxysilane, $H_2NCH_2CH_2NH\text{-}CH_2CH_2CH_2\text{-}Si(OMe)_3$ (aapts) — was used to complex with the Cu^{2+} template. The ethylenediamine group in this ligand forms strong bidentative interactions with the Cu^{2+} template. The imprinting complex precursor used in this study was $[Cu(aapts)_2]^{2+}$, synthesized according to standard literature procedures.[23] The $[Cu(aapts)_2]^{2+}$ complexes, CTAB, tetraethylorthosilicate (TEOS), water, and base (NaOH) were mixed and heated to imprint the two templates (Cu^{2+} and CTAB) into a silica matrix via the base-catalyzed hydrolytic condensation of TEOS. The blue solid products were recovered from filtration. Ethanol/HCl was used to reflux these products, and to extract the surfactant and the Cu^{2+} templates from the silica matrix. The final material was washed with large volumes of 1 N HNO_3 to ensure complete removal of the Cu^{2+} template. Control adsorbents, using the same procedure without the Cu^{2+} template, were also prepared. In this process, other surfactants — the anionic sodium dodecylsulfate (SDS) and the neutral dodecylamine (DDA) — were used to prepare Cu^{2+}-imprinted hierarchical structures and their corresponding control adsorbents.

These hierarchically imprinted silica gels and the corresponding control adsorbents have large surface areas (in the range of 200–600 m^2/g) in comparison with those of all other adsorbents imprinted only with $[Cu(aapts)_2]^{2+}$ ($< 30 \, m^2/g$). The UV-Vis spectrum of $[Cu(aapts)_2]^{2+}$ in methanol solution and that of $[Cu(aapts)_2]^{2+}$ covalently immobilized in a mesoporous silica gel prepared with CTAB show close agreement (Figure 10.3).

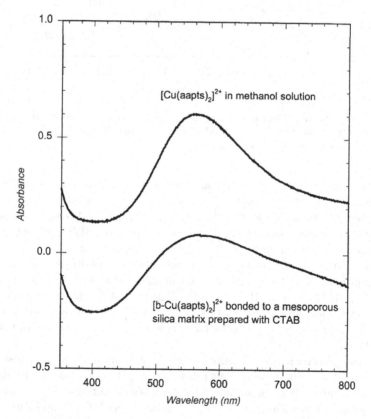

Figure 10.3.   UV-Vis spectra of $[Cu(aapts)_2]^{2+}$ in methanol solution and $[Cu(aapts)_2]^{2+}$ covalently immobilized in a mesoporous silica gel prepared with CTAB.[20]

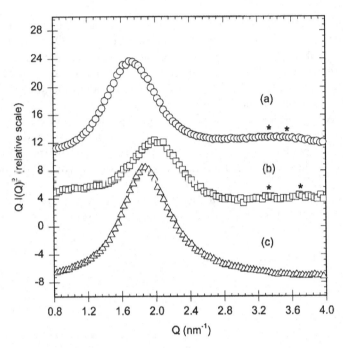

Figure 10.4. Small-angle scattering of hierarchically imprinted adsorbents prepared using (a) CTAB, (b) SDS, and (c) DDA, where $Q = (4\pi/\lambda)\sin(\theta/2)$, λ = X-ray wavelength (0.154 nm), and θ = scattering angle.[20]

This close match indicates that the environment of the copper ion is similar and that the $[\mathrm{Cu(aapts)_2}]^{2+}$ complexes are immobilized in the silica adsorbents.[24] Small-angle X-ray scatterings (Figure 10.4) from the hierarchically $\mathrm{Cu^{2+}}$-imprinted silica adsorbents with CTAB, SDS, and DDA show the common peak around $Q = 1.2 - 2.4$ nm, in agreement with the pore size independently measured in nitrogen adsorption experiments. Additional peaks present from the scattering on the silica adsorbents prepared either with CTAB (Figure 10.4(a)) or SDS (Figure 10.4(b)) indicate hexagonal structures, and the absence of this additional peak from the scattering on the silica adsorbents prepared by DDA (Figure 10.4(c)) indicates worm-like structures.

These hierarchically imprinted silica adsorbents and their corresponding control adsorbent were neutralized to a pH of 7 and dried using a vacuum oven at 50°C for six hours before adsorption tests were conducted. Competitive adsorption studies were conducted with $\mathrm{Cu^{2+}}$ and $\mathrm{Zn^{2+}}$ ions in order to measure the selectivity of these hierarchically $\mathrm{Cu^{2+}}$-imprinted adsorbents. The selectivity coefficient, k, for the binding of a specific ion in the presence of a competitor species is obtained from Equation (10.2):

$$M_1(solution) + M_2(adsorbent) \leftrightarrow M_2(solution) + M_1(adsorbent) \quad (10.1)$$

$$k = \frac{[M_2(solution)][M_1(adsorbent)]}{[M_1(solution)][M_2(adsorbent)]} = \frac{k_d[Cu^{2+}]}{k_d[Zn^{2+}]} \quad (10.2)$$

where M_1 and M_2 correspond to Cu^{2+} and Zn^{2+}, respectively.
The distribution coefficient, k_d, is calculated from Equation (10.3):

$$k_d = (C_i - C_f)/C_f \times volume\ of\ solution\ (mL)/mass\ of\ gel\ (g) \quad (10.3)$$

where C_i is the initial solution concentration and C_f is the final solution concentration.

A measure of the increase in selectivity due to the imprinting process of the Cu^{2+} template (k') is calculated by the ratio of the selectivity coefficient (k) of the imprinted adsorbents to that of the control adsorbents:

$$k' = k\ (imprinted)/k\ (control). \quad (10.4)$$

The k and k' values for these hierarchically imprinted silica adsorbents prepared with CTAB, SDS, or DDA range between 180 and 33,000, and between 5 and 280, respectively. These values are much higher than those obtained for imprinted silica adsorbents prepared without the surfactants ($k = 20$ and $k' = 1.3$). This finding can be attributed to the low surface areas obtained on the imprinted silica adsorbents without the surfactant ($26\,m^2/g$) as compared with those obtained on the hierarchically imprinted silica adsorbents (240–$520\,m^2/g$). These results demonstrate the importance of using surfactant templates to generate large surface areas in these hierarchically Cu^{2+}-imprinted silica adsorbents.

Further optimization to increase the selectivity was explored by adding various ratios of methyltrimethoxysilane ($CH_3Si(OCH_3)_3$; MTMOS) or phenyltrimethoxysilane ($C_6H_5Si(OCH_3)_3$; PTMOS) to the mixtures of $[Cu(aapts)_2]^{2+}$ and tetramethylorthosilicate ($Si(OCH_3)_4$; TMSO),[25] using adequate co-solvents,[24] and with bridged polysilsesquioxane matrices.[26] These results demonstrated that an enhanced imprinting effect and selectivity of these hierarchically Cu^{2+}-imprinted silica adsorbents toward the template ion can be obtained by enhancing the hydrophobicity of the sol-gel materials. This double-imprinting technique was also successfully used in creating Cd^{2+}-selective recognition sites on silica mesoporous adsorbents.[27]

10.2.2 *Surface-imprinted polymeric adsorbents*

The co-assembly approach to generate hierarchically imprinted adsorbents has been also explored in polymer matrices by creating microspheres with

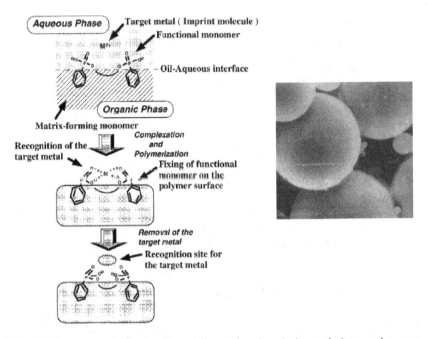

Figure 10.5.   Schematic illustration of the surface imprinting technique and a scanning electron microscopy micrograph of the resulting polymers using W/O/W emulsion polymerization.[34]

imprinted external cavities using emulsion polymerization. This method was originally pioneered by Uezu and co-workers to generate surface-imprinted polymers for Zn^{2+} templates in water-in-oil (W/O) emulsions.[28] After thorough optimization studies,[29–34] successful hierarchically imprinted microspheres for the Zn^{2+} template were synthesized in water-in-oil-in-water (W/O/W) emulsions (Figure 10.5).[34]

These microspheres have hierarchical structures composed of spherical surfaces and cavities scattered on the surfaces. The morphology of the microspheres was controlled by the following parameters: (1) Magnesium ion was used to control ionic strength to prevent shrinking of W/O emulsions during the polymerization. (2) The negatively charged surfactant SDS was used at optimum concentration to prevent aggregation of the spherical W/O emulsions. The position of the cavities on the surfaces of the microspheres was achieved using the bifunctional surfactant 1,12-dodecanediol-O-O'-diphenyl phosphonic acid (DDDPA). The bifunctional surfactant, which is amphiphilic in nature, complexes with the Zn^{2+} template and positions the complexes at the interface between the water and oil surfaces. This method also provides an additional advantage in imprinting water-soluble templates such as metal ions and biological components such as amino acids and proteins.

The steps in the synthesis of these surface-imprinted polymers are as follows: the bifunctional surfactant (DDDPA), the emulsion stabilizer (L-glutamic acid dioleylester; $2C_{18}\Delta^9GE$), and the cross-linking monomer (trimethylolpropane trimethacrylate; TRIM) were dissolved in organic solvents (i.e. toluene with 5 vol. % 2-ethylhexyl alcohol). This oil mixture was added to an aqueous solution of $Zn(NO_3)_2$ at pH 3.5, and the oil and water mixture was sonicated to produce a W/O emulsion. The W/O emulsion was then placed in an aqueous buffer solution (pH 3.5) that contained an ionic strength adjuster $[Mg(NO_3)_2]$ and the surfactant SDS to form the W/O/W emulsion. The W/O/W emulsion was polymerized with the initiator 2,2'-azobis(2,4'-dimethylvaleronitrile), ADMVN at 55°C for four hours. After the polymerization, the resulting microspheres were washed with 1 M hydrochloric acid to remove the Zn^{2+} template. These surface-imprinted polymers have been successfully used for the selective uptake of Zn^{2+} over Cu^{2+} in buffered aqueous solutions. These studies identified several critical parameters to optimize the selectivity of these surface-imprinted polymers. It was found that increasing rigidity of the phosphonic acid-functionalized molecule significantly increases the selectivity of the surface-imprinted polymers toward metal ions.[30] In the procedure using the bifunctional surfactant DDDPA, it was also found that the length of the spacer connecting the bifunctional phosphonic groups is an important factor in ensuring optimized recognition sites and rigidity of these sites on the surfaces of the polymer.[31]

This method of generating hierarchically imprinted polymers has been also extended to organic molecules, such as enantiomers of derivatized amino acids[35,36] and bifunctional amino acids,[37] as well as to nucleotides.[38] This idea of using a functionalized surfactant to generate hierarchically surface-imprinted polymers has been also applied with a core-shell morphology of the polymers for cholesterol,[39] caffeine, and tripeptides containing the Gly-Gly sequence.[40,41]

10.3 Stepwise Assembly Approach

The stepwise assembly approach to generate hierarchically imprinted adsorbents involves sequential generation of hierarchical structures. In this approach, various forms of inorganic hosts have been used as modules to control the structural parameters on the hierarchically imprinted adsorbents. The stepwise assembly approach has been used to generate surface-imprinted inorganic, polymeric, and inorganic-organic hybrid adsorbents.

10.3.1 Surface-imprinted sol-gel adsorbents

The stepwise assembly approach to generate hierarchically imprinted adsorbents was originally demonstrated by Dai and co-workers.[42] The key to this method is to coat the pore surface of an ordered mesoporous silica host (MCM-41) with the complex precursor $[Cu(aapts)_2]^{2+}$ via the hydrolysis and condensation reactions of silicon alkoxide groups in aapts (Figure 10.6).

A mesoporous ordered silica host was synthesized with optimum pore diameter to match the stereochemical requirements for the surface imprinting of the complexes. The prepared mesoporous silica samples have surface areas of over $1,000\,m^2/g$, mesopore volumes of $0.98\,cm^3/g$, and average pore sizes of $\sim 25\,\text{Å}$, as measured by powder X-ray diffraction (XRD) and small-angle X-ray scattering. This pore diameter of the mesoporous silica samples is ideal to fit the complexes $[Cu(aapts)_2]^{2+}$ (~ 16–$25\,\text{Å}$). These hierarchically imprinted adsorbents showed a k' value of 40 for the Cu^{2+} template over Zn^{2+}. When $[Cu(aapts)_2]^{2+}$ was coated on the pore surfaces of commercial amorphous silica gel (Aldrich; $\bar{d} = 60\,\text{Å}$; surface area $= 600\,m^2/g$), the resulting imprinted materials did not show any significant selectivity toward the Cu^{2+} template ($k' = 1.54$). This demonstrates the importance of the optimized pore diameter and ordered structure of the inorganic hosts for the optimized selectivity toward the metal ions.

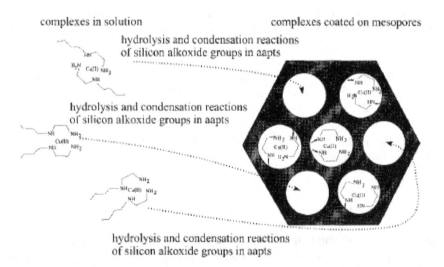

Figure 10.6.   Schematic diagram of the imprint-coating process. First, the complexes are introduced between target metal ions and bifunctional ligands. Then the siloxane groups in the bifunctional ligands are hydrolyzed, and finally the complexes are covalently coated on the mesopore surfaces.[42]

Liu and co-workers[43] have successfully explored the stepwise assembly strategy to imprint organic molecules inside mesopores, allowing control over both the binding properties and pore structures. The ordered mesoporous silica hosts (MCM-41) were also successfully used to imprint dopamine and glucosamine by creating a multifunctionalized mesopore that can interact with the template by covalent and non-covalent interactions.[44] The importance of the ordered mesopores in creating cavities was also demonstrated in this study by the lack of selectivity towards the template when an amorphous silica was used as the host.[44] Another successful application of the mesoporous silica hosts was demonstrated by creating the surface-imprinted sol-gel adsorbents for estrone in the inorganic hosts.[45]

Even though the mesoporous silica hosts used in the previous study have a perfect nanoscale order, the wall structure of the mesoporous materials is amorphous, which could collapse under hydrothermal conditions. Thus, the use of porous zeolitic crystalline hosts to imprint Cu^{2+} were tested for a possible improved imprinting effect, which may come from the generation of uniform imprints and the limited choices of coordination environments.[46] Due to the small pore opening of the L-zeolite used in this study (7.1–7.8 Å), a stepwise ship-in-a-bottle imprinted methodology was used (Figure 10.7). The Cu^{2+} template was first used to replace K^+ or Na^+ in the L-zeolite through the ion-exchange interaction. This step was then followed by complexations of the surface Cu^{2+} template with aapts ligands and the surface coupling of aapts with neighboring surface SiO^- or $-SiOH$ groups. By extracting the Cu^{2+} template via protonation of the amine groups on the complexes, a zeolite imprinted on the channel surface with the Cu^{2+} ion was prepared. Using the same procedure without the Cu^{2+}, control zeolite adsorbents were also prepared.

The exchange of the Cu^{2+} ion with the precursor complex $[Cu(aapts)_x]^{2+}$ in the zeolite-imprinted adsorbent was confirmed by the UV-Vis and Fourier transform infrared (FTIR) spectra. In the UV-Vis spectrum, the maximum band position moves from 681 nm to 585 nm in agreement with the change of the coordination environment of the Cu^{2+}

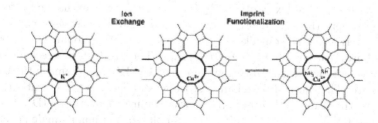

Figure 10.7.   Schematic diagram of the Cu^{2+} imprinting in L-zeolite hosts.[46]

ion from water to amine. In the FTIR spectrum, the C-H stretching around
$2,900 \, \text{cm}^{-1}$ can be attributed to the C-H groups of the aapts ligands. The
quantification of the C-H stretching peak also showed that a similar amount
of the aapts ligands was present in the imprinted and the control zeolite
adsorbents. Powder XRD was used to observe the crystallinity of the zeolite
structure during the various stages of the preparation. The XRD patterns of
the imprinted and the control zeolite adsorbents show that the crystallinity
of the zeolite adsorbents was maintained during the preparation process and
that the space distribution of the functional ligands in the imprinted zeolite
adsorbent was different from that in the control zeolite adsorbents, as would
be expected in the imprinting process. The selectivity of the imprinted zeolite adsorbents towards the Cu^{2+} template in aqueous Cu^{2+}/Zn^{2+} mixtures
was tested using a typical batch procedure. The imprinted zeolite adsorbents showed k and k' values of 29 and 12, respectively, for the imprinted
Cu^{2+} in aqueous Cu^{2+}/Zn^{2+} mixtures. These values for the imprinted zeolite adsorbents are higher than those obtained for the control adsorbents.
However, the values for the former are much lower than those obtained for
the imprinted mesoporous silica hosts ($k = 91$ and $k' = 40$).[42]

Several other forms of inorganic hosts, such as layered nanomaterials[47]
and nanomembranes,[48] have also been successfully used to produce surface-
imprinted sol-gel materials via the stepwise assembly approach. For example, Zhang et al. used polysilicate magadiite as a layered nanomaterial to
imprint Cu^{2+} ions at the interface of the layered magadiite (Figure 10.8).[47]

The layered magadiite is composed of one or more negatively charged
sheets of SiO_4 tetrahedra with abundant silanol-terminated surfaces where
these negative charges are balanced by either Na^+ or H^+ in the interlayer
spacing. The layered magadiite provides unique features of tunable gallery
spacing and stable crystalline forms for hierarchically imprinted structures.
The synthesis steps for the imprinting of Cu^{2+} ions in the layered magadiite are as follows: First, the Na^+-magadiite host was synthesized according to previously published methods.[49] Next, via ion exchange, the Na^+
in the galleries of the layered magadiite was replaced by a much larger
CTA^+ cation (cetyltrimethylammonium), which was then exchanged with
the precursor complex $[Cu(aapts)_2]^{2+}$. The complexes exchanged into the
galleries were then covalently attached to the surfaces through condensation reactions of silicon alkoxide groups in the bifunctional ligands with
neighboring surface SiO^- or $SiOH$ groups. The product was recovered by
filtration and washed with large volumes of 1 M HNO_3 to remove the Cu^{2+}
template. The successful exchange process from Na^+, to CTA^+ cation, and
then to the complex $[Cu(aapts)_2]^{2+}$ was confirmed by the XRD patterns
and the UV-Vis spectra. The ^{29}Si cross-polarization magic-angle spinning
(CP/MAS) NMR spectra of the imprinted magadiite adsorbents confirmed

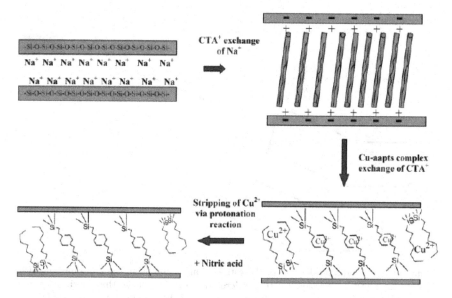

Figure 10.8. Schematic diagram of imprinting Cu^{2+} ions in magadiite silicate.[47]

that the crystallinity of the host was maintained during the preparation process. The amount of the aapts ligands incorporated in the galleries of the magadiite hosts was higher in the imprinted magadiite than in the control magadiite, as indicated by the basal spacing calculated from the corresponding XRD patterns and the resonance peaks from the corresponding ^{29}Si CP/MAS NMR spectra. The results from the adsorption studies on the magadiite adsorbents toward the Cu^{2+} template in aqueous Cu^{2+}/Zn^{2+} mixtures showed the preferential adsorption of the Cu^{2+} template on the imprinted magadiite adsorbents ($k = 140.03$) over that on the control magadiite adsorbents ($k = 45.47$), resulting in the imprinting factor (k') of 3.24.

Recently, Yang *et al.* applied a molecular imprinting technique to imprint estrone into the walls of the sol-gel nanotubes (Figure 10.9).[48] The silica nanotubes were synthesized within the pores of nanopore alumina template membranes using the sol-gel method.

First, silica monomer-template complexes were synthesized by reacting estrone with 3-(triethoxylsilyl)propyl isocyanate. The silica monomer-estrone complexes and TEOS were then added to a mixture of acetate buffer at pH 5.1 with ~10 vol. % ethanol. After the mixture was stirred for five minutes, the alumina membrane was immersed. After 0.5 hours at low pressure, the alumina membrane was removed by phosphoric acid, rinsed with ethanol, and cured at 150°C for one hour. To extract the estrone template, the imprinted silica nanotube membrane was heated at 180°C in a

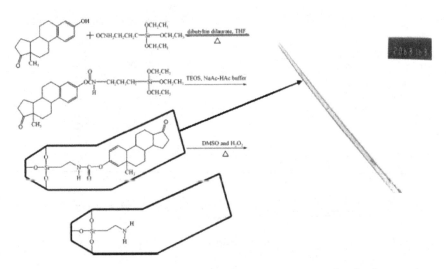

Figure 10.9. Schematic diagram of the estrone imprinting using the alumina template membrane. The transmission electron microscopy (TEM) shows the imprinted silica nanotubes (with 100 nm diameter) after the removal of the alumina template membranes by phosphoric acid.[48]

mixture of dimethyl sulfoxide (DMSO) and water and then washed with the mixture at room temperature. The amount of estrone that could not be removed via this process was $\sim$1.2%, as estimated from the carbonyl peak of the urethane group of the silica monomer-estrone complexes at 1,736 cm^{-1} in the infrared (IR) spectroscopy. After the alumina template had been removed by phosphoric acid, transmission electron microscopy (TEM) of the imprinted silica nanotubes showed pore diameters of 100 nm (Figure 10.9). Calculations from a steady-state binding method showed that the ratio of the selectivity of the imprinted silica nanotubes towards the estrone template to that of the corresponding control silica nanotubes was 6.94 in chloroform.

10.3.2 *Surface-imprinted polymeric adsorbents*

The surface-imprinted polymeric adsorbents using silica hosts were first demonstrated by Yilmaz *et al.* for theophylline templates (Figure 10.10).[21] The major objective in employing this method is to covalently immobilize the template in the silica hosts by reacting 8-carboxypropyl-derivatized theophylline with aminopropyl-derivatized silica hosts to form amide bonds using carbodiimide chemistry. Subsequently, the functional monomers (trifluoromethylacrylic acid; TFMAA) and the cross-linking monomers

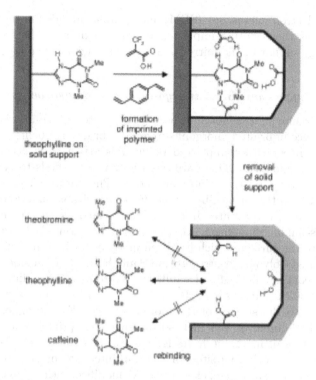

Figure 10.10. Schematic diagram of surface imprinting method using immobilized template on the silica hosts.[21]

(divinylbenzene; DVB) are added to the silica hosts with the immobilized template, and these mixtures are then polymerized. The removal of the silica hosts by aqueous hydrofluoric acid generates the surface-imprinted polymeric adsorbents. The successful imprinting effect on these surface-imprinted polymeric adsorbents for the theophylline templates was demonstrated by radioligand binding studies on toluene. This concept of immobilizing templates on silica hosts to independently control the structural parameters of the imprinted adsorbents was also demonstrated for 9-ethyladenine or triaminopyrimidine,[22] peptides,[50] and hemoglobin.[51] Yang *et al.*[52] used this concept to prepare imprinted nanowires for glutamic acid template using alumina membranes as inorganic hosts instead of the spherical silica hosts. Rather than immobilizing templates in the pores of the spherical silica hosts, Yilmaz *et al.*[53] generated (-)-isoproterenol-imprinted polymers by adding the mixtures of the pre-polymer components within the pores of spherical silica hosts, followed by polymerization of the mixtures within the pores and resolution of the spherical silica hosts. The scanning electron microscopy (SEM) micrographs provided in these studies showed that the resulting imprinted polymers exhibited a structure and morphology similar to the mirror image of the original silica hosts.

10.3.3 *Surface-imprinted inorganic-organic adsorbents*

Instead of removing the inorganic hosts to generate surface-imprinted polymers, surface-imprinted inorganic-organic hybrid adsorbents have been developed, in which an imprinted polymer is attached to the surface of porous inorganic hosts. There are essentially two methods to generate an imprinted surface on these inorganic hosts. The imprinted polymers can be grafted onto the surface by attaching polymerizable functional groups, which are then incorporated into a polymer matrix by co-polymerization with cross-linking monomers. Alternatively, an imprinted polymer can be grown onto the surface, which has been modified with an initiator.

Imprinted polymer grafts onto inorganic hosts have included methods involving coating the walls of capillary electrophoresis columns[54,55] and the use of silica particles.[56–59] Prior to the polymerization, a vinyl group is attached to the surface of the inorganic hosts. For example, Plunkett and Arnold[56] employed this method to generate imprinted polymer grafts on the surface of inorganic hosts for a series of mono- and bis-imidazole templates. The anchoring sites for the imprinted polymers were provided by $10\,\mu$m porous silica particles (1,000 Å) modified with propylmethacrylate using a silanizing reagent, 3-(trimethosylsilyl)-propylmethacrylate (Figure 10.11).

One of the templates, the functional monomer (copper(II) [N-(4-vinylbenzyl)imino]diacetate $\bullet$ 2H$_2$O) and the cross-linking monomer (ethylene glycol dimethacrylate; EGDMA) were mixed with the modified silica suspended in 80% aqueous methanol. The mixture was then polymerized at 40°C for 48 hours with ammonium peroxydisulfate (APS) as an initiator. The resulting adsorbents were washed extensively with solutions of EDTA and 1,4,7-triazacyclononane to remove the template and incorporated copper ions. The SEM morphology of the resulting adsorbents showed the continuous polymer nanospheres (in the 50–100 nm range) coated in a thin

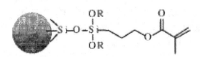

Figure 10.11.   Modification of the silica hosts using chemically attached polymerizable groups.

layer (on the order of 1 μm or less) on the silica hosts. The selectivity of these hybrid adsorbents towards the template, tested using high-performance liquid chromatography (HPLC), showed a strong imprinting effect even when the copper ion was replaced with zinc ion. However, this approach to graft the imprinted polymers on the surfaces of the hosts requires that the monomer mixture be applied as a thin liquid film on the surface prior to the polymerization because the initiator is in solution. Thus, control of the thickness of the polymer layer is difficult, and the maximum density of grafted polymer chains is limited due to kinetic and steric factors, as demonstrated by Brueggemann *et al.*[54]

A better control of the morphology of inorganic-organic hybrid adsorbents was obtained when the imprinted polymers were grown on the surface of the inorganic hosts modified with an initiator.[60,61] A free-radical initiator was attached to a surface either chemically (Figures 10.12(a) and 10.12(b)) or by physical adsorption (Figure 10.12(c)).[60]

The covalent attachment of the initiator on the surface of the silica hosts comprised two successive surface reactions. The free silanol groups ($\sim$8 μmol/m^2) on the silica surface, re-hydroxylated according to standard

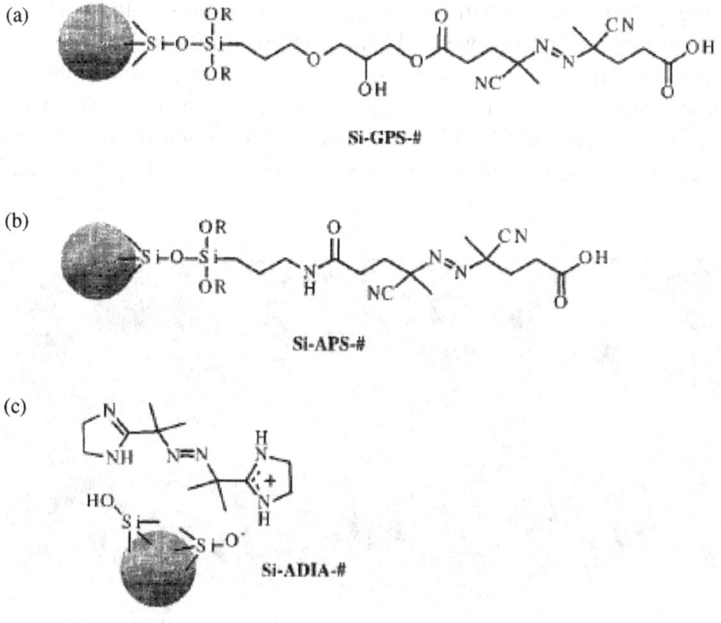

Figure 10.12. Modification of the silica hosts either (a, b) using covalently attached initiators or (c) non-covalently attached initiators.

procedures,[62] were reacted with trimethoxyglycidoxypropylsilane (GPA) or (3-aminopropyl)triethoxysilane (APS), followed by reaction of the epoxy or amino groups with an azo initiator such as azobis(cyanopentanoic acid) (ACPA), which led to the formation of an ester or amide bond between the surface and the azo initiator. The results from the elemental analysis after the sequential coupling to the silica hosts showed an overall conversion of the silanol groups of $\sim 10\%$ (i.e. 0.7–$0.8\,\mu\mathrm{mol/m}^2$). In the non-covalent approach to incorporating the initiator on the surface of the silica hosts, a strongly basic amidine-containing initiator ($2,2'$-azobis-(N,N'-dimethyleneisobutyramidine; ADIA) was used. This basic initiator was adsorbed to bare silica particles from chloroform, resulting in a reproducible and slightly higher surface coverage ($1\,\mu\mathrm{mol/m}^2$) than those generated by the covalent attachments. These initiator-modified particles were then immersed in the monomer mixture — the template (L-phenylalanine anilide; L-PA), the functional monomer (methacrylic acid; MAA), the cross-linking monomer (EGDMA), and the porogen (toluene or dichloromethane) — followed by photopolymerization. The SEM micrographs of the resulting particles (Figure 10.13) showed morphology changes by the imprinting process on the surface of the silica hosts (Figure 10.13(b)) in comparison to the unmodified silica hosts. At higher grafting levels with increasing polymerization time, aggregated silica hosts held together by a web-like polymer structure were observed (Figure 10.13(c)).

Systematic investigation of these imprinted inorganic-organic hybrid adsorbents was accomplished by measuring the separation and the resolution parameters in HPLC. The non-covalent approach resulted in a much lower separation and resolution factor compared with that of the adsorbents

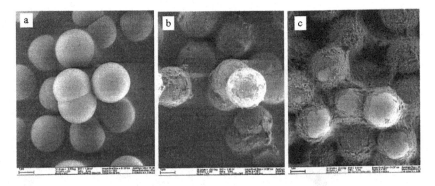

Figure 10.13. SEM micrographs of (a) unmodified silica hosts, (b) modified silica hosts with covalently attached APS initiator containing 3.6% carbon, and (c) modified silica hosts with covalently attached APS initiator containing 16% carbon.[60]

prepared by covalently attached initiators. Chromatographic performance of the adsorbents prepared by covalently attached initiators showed that the column efficiency depends on the thickness of the imprinted polymers on the silica hosts. Imprinted polymers grafted as thin films ($\sim$0.8 nm) showed a higher column efficiency, which decreased with increasing thickness of the polymer films. On the other hand, the sample loading capacity and separation factor were observed at a higher thickness of the imprinted polymers ($\sim$7.0 nm). These results show that these inorganic-organic hybrid adsorbents can be rendered tunable, by controlling the thickness of the imprinted polymers (e.g. via polymerization time), for different applications such as in HPLC stationary phases (i.e. analytical and preparative chromatography).

To gain further control over the morphology of these inorganic-organic hybrid adsorbents, Sellergren *et al.* used a so-called iniferter (Figure 10.14)[63] to prevent undesired polymerization in solution, which invariably accompanies the use of the covalently attached azo initiators.

An iniferter is an initiator for free-radical polymerization, which undergoes primary radical termination and ordinarily avoids irreversible biomolecular termination. Iniferters decompose to give two different radicals: one able to initiate polymerization and the other being stable but capable of terminating the growing polymer chains by recombination. Recently, Ruckert and Kolb used scanning-mode transmission electron microscopy (STEM) and energy dispersive X-ray spectroscopy (EDX) to monitor the distribution of the imprinted polymers grafted on the surface of silica hosts using iniferters.[64] This living nature of the iniferters was also applied to consecutively graft two polymer layers imprinted with two different templates, or one imprinted and one non-imprinted layer in any order.[65] Recent comparative studies of these different formats of hierarchically imprinted adsorbents showed better accessibility to substrates when thin films of imprinted polymers were grafted on the silica hosts than with other formats, generated either by filling the pores of the silica hosts with polymerization mixtures and subsequent polymerization (leading to silica-imprinted polymer composites) or spherical imprinted adsorbents after the dissolution of the silica hosts.[66] These different morphologies of the imprinted adsorbents were shown to affect the mass transfer kinetics of the substrates without any significant influences on the molecular recognition properties of these imprinted adsorbents.[67]

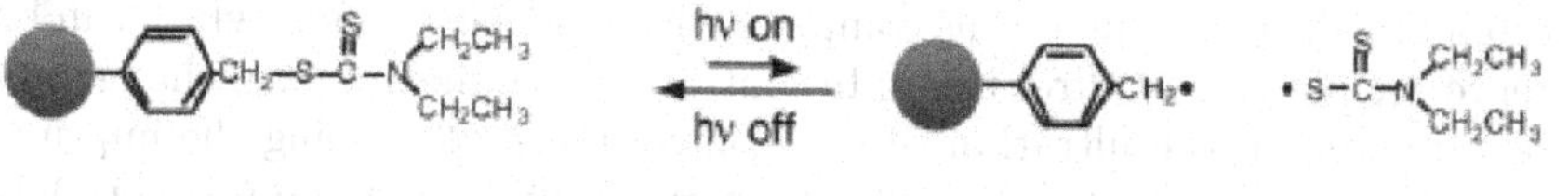

Figure 10.14.   Modifications of the silica hosts with a dithiocarbamate iniferter.

10.3.4 Surface-imprinted titania gel film

Even though the above methods to prepare hierarchically imprinted adsorbents by surface modifications provide satisfactory substrate selectivity, the molecular details of these modified surfaces are not clear. Forming ultrathin films on the surfaces provides an opportunity to study the molecular details of these modified surfaces and the practical application in sensor areas. In principle, the use of organized molecular films such as Langmuir–Blodgett multilayers and surface-bound monolayers is not compatible with the imprinting process, which requires adaptable structural modification. Well-characterized imprinted ultrathin films can be formed on metal oxides by chemical vapor deposition, as shown by Kodakari et al., using preadsorbed benzoate anion as a template.[68] However, this chemical vapor deposition method requires a high temperature, at which thermally labile organic compounds cannot be used as templates.

Recently, Kunitake et al.[69,70] developed a method to prepare imprinted ultrathin titania gel films by sequential chemisorption and activation, where individual metal oxide layers are formed with nanometer precision at mild preparatory conditions appropriate for thermally labile organic templates (Figure 10.15).

First the titanium alkoxide [Ti(O-nBu$_4$] was interacted with the azobenzene carboxylic acid template (C$_3$AzoCO$_2$H) to form the complexes. These complexes were added with water and aged for several hours to produce Ti$_4$O$_4$(OH)$_4$ (O-nBu)$_4$C$_3$AzoCO$_2$H. A gold-coated quartz crystal microbalance (QCM) electrode modified with mercaptoethanol was allowed to interact with the template solutions to form covalently bound surface monolayers of the titanium oxide-template complexes. The physisorbed complexes were removed by washing with toluene for one minute. The chemisorbed surface layers of the template complexes were then hydrolyzed in air to give a new hydroxylated surface. This process comprises one cycle, and multilayers on the surface can be formed by repeating this cycle several times with reproducible control of the thickness of each layer (0.8–2.7 nm). The templates in the ultrathin titania films were completely removed with 1% aqueous ammonia for 30 minutes as estimated by the UV absorbance and the QCM frequency measurements.

The imprinted ultrathin titania films (ten cycles) were tested for their adsorption kinetics and selectivity toward the template using in situ QCM experiments. Figure 10.16(a) shows that the mass increase due to the rebinding of the template was complete within 40–60 seconds, which can be reproducible at least three times. In contrast, no significant mass change can be observed on the ultrathin titania control films, confirming the imprinting effect. Figure 10.16(b) shows the highest binding of the C$_3$AzoCO$_2$H

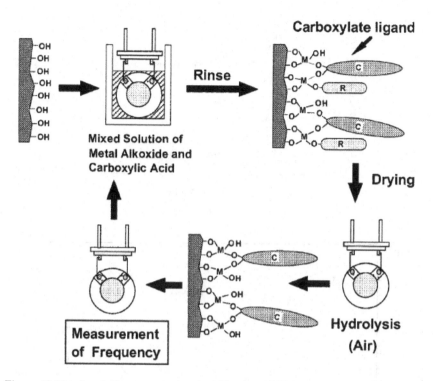

Figure 10.15. Imprinting process in ultrathin titania films using the surface sol-gel process. "C" and "R" denote the template molecule and the unhydrolyzed butoxide group of $Ti(O-nBu)_4$.[69]

template on the imprinted ultrathin titania films compared with other structural analogues. This method of generating well-controlled imprinted ultrathin titania films was extended for derivatized amino acids,[71] Mg^{2+} ions,[72] di- and tripeptides,[73] and (+)-glucose.[74] Further refinements of this approach were also shown in the development of self-supporting imprinted ultrathin films[75] and liquid-phase deposition to make the titania ultrathin films.[76]

10.4 Conclusions

This chapter reviews the known available methods to generate hierarchically imprinted adsorbents for separation purposes. These hierarchically imprinted adsorbents allow us to independently control structural parameters (to facilitate accessibility and mass transfer kinetics) in addition to the desired adsorption parameters (i.e. selectivity toward imprinted molecules)

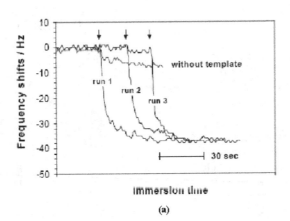

(a)

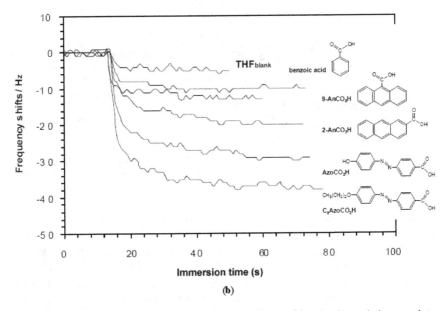

(b)

Figure 10.16.   *In situ* QCM frequency decrease due to (a) rebinding of the template molecules and (b) rebinding of a series of carboxylic acids.[70]

using co-assembly and stepwise assembly synthesis. Thus, these materials should allow us to independently optimize these parameters for efficient separation and detection of various metal ions and biological molecules in environmental applications. Further systematic studies on these hierarchically imprinted adsorbents for their mass transfer kinetics and adsorption properties are also desired to facilitate the development of high-performance imprinted adsorbents for practical applications in future.

References

1. E. Fisher, *Ber. Dtsch. Chem. Ges.* **27**, 2985 (1894).
2. S. Mudd, *J. Immunol.* **23**, 423 (1932).
3. L. Pauling, *J. Am. Chem. Soc.* **62**, 2643 (1940).
4. F. H. Dickey, *Proc. Natl. Acad. Sci. U.S.A.* **35**, 227 (1949).
5. G. Wulff and A. Sarhan, *Angew. Chem. Int. Ed. Engl.* **11**, 341 (1972).
6. G. Wulff, *Chem. Rev.* **102**(1), 1–27 (2002).
7. K. Mosbach, *Trends Biochem. Sci.* **19**(1), 9–14 (1994).
8. B. Sellergren, *Trends Anal. Chem.* **16**(6), 310–320 (1997).
9. B. Sellergren (ed.) *Molecularly Imprinted Polymers. Man-Made Mimics of Antibodies and their Applications in Analytical Chemistry* (Elsevier, Amsterdam, 2003).
10. D. Spivak, M. A. Gilmore and K.J. Shea, *J. Am. Chem. Soc.* **119**(19), 4388–4393 (1997).
11. H. Kim and G. Guiochon, *Anal. Chem.* **77**, 1708–1717 (2005).
12. H. Kim and G. Guiochon, *Anal. Chem.* **77**, 1718–1726 (2005).
13. H. Kim and G. Guiochon, *Anal. Chem.* **77**, 2496–2504 (2005).
14. B. Sellergren and K. J. Shea, *J. Chromatogr.* **115**, 3368–3369 (1993).
15. K. Miyabe and G. Guiochon, *Biotechnol. Prog.* **16**, 617 (2000).
16. H. Kim, K. Kaczmarski and G. Guiochon, *Chem. Eng. Sci.* **60**, 5425–5444 (2005).
17. H. Kim, K. Kaczmarski and G. Guiochon, *Chem. Eng. Sci.* **61**, 1122 (2006).
18. H. Kim and G. Guiochon, *J. Chromatogr.* **1097**, 84 (2005).
19. H. Kim, K. Kaczmarski and G. Guiochon, *Chem. Eng. Sci.* **61**, 5249 (2006).
20. S. Dai, M. C. Burleigh, Y. H. Ju, H. J. Gao, J. S. Lin, S. J. Pennycook, C. E. Barnes and Z. L. Xue, *J. Am. Chem. Soc.* **122**, 992 (2000).
21. E. Yilmaz, K. Haupt and K. Mosbach, *Angew. Chem. Int. Ed.* **39**(12), 2115 (2000).
22. M. M. Titirici, A. J. Hall and B. Sellergren, *Chem. Mater.* **14** (2002).
23. G. De, M. Epifani and A. Licciulli, *J. Non-Cryst. Solids* **201**, 250–255 (1996).
24. A. T. Baker, *J. Chem. Educ.* **75**, 98 (1998).
25. R. D. Makote and S. Dai, *Anal. Chim. Acta* **435**, 169–175 (2001).
26. M. C. Burleigh, S. Dai, E. W. Hagaman and J. S. Lin, *Chem. Mater.* **13**, 2537–2546 (2001).
27. Y. K. Lu and X. P. Yan, *Anal. Chem.* **76**, 453–457 (2004).
28. K. Uezu, R. Nakamura, M. Goto, M. Murata, M. Maeda, M. Takagi and F. Nakashio, *J. Chem. Eng. Jpn.* **27**, 436–438 (1994).
29. M. Yoshida, K. Uezu, M. Goto and F. Nakashio, *J. Chem. Eng. Jpn.* **29**, 174–176 (1996).
30. M. Yoshida, K. Uezu, M. Goto and S. Furusaki, *Macromolecules* **32**, 1237–1243 (1999).

31. M. Yoshida, K. Uezu, F. Nakashio and M. Goto, *J. Polym. Sci. Part A: Polym. Chem.* **36**, 2727–2734 (1998).

32. K. Uezu, H. Nakamura, J. Kanno, T. Sugo, M. Goto and F. Nakashio, *Macromolecules* **13**, 3888–3891 (1997).

33. K. Uezu, H. Nakamura, M. Goto, F. Nakashio and S. Furusaki, *J. Chem. Eng. Jpn.* **32**, 262–267 (1999).

34. M. Yoshida, K. Uezu, M. Goto and S. Furusaki, *J. Appl. Polym. Sci.* **73**, 1223–1230 (1999).

35. M. Yoshida, K. Uezu, M. Goto and S. Furusaki, *J. Appl. Polym. Sci.* **78**, 659–703 (2000).

36. M. Yoshida, Y. Hatate, K. Uezu, M. Goto and S. Furusaki, *Colloid. Surface. A: Physicochem. Eng. Aspect.* **109**, 259 269 (2000).

37. K. Araki, M. Goto and S. Gurusaki, *Anal. Chim. Acta* **469**, 173–181 (2002).

38. H. Tsunemori, K. Araki, K. Uezu, M. Goto and S. Furusaki, *Bioseparation* **10**, 315–321 (2001).

39. N. Perez, M. J. Whitcombe and E. N. Vulfson, *Macromolecules* **34**, 830–836 (2001).

40. S. Carter, L. Shui-Yu and S. Rimmer, *Supramol. Chem.* **15**, 213–220 (2003).

41. S. Carter and S. Rimmer, *Adv. Funct. Mater.* **14**, 553–561 (2004).

42. S. Dai, M. C. Burleigh, Y. Shin, C. C. Morrow, C. E. Barnes and Z. Xue, *Angew. Chem. Int. Ed.* **38**, 1235–1239 (1999).

43. Y. S. Shin, J. Liu, L. Q. Wang, Z. M. Nie, W. D. Samuels, G. E. Fryxell and G. J. Exarhos, *Angew. Chem. Int. Ed. Engl.* **39**, 2702–2707 (2000).

44. V. S. Lin, C. Lai, J. Huang, S. Song and S. Xu, *J. Am. Chem. Soc.* **123**, 11510–11511 (2001).

45. C. D. Ki, C. Oh, S. Oh and J. Y. Chang, *J. Am. Chem. Soc.* **124**, 14838–14839 (2002).

46. Z. Zhang, S. Dai, R. D. Hunt, Y. Wei and S. Qiu, *Adv. Mater.* **13**, 493–496 (2001).

47. Z. Zhang, S. Saengkerdsub and S. Dai, *Chem. Mater.* **15**, 2921–2925 (2003).

48. H. Yang, S. Zhang, W. Yang, X. Chen, Z. Zhuang, J. Xu and X. Wang, *J. Am. Chem. Soc.* **126**, 4054–4055 (2004).

49. R. A. Fletcher and D. M. Bibby, *Clays Clay Miner.* **35**, 318–320 (1987).

50. M. M. Titirici and B. Sellergren, *Anal. Bioanal. Chem.* **378**, 1913–1921 (2004).

51. T. Shiomi, M. Matsui, F. Mizukami and K. Sakaguchi, *Biomaterials* **26**, 5564–5571 (2005).

52. H. Yang, S. Zhang, F. Tan, Z. Zhuang and X. Wang, *J. Am. Chem. Soc.* **127**, 1378–1379 (2005).

53. E. Yilmaz, O. Ramstrom, P. Moller, D. Sanchez and K. Mosbach, *J. Mater. Chem.* **12**, 1577–1581 (2002).

54. O. Brueggemann, R. Freitag, M. J. Whitcombe and E. N. Vulfson, *J. Chromatogr. A* **781**, 43–53 (1997).

55. J. M. Lin, T. Nakagama, K. Uchiyama and T. Hobo, *J. Pharmaceut. Biomed. Anal.* **15**, 1351–1358 (1997).

56. S. D. Plunkett and F. H. Arnold, *J. Chromatogr. A* **708**, 19–29 (1995).

57. S. Vidyasankar, M. Ru and F. H. Arnold, *J. Chromatogr. A* **775**, 51–63 (1997).

58. T. H. Kim, K. C. Do, H. Cho, T. Y. Chang and J. Y. Chang, *Macromolecules* **38**, 6423–6428 (2005).

59. E. J. Acosta, S. O. Gonzalez and E. E. Simanek, *J. Polym. Sci. Part A. Polym. Chem.* **43**, 168–177 (2005).

60. C. Sulitzky, B. Ruckert, A. J. Hall, F. Lanza, K. Unger and B. Sellergren, *Macromolecules* **35**, 79–91 (2002).

61. L. Schweiz, *Anal. Chem.* **74**, 1192–1196 (2002).

62. K. K. Unger, in *Adsorbents in Column Liquid Chromatography*, ed. K. K. Unger, Vol. 47 (Marcel Dekker Inc., New York, 1990), pp. 331–470.

63. B. Ruckert, A. J. Hall and B. Sellergren, *J. Mater. Chem.* **12**, 2275–2280 (2002).

64. B. Ruckert and U. Kolb, *Micron* **36**, 247–260 (2005).

65. B. Sellergren, B. Ruckert and A. J. Hall, *Adv. Mater.* **14**, 1204–1208 (2002).

66. F. G. Tamayo, M. M. Titirici, A. B. Martin-Esteban and B. Sellergren, *Anal. Chim. Acta* **542**, 38–46 (2005).

67. C. Baggiani, P. Baravalle, L. Anfossi and C. Tozzi, *Anal. Chim. Acta* 125–134 (2005).

68. N. Kodakari, N. Katada and M. Niwa, *J. Chem. Soc. Chem. Commun.* **6**, 623–624 (1995).

69. S. Lee, I. Ichinose and T. Kunitake, *Langmuir* **14**, 2857–2863 (1998).

70. T. Kunitake and S. W. Lee, *Anal. Chim. Acta* **504**, 1–6 (2004).

71. S. Lee, I. Ichinose and T. Kunitake, *Chem. Lett.* **12**, 1193–1194 (1998).

72. J. H. He, I. Ichinose, S. Fujikawa, T. Kunitake and A. Nakao, *Chem. Mater.* **14**, 3493–3500 (2002).

73. I. Ichinose, T. Kikuchi, S. W. Lee and T. Kunitake, *Chem. Lett.* **1**, 104–105 (2002).

74. S. W. Lee and T. Kunitake, *Mol. Cryst. Liq. Cryst.* **371**, 111–114 (2001).

75. M. Hashizume and T. Kunitake, *Langmuir* **19**, 10172–10178 (2003).

76. L. Feng, Y. Liu and J. Hu, *Langmuir* **20**, 1786–1790 (2004).

Chapter 11

Functionalization of Periodic Mesoporous Silica and its Application to the Adsorption of Toxic Anions

Hideaki Yoshitake

Division of Materials Science and Chemical Engineering,
Yokohama National University, Yokohama, Japan

11.1 Introduction

11.1.1 *Why mesoporous materials?*

While the practical applications of porous solids are numerous and have been studied extensively, the pore structures of porous solids are generally grouped into three categories. According to the IUPAC Compendium, mesopores are pores with widths between those of micropores (pore size <2 nm) and macropores (pore size >50 nm). The difference in size does not simply lead to various molecular sieving effects, but also to a wide range of characteristic chemical phenomena occurring within the pores. To discover chemically interesting phenomena specific to a particular pore size, we need ordered porous material, which has a defined pore size and a well-developed porous structure.

Mesoporous silica with a periodic structure was first reported in the early 1990s.[1,2] Attention has been drawn to this material because of the unique way in which the structure is formed, the narrow pore size distribution on the mesoscale and the periodicity of mesostructure. Various kinds of surfactant micelle have been used as templates for directing the structure of mesoporous silicas, in which the size of the pores is almost the same as that of the micelle and the arrangement of the pore channels traces the self-organized pattern of the micelle.[3-5] Typical mesoporous silicas appearing most frequently in the literature are illustrated in Figure 11.1. The procedure for synthesis is as follows: a silica precursor (e.g. tetraethyl orthosilicate, TEOS) and a surfactant are mixed in a basic or acidic solution.

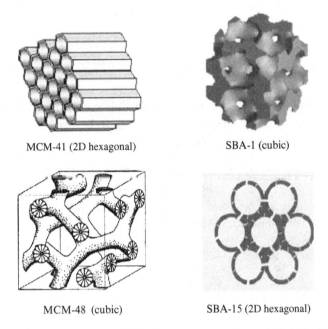

MCM-41 (2D hexagonal)   SBA-1 (cubic)

MCM-48 (cubic)   SBA-15 (2D hexagonal)

Figure 11.1.   Porous structures of MCM-41, MCM-48, SBA-1 and SBA-15. Typical pore sizes are shown. The pore size distribution is narrow in these mesoporous silicas though the diameter ($2R_P$) is usually controllable. Although the structure of the mesopore arrays of MCM-41 and SBA-15 (a cross-sectional view is shown) is similar, in the latter silica these are connected to each other with micropores.

The mixture is maintained at 273–373 K for several days, during which extensive hydrolysis and condensation occur, giving rise to the formation of siloxane (Si-O-Si) bonds. This polymerization leads to precipitation and the solid is filtered out, washed and collected. Finally the product is calcined at 600–900 K or washed with a suitable solvent, such as hydrochloric acid or ethanol, to remove the surfactant. The space occupied by the surfactant micelle is converted into mesopores at lattice points in a certain space group. The structural characteristics of mesoporous silica, therefore, arise from the micelle structure. From time to time, we call this synthesis "inverse replication." The pore size is distributed according to the micelle structure and the framework is usually amorphous. This is very different from the structure of a crystalline zeolite prepared with a molecular ion template, which gives a much narrower pore size distribution than a surfactant micelle. The choice of surfactant is quite important in obtaining particular mesostructures. Cetyltrimethylammonium bromide (CTMAB), cetyltriethylammonium bromide (CTEAB) and Pluronic® P123 are typically used for the synthesis of MCM-41 (and MCM-48, too), SBA-1 and SBA-15, respectively.

The application of this type of silica has been explored in most of the fields concerned with interfacial phenomena, such as molecular sieving, catalysis and sensors,[6-9] and a considerable number of studies have been contributed to the functionalization with reactive organic compounds, such as organosilanes. The adsorption of environmentally toxic cations and anions on such functionalized mesoporous silicas has been intensively investigated. This trend is not due to a simple coincidence of the rising importance of the chemistry of mesostructured materials and the increasing worldwide concern with environmental problems. If not, why have mesoporous silicas frequently been employed as platforms for the synthesis of functionalized solids in the development of environmental remediation tools?

Large surface areas are often required in the application of solid materials. This surface should be, needless to say, accessible to the molecules and ions in the particular application, while the value evaluated by the BET surface area measurement, using nitrogen adsorption, does not always characterize the surface properly. Thus, when we conceive or plan a certain environmental application of a solid, we should further consider the physical and chemical properties of the surface, such as hydrophilicity, porosity and pore size distribution. Since the surface energy increases with the specific surface area, fine particles tend to coagulate, resulting in a decrease in surface area and a deterioration in performance while they function. The porous structure is an important factor in extending the surface area while at the same time preventing such coagulation. This illustration is rather simplified, but explains why the attention of researchers concerned with applications requiring large surface areas has been drawn toward porous solids.

Uniform pore size is considered to be one of the characteristics of practical importance. Among porous solids, zeolites[10] and carbons[11] have been most frequently studied. In the zeolites, which are found naturally or synthesized artificially, the pore size is at most 1.3 nm. This is not large enough for aqueous ions to diffuse rapidly into the pores. Moreover, the chemical nature of the zeolite surface is not always the most favourable for aqueous ions for the following reason. The pore surface of pure silica zeolite is hydrophobic rather than hydrophilic due to its crystalline nature. Higher hydrophilicity is obtained by choosing zeolites containing Al, which gives rise to defects with acidity. Hydrophobicity in zeolites is sometimes a disadvantage for the adsorption of toxic materials in an aqueous environment. Water is also the most widely used medium in industry. Furthermore, it is difficult to decorate the pore surface with organic functionalities due to the small pore size. This means that there are not many ways by which the structure of the adsorption sites for a particular target adsorbate can be optimized.

Active carbons have been widely used for the removal of harmful or unpleasant substances, especially in closed environments. Although the experimental surface area often exceeds 2,000 m^2/g, they have a wide pore size distribution from micro- to mesopores. In active carbons the micropores usually contribute a significant fraction of the total pore volume and the BET surface area. This pattern of pore size distribution is not suitable for solid materials used in solution. In fact, considerable effort has been devoted to diminishing the number of small micropores in order to increase the capacitance of carbon electrodes.[12] On the other hand, the surface hydrophilicity can be controlled, more or less, by a chemical treatment, because the organic functional groups terminate the lattice at the carbon surface. For the development of active carbons as adsorbents or sensors in an aqueous environment, it should be noted that several kinds of surface group, which have quite different reactivities, exist at the same time and the density of these functional groups is usually much lower than that of silanols on the surface of amorphous silica. These properties make the formation of a uniform and dense layer of functional groups difficult. The reaction of silane at the surface of amorphous silica mainly occurs between the surface silanols (Si-OH) and the alkoxy (Si-OR) bond in the silane molecule. In addition, the density of silanol in mesoporous silica is high, say 2–3 nm^{-2}. These properties seem advantageous for the purpose of functionalization for synthesizing the adsorption sites of target ions.

In spite of the advantages of mesoporous silica for the adsorption of toxic substances, there are several defects *a priori*. Shape selectivity, such as that found in the molecular sieving effect of zeolites, is not to be expected for two reasons. One is that the pore sizes are distributed more than in zeolites. This is because the zeolites are crystalline solids while mesoporous silicas are amorphous, and their templates are a molecular ion and a micelle of a surfactant, respectively. The second reason is that, even if we could prepare silica with well-defined pore size as in crystalline solids, the molecular sieving effect is less clear-cut in a mesopore window than in a microporous one. The variation of conformational isomers increases exponentially with molecular size, and this effect makes the dimensions of the molecules ambiguous. Therefore, it is not always reasonable to expect to have precise control in simple molecular sieving with mesoporous silica and the materials on a mesoporous silica platform. Another disadvantage of mesoporous silica is the instability of the mesoporous structure under hydrothermal conditions, such as in boiling water.[13] The fragility in water raises doubts about its durability in practical applications. Once these disadvantages can be overcome, mesoporous silica will provide an excellent platform for synthesized interfaces for remediation in aqueous environments, making full use of the fast diffusion of molecules and ions in its pores as well as large accessibility of the surface.

11.1.2 *Environmentally toxic anions*

Drinking water is regulated as shown in Table 11.1; however, excessive contaminant anions, such as arsenate, chromate, selenate and molybdate, have often been found in the environment. It has been claimed in several different countries that chronic human health disorders caused by these toxins have been discovered. The arsenic crisis in Bangladesh[14,15] is considered to be due to drinking water polluted with natural arsenate and arsenite in groundwater. An examination into the quality of groundwater

Table 11.1.   Guidelines for the maximum amount of inorganic pollutants in drinking water (mg/L).

	WHO Guideline	US EPA Regulation	EU Direction	Japan Regulation
Antimony	0.005[a]	0.006	0.005	0.002[a]
Arsenic	0.01[a]	0.05	0.01	0.01
Barium	0.7	2		
Beryllium		0.004		
Boron	0.5[a]		1	1
Cadmium	0.003	0.005	0.005	0.01
Chloride			250	
Chromium	0.05[a]	0.1	0.05	0.05[b]
Copper	2[a]		2	
Cyanide	0.07	0.2[c]	0.05	0.01
Fluoride	1.5	4.0	1.5	0.8
Iron			0.2	
Lead	0.01		0.01	0.01
Manganese	0.5[a]		0.05	
Mercury	0.001	0.002[d]	0.001	0.0005
Molybdenum	0.07			0.07
Nickel	0.02[a]		0.02	0.01
Selenium	0.01	0.05	0.01	0.01
Sodium			200	
Sulfate			250	
Uranium	0.002[a]	0.03		0.002[a]
Thallium		0.002		

[a] Temporary value.
[b] As hexavalent Cr.
[c] As free cyanide.
[d] As inorganic mercury.

in the Chikugo Plain (Japan) demonstrated that about 25% of the 11,673 wells are contaminated with arsenic above the regulation limit for drinking water.[16] Chronic toxicity is also remembered for triggering the emergence of blackfoot disease on the south-western coast of Taiwan.[17,18] Exposure to arsenic from drinking water has also been reported from the American continents.[19] A major source of chromium pollution is industrial, such as from leather tanning and electroplating factories, where proper waste treatment facilities may be lacking.[20,21] These observations show that although access to safe water is a fundamental human need, it is often threatened in societies that depend on wells.

The exposure to anionic pollutants in groundwater is a universal phenomenon and the development of remediation methods is an urgent issue for environmental technology and materials chemistry. Building a water purification plant is normally an inappropriate measure for societies dependent on wells, and more cost-effective and environmentally friendly methods are needed. Chemical separation of harmful anions, such as chemisorption, can be a good solution; however, the difficulties will be encountered in the selectivity of such anions.

In the toxic oxyanions referred to earlier, the central element is penta- or hexavalent in the oxygen tetrahedron. These anions have a quite similar structure; the central atom-oxygen distances in AsO_4^{3-}, CrO_4^{2-}, SeO_4^{2-} and MoO_4^{2-} are 0.169, 0.165, 0.166 and 0.176 nm, respectively. This similarity in structure implies that the chemical behaviour is almost the same for adsorption on solid surfaces. From the environmental point of view, the selectivity against chloride, sulfate, nitrate, carbonate and phosphate will be more important in the removal of toxic oxyanions, because their typical concentration in fresh water is from several ppm to several tenths of ppm. Considering the environmental regulation, the adsorption of toxic oxyanions needs to be 100–1,000 times more selective than that of these abundant natural anions. Sulfate and phosphate are also tetrahedral oxyanions, even though the bond length between the central element and oxygen is smaller than in the pollutants: $r(S\text{-}O) = 0.148$ nm (in SO_4^{2-}) and $r(P\text{-}O) = 0.154$ nm (in PO_4^{3-}). The structural nature of such aqueous anions will cause difficulties in the separation of toxins. This problem would emerge as a severe inhibition of adsorption, because the number of adsorption sites would be limited. Fine control of the structure of the adsorption site is thus usually necessary for preparing a selective adsorbent.

11.1.3 Hierarchy of solid structures and adsorption

It has been demonstrated that solids synthesized by conventional methods, such as graining and impregnation, have unsatisfactory adsorption strengths and capacities for toxic oxyanions.[22] To synthesize effective adsorbents for a particular target ion, the structural factors of the solid should be considered both at the micro- and mesoscales. These views on different scales cannot always be separated but sometimes interfere with each other. Fumed silica with a large surface area on which organic functional groups are grafted can be used as a synthetic adsorbate. In such a solid, the strength of the adsorption is controlled mainly by the microstructure of the adsorption site. This is critically important for oxyanion adsorbents, because a nearly complete removal (= a huge distribution coefficient) is necessary in order to meet the environmental regulations and high selectivity to other bulk anions with a similar structure is also needed. On the other hand, when using porous silica, the rate of diffusion during adsorption depends on the porous structure, which is to be evaluated at the micro- and mesoscales. The preparation of the surface, which is called functionalization or decoration of the surface, also depends considerably on the micro- and mesostructures. One of the important examples of the study carried out on microstructures is the reactivity of silanol groups on a silica surface, which has, in fact, been argued for a long time.[23] On the other hand, it has been pointed out that uniform dispersion of organic functional groups is difficult in the functionalization of a mesoporous silica.[24] The structure of the surface organic groups and the results of adsorption experiments can vary considerably, even when the same kind of functional group is grafted on silica with a similar pore size and surface area, since they depend on the preparation method and the porous structure. We will see the examples later in this chapter.

In functionalized mesoporous silicas, the microstructure revealed by conventional physicochemical techniques, such as IR, NMR and UV-Vis, is less like that predicted by the adsorption characteristics than in functionalized non-porous silicas. This is due to the interference by mesopores on the microscopic structure of the organic groups during preparation as well as in the working state. The adsorption is sometimes influenced more by the distribution of adsorption sites in the mesopores and the interactions between organic groups and the framework silica surface than the microstructure of adsorption sites. This cannot only be true for the case of functionalized mesoporous silica but also for functionalized fumed silica, polymers and natural substances. It is proper to mention that the adsorption properties sensitive to the mesostructure will appear explicitly only due to the usage of a defined and rigid mesoporous structure of the adsorbates. In this sense, choosing mesoporous silica with a periodically defined structure as a platform for synthetic adsorbate is a good way to obtain an understanding of mesostructural factors in the disturbance of adsorption. The features of oxyanion adsorption under mesostructural influence are described in the following sections of this chapter.

11.2 Functionalization of Mesoporous Silicas

11.2.1 *Decoration of the pore surface with an organic layer*

The three methods described in this section are illustrated in Scheme 11.1.

One of the simplest preparation methods of organic layers in mesopores is grafting, in which reactive silane with a desired functionality is grafted via the reaction mainly with isolated silanols on the mesoporous silica. The loading of the organic groups and the uniformity of the organic layer depend on the density of the silanols and adsorbed water on the silica surface. Although this preparation method can be applied to all kinds of amorphous silicas, specific problems arising from the mesoporous structure have been recognized, with an increasing number of research papers on grafting. One of the prominent problems is non-uniform distribution of the functional groups.

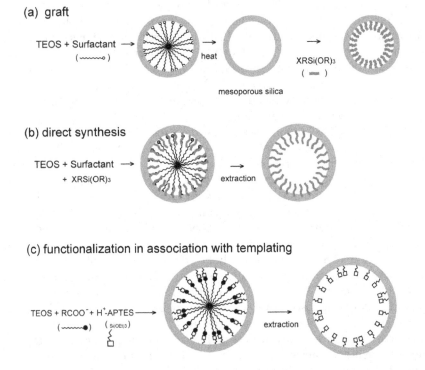

Scheme 11.1.   Preparation of functionalized mesoporous silica, illustrated with the formation of the silica framework: grafting, direct synthesis and functionalization in association with templating.

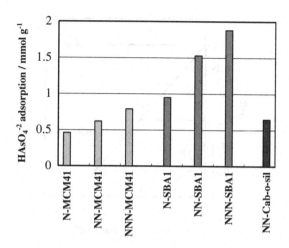

Figure 11.2. Adsorption capacity of arsenate on 3-aminopropyltrimethoxylsilane (N-), [1-(2-aminoethyl)-3-aminopropyl]trimethoxysilane (NN-), 1-[3-(trimethoxysilyl)propyl] diethylenetriamine (NNN-) grafted on MCM-41 and SBA-1. The BET specific surface area of these mesoporous silica frameworks is 1,280 and 1,220 m^2/g, respectively, and the pore size is 2.9 and 3.0 nm, respectively. The result on [1-(2-aminoethyl)-3-aminopropyl]trimethoxysilane grafted on non-porous silica (Cab-O-sil M-7D, BET surface area: 200 m^2/g) is shown for comparison.

Figure 11.2 shows the adsorption capacity of arsenate on amino-functionalized MCM-41 and SBA-1 prepared by grafting 3-aminopropyl trimethoxylsilane ($NH_2CH_2CH_2CH_2Si(OCH_3)_3$), [1-(2-aminoethyl)-3-aminopropyl]trimethoxysilane ($NH_2CH_2CH_2NHCH_2CH_2CH_2Si(OCH_3)_3$, which can be called EDA-propylsilane), and 1-[3-(trimethoxysilyl)propyl] diethylenetriamine ($NH_2CH_2CH_2NHCH_2CH_2NHCH_2CH_2CH_2Si(OCH_3)_3$, which can be called DETA-propylsilane). The adsorption increases with the number of amino groups in the silane molecule, though, with careful observation, the lack of linearity can be found on the data of MCM-41. In addition, in spite of the similar surface area and pore size, the adsorption is considerably different between these two mesoporous frameworks. This is certainly because the amino groups in a large organic group or in highly loaded organic groups are bound to fewer arsenates than those in small or dispersed organic groups.[25] The difference between the frameworks is more significant in a plot of the As/N ratio versus the amount of amino groups loaded on the solid (Figure 11.3). This apparent stoichiometry is constantly found in EDA-propylsilane-grafted SBA-1, NN-SBA-1, while it decreases with the loading of amino groups in EDA-propylsilane-grafted MCM-41, NN-MCM-41. However, the extrapolation to zero gives us a common stoichiometry: As/N = 0.5. These plots suggest that the stable

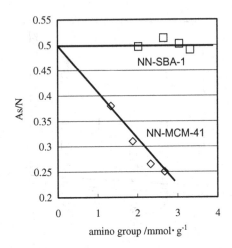

Figure 11.3. Arsenate adsorption capacity of [1-(2-aminoethyl)-3-aminopropyl] trimethoxysilane-grafted MCM-41 and SBA-1. The ratio of arsenic to nitrogen is plotted against the amount of amino groups.

structure at adsorption saturation is two amine groups in the EDA-like ligand binding one arsenate anion, unless there is inhibition by neighbouring organic groups or by the pore wall, and that a certain proportion of the organic groups in NN-MCM-41 do not work, probably due to mutual interference in the adsorption. It has often been claimed but not sufficiently demonstrated that, in grafting silane on mesopore walls, dispersion of the silanes is hampered by the competition between the diffusion into the pores and the reaction with the pore surface.[24] As a result, the grafted silane molecules tend to be densely populated near the pore windows. Although this hypothetical mechanism might be closely related to the decrease of As/N in Figure 11.3, the difference between the framework structures cannot be explained. It could be caused by the pore-connecting geometry of MCM-41, which is a straight channel and the entrance exists only on the ends of the tubes. In contrast, the pores are connected three-dimensionally in the cubic Pm3n structure of SBA-1 and self-blocking by grafted silanes gradually clogs the pores.

Since uniform dispersion is often unlikely when grafting silanes onto mesoporous silica, the co-condensation of TEOS and silane with a desired function is proposed for the preparation of functionalized mesoporous silica. This method is also called direct synthesis.[24] Although several problems have been pointed out in direct synthesis, the organic groups are much more uniformly dispersed than in grafting when TEOS and silane are well mixed before hydrolysis. A distinct difference between amino-functionalized

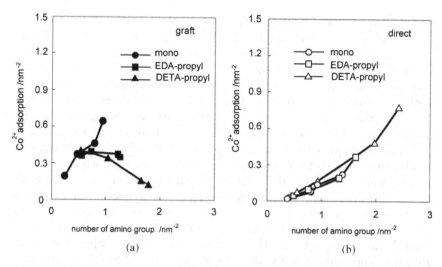

Figure 11.4. Adsorption of Co^{2+} on (a) 3-aminopropyltrimethoxylsilane (mono-), [1-(2-aminoethyl)-3-aminopropyl]trimethoxysilane (EDA-), 1-[3-(trimethoxysilyl)propyl] diethylenetriamine (DETA-) grafted MCM-41 and (b) those equivalent mesoporous functionalized silicas prepared by the direct co-condensation method. These are plotted against the amount of amino groups.

mesoporous silicas prepared by grafting and co-condensation has been found in cobalt(II) adsorptions (Figure 11.4).[26] The adsorption capacity is expected to increase with the number of adsorption sites grafted, if the functional groups are grafted uniformly on the surface. On the contrary, the variation of adsorption depends on the type and the surface density of silanes. In grafted MCM-41 ($2R_p = 2.6$ nm), the amount of Co^{2+} adsorption is not proportional to the amount of amino groups. The adsorption is even unchanged or decreased with the loading, when EDA-propylsilane or DETA-propylsilane is applied, respectively. These non-linearities can be explained by the following mechanism. A large silane molecule can suffer from interference with its diffusion into the pores, while the reactivity of methoxy groups is almost the same as in short silanes, and, consequently, the silane may settle near the pore window. This settling can inactivate the amino groups by the congestion of organic chains as well as inhibit the diffusion of Co^{2+}.[26] In contrast, all the adsorption data for the functionalized MCM-41-motif prepared by direct synthesis appear almost on the same line, implying a homogeneous dispersion of amino groups in the pores, regardless of the surface density of the amino groups and the structure of the organic chains in the silane molecule.

Direct synthesis can inactivate a part of the functional groups through strong interaction with the surface or occlusion of the organic groups into the framework.[24] This problem appears considerably in aminopropyl-functionalization, which may be closely related to the significant distur-bance in the formation of ordered mesopore walls (suggesting a strong interaction with the silica framework).[27] A solution for avoiding the dis-turbance of amino groups during the siloxane bond formation is functional-ization in association with templating (FAT).[24,28,29] In this modified direct synthesis, the template molecule and the functional group are strongly bonded during hydrothermal synthesis. The functional group is a part of the silane molecule and the other silane bond, Si-OR, is hydrolysed with a silica precursor to form a siloxane network. After the mesostructure is formed, the solid is treated with a suitable solvent to dissociate the bond in the surfactant-functional group complex. To prepare 3-aminopropyl-functionalized mesoporous silica, 3-aminopropyltriethoxysilane (APTES) and TEOS are hydrolysed in the presence of a carboxylate with a long alkyl chain. Under suitable pH conditions (pH $=$ $ca.$ 10), where the amino groups are protonated, while the rate of hydrolysis of Si-OC$_2$H$_5$ bonds is still high enough, amino groups and carboxylate are bound together dur-ing hydrothermal synthesis. The hydrolysis of Si-OC$_2$H$_5$ in APTES and TEOS results in polymerization and a solid formation around the carboxy-late micelles. The template molecules are removed by washing with acidi-fied CH$_3$CN to liberate the amino groups on the pore surface. Considering the lack of reports on the successful synthesis of mesoporous silica using TEOS and an anionic surfactant, the interaction between amino and car-boxyl groups is essential for the mesostructure formation. (In fact, under pH conditions higher than 11, no mesoporous solid is formed.) It may also be claimed that all amino groups exist, theoretically, only on the pore surface and that their dispersion is uniform.

Figure 11.5 shows a comparison between argentometric titration and elemental analysis of nitrogen in functionalized mesoporous silica. The for-mer measurement is based on the reaction Ag$^+$ $+$ Cl$^-$ $=$ AgCl (precipi-tated), where the Cl$^-$ neutralizes the –NH$_3$$^+$ of the aminopropyl groups, and, hence, is a quantification of the –NH$_3$$^+$ exposed on the surface. The ratio of nitrogen measured by titration to that by elemental analysis, there-fore, indicates the surface exposure of the amino groups. This is independent of the loading of amino groups in 3-aminopropyl-functionalized mesoporous silica prepared by the FAT method and the titration records 95% of nitro-gens detected by bulk elemental analysis. On the other hand, the equiva-lent mesoporous silica (MCM-41-motif) prepared by direct synthesis using CTMAB and the same silane (3-aminopropyltrimethoxysilane) provides a linearly decreasing function against loading. This distinct difference clearly demonstrates the advantage of the FAT method in exposing the func-tional groups to the surface. The fact that almost all the amino groups

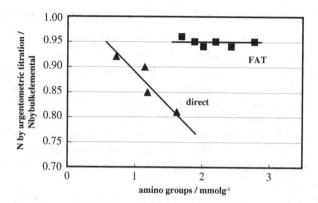

Figure 11.5. Ratio of nitrogen measured by argentometric titration to that measured by bulk elemental analysis in 3-aminopropyl-functionalized mesoporous silicas. The solid samples are prepared by functionalization in association with templating (FAT) and by direct synthesis. The template surfactants are sodium laurate (LAS) and CTMAB, respectively.

are exposed to the surface over all loadings reveals that the attractive interaction between the amino and carboxyl groups really works during the mesostructure formation. The head group of CTAB that is used for direct synthesis can be repulsive to the amino groups during the mesostructure formation. This mode of interaction promotes the interaction between amines and silanol groups or expels the organic chain into the micropores in the wall.

Another advantage of the FAT method is an easy achievement of high loading of the functional group. More than 5.4 mmol/g of 3-aminopropyl groups can be found in functionalized mesoporous silica without loss of periodicity, when it is prepared by the FAT method.[24] On the other hand, in direct synthesis, 3-aminopropyltrimethoxysilane and CTAB provides a loading of 1.7 mmol/g.[26] Although the density of amino groups depends considerably on the pre-treatment of the MCM-41, the grafting provides around 1.5–2 mmol/g after the removal of physisorbed water to activate Si–OH groups on MCM-41.[25,26] The comparison with simple direct synthesis and grafting reveals the outstanding loading of functional groups by FAT.

Other valuable techniques, such as on-surface synthesis, grafting on as-synthesized mesoporous silica, multiple functionalization of specific sites (in the pores or on the exterior surface), in the preparation of organic functionalities on mesoporous silicas have been described in recent reviews.[24]

11.2.2 *Transition metal cation-incorporated adsorption sites*

The amino groups, especially chelating ligands such as EDA, coordinate transition metal cations. Many of these kinds of complex are usually stable over a wide pH range. Considering that the interaction between transition metal ions and oxyanions is often found in insoluble minerals, such as scorodite (iron(III) arsenate dihydrate, $FeAsO_4$ $2H_2O$), much stronger chemical interactions with oxyanions can be obtained at the transition metal cation coordinate sites than the protonated amino group. Iron(III)-, cobalt(II)-, nickel(II)- and copper(II)-coordinated EDA-propyls have been prepared with [1-(2-aminoethyl)-3-aminopropyl]trimethoxysilyl mesoporous silicas.[30,31]

This is a typical on-surface synthesis (post-modification), where the coordination of the cation is carried out at the grafted functional groups, and, hence, attention should be paid to the uniformity. This is because the conversion of surface organic groups by substitution, oxidation, reduction, etc., has been usually small in the studies on on-surface synthesis of mesoporous silicas. The examples have been documented in Ref. 24(a). The cation coordination also provides unsatisfactory results for a uniform conversion, as shown in Table 11.2. The nitrogen/metal cation ratios in the cation-coordinated amino-functionalized mesoporous silicas quite often have an unreasonable stoichiometry for the coordination complexes expected.[31,32] The ratio of less than unity found in Ni^{2+}-coordinated mesoporous silicas suggests the adsorption of nickel(II) on the silica surface.

Table 11.2. Uptake of transition metal cations by amino-functionalized mesoporous silicas. N/M^{n+} ratios measured by elemental analyses.

	N density* mmol/g	N/Fe^{3+}	N/Co^{2+}	N/Ni^{2+}	N/Cu^{2+}
N-MCM-41	1.5	3.2	2.5	0.48	2.6
N-MCM-48	1.5	2.0	1.8	0.43	2.0
NN-MCM-41	2.8	3.8	3.3	0.55	3.6
NN-MCM-48	2.7	2.7	2.3	0.40	2.4
NNN-MCM-41	3.5	2.8	7.9	1.1	9.4
NNN-MCM-48	4.2	2.8	7.0	1.0	9.0

N.B. MCM-41 and MCM-48 functionalized by grafting 3-aminopropyl-trimethoxylsilane (N-), [1-(2-aminoethyl)-3-aminopropyl] trimethoxysilane (NN-) and 1-[3-(trimethoxysilyl)propyl] diethylenetriamine (NNN-) are stirred in chlorides of these cations at room temperature.

*The specific amount of nitrogen in the solid before coordination of metal cations. A part of it is lost after coordination, accompanied with a decrease of the BET surface area.

In contrast, $N/M^{n+} = 7-9$ in NNN-functionalized mesoporous silicas indicates a large number of uncoordinated amino groups in the pores. Table 11.2 shows only the average compositions of the bulk solids and this may be the results from mixtures of several sites with different structures. However, the N/M^{n+} ratios for $Fe^{3+}/$ and Cu^{2+}/NN-MCM-41 suggest a structure in tetra-coordination of the amino groups.

The N/M^{n+} ratio depends on the loading of the EDA-propyl groups. It has been claimed from the results of EDS-SEM that, at an extremely high loading of the EDA-propyl groups, a $[Cu(EDA)_3]^{2+}$-type complex is formed in NN-MCM-41.[30,33]

In adsorptions on real solids, since the adsorption sites are not usually uniform and interference between the sites occurs, Langmuir-type isotherms are not obtained. Furthermore, high-quality plots are hardly obtained due to the low concentration of oxyanions in the experimental solution. Instead, the strength of the adsorption should be evaluated by the distribution coefficient, K_d, between the solid and the solution. A typical K_d-adsorption plot is shown in Figure 11.6. The detection limit of the chemical analysis gives the maximum distribution coefficient, which is equal to 2×10^5 in Figure 11.6. In all the adsorbents, K_d decreases with adsorption, implying the distribution of the adsorption strength or interference between the adsorption sites. The vertical drops that are observed with relatively low K_d

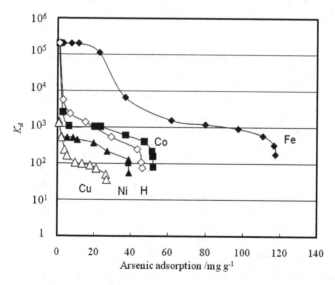

Figure 11.6. Distribution coefficient versus adsorption of arsenate on Fe^{3+}-, Co^{2+}-, Ni^{2+}-, Cu^{2+}- and H^+-coordinated NN-MCM-41.

indicate adsorption saturation. The adsorption capacity is given by these drops. Both the strength and capacity of the adsorption are highest in Fe^{3+}/NN-MCM-41. The value of K_d for this adsorbent reaches more than 10^5 in the region where that for Cu^{2+}/NN-MCM-41 is 10^2; the strength of arsenate adsorption differs by three orders of magnitude. The adsorption saturation of this iron-activated adsorbent is four times higher than the copper-activated one. Thus the adsorption strength and capacity depend considerably on the cations in the adsorption site.

The adsorption capacity of these transition metal cation-activated amino-functionalized mesoporous silicas is summarized in Table 11.3.[25,31,32] Although in practice the uptake of arsenate per unit amount of adsorbent has a large impact, and is, in fact, superior to that found in natural or nearly natural solids, such as goethite $(20\,\text{mg/g})$,[34] natural iron ore $(0.4\,\text{mg/g})$,[35] activated alumina $(15.9\,\text{mg/g})$,[36] anatase with a BET surface area of $251\,\text{m}^2/\text{g}$ $(41.4\,\text{mg/g})$,[37] etc., the molar ratio of arsenic to cation is more important in exploring the mechanism and the structure of the adsorption site. Many of the As/M^{n+} values are irrational for a stoichiometric reaction, just as found in Table 11.2; the adsorption structures are likely to be mixtures. The As/M^{n+} ratio will contain structural information of the adsorbent only when the molar ratio of nitrogen to cation is nearly stoichiometric: $N/M^{n+} \sim 4$ for Fe^{3+}/NN-MCM-41 and Cu^{2+}/NN-MCM-41 (Table 11.2). The As/M^{n+} values for these adsorbents are 2.8 and 0.7, respectively, suggesting that the stoichiometric ratio at saturation is As : $M^{n+} = 3 : 1$ and 1:1 for iron and copper, respectively.

Table 11.3. The adsorption capacity of arsenate on cationized amino-functionalized mesoporous silicas.

		H^+	Fe^{3+}	Co^{2+}	Ni^{2+}	Cu^{2+}
N-MCM-41	C_{max}	64	122	84		
	As/M^{n+}	0.35	2.1	1.1		
N-MCM-48	C_{max}	71	146	124		
	As/M^{n+}	0.37	1.6	1.2		
NN-MCM-41	C_{max}	86	226	97	73	51
	As/M^{n+}	0.22	2.8	1.2	0.25	0.69
NN-MCM-48	C_{max}	122	353	141	121	71
	As/M^{n+}	0.38	2.7	1.2	0.32	0.65
NNN-MCM-41	C_{max}	110	226	74		
	As/M^{n+}	0.25	1.6	1.3		
NNN-MCM-48	C_{max}	182	251	88		
	As/M^{n+}	0.37	1.4	1.1		

N.B. C_{max} in $\text{mg}(\text{g-adsorbent})^{-1}$. The blanks indicate that the capacity was not measured.

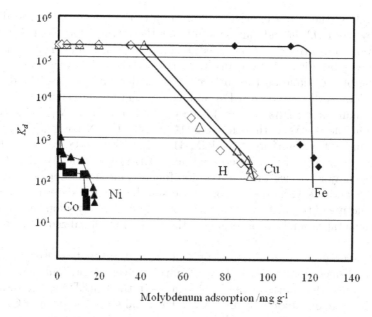

Figure 11.7. Distribution coefficient versus adsorption of molybdate on Fe^{3+}-, Co^{2+}-, Ni^{2+}-, Cu^{2+}- and H^{+}-coordinated NN-MCM-41.

The K_d-adsorption plot is useful for an appraisal of the adsorption properties of environmentally toxic oxyanions.[31(b)] We can easily tell the behaviour of an ideal adsorbent; K_d is immeasurably high (= below the detection limit in the analysis of the concentration of anions remaining in solution) till adsorption saturation is reached in addition to a large capacity. This is observed in the adsorption of molybdate on Fe^{3+}/NN-MCM-41 as shown in Figure 11.7. The profiles are completely different from those in arsenate adsorption. The strength and capacity of Cu^{2+}/ and H^{+}/NN-MCM-41 are larger than Ni^{2+}/ and Co^{2+}/NN-MCM-41, while Co^{2+}, H^{+} and Ni^{2+} show larger K_d and capacities than Cu^{2+} in arsenate adsorption.

11.2.3 *Structure of the adsorption sites*

In solution, tris(ethylenediamine)iron(III) sulfate has been known as an Fe(EDA)$_3$ complex.[38] The complexation constants, log K_1, log K_2 and log K_3, for the EDA ligand with Co^{2+} are 5.97, 4.91 and 3.18 (in an ionic strength of 1.4 mol/dm^3), respectively, while those with Cu^{2+} are 10.72, 9.31 and −1.0. The third constant for Cu^{2+} suggests that [Cu(EDA)$_3$]$^{2+}$ is possible, only if the concentration of EDA is high.[39,40] These equilibrium

constants in the homogenous chemistry imply that the threefold coordination by the EDA ligand can be stable for these transition metal cations.
It is well known that, in d^9 octahedral complexes, the number of electrons
in the e_g orbital is odd and trans-$[Cu(EDA)_2(H_2O)_2]^{2+}$ is very stable. The
fixation of the silane on the surface will considerably decrease the degree
of freedom in the motion of EDA and, consequently, a high surface density
(more than three EDAs per nm^2) is necessary to obtain an $M(EDA)_3$ complex. On the contrary, the surface EDA density (= (N content)/2 S_{BET})
is 0.9 and 1.4 per nm^2 for NN-MCM-41 and NN-MCM-48, respectively.[30]
These values support the contention that $M(EDA)_2Cl^-_n$ and $M(EDA)Cl^-_n$
are the major species on these amino-functionalized mesoporous silicas.
XANES and EXAFS spectroscopies are suitable for the determination of
the structure of the surface sites, both before and after oxyanion adsorption,
when a metal cation is incorporated in the amino-functionalized mesoporous
silica.

A structural determination of sulfate and arsenate adsorption on
Cu^{2+}/NN-mesoporous silica ($2R_p = 6$ nm) has been carried out in detail.[33]
It is claimed that parts of the EDA ligands in the $Cu(EDA)_3$ site dissociate when the oxyanion is coordinated. The local structure around Cu and
As indicates direct Cu-O bond formation, a monodentate linkage between
the anion and Cu^{2+} and a trigonal bipyramidal geometry of the Cu centre (Figure 11.8). With this adsorbent, however, the specific adsorption
reached 142 mg/g.[30] This is significantly larger than the data in Table 11.3:
51 and 71 mg/g for NN-MCM-41 and NN-MCM-48, respectively. This difference is likely to be due to a larger number of silanes being grafted and,
hence, Cu^{2+} coordinated: 1.0–1.4 mmol/g in the study in Ref. 30, which

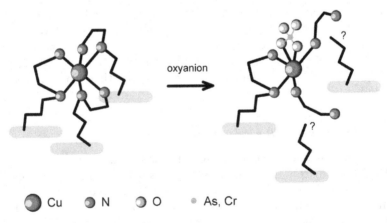

oxyanion

● Cu   ● N   ○ O   • As, Cr

Figure 11.8. $[Cu(EDA)_3]$ centre for arsenate adsorption. A possible adsorption model.

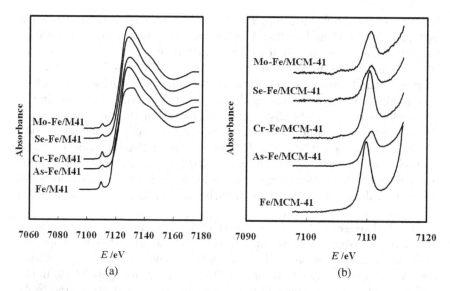

Figure 11.9. (a) Fe K-edge XANES spectra of Fe^{3+}/NN-MCM-41 and that after adsorbing arsenate, chromate, selenate and molybdate. (b) The pre-edge region expanded for comparison.

is almost twice as large as the density in Cu^{2+}/NN-MCM-41 (Table 11.2). Interestingly, the As/Cu ratio in Table 11.3, 0.69, agrees remarkably with the adsorption on highly dense Cu^{2+}/NN-mesoporous silica, 0.71. These comparisons suggest that the capacity is explained simply by that of the loading of Cu^{2+} on amino-functionalized mesoporous silica and the mode of interaction of arsenate is not very sensitive to the coordination environment of Cu^{2+}.

The oxyanion adsorbed on Fe^{3+}/NN-MCM-41 has been investigated also by XAFS spectroscopies.[31,32] Figure 11.9 shows the Fe K-edge XANES of Fe^{3+}/NN-MCM-41 and those adsorbing arsenate, chromate, selenate and molybdate. Two unresolved peaks appear in the pre-edge region of the spectra for arsenate- and selenate-adsorbed Fe^{3+}/NN-MCM-41. The first component appears at *ca.* 7,110.5 eV, which has a smaller area than the second one at *ca.* 7,110.8 eV. These combined peaks are a typical feature of an O_h-like Fe(III) centre. Theoretical and simulative studies of the Fe K-edge 1s–3d transition of iron complexes revealed that the absorption at the lower energy is related to electric quadrupole transitions, whereas that at the higher energy is due to both quadrupole and dipole transitions.[41] In molybdate-adsorbed Fe^{3+}/NN-MCM-41, the intensity is almost the same as in the selenate-adsorbed Fe, but the peaks are not distinguished from each other. The enhanced pre-edge peak is caused by the dipole-allowed transition by p-d hybridization. This feature is observed in the spectrum of Fe^{3+}/NN-MCM-41 and in that binding with chromate. After the adsorption of arsenate, selenate and probably molybdate, the structure of Fe^{3+} becomes more symmetric than the initial Fe^{3+}/NN-MCM-41, while no such structural change is observed in chromate adsorption. The difference in the pre-edge peak among the oxyanions suggests a variation in the adsorption strengths. The distribution coefficient in chromium adsorption is smaller than the other three oxyanions (e.g. comparing the adsorption of 20 mg/g, K_d for CrO_4^{2-} is 3,000 while those for $HAsO_4^{2-}$, SeO_4^{2-} and MoO_4^{2-} are 11,0000, >200,000 and >200,000, respectively). This is not a simple coincidence but suggests a mechanism where the relatively weak interaction does not significantly disturb the coordination environment of Fe^{3+}.

The Fe K-edge EXAFS spectra measured before and after the adsorption of 1, 2 and 2.8 equivalent arsenates on Fe^{3+}/NN-MCM-41 follow the growth of the adsorption structure.[31] $k^3\chi(k)$ EXAFS oscillations with curve-fitting results and their Fourier transforms (FT) are shown in Figure 11.10. In the radial distribution functions, the peak around 2.9–3.0 Å is attributed to As scatterers. This peak appears in the FT for the adsorption of one equivalent arsenate and it grows for the adsorption of two equivalent arsenates. However, the intensity for three equivalent adsorptions is almost the same as that for two. The coordination number of this bond is 0, 1.0 ± 0.4, 1.6 ± 0.5 and 1.6 ± 0.5 for before adsorption, adsorption of one, two and three equivalent arsenates, respectively. In contrast, as the adsorption progresses the coordination number for the Cl shell decreases; found to be 2.1 ± 0.4 prior to adsorption, 1.2 ± 0.4, 0.4 ± 0.2 and 0.2 ± 0.2, for one, two and three equivalent arsenates. The distance is unchanged at 0.227 nm. The Fe–As bond length is constant at 0.327 nm. These fitting results suggest that two of the arsenate ions can be bound in the inner-shell environment of Fe^{3+} even with full coverage of arsenate, where the stoichiometry is almost As : Fe = 3:1. Unlike the change in the Fe^{3+} centre, the EXAFS of the As K-edge showed no significant change in the As-O bond after adsorption, both in the coordination number and the bond length, as shown in the $k^3\chi(k)$ EXAFS oscillations and their Fourier transforms in Figure 11.11. The result of the curve-fitting calculation is $N_{As-O} = 3.9 \pm 0.4$ and $r_{As-O} = 0.169$ nm. The tetrahedral coordination environment of arsenic is unlikely to be disturbed by the adsorption on Fe^{3+}/NN-MCM-41. The As-Fe length and the coordination number in the solid with As : Fe = 1:1 are 0.327 nm and 0.95, respectively, and these parameters are not changed in the adsorption with As : Fe = 2:1. The agreement of Fe-As and As-Fe distances from the Fe and As K-edge EXAFS spectra, respectively, provides the validity of the analysis.

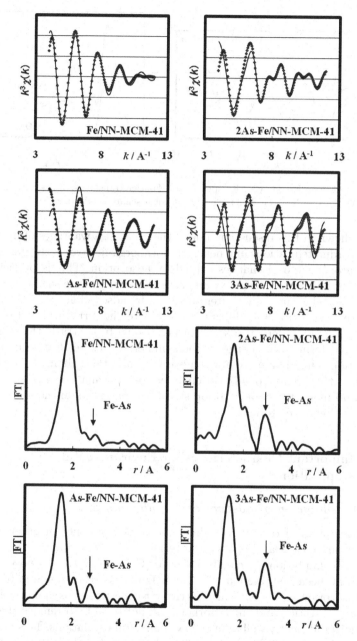

Figure 11.10. Fe K-edge EXAFS spectra and Fourier transforms of Fe^{3+}/NN-MCM-41 and that after adsorbing arsenate. The amount of arsenate adsorption was changed to control the coverage.

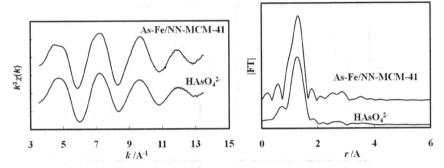

Figure 11.11.   As K-edge EXAFS spectra and Fourier transforms of arsenate in water and on Fe^{3+}/NN-MCM-41. The molar equivalent arsenate is adsorbed.

As shown in this discussion, one of the greatest advantages of mononuclear coordinated cations in a nearly uniform organic layer is the possibility to monitor structural changes caused by adsorption. This is particularly important in the analysis of oxyanion adsorptions, because the oxygen shell of the toxic elements is not often distorted by adsorption, just like that of arsenate, and hence it is difficult to elucidate the structural factors that can explain the different adsorption behaviour. In contrast, the dissociation, bond formation, change in the symmetry, etc., readily occur in the adsorption centres after oxyanion adsorption and their changes are closely related to the strength of adsorption. Spectroscopic analysis will be applied, more and more, to the adsorption structures on cation-anchored organic-inorganic hybrid materials.

11.3 Important Characteristics for Environmental Applications

11.3.1 *Inhibition of adsorption by other anions*

The adsorption of oxyanions is hindered in the presence of other anions in solution. This is an important criterion for the environmental applications. The inhibition by co-adsorption of Cl^- and SO_4^{2-} is particularly important, because these ions are widely found at 10^0–10^2 ppm in freshwater. Adsorption with a similar selectivity to the target oxyanion will lead to rapid saturation and the adsorbent will not be able to remove the toxic oxyanions sufficiently well to meet environmental regulations. This is a realistic problem, since it has been, in fact, demonstrated that the retention of chromate is weaker than sulfate on γ-Al_2O_3.[42] The influence of the presence of phosphate and carbonate on the adsorption of arsenate on goethite has

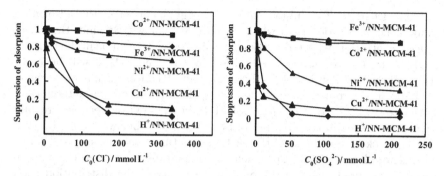

Figure 11.12. Inhibition of arsenate adsorption by the presence of chloride (upper) and sulfate (lower) evaluated by the change of adsorption capacity. The ordinates are the adsorption capacity in the presence of the inhibitor at an initial concentration of C_0 divided by that without the inhibitor. The adsorptions were carried out under the following conditions: initial concentration of arsenate: 12.2 mmol/L, adsorbent: 50 mg, volume of solution: 10 mL, and reaction temperature: 298 K.

been analysed in detail in order to understand the adsorption mechanisms of arsenic species in the environment in a more realistic interface model than in a pure arsenate solution.[43–45]

The inhibition of arsenate adsorption on the EDA-propyl-functionalized MCM-41 (NN-MCM-41) that incorporates a transition metal cation has been explored by following the change in the uptake of arsenate in the presence of Cl^- and SO_4^{2-}.[31(a)] The ratio of the adsorption capacity to that without inhibitive anions is plotted against the initial concentration of the inhibitor, C_0, in Figure 11.12. Co^{2+}/NN-MCM-41 shows a nearly constant adsorption capacity with respect to the concentration of Cl^-. The persistence of Fe^{3+}/NN-MCM-41 is ranked between those of Co^{2+}/ and Ni^{2+}/NN-MCM-41: a loss of 19% of adsorption when $C_0(Cl^-) = 342$ mmol/L. The suppressions for Cu^{2+}/ and H^+/NN-MCM-41 are considerably larger than the other three cation-incorporated NN-MCM-41; with $C_0(Cl^-) = 342$ mmol/L, 89 and 98% of the adsorption is lost in Cu^{2+}/ and H^+/NN-MCM-41, respectively. A similar result is obtained as for the inhibition of arsenate by the presence of sulfate; the adsorption on Fe^{3+}/ and Co^{2+}/NN-MCM-41 is almost unaffected, whereas the Cu^{2+}- and H^+-based adsorbents suffer from considerable inhibition.

Although the degree of inhibition by Cl^- and SO_4^{2-} varies among the cations, the concentration of Cl^- and SO_4^{2-} in rivers is generally 0.16–0.23 and 0.086–0.18 mmol/L, respectively.[46,47] When these concentrations are applied to the plots in Figure 11.12, the inhibition due to Cl^- and SO_4^{2-} is negligible, implying that these anions have little effect in the environment. The strength of adsorption is another important factor as discussed earlier. The inhibition is more significant when the adsorption approaches saturation than when the coverage is low.[30,32] Since the arsenate concentration expected in the contaminated water will be in the range of 1×10^2 to 1×10^3 ppb, the inhibition by Cl^- and $SO_4{}^{2-}$ will be much lower than in Figure 11.12, where the initial concentration of arsenate is 1,700 ppm. From these considerations, the adsorption characteristics of arsenate on $Fe^{3+}/$ or Co^{2+}/NN-MCM-41 in the environment will remain unchanged, even in the presence of $SO_4{}^{2-}$ and Cl^- below 1 mmol/L.

The lack of uniformity of the surface structure often causes problems in the extrapolation of the experimental results, especially when the adsorption is measured near saturation. The decrease in the distribution coefficient with the progress of adsorption in Figure 11.6 suggests that the surface contains weak and strong adsorption sites. In addition, we can see that the release of arsenate occurs more easily on H^+ sites than on Fe^{3+} and Co^{2+} sites. If a meaningful number of H^+ sites exist in $Fe^{3+}/$ and Co^{2+}/NN-MCM-41, the exchange of adsorbed arsenate with Cl^- or $SO_4{}^{2-}$ anions would occur in the same manner as in H^+/NN-MCM-41 in Figure 11.12. On the contrary, the absence of a rapid decrease in (arsenate adsorption)/(arsenate adsorption at $C_0 = 0$) near $C_0 = 0$ in $Fe^{3+}/$ and Co^{2+}/NN-MCM-41 demonstrates that the contribution of H^+ sites to the adsorption capacity in M^{n+}/NN-MCM-41, $(M^{n+} = Fe^{3+}$ and $Co^{2+})$ is negligible, even if such sites exist.

The separation of $SeO_4{}^{2-}$ from $SO_4{}^{2-}$ and of $AsO_4{}^{3-}$ from $PO_4{}^{3-}$ has been considered difficult and can be benchmarks for selective adsorption.[31b,32] Figure 11.13 shows the suppression of the adsorption capacity in the co-adsorption systems of these anion combinations. The effect of the addition of sulfate on the adsorption of selenate is considerably greater than the suppressions in Figure 11.12. Even in the best adsorbent, Fe^{3+}/NN-MCM-41, more than 50% of adsorption capacity is lost with co-existence of 50 mmol/L of sulfate. Nevertheless, inhibition by the sulfate can be considered to be still limited for the following reason. As shown in Table 11.4, with the addition of 202 ppm of sulfate, suppression cannot be detected and the distribution coefficient remains beyond the detection limit (i.e. $K_d > 200,000$) in the adsorption of selenate on Fe^{3+}/NN-MCM-41. With 1,014 ppm of sulfate, about 8% suppression is observed though K_d diminishes to 2,381, probably becoming less than 1% of the K_d observed in the adsorption without an inhibitor. However, since water containing 1,000 ppm of $SO_4{}^{2-}$ is not generally suitable for drinking as shown in Table 11.1, Fe^{3+}/NN-MCM-41 is qualified for the adsorbent that removes selenate from sulfate. On the other hand, 20% of the adsorption of selenate is lost and the distribution coefficient diminishes to 800 in the presence of 202 ppm sulfate, when the adsorption is carried out

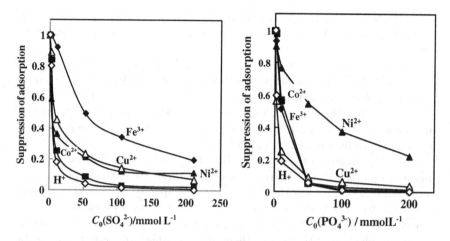

Figure 11.13.   Inhibition of selenate adsorption by the presence of sulfate (upper) and inhibition of arsenate adsorption by the presence of phosphate (lower). The ordinates are the adsorption capacity in the presence of the inhibitor at the initial concentration of C_0 divided by that without the inhibitor. The central cations are indicated in the plots and the substrate is NN-MCM-41. The adsorptions were carried out under the following conditions: initial concentration of selenate or arsenate: 12.2 mmol/L^1, adsorbent: 50 mg, volume of solution: 10 mL, and reaction temperature: 298 K.

Table 11.4.   Suppression of selenate adsorption (= [adsorption of selenate in the presence of $SO_4{}^{2-}$]/[adsorption of selenate in the absence of $SO_4{}^{2-}$]) on cation-incorporated NN-MCM-41.

Cation	$SO_4{}^{2-}$ Initial ppm*	$SeO_4{}^{2-}$ Initial ppm*	$SeO_4{}^{2-}$ Final ppm*	$SeO_4{}^{2-}$ Adsorption mmol/g	Suppression**	K_d***
Fe^{3+}	0	71.2	0	0.0997	—	$>2 \times 10^5$
	202		0	0.0997	1.00	$>2 \times 10^5$
	1,014		5.5	0.0920	0.922	2,381
	5,071		36.1	0.0492	0.494	195
H^+	0	71.2	0	0.0997	—	$>2 \times 10^5$
	202		14.2	0.0797	0.800	800
	1,014		58.4	0.0180	0.181	44
	5,071		68.2	0.00427	0.0428	9

* w/w ppm of oxyanions.
** = [adsorption of selenate in the presence of $SO_4{}^{2-}$]/[adsorption of selenate in the absence of $SO_4{}^{2-}$].
*** Distribution coefficient of selenate.

with H^+/NN-MCM-41. This result implies that protonated NN-MCM-41 is possibly disqualified as an adsorbent of selenate used in sulfate-containing water.

The order of resistance against inhibition by sulfate is $Fe^{3+} > Cu^{2+} \sim Ni^{2+} > Co^{2+} \sim H^+$ in selenate adsorption, which differs slightly from that in arsenate adsorption, $Fe^{3+} \sim Co^{2+} > Ni^{2+} > Cu^{2+} \sim H^+$ (Figure 11.12). The difference in suppression implies that the transition metal cation at the adsorption site has a key importance for the prevention of interference by co-existing anions.

The suppression of arsenate adsorption by the inhibition of phosphate (Figure 11.13) is clearly larger than those by chloride and sulfate (Figure 11.12). Ni^{2+} has the most resistant capacity against this adsorption inhibitor. However, natural waters usually contain phosphorus whose concentration is lower than sulfate, as low as 20 ppb, which is a limiting factor for plant growth,[47] and, therefore, the effect of phosphate will be limited.

11.3.2 Recyclability of used adsorbents

Mesoporous silica frameworks, surface functional groups and adsorption sites are synthesized with more costly reagents than simple or modified natural adsorbents, such as grained goethite, pumice and γ-Al_2O_3. The reactivation of used adsorbent is critically important for the demand in the potential market for reducing cost as well as emissions. When the adsorption is strong, reuse of the adsorbent is usually difficult because the regeneration of active sites inevitably includes desorption of the adsorbates. The removal may be accompanied by the destruction of the surface structure, resulting in the loss of adsorption capacity. It was demonstrated, in fact, that the desorption of arsenate from active carbon by treatment with a strong acid or base results in a significant loss of adsorption capacity.[49]

The arsenate adsorbed on Fe^{3+}-incorporated NN-MCM-41 can be easily removed by washing with 1 M HCl at room temperature for 10 hours.[50] The results of the elemental analysis are summarized in Table 11.5. More than 99% of arsenate together with 94% of Fe are removed by this acid treatment, while the loss of amino groups is held at 23%. These percentages demonstrate that most of the arsenate is washed out with Fe^{3+} cation. The loss of iron signifies the destruction of the adsorption site, but does not mean the loss of a recovery method. Although the damage to the organic layer is not negligible, the re-incorporation of Fe^{3+} into the mesoporous silica after the removal of As recovers up to 82% of Fe^{3+} of the initial adsorbent level. The N/Fe ratio was nearly equal to 4 in this regenerated Fe^{3+}/NN-MCM-41, implying the recovery of the same coordination structure.

Table 11.5. Elemental analysis of arsenate-adsorbed Fe^{3+}/ NN-MCM-41, that followed by treatment with hydrochloric acid and that by successive Fe^{3+} incorporation in addition.

	As* mmol	Fe^{3+}* mmol	N* mmol	N/Fe
After the adsorption**	1.50	0.51	1.98	3.9
After 1 M HCl treatment	0.01	0.03	1.53	—
Fe^{3+} re-coordination	0.01	0.42	1.51	3.6

* The amount of the element per g-solid.
** Nearly full coverage.

Table 11.6. Arsenate adsorptions on fresh Fe^{3+}-incorporated NN-MCM-41 and regenerated Fe^{3+}-incorporated NN-MCM-41.

	Arsenate		Specific Adsorption mmol/g	K_d
	Initial mg/L	Final mg/L		
Fresh adsorbent	10.1	n.d.	0.01	$> 2 \times 10^5$
Regenerated adsorbent	10.1	n.d.	0.01	$> 2 \times 10^5$
Fresh adsorbent	1,520	750	1.10	205
Regenerated adsorbent	1,560	858	1.00	164
	As/Fe at the saturation			
Fresh adsorbent	2.8			
Regenerated adsorbent	3.1			

N.B. The adsorptions were carried out under the following conditions: initial concentration of arsenate: 12.2 mmol/L, adsorbent: 50 mg, volume of solution: 10 mL, and reaction temperature: 298 K.

This recycling procedure works as well as or even better for the other oxyanions than for arsenate. The recovery of Fe^{3+} re-incorporation is 83, 90 and 96% using the same recycling treatment applied to chromate-, selenate- and molybdate-adsorbed Fe^{3+}/NN-MCM-41, respectively. The difference is probably due to the amount of amino groups that remain after acid treatment: 83, 79 and 93% for chromate-, selenate- and molybdate-adsorbed Fe^{3+}/NN-MCM-41, respectively.[50]

The arsenate adsorptions on the fresh and regenerated Fe^{3+}/NN-MCM-41 are compared in Table 11.6. It is clearly demonstrated in the table that the specific adsorption and distribution coefficient are sufficiently (80–100%) recovered. The adsorption capacity of regenerated adsorbent reaches 1.31 mmol/g,[50] which is equal to *ca.* 81% of the capacity of fresh adsorbent (226 mg/g = 1.61 mmol/g in Table 11.3), and, therefore, the

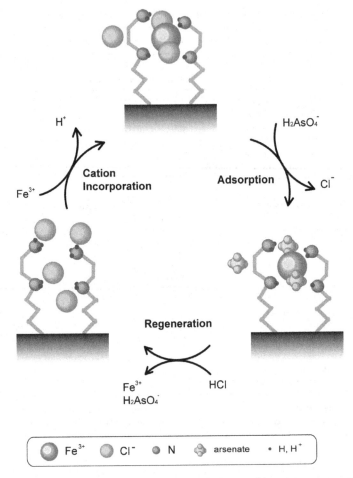

Figure 11.14.   A schematic illustration of the recycling of Fe^{3+}/NN-MCM-41. This cycle includes the incorporation of Fe^{3+} by amino ligands, the adsorption of oxyanions, and the simultaneous removal of oxyanions and Fe^{3+} by acid treatment.

As/Fe ratio does not change significantly, as shown in Table 11.6, suggesting that the same kinds of adsorption site and adsorption mechanism are obtained in this regenerated adsorbent.

Together with these results, the recycling of Fe^{3+}-incorporated NN-MCM-41 is illustrated in Figure 11.14. Nearly 81% of the initial capacity was regained through one cycle and the high efficiency of regeneration may serve to meet the cost and environmental demands in the future market. In the filtration solution of the regeneration process (washing with

HCl to remove arsenate and Fe^{3+}), the arsenate is probably dissolved due to the low pH. This species will be precipitated in the form of an insoluble ferric arsenate oxide, such as scorodite, by increasing the solution pH. In the other oxyanion adsorptions, such as chromate, selenate and molybdate adsorptions, 82, 86 and 85%, respectively, of adsorption capacity was recovered using regenerated Fe^{3+}/NN-MCM-41. The oxyanion/Fe ratio remained almost the same as in the fresh material (1.9, 1.5 and 2.2, respectively).

11.4 Recent Progress in Analysis and Synthesis of Functionalized Mesoporous Silica

11.4.1 *Distributions of organic functions*

As we have already recognized in the plots in Figure 11.4, the distributions of organic functions on the mesopores can become non-uniform when they are grafted. When the diffusion of reactive silane into the pores competes with the surface reaction, the distribution will become more uniform in large mesopores than in small mesopores. This hypothesis has been proved experimentally by analysis using a reaction probe.

When the grafted tethers are distributed randomly on the surface, the pair distribution function can explain the distance between adjacent tethers (or the distance between the nearest surface sites bound to the tether). However, there are no suitable physicochemical techniques or instrumental analyses with which to determine such distribution functions. This is mainly because the degree of freedom of organic tethers is generally large and the conformations are constantly changing at room temperature, causing a huge disorder in the atomic positions. Miyajima *et al.* have proposed the use of a probe reaction for experimental determination of the pair distance distribution functions of adjacent 3-bromopropyl groups on a silica surface.[51] They used a series of diamines, $RNH(CH_2)_nNHR$, where $R = CH_3$ or H and $n = 2, 3, 4, 5$ and 6, as the substituent for the Br atom of the 3-bromopropyl group on the silica surface and proved that both of the amine groups react with bromopropyl tethers when they are within the span of the two nitrogen atoms (Scheme 11.2). Elemental analysis and ^{13}C-CP-MAS-NMR spectroscopy have been used to determine the composition and the structure of surface species after the substitution reaction in Scheme 11.2. The compositions of C and N provide the populations of the linear species $RNH(CH_2)_nNH(R)CH_2CH_2CH_2*$ and of the bridge species $*CH_2CH_2CH_2(R)NH(CH_2)_nNH(R)CH_2CH_2CH_2*$. The former and latter species are generated from the two nearest bromopropyl

a larger spacing than N-N length of diamine

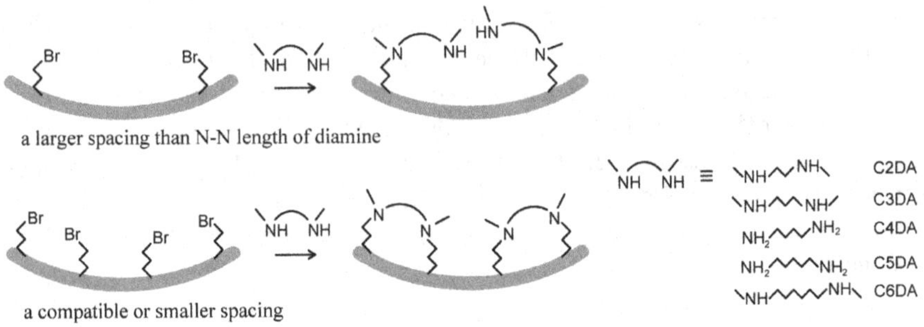

a compatible or smaller spacing

$\searrow$NH　NH$\swarrow$ ≡	$\searrow$NH$\sim$NH$\searrow$	C2DA
	$\searrow$NH$\sim$NH$\diagup$	C3DA
	NH$_2\sim$NH$_2$	C4DA
	NH$_2\sim\sim$NH$_2$	C5DA
	$\searrow$NH$\sim\sim$NH$\searrow$	C6DA

Scheme 11.9. Determination of pair distance of surface tethers by the probe reaction with diamines of various molecular lengths.

tethers, which are positioned such that they have larger and smaller spacings, respectively, than the span of two nitrogen atoms of the diamine molecule. When using a long diamine, the ratio of the bridge species becomes large. The pair distance distributions were determined after reactions with six different diamines and have been compared with the integral of the theoretical random pair distribution functions assuming a plane surface (Figure 11.15). A comparison of three 3-bromopropyl-functionalized silicas (MCM-41, SBA-15 and fumed silica) reveals that the spacing of 3-bromopropyl tethers is ordered such that fumed silica (non-mesoporous, $A_{BET} = 200 \, m^2/g$) > SBA-15 ($2R_P = 7.6 \, nm$, $A_{BET} = 687 \, m^2/g$) > MCM-41 ($2R_P = 2.4 \, nm$, $A_{BET} = 1{,}289 \, m^2/g$). On the other hand, the densities of organic functions as determined by analysis of the whole solids are 1.2, 1.1 and 0.8 Br-propyl/nm^2 for fumed silica (most populated), SBA-15 and MCM-41 (least populated), respectively. These values are calculated from data derived from elemental analysis and nitrogen adsorption experiments. The discrepancy between these microscopic and macroscopic data arises from local and whole analyses, respectively, implying that the distributions of bromopropyl tethers are most uniform on fumed silica and least uniform on MCM-41. The order of uniformity (fumed silica > SBA-15 > MCM-41) follows the pore size, if that of fumed silica is assumed to be ∞, and is consistent with the hypothesis that the graft reaction occurs preferentially at the pore openings when the mesopores are small.

Although providing absolute values for the pair distance distribution functions is still difficult due to conformational changes of the organic tethers and the curvature of the mesopores, comparison can be carried out among different kinds of substrate silicas. The method that Miyajima *et al.* developed is basically applicable to various tethers such as 3-aminopropyl by the use of suitable coupling reactions.

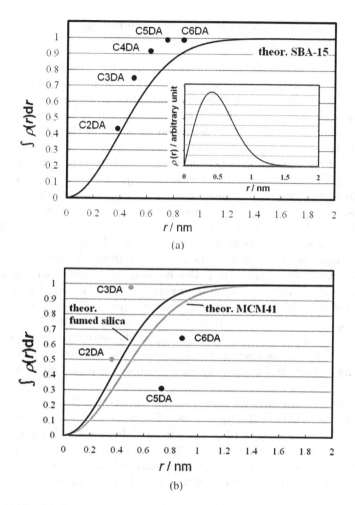

Figure 11.15. (a) Integration of the pair distance distribution function of tethers expected by the loading and the surface area of SBA-15 and the experimental values from the probe reactions with diamines of various molecular lengths. Inset is the pair distance distribution function of tethers. (b) Integrations of the pair distance distribution function of tethers expected by the loading and the surface area of MCM-41 (in gray) and fumed silica (in black) and experimental values from the probe reactions with diamines of various molecular lengths. The molecular length of diamines is represented with the distance between two amine functions in an all-trans conformer and the curvature of the silica surface is ignored. These assumptions explain the reason why the theoretical curve becomes larger than the experimental data for bromopropyl-functionalized fumed silica, where the organic tethers are expected to be distributed uniformly. However, the larger difference between the theory and the experiment (the theoretical curves smaller than the experimental data) in MCM-41 than in SBA-15 implies that the dispersion of organic tethers is more heterogeneous in the former than in the latter.

11.4.2 Functionalization at specific structure of mesoporous silica

The structural characteristics of mesoporous solids can be described not only by the periodicity of mesopores but also by uniform structural features on the mesoscale, such as the curvatures of mesopore walls, inner and outer surfaces, and different kinds of pores in a bimodal structure. Considerable efforts have been dedicated to elucidate the detailed geometries of these structures.[52-53] In contrast, the number of successful works that report the selective functionalization at such specific structures is not large. The simplest and most important concept of these works is probably the selective grafting of the inner and outer surfaces of mesoporous particles, i.e. one kind of silane on the outer surface and another on the pore wall (Figure 11.16).[54-57] A wide range of applications (e.g. separation and chromatography) are expected to become available for mesoporous silicas functionalized by this technique, because it can define the interaction of the silica particles with their surrounding medium separately from that of ions and molecules penetrating from the medium into the mesopores. Recently, the use of confocal laser scanning microscopy has made possible the selective imaging of the outer surface by detecting fluorescence from fluorescein isothiocyanate captured by 3-aminopropyl groups that are expected to exist only on the outer surface.[58] This study has implied that the size of silane molecules can be a key for successful selective grafting of the outer surfaces.

Another promising method for the bi-functionalization of specific mesostructures is provided by the use of periodic mesoporous organosilica (PMO). This material is normally prepared by using a bridged bisilane $(RO)_3Si-X-Si(OR)_3$, instead of TEOS $Si(OC_2H_5)_4$ for mesoporous silicas, or a mixture of a bridged bisilane and TEOS. Trisilane and tetrasilane can also be used in the mixture with TEOS. The mesoporous solid

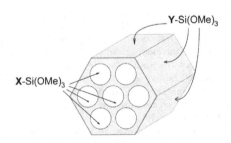

Figure 11.16.   Selective functionalization of the outer and inner surfaces of mesoporous silica particles.

obtained by hydrothermal synthesis contains the organic group -X- (-X- = -C_2H_2-, -C_6H_4-, -1,4-$C_2H_2C_5H_3NC_2H_2$-, -C_4H_2S-, etc.) in a framework structure. On the other hand, the surface contains significant amounts of silanols, onto which the organic group Y- can be grafted by a reaction with a reactive silane Y-Si(RO)$_3$. As a result, organic pendant group Y- is fixed on the surface of the pore walls, while the pore walls are built up with the bridging organic group -X- and silsesquioxane unit. A number of reviews[60] have been written on PMOs in which the various applications are described.

11.4.3 *Physical chemistry of aqueous solutions in mesopore confinement*

Practical adsorbents might be used in cold water, so the freezing point should always be kept in mind in order to know the minimum applicable temperature and not to destroy the porous structure due to the expansion of volume by solidification of water. In addition, it is necessary to remember that the properties of liquids change considerably when they are confined in mesopores. One good example is phase transitions. The decrease of the vapour pressure of liquid nitrogen in mesopores is frequently observed in the nitrogen adsorption experiment at 77 K, which is used for the determination of pore size distributions. It is often said that the adsorption of nitrogen shows an abrupt increase due to condensation in the mesopores when the pressure of nitrogen reaches around 0.4 atm. Similar phenomena for the freezing of water have long been investigated,[61] though great concerns arose after the discovery of MCM-41.[62] The core water that exists in the centre of the pores freezes at a higher temperature than the normal freezing point (=273 K) by $\Delta T = K/(D - l)$, where D is the pore diameter, l is the thickness of the bound water near the pore wall surface and K is a constant. In this model, water in the mesopores is classified into two categories: core and bound water. ^{1}H-NMR and XRD studies have demonstrated that core water shows hysteresis in its melting/freezing cycle, while bound water is non-freezable and undergoes a continuous transition to a more rigid structure. The screening length of this unfrozen water $(= l)$ is assumed to be 0.4–0.8 nm, which depends on the composition and topology of the mesopores.

Another important chemical property that is significantly influenced by confinement is the pH of aqueous solutions under equilibrium between the mesopores and the exterior.[63] The volume of water confined in a hollow sphere with an inner diameter of 2 nm would contain only 140 molecules. It is obvious that this number is not large enough to exhibit properties of bulk water. The ionization of water ($[H^+][OH^-] = 10^{-14}$), for example, is a probability event, since 140 is much smaller than 10^7. From a thermodynamic point of view, the number of such spheres could be 1×10^{20} in a volume of $1\,cm^3$, some of which would contain H^+ and OH^- ions. However, a coordination experiment involving $AuCl_4^-/AuCl_3$ to an aminopropyl-grafted silica cage of $2\,nm^{63(a)}$ implies that amino groups inside such a nanocage do not become protonated even when the pH of the solution (when the exterior is under equilibrium with the inside of the cages) reaches 4.0. It should be noted that pKa of primary amine is $ca.$ 10.5 and the aminopropyl group in the cage is stabilized by the order of 6.5. This result suggests that the coordination structure M^{n+}/NN-MCM-41 ($M^{n+} - Fe^{3+}$, Cu^{2+} and Cu^{2+}) is more stable in acidic solution than would be expected from the equilibrium constants of M^{n+}-ethylenediamine and H^+-ethylenediamine in aqueous solution. (N.B. The lone pair that is necessary for the formation of a coordination bond with M^{n+} is lost after the conversion from NH_2-$CH_2CH_2CH_2$- to NH_3^+-$CH_2CH_2CH_2$-.) The result also implies that the adsorbents M^{n+}/NN-MCM-41 can be used under more acidic conditions than M^{n+}/NN-non-porous silica.

11.5 Concluding Remarks

This chapter describes how amino-functionalized mesoporous silicas work as an adsorbent of toxic oxyanions, such as arsenate, chromate, selenate and molybdate. Their properties depend on the molecular structure of organic groups as well as the mesoporous structure. It is likely that the latter factor is closely related to the uniformity or the "real" density of the organic layer. The adsorption properties are influenced considerably by the cation on the amino group. An intense and large adsorption of arsenate is obtained on Fe^{3+}-incorporated EDA-propyl-functionalized MCM-41. The adsorption is selective even in the presence of chloride and sulfate. This adsorbent can be recycled by washing with hydrochloric acid and re-incorporation of Fe^{3+}.

The mesoporous adsorbents without organic functionalities, Fe, Al, Zn and La supported on SBA-15,[64] have been used for oxyanion adsorptions. When the element interacting with the oxyanions is highly dispersed and highly dense in a particular solid, a strong and capacious adsorption will be obtained. The platform used for the adsorption sites can be clay.[65]

The adsorption of cations on an adsorbent derived from mesoporous silica in a periodic structure has been investigated much more than that of anions. The typical example is the adsorption of mercury(II) on functionalized mesoporous silicas where 3-mercaptopropylsilane is used as a surface modifier.[66] It has been demonstrated that the Hg/S ratio depends on the kind of mesoporous silica,[67] being like arsenate adsorption on EDA-propyl functionalized mesoporous silicas as shown in Figures 11.2 and 11.3. The stoichiometry of the adsorption site relative to the adsorbate ion is clearly important in the design of adsorbents, because the capacity is, roughly speaking, the product of the surface area, the surface density of the adsorption sites and this stoichiometry. The large capacity of Fe^{3+}/NN-MCM-41 for arsenate adsorption stems from the stoichiometry, Fe:As = 3:1, when the adsorption is saturated.

The reduction in the number of synthetic processes may be important. An interesting approach for reducing the processes is the combination of functionalization with amines, such as N, N-dimethyldecylamine (DMDA), with a pore expansion technique,[68] which is one of the most important techniques in the preparation of mesoporous silica. Since amine works as an anchoring site for transition metal cations, this pore-expanded mesoporous silica adsorbs toxic cations before extracting the surfactants. The adsorbent exhibits a high capacity: 106 mg/(g-solid) of Cu^{2+}. Further development and application of this method are expected. The functionalization in association with templating, described earlier, is also a method combining templating and functionalization. These combined syntheses will become more and more important and so will be control and analysis of the distributions of organic functional groups.[69]

Acknowledgements

The author thanks his collaborators, Prof. Toshiyuki Yokoi (Tokyo Institute of Technology), Prof. Takashi Tatsumi (Tokyo Institute of Technology), Prof. Yasunori Oumi (Gifu University) and Prof. Tsuneji Sano (Hiroshima University). He is also grateful to the Ministry of Education, Culture, Sports, Science and Technology, the Japan Society for the Promotion of Science, the Japan Securities Scholarship Foundation and the Japan Science and Technology Agency for financial support.

References

1. T. Yanagisawa, T. Shimizu, K. Kuroda and C. Kato, *Bull. Chem. Soc. Jpn.* **63**, 988–992 (1990).
2. C. T. Kresge, M. E. Leonowicz, W. J. Roth, J. C. Vartuli and J. S. Beck, *Nature* **359**, 710–712 (1992).
3. U. Ciesla and F. Schüth, *Micropor. Mesopor. Mater.* **27**, 131–149 (1999).
4. J. Y. Ying, C. P. Mehnert and M. S. Wong, *Angew. Chem. Int. Ed.* **38**, 56–77 (1999).

5. (a) Q. Huo, D. I. Margolese, U. Ciesla, P. Feng, T. E. Gier, P. Sieger, R. Leon,
    P. M. Petroff, F. Schüth and G. D. Stucky, *Nature* **368**, 317–321 (1994).
    (b) Q. Huo, D. I. Margolese, U. Ciesla, D. G. Demuth, P. Feng, T. E. Gier,
    P. Sieger, A. Firouzi, B. F. Chmelka, F. Schüth and G. D. Stucky, *Chem.
    Mater.* **6**, 1176–1191 (1994). (c) Q. Huo, R. Leon, P. M. Petroff and G. D.
    Stucky, *Science* **268**, 1324–1327 (1995). (d) Q. Huo, D. I. Margolese and G.
    D. Stucky, *Chem. Mater.* **8**, 1147–1160 (1996). (e) S. A. Bagshaw, E. Prouzet
    and T. J. Pinnavaia, *Science* **269**, 1242–1244 (1995).
6. A. Tuel, *Micropor. Mesopor. Mater.* **27**, 151–169 (1999).
7. (a) A. Corma, *Topics Catal.* **4**, 249–260 (1997). (b) A. Corma, *Chem. Rev.*
    97, 2373–2419 (1997).
8. G. E. Fryxell and J. Liu, in *Adsorption at Silica Surface*, eds. E. Papirer
    (Marcel Dekker, New York, 2000) p. 665.
9. A. Sayari and S. Hamoudi, *Chem. Mater.* **13**, 3151–3168 (2001).
10. For general information on zeolites, see H. van Bekkum, E. M. Flanigen, P. A.
    Jacobs and J. C. Jansen (eds.), *Introduction to Zeolite Science and Practice*
    (Elsevier, Amsterdam, 2001).
11. For general information on active carbon, see K. Kinoshita, *Carbon* (Wiley-
    Interscience Publication, New York, 1988).
12. S. Shiraishi, in *Carbon Alloys*, eds. E. Yasuda, M. Inagaki, K. Kaneko, M.
    Endo, A. Oya and Y. Tanabe (Elsevier, Amsterdam, 2003) Chapter 27.
13. (a) C. Y. Chen, H. X. Li and M. E. Davis, *Microporous Mater.* **2**, 17–26
    (1993). (b) J. J. M. Kim, J. H. Kwak, S. Jun and R. Ryoo, *J. Phys. Chem.*
    99, 16742–16747 (1995).
14. R. Nickson, J. McArthur, W. Burgess, K. M. Ahmed, P. Ravenscroft and
    M. Rahman, *Nature* **395**, 338 (1998).
15. M. Karim, *Water Res.* **34**, 304–310 (2000).
16. H. Kondo, *Mizukankyo Gakkaishi* **20**, 6–10 (1997).
17. Y. S. Sen and C. S. Shen, *J. Water Pollut. Control Fed.* **36**, 281 (1964).
18. S. L. Chen, S. R. Dzeng, M. H. Yang, K. H. Chiu, G. M. Shieh and C. M.
    Wai, *Environ. Sci. Technol.* **28**, 877–881 (1994).
19. (a) R. D. Foust, Jr., P. Mohapatra, A.-M. Compton-O'Brien and J. Reifel,
    Appl. Geochem. **19**, 251–255 (2004). (b) A. E. Grosz, J. .N. Grossman,
    R. Garrett, P. Friske, D. B. Smith, A. G. Darnley and E. Vowinkel, *Appl.
    Geochem.* **19**, 257–260 (2004). (c) J. R. Lockwood, M. J. Schervish, P. Gurian
    and M. J. Small, *J. Am. Stat. Assoc.* **96**, 1184–1193 (2001). (d) J. M. Har-
    rington, J. P. Middaugh, D. L. Morse, J. Housworth, *Am. J. Epidemiol.*
    108, 377–385 (1978). (e) J. Bundschuh, B. Farias, R. Martin, A. Storniolo,
    P. Bhattacharya, J. Cortes, G. Bonorino and R. Albouy, *Appl. Geochem.* **19**,
    231–243 (2004).
20. M. M. Lawrence, *Environ. Sci. Technol.* **15**, 1482–1484 (1981).
21. E. L. Tavani and C. Volzone, *J. Soc. Leather Technol. Chem.* **81**, 143–148
    (1997).
22. H. Yoshitake, in *Bottom-Up Nanofabrication: Supramolecules, Self-
    Assemblies, and Organic Film*, Vol. 6, eds. K. Ariga and H. S. Nalwa (Amer-
    ican Scientific Publishers, Stevenson Ranch, CA, 2009) Chapter 10.

23. A. P. Legrand (ed.), *The Surface Properties of Silicas* (John Wiley & Sons, Chichester, 1998) Chapter 3.
24. H. Yoshitake, *New J. Chem.* **29**, 1097 (2005).
25. H. Yoshitake, T. Yokoi and T. Tatsumi, *Chem. Mater.* **14**, 4603–4610 (2002).
26. T. Yokoi, H. Yoshitake and T. Tatsumi, *J. Mater. Chem.* **14**, 951–957 (2004).
27. A. S. M. Chong, X. S. Zhao, A. T. Kustedjo and S. Z. Qiao, *Micropor. Mesopor. Mater.* **72**, 33–42 (2004).
28. T. Yokoi, H. Yoshitake and T. Tatsumi, *Chem. Mater.* **15**, 4536–4538 (2003).
29. Q. Zhang, K. Ariga, A. Okabe and T. Aida, *J. Am. Chem. Soc.* **126**, 988–989 (2004).
30. G. E. Fryxell, J. Liu, T. A. Hauser, Z. Nie, K. F. Ferris, S. Mattigod, M. Gong and R. T. Hallen, *Chem. Mater.* **11**, 2148–2154 (1999).
31. (a) H. Yoshitake, T. Yokoi and T. Tatsumi, *Chem. Mater.* **15**, 1713–1721 (2003). (b) H. Yoshitake, T. Yokoi and T. Tatsumi, *Bull. Chem. Soc. Jpn.* **76**, 2225–2232 (2003).
32. T. Yokoi, M.S. Engineering thesis, Yokohama National University, Yokohama, Japan, 2001.
33. S. D. Kelly, K. M. Kemner, G. E. Fryxell, J. Liu, S. V. Mattigod and K. F. Ferris, *J. Phys. Chem. B* **105**, 6337–6346 (2001).
34. Y. Gao and A. Mucci, *Geochim. Cosmochim. Acta* **65**, 2361–2378 (2001).
35. W. Zhang, P. Singh, E. Paling and S. Delides, *Minerals Eng.* **17**, 517–524 (2004).
36. T. F. Lin and J. K. Wu, *Water Res.* **35**, 2049–2057 (2001).
37. S. Bang, M. Patel, L. Lippincott and X. Meng, *Chemosphere* **60**, 389–394 (2005).
38. A. N. Garg and P. N. Shukla, *Indian J. Chem.* **12**, 996 (1974).
39. R. L. Pecsok and J. Bjerrum, *Acta Chem. Scand.* **11**, 1419–1421 (1957).
40. G. Gordon and R. K. Birdwhistell, *J. Am. Chem. Soc.* **81**, 3567 (1959).
41. T. E. Westre, P. Kennepohl, J. P. Dewitt, B. Hedman, K. O. Hodgson and E. I. Solomon, *J. Am. Chem. Soc.* **119**, 6297–6314 (1997).
42. C. H. Wu, S. L. Lo and C. F. Lin, *Colloid Surf. A* **116**, 251–259 (2000).
43. Z. Hongshao and R. Stanforth, *Environ. Sci. Technol.* **35**, 4753–4757 (2001).
44. J. Antelo, M. Avena, S. Fiol, R. Lopez and F. Arce, *J. Colloid Interf. Sci.* **285**, 476–486 (2005).
45. Y. Arai, D. L. Sparks and J. A. Davis, *Environ. Sci. Technol.* **38**, 817–824 (2004).
46. M. Meybeck, *Rev. Geol. Dyn. Geogr. Phys.* **21**, 215–246 (1979).
47. E. J. R. Conway, *Proc. Roy. Irish Acad. B* **48**, 119–160 (1942).
48. (a) C. D. Rail, *Groundwater Contamination: Sources, Control, and Preventative Measures* (Technomic Publishing Co., Inc., Lancaster, PA, 1989) p. 17. (b) E. A. Laws, *Aquatic Pollution,* 2nd edn. (John Wiley & Sons, Inc., New York, NY, 1993) pp. 148–149.
49. C. P. Huang and P. L. K. Fu, *J. Water Pollut. Control. Fed.* **56**, 233–242 (1984).
50. T. Yokoi, T. Tatsumi and H. Yoshitake, *J. Colloid Interf. Sci.* **274**, 451–457 (2004).

51. T. Miyajima, S. Abry, W. Zhou, B. Albela, L. Bonneviot, Y. Oumi, T. Sano and H. Yoshitake, *J. Mater. Chem.* **17**, 3901–3909 (2007).

52. J. M. Thomas, O. Terasaki, P. L. Gai, W. Z. Zhou and J. Gonzalez-Calbet, *Acc. Chem. Res.* **34**, 583–594 (2001).

53. I. Diaz, V. Alfredsson and Y. Sakamoto, *Curr. Opin. Colloid Interface Sci.* **11**, 302–307 (2006).

54. D. S. Shepard, W. Zhou, T. Maschmeyer, J. M. Matters, C. L. Roper, S. Parsons, B. F. G. Johnson and M. J. Duer, *Angew. Chem. Int. Ed.* **37**, 2719–2723 (1998).

55. F. Juan and E. Ruiz-Hitzky, *Adv. Mater.* **12**, 430–432 (2000).

56. J. Kecht, A. Schlossbauer and T. Bein, *Chem. Mater.* **20**, 7207–7214 (2008).

57. J. M. Rosenholm, A. Duchanoy and M. Lindón, *Chem. Mater.* **20**, 1126–1133 (2008).

58. J. D. Lunn and D. F. Shantz, *Chem. Commun.* 2926–2928 (2010).

59. N. Gartmann and D. Brühwiler, *Angew. Chem. Int. Ed.* **48**, 6354–6356 (2009).

60. (a) S. Inagaki, S. Guan, T. Ohsuna and O. Terasaki, *Nature* **416**, 304–307 (2002). (b) G. J. D. Soler-Illia, C. Sanchez, B. Lebeau and J. Patarin, *Chem. Rev.* **102**, 4093–4138 (2002). (c) A. Taguchi and F. Schuth, *Micropor. Mesopor. Mater.* **77**, 1–45 (2005). (d) F. Hoffmann, M. Cornelius, J. Morell and M. Fröba, *Angew. Chem. Int. Ed.* **45**, 3216–3251 (2006). (e) B. Hatton, K. Landskron, W. Whitnall, D. Perovic and G. A. Ozin, *Acc. Chem. Res.* **38**, 305–312 (2005). (f) A. Corma and H. Garcia, *Adv. Synth. Catal.* **348**, 1391–1412 (2006). (g) K. Ariga, J. P. Hill, M. V. Lee, A. Vinu, R. Charvet and S. Acharya, *Sci. Technol. Adv. Mater.* **9**, 014109 (2008). (h) C. Sanchez, C. Boissiere, D. Grosso, C. Laberty and L. Nicole, *Chem. Mater.* **20**, 682–737 (2008). (i) S. Fujita and S. Inagaki, *Chem. Mater.* **20**, 891–908 (2008). (j) M. P. Kapoor and S. Inagaki, *Bull. Chem. Soc. Jpn.* **79**, 1463–1475 (2006). (k) Y. Yamauchi and K. Kuroda, *Chem. Asian J.* **3**, 664–676 (2008). (l) P. van der Voort, C. Vercaemst, D. Schaubroeck and F. Verpoort, *Phys. Chem. Chem. Phys.* **10**, 347–360 (2008). (m) Q. H. Yang, J. Liu, L. Zhang and C. Li, *J. Mater. Chem.* **19**, 1945–1955 (2009). (n) L. D. Carlos, R. A. S. Ferreira, V. D. Bermudez and S. J. L. Ribeiro, *Adv. Mater.* **21**, 509–534 (2009). (o) Z. Q. Wang and S. M. Cohen, *Chem. Soc. Rev.* **38**, 1315–1329 (2009).

61. (a) D. Everett, *Trans. Faraday Soc.* **57**, 1541–1551 (1961). (b) B. R. Puri, Y. P. Myer and D. D. Singh, *Trans. Faraday Soc.* **53**, 530–534 (1957). (c) A. A. Antoniou, *J. Phys. Chem.* **68**, 2754–2764 (1964). (d) T. Takamuku, M. Yamagami, H. Wakita, Y. Masuda and T. Yamaguchi, *J. Phys. Chem. B* **101**, 5730–5739 (1997). (e) G. Papavassiliou, M. Fardis, A. Leventis, F. Milia, J. Dolinsek, T. Apih and M. U. Mikac, *Phys. Rev. B* **55**, 12161–12174 (1997). (f) R. Schmidt, E. W. Hansen, M. Stöcker, D. Akporiaye and O. H. Ellestad, *J. Am. Chem. Soc.* **117**, 4049–4056 (1995). (g) D. D. Awschalom and J. Warnock, *Phys. Rev. B* **35**, 6779–6785 (1987).

62. (a) E. W. Hansen, M. Stöcker and R. Schmidt, *J. Phys. Chem.* **100**, 2195–2200 (1996). (b) D. Akporiaye, E. W. Hansen, R. Schmidt and M. Stöcker,

J. Phys. Chem. **98**, 1926–1928 (1994). (c) K. Morishige and K. Nobuoka, *J. Chem. Phys.* **107**, 6965–6969 (1997). (d) K. Morishige and K. Kawano, *J. Chem. Phys.* **110**, 4867–4872 (1999). (e) M. Sliwinska-Bartkowiak, J. Gras, R. Sikorski, R. Radhakrishnan, L. Gelb and K. E. Gubbins, *Langmuir* **15**, 6060–6069 (1999). (f) S. Sklari, H. Rahiala, V. Stathopoulos, J. Rosenholm and P. Pomonis, *Micropor. Mesopor. Mater.* **49**, 1–13 (2001).

63. (a) J. D. Henao, Y. W. Suh, J. K. Lee, M. C. Kung and H. H. Kung, *J. Am. Chem. Soc.* **130**, 16142–16143 (2008). (b) A. Walcarius and C. Delacôte, *Anal. Chim. Acta* **547**, 3–13 (2005).

64. (a) M. Jang, J. K. Park and E. W. Shin, *Micropor. Mesopor. Mater.* **75**, 159–168 (2004). (b) M. Jang, E. W. Shin, J. K. Park and S. I. Choi, *Environ. Sci. Technol.* **37**, 5062–5070 (2003).

65. Y. Izumi, K. Masih, K. Aika and Y. Seida, *J. Phys. Chem. B* **109**, 3227–3232 (2005).

66. X. Feng, G. E. Fryxell, W.-Q. Wang, A. Y. Kim, J. Liu and K. M. Kemner, *Science* **276**, 923–926 (1997).

67. L. Mercier and T. J. Pinnavaia, *Environ. Sci. Technol.* **32**, 2749–2754 (1998).

68. (a) A. Sayari, Y. Yang, M. Kruk and M. Jaroniec, *J. Phys. Chem. B* **103**, 3651–3658. (b) A. Sayari, *Angew. Chem. Int. Ed. Engl.* **39**, 2920–2922 (2000).

69. H. Yoshitake, *J. Mater. Chem.* **20**, 4537–4550 (2010).

Chapter 12

Layered Semi-crystalline Polysilsesquioxane: A Mesostructured and Stoichiometric Organic-Inorganic Hybrid Solid for the Removal of Environmentally Hazardous Ions

Hideaki Yoshitake

Division of Materials Science and Chemical Engineering
Yokohama National University, Yokohama, Japan

12.1 General Introduction

Polysilsesquioxane is a siliceous polymer composed of silsesquioxane units
R-Si-O- with ,O- and `O- substituents, and its formula is given as $RSiO_{1.5}$. This is formally categorized between polysiloxane (linear chain polymer, silicone, R_1R_2SiO) and silicon dioxide (inorganic crystal, silica, SiO_2), though this point of view may be also useful for understanding the physical and chemical properties of this solid. The various studies reviewed in this chapter have revealed the characteristics of polysilsesquioxane, which are between those of a polymer-like material and a typical inorganic solid. It is prepared by the dehydration-condensation of a monoorganosilane such as $RSi(OR')_3$ or organotrichlorosilicon $RSiCl_3$ under ambient pressure. It is generally obtained in the form of a lamellar solid in which inorganic layers built up with siloxane bonds and organic interlayers are accumulated alternately. The periodic order becomes clear when R is a self-assembling alkyl chain. The lamellar pattern is smeared for small values of R, though the positional correlation is still found in X-ray diffraction. The structure of polysilsesquioxane is illustrated schematically in Figure 12.1.

The index of the condensation level of polysilsesquioxane, $I(T^3)/(I(T^1) + I(T^2) + I(T^3))$, which is calculated from the peak intensities given by ^{29}Si NMR, is usually much higher than the equivalent index,

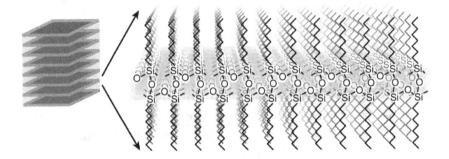

Figure 12.1.   Lamellar polysilsesquioxane.

$I(Q^4)/(I(Q^1)+I(Q^2)+I(Q^3)+I(Q^4))$, of amorphous silica from tetraethyl orthosilicate, TEOS. In many cases, $I(T^3)/(I(T^1) + I(T^2) + I(T^3))$ for polysilsesquioxane is nearly 100%. The polysilsesquioxane lamellar structure, which is composed of sheets of a rigid network of siloxane bonds and self-assembled organic interlayers, can be classified between the flexible linear chain of polysiloxane and the hard three-dimensional network of siloxane bonds that are found in silica.

Since the organic part R can contain functional groups such as -NH$_2$, -COOH and -SH, polysilsesquioxane can be used as a host for other ions and molecules, and is especially applicable to the removal of harmful pollutants from aqueous environments. Furthermore, the density of such organic functions can be extremely large and, therefore, good performance is expected if it is used for adsorptions and separations for environmental remediation.

In various industrial usages of silicon dioxide, its surface properties are modified, and chemical modification is often achieved by so-called functionalization with organic groups. Adsorptions have been clearly one of the most investigated benchmarks for functionalized silicas with a high surface area. With regard to applications such as the removal of aqueous pollutants, the adsorption capacity per unit weight is particularly important and should even be competitive with other kinds of adsorbents such as functionalized polymers and conventional inorganic powdery solids. Recent developments in the synthesis of mesoporous solids[1,2] have strengthened the advantages of silica because their large surface area (say, 500–1,200 m^2/g) is mainly due to the mesopores, which will facilitate the diffusion and adsorption of aqueous ions when compared with microporous materials. In fact, functionalized mesoporous silicas often show high adsorption capacities for cations and anions from aqueous solutions.[3,4] In this situation, the population of functional groups is 1–2/nm^2 or 1.6–3.3 mmol/g, which means that the ratio (organic function)/Si $= 0.1$–0.2. If grafting is carried out with some water remaining on the surface, the number tends to become even larger, say $5\,mmol/g$. On the other hand, (organic function)/Si $= 1$ can usually be realized in polysilsesquioxanes derived from monoorganosilane $RSi(OR')_3$. This high level of organic functionality could potentially enhance the uptake of aqueous pollutants.

We may find some resemblance between polysilsesquioxane and functionalized clays, which also have a layer-by-layer periodic structure composed of thin inorganic oxides and self-assembled organic molecules.[5–15] However, this class of solids is clearly different from polysilsesquioxane for the following three reasons. First, the inorganic layer is crystalline and multi-atomic; second, organic molecules are often inserted in the interlayer as a result of an exchange reaction with a counter anion, and, therefore, the inorganic-organic junction has an ionic nature; finally, the organic-inorganic stoichiometry of the surfactant and the SiO_4 units is variable and lower than 1. Properties that are characteristic of the structure of two-dimensional polymers are not found in functionalized clays, even though these materials have low-dimensional properties. Exploitation of the unique properties of polysilsesquioxane will therefore extend our understanding of the structure, the reactions and the functionality of organic-inorganic hybrid materials in general. In this chapter, current knowledge about the structure and chemical reactivities of polysilsesquioxane are reviewed and restructured for the future development of new functionalized materials.

Many small bisilane $(R'O)_3Si$-X-$Si(OR')_3$ can be converted into periodic mesoporous organosilica (PMO) when they are treated hydrothermally with a surfactant in the form of rod-like or spherical micelles.[16] The conformational degree of freedom of -X- is normally very small in PMO and the organic group is occluded in the framework. Therefore only a small part of the -X- is active for various chemical reactions. Although PMO is also a polysilsesquioxane and has been studied both intensively and extensively, it is outside the scope of this review. Several excellent reviews can be found in the literature.[16]

12.2 Synthesis and Variation of Materials

The basic scheme of polysilsesquioxane is simple and common for all precursors: the hydrolysis of a monoorganotrialkoxysilane, $RSi(OR')_3$, or a monoorganotrichlorosilane $RSiCl_3$ in an acidic or basic solution followed by drying. The pH of the solvent is normally monitored after mixing with the silane, since it influences not only the rate of hydrolysis of the silane and the subsequent dehydration reaction but it also changes the ionization state of functional groups such as NH_2-, if they exist within R-. The latter effect also explains why the pH is considerably different among the recipes that have been reported in the literature. An organic liquid such as ethanol has often been added. The reaction conditions are varied on the silane, as shown in Scheme 12.1. A simple polysilsesquioxane (which ideally becomes $[RSiO_{1.5}]$ although the condensation level has not always been as high as in this formula) is obtained when R is a long alkyl chain such as $C_{12}H_{25}$-, $C_{14}H_{29}$-, $C_{16}H_{33}$- or $C_{18}H_{37}$- (Scheme 12.1(a)).[17,18] When using 3-aminotrimethoxysilane $NH_2CH_2CH_2CH_2Si(OCH_3)_3$, APTMS, instead of alkylsilane, chloride is occluded into the solid product, in which case the positive charge on the $^+NH_3$- is expected to be neutralized. This neutralization occurs during the hydrolysis of APTMS in hydrochloric acid (Scheme 12.1(b)).[19] Although the self-assembly mechanisms that have been elucidated for R = $C_{12}H_{25}$-, $C_{14}H_{29}$-, $C_{16}H_{33}$- and $C_{18}H_{37}^{18}$ are likely to be absent in the synthesis shown in Scheme 12.1(b), the hydrolysis-condensation of APTMS or 3-aminotriethoxysilane $NH_2CH_2CH_2CH_2Si(OCH_3)_3$, APTES, in the presence of monoalkyl-sulfate (Scheme 12.1 (c)) or carboxylate (Scheme 12.1(d)) results in the formation of a self-assembled solid without any chloride in the interlayer.[20,21]

This synthetic principle is effective when using CO_2 instead of an organic acid. The CO_2 bridges aminosilane molecules such as 3-aminopropyltrimethoxysilane $NH_2C_3H_6Si(OCH_3)_3$, 11-aminoundecyltrimethoxysilane $NH_2C_{11}H_{22}Si(OCH_3)_3$, N-aminoethyl-3-aminopropyltrimethoxysilane $NH_2C_2H_4NHC_3H_6Si(OCH_3)_3$ and N-aminohexyl-3-aminopropyltrimethoxysilane $NH_2C_6H_{12}NHC_3H_6Si(OCH_3)_3$ during the condensation reaction. The polysilsesquioxane thus formed occludes CO_2 and has a good lamellar structure (Scheme 12.1(e)). This structure remains relatively intact after the thermal removal of CO_2.[22] These decarbonated poly-amino-alkylsilsesquioxanes are considered to contain no anions in the interlayer, and, consequently, the amines are neutral. Other polysilsesquioxanes that can be prepared by anion-assisted synthesis are those from 4-amino-phenyltrimethoxysilane $NH_2C_6H_4Si(OCH_3)_3$ hydrolysed with Cl and $C_{12}H_{25}OSO_3H$, which is followed by the reaction with chloroacetylchloride (Scheme 12.1(f)).[23]

The hydrolysis of bridged bisilane, $(R'O)_3Si$-R-$Si(OR')_3$, also provides a layered silsesquioxane when -R- is a relatively long chain such as -$C_nH_{2n}N$ $HCONHC_6H_4NHCONHC_nH_{2n}$-, ($n = 3, 10$, Scheme 12.1(g)),[24] -$C_{11}H_{22}NH$ $COC_6H_4CONHC_{11}H_{22}^{25}$ and -$C_{11}H_{22}OC_6H_4OC_{11}H_{22}$- (Scheme 12.1(h)).[25] An organic solvent such as DMSO has, if necessary, been used for the dissolution of these large bridged bisilanes. Substituted alkylsilanes $XC_nH_{2n}Si(OR')_3$, where X is HS-[26] or NC-[27] have been used as precursors

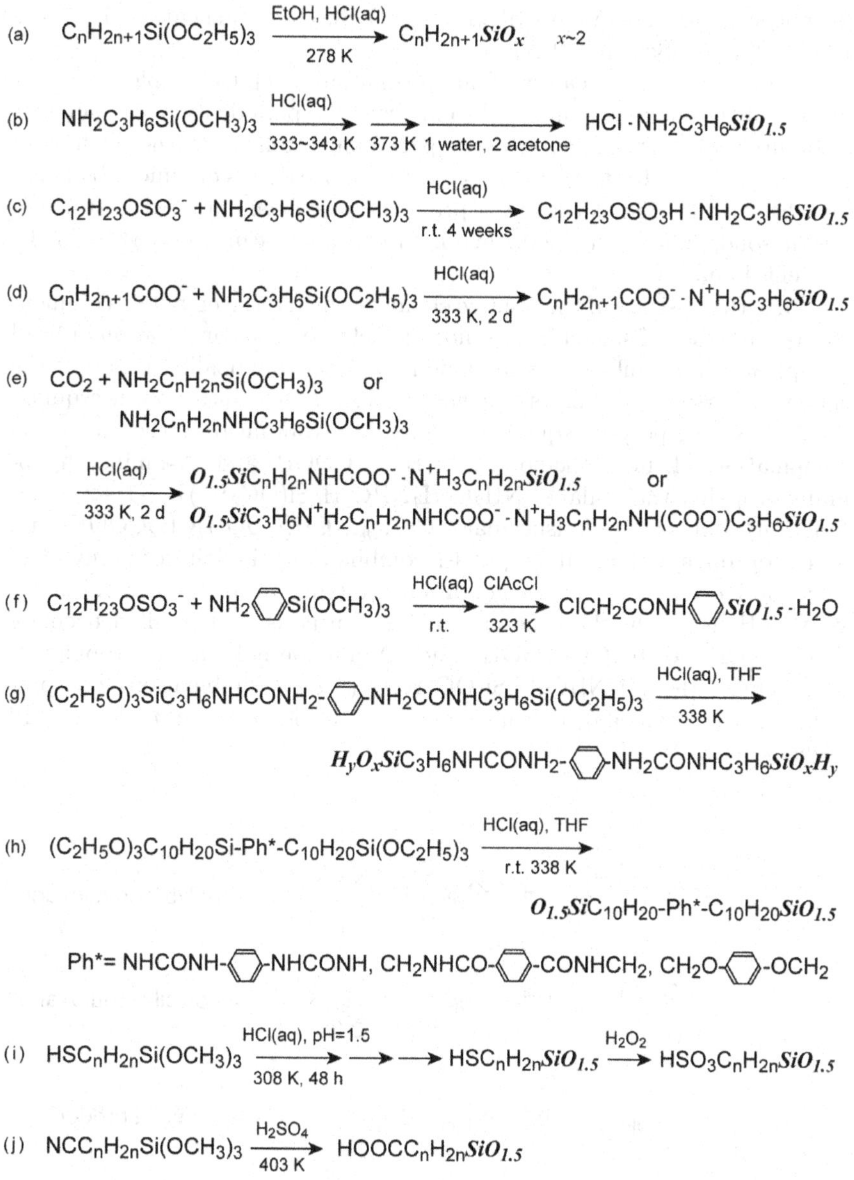

(a) $C_nH_{2n+1}Si(OC_2H_5)_3 \xrightarrow[278\ K]{EtOH,\ HCl(aq)} C_nH_{2n+1}SiO_x \quad x\sim2$

(b) $NH_2C_3H_6Si(OCH_3)_3 \xrightarrow[333\sim343\ K]{HCl(aq)} \xrightarrow{373\ K} \xrightarrow{1\ water,\ 2\ acetone} HCl\cdot NH_2C_3H_6SiO_{1.5}$

(c) $C_{12}H_{23}OSO_3^- + NH_2C_3H_6Si(OCH_3)_3 \xrightarrow[r.t.\ 4\ weeks]{HCl(aq)} C_{12}H_{23}OSO_3H\cdot NH_2C_3H_6SiO_{1.5}$

(d) $C_nH_{2n+1}COO^- + NH_2C_3H_6Si(OC_2H_5)_3 \xrightarrow[333\ K,\ 2\ d]{HCl(aq)} C_nH_{2n+1}COO^-\cdot N^+H_3C_3H_6SiO_{1.5}$

(e) $CO_2 + NH_2C_nH_{2n}Si(OCH_3)_3$ or

    $NH_2C_nH_{2n}NHC_3H_6Si(OCH_3)_3$

$\xrightarrow[333\ K,\ 2\ d]{HCl(aq)} O_{1.5}SiC_nH_{2n}NHCOO^-\cdot N^+H_3C_nH_{2n}SiO_{1.5}$ or

$O_{1.5}SiC_3H_6N^+H_2C_nH_{2n}NHCOO^-\cdot N^+H_3C_nH_{2n}NH(COO^-)C_3H_6SiO_{1.5}$

(f) $C_{12}H_{23}OSO_3^- + NH_2\langle\bigcirc\rangle Si(OCH_3)_3 \xrightarrow[r.t.]{HCl(aq)} \xrightarrow{ClAcCl}_{323\ K} ClCH_2CONH\langle\bigcirc\rangle SiO_{1.5}\cdot H_2O$

(g) $(C_2H_5O)_3SiC_3H_6NHCONH_2\text{-}\langle\bigcirc\rangle\text{-}NH_2CONHC_3H_6Si(OC_2H_5)_3 \xrightarrow[338\ K]{HCl(aq),\ THF}$

    $H_yO_xSiC_3H_6NHCONH_2\text{-}\langle\bigcirc\rangle\text{-}NH_2CONHC_3H_6SiO_xH_y$

(h) $(C_2H_5O)_3C_{10}H_{20}Si\text{-}Ph^*\text{-}C_{10}H_{20}Si(OC_2H_5)_3 \xrightarrow[r.t.\ 338\ K]{HCl(aq),\ THF}$

    $O_{1.5}SiC_{10}H_{20}\text{-}Ph^*\text{-}C_{10}H_{20}SiO_{1.5}$

$Ph^* = NHCONH\text{-}\langle\bigcirc\rangle\text{-}NHCONH,\ CH_2NHCO\text{-}\langle\bigcirc\rangle\text{-}CONHCH_2,\ CH_2O\text{-}\langle\bigcirc\rangle\text{-}OCH_2$

(i) $HSC_nH_{2n}Si(OCH_3)_3 \xrightarrow[308\ K,\ 48\ h]{HCl(aq),\ pH=1.5} \xrightarrow{} \xrightarrow{} HSC_nH_{2n}SiO_{1.5} \xrightarrow{H_2O_2} HSO_3C_nH_{2n}SiO_{1.5}$

(j) $NCC_nH_{2n}Si(OCH_3)_3 \xrightarrow[403\ K]{H_2SO_4} HOOCC_nH_{2n}SiO_{1.5}$

Scheme 12.1.   Various polysilsesquioxanes reported in the literature.

for polysilsesquioxane with HSC_nH_{2n}-, $HO_3SC_nH_{2n}$- (Scheme 12.1(i)) and $HOOCC_nH_{2n}$- (Scheme 12.1(j)).

The drying process can need more than one week to complete the formation of the siloxane bonds and to improve the periodicity of the lamellar structure.[21,28] It is necessary to pay particular attention to the washing of the powder just after filtration in order to remove excess organic substances attached to the solid. Most of the products have been obtained as stoichiometric solids (*vide infra*) and such organic residues can cause errors in the elemental analysis.

Since the position of the organic chains in the interlayer will be commensurate with that of silicon in the siloxane network, the combination of head groups of self-assembled organic anions with the aminoalkyl group of the silane precursor is an important subject in research into polysilsesquioxanes. However, polysilsesquioxanes have been obtained neither from the combination of monoalkanoates $C_nH_{2n+1}COOH$ and N-aminoethyl-3-aminopropyltriethoxysilane $NH_2C_2H_4NHC_3H_6Si(OC_2H_5)_3$ (-1:+2 combination) nor of n-alkylsuccinate $C_nH_{2n+1}CH(COOH)CH_2COOH$ and 3-aminopropyltriethoxysilane (−2:+1 combination), in spite of many attempts under various reaction conditions (mixing ratio, temperature, solvents, pH, etc.) On the other hand, the synthesis from n-alkylsuccinate $C_nH_{2n+1}CH(COOH)CH_2COOH$ and N-aminoethyl-3-aminopropyltriethoxysilane $NH_2C_2H_4NHC_3H_6Si(OC_2H_5)_3$ (−2:+2 combination) has provided a good lamellar polysilsesquioxane, at least when $n = 8 - 18$ (Scheme 12.2).[29]

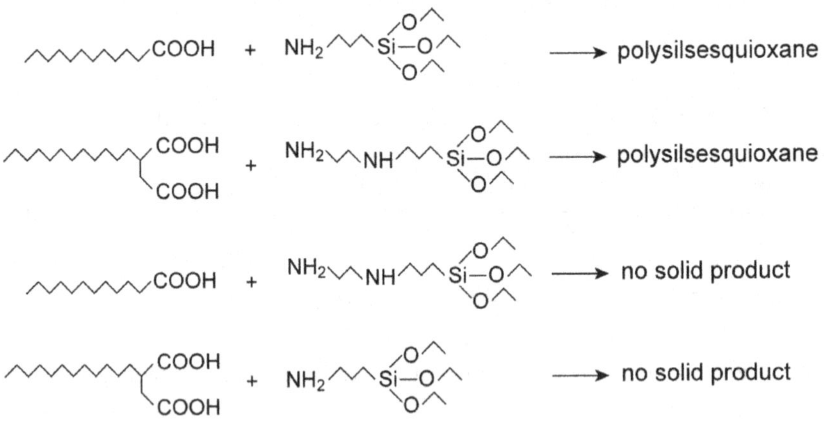

Scheme 12.2. Combinations of surfactant and silane in the preparation of polysilsesquioxane.

12.3 Structure

12.3.1 *Composition*

As with other inorganic-organic hybridized solids, determination of the elemental composition is particularly important for understanding the structure of polysilsesquioxane. It provides the density of the organic functional group as well as a key to check the structural model proposed from the results of spectroscopic analyses. For example, 70.15% C, 11.84% H, 7.28% Si and 10.73% O have been obtained as the elemental composition for a solid prepared from octadecyltrichlorosilane $C_{18}H_{37}SiCl_3$.[18] These percentages agree with the formula of $C_{18}H_{36.15}Si_{0.797}O_{2.065}$, while the expected polysilsesquioxane would be $C_{18}H_{37}SiO_{1.5}$. The observation that the oxygen content is larger than it should be may be attributed to incomplete condensation and the presence of physisorbed water, which should be supported by ^{29}Si NMR (*vide infra*). When the author of this chapter recalculated the Si/O ratio from the intensities of ^{29}Si NMR due to the T^1, T^2 and T^3 species in Ref. 18, Si : O = 1:1.9 was obtained. It is to be noted that disagreement as large as that seen in this case is often found in the elemental analyses of polysilsesquioxanes, probably due to intrinsic errors in the CHN elemental analysis.

Nearly stoichiometric results have been obtained in the elemental analysis of C and N in polysilsesquioxane synthesized from APTES and a series of alkanoic acids $C_nH_{2n+1}COOH$, where $n = 11, 13, 15$ and 17. Their weight percentages are consistent with the respective formulae of $C_nH_{2n+1}COOHNH_2C_3H_6SiO_{1.5}$.[21,28] Apart from these studies, good agreement with the structure of the silane precursor has been obtained in the elemental analysis of polysilsesquioxane in Refs. 24–27, 29 and 30.

Thermogravimetry (TG) has been used to determine the amount of physisorbed water in the solid.[28] Furthermore, if CHNS elemental analysis is not available, the combination of ^{29}Si NMR, ^{13}C NMR and TG will provide approximately complementary information about the composition of the solid.

12.3.2 *Periodicity measured by X-ray diffraction*

X-ray diffraction (XRD) has usually been applied as the first-choice technique for the structural analyses of polysilsesquioxanes, since the material is normally obtained as a layered solid.[17–27] A typical lamellar pattern has been obtained by the measurement of samples where the self-assembly of organic chains is expected. For example, the use of the precursor

$C_nH_{2n+1}Si(OC_2H_5)_3$, where $n = 12, 14, 16$ and 18, provides solids that display a lamellar pattern with three or more prominent peaks.[17] Furthermore, for this series of solids, the interlayer distance d increases linearly with n, with a slope of 0.25 nm per carbon, which is consistent with a structural model where all of the alkyl chains have the trans conformation and are perpendicular to the layers. All these results imply that the formation of a layered structure is promoted by the self-assembly of the organic part of the silane. However, such a clear XRD pattern has not been obtained in a study using $C_nH_{2n+1}SiCl_3$,[18] probably due to the different amount of water present in the solid.

Chujo et al. optimized the drying time (between 1 and 2 weeks) by observation of the XRD patterns of the lamellar solids.[28] The drying is considerably longer than the synthesis from $C_nH_{2n+1}Si(OC_2H_5)_3$ (1 day). This is because the former solid is not composed of the hydrolysed product from alkylsilane, but of those from $NH_2CH_2CH_2CH_2Si(OCH_3)_3$ and $C_nH_{2n+1}COOH$, $n = 11, 13, 15$ and 17, which are more hydrophilic than alkyl chains. The study by Chujo et al. has, nevertheless, demonstrated that the drying conditions are important to obtain a periodic structure. In this series, however, the interlayer distance d (in nm) increases linearly as $d = 0.166n + 1.082$ where n is the number of carbon atoms in the carboxylic acid, as shown in Figure 12.2. The observation that the slope is clearly less than 0.25 nm per carbon indicates either a common mixing ratio of trans and gauche conformations for all n, or an inclination of the molecular axis of the surfactant in a nearly all-trans conformation with respect to the basal plane of the siloxane network. They concluded that the latter model was more likely. The possibility of interdigitation can be excluded by the results obtained by IR and ^{13}C NMR spectroscopies (vide infra).

12.3.3 Silica layer

Considering the elemental composition and the self-assembled interlayer structure, layers built up with a Si-O-Si network are extremely thin and the position of the Si and O atoms attracts special interest. Historically, the bond angle distributions of Si-O-Si have long been a subject of interest for amorphous silica and other silicates. However, the structural data obtained by physicochemical analyses rarely provide sufficient evidence to elucidate the bond angle in polysilsesquioxane. A hexagonal pattern has been obtained in the electron diffraction of poly-3-aminopropylsilsesquioxane, which agrees with the pseudo-hexagonal structure in the sheet of siloxane bonds in the network with lattice parameters of 0.52 and 0.90 nm (Figure 12.3).[20] These parameters suggest a Si-O-Si bond angle near

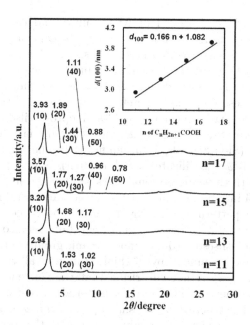

Figure 12.2. X-ray diffractions by polysilsesquioxane $C_nH_{2n+1}COOH$-$NH_2C_3H_6SiO_{1.5}$ prepared with 3-aminopropyltriethoxysilane and alkanoic acid. The alkanoic acids are (a) lauric acid $C_{11}H_{23}COOH$, (b) myristic acid $C_{13}H_{27}COOH$, (c) palmitic acid $C_{15}H_{31}COOH$ and (d) stearic acid $C_{17}H_{35}COOH$. The interlayer distance calculated with the position of the 100 diffraction is plotted against the number of carbon atoms in the alkyl chain of the alkanoic acid.

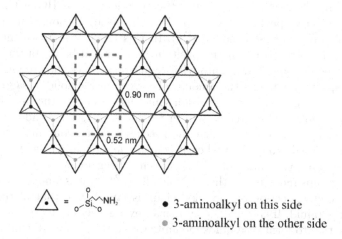

Figure 12.3. Model of a silica layer in the polysilsesquioxane $NH_2C_3H_6SiO_{1.5}$, which is based on the electron diffraction pattern in TEM observations.

to 2π if the solid is composed of $RSiO_3$ units and contains no terminal SiOH. Studies using ^{29}Si magic angle nuclear magnetic resonance (MAS NMR) have demonstrated negligible signals due to the T^1 (around -49 ppm from TMS) and T^2 (around -57 ppm) species, and a clear resonance due to the T^3 (around -68 ppm) species for the Si atom in poly-3-aminopropylsilsesquioxane.[20] Therefore, it is concluded that most of the Si-O-Si bonds are nearly linear in poly-3-aminopropylsilsesquioxane. However, experimental and theoretical studies showed that the contribution from the angle (Si-O-Si) $= 2\pi$ is insignificant in the angle distributions of amorphous silica and silicate.[31,32] A comparison of a two-dimensional $RSiO_{1.5}$ layer with three-dimensional silica solids suggests that the chemical bond Si-O-Si in polysilsesquioxane is considerably strained.

The spectral pattern in ^{29}Si NMR has not only been reproduced in other studies[19] on almost the same material, but has also been found in those on other polysilsesquioxanes whose organic group contains thymine,[24] carboxylate-bound 3-aminopropyl,[28] thiol-terminated alkyl[26] and carboxyl-terminated alkyl.[27] It is not surprising that the elemental compositions determined experimentally in all of these materials are consistent with those expected from the structure of the silane precursors, considering that the CHN compositions by elemental analysis agree with the theoretical formula deduced from the structure of silane precursors in the absence of terminal silanol.

The complete condensation without the formation of silanols implies that the linear Si-O-Si bond is stable in layered silsesquioxane, unlike the situation in amorphous silica and silicate. The same pattern has also been obtained in ^{29}Si NMR spectra when the organic group is $HOOCC_nH_{2n}$.[27] In contrast, the peak assigned to the T^2 species is dominant in ^{29}Si NMR of poly-alkylsilsesquioxane.[17,18] In this case, no experimental evidence has been provided for the positions or chemical bonds of the Si and O atoms, unlike poly-3-aminopropylsilsesquioxane. Since the presence of T^2 and T^1 species augments the amount of oxygen in the solid, disagreement in terms of the elemental composition sometimes occurs in the case of poly-alkylsilsesquioxane, as already seen in Section 12.3.1. The condensation is probably much more rapid when the precursor silane contains a basic or acidic functional group than when it is a simple monoalkylsilane.

Considering that only the T^3 species has been found in ^{29}Si NMR of organic groups more bulky than simple alkyls,[24-27] it is unlikely that the projective cross-section of the alkyl chain is larger than that of the SiO_3 trigonal pyramid. If the silane contains an excessively large organic group, condensation to form silsesquioxane units would be prevented, while a considerable number of silanols would be generated after the hydrolysis of silane.

In all cases where ^{29}Si NMR is applied, the Q^n peaks are negligible in the spectra, and, therefore, we can conclude that little bond scission occurs between C and Si atoms in the formation of the solid.

12.3.4 *Organic interlayer*

Since most of the polysilsesquioxane lamellar solids are formed with a self-assembled organic interlayer (and without such an interlayer, hardly any high-index diffractions appear in XRD), understanding the mode of interaction among the organic chains is important for elucidating the conditions for the formation of a good periodic order. We could find similarities with Langmuir or Langmuir–Blodgett films in many aspects of structural analyses. For example, when using Cu K_α radiation ($\lambda = 0.154$ nm), a distinct diffraction is observed at $2\theta = 21.44°$ from octadecylpolysilsesquioxane,[18] which agrees with $d = 0.414$ nm. This spacing is roughly in accordance with the chain-chain spacings found in Langmuir monolayers, 0.415–0.430 nm, which depend on the lateral pressure.[33-35] The same kinds of diffraction are found in the measurement of solids prepared from bridge bisilanes, though these are composed of two or more peaks and have been attributed to hydrogen bonds.[24] The peaks become broader than these two kinds of silsesquioxanes when functional groups are present within the alkyl chain.[27-28]

Historically speaking, the solid products prepared by the hydrolysis of organosilanes were studied with respect to derivative materials from Langmuir films and the structures were analysed by extending the concept of self-assembly.[36-38] The formation and structure of such self-assembled monolayers and multilayers are reviewed in Ref. 39.

Since the vibrational spectra are influenced by the conditions of self-assembly, infrared (IR) spectroscopy has been a powerful tool for the analysis of organic crystals and semi-crystalline materials such as Langmuir films.[40-49] Blue-shifts of methylene stretching vibrations are attributed to highly crystalline n-alkyl chains with large populations of the trans conformer; two impressive peaks positioned at *ca.* 2,915–2,917 cm^{-1} (d$^-$) and 2,846–2,848 cm^{-1} (d$^+$) appear in the C-H stretching region, whereas liquid decane shows bands at *ca.* 2,925.3 cm^{-1} (d$^-$) and 2,960.0 cm^{-1} (d$^+$), respectively.[50,51] The structure of polysilsesquioxane has also been analysed from the same point of view. The positions of these bands of polysilsesquioxanes are 2,917.5 and 2,850.0 cm^{-1},[18] 2,918 and 2,849 cm^{-1},[28] and 2,920 and 2,951 cm^{-1}.[30] The material in the last case (poly-C$_{12}$H$_{25}$OSO$_3$H-NH$_2$C$_6$H$_4$SiO$_{1.5}$H$_x$) contains not only a dodecyl chain but also a phenyl group bound to a silicon atom, which may provide space for a conformational change in the dodecyl chain. This structure of the organic chain can explain the relatively small shift of the IR bands. Figure 12.4 shows the IR

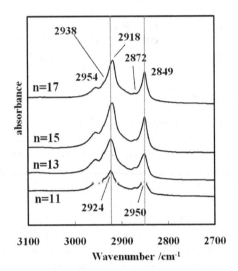

Figure 12.4. Infrared absorptions due to C-H stretching vibrations in the poly-silsesquioxane $C_nH_{2n+1}COOH-NH_2C_3H_6SiO_{1.5}$ prepared from 3-aminopropyltrie-thoxysilane and alkanoic acid. The vertical lines are added for comparison of the peak position.

absorptions for $C_nH_{2n+1}COOH-NH_2C_3H_6SiO_{1.5}$ ($n = 11$, 13, 15 and 17).[28] The position of the d^- band ($2{,}924$–$2{,}918$ cm^{-1}) is considerably lower than that of liquid decane ($2{,}925.3$ cm^{-1}) and it gradually decreases with increasing values of n. At the same time, the FWHM is narrowed according to n. These observations imply that the increase in carbon chain length promotes the self-assembly of alkanoic acid molecules and restricts them to change from trans conformation. These spectral transformations are much slighter in the d^+ band than in the d^- band.

For the measurements at room temperature, the pair of asymmetric stretching modes due to methyl r_a^- ($\sim 2{,}962$ cm^{-1}) and r_b^- ($\sim 2{,}952$ cm^{-1}) are not usually separated and the symmetric stretch r^+ ($\sim 2{,}870$ cm^{-1}) is not resolved from the Fermi resonance due to the methylene symmetric stretch ($\sim 2{,}890$ cm^{-1}).[51] In addition, these absorption bands of methyl are much weaker than those of methylene d^- and d^+. For these reasons, the bands due to methyl stretching vibrations are not as informative as those of methylene (d^+ and d^-) for characterizing the interlayer self-assembly of polysilsesquioxane.

Another characteristic spectral feature potentially important is the band progressions due to the wagging, twisting and rocking modes of the -CH$_2$- unit of the all-trans conformational sequence. Although several conformers can be distinguished in n-alkane solids,[43] this feature becomes smeared out in the IR absorptions of slightly disordered solids[44,45] or of self-assembled and layered solids.[46,48] The wagging progression at *ca.* 1,170–1,350 cm^{-1} appears most clearly in poly-alkylsilsesquioxane.[18] This band structure is also found in the spectra of other complex polysilsesquioxanes,[42,52] though the data are not discussed explicitly in these papers. The portion of the trans conformations in the alkyl chain is roughly estimated with the inter-band energy $\Delta\nu$ (in cm^{-1}) using the relationship with the number of -CH$_2$- units in an all-trans alkyl chain CH$_3$(CH$_2$)$_n$-: $\Delta\nu = 326/(n+1)$.[18]

Since the structures of alkyl chains in solid and self-assembled layers have long been analysed by IR spectroscopy, this technique should be applied more frequently for the structural characterization of polysilsesquioxane, considering that the conformational and self-assembled structure is definitively influential to its reactivities such as the absorptions of aqueous ions/molecules, delamination and the exchange of inter-layer organic substances.

^{13}C MAS NMR has been applied not only to confirm the conservation of the organic part of the silane under hydrothermal conditions, but also to explore the detailed structure of organic chains of polysilsesquioxane.[18,20,24–27] For a long linear alkyl chain in the solid state, the carbon bound directly to an Si atom and the terminal carbon have nearly the same chemical shift at *ca.* 15 ppm from TMS and the second neighbour of the Si atom and the neighbour of terminal carbon also provide unresolved resonances at around 24 ppm. On the other hand, the resonances of the other carbons are apparently found as a single peak at *ca.* 34 ppm.[53] The carbon atom in the internal methylenes of a linear alkane resonates at 30 ppm in the liquid phase.[54] Together with the band shift and the wagging progression in the FT-IR spectra, this upfield shift of the methylene resonance has been used for the analysis of self-assembled monolayers.[54,55] The shifts have been recorded from 30.5 to 33.5 ppm[54] and from 33 to 35 ppm[55] when the alkyl chains change into a more solid state. The resonances of poly-octadecylsilsesquioxane have been observed at 15, 25 and 34 ppm,[18] which agrees with the solid-state alkyl chain and is consistent with the structure of an all-trans conformational assembly. However, peaks are found at 31.0 and 30.5 ppm respectively when the organic part of polysilsesquioxane is C$_{13}$H$_{27}$COOH-NH$_2$C$_3$H$_6^{2-}$ and HOOCC$_{11}$H$_{22-}$,[27] suggesting that these complex organic groups are nearer to liquid than the alkyl chain in poly-octadecylsilsesquioxane. The presence of a carboxyl group can disturb the self-assemblage of linear alkyl chains. In spite of the observation of the wagging mode progression and the significant blue-shift in the IR absorption due to the ν(C-H) of -CH$_2$-, linear alkanoate-aminopropyl may contain a considerable quantity of gauche conformers in the alkyl chains of the alkanoate.[28]

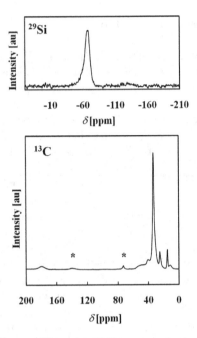

Figure 12.5. Typical ^{29}Si and ^{13}C MAS-NMR spectra of polysilsesquioxane. The sample is $C_{16}H_{33}C(COOH)CHCOOH-NH_2C_2H_4NHC_3H_6SiO_{1.5}$. The peaks marked with an asterisk are the spinning sidebands.

Typical ^{29}Si and ^{13}C MAS NMR spectra of polysilsesquioxane containing a functional group are depicted in Figure 12.5.

12.4 Reactions and Applications

12.4.1 *Conversions into other assembled structures*

Polysilsesquioxanes that contain surfactant molecules in the organic interlayer can be converted by an exchange reaction. For example, the substitution of carboxylic acid occurs when polysilsesquioxane $C_{17}H_{35}COOH-NH_2C_3H_6SiO_{1.5}$ is reacted with $C_{15}H_{31}COONa$ under the same hydrothermal conditions as those applied for the synthesis. The termination of the reaction can be confirmed by the elemental composition and the shift of peaks in XRD as shown in Figure 12.6. The shrinking interlayer distance agrees well with that of $C_{15}H_{31}COOH$-$NH_2C_3H_6SiO_{1.5}$ freshly prepared with $C_{15}H_{31}COONa$ and APTES.[28] In contrast, the alkanoic acid (laurate) in $C_{11}H_{23}COOH-NH_2C_3H_6SiO_{1.5}$ is

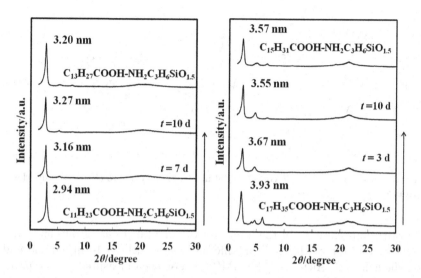

Figure 12.6. Exchange of interlayer surfactant alkanoic acid to expand and to shrink the interlayer distance in the lamellar structure. (Left) The reaction of $C_{11}H_{23}COOH-NH_2C_3H_6SiO_{1.5}$ with $C_{13}H_{27}COOH$. (Right) The reaction of $C_{17}H_{35}COOH-NH_2C_3H_6SiO_{1.5}$ with $C_{15}H_{31}COOH$. The reaction conditions are identical to those used for the preparation of the original polysilsesquioxanes.

exchanged in a stoichiometric manner with $C_{13}H_{27}COOH$ by the same method or even with CH_3COOH.[28,56] These results indicating successful substitutions suggest that the interlayer surfactant of poly(alkanoate-3-aminopropyl)silsesquioxane can be exchanged with all kinds of linear chain alkanoates and that the interlayer distance of this polysilsesquioxane can be elongated and shortened according to the number of carbons in the alkanoic acid, $d = 0.166n + 1.052$ (Scheme 12.3). Poly-3-aminopropylsilsesquioxane, whose composition is $Cl^-N^+H_3C_3H_6SiO_{1.5}\cdot H_2O$, can be obtained by surfactant removal from $C_{12}H_{25}SO_4H-NH_2C_3H_6SiO_{1.5}$, after washing with hydrochloric acid.[20] This process can be considered as an anion exchange reaction. In order to prepare a stoichiometric solid with a well-defined structure, it is important to complete the exchange reaction. However, partial substitutions of various organic anions such as ethanesulfonate, 1-hexanesulfonate, p-toluenesulfonnate and octylsulfonate have been reported[20] as insertion reactions into poly-3-aminopropylsilsesquioxane. It is not clear if a complete anion exchange is possible simply by optimization of the reaction conditions.

Oxidations of the terminal functional group of polysilsesquioxane have been attempted in the preparation of $HSO_3C_nH_{2n}SiO_{1.5}$ and $HOOCC_nH_{2n}SiO_{1.5}$ (from $HSC_nH_{2n}SiO_{1.5}$ and $NCC_nH_{2n}SiO_{1.5}$, respectively).[26,27]

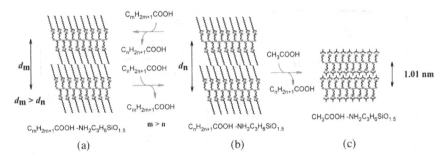

Scheme 12.3.   Stoichiometric conversion of interlayer molecule to change the interlayer distance.

Substitution reactions in the interlayer organic functional group have also been carried out with small organic molecules such as 1-butylamine and acrylic acid.[57,58] Although ^{29}Si NMR has not been measured for the products of these substitution reactions (which makes determination of the composition difficult), the C/N/Cl ratios imply that the conversion attains 96% in the latter substitution reaction.

The capture of ions and molecules can also change the interlayer distance of polysilsesquioxane even if the lamellar structure is well preserved. This is discussed in the following section.

Delamination and exfoliation are generally important subjects in research into layered solids. In the case of polysilsesquioxane, the highly dense functional group may be made more easily accessible for guest ions or small molecules when the solid is exfoliated than in the form of the lamellar semi-crystalline solid. However, the layered structure of polysilsesquioxane is mostly retained (as confirmed by XRD, even though only one peak appears) after washing with organic solvent or electrolytic solutions. By washing $C_{11}H_{23}COOH-NH_2C_3H_6SiO_{1.5}$ with acetic acid, the lamellar pattern ($2\theta = 2.09°$, $5.67°$ and $8.47°$, $d = 3.05$ nm) in XRD of the solid turns into a single peak at $2\theta = 8.74°$ ($d = 1.01$ nm). Although diffractions due to higher indices are not clearly observed, the positional correlation of the silsesquioxane sheets is still retained[56] as with other polysilsesquioxane prepared from APTMS.[19,20] On the other hand, when treated with Fe^{3+} in hydrochloric acid, the XRD pattern of $C_{11}H_{23}COOH-NH_2C_3H_6SiO_{1.5}$ completely disappears. This change implies that delamination happens but does not mean that the layers are exfoliated into separate sheets. The mechanism for the removal of surfactant molecules by Fe^{3+} is different from washing with acetic acid or hydrochloric acid; the former reaction involves the neutralization of the aminopropyl group followed by coordination of a lone pair of nitrogen atoms to the unoccupied d-state

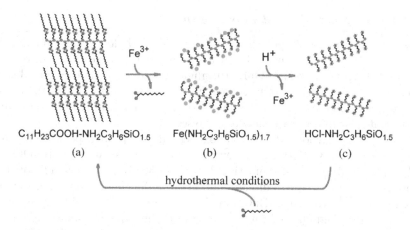

$C_{11}H_{23}COOH\text{-}NH_2C_3H_6SiO_{1.5}$ $Fe(NH_2C_3H_6SiO_{1.5})_{1.7}$ $HCl\text{-}NH_2C_3H_6SiO_{1.5}$

(a)   (b)   (c)

hydrothermal conditions

Scheme 12.4. Delamination of polysilsesquioxane.

of Fe^{3+}, while the latter is considered to be an anion exchange reaction. The interactions between the layers after the former reaction are likely to be weaker than those in the latter case. The composition after the reaction with Fe^{3+} is $Fe(NH_2C_3H_6SiO_{1.5})_{1.7}$, implying the coexistence of several coordination structures. This Fe^{3+}-coordinated polysilsesquioxane can be transformed into a protonated polysilsesquioxane, whose composition is $Cl^-\text{-}N^+H_3C_3H_6SiO_{1.5}$, by washing with hydrochloric acid (Scheme 12.4). Although this solid provides no diffraction patterns, in spite of having the same composition as layered $Cl^-\text{-}N^+H_3C_3H_6SiO_{1.5}$, a well-developed lamellar pattern reappears after treatment with $C_{11}H_{23}COONa$ under hydrothermal conditions.[56] The peak positions in XRD are the same as those for $C_{11}H_{23}COOH\text{-}NH_2C_3H_6SiO_{1.5}$, the starting polysilsesquioxane.

Since $Fe(NH_2C_3H_6SiO_{1.5})_{1.7}$ and $Cl^-\text{-}N^+H_3C_3H_6SiO_{1.5}$ are obtained in the form of a white powder that apparently shows no sign of exfoliation, the sheets are likely to be agglomerated randomly, though the lamellar structure is destroyed. This structure may facilitate the accommodation of guest molecules and ions at the functional sites of the organic part of the solid.

A little note of caution that should be mentioned is that the conversions shown in Scheme 12.3 are accompanied by a considerable change in hydrophilicity. The desorption of water monitored by TG between room temperature and 423 K is enhanced from 0.5 wt.% in $C_{11}H_{23}COOH\text{-}NH_2C_3H_6SiO_{1.5}$ to 3.4 wt.% in $CH_3COOH\text{-}NH_2C_3H_6SiO_{1.5}$. The factor is therefore 6.8, which is larger than that due to the decrease in the molecular weight of alkanoic acid. The enhancement is probably attributed to the shortening of the hydrophobic alkyl chain from $C_{11}H_{23}-$ to CH_3-.

12.4.2 Capture of guest ions and molecules

As already discussed, polysilsesquioxane can include various functional groups with an enormously high density and at the same time can retain a good lamellar structure. Only a negligible number of silanols have been demonstrated experimentally, suggesting the chemical stability of the inorganic sheet. These two features imply that this material is potentially an excellent host in adsorptions and catalyses.

Since poly-carboxyalkylsilsesquioxanes (alkyl $=$ propyl, pentyl and undecyl) contain -COOH at the end of an organic chain (i.e. $HOOCC_nH_{2n+1}$-, $n = 3, 5, 11$), the functional sites are located in the middle of the interlayer and form a two-dimensional hydrophilic region. Once a proton is exchanged with K^+ using ^tBuOK or $K(acac)$, this lamellar solid can capture a considerable number of Eu^{3+} from chloride in ethanoic solution.[27] The ion exchange proceeds almost stoichiometrically, i.e. $K/Si = 0.90$, 0.99 and 0.93, while the capture of europium varies as $Eu/Si = 0.34$, 0.32 and 0.51, for propyl, pentyl and undecyl chains in polysilsesquioxane. Stoichiometries of 1:3 and 1:2 can be expected for these stable coordination structures. The important point might be that the specific capacity for Eu^{3+} becomes 2.0–2.4 mmol/g, which is considerably large for an adsorbent for lanthanide cations.

By bridging with ethylenediamine, ((triethoxy)silylpropylcarbamoyl) butyric acid turns into bis-zwitterionic polysilsesquioxane (Scheme 12.5). The ethylenediamine in the interlayer can be substituted with transition metal cations such as Co^{2+}, Cu^{2+}, Ni^{2+} and Eu^{3+}, where the amount of substituent per Si atom is 0.2, 0.5, 0.5 and 0.5.[59] The N/Si ratio measured after the adsorption of these transition metals is 1, except in the case of Co^{2+}. Thus, it is likely that ethylenediamine is exchanged completely with these ions. A coordination model where one cation is bound to two carboxyl functional groups has been proposed as shown in Scheme 12.5. The capacity is 2.1 mmol/g in these absorptions of transition metal ions. Poly-aminoalkylsilsesquioxanes generated by a CO_2 bridge-assisted synthesis also absorb Cu^{2+}, Eu^{3+} and Gd^{3+}.[22] In this case, the N/M^{n+} ratio varies from 1.67 to 6.44; the absorption of Cu^{2+} nearly agrees with $N/Cu^{2+} = 2$, while those of the lanthanides require five or six amine functions. Since the molecular weight of the organic group is small in these polysilsesquioxanes, the absorption capacity often exceeds even 6 mmol/g. These capacities are comparable with, or even higher than, those of functionalized amorphous silica solids with high surface area.[3,4,60-64]

The exfoliation of $C_{11}H_{23}COOH-NH_2C_3H_6SiO_{1.5}$ with Fe^{3+} as described can be regarded as an absorption of Fe^{3+} and the capacity in this case is 1.9 mmol/g.

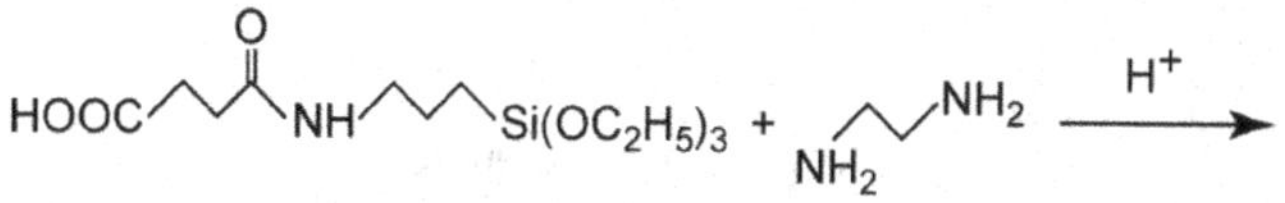

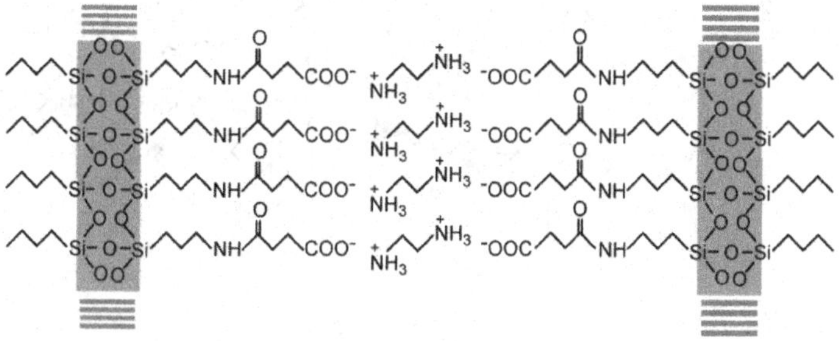

Scheme 12.5. Zwitterion-linked polysilsesquioxane.

12.4.3 Host-guest chemistry in confined space

The density of functional groups is generally considered to be closely related with the capacity. Therefore, high density is preferable for practical use, as long as secondary effects are negligible. In this sense, functionalized polymers can potentially be the best materials, because they can include much more functional sites due to their molecular frameworks than other adsorbents such as functionalized silicas. For example, the density of the amino group in a chelate fibre $[CH_2CH(C(=NH)NHCH_2CH_2NH_2)]_n$ is calculated to be 26.5 mmol/g, which is considerably larger than the density of aminopropyl functions grafted onto a mesoporous silica, typically 1–5 mmol/g.[3,60−66] In spite of the large difference in density, the sorption capacity of chromate CrO_4^{2-} on these two kinds of materials is 0.067 mmol/g and –2 mmol/g, respectively.[3,65] The uptake of chromate by this polymer (0.067 mmol/g) corresponds to $Cr/N = 2.6 \times 10^{-3}$, suggesting that a large part of the amino group is inactive and does not interact with aqueous chromate. On the other hand, much more stoichiometric (e.g. $Cr/N = 0.5$) adsorptions have been observed on mesoporous silica-based solids.[65,66] This difference may arise from the nature of the frameworks in a chain polymer and in silica (flexible/rigid, three-dimensional/low-dimensional, organic/inorganic, etc.). The characteristics of polysilsesquioxane in the sorptions of cations and anions are, roughly speaking, regarded as a melange of those of functionalized polymers and grafted mesoporous silicas (Figure 12.7) and particularly attract interest.

The captures of transition metal cations, as described in the previous section, are not generally stoichiometric, though alkaline metal cations such

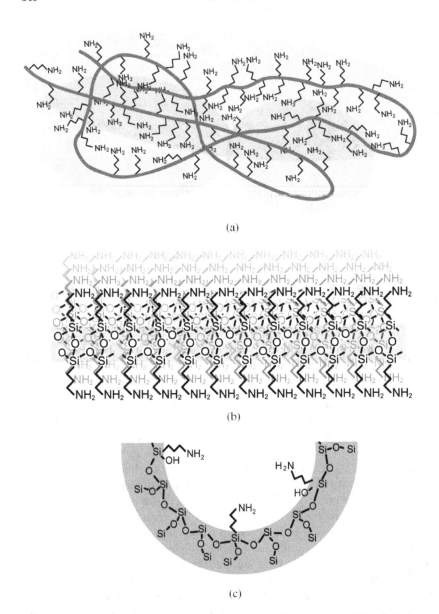

(a)

(b)

(c)

Figure 12.7.　Schematic illustrations of (a) functionalized polymer, (b) polysilsesquioxane and (c) functionalized silica.

as K^+ are exchanged stoichiometrically in poly-carboxyalkylsilsesquioxane. Differences among these guest ions certainly arise from the nature of the chemical bonds; the coordination bond for binding transition metal cations requires certain orientations due to the d orbitals, while the ionic interaction for binding alkaline cations is not anisotropic. In addition, although it is still not clear even today how the cations and anionic sites are hydrated by water molecules (or solvated by solvent molecules) in polysilsesquioxane, the host-guest chemistry of such a highly dense organic group is likely to be sensitive to its density. This is not only because of hindrance of the rotation of single bonds involved in the organic chains, but also because of the facility for the diffusion of ions.

The relationship between the absorption capacity and the lamellar structure has been investigated for the arsenate absorption by poly-3-aminopropylsilsesquioxanes.[56] This study revealed that the number of arsenate anions per amino function is As/N = 0.23, 0.13, 0.41 and 0 for the lamellar form with a large interlayer distance, $C_{11}H_{23}COOH$-$NH_2C_3H_6SiO_{1.5}$, (Scheme 12.4(a)) the lamellar form with a small interlayer distance, CH_3COOH-$NH_2C_3H_6SiO_{1.5}$ (Scheme 12.3(c)), the defoliated and Fe^{3+}-coordinated form, $Fe(NH_2C_3H_6SiO_{1.5})_{1.7}$ (Scheme 12.4(b)) and the defoliated Cl-$NH_2C_3H_6SiO_{1.5}$ (Scheme 12.4(c)), respectively. Thus, an exfoliated and Fe^{3+}-terminated surface is most effective for absorbing a large amount of arsenate anion, while, when Fe^{3+} is exchanged with H^+, the adsorption capacity is completely lost. The loss is attributed to the assemblage of polysilsesquioxane sheets. The low As/N ratios for $C_{11}H_{23}COOH$-$NH_2C_3H_6SiO_{1.5}$ and CH_3COOH-$NH_2C_3H_6SiO_{1.5}$ suggest the inhibition of sorption by the size of the arsenate anion, which is a tetrahedron with a side of 0.276 nm, and makes a clear contrast with the elemental ratio at the adsorption saturation on 3-aminopropyl-MCM-41 and 3-aminopropyl-SBA-1, As/N = 0.35 or 0.55, respectively.[66] The adsorption capacity per unit weight among these four polysilsesquioxane derivatives is largest in the case of $Fe(NH_2C_3H_6SiO_{1.5})_{1.7}$, which appears to be 2.9 mmol/g and is larger than most of the conventional adsorbents.[67]

As already mentioned, it is likely to be due to steric effects that fewer transition metal cations are captured by the terminal amino or carboxyl groups of polysilsesquioxane than alkaline metal cations. On the other hand, the nearest neighbour interaction in a well-defined structure often induces a cooperative effect such as a phase transition. This is rarely observed in the adsorption/absorption onto/into amorphous solids, while it is a rather common phenomenon in the adsorption of metal atoms or small molecules onto single crystal surfaces.[68] Gonda and Yoshitake proved that, in the capture of Cu^{2+} by polysilsesquioxane, the absorption occurs with two steps as shown in the isotherm in Figure 12.8.[69] The intensity of the pre-edge

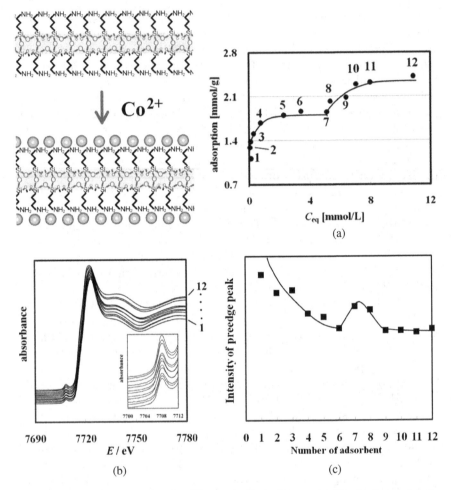

Figure 12.8.   The absorption of Co^{2+} in polysilsesquioxane. (a) The isotherm. (b) Co K-edge XANES spectra of the solids after absorption of Co^{2+}. The inset is an expansion of the pre-edge. (c) The intensity of the pre-edge peak of Co K-edge. The numbers from 1 to 12 are common between (a), (b) and (c).

peak of the Co K-edge spectrum gradually decreases with the coverage, though it suddenly increases just after the first absorption step is finished. Considering that this peak is assigned to the 1s-3d inhibited transition and declines with the stereochemical relaxation of the Co^{2+} centre, the sudden increase agrees with the emergence of space around the Co^{2+} and easing of the steric hindrance at Co^{2+} where the amine ligands are coordinated more freely than in the termination of the first step. All of these

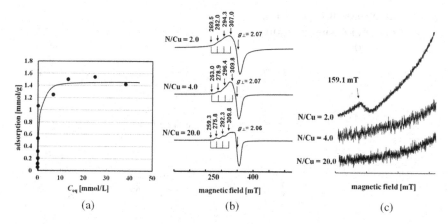

Figure 12.9. The absorption of Cu^{2+} in polysilsesquioxane. (a) The isotherm. (b) ESR spectra of Cu^{2+} in the course of the absorption. (c) The low-field region where the resonances due to the spin coupling of Cu^{2+}-Cu^{2+} appear.

data are consistent with a phase transition of the absorption layer of Co^{2+}, though it is not observed visually. The nearest neighbour interaction has also been observed in Cu^{2+} absorption as a form of spin-spin coupling of the Cu^{2+}-Cu^{2+} pair in the ESR spectra (Figure 12.9).[69] The peak appears at 159.1 mT, even when the measurement is carried out at room temperature, together with a change in the hyperfine splitting constants. These physicochemical data imply that transition metal cations with high coverage in poly-3-aminopropylsilsesquioxane interact with their nearest neighbours (even though they are captured by the amine ligand) and that the interaction can induce a structural transition. This is probably one of the most unique aspects of polysilsesquioxane that rarely if ever occurs on silica surfaces functionalized with the same organic tethers.

The absorptions of Cu^{2+} and Co^{2+} are finally saturated at $Co/N = 0.73$ and $Cu/N = 0.45$, respectively.[69] These ratios can be compared with those of 3-aminopropyl-grafted MCM-41, where $Co/N = 0.40$ and $Cu/N = 0.38$.[70] Although the capacity for Cu^{2+} is similar between the two sorbents, that for Co^{2+} is considerably different. This may be attributed to the flexibility of the coordination structure of Co^{2+}.

If a steric effect working between the absorbates can cause a decrease in the capacity or strength of the absorptions, it is possible to optimize these properties by expanding the spacing between the functional ligands. For poly-aminoalkylsilsesquioxane with alkanoate in the interlayer, this modification is easily achieved by partially substituting the alkanoate with alkanol and partially substituting the aminoalkyl unit with a hydroxyl unit, though the solid becomes no longer a polysilsesquioxane and the

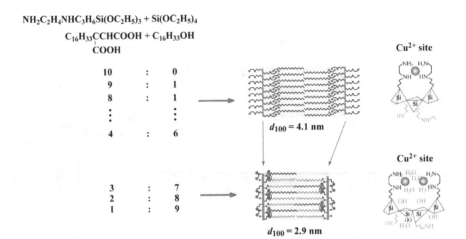

Scheme 12.6. Polysilsesquioxane derivatives.

density of the aminoalkyl functions is decreased.[71] The material can be synthesized by mixing alkylsuccinate and alkanol (additive) as the self-assembling surfactants and N-aminoethyl-3-aminopropyltriethoxysilane and tetraethyl orthosilicate (TEOS, additive) as the silica source with common mixing ratios $x = $ (functional agent) : (additive) from 10:0 to 1:9. The alkanol has the same alkyl group with the alkylsuccinate (Scheme 12.6). For a ratio of 10:0, i.e. containing neither alkanol nor TEOS, a poly-alkanoate-aminoalkylsilsesquioxane in a 2^-:2^+ combination is formed. From the elemental composition and the XRD pattern, a mixing ratio of 1:2 is the critical point for the structural transition; the lattice constant and nitrogen content changes from 4.1 nm and 4 mmol/g into 2.8 nm and 2.5 mmol/g, respectively. The IR absorption band in the carbonyl region correspondingly changes from 1,701 cm^{-1} to 1,574 cm^{-1}, which can be reasonably attributed to non-dissociative and dissociative carboxyls, respectively (Figure 12.10). This change in the IR absorption band can be reasonably explained by the creation of space for the amino functions to become dissociated, since the ionic interaction is a long-range interaction and requires alternate positioning of the cations and anions. The Cu^{2+} coordination to this diluted polysilsesquioxane derivative also changes with the mixing ratio from Cu/N $= 0.25$ ($x > 1$:3) to Cu/N $= 0.43$ ($x < 1$:3), implying the transformation of Cu(en)$_2$ into Cu(en)$_1$ where en $= $ NH$_2$CH$_2$CH$_2$NH-. The square planar and elongated octahedron structures are in agreement with the ESR spectra of the former and latter solids. The strength of the chromate absorption is quite different between these two kinds of Cu^{2+}-polysilsesquioxane derivatives; the latter structure provides

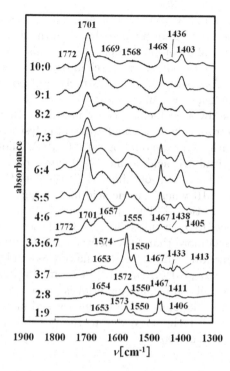

Figure 12.10. Infrared absorptions of polysilsesquioxane-derivatives prepared with the silane precursors (N-aminoethyl-3-aminopropyltriethoxysilane and tetraethyl orthosilicate) and the surfactants (hexadecylsuccinate and hexadecanol). The mixing ratio of (N-aminoethyl-3-aminopropyltriethoxysilane) : (tetraethyl orthosilicate) = (hexadecylsuccinate) : (hexadecanol) is shown.

weaker absorption sites than the former, which is verified by EXAFS and XANES spectroscopies, and the difference is probably due to the coordination structure of Cu^{2+}. These results reveal the aspect of dense packing of the functional group; $NH_2C_2H_4CHNHC_3H_6$- provides a Cu^{2+} coordination site to become $Cu(en)_2$, even though $NH_2C_2H_4CHNHC_3H_6$- is closely packed as $NH_2C_2H_4CHNHC_3H_6$-/$Si = 1$, and when the functional chains are separated from each other, the coordination structure becomes $Cu(en)_1$.

When we graft N-aminoethyl-3-aminopropyltriethoxysilane on MCM-41 (pore size: 2.9 nm) or SBA-1 (pore size: 3.0 nm) and then coordinate Cu^{2+}, we will obtain a functionalized mesoporous silica with Cu/N $\sim$0.25, though the surface density is only 2.8–3.3 N/nm^2.[72] Considering the density of the amino group in the aforementioned polysilsesquioxane derivative and the coordination structure of Cu^{2+} revealed by the Cu/N ratio and the ESR spectra, the organic functional tethers are clearly not sufficiently dense to provide Cu(en)$_2$ sites in MCM-41 and SBA-1. This implies that the amino functional tethers are not uniformly distributed in these mesoporous silicas.

The final example of a potential application of this material is as a proton conductive electrolyte, namely, research into the transport of protons in polysilsesquioxane. As shown above, the structure of polysilsesquioxane is often tightly assembled, containing almost no water. However, it is possible to swell it with water by introducing a certain mediator that can be strongly bound to both water and the functional group of polysilsesquioxane. When APTES is mixed with a large amount of water (1:100), followed by the addition of concentrated sulfuric acid, a siliceous polymer containing significant amounts of water and sulfate is obtained.[73] T^3 becomes the only peak that is observed in ^{29}Si NMR when sulfuric acid molar-equivalent with APTES is added to the synthetic gel. The application of IR spectroscopy reveals that sulfate becomes HSO$_4$ in this solid. The weight loss till 300°C is about 5%, whilst it exceeds more than 50% between 300°C and 500°C. Considering these experimental results, the formula of this solid is deduced to be HSO$_4^-$-N$^+$H$_3$C$_3$H$_6$SiO$_{1.5}$-nH$_2$O ($n > 12$) at room temperature. A notable property of this polysilsesquioxane material is its proton conductivity at 473 K: 2×10^{-3} S/cm. The application of this swelled material is especially interesting for fuel cell technology, due to the possibility of operation at temperatures higher than 100°C.

12.5 Conclusion and Outlook

Layered semi-crystalline polysilsesquioxane is prepared simply by the hydrolysis and dehydration-condensation of monoorganosilane RSiX$_3$ (X = OCH$_3$, OC$_2$H$_5$ or Cl), which is often accompanied by self-assembly due to R or by the aid of a surfactant. The occlusion of the surfactant is stoichiometric and is usually carried out in the course of the synthesis. The siloxane network is completed producing only a few silanols and the solid is usually obtained in the form of RSiO$_{1.5}$. Analysis of the periodic structure has been carried out by XRD and a good lamellar structure has often been observed. In several cases, a linear relationship has been found between d_{100} and the number of carbon atoms in the organic chain. Infrared spectroscopy and ^{13}C NMR have been applied to prove that the interlayer organic chain is densely packed. Transformation of the periodic structure is carried out and various substitutions of the interlayer organic molecules can be found in the literature. The sorption of hazardous cations and anions occurs easily with a high capacity. The relationship between the self-assembled structure of polysilsesquioxane and its sorption capacity has been studied for some transition metal cations. Unique physical chemistry has been unveiled for low-dimensional as well as extremely densely populated functional groups in polysilsesquioxane.

Due to the simple preparation that is required, various polysilsesquioxanes without any particular mesostructure have been synthesized and the number of publications has increased particularly since the year 2000.[74-80] The chemical properties of these materials differ considerably, partly because these solids can be non-stoichiometric with no hierarchical structure. Such ill-defined structures often conceal the mechanisms of their excellent properties in certain applications. It is desirable that model solids with a defined structure are prepared for analysing the mode of working in their applications.

References

1. T. Yanagisawa, T. Shimizu, K. Kuroda and C. Kato, *Bull. Chem. Soc. Jpn.* **63**, 988–992 (1990).
2. C. T. Kresge, M. E. Leonowicz, W. J. Roth, J. C. Vartuli and J. S. Beck, *Nature* **359**, 710–712 (1992).
3. (a) H. Yoshitake, *New J. Chem.* **29**, 1107–1117 (2005). (b) J. Liu, X. Feng, G. E. Fryxell, L. Q. Wang, A. Y. Kim and M. Gong, *Adv. Mater.* **10**, 161–165 (1998). (c) G. E. Fryxell and J. Liu, in *Adsorption on Silica Surfaces*, ed. E. Papirer Marcel Dekker, New York, 2000) p. 665. (d) A. Sayari and S. Hamoudi, *Chem. Mater.* **13**, 3151–3168 (2001). (e) J. H. Clark and D. J. Macquarrie, *Chem. Commun.* 853–854 (1998). (f) K. Moller and T. Bein, *Chem. Mater.* **10**, 2950–2963 (1998). (g) A. Stein, B. J. Melde and R. C. Schroden, *Adv. Mater.* **12**, 1403–1419 (2000). (h) R. Anwander, *Chem. Mater.* **13**, 4419–4438 (2001). (i) G. Kickelbick, *Angew. Chem. Int. Ed.* **43**, 3102–3104 (2004).
4. (a) H. Yoshitake, *J. Mater. Chem.* **20**, 4537–4550 (2010). (b) A. Jentys, N. H. Pham and H. Vinek, *J. Chem. Soc., Faraday Trans.* **92**, 3287–3291 (1996). (c) X. S. Zhao and G. Q. Liu, *J. Phys. Chem. B* **102**, 1556–1561 (1998). (d) C. P. Jaroniec, M. Kruk and M. Jaroniec, *J. Phys. Chem. B* **102**, 5503–5510 (1998). (e) T. Ishikawa, M .Matsuda, A. Yasukawa, K. Kandori, S. Inagaki, T. Fukushima and S. Kondo, *J. Chem. Soc., Faraday Trans.* **92**, 1985–1989 (1996). (f) J. Chen, Q. Li, R. Xu and F. Xiao, *Angew. Chem. Int. Ed. Engl.* **34**, 2694–2696 (1995). (g) X. S. Zhao, G. Q. Lu, A. K. Whittaker, G. J. Millar and H. Y. Zhu, *J. Phys. Chem. B* **101**, 6525–6531 (1996).
5. G. Lagaly and K. Beneke, *Colloid Polym. Sci.* **269**, 1198–1211 (1991).
6. R. Krishnamoorti, R. A. Vaia and E. P. Giannelis, *Chem. Mater.* **8**, 1728–1734 (1996).
7. M. Ogawa and K. Kuroda, *Chem. Rev.* **95**, 399–438 (1995).
8. M. Ogawa and K. Kuroda, *Bull. Chem. Soc. Jpn.* **70**, 2593–2618 (1997).
9. J. H. Fendler, *Chem. Mater.* **8**, 1616–1624 (1996).
10. S. P. Newman and W. Jones, *New J. Chem.* **22**, 105–115 (1998).

11. A.C. Balazs, C. Singh, E. Zhulina and Y. Lyatskaya, *Acc. Chem. Res.* **32**, 651–657 (1999).
12. M. Zanetti, S. Lomakin and G. Camino, *Macromol. Mater. Eng.* **279**, 1–9 (2000).
13. K. A. Carrado, *Appl. Clay Sci.* **17**, 1–23 (2000).
14. S. S. Ray and M. Okamoto, *Prog. Polym. Sci.* **28**, 1539–1641 (2003).
15. (a) C. Sanchez, G. J. A. A. Soler-Illia, F. Ribot and D. Grosso, *C. R. Chimie* **6**, 1131–1151 (2003). (b) C. Sanchez, C. Bossière, D. Grosso, C. Laberty and L. Nicole, *Chem. Mater.* **20**, 682–737 (2008).
16. (a) S. Inagaki, S. Guan, T. Ohsuna and O. Terasaki, *Nature* **416**, 304–307 (2002). (b) G. J. D. Soler-Illia, C. Sanchez, B. Lebeau and J. Patarin, *Chem. Rev.* **102**, 4093–4318 (2002). (c) A. Taguchi and F. Schuth, *Micropor. Mesopor. Mater.* **77**, 1–45 (2005). (d) F. Hoffmann, M. Cornelius, J. Morell and M. Fröba, *Angew. Chem. Int. Ed.* **45**, 3216–3251 (2006). (e) B. Hatton, K. Landskron, W. Whitnall, D. Perovic and G. A. Ozin, *Acc. Chem. Res.* **38**, 305–312 (2005). (f) A. Corma and H. Garcia, *Adv. Synth. Catal.* **348**, 1391–1412 (2006). (g) K. Ariga, J. P. Hill, M. V. Lee, A. Vinu, R. Charvet and S. Acharya, *Sci. Technol. Adv. Mater.* **9**, 014109 (2008). (h) C. Sanchez, C. Boissiere, D. Grosso, C. Laberty and L. Nicole, *Chem. Mater.* **20**, 682–737 (2008). (i) S. Fujita and S. Inagaki, *Chem. Mater.* **20**, 891–908 (2008). (j) M. P. Kapoor and S. Inagaki, *Bull. Chem. Soc. Jpn.* **79**, 1463–1475 (2006). (k) Y. Yamauchi and K. Kuroda, *Chem. Asian J.* **3**, 664–676 (2008). (l) P. van der Voort, C. Vercaemst, D. Schaubroeck and F. Verpoort, *Phys. Chem. Chem. Phys.* **10**, 347–360 (2008). (m) Q. H. Yang, J. Liu, L. Zhang and C. Li, *J. Mater. Chem.* **19**, 1945–1955 (2009). (n) L. D. Carlos, R. A. S. Ferreira, V. D. Bermudez and S. J. L. Ribeiro, *Adv. Mater.* **21**, 509–534 (2009). (o) Z. Q. Wang and S. M. Cohen, *Chem. Soc. Rev.* **38**, 1315–1329 (2009).
17. A. Shimojima, Y. Sugahara and K. Kuroda, *Bull. Chem. Soc. Jpn.* **70**, 2847–2853 (1997).
18. A. N. Parikh, M. A. Schivley, E. Koo, K. Seshadri, D. Aurentz, K. Mueller and D. L. Allara, *J. Am. Chem. Soc.* **119**, 3135–3143 (1997).
19. Y. Kaneko, N. Iyi, T. Matsumoto, K. Fujii, K. Kurashima and T. Fujita, *J. Mater. Chem.* **13**, 2058–2060 (2003).
20. K. Yao, Y. Imai, L. Y. Shi, A. M. Dong, Y. Adachi, K. Nishikubo, E. Abe and H. Tateyama, *J. Colloid. Interf. Sci.* **285**, 259–266 (2005).
21. T. Chujo, Y. Gonda, Y. Oumi, T. Sano and H. Yoshitake, *Chem. Lett.* **35**, 1198–1199 (2006).
22. J. Alauzun, E. Besson, A. Mehdi, C. Reyé and R. J. P. Corriu, *Chem. Mater.* **20**, 503–513 (2008).
23. K. Yao, Y. Imai, L. Y. Shi, E. Abe, Y. Adachi, K. Nishikubo and H. Tateyama, *Chem. Lett.* **33**, 1112–1113 (2004).
24. J. J. E. Moreau, B. P. Pichon, M. W. C. Man, C. Bied, H. Pritzkow, J. L. Bantignies, P. Dieudonné and J. L. Sauvajol, *Angew. Chem. Int. Ed.* **43**, 203–206 (2004).

25. J. J. E. Moreau, B. P. Pichon, G. Arrachart, M. W. C. Man and C. Bied, *New J. Chem.* **29**, 653–658 (2005).
26. J. Alauzun, A. Mehdi, C. Reyé and R. J. P. Corriu, *Chem. Commun.* 347–349 (2006).
27. R. Mouawia, A. Mehdi, C. Reyé and R. J. P. Corriu, *J. Mater. Chem.* **18**, 2028–2035 (2008).
28. T. Chujo, Y. Gonda, Y. Oumi, T. Sano and H. Yoshitake, *J. Mater. Chem.* **17**, 1372–1380 (2007).
29. H. Nakajima, Y. Oumi, T. Sano and H. Yoshitake, *Bull. Chem. Soc. Jpn.* **82**, 1313–1321 (2009).
30. K. Yao, Y. Fu, L. Shi. W. Wan, X. Q. You and F. Yu, *J. Colloid Interf. Sci.* **315**, 400–404 (2007).
31. J. Neuefeind and K. D. Liss, *Ber. Bunsenges. Phys. Chem.* **100**, 1341–1349 (1996).
32. T. M. Clark and P. J. Grandinetti, *Phys. Rev. B* **70**, 064202 (2004).
33. S. W. Barton, B. N. Thomas, E. B. Flom, F. Novak and S. A. Rice, *Langmuir* **4**, 233–234 (1988).
34. K. Kjaer, J. Als-Nielsen, C. A. Helm, L. A. Laxhauber and H. Mohwald, *Phys. Rev. Lett.* **158**, 2224–2227 (1987).
35. P. Tippmann-Krayer and H. Möhwald, *Langmuir* **7**, 2303–2306 (1991).
36. W. R. Thompson and J. E. Pemberton, *Langmuir* **11**, 1720–1725 (1995).
37. J. Sjöblom, G. Stakkestad and H. Ebeltoft, *Langmuir* **11**, 2652–2660 (1995).
38. R. Maoz, S. Matlis, E. DiMasi, B. M. Ocko and S. Sagiv, *Nature* **384**, 150–153 (1996).
39. A. Ulman, *Chem. Rev.* **96**, 1533–1554 (1996).
40. P. G. Snyder, *J. Chem. Phys.* **47**, 1316–1360 (1967).
41. J. H. Schachtschneider and P. G. Snyder, *Spectrochim. Acta* **19**, 117–168 (1963).
42. M. Tasumi, T. Shimanouchi and T. Miyazawa, *J. Mol. Spectrosc.* **9**, 261–287 (1962).
43. M. Maroncelli, S. P. Qi, H. L. Strauss and R. G. Snyder, *J. Am. Chem. Soc.* **104**, 6237–6247 (1982).
44. Y. Kim, H. L. Strauss and R. G. Snyder, *J. Phys. Chem.* **93**, 485–490 (1989).
45. L. Senak, D. Moore and R. Mendelsohn, *J. Phys. Chem.* **96**, 2749–2754 (1992).
46. N. V. Venkataraman, S. Barman, S. Vasudevan and R. Seshadri, *Chem. Phys. Lett.* **358**, 139–143 (2002).
47. N. V. Venkataraman and S. Vasudevan, *J. Phys. Chem. B* **106**, 7766–7773 (2002).
48. K. R. Rodriguez, S. Shah, S. M. Williams and S. Teeters-Kennedy, *J. Chem. Phys.* **121**, 8671–8675 (2004).
49. S. H. Park and C. E. Lee, *Chem. Mater.* **18**, 981–987 (2006).
50. I. R. Hill and I. W. Levin, *J. Chem. Phys.* **70**, 842–851 (1979).
51. R. A. MacPhail, H. L. Strauss, R. G. Snyder and C. A. Elliger, *J. Phys. Chem.* **88**, 334–341 (1984).

52. H. Nakajima, Y. Oumi, T. Sano and H. Yoshitake, *Bull. Chem. Soc. Jpn.* **82**, 1313–1321 (2009).
53. W. L. Earl and D. L. VanderHart, *Macromolecules* **12**, 762–767 (1979).
54. W. Gao and L. Reven, *Langmuir* **11**, 1860–1863 (1995).
55. R. Suresh, N. Venkataraman, S. Vasudevan and K. V. Ramanathan, *J. Phys. Chem. C* **111**, 495–500 (2007).
56. Y. Gonda, Y. Oumi, T. Sano and H. Yoshitake, *Colloid Surf. A* **360**, 159–166 (2010).
57. F. T. Yu, K. Yao, L. Y. Shi, H. Z. Wang, Y. Fu and X. Q. You, *Chem. Mater.* **19**, 335–337 (2007).
58. K. Yao, Y. Fu, L. Y. Shi, W. Wan, X. Q. You and F. T. Yu, *J. Colloid Interf. Sci.* **015**, 400 404 (2007).
59. E. Besson, A. Mehdi, H. Chollet, C. Réye, R. Guilard and R. J. P. Corriu, *J. Mater. Chem.* **18**, 1193–1195 (1998).
60. A. Walcarius and L. Mercier, *J. Mater. Chem.* **20**, 4478–4511 (2010).
61. J. Liu, X. Feng, G. E. Fryxell, L. Q. Wang, A. Y. Kim and M. Gong, *Adv. Mater.* **10**, 161–165 (1998).
62. A. Sayari and S. Hamoudi, *Chem. Mater.* **13**, 3151–3168 (2001).
63. K. Moller and T. Bein, *Chem. Mater.* **10**, 2950–2963 (1998).
64. A. Stein, B. J. Melde and R. C. Schroden, *Adv. Mater.* **12**, 1403–1419 (2000).
65. H. Yoshitake, T. Yokoi and T. Tatsumi, *Chem. Lett.* **31**, 586–587 (2002).
66. H. Yoshitake, T. Yokoi and T. Tatsumi, *Chem. Mater.* **14**, 4603–4610 (2002).
67. H. Yoshitake, in *Bottom-up Nanofabrication: Supramolecules, Self-Assemblies, and Organic Film*, Vol. 6, eds. K. Ariga and H. S. Nalwa (American Scientific Publishers, Stevenson Ranch, CA, 2009) Chapter 10.
68. G. A. Somorjai, *Introduction to Surface Chemistry and Catalysis* (John Wiley and Sons, New York, 1994), Chapter 2.
69. Y. Gonda and H. Yoshitake, *J. Phys. Chem. C* **114**, 20076–20082 (2010).
70. T. Yokoi, T. Tatsumi and H. Yoshitake, *Bull. Chem. Soc. Jpn.* **76**, 2225–2232 (2003).
71. H. Yoshitake, H. Nakajima, Y. Oumi and T. Sano, *J. Mater. Chem.* **20**, 2024–2032 (2010).
72. H. Yoshitake, T. Yokoi and T. Tatsumi, *Chem. Mater.* **15**, 1713–1721 (2003).
73. T. Tezuka, K. Tadanaga, A. Hayashi and M. Tatsumisago, *Solid State Ionics* **178**, 705–708 (2007).
74. D. A. Loy and K. J. Shea, *Chem. Rev.* **95**, 1431–1442 (1995).
75. G. Cerveau and R. J. P. Corriu, *Coord. Chem. Rev.* **178–180**, 1051–1071 (1998).
76. R. H. Baney and X. Cao, in *Silicon-Containing Polymers*, eds. R. G. Johns, W. Ando and J. Chojnowski (Kluwer, Dordrecht, 2000) pp. 157–184.
77. (a) K. J. Shea and D. A. Loy, *Acc. Chem. Res.* **34**, 707–716 (2001). (b) K. J. Shea and D. A. Loy, *Chem. Mater.* **13**, 3306–3319 (2001). (c) K. J. Shea, J. Moreau, D. A. Loy, R. J. P. Corriu and B. Boury, *Funct. Hybrid Mater.* 50–85 (2004).

78. Y. Kaneko, N. Iyi and Z. Kristall, *Zeitschrift für Kristallographie* **222**, 656–662 (2007).
79. Q. L. Zhou, S. Yan, C. C. Han, P. Xie and R. B. Zhang, *Adv. Mater.* **20**, 2970–2976 (2008).
80. K. Kanamori and K. Nakanishi, *Chem. Soc. Rev.* **40**, 754–770 (2011).

Chapter 13

A Thiol-functionalized Nanoporous Silica Sorbent for Removal of Mercury from Actual Industrial Waste

Shas V. Mattigod, Glen E. Fryxell and Kent E. Parker

Pacific Northwest National Laboratory
Richland, WA, USA

13.1 Introduction

There are a number of existing technologies for mercury removal from water and wastewater. These include sulfide precipitation, coagulation/co-precipitation, adsorption, ion exchange, and membrane separation. Recent reviews have included detailed discussions of the performance characteristics, advantages, and disadvantages of these treatment methods.[1,2] Some of the proposed treatment technologies for RCRA metal and mercury removal include sulfur-impregnated carbon,[3] microemulsion liquid membranes,[4] ion exchange,[5] and colloid precipitate flotation.[6] The sulfur-impregnated carbon weakly bonds a heavy metal; therefore, the adsorbed heavy metal needs secondary stabilization. In addition, a large portion of the pores in the active carbon is large enough for the entry of microbes that solubilize the mercury-sulfur compounds. Additionally, the RCRA metal loading on the carbon substrate is extremely limited.

The microemulsion liquid membrane technique consists of an oleic acid microemulsion liquid membrane containing sulfuric acid as the internal phase for mercury adsorption. However, this removal technology uses large volumes of organic solvent and involves multiple steps of extraction, stripping, de-emulsification, and finally mercury recovery by electrolysis. Also, the swelling of the liquid membrane reduces the efficiency of mercury extraction, and the slow kinetics of the metal adsorption requires long contact times.

Typically, the ion-exchange organic resins used for mercury removal are prone to oxidation, leading to substantial reduction of resin life and their inability to reduce the mercury level to below the permitted level. Additionally, the mercury loading on organic resins is limited and the adsorbed mercury can be released into the environment if it is disposed of as a waste form because the organic substrates do not have the ability to resist microbial attack.

The removal of RCRA metal from water by colloid precipitate flotation has been reported to reduce mercury concentration from 0.16 ppm to about 0.0016 ppm. However, to remove colloids, this process involves adding hydrochloric acid (to adjust the wastewater pH), sodium sulfide, and oleic acid solutions to the wastewater. Consequently, the treated water from this process needs additional processing to remove the residual additives. Also, the separated mercury needs further treatment to be stabilized as a permanent waste form.

During the last few years, we have designed and developed a new class of high-performance nanoporous sorbent materials for heavy metal removal that overcomes the deficiencies of existing technologies. These novel materials are created from a combination of synthetic mesoporous ceramic substrates that have specifically tailored pore sizes (2–10 nm) and very high surface areas ($\sim$1,000 m^2/g) with self-assembled monolayers of well-ordered functional groups that have high affinity and specificity for specific types of free or complex cations or anions. Typically, the nanoporous supporting materials are synthesized through a templated assembly process using oxide precursors and surfactant molecules. The synthesis is accomplished by mixing surfactants and oxide precursors in a solvent and exposing the solution to mild hydrothermal conditions. The surfactant molecules form ordered liquid crystalline structures, such as hexagonally ordered rod-like micelles, and the oxide materials precipitate on the micellar surfaces to replicate the organic templates formed by the rod-like micelles. Subsequent calcination to 500°C removes the surfactant templates and leaves a high-surface-area nanoporous ceramic substrate. Using surfactants of different chain lengths produces nanoporous materials with different pore sizes.

These nanoporous materials can be used as substrates for self-assembled monolayers of adsorptive functional groups that are selected to specifically adsorb heavy metals. Molecular self-assembly is a unique phenomenon in which functional molecules aggregate on an active surface, resulting in an organized assembly having both order and orientation. In this approach, bifunctional molecules containing a hydrophilic head group and a hydrophobic tail group adsorb onto a substrate or an interface as closely packed monolayers. The driving forces for the self-assembly are the intermolecular interactions (van der Waals forces) between the functional molecules and the substrate. The tail group and the head group can be chemically modified to contain certain functional groups to promote covalent bonding between the functional organic molecules and the substrate on one end, and the molecular bonding between the organic molecules and the metals on the other. Populating the head group with alkylthiols (which are well known to have a high affinity for various heavy metals, including mercury) results in a functional monolayer which specifically adsorbs heavy metals. Using this technology, we synthesized a novel sorbent (thiol-SAMMS — thiol-self-assembled monolayers on mesoporous silica) for efficiently scavenging heavy metals from waste streams. Detailed descriptions of the synthesis, fabrication, and adsorptive properties of these novel materials have been published previously.[7–11] This chapter will summarize the effectiveness of thiol-SAMMS for mercury removal from two waste streams generated from a pilot-scale radioactive waste vitrification process. The treatment goal was to achieve residual mercury concentrations of ≤ 0.2 ppm as specified by Universal Treatment Standards 40 CFR §268.38 (U.S. EPA).

13.2 Experimental

The samples of waste streams tested were obtained from a pilot-scale study conducted at the Pacific Northwest National Laboratory. These waste streams result from a process designed to vitrify radioactive sludges that result from chemical separation of targeted actinide species. One of the waste streams (melter condensate) results from condensed liquid that is collected during the formation of waste glass in a melter. During this process, mercury contained in the radioactive sludge vaporizes and is collected in a condensate waste stream. A second waste stream results from periodic rinsing of a high-efficiency mist eliminator (HEME) filter with concentrated nitric acid. This highly acidic waste stream (HEME waste) typically contains concentrations of mercury that are over two orders of magnitude higher than the melter condensate waste stream.

Following filtration (0.45 μm), aliquots of each waste stream were analyzed using an inductively coupled plasma optical emission spectrometer (ICP-OES), and an ion chromatograph (IC). Mercury concentrations in these waste streams were measured by an inductively coupled plasma mass spectrometer (ICP-MS). The pH of the melter condensate sample was measured with a glass electrode and the acid content of HEME waste was determined by titrating an aliquot with 2.5 M sodium hydroxide solution.

Thiol-SAMMS used in these tests consisted of 3.5 nm MCM-41 material that was functionalized as described previously.[7–9] Because of the very high acidity, the HEME waste was preprocessed by neutralization with sodium hydroxide followed by filtration. This neutralization step also resulted in five-fold dilution of the sample. About a 250 mL aliquot of this pretreated

HEME waste was initially equilibrated with 0.5 g of thiol-SAMMS material for about four hours. Following equilibration and the solid-liquid separation by filtration, the residual mercury concentration was measured by ICP-MS. Additional treatments were carried out by adding 0.5 g portions of thiol-SAMMS and sampling the solution until the residual concentration of mercury was ≤ 0.2 ppm. For melter condensate tests, we added 0.02 g of thiol-SAMMS to an aliquot of 50 mL of the waste solution and equilibrated the mixture for four hours. After equilibrating, the solid and liquid were separated by filtration, and the residual mercury concentration was measured by ICP-MS. Additional treatment with 0.02 g portions of thiol-SAMMS was carried out until a residual mercury concentration of ≤ 0.2 ppm was achieved.

13.3 Results and Discussion

The data showed the principal component in HEME waste solution to be ~9.15 M nitric acid. The mercury concentration in this waste stream was 732.7 mg/L (Table 13.1). Other dissolved components in the waste solution were Al, B, Ca, Fe, K, Na Mg, Na, Si, and Zn in concentrations ranging from 6 to 500 mg/L. Minor dissolved components such as, Ba, Cd, Co, Cr, Cu, Mo, Ni, P, and Pb, were present in trace concentrations (< 3 mg/L).

Following neutralization (pH 6.9 SU), the principal component in HEME waste was ~1.83 M NaNO$_3$ with a minor amount (~0.002 M) of sodium borate. The waste after pretreatment was relatively concentrated with an ionic strength of 1.83 M. After applying dilution correction, the results (Table 13.2) indicate that the neutralization step removed a significant fraction (~58%) of the mercury originally present in the HEME waste. The residual concentration of mercury after pretreatment was 61 mg/L. Analysis of the neutralized and filtered solutions indicate that this pretreatment step also removed substantial fractions of Al (~90%), Cr (~87%), Fe (~99%), Mn (~81%), and Zn (~82%) from the waste solution.

The melter condensate waste stream was alkaline in nature (pH 8.5 SU) with the main dissolved components being ~0.03 M sodium borate, ~0.009 M sodium fluoride, and ~0.003 M sodium chloride with minor amounts of sodium sulfate, sodium nitrate, sodium nitrite, and sodium iodide (Table 13.1). The ionic strength of this solution was ~0.15 M which is an order of magnitude more dilute than the HEME waste. The mercury concentration in this waste stream was measured to be 4.64 mg/L. Other dissolved components such as Al, Ba, Ca, Cd, Co, C$_2$O$_4$ (oxalate), Cr, Cu, Fe, Mg, Mn, Mo, Ni, P, Pb, and Zn were present in trace concentrations (< 2 mg/L).

Table 13.1.   Composition of mercury-containing waste streams.

Constituent	HEME Waste (mg/L)	Neutralized HEME (mg/L)	Melter Condensate (mg/L)	Constituent	HEME Waste (mg/L)	Neutralized HEME (mg/L)	Melter Condensate (mg/L)
Al	15.63	0.33	0.86	K	7.53	134.80	3.12
B	94.42	17.33	329.70	Mg	10.83	1.82	0.85
Ba	1.02	0.22	0.02	Mn	3.01	0.11	0.02
Ca	34.06	7.72	0.57	Mo	1.18	0.19	0.62
Cd	0.07	0.01	0.04	Na	510	42,146	522.00
Cl	—	—	105.00	Ni	1.03	0.16	0.08
Co	0.03	0.02	0.03	NO_3	567,300	113,460	39.00
C_2O_4	—	—	1.30	NO_2	—	—	40.00
Cr	2.26	0.06	1.67	P	0.39	0.13	0.65
Cu	0.28	0.07	0.08	Pb	0.78	0.13	0.23
Fe	26.55	0.07	0.68	Si	6.39	0.80	8.01
F	—	—	169.00	SO_4	—	—	75.00
Hg	732.7	61.0	4.64	Zn	5.99	0.21	0.25
I	—	—	25.00	pH	—	6.9 SU	8.5 SU

Table 13.2.   Mercury removal from HEME waste solution.

Treatment	Residual Conc. (mg/L)	Cumulative Removal (%)
pH adjusted & filtered	61.00	—
0.5 g thiol-SAMMS	8.10	86.7
0.5 g thiol-SAMMS	4.20	93.1
0.5 g thiol-SAMMS	0.20	99.7

Results of the treatment of the HEME waste with thiol-SAMMS showed (Table 13.2) that the initial treatment removed ~87% of mercury in solution, resulting in a residual concentration of 8.1 mg/L. Subsequent thiol SAMMS addition reduced the mercury concentration to 4.2 mg/L. Adding an additional 0.5 g of thiol-SAMMS, for a total of 1.5 g per 250 mL of waste solution, resulted in a cumulative reduction of mercury to 99.7% with the residual concentration of 0.2 mg/L. The cumulative mercury loading on the solid phase was calculated to be ~10.1 mg/g of thiol-SAMMS.

In the absence of strongly complexing ligands, mercury in neutralized HEME waste would exist mainly as hydrolytic species $[Hg(OH)_2^0]$. Mercury therefore can be adsorbed by the thiol functionality of SAMMS through two different mechanisms, namely:

$$R\text{-}SH + Hg(OH)_2^0 = R\text{-}SHgOH + H_2O \qquad (13.1)$$

$$R\text{-}(SH)_2 + Hg(OH)_2^0 + R\text{-}S_2Hg + 2H_2O. \qquad (13.2)$$

The adsorption reaction (13.1) indicates a monodentate bonding of mercury whereas reaction (13.2) suggests a bidentate bonding of mercury to the thiol sites. Extended X-ray adsorption fine structure (EXAFS) data obtained on mercury-loaded thiol-SAMMS has shown that bonding is typically bidentate in nature as indicated by the second reaction.[12]

We also calculated the selectivity (affinity) of thiol-SAMMS for adsorbing mercury from a high-ionic-strength waste solution. The selectivity (affinity) of a sorbent for a contaminant is typically expressed as a distribution coefficient (K_d) which defines the partitioning of the contaminant between the sorbent and solution phases at equilibrium. The distribution coefficient is the measure of an exchange substrate's selectivity or specificity for adsorbing a specific contaminant or a group of contaminants from matrix solutions, such as waste streams. The distribution coefficient (sometimes referred to as the partition coefficient at equilibrium) is defined as a ratio of the adsorption density to the final contaminant concentration in solution at equilibrium. This measure of selectivity is defined as:

$$K_d = \frac{(x/m)_{eq}}{c_{eq}} \qquad (13.3)$$

where K_d is the distribution coefficient (mL/g), $(x/m)_{eq}$ is the equilibrium adsorption density (mg of mercury per gram of thiol-SAMMS), and c_{eq} is the mercury concentration (mg/mL) in waste solution at equilibrium.

We computed a distribution coefficient (K_d) value of $\sim 5 \times 10^4$ mL/g for mercury adsorption by thiol-SAMMS from HEME solution. These data showed that the combination of pretreatment (neutralization of the acid waste) and thiol-SAMMS treatment removed almost all ($\sim 99.9\%$) mercury originally present in very high concentrations (~ 730 mg/L) in this waste. This bench-scale test demonstrated the feasibility of effectively removing very high concentrations of mercury from strong acidic wastes by first neutralizing the waste, and then using the highly selective adsorptive properties of thiol-SAMMS to scavenge the remaining mercury from the waste to meet the UTS limit of ≤ 0.2 mg/L.

Data from the treatment of the melter condensate waste showed (Table 13.3) that the initial addition of 0.02 g of thiol-SAMMS removed nearly 85% of mercury in solution with a residual concentration of 0.72 mg/L. Adding an additional 0.02 g of thiol-SAMMS, for a total of 0.04 g per 50 mL of waste solution, resulted in a cumulative reduction of mercury to 98.9% with the residual concentration of 0.05 mg/L. The cumulative mercury loading on the solid phase was calculated to be ~ 5.6 mg/g of thiol-SAMMS. As compared to the HEME waste, the melter condensate waste contained halide ligands such as Cl^- and I^- which form strong complexes with dissolved mercury.[10,13] Based on the complexation constants listed,[10] we calculated that dissolved mercury in this waste existed mainly as $HgOHI^0$ complex. Therefore, the bidentate adsorption mechanism for this complex can be represented by:

$$R\text{-}(SH)_2 + HgOHI^0 + R\text{-}S_2Hg + H_2O + H^+ + I^-. \quad (13.4)$$

We computed a distribution coefficient (K_d) value of $\sim 1 \times 10^5$ mL/g for mercury adsorption by thiol-SAMMS from melter condensate solution. This test demonstrated that treatment of this alkaline waste solution with thiol-SAMMS can effectively reduce the level of mercury to an order of magnitude below the UTS threshold of ≤ 0.2 mg/L waste.

The data from these experiments showed that the presence of competing cations in the waste solutions did not significantly affect the high affinity

Table 13.3. Mercury removal from melter condensate waste solution.

Treatment	Residual Conc. (mg/L)	Cumulative Removal (%)
Untreated	4.64	—
0.02 g thiol-SAMMS	0.72	84.5
0.02 g thiol-SAMMS	0.05	98.9

of thiol-SAMMS for mercury in solution (K_d : 5×10^4 and 1×10^5 mL/g). Such selectivity and affinity in binding mercury by thiol-SAMMS can be explained on the basis of the hard and soft acid-base theory (HSAB),[14,16] which predicts that the degree of cation softness directly correlates with the observed strength of interaction with soft base functionalities such as thiols (-SH groups). According to the HSAB principle, soft cations and anions possess relatively large ionic size, low electronegativity, and high polarizability (highly deformable bonding electron orbitals), and therefore mutually form strong covalent bonds. In these waste streams, all cations except lead and cadmium are either hard or borderline cations and therefore were less likely to interact significantly with the soft base moiety (-SH) of thiol-SAMMS. Because the relative degree of softness is in the order Hg $\gg$ Pb $\gg$ Cd,[17] mercury would preferentially bind to the thiol functionality. These tests showed that thiol-SAMMS can very selectively bind dissolved mercury from waste solutions containing varying concentrations of major and minor cations and anions. Data also showed that thiol-SAMMS sorbent can very effectively scavenge strongly complexed mercury from dilute to relatively concentrated waste matrices to meet the UTS limits for effluents.

13.4 Conclusions

Tests were conducted using a novel sorbent, thiol-functionalized nanoporous silica material, demonstrating its effectiveness in removing mercury from two waste streams. These waste streams originated from pilot-scale tests being conducted to refine the process of vitrifying radioactive sludges that result from chemical separation of targeted actinide species. Two waste streams resulting from this process (HEME and melter condensate) contain mercury concentrations that ranged from ~700 ppm to ~5 ppm respectively. The data showed that thiol-SAMMS was effective in reducing mercury concentrations in these two waste streams to meet a treatment limit of ≤ 0.2 ppm. These tests demonstrated that thiol-SAMMS can very selectively (K_d : 5×10^4 and 1×10^5 mL/g) and effectively scavenge strongly complexed mercury from dilute to relatively concentrated waste matrices to meet the UTS limits for effluents.

Acknowledgments

The Pacific Northwest National Laboratory is a multiprogram national laboratory operated for the U.S. Department of Energy by Battelle Memorial Institute under Contract DE-AC06-76RLO 1830.

References

1. M. A. Ebadian, M. Allen, Y. Cai and J. F. McGahan, *Mercury Contaminated Material Decontamination Methods: Investigation and Assessment*, Final Report. Hemispheric Center for Environmental Toxicology, Florida International University, Miami (2001).

2. United States Environmental Protection Agency, *Aqueous Mercury Treatment*, Capsule Report, EPA/625/R-97/004. Office of Research and Development, Washington, D.C. (1997).

3. Y. Otani, H. Eml, C. Kanaoka and H. Nishino, *Environ. Sci. Technol.* **22**, 708 (1988).

4. K. A. Larson and J. M. Wiencek, *Environ. Prog.* **13**, 253–262 (1994).

5. S. E. Ghazy, *Separ. Sci. Technol.* **30**, 933–947 (1995).

6. J. A. Ritter and J. P. Bibler, *Water Sci. Technol.* **25**, 165–172 (1992).

7. G. E. Fryxell, J. Liu, A. A. Hauser, Z. Nie, K. F. Ferris, S. V. Mattigod, M. Gong and R. T. Hallen, *Chem. Mater.* **11**, 2148–2154 (1999).

8. G. E. Fryxell, J. Liu and S. V. Mattigod, *Mat. Tech. Adv. Perf. Mat.* **14**, 188–191 (1999).

9. G. E. Fryxell, J. Liu, S. V. Mattigod, L. Q. Wang, M. Gong, T. A. Hauser, Y. Lin and K. F. Ferris and X. Feng, Environmental Applications of Interfacially-Modified Mesoporous Ceramics, *Proceedings of the 101st National Meetings of the American Ceramic Society* (1999).

10. S. V. Mattigod, X. Feng, G. Fryxell, J. Liu and M. Gong, *Separ. Sci. Tech.* **34**, 2329–2345 (1999).

11. J. Liu, G. E. Fryxell, S. V. Mattigod, T. S. Zemanian, Y. Shin and L. Q. Wang, *Stud. Surf. Sci. Catal.* **129**, 729–738 (2000).

12. K. M. Kemner, X. Feng, G. E. Fryxell, L.-Q. Wang, A. Y. Kim and J. Liu, *J. Synchrotron Rad.* **6**, 633–635 (1999).

13. D. Sarkar, M. E. Essington and K. C. Mishra, *Soil Sci. Soc. Am. J.* **64**, 1968–1975 (2000).

14. R. G. Pearson, *J. Chem. Educ.* **45**, 581–587 (1968).

15. R. G. Pearson, *J. Chem. Educ.* **45**, 643–648 (1968).

16. R. D. Hancock and A. E. Martell, *J. Chem. Educ.* **74**, 644 (1996).

17. M. Misono, E. Ochiai, Y. Saito and Y. Yoneda, *J. Inorg. Nucl. Chem.* **29**, 2685–2691 (1967).

Chapter 14

Functionalized Nanoporous Silica for Oral Chelation Therapy of a Broad Range of Radionuclides

Wassana Yantasee*, Wilaiwan Chouyyok[†], Robert J. Wiacek[†],
Jeffrey A. Creim[†], R. Shane Addleman[†], Glen E. Fryxell[†]
and Charles Timchalk[†]

*Department of Biomedical Engineering, Oregon Health & Science
University School of Medicine, Portland, Oregon 97239, USA
[†]Pacific Northwest National Laboratory
Richland, Washington 99352, USA

14.1 Introduction

In relation to the serious national threats posed by radiological and nuclear materials, the field of radionuclide decorporation research is lagging significantly (e.g. most work on radionuclide decorporation was done during the Cold War period). Current chelation therapies are effective only for a selected class of radionuclides, and some radionuclides do not have recommended treatment. In the event of a nuclear or dirty bomb explosion, it is very likely that the public will be exposed to multiple radionuclides at once. For example, due to their relatively prevalent use in the nuclear industry, for medical purposes, and academic applications, ^{241}Am, ^{137}Cs, ^{60}Co, ^{192}Ir, ^{238}Pu, ^{210}Po, and ^{90}Sr may be found together as the components of dirty bombs. In such events, identification of the exact radionuclide(s) may not be accomplished in a timely manner (normally done by blood tests and monitoring of radiation levels in hospitals) prior to starting the treatment. Thus, a cocktail of chelating materials that decorporates a broad range of radionuclides is highly desirable as a first response measure to these events. In addition to an increased range of radionuclide efficacy, another important desired characteristic of such a first response measure is oral administration

W. Yantasee et al.

Table 14.1.   Fractional intestinal absorption
values for the selected radionuclides.

Radionuclides	Adult	Infant
Co*	0.1	0.6
Sr*	0.3	0.6
Cs	1.0	1.0
I	1.0	1.0
Po	0.5	1.0

*Intermediate values for 1–15-year-old children: 0.3 for Co, and 0.4 for Sr.

(as opposed to intravenous DTPA injection currently used for actinides), making it much more amenable to the treatment of large numbers of people.

To this end, self-assembled monolayers on mesoporous supports (SAMMS) is a notable class of sorbent materials that were originally created at the Pacific Northwest National Laboratory (PNNL) for nuclear waste cleanups. SAMMS are built on silica with an extremely large surface area (up to $1,000\,m^2/g$) and contain organic functionality that has been fine-tuned to selectively capture heavy metals,[1-3] actinides,[4,5] lanthanides,[6,7] radioiodine,[8] monovalent cesium and thallium,[9,10] transition metals,[11,12] and oxometallate anions.[13,14]

Oral chelating materials like SAMMS provide beneficial intervention for cesium, iodine, strontium, and polonium, which are orally absorbed at a very high percentage, as shown in Table 14.1.[15] By the inhalation route, insoluble particulates of Co, U, Pu, Ir, Cm, and Am would be cleared from the respiratory system via mucociliary clearance and transported to the gastrointestinal (GI) system where they could then be captured by SAMMS. Although these radionuclides, depending on their chemical forms, may not be absorbed from the gut efficiently (e.g. 2% for U), their level of toxicity is significant (e.g. the tolerable daily intake (TDI) for uranium established by the World Health Organization (WHO) is only $0.6\,\mu g/kg$ body weight per day). If exposure is high (e.g. acute exposure), blocking even 2% of U uptake is important. For long-term treatment, individual SAMMS materials could reduce the total body burden of specific radionuclides (once identified) from all routes of exposure that undergo enterohepatic recirculation.

14.2 Selection of SAMMS Materials

For each class of radionuclides to be captured, the best SAMMS was selected after a thorough screening of five to ten potential candidates based on metal and ligand chemistry as well as our 10+ years of experience in environmental

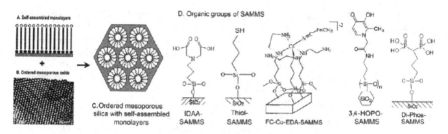

Figure 14.1.   Chemical structures of self-assembled monolayers on mesoporous supports (SAMMS).

cleanups of these metals. Studies on the material synthesis and some screening via batch contact experiments have been published elsewhere.[1-14] The chemical structures of various chosen SAMMS are shown in Figure 14.1. For illustration, actinides (e.g. U, Pu, Cm) are hard Lewis acids and are considerably larger than the typical transition-metal cations, thus both hard anionic ligands and ligand synergy are important attributes in designing effective complexing agents for these metals. For capturing actinides, the best SAMMS was selected from those materials built around the following ligands: three isomers of hydroxypyridinone (HOPO), diphosphonic acid, acetamide phosphonic acid, glycinyl urea, or a diethylenetriamine pentaacetate (DTPA) analog. 3,4-HOPO-SAMMS was found to be best at capturing U, Am, Pu, and Th in water and blood.[4] SH-SAMMS has been the most outstanding material for selectively capturing soft heavy metals like Pb, Hg, and Cd,[1] and thus will be selected for chelating Po. An intermediate ligand like IDAA-SAMMS (an analog of EDTA) was found to be most effective at capturing intermediate transition metals like cobalt following Pearson's hard-soft acid-base principle.[16] Ag-SH-SAMMS has been found to be very effective in glass leachate containing radioactive iodine.[8] FC-Cu-EDA-SAMMS has shown superior performance over insoluble Prussian blue for Cs capture *in vitro*[9] and comparable performance *in vivo* even when the SAMMS material had not yet been optimized.[10] Decorporating alkaline earth metal strontium is known to be very challenging due to its similar chemical properties to the large excess of calcium found in the body. Diphosphonic acid (Di-Phos) SAMMS was evaluated for Sr capture due its known affinity for alkaline earth metals.

14.3 Efficacy of SAMMS in Capturing a Broad Range of Radionuclides

Capture of radionuclides of interest was performed *in vitro* using the best SAMMS selected for actinides (U, Pu, Am), heavy metals (Pb used as a

Table 14.2. Radionuclide removal[a] in synthetic gastric fluid (SGF) and synthetic intestinal fluid (SIF).

Target	Selected SAMMS	% Removal in SGF[b]	% Removal in SIF[c]
U[d]	3,4-HOPO-SAMMS	87	100
Pb[e]	SH-SAMMS	0	94
I	Ag-SH-SAMMS	91	90
Cs	FC-Cu-EDA-SAMMS	100	99
Co	IDAA-SAMMS	0	94
U, Pb, I, Cs, Co, Ir, Sr	All SAMMS	88 (U), 11(Pb), 64 (I), 100 (Cs), 0 (Co), 55 (Ir), 0 (Sr)	100 (U), 92 (Pb), 75 (I), 99 (Cs), 100 (Co), 29 (Ir), 20 (Sr)

[a] All measured in replicates (S.D. < 5%), initial metal conc. ~50 ppb, 1 g SAMMS/L of solution, one hour of contact.
[b] SGF (pH 1.1) contained 0.03 M NaCl, 0.085 M HCl, and 0.32% (w/v) pepsin.[19]
[c] SIF (pH 7.9) contained 0.14 M NaCl, 0.005 M KCl, and 0.008 M NaHCO$_3$.[20]
[d] U is a representative of the actinide class.
[e] Pb was used as a surrogate for Po because (1) studies in rats indicate that Po binds to the same binding sites such as metallothionein in the body as does Pb,[21] (2) in the presence of reducing chelating agents, Po exists as Po^{2+} which is similar to Pb^{2+}, and (3) Po has been found to be effectively complexed by chelating agents that contain sulfhydryl thiol (RSH) groups, similar to those clinically used for decorporation of Pb.[21]

surrogate for Po), iodine, cesium, and transition metal cobalt, as shown in Table 14.2. Nonradioactive forms of the radionuclides were used in this initial evaluation since binding by SAMMS is dependent on the chemistry and not the isotope of radionuclides. Table 14.2 suggests that some radionuclides such as U, I, and Cs could be removed effectively (>90%) from both gastric and intestinal fluids by the selected SAMMS. Uranium could be removed by 87% in synthetic gastric fluid and by 100% in synthetic intestinal fluid using 3,4-HOPO-SAMMS. Our previous study indicated that 3,4-HOPO-SAMMS was also best at the removal of other actinides like Am, Pu, and Th, and outperformed the DTPA analog generally used for actinide chelation therapy.[4] Over 90% of I could be removed from both fluids using Ag-SH-SAMMS. Likewise, over 99% of Cs could be removed from both fluids using FC-Cu-EDA-SAMMS. Some metals such as Pb (a Po surrogate) and Co could be removed effectively (>94%) from synthetic intestinal fluid (but not in stomach fluids due to hydrogen competition) using SH-SAMMS and IDAA-SAMMS, respectively.

Table 14.2 also shows the percentage removal of all radionuclides together using mixed SAMMS materials. Removal of a given radionuclide using mixed SAMMS was mostly as effective as using individual SAMMS, indicating no negative interaction of the SAMMS materials. The exception to this was I; a 20–30% lower removal was found in both fluids in the mixture system compared to a single Ag-SH-SAMMS system. Reducing such negative interaction will be a component of our optimization studies. On the other hand, additional removal was found with Pb in synthetic gastric fluid and Co in synthetic intestinal fluid using mixed SAMMS, which is attributed to IDAA-SAMMS and SH-SAMMS working together. The mixed SAMMS (most likely by IDAA-SAMMS) also removed a respectable percentage (30–50%) of Ir from both fluids. Although the SAMMS in Table 14.2 were not specifically designed for Sr, about 20% of Sr was removed from synthetic intestinal fluid. In a separate study using Di-Phos-SAMMS, Sr could be removed to a greater extent (66% in SIF) even in the presence of a sixfold higher concentration of competing calcium. Since Sr is substantially (~30%) absorbed following oral exposure, this level of Sr decorporation by the SAMMS mixture to limit the gut absorption is very beneficial.

14.4 Decorporation Rate

It is important that the chelating sorbent offers rapid capture of radionuclides in the GI tract in order to minimize the absorption of radionuclides from the GI tract to the blood. To determine the rates of radionuclide removal, the concentration profiles of radionuclides were obtained in synthetic gastric fluid and synthetic intestinal fluid as a function of the time they were in contact with the mixed five SAMMS (as in Table 14.2). The profiles are shown in Figure 14.2. The removal rate of radionuclides by SAMMS was very rapid. In synthetic gastric fluid, 56% of I, 99% of Cs, 50%

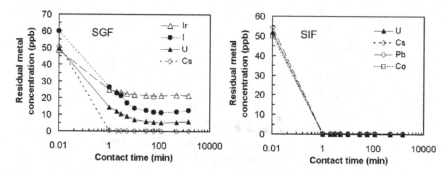

Figure 14.2. Removal rate of metal mixture from synthetic gastric fluid (SGF) and synthetic intestinal fluid (SIF) using mixed SAMMS materials (as in Table 14.2), 1 g SAMMS per L of solution.

of Ir, and 71% of U were removed within the first minute. In synthetic intestinal fluid, over 99% of U, Co, Cs, and Pb were removed within the first minute. The commercial product counterparts, Chelex-100 (IDAA on resins) and GT-73 (SH on resins), took over 10 minutes and 120 minutes to remove 96% of Pb, respectively.[17] The rapid removal rate by SAMMS was attributed to the rigid, open-pore structure of SAMMS as well as pores in the right size range (in the mesopore range), which promote mass transport of the radionuclides to the binding sites inside the pores.

14.5 Chemical Stability

The stability of SAMMS materials after being vigorously stirred for two hours in pH-adjusted seawater (pH 1.0–8.1) was measured as percent Si dissolved. Most SAMMS remained stable across the whole pH range; the wt% Si dissolved (per weight of material) was 0.03% for SH-SAMMS, 0.2% for Cu-FC-EDA-SAMMS, and 0.4% for 3,4-HOPO-SAMMS.[4] In comparison, FC-Cu-EDA-SAMMS (best for Cs capture) leaches a lot less Fe than the FDA-approved insoluble Prussian blue (as shown in Figure 14.3), suggesting that SAMMS could be safer to use than insoluble Prussian blue. After seven cycles of regeneration by 0.5 M HCl washing and reuse of HOPO-SAMMS, there was no loss in the binding affinity of the material.[6] The bond between Ag and SH is extraordinarily stable even at pH < 1, and

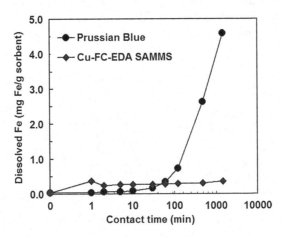

Figure 14.3.   The dissolved Fe from FC-Cu-EDA-SAMMS and Prussian blue as a function of contact time between the materials and seawater (pH 7.7), sorbent per liquid of 1 g/L.

this material would not leach Ag out in stomach acid, suggesting that its stability will not be affected by the acidity of the stomach environment.

These SAMMS were prepared by a standard method that relies on toluene for monolayer deposition on SAMMS. For use in humans, SAMMS materials can be made much more stable; we have developed a synthetic strategy using supercritical carbon dioxide in place of toluene, which not only significantly enhances the stability of the SAMMS materials (by high pressure-enhancing monolayer deposition which leads to a higher degree of cross-linking and minimized defect density), but also increases the number of binding sites on SAMMS.[18]

14.6 Permeability and Uptake of SAMMS Across Gut Cells

The potential transport of SAMMS across gut epithelium was evaluated *in vitro* using Caco-2 cells (human small intestinal epithelium).[1] The cell-based systems facilitate initial broad-based screening of a variety of SAMMS materials, leading to more focused validation studies in whole animals. After 30 minutes of suspension of SH-SAMMS (that was prebound with 1.0 mg (each) of Cd, Pb, and Hg) in the transport buffer (pH 7.4), there was no detectable leaching of Cd, Pb, and Hg. Thus most metals remained bound to the SAMMS material prior to adding it to the Caco-2 cells. The metal-bound SAMMS suspension was then added to the Caco-2 Transwell insert (apical). After two hours of incubation, there was no difference in the concentrations of the metals and Si in the receiver (basolateral) between the test and the control groups (with no metal-bound SAMMS material added). Thus, the percentage transport of the metals across the Caco-2 monolayers was negligible, indicating that once bound with SH-SAMMS the metals were not released into the transport buffer, which is due to the strong affinity between the SH functionality and the metal ions. The lack of detectable Si also indicated that the SAMMS was not taken up by the Caco-2 cells, which is due to the relatively large particle size of SAMMS (most SAMMS materials have a mean particle size of $\sim$20 μm).

A series of differential interference contrast (DIC) and fluorescence images, taken through the z-axis of the cells after exposure to fluorescent dye-tagged SH-SAMMS for three hours, followed by fluorescence quenching by Trypan blue, revealed that large particles ($>$5 μm) remained on the cell surface, while smaller particles (1–2 μm) could enter the cell cytoplasm.[1] No change in the morphology of the cells was detected in the presence of the larger particles, when compared with control cells. Thus for oral drug candidates, SAMMS that are larger than 5 μm in size are recommended. To be very safe, SAMMS that are much larger (e.g. $>$20 μm) may be used in case there is injury of cells in the GI tract (e.g. due to radioactivity). It is worth noting that although their particle size is large, SAMMS materials achieve a high surface area (equivalent to two tennis courts) as a result of their high porosity.

Last, there was no decrease in the transepithelial electrical resistance (TEER) across the cell monolayers when comparing those that were not exposed to metal-bound SH-SAMMS ($792 \pm 19 \, \Omega$-cm^2) to those that were ($798 \pm 16 \, \Omega$-cm^2).[1] Thus the SAMMS material did not alter the monolayer integrity and no cell damage was observed based on the cell resistivity measurement.

14.7 *In Vivo* Decorporation with SAMMS

Once the binding affinity, removal rate, material stability, gut cell permeability, and uptake of SAMMS were evaluated *in vitro*, a limited *in vivo* decorporation efficacy study was performed using a rodent animal model. SAMMS functionalized with FC-Cu-EDA were evaluated against FDA-approved insoluble Prussian blue for ^{137}Cs decorporation in rat. The detailed study and results were reported elsewhere.[10] In brief, we established that SAMMS could rapidly decorporate ^{137}Cs following oral administration and the SAMMS-^{137}Cs complex is very stable in the GI tract. Following the SAMMS administration, less than 1.5% of the administered dose of ^{137}Cs was accounted for in the urine of rats (through 72 hours postdosing); whereas for the ^{137}Cs-only treatment, more than 11% of the administered dose was accounted for in the urine. In contrast, the prebound (with Cs) and sequential SAMMS (given after Cs intake) treatments resulted in substantially more fecal excretion of ^{137}Cs, particularly in the first 24 hours where prebound and sequential administration accounted for 70% and 39% of the dose, respectively. In comparison, less than 0.5% of the ^{137}Cs-only dose was accounted for in the feces over the same collection interval. These results suggest that SAMMS binds rapidly with available ^{137}Cs in the gut and once the ^{137}Cs is bound, it is stable and readily excreted in the feces.

Figure 14.4 shows that given orally after ingesting ^{137}Cs, FC-Cu-EDA-SAMMS was approximately as effective as insoluble Prussian blue for reducing the tissue content of Cs. This result was from our first *in vivo* study and we believe the FC-Cu-EDA-SAMMS material can be further optimized to exceed the performance of Prussian blue. In this regard, optimizing SAMMS materials may be accomplished by increasing the pore size to maximize the access of the functional groups to ^{137}Cs in the GI tract content matrix, while still preventing food particulates from clogging the pores. Results will be reported in due course.

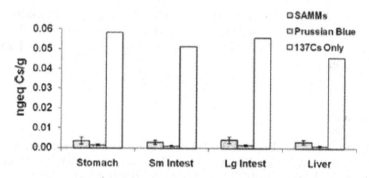

Figure 14.4. Concentration of ^{137}Cs in selected tissues obtained from rats at 72 hours postdosing; ^{137}Cs was orally administered, followed by equal amounts (0.1 g) of FC-Cu-EDA-SAMMS or Prussian blue administered by oral gavage.

14.8 Conclusion

SAMMS materials are nontoxic solid particles that can be mixed together without interaction or loss of activity. A small quantity of SAMMS materials will be needed due to the high efficacy of the materials being used. The SAMMS cocktail can be orally delivered (by pill or perhaps as a "milkshake") and could be taken in response to suspected exposure or as prevention, mitigation or treatment for first responders. For short-term treatment, prompt administration of the cocktail will limit gut absorption and facilitate fecal excretion of these radionuclides entering the body via ingestion, which is one of the primary exposure routes. Since SAMMS are not systemically absorbed and are not cleared by the kidney (in contrast to chelating agents like DTPA), SAMMS are anticipated to be safe for long-term use. In addition to efficacy and safe use, SAMMS will be viable as an emergency medical countermeasure because they are easily administered, have a long shelf-life (~10 years) for stockpiling in emergency preparedness, have the ability to be manufactured on a large scale (e.g. SAMMS have been licensed by PNNL to Steward Advanced Materials, Chattanooga, TN, which has manufactured multiple grades of SAMMS including pharmaceutical grade), and are cost-competitive.

Acknowledgments

This research was supported by the National Institute of Allergy and Infectious Diseases (NIAID), grant# R01 AIO74064, National Institute of Environmental Health Sciences (NIEHS), grant# R21 ES015620, and National

Institute of General Medical Sciences (NIGMS), grant# R01 GM089918. The authors thank Dr Jide Xu, Dr Kenneth N. Raymond, Dr Galya Orr, Dr Marvin G Warner, Dr Rafal M. Grudzien, and Cynthia Warner for their contribution.

References

1. W. Yantasee, R. D. Rutledge, W. Chouyyok, V. Sukwarotwat, G. Orr, C. L. Warner, M. G. Warner, G. E. Fryxell, R. J. Wiacek, C. Timchalk and R. S. Addleman, *ACS Appl. Mater. Interfaces* **2**, 2749 2758 (2010).
2. W. Yantasee, Y. Lin, G. E. Fryxell, B. J. Busche and J. C. Birnbaum, *Sep. Sci. Technol.* **38**, 3809–3825 (2003).
3. X. Feng, G. E. Fryxell, L. Q. Wang, A. Y. Kim, J. Liu and K. M. Kemner, *Science* **276**, 923–926 (1997).
4. W. Yantasee, T. Sangvanich, J. A. Creim, K. Pattamakomsan, R. J. Wiacek, G. E. Fryxell, R. S. Addleman and C. Timchalk, *Health Phys.* **99**, 413–419 (2010).
5. G. E. Fryxell, Y. Lin, S. Fiskum, J. C. Birnbaum, H. Wu, K. M. Kemner and S. Kelly, *Environ. Sci. Technol.* **39**, 1324–1331 (2005).
6. W. Yantasee, G. E. Fryxell, G. A. Porter, K. Pattamakomsan, V. Sukwarotwat, W. Chouyyok, V. Koonsiripaiboon, J. Xu and K. N. Raymond, *Nanomed. Nanotech. Biol. Med.* **6**, e1–e8 (2010).
7. W. Yantasee, G. E. Fryxell, Y. Lin, H. Wu, K. N. Raymond and J. Xu, *J. Nanosci. Nanotechnol.* **5**, 526–529 (2005).
8. S. V. Mattigod, G. E. Fryxell, R. J. Serne and K. E. Parker, *Radiochim. Acta* **91**, 539–546 (2003).
9. T. Sangvanich, V. Sukwarotwat, R. J. Wiacek, R. M. Grudzien, G. E. Fryxell, R. S. Addleman, C. Timchalk and W. Yantasee, *J. Hazard. Mater.* **182**, 225–231 (2010).
10. C. Timchalk, J. A. Creim, V. Sukwarotwat, R. Wiacek, R. S. Addleman, G. E. Fryxell and W. Yantasee, *Health Phys.* **99**, 420–429 (2010).
11. W. Chouyyok, W. Yantasee, Y. Shin, R. M. Grudzien and G. E. Fryxell, *Inorg. Chem. Commun.* **12**, 1099–1103 (2009).
12. W. Chouyyok, Y. Shin, J. Davidson, W. D. Samuels, N. H. LaFemina, R. D. Rutledge, G. E. Fryxell, T. Sangvanich and W. Yantasee, *Environ. Sci. Technol.* **44**, 6390–6395 (2010).
13. G. E. Fryxell, J. Liu, T. A. Hauser, Z. Nie, K. F. Ferris, S. V. Mattigod, M. Gong and R. T. Hallen, *Chem. Mater.* **11**, 2148–2154 (1999).
14. W. Chouyyok, R. J. Wiacek, K. Pattamakomsan, T. Sangvanich, R. M. Grudzien, G. E. Fryxell and W. Yantasee, *Environ. Sci. Technol.* **44**, 3073–3078 (2010).
15. ICRP, *Ann. ICRP Pub.* **100**, 41–59 (2006).
16. B. Busche, R. Wiacek, J. Davidson, V. Koonsiripaiboon, W. Yantasee, R. S. Addleman and G. E. Fryxell, *Inorg. Chem. Commun.* **12**, 312–315 (2009).

17. W. Yantasee, C. L. Warner, T. Sangvanich, R. S. Addleman, T. G. Carter, R. J. Wiacek, G. E. Fryxell, C. Timchalk and M. G. Warner, *Environ. Sci. Technol.* **41**, 5114–5119 (2007).

18. T. S. Zemanian, G. E. Fryxell, J. Liu, S. Mattigod, J. A. Franz and Z. Nie, *Langmuir* **17**, 8172–8177 (2001).

19. USP, *The United States Pharmacopeia*, 22nd ed. (United States Pharmacopeial Convention Inc., Rockville, MD, 1990).

20. USP, *The United States Pharmacopeia*, 26th ed. (United States Pharmacopeial Convention Inc., Rockville, MD, 2003).

21. J. Rencová, A. Vlková, R. Čuřík, R. Holuša and G. Veselá, *Int. J. Radiat. Biol.* **80**, 769–776 (2004).

22. J. Rencová, V. Volf, M. M. Jones and P. K. Singh, *Radiat. Protect. Dosim.* **53**, 311–313 (1994).

Chapter 15

Amine-functionalized Nanoporous Materials for Carbon Dioxide (CO_2) Capture

Feng Zheng, R. Shane Addleman, Christopher L. Aardahl,
Glen E. Fryxell, Daryl R. Brown and Thomas S. Zemanian

Pacific Northwest National Laboratory

Richland, WA, USA

15.1 Introduction

Increasing levels of CO_2 concentration in the earth atmosphere and rising average global temperatures have raised serious concerns about the effects of anthropogenic CO_2 on global climate change. Meanwhile, most analyses project that fossil fuels will continue to be the dominant energy source worldwide until at least the middle of the 21st century.[1] A significant reduction in the current level of CO_2 emission from the consumption of fossil fuels is necessary to stabilize the atmospheric concentration of CO_2. To this end, it is vital to capture and store CO_2 from large point sources of emission such as fossil-fuel power plants, cement production, oil refineries, iron and steel plants, and petrochemical plants. The focus of this chapter will be on the application of CO_2 capture technologies in relation to energy production from fossil fuels, as over one third of the world's CO_2 emissions from fossil-fuel use are attributed to fossil-fuel electric power-generation plants.[1] However, the same set of technologies will also be applicable to the other industrial processes mentioned. In addition, the CO_2 capture technologies are also relevant to applications in space exploration and submarines, where the separation of CO_2 from air in contained breathing atmospheres is necessary.

In fossil-fuel-based power plants, the chemical energy stored in coal, natural gas, oil, etc., is converted to thermal and mechanical energy by combustion and subsequently converted to electricity, with CO_2 as a primary by-product of fossil-fuel combustion. There are three competitive CO_2

capture technologies for fossil-fuel-fired power plants: post-combustion capture, pre-combustion capture, and oxyfuel combustion. In post-combustion capture, CO_2 is captured from the flue gases after the fuel is burned in air. This system is applicable to retrofit existing power plants as well as new systems. However, the CO_2 concentration in flue gases is relatively low and the total flow rate is large because the nitrogen gas from combustion air significantly dilutes the flue gas stream. This makes it challenging to capture CO_2 from flue gas streams efficiently and economically. In pre-combustion capture, there is an initial fuel conversion step (usually gasification) and CO_2 is captured from the fuel gas. The advantage of this system is that the pressure and the CO_2 concentration of the fuel gas is higher, which makes separation easier. The flow rate of fuel gas streams is also smaller than that of flue gas streams as no nitrogen is introduced prior to combustion in air. The drawbacks of pre-combustion capture are twofold: the flowsheet of the gasification and capture plant are inherently complex, and the technology is not nearly as proven as post-combustion capture. In oxyfuel combustion, fossil fuels are combusted in high-purity oxygen to produce energy and a flue gas of essentially CO_2 and water, from which the CO_2 gas can be easily separated. The disadvantage here is the high energy cost associated with the air separation process that supplies the oxygen stream for consumption.

The selection of the separation processes for CO_2 capture from energy production depends on the gas stream conditions, which in turn depend on the selection of the overall CO_2 capture strategy. The CO_2 separation technologies currently available include gas/liquid scrubbing using physical or chemical absorbents as solvents, gas/solid scrubbing using sorbents by adsorption or absorption, cryogenic separation, and membrane separation. The topics covered in this chapter are those related to solid absorbents. Cryogenic separation and gas separation membranes are not discussed here as both of them would certainly deserve separate monographs of their own.

Physical solvents are nonreactive and the absorption and release of CO_2 are governed by Henry's law; in that, the concentration of CO_2 in the solvent is proportional to CO_2 partial pressure in the gas phase. Physical solvents thus work best when CO_2 partial pressure is high and more applicable to pre-combustion CO_2 capture. Commercial solvent processes such as Selexol^{TM} (polyethylene glycol dimethyl ethers) have an existing installation base in ammonia production, natural gas processing, and refinery operations. In the Rectisol^{TM} process, refrigerated methanol is used as solvents.

Unlike physical solvents where the absorption is driven by relatively weak intermolecular forces between the solvent and CO_2, chemical bonds are formed in chemical solvents when CO_2 is absorbed. Liquid amines such as monoethanolamine (MEA), diethanolamine (DEA), and methyldiethanolamine (MDEA) are chemical solvents that have been used in commercial plants to capture CO_2 from flue gases. Chemical solvents are more attractive for CO_2 capture at lower partial pressure. High-energy consumption required for regeneration is one of the main disadvantages of liquid amine solvents.

Although liquid amine absorption is a mature technology for industrial CO_2 separation, significant improvements are needed in order to apply this concept to the extremely large-scale CO_2 capture in energy production. The major drawbacks of the current liquid amine technologies are the large amount of energy required for regeneration, corrosiveness of the amine solvents, and solvent degradation problems in the presence of oxygen. For example, the amine concentration in the commercial amine process is limited to lower than 30% due to the increasing corrosiveness of concentrated amine towards carbon steel. Therefore significant energy is consumed in the solvent regeneration step to heat and partially vaporize the water that makes up a large portion of the aqueous solvent, resulting in the low energy efficiency of the overall process.

Solid sorbents include physical and chemical adsorbents. Examples of physical adsorbents are zeolitic materials and activated carbons. Solid chemical absorbents include alkali carbonates, alkaline earth metal oxides, zirconate and silicate salts. These chemical absorbents can be used in bulk powder or supported forms. A different kind of solid absorbent incorporates organic amines in a porous support. This can be achieved by impregnating the pores with a bulk liquid or grafting ligands to the pore surfaces through chemical bonds. In this chapter, these two approaches are referred to as supported amine and tethered amine sorbents, respectively. The CO_2 capture mechanism for both types of amine sorbents is similar to that of liquid amine solvents. Chemically active solid sorbents have not been previously utilized commercially because the physical and economic performance was inferior. However, new materials under development may produce solid sorbents with performance superior to liquid systems with desirable economics.

Solid amine sorbents have the promise to overcome some of the shortcomings of liquid amines. The dispersion, immobilization, and confinement of the amine functional groups into a porous solid support can result in a more stable, more mass-transfer-efficient, less toxic, and less corrosive material than the corresponding liquid amines. Solid amine sorbents allow a dry scrubbing process where the energy penalty associated with the evaporation of a large amount of water is avoided. Further, the amine functional groups can be tailored for lower regeneration energy requirement. The supports can be tailored independently for high stability and low mass transfer resistance.

In this chapter we will focus on solid amine sorbents (both supported and tethered amines) as the functional materials for CO_2 separation, and the design and performance of these structures is discussed. A detailed discussion of this topic is warranted and timely as increasingly more researchers are attracted to this area, as evidenced by the rapid growth of papers published in the last few years.[2-24]

15.2 Rational Design of Solid Amine Sorbents for CO_2 Capture

15.2.1 Desired attributes of an ideal CO_2 sorbent

15.2.1.1 Adsorption capacity

In order to apply the solid amine sorbent approach to CO_2 capture in energy production processes, a high CO_2 capture capacity is essential because the amount of CO_2 to be captured is extremely large and the capacity of the sorbents is one of the key drivers influencing the overall economic feasibility of the system.

Adsorption capacity is typically reported as the mass or moles of CO_2 adsorbed per mass of the sorbent, on a dry sorbent basis, at equilibrium with a certain CO_2 concentration in the gas phase at a specific temperature and pressure. When comparing literature data, care should be taken to compare capacities at the same gas-phase conditions.

The equilibrium adsorption capacity is determined by the amine functionality, the type of supports, and the loading level of amine on the support. Depending on the method by which the amine functionality is imparted, there is generally an upper limit on the amine loading. For example, impregnation with liquid amine is limited by the pore volume while grafting amine groups on the surface is limited by the surface area of the support and the density of the anchor groups on the surface. In general we have found that the CO_2 capacity generally increases with higher amine loading; therefore, maximizing amine loading is desirable.

15.2.1.2 Selectivity

Bulk gas streams in a power plant contain N_2 and O_2 in the case of flue gases and H_2, CO, and CH_4 in the case of synthesis gas. Good solid sorbents should have high selectivity towards CO_2 over these bulk gas components. Additionally, there is a large presence of moisture or steam in both flue gas and synthesis gas. It is critical that the solid sorbents maintain a high capacity for CO_2 in the presence of a significant amount of water vapor.

Some physical sorbents such as certain zeolites have a high CO_2 capacity, but the selectivity over water is poor and the sorbents need to be regenerated more frequently and at a higher temperature than desired, unless the feed gas is dehumidified prior to CO_2 separation.[13,25] This increases the energy cost and severely limits the use of such physical sorbents for CO_2 capture.

Gas streams from fossil-fuel power plants also contain other acid gas contaminants such as SO_2, NO_2, and H_2S. An ideal CO_2 sorbent should exhibit minimal degradation of CO_2 capacity in the presence of these acid gases. Although the effect of the presence of water vapor on CO_2 capacity is commonly reported in the literature, data on the selectivity of solid amine sorbents over acid gas contaminants are few.[18,21,26]

15.2.1.3 *Kinetics*

The overall kinetics of CO_2 adsorption on a solid amine sorbent is influenced by the intrinsic reaction kinetics of the amine functional groups with CO_2 as well as by the diffusion of the gas phase through the sorbent structures. The amine functional groups can be selected to have reasonably fast kinetics. For example, it is known that better liquid amine absorbents can be formulated by having a fast-reacting amine such as piperazine blended with another more strongly CO_2-binding amine.[27-30] The addition of the "promoter" amine improves the overall kinetics. A similar approach may be developed for solid amine sorbents but has not been reported. The porous supports of the solid amine sorbents can also be tailored to reduce mass transfer resistance. Large pore diameters, hierarchical pore structures, and cross connections of pore channels are all favorable for the enhancement of mass transport.

Adsorption is an exothermic process and is temperature-dependent. Large sorbent beds typically operate adiabatically and therefore the overall kinetics may also be influenced by the development of a thermal front traveling with the mass transfer zone. Ideally, high thermal conductivity of the sorbent materials is desirable. However, the choice of sorbent supports is usually limited by other considerations such as porosity, stability, ability to functionalize/impregnate, and of course cost.

15.2.1.4 *Durability*

Another key driver for the economics of CO_2 capture using solid sorbents is the durability of the sorbent. As the amount of sorbent required will be very large for fossil-fuel power plants, sorbent durability is critical to minimize the sorbent makeup rate and to keep the process economically feasible.

While synthesis gas in pre-combustion capture is highly reducing, flue gas in post-combustion capture contains oxygen. The solid amine sorbents must be chemically stable in the flue gas. The operation temperature of the solid amine sorbents is limited by the regeneration temperature, typically from 100–140°C. At such temperatures, amine degradation is usually not a concern. However, amine loss to evaporation in supported amine sorbents is likely to occur with moderate heating. Therefore, the thermal stability of the sorbents should be well characterized.

Porous silica is among the commonly used solid supports for amine sorbents. Silica gel and mesoporous silica materials can undergo destruction of their pore structures under aggressive hydrothermal conditions. The long-term effects of the presence of moisture or steam in the feed gas on the silica support for solid amine sorbents must be understood and minimized. Porous substrates with high hydrothermal stability are essential from a cost perspective.

Gas streams from fossil-fuel conversion are quite dirty and typically contain acid gas contaminants such as SO_2 and NO_x in flue gases or H_2S in synthesis gas. If solid amine sorbents irreversibly bond to these contaminants, the CO_2 capture capacity is lost and the economical viability of using such solid amine sorbents is significantly reduced. However, if the acid gas contaminants are reversibly co-captured with CO_2, significant savings in cost can be realized. A good solid amine sorbent should be at least inert to the presence of relevant sulfur species.

15.2.1.5 Energy for regeneration

During sorbent regeneration, energy equivalent to the heat of adsorption needs to be supplied as desorption is an endothermic process. Thus, sorbents that bond with CO_2 weakly are desirable from an energy consumption perspective. In a temperature swing adsorption (TSA) process, the sorbent bed needs to be cycled between adsorption and desorption temperatures. Therefore a sorbent material that requires a small temperature differential is also advantageous. Chemisorption-based adsorbents typically require higher energy input to regenerate due to their higher heat of adsorption than that of the physical adsorbents such as zeolites or activated carbons. However, physical adsorbents typically have poor selectivity for CO_2 in the presence of water. Thus, the overall economics will be a trade-off between the above considerations.

15.2.2 Tailoring amine functionalities

It has long been recognized that CO_2 reacts with ammonia and certain organic amines.[31-34] However, not all amines enter into this reaction with equal ease, and in fact, many do not react with CO_2 at all. These reactivity trends are a function of the nucleophilicity of the nitrogen lone pair, the steric environment surrounding the N atom, and the presence (or absence) of a proton on the amine. For example, typical primary alkyl amines (e.g. butyl amine) enter into this reaction with ease, while aniline does not. In the case of aniline, a combination of the inductive effect of the phenyl ring and conjugation of the π-system with the N atom's lone pair causes the amine to be insufficiently nucleophilic. Similar effects are in operation with various aryl amines (e.g. pyrroles, indoles), silazanes, and cyanamides. The steric environment surrounding the nucleophilic N atom is also important. Even though CO_2 is a small, linear molecule, steric congestion can impede the nucleophilic addition of the amine to the "carbonyl". This kinetic barrier also hinders the reversibility of the reaction. Thus, sterically congested amines (e.g. di-isopropylamine) are less desirable than are amines of similar molecular structures that are not so congested (e.g. *n*-hexylamine). Third, it is important that the amine nucleophiles have a proton to lose as the initially formed CO_2 adduct is not stable, and loss of a proton from the ammonium center creates a stable ureate salt. In the absence of this proton loss (e.g. with tertiary amines like triethylamine or pyridine), the CO_2 adducts are unstable and the equilibrium lies strongly on the side of the neutral amine and free CO_2. Thus, when tailoring the ideal nucleophilic center for this chemistry, alkyl amines are superior to aromatic amines, and primary alkyl amines are better than secondary alkyl amines.

However, it is also important to consider this chemistry from the perspective of the proton transfer reaction as well. Because of the spherical symmetry of the 1s orbital, and the lack of congesting substituents in the forward hemisphere of the proton's ligand field, proton transfer reactions tend to be quite facile and less susceptible to steric hindrance than other reactions. This means that the rate at which this proton is transferred (i.e. conversion of the unstable initial adduct to the stable ureate salt), and the position of this equilibrium will depend heavily on the basicity of the amine N atom that is removing the proton (assuming a primary amine of fixed identity for the initial nucleophilic addition). Alkyl amines are more basic than are aromatic amines, and tertiary amines are more basic than are secondary, which in turn are more basic than primary amines. Thus, basicity is a parameter that can be manipulated in order to "tune" the reversibility of CO_2 sorption.

Traditionally, amines have been used to absorb CO_2 in the liquid or amorphous polymer phases. Most commonly, a single amine is used to serve as both the nucleophile and the base in this chemistry. By employing an ordered self-assembled monolayer in a nanoporous ceramic architecture we have the luxury of being able to spatially organize these components within our system, and the ability to use different species for coordinating the

CO_2 nucleophile with the proton and the base. We also have the ability to incorporate both of these moieties into the same molecule, which allows us to "pre-organize" the transition state for the proton transfer reaction, facilitating both the forward and reverse reactions. Thus, suitably tethered diamines are particularly well-suited to this chemistry. For example, tethered ethylenediamine functional groups have two amine sites per group, with the primary amine functioning as the nucleophile and the secondary amine as the base. This allows the formation of an intramolecular carbamate when a CO_2 molecule is captured by the ethylenediamine moiety, as shown in Figure 15.1. The self-assembled monolayer also provides the ability to attach the greatest number of active sites per unit surface area, and still have all of them at the interface and easily accessible to the gas-phase CO_2.

15.2.3 *Tailoring nanoporous substrates*

The ability to tailor the porosity and chemical functionality of a material is of critical importance to a range of technologies including microelectronics, catalysis, sensing, and separations. Since the chemical activity of a substance is frequently proportional to surface area, materials with high relative surface areas can be very valuable. However, the surface area is only useful to chemical processes if it is accessible and provides the desired functionality. The porosity of a material is therefore of importance since it provides access to the surface for chemical processes to occur. Many porous architectures prevent or retard mass transfer in a material, nullifying the advantages of high relative surface areas. Porous silica is a frequently utilized high-surface-area material since both the porosity and surface chemistry can be dictated to create the desired material functionality.

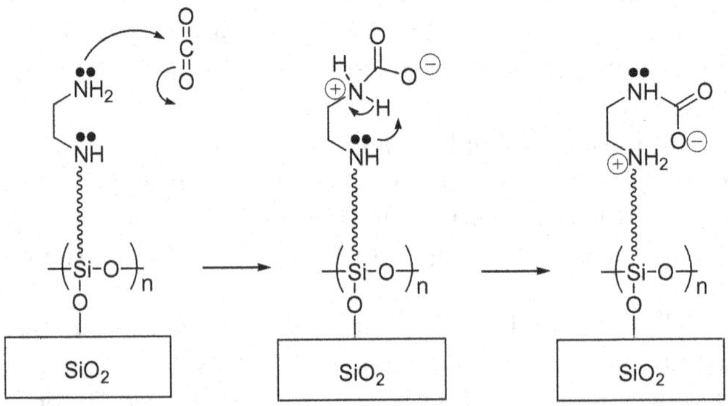

Figure 15.1. Formation of intramolecular carbamate salt through the reaction of CO_2 with the surface-tethered EDA.

In particular, functionalized nanoporous (or mesoporous) silica has recently been a material of significant interest.[35-45] The advantages of nanoporous silica over microporous silica structures stem from its higher surface area and its open, frequently ordered porosity.

Nanoporous silica materials can be created with large uniform pore structures, high specific surface area, and specific pore volumes. The nanostructure is imposed through the use of surfactants in siliceous sol-gel solutions that are used to create these materials. The surfactants can form ordered micellar structures in the sol-gel solution and as the sol-gel solution undergoes condensation, the solution's micellar structure is retained in the final solid form. Solution conditions can be controlled to produce cubic, hexagonal, lamellar, amorphous, and other material phases. Selection of the surfactant and sol-gel condensation conditions enable tuning of the final silica material properties such as percentage of porosity, pore size, surface area, pore connectivity, and structural phase (i.e. hexagonal, cubic, amorphous). Selection of these parameters will dictate the material performance and necessitate trade-offs. Detailed discussions of sol-gel chemistry and mesoporous silica synthesis are available.[35-39]

For CO_2 capture, a material with the highest surface area possible is desired to provide the best capacity. However, this must be balanced with material stability and cost. As previously discussed, durability is a critical requirement for a CO_2 sorbent. Wall thickness must be maintained to a level that provides sufficient thermal and chemical stability. Pores should be large enough to allow facile mass transport (after functionalization) but not so large as to unacceptably reduce the surface area. Interpore connectivity is advantageous since it improves mass transport throughout the material. Ordered porosity can provide faster mass transfer (sorbent kinetics) but such materials are typically more expensive due to increased synthesis costs. Mesoporous materials that have been explored as nanoporous silica supports for CO_2 sorbents include MCM-41, MCM-48, and SBA-15. Templated silica pore diameters have been demonstrated from 2 to over 200 nm.[39,46] In some materials the pore diameter can be tuned by changing the surfactants and using pore expanders without the loss of ordered structure. For example, MCM-41 has a stable hexagonal structure with no pore interconnectivity, and the pore diameter can be tuned from 2 to 30 nm using established methods.[46] As one of the first mesoporous materials, MCM-41 has been shown to be very effective as a catalyst and sorbent support.[12,22,23,39,47,48] Templating with neutral amines instead of long-chain cationic surfactants produces a more disordered hexagonal mesoporous silica (HMS) with somewhat better stability and pore connectivity than MCM-41.[5,47] MCM-48 has a three-dimensional cubic-phase structure with excellent pore interconnectivity that provides effective mass transport.[49,50] SBA-15 has a hexagonal structure with significant pore interconnectivity. The thicker walls provide SBA-15 with greater stability than many nanoporous silica structures and the interconnected channels provide excellent mass transfer kinetics as well as resistance to pore plugging.[51,52] These properties make SBA-15 an excellent candidate structure for the development of materials for chemical separations or transform processes.

A weakness with all silica structures is their chemical instability at higher and lower pH conditions. At lower pH, the silica structure is typically stable but the alkyl silane bridge that allows functional groups to be attached is subject to acidic attack and degradation. At higher pH, the silica structure itself can undergo hydrolysis and even complete dissolution. The stability of mesoporous silica materials are very dependent upon their structure and any subsequent material modifications.[36,53,54] Typically, mesoporous silica exhibits stability between pH 3 and 7.5. The stability range can be extended by utilizing mesoporous structures with thicker walls, additives such as alumina to the silica, or coating with a high-density cross-linked organosilane surface layer.[53,55−60] The typical thermal stability of calcined mesoporous structures is typically over 500°C. However, the stability of any organosilane functionalization is much lower, 100–300°C depending upon the ambient constituents and the chemistry employed on the surface.

15.2.4 Amine functionalization

One of the principal advantages of nanoporous silica is its amenability to surface functionalization with different chemical moieties via a number of direct and indirect methods.[35,37,39,41] Silica functionalization methods include co-condensation, simple impregnations or surface deposition, and post-condensation covalent grafting.

Silica functionalization with co-condensation is accomplished by incorporation of the silanes with the desired chemical moieties into the original sol-gel solution before polymerization is initiated. During the sol-gel condensation reaction, the functional silane is covalently bound into the silica structure in a convenient one-pot synthesis approach. Because the functional organosilane is present during the sol-gel reaction, organic groups which could interfere with the condensation process or disrupt the micellar structure (such as organic acids) must be avoided. If such a group is desired, incorporation must be performed post-condensation via chemical conversion with established synthetic methods. The templating surfactant can be removed chemically, but this is not ideal for many applications, because the mesoporous structure has not been calcined (which would destroy the intercollated functional organosilane) and the silica continues to slowly undergo condensation, resulting in an unstable structure. These structures are particularly vulnerable to hydrothermal degradation due to hydrolysis of the partially cross-linked silica structure.

More popular silica functionalization methods are the impregnation or surface deposition of the desired material directly into a fully calcined mesoporous material. Since the calcined mesoporous support material is fully condensed and cross-linked, many of the structural stability issues that arise with co-condensation functionalization are eliminated. This approach has been effectively applied for the functionalization of mesoporous materials with heterogeneous catalysts and absorbing materials, including amines for CO_2.[48]

Effective CO_2 sorbents have been created by deposition of polyethyleneimine (PEI) in the pores of mesoporous structures.[22,23] The synthetic methods and materials selected attempt to maximize amine loading (thereby increasing capacity) without plugging the pores (thereby reducing capacity and kinetics) within the materials. For polyamine surface deposition, materials with larger pores and good interpore connectivity have demonstrated superior performance.[7,49] CO_2 sorbents have also been created by impregnation of liquid amines into the pores of the mesoporous structures.[10,13] The high porosity of the mesoporous supports (typically over 50%) enable significant liquid loading of the sorbent. The open pore structure and short diffusion distances in the liquid phase provide excellent kinetics. However, the existence of a liquid phase places fundamental mass transfer limitations upon the system. Further, silica functionalization with liquid amine impregnation or surface deposition of polyamines produces a material where the active component is not covalently bound to the substrate. Consequently, the functional component of the material can be removed from the sorbent via evaporative or elution processes, resulting in a reduction of sorbent performance. The operating conditions for such sorbents will be bound by the thermal, pressure, and chemical conditions required to retain the active components within the solid support.

Covalent grafting of the chemically active moieties to the surface of calcined nanoporous silica is an attractive functionalization approach. It avoids the structural instabilities and chemical incompatibility issues inherent in the co-condensation approach since the silica structure is completely calcined before the desired organosilane species are introduced. Since the active moieties are covalently bound, grafting avoids the issues with evaporative or elution processes that may degrade sorbents materials created via impregnation or simple surface deposition. Consequently, covalent grafting of organosilanes post-calcination is a very popular method for functionalization of nanoporous silica and has been demonstrated with a vast array of active groups for a range of applications.[35,37,41,43]

A number of groups have demonstrated that aminoalkoxysilanes grafted onto mesoporous silica supports result in effective CO_2 sorbents.[5−7,9,15,17−19,24,49] The selection of the terminal amine structure determines the complexation mechanisms and binding chemistries. The selection of the mesoporous silica support can strongly affect factors such as kinetics and capacity. Aminoalkoxysilanes grafted onto MCM-48 and SBA-15 are particularly effective CO_2 sorbent materials. MCM-48 with monomeric aminopropyl grafting has excellent pore interconnectivity and has been demonstrated to be an effective support for CO_2 sorbents. While polymeric amines in MCM-48 showed high concentrations of sorbent sites, much of the pore structure was plugged, resulting in reduced adsorption rates and capacities.[49] The thicker walls of SBA-15 sorbents provide greater stability than MCM-48-based materials while retaining good interconnected porosity. Several research groups have studied monomeric aminopropyl-functionalized SBA-15 and found it to be an effective CO_2 sorbent.[6,7,15,17] Monomeric aminopropyl-functionalized materials react at a 2:1 ratio with CO_2 (in the absence of water) and at a 1:1 ratio (captured as a bicarbonate salt) if sufficient water is present. EDA can react at a 1:1 mole ratio with CO_2, providing improved capacity (on a per ligand basis in the absence of water). Further, the principal complexation reaction for EDA-SBA-15 is believed to be the formation of a carbamate salt which does not require the presence of water and has inherently fast reaction kinetics.[24,61] The capacity of EDA-SBA-15 reported to date compares favorably with previously reported aminopropyl-modified mesoporous silica. Furthermore, the adsorption capacity is not dependent upon the presence of water and the complexation kinetics is better.

Additional increases in sorbent capacity and kinetics are desirable and work is under way to provide materials with improved performance and economics. One way to improve performance is to increase the surface area of the support material. Undoubtedly there is room for incremental improvements but the specific surface area(s) of mesoporous silica cannot be significantly improved beyond the structures already used. However, many researchers have reported the loss of significant portions of the material surface area after installation of the amine moieties onto the surface, presumably resulting from pore plugging due to polymerization across the pores and not over the surface. Refinement of covalent grafting methods of relevant aminoalkoxysilanes to prevent vertical polymerization across the nanopore(s) should result in substantial improvements in both the capacity and kinetics. A direct approach to improving capacity is to increase the number of complexation sites per square nanometer of material. Such improvements also result in increased stability because the underlying substrate is protected by the functional monolayer. Ligand surface densities for EDA-SBA-15 have been estimated at or below 2.3 ligands per square nanometer.[15,17,24] Thiol-terminated monolayers on mesoporous silica have been reported as high as six silanes per square nanometer.[60] This suggests there can be significant improvements in CO_2 sorbent performance with increased amine site density. Refinement of covalent grafting methods to increase the density of ligands on the surface of mesoporous silica should result in improvements in both the capacity and stability. Improved surface ligand densities may be achieved with better silica surface preparation, multistep silanization processes, longer treatment times to enhance diffusion of precursors into pores, or more aggressive deposition conditions, i.e. intensive refluxing or the use of supercritical fluid solvents. Another approach to improving the performance of CO_2 sorbents may be to optimize the molecular environment around each ligand site to prevent competitive or interfering processes from reducing the chemical activity of complexation sites.

It should noted that the performance of aminoalkoxysilane-modified nanoporous silica has met or exceeded the performance characteristic of commercially available solid-phase CO_2 sorbents.[5−7,9,15,17−19,24,49] Furthermore, it should be clear that these new materials are not optimized and additional research will result in further improvements.

15.2.5 *Sorbent characterization and performance evaluation*

Any sorbent materials must be well characterized because their physico-chemical properties directly impact their performance in CO_2 adsorption and desorption. In addition, physical and chemical characterizations of the functionalized and unfunctionalized materials are essential for the development of improved synthesis techniques. The performance of the CO_2 sorbents is evaluated based on direct measurements of CO_2 uptake and release properties using various analytical methods.

One of the most critical properties of solid amine sorbents is the amine loading. This is best defined by the number of moles of amine functional groups per unit mass of the sorbent. There are several ways to determine the amine loading. The mass increase due to amine functionalization can be determined by gravimetric measurements.[12,13,15,18] Provided that the sorbent materials are properly outgassed to remove water and CO_2 that may have been adsorbed during synthesis or storage, the mass fraction of the amine moieties can be determined accurately. The amine loading can be calculated from the amine mass fraction, the composition, and molecular weights of the amine compounds. Direct elemental analysis of the functionalized sorbents can also provide the nitrogen mass fraction from which the amine loading is calculated.[4,5,19,49,62] X-ray photoelectron spectroscopy (XPS) can also be used to determine the atomic composition on the surface of amine-modified silica sorbents.[6-8] Because only the surface layer is probed by XPS, the nitrogen composition from XPS data cannot be converted to amine loading or surface amine concentration easily. The amine loading of porous silica sorbents can also be estimated from ^{29}Si solid-state nuclear magnetic resonance (NMR) data.[24,63] The chemical environment of the silicon atoms is probed by ^{29}Si NMR and the abundance of Si-C bonds can be determined.

For tethered amine sorbents, the concentration of the amine groups on the sorbent surface, measured by the number of molecules per square nanometer, is useful when comparing the effectiveness of various synthesis methods. The surface amine concentration can be estimated from amine loading and the specific surface area of the sorbent material. The surface area of the sorbents is typically determined by nitrogen adsorption measurements using the Brunauer–Emmett–Teller method. The surface amine concentrations reported in the literature are typically between 1.0 to 3.0 molecules per square nanometer on porous silica supports.[5,15,19,24,49,63,64] Amine surface concentrations based on the surface area of both the unmodified and the functionalized substrates can be found in the literature. However, the surface area does change when the sorbent material is functionalized due to pore partial filling as well as plugging. Therefore one needs to be careful when comparing amine surface concentration values in the literature. It was proposed that the surface concentration based on the surface area of the unmodified substrate is a better indicator of the degree of functionalization.[5] In any case, it is important to characterize the surface area of the sorbent material both before and after the functionalization so that the amine surface concentration can be compared with other published data consistently.

The CO_2 adsorption and desorption properties such as capacity can be measured gravimetrically, typically using a thermogravimetric analysis (TGA) instrument.[4,5,12,13,18,22,23,26,49,65] When using TGA instruments that have differential scanning calorimetry capability, the heat of adsorption and the specific heat capacity of the sorbent can also be measured. Both properties are important for the analysis of the economics of the CO_2 capture process as they affect the energy consumption during the sorbent regeneration step. The thermal stability of the solid amine sorbents is also usually determined by thermogravimetric methods. TGA coupled with mass spectrometry of the effluent gas is particularly useful for measurements of the desorption temperature of water and CO_2, as well as the decomposition temperature of the incorporated amine moieties. Because the TGA method can be coupled with calorimetry and effluent gas analysis, and the typical sample size required for TGA analysis is only about a few tens of milligrams, it is a rather useful characterization tool for CO_2 sorbent studies. Static volumetric measurements similar to the nitrogen adsorption method for surface area measurements were also reported to gather CO_2 adsorption isotherm data.[18,19]

Another popular technique for studying CO_2 sorption is to use a fixed-bed reactor to make flow-through-type measurements, where the sorption rate is determined from a mass balance based on the gas-phase flow rate and concentrations.[6–11,15,17–21] In a typical fixed-sorbent-bed experiment, a step change in the inlet CO_2 concentration of the sorbent bed results in a mass transfer front traveling down the bed during the adsorption step. The amount of CO_2 adsorbed by the sorbent bed at any moment equals the difference in the accumulated amount of CO_2 between what has entered and what has exited the bed up to that moment. Both quantities can be easily calculated by integration of the product of the total flow and the CO_2 concentration versus time. The variation in the total flow rate of the effluent stream due to adsorption or desorption can be compensated by simple mass balance considerations. Compared to the TGA method, the flow-through method allows better control of gas flow since TGA instruments typically have quite significant dead volumes. The data by flow-through methods can be fitted to an adsorption column model for kinetics studies. The ability to obtain sorption kinetics data at conditions relevant to the actual adsorption process is one of the main advantages of the flow-through-type measurements. The flow-through method also allows testing of different adsorption processes such as temperature swing adsorption (TSA) and pressure swing adsorption (PSA) and so on. One drawback of flow-through measurements is that the mass change of the sorbent is determined indirectly from concentration and flow rate measurements. Therefore, proper calibration of the concentration measurement instrument is critical, whether it is a gas chromatogram, a mass spectrometer, or some other CO_2 detector. Baseline measurements with an empty sorbent bed are also necessary in order to eliminate the error in CO_2 adsorption calculation caused by the dead volume of the flow system.

Quantitative comparison of the CO_2 adsorption capacities reported in the literature is difficult as researchers have used different measurement techniques and different test conditions for CO_2 inlet concentration and adsorption temperature. Conversion of these capacity data to equivalent values at identical concentration and temperature is not practical as the adsorption isotherms are generally unknown other than for those discrete data points reported. Additionally, some capacity data were reported in units based on actual sorbent mass while others based on the mass of unfunctionalized starting materials. However, a qualitative comparison of

Table 15.1. CO_2 capacity of solid amine sorbents reported in literature.

Amine Groups	Solid Supports	Capacity (mol/kg)	Amine Efficiency	Measurement Conditions[§]	Reference
NH_2[†]	Silica gel	0.62	0.67[€]	100% CO_2, 23°C	19
NH_2[†]	SBA-15	1.75	—	10% CO_2/He 20°C	7
NH_2[†]	SBA-15	0.40	—	4% CO_2/He 25°C	6
NH_2[†]	HMS	1.59	0.69	90% CO_2/Ar 20°C	5
NH_2[†]	MCM-48	2.05	0.89	100% CO_2 25°C	18
	Silica xerogel	1.14	0.67		
NH_2[†]	MCM-48	0.80	0.33	100% CO_2 25°C	49
Pyrrolidine[†]		0.30	0.20		
PEI[†]		0.40	0.08		
NH_2[†]	SBA-15	0.52	0.19	15% CO_2/N_2 60°C	15
EDA[†]		0.87	0.21		
DETA[†]		1.10	0.22		
EDA[†]	SBA-15	0.79	—	50% CO_2/He	9
EDA[†]	SBA-15	0.45	0.08	15% CO_2/N_2 25°C	24
PEI[‡]	MCM-41	4.89	—	100% CO_2 75°C	22
PEI[‡]	PMMA	0.84	—	2% CO_2/N_2	11
DEA[‡]	MCM-41	2.81	0.40	5% CO_2/N_2 25°C	13
MEA[‡]	PMMA	0.83	0.14	1% CO_2/N_2 20°C	10

[§]At 1 atm or ambient pressure and room temperature unless noted otherwise.
[€]Average amine loading used based on a reported range.
[†]Anchored covalently to surface as propyl silanes derivatized with the following amine functional groups: PEI, polyethyleneimine; EDA, ethylenediamine; DETA, diethylenetriamine.
[‡]Immobilized by impregnation of the following liquid amines: PEI, polyethyleneimine; DEA, diethanolamine; MEA, monoethanolamine.

the adsorption capacity is still valuable. In Table 15.1, CO_2 adsorption capacities reported in the literature have been compiled. The raw data were expressed in various units such as weight percentage, milligrams, standard cubic centimeters, or millimoles of CO_2 per gram of the dry adsorbent. For the purpose of comparison the literature CO_2 capacity data were all converted to moles per kilogram.

The CO_2 capacity of the tethered amine sorbents at near ambient temperature is generally 0.5–2.0 mol/kg. Comparable or higher capacities have been reported for the impregnated amine sorbents but their adsorption kinetics is slower than tethered amine sorbents, if the same solid support is used.[49] Although high amine loading, and therefore high CO_2 capacity, can be achieved by filling almost all available pore volume with a liquid amine, the adsorption rate is reduced as diffusion through the liquid amine phase is much slower than through the gas phase in open pores. A thorough analysis of the kinetic advantages of the different types of solid amine sorbents is not possible at this time because there are too few adsorption rate data available in the literature.

It is worth comparing the various solid amine sorbents in terms of amine efficiency, which is defined as the number of CO_2 molecules adsorbed for every nitrogen atom in the amine function groups. Therefore, the theoretical maximum of amine efficiency for carbamate formation is 0.5, i.e.

$$2R - NH_2 + CO_2 \longrightarrow R - NH_3^+ + R - NHCOO^-. \qquad (15.1)$$

When CO_2 is captured as bicarbonates, the theoretical maximum amine efficiency is 1, i.e.

$$R - NH_2 + CO_2 + H_2O \longrightarrow R - NH_3^+ + HCO_3^-. \qquad (15.2)$$

As can be seen in Table 15.1, the amine efficiency of the solid amine sorbents in the literature is typically lower than the theoretical values. Most synthetic amine sorbents contain defects that render a certain fraction of functional groups unavailable for reaction with CO_2. For example, polymerization of the amine silane precursors may result in plugged pores. The tethered amine groups can also form hydrogen bonds with surface silanol groups and become deactivated. Therefore, there are opportunities for improvements in sorbent performance through better synthesis techniques.

Generally, higher amine loading results in higher CO_2 adsorption capacity. For impregnated amine sorbents, the capacity increases with the amine loading until the loading is slightly over the level corresponding to complete pore filling.[13,22,23] Further increases in amine loading result in amine deposition outside the pores, which leads to reduced adsorption rates. At such high amine loading, the effective adsorption capacity after a fixed adsorption time also decreases due to the reduced adsorption rate even though the equilibrium capacity is expected to be higher with the added amine content. It was proposed that the liquid amine incorporated within the pores reacts with CO_2 faster than in the bulk liquid phase,[22,23] although the mechanism for this phenomenon has not been analyzed in detail. For tethered amine sorbents, it has been shown that the CO_2 concentration on the surface increased with surface amine concentration.[5] When the surface concentration of the adsorption sites was lower than approximately one nitrogen atom per square nanometer, the data suggested that physical adsorption was dominant.[5]

The effect of physical adsorption on the overall adsorption capacity of the solid amine sorbents has not been well studied. The physical adsorption on unmodified supports is typically small compared to the capacity of the solid amine sorbents.[15,17,18,22,23] The heats of adsorption reported in the literature are consistent with strong adsorbate-surface interactions,

Table 15.2.   The heat of CO_2 adsorption of solid amine sorbents reported in literature.

Amine Groups	Solid Supports	Heat of Adsorption (kJ/mol)	Method	Reference
NH_2[†]	Silica gel	58.8 ± 4.2	Calculated from isosteres	19
NH_2[†]	HMS	60 ± 2	Measured by differential thermal analysis	5
EDA[†]	SBA-15	47.8	Calculated from desorption peak temperature and half width	9
PEI[‡]	PMMA[§]	94.6 ± 8.2	Measured by isothermal flow microcalorimetry	11

[†]Anchored covalently to surface as propyl silanes; EDA = ethylenediamine.
[‡]Impregnated polyethyleneimine.
[§]Polymethyl methacrylate.

i.e. greater than two to three times the heat of evaporation (see Table 15.2). For CO_2 the heat of evaporation at 0°C is 10.3 kJ/mol.[66] However, it is oversimplified to treat the interaction between CO_2 molecules and amine-modified surfaces as purely chemisorption. All of the reported CO_2 adsorption isotherms by solid amine sorbents show that the CO_2 capacity increases significantly with the CO_2 partial pressure.[9,13,18,19,22,24,49] This behavior is similar to that of physical adsorption. In strict chemisorption the sorbent surface is saturated rapidly even at low partial pressure and the equilibrium capacity is essentially independent of adsorbate partial pressure. Therefore contributions from intermolecular interactions between the CO_2 and amine-modified surfaces are also important to the overall CO_2 uptake/release behavior of solid amine sorbents. Hydrogen bonds of CO_2 molecules with tethered amine and residual surface silane groups are one example of the weaker interactions. A molecular dynamics study showed that there are distinctive structural "pockets" on amine-modified hexagonal mesoporous silica surfaces where the motion of CO_2 molecules is restricted.[3]

There are conflicting results in the literature regarding the effect of water on the CO_2 adsorption by solid amine sorbents. Some reported that when water was present the adsorption capacity increased significantly and the reason was thought to be the shift from carbamate formation to bicarbonate formation, which only requires 1 mole of amine to react with 1 mole of CO_2.[5,11,19,20] Others reported that the CO_2 adsorption capacity was not enhanced by the presence of water.[13,15,17,18,24] One group also reported that although the CO_2 adsorption capacity increased in the presence of water, the rate of desorption decreased.[5] Obviously, more studies are needed to elucidate the role of water and to understand how sorbent synthesis can be improved to better control adsorption behavior. All existing studies show that the presence of water does not diminish the CO_2 adsorption capacity, so the presence of moisture in gas streams from energy production is not a problem for solid amine sorbents from a capacity point of view.

Gas streams from fossil-fuel-fired power plants contain other acid gases such as SO_2, NO_x, or H_2S, depending on the fuel conversion process. Strong tolerance of solid amine sorbents to these acid gas contaminants is highly desirable. If solid amine sorbents are also reactive towards these acid gases, whether they can be regenerated easily after exposure is also critical. There are very few reports on the performance of solid amine sorbents with acid gas contaminants. One group of researchers showed that H_2S was adsorbed reversibly on amine-grafted MCM-48 mesoporous silica.[18] The regeneration temperature for H_2S was 60°C, which was lower than that of CO_2 at 75°C. The formation of the $NH_3^+HS^-$ group was evidenced by characteristic infrared absorption bands and the adsorption of H_2S was not affected by the presence of water. Another group reported that the adsorption of NO_x on polyethyleneimine-impregnated MCM-41 sorbents was irreversible under flue gas conditions.[21] In another study,[26] it was found that copolymers of styrene with pendant ethylenediamine, N, N, N'-trimethylethylenediamine, and 1-methylpiperazine groups absorbed NO and NO_2 but not N_2O. The amine/NO adducts oxidized to ammonium nitrites and nitrosamines in the presence of oxygen, leading to the loss of active amine sites. The NO_2/amine adducts underwent similar oxidative degradation and deactivation. On the other hand, the same amino-functional polymers exhibited thermally reversible adsorption of SO_2. The affinity for SO_2 was much higher than for CO_2 as SO_2 is a stronger Lewis acid.

Based on the above limited literature results, it appears that the utility of solid amine sorbents for CO_2 capture from synthesis gas will not be affected by the presence of trace H_2S contaminants. It is possible that H_2S can even be captured simultaneously with CO_2. For CO_2 capture from flue gas, the presence of NO_x will likely cause sorbent degradation and thus removal of NO_x prior to CO_2 capture is recommended. Simultaneous capture of SO_2 with CO_2 from flue gas seems possible using solid amine sorbents.

15.3 Economics

Solid-based CO_2 sorbent systems for the most part have not been examined for cost-effectiveness in large-scale applications. Such analysis is difficult because the unit operations required and their configuration are considerably different from typical liquid-based scrubbing systems. Here, a preliminary analysis of the economics of using solid sorbents is presented to understand where the current level of development stands in comparison to commercial technologies such as liquid amine systems.

15.3.1 *Methods and assumptions*

The U.S. EPA has published costing guidelines for carbon adsorbers used in the application of volatile organics capture.[67] This model has been used in combination with the TEAM model developed by Brown *et al.*[68] to make a rough estimate of levelized costs of carbon capture using amine-functionalized solids. Several assumptions had to be made to adapt these models to the problem at hand:

- It is envisioned that the desorbing period may be less than that required for adsorption; therefore, it is possible to be capturing CO_2 with several columns, while regenerating a lesser number of columns at any given time on a rotating basis.
- Flow of CO_2 through the adsorbing bed is terminated once a critical breakthrough concentration is reached. In this case, the adsorption cycle is ended when the integrated average capture fraction drops to 90%.
- Steam is used for desorption of CO_2 because it can quickly and more uniformly heat the bed to a given desorption temperature. It is also in good supply in a coal-fired power plant. Steam is also used to carry the CO_2 out of the adsorber zone.
- Once desorption is complete, ambient air is blown through the bed to cool for another adsorption cycle. This air also serves to remove excess moisture from any steam condensation.
- Working capacity for CO_2 adsorption is assumed to be 5%. This is consistent with laboratory results for a feed of 15% CO_2 in a simulated coal flue gas.
- Adsorption is performed at 95°C and desorption is performed at 135°C. This temperature swing results in an estimated sensible heat of 325 Btu/lb CO_2.
- 1,177 Btu/lb CO_2 is the desorption energy for a CO_2-liquid EDA complex; whereas the desorption energy is 98 Btu/lb CO_2 for physical adsorption on zeolitic materials. Laboratory results show a combination of chemi- and physisorption, so it was assumed that the desorption energy was 357 Btu/lb CO_2 based on the ratio of chemi- to physisorption observed in lab tests.
- Steam consumption in the carrier role could dominate the steam demand. A rule of thumb for carbon beds is 3–5 lb of steam per pound of adsorbate or 3,000–5,000 Btu/lb CO_2.[67] It is possible to recover up to

approximately 70% of this energy through recovery processes depending on the plant configuration; therefore, steam regeneration energy assumptions of 1,000, 2,500, and 4,000 Btu/lb CO$_2$ were used.

- Steam cost is estimated at $1.57/MMBtu based on the foregone electricity production.[69]
- The cost of large-scale production of the adsorbent material is unknown; therefore, costs of $5, $20, and $40/lb were used to understand the sensitivity to this parameter.
- Economic life of the equipment assumed to be 30 years.
- Depreciable life of the equipment of 20 years.
- Economic life of the sorbent assumed to be 5 or 30 years.
- Depreciable life of the sorbent of 3 or 20 years.
- 9.3% discount rate.
- 3.1% inflation rate.
- 39.1% income tax rate.
- 2% property tax rate.

15.3.2 Results

The results of the analysis are presented in the form of two tables where Table 15.3 corresponds to a five-year sorbent life and Table 15.4 corresponds to a 30-year sorbent life.

From Tables 15.3 and 15.4, it is evident that for short sorbent lifetimes the cost of the sorbent is the dominant factor, and for long sorbent lifetimes

Table 15.3. Levelized capture cost for five-year sorbent lifetime ($/ton).

Sorbent Cost/Lb	Regeneration Energy Cost (Btu/lb CO$_2$)		
	1,000	2,500	4,000
$5	$20.00	$26.60	$33.10
$20	$52.40	$58.90	$65.40
$40	$95.40	$101.90	$108.40

Table 15.4. Levelized capture cost for 30-year sorbent lifetime ($/ton).

Sorbent Cost/Lb	Regeneration Energy Cost (Btu/lb CO$_2$)		
	1,000	2,500	4,000
$5	$15.90	$22.40	$28.90
$20	$35.30	$41.80	$48.30
$40	$61.30	$67.80	$74.30

the regeneration energy cost is a more significant factor. The relative order of magnitudes of cost per ton of CO_2 captured are in the neighborhood of estimates for liquid amine systems (\$20 to \$70/ton) or somewhat higher.

15.3.3 Conclusions

With the sorbent cost being a significant factor in the overall CO_2 capture cost, it is important to explore inexpensive substrates, e.g. silica gel, as the starting material for amine modifications. Replicating the amine-modified structures found in the mesoporous silica with tethered amine groups is possible when silica gel is used instead. In our own laboratory, using novel supercritical fluid processing techniques, we have successfully modified silica gels with amine functional groups. The above preliminary economics analysis also illustrates the importance of the lifetime of the solid sorbents. In currently published work, the durability of solid amine sorbents have not been tested adequately. Additional investigation of the stability of solid amine sorbents in CO_2 capture should be undertaken to obtain a better understanding of the deactivation process and its prevention. Provided that the uncertainty in sorbent cost and sorbent lifetime can be narrowed, a more detailed comparison of the solid amine sorbents with existing technologies is certainly highly desirable. While that is beyond the scope of this book chapter, it is our hope that this chapter can serve as a useful resource for practitioners in this rapidly developing research field.

References

1. Working Group III of the Intergovernmental Panel on Climate Change, *IPCC Special Report on Carbon Dioxide Capture and Storage* (Cambridge University Press, New York, 2005).
2. P. J. Branton, P. G. Hall, M. Treguer and K. S. W. Sing, *J. Chem. Soc., Faraday Trans.* **91**(13), 2041–2043 (1995).
3. A. L. Chaffee, *Fuel Process. Technol.* **86**(14–15), 1473–1486 (2005).
4. G. P. Knowles, S. W. Delaney and A. L. Chaffee, *Nanoporous Materials IV: Stud. Surf. Sci. Catal.* **156**, 887–896 (2005).
5. G. P. Knowles, J. V. Graham, S. W. Delaney and A. L. Chaffee, *Fuel Process. Technol.* **86**(14–15), 1435–1448 (2005).
6. A. C. C. Chang, S. S. C. Chuang, M. Gray and Y. Soong, *Energ. Fuel.* **17**(2), 468–473 (2003).
7. M. L. Gray, Y. Soong, K. J. Champagne, H. Pennline, J. P. Baltrus, R. W. Stevens, R. Khatri, S. S. C. Chuang and T. Filburn, *Fuel Process. Technol.* **86**(14–15), 1449–1455 (2005).

8. M. L. Gray, Y. Soong, K. J. Champagne, H. W. Pennline, J. Baltrus, Jr. R. W. Stevens, R. Khatri and S. S. C. Chuang, *Int. J. Environ. Techn. Manag.* **4**(1/2), 82–88 (2004).

9. R. A. Khatri, S. S. C. Chuang, Y. Soong and M. Gray, *Ind. Eng. Chem. Res.* **44**(10), 3702–3708 (2005).

10. T. Filburn, J. J. Helble and R. A. Weiss, *Ind. Eng. Chem. Res.* **44**(5), 1542–1546 (2005).

11. S. Satyapal, T. Filburn, J. Trela and J. Strange, *Energ. Fuel.* **15**(2), 250–255 (2001).

12. R. Franchi, P. J. E. Harlick and A. Sayari, *Nanoporous Materials IV: Stud. Surf. Sci. Catal.* **156**, 879–886 (2005).

13. R. S. Franchi, P. J. E. Harlick and A. Sayari, *Ind. Eng. Chem. Res.* **44**(21), 8007–8013 (2005).

14. N. Hiyoshi, K. Yogo and T. Yashima, *Recent Advances in the Science and Technology of Zeolites and Related Materials, Pts A–C* **154**, 2995–3002 (2004).

15. N. Hiyoshi, K. Yogo and T. Yashima, *J. Jpn. Petrol. Inst.* **48**(1), 29–36 (2005).

16. N. Hiyoshi, K. Yogo and T. Yashima, *Carbon Dioxide Utilization for Global Sustainability* **153**, 417–422 (2004).

17. N. Hiyoshi, K. Yogo and T. Yashima, *Chem. Lett.* **33**(5), 510–511 (2004).

18. H. Y. Huang, R. T. Yang, D. Chinn and C. L. Munson, *Ind. Eng. Chem. Res.* **42**(12), 2427–2433 (2003).

19. O. Leal, C. Bolivar, C. Ovalles, J. J. Garcia and Y. Espidel, *Inorg. Chim. Acta* **240**(1–2), 183–189 (1995).

20. X. Xu, C. Song, B. G. Miller and A. W. Scaroni, *Ind. Eng. Chem. Res.* **44**(21), 8113–8119 (2005).

21. X. C. Xu, C. S. Song, B. G. Miller and A. W. Scaroni, *Fuel Process. Technol.* **86**(14–15), 1457–1472 (2005).

22. X. C. Xu, C. S. Song, J. M. Andresen, B. G. Miller and A. W. Scaroni, *Energ. Fuel.* **16**(6), 1463–1469 (2002).

23. X. C. Xu, C. S. Song, J. M. Andresen, B. G. Miller and A. W. Scaroni, *Micropor. Mesopor. Mat.* **62**(1–2), 29–45 (2003).

24. F. Zheng, D. N. Tran, B. J. Busche, G. E. Fryxell, R. S. Addleman, T. S. Zemanian and C. L. Aardahl, *Ind. Eng. Chem. Res.* **44**(9), 3099–3105 (2005).

25. E. S. Kikkinides, V. I. Sikavitsas and R. T. Yang, *Ind. Eng. Chem. Res.* **34**(1), 255–262 (1995).

26. A. Diaf, J. L. Garcia and E. J. Beckman, *J. Appl. Polym. Sci.* **53**(7), 857–875 (1994).

27. G. W. Xu, C. F. Zhang, S. J. Qin, W. H. Gao and H. B. Liu, *Ind. Eng. Chem. Res.* **37**(4), 1473–1477 (1998).

28. G. W. Xu, C. F. Zhang, S. J. Qin and B. C. Zhu, *Ind. Eng. Chem. Res.* **34**(3), 874–880 (1995).

29. G. W. Xu, C. F. Zhang, S. J. Qin and Y. W. Wang, *Ind. Eng. Chem. Res.* **31**(3), 921–927 (1992).

30. H. Y. Dang and G. T. Rochelle, *Separ. Sci. Tech.* **38**(2), 337–357 (2003).

31. E. Schering, *Chem. Ztg.* **72**(II), 519 (1911).
32. E. Schering, German Patent 123,138, 1900.
33. F. Fichter and B. Becker, *Chem. Ber.* **44**, 3481 (1911).
34. V. Meyer and P. Jacobson, in *Lehrbuch der Organischen Chemie* (Veit, Leipzig, 1913), p. 1370.
35. A. Vinu, K. Z. Hossain and K. Ariga, *J. Nanosci. Nanotechnol.* **5**(3), 347–371 (2005).
36. C. J. Brinker and G. W. Scherer, *Sol-Gel Science: The Physics and Chemistry of Sol-Gel Processing* (Academic Press, Boston, 1990).
37. G. E. Fryxell, Y. H. Lin, H. Wu and K. M. Kemner, in *Nanoporous Materials III*, eds. A. Sayari and M. Jaroniec (Elsevier Science, Amsterdam, 2002), pp. 583 500.
38. M. S. Morey, A. Davidson and G. D. Stucky, *J. Porous Mater.* **5**(3–4), 195–204 (1998).
39. J. Y. Ying, C. P. Mehnert and M. S. Wong, *Angew. Chem. Int. Ed.* **38**(1–2), 56–77 (1999).
40. A. Sayari, *Chem. Mater.* **8**(8), 1840–1852 (1996).
41. K. Moller and T. Bein, *Chem. Mater.* **10**(10), 2950–2963 (1998).
42. U. Ciesla and F. Schuth, *Micropor. Mesopor. Mat.* **27**(2–3), 131–149 (1999).
43. A. Stein, B. J. Melde and R. C. Schroden, *Adv. Mater.* **12**(19), 1403–1419 (2000).
44. A. Sayari and S. Hamoudi, *Chem. Mater.* **13**(10), 3151–3168 (2001).
45. A. Taguchi and F. Schuth, *Micropor. Mesopor. Mat.* **77**(1), 1–45 (2005).
46. A. Sayari, Y. Yang, M. Kruk and M. Jaroniec, *J. Phys. Chem. B* **103**(18), 3651–3658 (1999).
47. M. Dubois, T. Gulikkrzywicki and B. Cabane, *Langmuir* **9**(3), 673–680 (1993).
48. B. Zhou, S. Hermans and G. Somorjai, *Nanotechnology in Catalysis*, Vol. 2 (Kluwer Academic/Plenum Publishers, New York, 2004).
49. S. Kim, J. Ida, V. V. Guliants and J. Y. S. Lin, *J. Phys. Chem. B* **109**(13), 6287–6293 (2005).
50. J. C. Vartuli, K. D. Schmitt, C. T. Kresge, W. J. Roth, M. E. Leonowicz, S. B. Mccullen, S. D. Hellring, J. S. Beck, J. L. Schlenker, D. H. Olson and E. W. Sheppard, *Chem. Mater.* **6**(12), 2317–2326 (1994).
51. D. Y. Zhao, Q. S. Huo, J. L. Feng, B. F. Chmelka and G. D. Stucky, *J. Am. Chem. Soc.* **120**(24), 6024–6036 (1998).
52. D. Y. Zhao, J. L. Feng, Q. S. Huo, N. Melosh, G. H. Fredrickson, B. F. Chmelka and G. D. Stucky, *Science* **279**(5350), 548–552 (1998).
53. M. V. Landau, S. P. Varkey, M. Herskowitz, O. Regev, S. Pevzner, T. Sen and Z. Luz, *Micropor. Mesopor. Mat.* **33**(1–3), 149–163(1999).
54. A. Doyle and B. K. Hodnett, *Micropor. Mesopor. Mat.* **63**(1–3), 53–57 (2003).
55. Y. D. Xia and R. Mokaya, *J. Phys. Chem. B* **107**(29), 6954–6960 (2003).
56. J. L. Zheng, Y. Zhang, Z. H. Li, W. Wei, D. Wu and Y. H. Sun, *Chem. Phys. Lett.* **376**(1–2), 136–140 (2003).
57. M. Luechinger, L. Frunz, G. D. Pirngruber and R. A. Prins, *Micropor. Mesopor. Mat.* **64**(1–3), 203–211 (2003).

58. D. R. Dunphy, S. Singer, A. W. Cook, B. Smarsly, D. A. Doshi and C. J. Brinker, *Langmuir* **19**(24), 10403–10408 (2003).
59. Y. S. Ooi, R. Zakaria, A. R. Mohamed and S. Bhatia, *Catal. Commun.* **5**(8), 441–445 (2004).
60. T. S. Zemanian, G. E. Fryxell, J. Liu, S. Mattigod, J. A. Franz and Z. M. Nie, *Langmuir* **17**(26), 8172–8177 (2001).
61. A. Dibenedetto, M. Aresta, C. Fragale and M. Narracci, *Green Chem.* **4**(5), 439–443 (2002).
62. A. R. Cestari and C. Airoldi, *Langmuir* **13**(10), 2681–2686 (1997).
63. S. Huh, J. W. Wiench, J. C. Yoo, M. Pruski and V. S. Y. Lin, *Chem. Mater.* **15**(22), 4247–4256 (2003).
64. G. S. Caravajal, D. E. Leyden, G. R. Quinting and G. E. Maciel, *Anal. Chem.* **60**(17), 1776–1786 (1988).
65. A. Diaf and E. J. Beckman, *React. Funct. Polym.* **27**(1), 45–51 (1995).
66. Compressed Gas Association, *Handbook of Compressed Gases*, 4th edn. (Kluwer Academic Publishers, Boston, 1999).
67. U.S. Environmental Protection Agency, *EPA Air Pollution Control Cost Manual*, EPA/452/B-02-001. US EPA Office of Air Quality Planning and Standards, Research Triangle Park, NC, Jan, 02.
68. D. R. Brown, D. A. Dirks, M. K. Spanner and T. A. Williams, *An Assessment of Methodology for Thermal Energy Storage Evaluation*, PNNL-6372. Pacific Northwest National Laboratory, Richland, WA, 87.
69. *Integrated Environmental Control Model*, version 5.01, URL: http://www.iecm-online.com (Carnegie Mellon University: Pittsburgh, PA, 2004).

Chapter 16

Carbon Dioxide Capture from Post-combustion Streams Using Amine-functionalized Nanoporous Materials

Rodrigo Serna-Guerrero* and Abdelhamid Sayari†

*Department of Chemical and Biological Engineering
†Department of Chemistry
University of Ottawa, Ottawa, ON, Canada

16.1 Introduction

Emission of carbon dioxide (CO_2) into the atmosphere is a major environmental concern. According to recent studies, the atmospheric concentration of CO_2 has increased dramatically from 280 ppm before the industrial age,[1] up to 385 ppm in 2009,[2,3] mainly due to extensive combustion of fossil fuels. Such an increase in CO_2 concentration has been directly correlated to the rise in atmospheric temperature, a phenomenon commonly referred to as global warming,[3,4] which is believed to have a negative impact on our ecosystem. However, it is generally accepted that, although there are efforts toward developing alternative energy sources, fossil fuels will remain a key source of energy in decades to come, both for power generation and vehicle transportation. Consequently, CO_2 capture technologies are expected to provide a solution to this crisis, at least until clean energy technologies are fully mature and implemented. Therefore, it is critical to develop effective methods for the capture and sequestration of CO_2 from post-combustion effluents, such as flue gas.

Currently, the most common technology used for the treatment of CO_2-containing streams is scrubbing using alkanolamine solutions. The removal of CO_2 by amines occurs via the widely accepted formation of carbamate and bicarbonate species, as represented in Scheme 16.1.[5] These are reversible reactions that permit the regeneration of amines, typically by heating the CO_2-rich solution.

$$2(RNH_2) + CO_2 \longleftrightarrow RNHCO_2^- RNH_3^+$$
$$\text{carbamate}$$

$$RNH_2 + CO_2 + H_2O \longleftrightarrow RNH_3^+ HCO_3^- \xleftrightarrow{\text{RNH}_2} (RNH_3^+)_2 CO_3^{2-}$$
$$\text{bicarbonate} \qquad\qquad \text{carbonate}$$

Scheme 16.1.   Typical reaction pathways between CO_2 and amines.[5]

The basic nature of the amine groups renders them an effective medium to capture acid gases, such as CO_2. However, some drawbacks have been associated with the use of amine solutions, including a high demand of energy for operation and regeneration of the absorbent,[6] corrosion of the equipment[7] and, recently, the environmental impact associated with the disposal of depleted amine solutions.[8] For these reasons, alternative CO_2 capture and separation technologies are being investigated, including absorption using ionic liquids,[9] membrane separation[10] and chemical reaction with solids at high temperature,[11] among others.[12,13] Adsorption separation has also drawn particular attention for being a technology generally regarded as consuming low amounts of energy while having the potential to produce high-purity adsorbates.[14,15] A major challenge with respect to adsorption technologies is the development of a suitable adsorbent for the particular application pursued. As a result, efforts have been reported in the literature with respect to CO_2 capture exploring commercially available adsorbents such as zeolites[16-20] and carbonaceous materials,[21-25] among others.

In the search for more efficient adsorbents, the liquid amine scrubbing process inspired researchers to use amines supported on solid materials as CO_2 adsorbents. As far as flue gas treatment is concerned, it was anticipated that supported amines would not present the aforementioned drawbacks associated with aqueous amine solutions while maintaining a high selectivity towards CO_2, particularly over N_2. Although the early attempts to produce amine-functionalized adsorbents were not particularly successful in terms of adsorption capacity, in recent years the collective effort of several research groups has resulted in significant performance improvements, leading to increasing interest in this subject matter. In particular, the exploration of nanostructured supports represented a definite turning point in the development of attractive amine-functionalized adsorbents, as they produced materials with suitable characteristics, previously unattainable with either amorphous or microporous supports.

The present chapter provides a general overview of recent advances in the development of amine-functionalized adsorbents for CO_2 capture. To determine whether an adsorbent is appropriate, it must be endowed with the following attributes:

(i) High adsorption capacity: Among the main properties used to screen new adsorbents, the adsorption capacity is of major interest as it is a measure of the potential of the adsorbent to capture particular species. Adsorption capacity at equilibrium measured at constant temperature over a range of partial pressure produces adsorption isotherms whose shape is of prime importance for early evaluation of potential adsorbents and can be used to model adsorptive behavior. The slope of the isotherm at low pressures can also be used to calculate thermodynamic properties. With respect to CO_2 capture for flue gas treatment, it typically occurs at low CO_2 partial pressures (i.e. less than 0.4 bar) and moderate temperature (i.e. below 70°C). Thus, an attractive adsorbent should exhibit high capacity within this range of conditions. As a point of reference, it has been suggested that, in order to be competitive against liquid amine scrubbing, an adsorbent should exhibit a capacity of 2–4 mmol/g.[26,27]

(ii) Fast kinetics: In dynamic processes such as adsorption in a fixed-bed column, the rates of adsorption and desorption have a direct impact on working adsorption capacity. A suitable CO_2 adsorbent will have a high rate of adsorption, resulting in a working capacity close to the equilibrium uptake, even in the presence of high gas flow. It should be mentioned, however, that determination of kinetic properties such as diffusion is one of the most challenging issues in adsorption science, as it involves measurements that may result in complicated experimental set-ups or the use of parameters not always readily available, such as the particle size of the adsorbent. A reported advantage offered by mesoporous materials over other more common adsorbents is their open pore structure, which has been associated with fast CO_2 adsorption kinetics.

(iii) High CO_2 selectivity: The adsorbent selectivity toward CO_2 has a direct impact on the degree of purity of the product. This in turn plays a major role in the economics of the CO_2 adsorption process.[26] Ideally, an adsorbent for flue gas treatment will not adsorb any nitrogen. As will be discussed later, unlike physical adsorbents such as zeolites, activated carbons and metal-organic frameworks, amine-functionalized mesoporous materials reportedly present a high selectivity for CO_2 over N_2.

(iv) Mild conditions for regeneration: One of the most important characteristics of an adsorbent from the economic point of view is its ease of regeneration, as that determines the energy input required to operate the system as well as the size required to achieve a certain productivity. The ease of regeneration is strongly dependent on the type of bonding formed between the adsorbent and the adsorbate, whether it is weak (e.g. van der Waals, hydrogen bonding) or strong (e.g. acid-base interactions). In general terms, an optimum interaction should be neither too weak nor too strong, as the former would result in an effortless desorption but low adsorption capacity, while the latter represents a costly regeneration. Surface modification of the adsorbent by functionalization can be used to modify such interactions. For example, ordered mesoporous silicas have been engineered to produce adsorbents with high capacity and ease of regeneration. Depending on the nature of the adsorbent-adsorbate interactions, adsorption-desorption cycling may be achieved via temperature, pressure (or vacuum) or concentration swing adsorption, or a combination thereof.

(v) Long-term stability: The lifetime of adsorbents, which determines the frequency of their replacement, is a critical property because of its direct impact on the economics of any commercial-scale operation. Thus, suitable adsorbents should exhibit high stability during extensive adsorption-desorption cycling.

(vi) Tolerance to moisture and other impurities in the feed: In addition to CO_2 and N_2, flue gas contains water vapor and other impurities such as H_2O and SO_2. The degree of tolerance and the affinity of the adsorbent to such impurities may affect significantly the strategy to be used, with a direct impact on the overall economics of the CO_2 separation process. One of the major drawbacks of microporous adsorbents, such as zeolites, is their preferential and competitive uptake of water, severely decreasing their CO_2 adsorption capacity. According to the reactions presented in Scheme 16.1, the presence of moisture is not expected to affect negatively the CO_2 adsorption on amine-functionalized materials, as it would promote the formation of bicarbonate species. On the other hand, it is generally established that CO_2 adsorbents have high affinity toward SO_2 and even some affinity toward NO_X, which may adversely affect the CO_2 adsorption capability of the material. Thus, abatement of SO_2 and NO_X from flue gas prior to CO_2 capture is required in most cases.

(vii) Other attributes: In addition to the aforementioned properties, it is highly desired that an adsorbent be low-cost, mechanically robust and easy to handle.

The present chapter is broadly organized according to the type of interactions between amine groups and the support, namely (i) amine-impregnated materials where mostly weak interactions occur, and

(ii) covalently bonded amine-containing species, typically obtained via surface-grafting of aminosilanes. The rationale behind such classification is that materials with either strong or weak interactions exhibit a number of common characteristics. In addition, our discussion will be mainly based on literature reports dealing with CO_2 adsorption using 5–20% CO_2-containing mixtures with a total pressure of 1–2 bar and temperatures between 25 and 70°C, as these represent the range of flue gas operating conditions after SO_2 and NO_x abatement.[28,29]

16.2 Amine-impregnated Materials

Possibly the simplest approach to introduce amine functionality into porous solids is by physically dispersing amine-bearing molecules on its surface. Typically, impregnation is performed by first dissolving the amine-containing species into a volatile organic solvent, usually methanol, followed by addition of the support to the solution as shown in Figure 16.1. The mixture is subsequently stirred and the solvent evaporated, with the resulting precipitate being an amine-functionalized adsorbent. In this case, although the amine content can be varied at will, it has been observed that beyond an amine content threshold, the product becomes a viscous material, probably due to deposition of organic molecules on the external surface of the support after filling its pores. This method of functionalization produces weakly bonded amine-containing groups on the support, and so, one of the main concerns associated with this type of materials is their thermal stability, particularly since regeneration may be performed using heat. However, since no chemical bonding with the support is required, a wide variety of

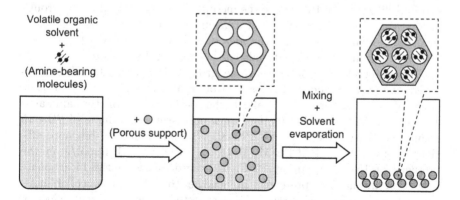

Figure 16.1. Schematic representation of amine wet impregnation.

supports can be applied. The following sections review some of the reported findings using amine-impregnated materials.

16.2.1 Ordered mesoporous supports

A vast number of literature contributions devoted to amine-impregnated materials have used ordered mesoporous silica as support. In general, this can be attributed to the large pore volume offered by such materials, permitting high organic loadings, which in turn are expected to produce significant CO_2 removal capacities. The first reports of such materials involved MCM-41 silica impregnated with polyethylencimine (PEI),[30] coining the term "molecular basket", as it was envisioned that the cross-linked organic chains of PEI would resemble a network inside the mesopores capable of retaining CO_2. It was found that the adsorption capacity of PEI-impregnated MCM-41 improved at increased loading. The highest adsorption capacity of 3.02 mmol/g reported in this work was obtained for a sample with 75 wt% PEI and using a stream of pure CO_2 at 75°C. However, the maximum efficiency, i.e. CO_2/PEI molar ratio, was obtained in the presence of a material containing 50 wt% PEI, and decreased steadily at higher loadings. As discussed later, this behavior was confirmed by other researchers. Evaluation under conditions relevant to flue gas treatment showed that the 50% PEI on MCM-41 silica exhibited an adsorption capacity of ca. 2.1 mmol/g in the presence of 10% CO_2/N_2 at 75°C. A particularly interesting behavior of PEI-impregnated MCM-41 materials was the fact that, unlike other adsorbents, the CO_2 uptake improved as the temperature increased from 25 to 75°C. Since the actual adsorption event is exothermic in nature, the increasing adsorption uptake with temperature was attributed to a diffusion-limited process associated with the occurrence of a bulk-like state of PEI inside the mesopores, with internal amine groups hardly accessible at low temperature. Since then, other researchers have reported similar findings and the idea of a diffusion-limited CO_2 adsorption on PEI-impregnated materials has been generally accepted.

In a later contribution, the same group[31] investigated PEI impregnated on SBA-15 under the assumption that the structural characteristics of the support would affect the performance of the aminated adsorbents. Allegedly, due to the larger pore size and volume of SBA-15 compared to MCM-41, 50 wt% PEI-SBA-15 was found to be more efficient than 50 wt% PEI-PE-MCM-41, i.e. 3.18 vs. 2.1 mmol/g in the presence of 15% CO_2/air at 75°C. In accordance with the bicarbonate formation mechanism presented in Scheme 16.1, the presence of moisture in the CO_2-containing streams was reported to have a positive effect on PEI-impregnated MCM-41[32] in terms of increased capacity. Interestingly, the increase was only observed when the molar concentration of water was equal or lower to that of CO_2 and not for streams with higher moisture content. Accordingly, the CO_2 uptake in a dry gas containing 15% CO_2 was 2.01 mmol/g vs. 2.84 mmol/g when the stream contained 10% moisture.

Furthermore, other researchers performed comparative studies using various mesoporous supports. It was found that at constant PEI loading, KIT-6-, SBA-16-, SBA-15-, MCM-48- and MCM-41-type silicas afforded different adsorption capacities.[33] Under otherwise the same conditions, the adsorption capacity appeared to be dependent on pore diameter (d_p). Using a 50% loading of PEI on KIT-6 silica $(d_p = 6\,nm)$ afforded a capacity of 3.07 mmol/g in a stream of pure CO_2 at 75°C, compared to 2.52 mmol/g for the smaller-pore MCM-41 $(d_p = 2.8\,nm)$. The adsorption capacity of 50% PEI-loaded KIT-6 in conditions closer to flue gas was 1.95 mmol/g in the presence of 5% CO_2/N_2 at 75°C. In addition, the pore size reportedly affected the rate of adsorption as the time required to achieve 90% of the total capacity was in the order of KIT-6 < SBA-16 = SBA-15 < MCM-48 < MCM-41. Following the hypothesis that large pore sizes afford enhanced adsorption capacity, PEI and tetraethylenepentamine (TEPA) were further impregnated on a silica monolith with a hierarchical pore structure.[34] Due to its larger d_p, with mean values at 3, 17 and 120 nm, the optimum loading of PEI was 65 wt%, with a CO_2 uptake of 3.75 mmol/g for a stream of 5% CO_2 in N_2 at 75°C, much higher than that obtained using conventional MCM-41 mesoporous silica as support. PEI impregnated on HMS, a mesoporous silica material with a disordered three-dimensional channel structure, was also studied for CO_2 capture.[35] An adsorption capacity of 4.18 mmol/g was reported for a sample with 60% PEI loading under a stream of pure CO_2 at 75°C.

Having established that at adequate temperatures the adsorption capacity of amine-impregnated materials is proportional to the amine loading, alternative supports with vast empty spaces used to accommodate amines have been explored. PEI and TEPA were loaded on hollow mesoporous capsules produced by growing a silica shell around polystyrene microcapsules in the presence of cetyltrimethylammonium bromide as a mesopore structure-directing agent.[36] The empty particle core permitted an unprecedented loading of >80% while it was believed the mesoporous shell would permit an easy access of adsorbate to its interior. Thus, the better performing adsorbent contained 83% TEPA and presented a capacity of 7.93 mmol/g for a wet stream (8% RH) of 10% CO_2/N at 75°C.

Besides PEI and TEPA, other typical amine molecules used in liquid scrubbing have been studied. Diethanolamine (DEA), for example, has been impregnated on a variety of supports, the most promising being MCM-41 silica with pores enlarged by post-synthesis treatment

(PE-MCM-41, $d_p = 9.7$ nm). This material afforded an adsorption capacity of 3 mmol/g at 25°C in the presence of 10% CO_2 in N_2.[37] The reported adsorption capacity of PEI-impregnated PE-MCM-41 was comparatively higher than that reported for 13X zeolite (i.e. 2.8 mmol g^{-1}) under the same conditions. Similarly to supported PEI, the adsorption capacity increased with DEA loading, but the adsorption efficiency (CO_2/N) decreased, leading to the conclusion that beyond 6 mmol of DEA per gram, the adsorbent becomes unattractive.

In addition to the purely siliceous mesoporous supports mentioned, as-synthesized mesoporous materials, i.e. with the supramolecular template occluded in the pores, have also been explored.[37] When TEPA (50 wt%) was impregnated on as-synthesized SBA-15, the adsorption capacity in the presence of 10% CO_2 in N_2 was ca. 3.25 mmol/g at 75°C. This was 10% higher than the corresponding TEPA on calcined support, with the added advantage that no steps are required to remove the organic template. This finding was corroborated in later reports using as-synthesized supports.[39,40] The proposed explanation was that the polymeric template in the pores of as-synthesized supports interacts with the TEPA, forming a more even distribution of the functional groups, and preventing TEPA from aggregating into a micellar-like form, which is believed to be its naturally occurring form (Figure 16.2). As-synthesized SBA-15 has also been impregnated using mixtures of TEPA and DEA.[40] In this case, the maximum CO_2/N ratio was found to be ca. 0.4 at a loading of ca. 30% TEPA and 20% DEA. In the case of TEPA-DEA-SBA-15 at 75°C, the adsorption capacity ranged from 3.77 to 3.61 mmol/g for 5% CO_2 in N_2 throughout six adsorption-desorption cycles. The good performance of this adsorbent was attributed to the hydroxyl groups present in DEA. This is analogous to the

| Silica |
| Supramolecular template |
| TEPA |

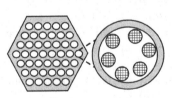
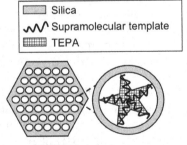

TEPA loaded on
Calcined mesoporous silica

TEPA loaded on
as-synthesized mesoporous silica

Figure 16.2. Schematic representation of amine distribution on calcined and as-synthesized mesoporous supports.

effect of water vapor, which is associated with a more favorable CO_2 to N stoichiometry, as shown in Scheme 16.1. The use of as-synthesized supports was also extended to MCM-41.[39] A TEPA-MCM-41 sample with 50 wt% loading had a capacity of 4.54 mmol/g for 5% CO_2/N_2 at 75°C, outperforming the aforementioned TEPA-SBA-15.

Besides ordered mesoporous silica, mesoporous alumina (MA) has also been impregnated with a series of amine-containing molecules such as diisopropanolamine, triethanolamine, 2-amino-2-methyl-1,3-propanediol, diethylenetriamine (DETA) and PEI.[41] The most attractive materials were those containing DETA with 40 wt% loading, with a capacity of *ca.* 1 mmol/g at 25°C for pure CO_2, increasing to *ca.* 1.4 mmol/g at 57°C. For PEI-MA, the adsorption capacity at 57°C was *ca.* 1.14 mmol/g in the presence of pure CO_2. It is worth noting that, according to this work, only when amine loading was high, the adsorption capacity was enhanced at higher temperatures, supporting the hypothesis that this behavior is associated with diffusion limitations within the amine phase at low temperatures.

Because an economically attractive adsorbent has to be easily regenerated for cyclic use, a number of literature reports explored the regeneration behavior of amine-impregnated mesoporous materials. In a work devoted to the influence of regeneration temperature on a proprietary mesoporous silica impregnated with PEI (40 wt%), it was observed that desorption at a temperature of less than 140°C resulted in incomplete regeneration.[29] This work analyzed the effect of regeneration temperature using pure CO_2 as stripping gas. However, some concerns were raised regarding the use of such high temperatures, mainly because of the following problems: (i) evaporation of PEI may occur and (ii) a secondary reaction between CO_2 and amine groups forms a stable product, most likely urea, resulting in a decreasing number of adsorption sites. The suggested alternative was to use a different stripping gas or lower desorption temperature although this would sacrifice some working adsorption capacity. As discussed later, a strategy to prevent the formation of urea during extensive cycling even at high temperature has been proposed recently by Sayari and Belmabkhout.[42] Another work dealing with regeneration of PEI-impregnated SBA-15 explored the use of pressure-swing adsorption (PSA).[43] The highest working capacity reported at 75°C was 1.36 mmol/g for 12% CO_2 in N_2. A steady state was obtained after 15 to 20 cycles, and the productivity was better compared to a similar PSA procedure using 13X zeolite at 75°C.

16.2.2 Zeolite supports

In addition to mesoporous materials, zeolites have been explored as supports of amine groups. In a study using monoethanolamine (MEA) dispersed on 13X zeolite with different loadings, an adsorbent with only 2.9 wt% MEA presented the highest capacity at low temperature (i.e. 35°C), with 1.96 mmol/g for 15% CO_2 in N_2, while the best capacity at 75°C (i.e. 0.45 mmol/g) was reported using a sample with 25 wt% amines.[44] These results suggest that the diffusion limitations observed in amine-impregnated mesoporous materials also occur when microporous supports are used, making it attributable to the bulk state of the amine at high loadings rather than the structural properties of the solid support. The adsorption capacity of the amine-functionalized zeolite was higher than unmodified 13X, which adsorbed 0.64 and 0.36 mmol/g at 35 and 75°C, respectively. An interesting advantage of amine-containing 13X was a significant improvement in tolerance to moisture. While it is generally accepted that preferential adsorption of water on 13X results in a drastic reduction of CO_2 uptake, the adsorption capacity for CO_2 in the presence of 100% RH decreased by only ca. 13% with respect to dry conditions.

Another type of zeolite, namely beta-zeolite, was also used to support TEPA and compared with TEPA impregnated on amorphous alumina (Al_2O_3) and silica (SiO_2).[45] The results clearly showed the advantages of using a support with good structural properties since beta-zeolite was loaded with up to 38.4 wt% compared to only 14.6 and 8.3 wt% for SiO_2 and Al_2O_3, respectively, reflecting their different values of pore volume. Such loading translated into a significantly higher adsorption capacity for TEPA-beta-zeolite compared to the other samples, being 2.08 mmol/g for 10% CO_2, balance nitrogen, at 30°C, versus 0.19 and 0.68 mmol/g for TEPA-Al_2O_3 and TEPA-SiO_2, respectively.

16.2.3 Other supports

While ordered porous supports have proven to be suitable substrates for the dispersion of amines, the use of other commercially available supports has also been explored. Extensive work has been reported on high-surface-area polymeric supports, mainly polymethylmethacrylate (PMMA) functionalized with a variety of amines, such as PEI, monoethanolamine, diethanolamine, triethanolamine and TEPA.[46-48] Although these adsorbents were originally conceived for air purification in confined environments, the results have proven to be of interest for other applications such as flue gas treatment. In particular, a TEPA-impregnated PMMA exhibited remarkably high CO_2 adsorption capacities, being 21.45 and 13.88 mmol/g at 20 and 70°C, respectively, in the presence of 15% CO_2 and 2.6% H_2O, balance N_2.[48] It was also observed that, contrary to amine-impregnated mesoporous inorganic supports, the CO_2 adsorption capacity decreased at a higher temperature for the polymer-based adsorbents.

In the same work,[48] TEPA was reacted with acrylonitrile to selectively transform primary amines into secondary amines before impregnation, with the intention of exploring whether a molecule with only secondary amines would be advantageous. However, this was not the case, since the sample produced by impregnation of the modified TEPA, referred to as TEPAN, presented adsorption capacities of 14.22 and 4.01 mmol/g at 20 and 70°C respectively, underperforming the aforementioned values for TEPA-PMMA. Furthermore, TEPAN-PMMA was not stable as its capacity decreased considerably, e.g. from 4.01 to 1.68 mmol/g after only three regeneration cycles, with adsorption at 25°C and regeneration at 70°C.

The hypothesis that secondary amines perform better than primary amines had also been proposed in a comparative study between ethyleneamine-impregnated polymer and aminopropyl-grafted SBA-15.[49] The latter had an amine loading of 5.07 mmol/g and a capacity of 2.01 mmol/g under a humid stream of 10% CO_2 in N_2 at 25°C, a higher capacity value than the ethyleneamine-containing adsorbent, i.e. 1.92 mmol/g. Nonetheless, a sample produced by loading ethyleneamine after reaction with acrylonitrile had an enhanced adsorption capacity of 4.18 mmol/g, although it exhibited poor stability, as capacity decreased to 2.69 mmol/g after regeneration.

Tertiary amines are expected to be advantageous, based on the rationale that they react with CO_2 to form bicarbonate, a reaction with a more favorable CO_2/N stoichiometry compared to the formation of carbamate (Scheme 16.1) with the caveat that water must be present in the adsorbate stream. In an attempt to explore the use of tertiary amines, 1,8-diazabicyclo-[5.4.0]-undec-7-ene (DBU) was dispersed on PMMA beads.[27] Since the reaction between CO_2 and tertiary amines can take place only in the presence of humidity, the simulated flue gas used in that study contained 10% CO_2 and 2% H_2O (i.e. 100% RH at 25°C), balance N_2. The sample with the highest loading (*ca.* 30% DBU) was also the best in terms of adsorption capacity, with an average of 3.02 mmol/g over four cycles at 25°C, and 2.34 mmol/g at 65°C, corresponding to a CO_2/N ratio of 0.76 and 0.59, respectively.

Other supports used for PEI impregnation include polystyrene (Macronet), silicon dioxide (CARiACT) and PMMA (Diaion).[50] While the polystyrene-supported amines performed poorly, the other two adsorbents were promising, with PEI-CARiACT and PEI-Diaion adsorbing 2.55 and 2.40 mmol/g, respectively, in a dry stream of 10% CO_2, balance He, at 40°C. The enhancement of the adsorption capacity due to moisture was quite evident in this work, with capacity increasing to 3.65 and 3.53 mmol/g for PEI-CARiACT and PEI-Diaion, respectively, when the stream contained 7% water vapor (100% RH).

Carbon-based materials as supports, including products from sewage sludge and air-oxidized olive stones (AOS), were also functionalized with PEI.[51,52] The resulting adsorbents, however, did not offer any particular advantage with an adsorption capacity of 1.98 mmol/g for PEI-AOS in the presence of pure CO_2 at 25°C, a value lower than other non-PEI-containing activated carbons included in the same work. It should be noted that the PEI-AOS sample had an amine loading of only 5 wt%, a value significantly smaller than other amine-impregnated materials with competitive adsorption capacities.

As summarized in Table 16.1, this review of literature data provides evidence of the great variety of amines impregnated on solid materials. Although adsorbents were produced using a variety of supports, including polymers, zeolites and mesoporous oxides, supports with large d_p seem to be more appropriate. Despite their diversity, some general properties can be mentioned about these adsorbents. It was found that high amine loadings result in enhanced adsorption capacity; however, this was usually accompanied by a decrease in the rate of adsorption and CO_2/N ratio. Another trend observed is that at a certain amine content, the optimum adsorption capacity may occur at high temperature, making them inappropriate for applications where lower temperature is required. Furthermore, because of the weak interactions between the active phase and the support, amine-impregnated adsorbents can be unstable. Thus, regeneration of such materials has to be performed under strictly controlled conditions, since low temperatures may result in incomplete regeneration, whereas higher temperatures induce evaporation and/or degradation of the supported amines.

16.3 Grafted Materials

Functionalization of silica supports using amine-containing silanes has been studied to produce materials with covalently bonded surface amine groups. Grafting is usually performed by dispersing an amine-containing silane in an organic solvent, followed by the addition of a siliceous support and, in some cases, water. Reaction proceeds at moderate temperatures for a few hours, depending on the type of silane and support. After reaction, the solids are filtered and dried. A schematic representation of a typical grafting process is presented in Figure 16.3. Because grafting is a result of the chemical reaction between silane and surface silanol groups, the amine loading depends strongly on the surface area of the support and the availability of surface silanol groups. Consequently, ordered mesoporous silicas have proven particularly attractive as their combination of high surface area and large pore

Table 16.1. Literature data on CO_2 adsorption capacity of amine-impregnated adsorbents.

Support	Amine	Amine Loading (wt%)	Capacity (mmol/g)	CO_2/N	Experimental Conditions		Reference
					CO_2 Concentration (%)	T (°C)	
PMMA	DBU	30	2.34	0.59	10 (2% H_2O)	65	27
MCM-41	PEI	50	2.1	0.18	10	75	30
SBA-15	PEI	50	3.18	0.27	15	75	31
MCM-41	PEI	50	2.84	0.27	13 (13% H_2O)	75	32
KIT-6	PEI	50	1.95	0.17	5	75	33
Monolith	PEI	65	3.75	0.25	5	75	34
PE-MCM-41	DEA	76	3	0.41	5	25	37
As-synthesized SBA-15	TEPA	50	3.25	0.28	10	75	38
As-synthesized MCM-41	TEPA	50	4.54	0.34	5	75	39
As-synthesized SBA-15	TEPA + DEA	(30% TEPA, 20% DEA)	3.77	0.38	5	75	40
Mesoporous SiO2	PEI	40	2.4	0.26	15	70	41
Mesoporous Al2O3	DETA	40	1.4	0.12	100	57	41
SBA-15	PEI	50	1.36	0.12	12	75	43
13X	MEA	25	0.45	0.11	15	75	44
Beta-zeolite	TEPA	38	2.08	0.21	10	30	45
PMMA	TEPA	41	13.88	1.28	15 (2.6% H_2O)	70	48
PMMA	Ethyleneamine + acrylonitrile	Proprietary information	4.18	Proprietary information	10 (humid)	25	49
SiO2 (CARiACT)	PEI	40	3.95	0.42	10 (2% H_2O)	40	50
PMMA (Diaion)	PEI	40	3.60	0.39	10 (2% H_2O)	40	50
AOS carbon	PEI	5	1.98	1.70	100	25	51

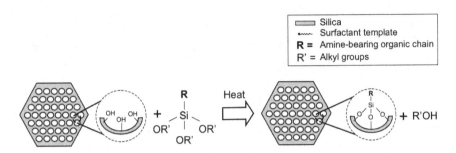

Figure 16.3.   Schematic representation of amine functionalization by grafting.

size and volume can be exploited to host large quantities of organic groups. In the particular case of ordered mesoporous materials, another potential route for surface grafting consists of co-condensation of the amine-bearing silanol species with silica during the hydrothermal preparation of the material, resulting in the direct incorporation of amine groups on the silica walls. This functionalization approach, however, has not been widely explored in the case of amine-grafted adsorbents for CO_2 capture.

To the best of our knowledge, the earliest studies dealing with amine-grafted materials for adsorption of CO_2 were those of Tsuda and Fujiwara[53] and Leal et al.[54] In the former, amorphous silica gel adsorbents were produced by co-condensation of silanes and a number of polyethyleneimines and macrocyclic polyamines, but the produced materials required long contact times to reach equilibrium (up to 3 hours) and had relatively low capacities, ranging from 0.3 to 0.7 mmol/g for ca. 7% CO_2 in N_2 at 30°C. In the work by Leal et al.,[54] silica gel was decorated with amine groups via grafting of (3-aminopropyl)triethoxysilane (AP) under anhydrous conditions, resulting in an adsorbent with a capacity of 0.41 mmol/g ($CO_2/N = 0.33$) under a stream of pure CO_2 at 23°C.[54] Although such adsorption capacity was lower compared to other benchmark adsorbents such as zeolites and activated carbons, the relevance of this pivotal work lies in the actual idea of using supported amines for CO_2 capture that was subsequently adopted by other researchers. In this work, infrared spectroscopy was used to substantiate the proposed reaction mechanisms between CO_2 and supported amine groups producing carbamate and bicarbonate in dry and humid streams, respectively (Scheme 16.1). Thus, the more favorable stoichiometry expected in the presence of moisture was confirmed by the experimental CO_2/N values, increasing from 0.41 to 0.89 mmol/g in dry and water-saturated CO_2 streams, respectively. For several years thereafter, no further CO_2 adsorption studies on amine-grafted materials appeared, most likely because of the lack of interest in this topic. However, with the rapid development of ordered mesoporous materials, and increasing awareness of the effects of global warming, new studies began to appear in the literature.

The advantages offered by ordered mesoporous supports were evidenced in a comparative investigation of propylamine grafted on MCM-48 and silica xerogel.[55] Using 10% CO_2 in N_2, a significantly higher capacity of *ca.* 1.42 mmol/g was obtained for AP-MCM-48 at room temperature versus *ca.* 0.58 mmol/g for silica xerogel under the same conditions. Since the amine loading was 2.3 and 1.7 mmol/g for MCM-48 and silica xerogel, respectively, it was shown that the higher adsorption capacity of AP-MCM-48 was accompanied by improved efficiency. Further, using a CO_2-containing stream with 100% RH gave rise to an adsorption capacity twice as high with a CO_2/N value of 1, corresponding to quantitative transformation of amine groups into ammonium bicarbonate.

Further evidence on the advantages of periodic mesoporous supports was reported in a study where AP was grafted onto hexagonal mesoporous silica (HMS) and amorphous silica gel.[56] A higher amine loading of 2.3 mmol/g was obtained versus only 1.1 mmol/g for amorphous silica. These loadings mirrored the difference in surface areas of HMS $(1,198\,m^2/g)$ compared to amorphous silica $(567\,m^2/g)$. The adsorption capacity at 20°C in the presence of 90% CO_2/Ar was 1.59 mmol/g for AP-HMS compared to 0.68 mmol/g for AP-grafted amorphous support. Later, (3-trimethoxysilylpropyl) diethylenetriamine (TRI) was grafted on HMS to a loading of 4.6 mmol/g.[57] The corresponding adsorption capacity was 1.34 mmol/g in the presence of 90% CO_2 balance Ar at 20°C. The material was found to be thermally stable up to 170°C under pure N_2 or mildly oxygenated environments, a comparatively higher temperature than amine-impregnated adsorbents.

Sayari's group made significant contributions to the area of CO_2 capture by amine-containing nanoporous materials. They demonstrated the beneficial effect of using materials with larger pore diameter and pore volume than typical MCM-41 silica.[58–60] To do so, they used a post-synthesis pore-expansion method developed earlier in which, using as-synthesized MCM-41 as the starting material, they generated PE-MCM-41 with a pore size and pore volume of up to 20 nm and 3.5 cm³/g, vs. typically *ca.* 3–4 nm and *ca.* 0.7–1 cm³/g for regular MCM-41, with hardly any change in surface area.[61,62] As shown in Figure 16.4, grafting MCM-41 and PE-MCM-41 with TRI led to comparable amine loadings because of similar surface areas. However, as shown in Figure 16.5, using 5% CO_2 in N_2 at 25°C, the CO_2 uptake was *ca.* 50% higher for TRI-PE-MCM-41 than TRI-MCM-41, at all amine loadings. Moreover, TRI-PE-MCM-41 adsorbed CO_2 about 30% faster than MCM-41-based material, showing the importance of pore size and volume for amine-grafted adsorbents.

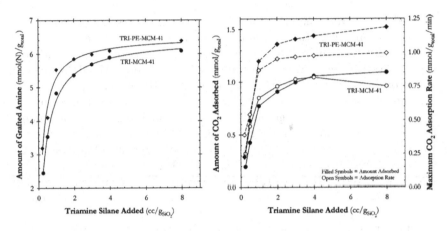

Figure 16.4. Amine loading (left) and adsorption capacity (right) versus TRI/SiO₂ ratio on MCM-41 and PE-MCM-41.[58]

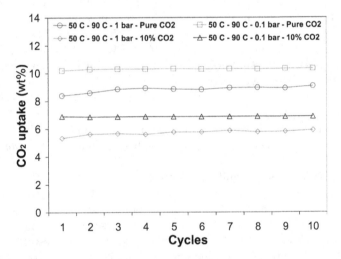

Figure 16.5. Adsorption-desorption cycles using TS and TVS regeneration configurations for pure and 10% CO₂ in N₂ on TRI-PE-MCM-41.

Another contribution of Sayari's group was the optimization of the grafting conditions, leading to dramatic improvement of amine loading and adsorptive properties. Grafting is traditionally practised under reflux, in dry solvent (typically toluene at 110°C) with a large excess of silane. Harlick and Sayari[59] found that the optimum grafting conditions of TRI on PE-MCM-41 in toluene were as follows: T $= 85°C$; water added: $0.3\,mL$ per gram of support; aminosilane added: $3\,mL$ per gram of support. Under such conditions, the amine content increased by *ca.* 30% (i.e. 7.98 vs. $6.11\,mmol/g$ for conventional dry grafting), whereas the adsorption capacity using 5% CO_2/N_2 at 25°C increased by *ca.* 70% from $1.55\,mmol/g$ for conventional dry grafting to $2.65\,mmol/g$. Thus, under these CO_2 adsorption conditions, the combination of pore expansion and optimization of grafting conditions improved the adsorption capacity by close to 300% compared to the adsorbent produced via anhydrous grafting on conventional MCM-41, in addition to a significant increase in the rate of adsorption. The advantage of using amine-functionalized mesoporous materials was further evidenced when a stream of humid CO_2 was used. In the presence of 5% CO_2 in N_2 with 27% RH, the adsorption capacity for TRI-PE-MCM-41 increased from 2.64 to $2.94\,mmol/g$, in contrast to a dramatic decrease reported for 13X, down to $0.09\,mmol/g$.

Further studies on TRI-PE-MCM-41[60] demonstrated that, besides an enhanced capacity at low CO_2 partial pressure, incorporation of amines on mesoporous silica increased the selectivity towards CO_2 over N_2. Using conditions directly related to flue gas, i.e. 10% CO_2, balance N_2, at 50°C, Serna-Guerrero *et al.*[63] obtained a stable working capacity of $1.59\,mmol/g$ over ten adsorption-desorption cycles with regeneration under vacuum at 90°C (Figure 16.5).

To further address the long-term stability of amine-containing adsorbents for CO_2 capture, Sayari and Belmabkhout[42] carried an in-depth investigation using extensive adsorption-desorption cycling under different conditions in the presence of monoamine- and triamine-functionalized PE-MCM-41 as well as PEI-impregnated PE-MCM-41. They found that under dry adsorption and desorption conditions, all materials deactivate at different rates depending on the nature of the materials and the experimental conditions. As an example, Figure 16.6 shows that under dry conditions, the adsorbent would ultimately deactivate, even under mild conditions. Based on ^{13}C NMR and *in situ* FTIR data, it was convincingly demonstrated that the adsorbent deactivation was associated with the gradual formation of urea groups, which are stable under the desorption conditions. To prevent the formation of urea and drastically improve the stability of amine-functionalized adsorbents, the use of humid streams was proposed. As illustrated in Figure 16.6, the adsorbent underwent more than 700 cycles without any loss in adsorption capacity when the adsorption and desorption gases contained 7% RH at 70°C. Another interesting finding was that by treating deactivated AP-grafted PE-MCM-41 in the presence of water vapor at *ca.* 200°C (as low as 0.15% RH), it was possible to hydrolyze the urea groups and fully regenerate the adsorbent.

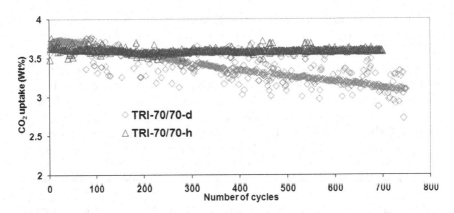

Figure 16.6. Working adsorption capacity of TRI-PE-MCM-41 over various adsorption-desorption cycles in dry (TRI-70/70-d) and humid (TRI-70/70-h 7% RH) streams with adsorption and desorption at 70°C.[42]

The advantages of using TRI were also discussed by other researchers in a thorough comparative analysis of monoamine-, diamine- and triamine-grafted SBA-15.[64] The higher amine density achieved through the use of TRI resulted in the best-performing adsorbent. The reported adsorption capacity for TRI-SBA-15 at 60°C in the presence of 15% CO_2 was 1.58 mmol/g under dry conditions and 1.80 mmol/g in a stream containing 60% RH. Furthermore, it was reported that TRI-SBA-15 was stable over 50 cycles of adsorption at 60°C and desorption at 100°C in flowing He.

Another study comparing mesoporous silica (MS) functionalized with molecules containing 1–3 amine groups produced by anhydrous grafting and by co-condensation has been reported as well.[65] In general, samples prepared by co-condensation presented higher amine contents, that reportedly promoted better amine efficiency and adsorption capacity. In line with the observations mentioned, the adsorbent with the highest capacity was TRI-MS, with 1.74 mmol/g for pure CO_2 at 25°C, while the most efficient under the same conditions was the AP-MS sample with a CO_2/N ratio of 0.43 and a capacity of 1.14 mmol/g. This is consistent with findings by Serna-Guerrero et al.,[66] who obtained the maximum efficiency of $CO_2/N = 0.5$ using AP-grafted PE-MCM-41, whereas the CO_2/N ratio for TRI-PE-MCM-41 never exceeded 0.34.[63]

The use of diamine-bearing molecules has been investigated under the hypothesis that the occurrence of two amine groups in close proximity will lead to enhanced formation of carbamate, and thus higher CO_2/N efficiency. In a work dealing with N-[3-(trimethoxysilyl)propyl] ethylene-diamine (PEDA) grafted on SBA-16 silica, it was clearly shown that incorporation of amine groups resulted in improved capacity at CO_2 partial pressures below 1 bar, although this work was mainly focused on CO_2 adsorption at high pressure (up to 35 bar).[67] The reported capacity for pure CO_2 at 1 bar was 1.4 mmol/g at 27°C for the best-performing sample. It was observed, however, that at high pressure (*ca.* 4 bar or more), the non-aminated samples exhibited higher adsorption capacity. A possible explanation is that physical adsorption predominates at high pressure, and so, the higher surface area and pore volume of the unmodified support offers a comparative advantage in terms of adsorption capacity.

Further, efforts to optimize the grafting process have been made, with the aim of enhancing the efficiency and capacity of aminated silicas. A proposed approach to incorporate AP-functionality consisted in the simultaneous extraction of the structure-directing agent and amine-grafting on as-synthesized SBA-15.[68] The CO_2 adsorption capacity of the adsorbent obtained by such a method (1.22 mmol/g) outperformed a sample synthesized using the typical grafting procedure on calcined SBA-15 (1 mmol/g), for CO_2 at 1 bar and 25°C. For conditions more representative of flue gas, the sample using as-synthesized support produced a material with an adsorption capacity of *ca.* 0.45 mmol/g at 65°C at a CO_2 partial pressure of 0.1 bar, corresponding to a CO_2/N efficiency of 0.44, close to the stoichiometric ratio of 0.5. It was suggested that the extraction of the surfactant template with ethanol was a milder treatment than calcination and thus preserved the surface silanol groups, which in turn translated into a better distribution of surface amines with a subsequent improvement of adsorption capacity.

Another proposed approach to tackle the drawback of surface silanol group removal during calcination of the support consisted of rehydration of the silica by contact with water. This was studied by soaking SBA-15 in water at 97°C before grafting with PEDA.[69] The obtained material had an amine content of 3.06 mmol/g, and a capacity of 0.73 mmol/g for 0.15 bar CO_2 at 60°C. In comparison, a similar material prepared using non-hydrated SBA-15 had an amine loading of 2.59 mmol/g and an adsorption capacity of 0.59 mmol/g.

Some interesting studies were devoted to establish the influence of the characteristics of the supports on amine-grafted materials. For instance, the impact of the pore size of the support on CO_2 capture has been studied.[70] Similarly to amine-impregnated materials, a larger pore size was considered beneficial. However, unlike amine-impregnated materials, this was not a result of higher amine loadings, but was rather associated with a more efficient use of amine groups. Indeed, when grafting AP on MCM-41 with a pore size of 3.3 nm, a high amine content was obtained (i.e. 3 mmol/g), but the adsorption capacity was only 0.57 mmol/g for 10% CO_2 at 25°C.

In contrast, a capacity of 1.54 mmol/g was obtained when SBA-15 with a pore size of 7.1 nm was used as support, despite a slightly lower amine loading of 2.7 mmol/g. In this case, the lower efficiency of the MCM-41-based material was not only attributed to a difference in pore sizes. The amine surface density reported for AP-MCM-41 was lower, having only 1.1 amine groups per nm^2 compared to 2.4 amines per nm^2 for AP-SBA-15. Since admittedly two amine molecules in close proximity are required for reaction with CO_2, this would be a disadvantage for the sample with lower amine density.

The influence of the chemical composition of the support was also analyzed in a comparative study between AP-grafted mesoporous silica and mesoporous titania (MT).[71] The highest adsorption capacity of ca. 0.24 mmol/g for 10% CO_2 at 30°C was obtained with AP-MS. This is a low capacity compared to other materials reported in the literature, but the main finding of this work was that the chemical properties of some supports might influence the behavior of the functionalized adsorbent. While no additional chemisorption to that between CO_2 and amine groups was detected when silica was used as support, CO_2 was chemically adsorbed on the surface of MT. This was reflected in a higher capacity when expressed in terms of surface area, i.e. 1 $\mu mol/m^2$ and 0.6 $\mu mol/m^2$ for AP-MT and AP-MS, respectively.

In addition, the use of support with defined morphology was explored by grafting PEDA on mesoporous spherical particles.[72] The adsorption capacity at 60°C in the presence of 10% CO_2 in air was 0.73 mmol/g. These adsorbents showed remarkable stability after the second adsorption-desorption cycle when regenerated using a TSA procedure at 120°C or under VSA. In addition, it was suggested that the combination of heat and vacuum resulted in an improvement in the desorption rate.

A different method of functionalization, introduced recently, consisted of iterative building of amine-containing dendrimers inside the porous supports.[73,74] Highly branched dendrimers were produced by stepwise reaction between diisopropylethylamine and cyanuric chloride inside the pores of SBA-15[73] or mesocellular siliceous foams.[74] The optimum adsorbent produced with this approach was obtained after three reaction cycles, with a capacity of ca. 1 mmol/g for 90% CO_2 in Ar at 20°C. However, as a larger number of synthesis reaction steps were performed to obtain higher-generation dendrimers, the adsorbent lost its structural properties, allegedly due to space limitations, which had a negative impact on the adsorptive properties.

Another innovative amine loading approach involved the polymerization of aziridine by ring opening inside the pores of SBA-15, producing a "covalently tethered hyperbranched aminosilica", as represented in

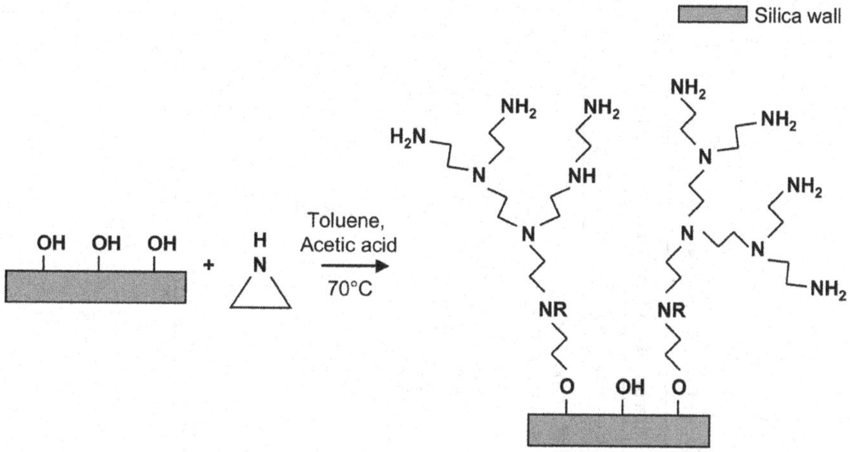

Figure 16.7. Schematic representation of the synthesis of hyperbranched aminosilica according to Jones *et al.*[75]

Figure 16.7.[75,76] This material exhibited a capacity of 3.11 mmol/g under a flow of water-saturated 10% CO_2/Ar at 25°C. The CO_2/N efficiency was as high as 0.44 at room temperature, close to the theoretical value of 0.5. With respect to its performance at 75°C, and 10% CO_2/Ar, the hyperbranched SBA-15 was stable, presenting an average adsorption capacity of *ca.* 1.98 mmol/g over 12 cycles with regeneration at 130°C. In a later contribution,[77] it was shown that higher loading of hyperbranched amines afforded a better capacity. The best reported adsorbent had an amine loading of 9.78 mmol/g and adsorbed *ca.* 4 mmol/g at 10% CO_2/ N_2 at 75°C in the presence of humidity.

As summarized in Table 16.2, similarly to amine-impregnated adsorbents, the covalently bonded aminated adsorbents span materials with a wide variety of characteristics and performances. However, a number of common advantages and limitations of amine-grafted materials can be outlined. For instance, only supports that exhibit surface hydroxyl groups can be used to produce amine-grafted materials. It was observed that high amine loading is a result of high surface area and the density of surface silanol groups, but the efficient use of functional groups is observed mainly in supports with large pores. While the equilibrium adsorption capacities are certainly not as high as those reported with some amine-impregnated adsorbents, properly designed amine-grafted materials do not exhibit the strong diffusion limitations observed in impregnated adsorbents. Therefore, high adsorption rates are not restricted upon operating at high temperature. A particular advantage offered by amine-grafted adsorbents is their high

Table 16.2. Literature data on CO$_2$ adsorption capacity of amine-grafted adsorbents.

Support	Amine	Capacity (mmol/g)	Amine Loading (mmol/g)	CO$_2$/N	Experimental Conditions		Reference
					CO$_2$ Concentration (%)	T (°C)	
Silica gel	AP	0.89	1.26	0.71	100% (100% RH)	50	54
MCM-48	AP	2.3	2.3	1	10% (100% RH)	25	55
HMS	AP	1.59	2.29	0.69	90%	20	56
HMS	TRI	1.34	4.57	0.29	90%	20	57
PE-MCM-41	TRI	1.59	7.9	0.20	10%	50	63
SBA-15	TRI	1.80	5.80	0.31	15% (humid)	60	64
MS	TRI (co-cond)	1.74	5.18	0.34	100%	25	65
SBA-16	EDA	1.4	0.76	1.84	100%	27	67
SBA-15	AP	0.45	2.56	0.18	10%	65	68
SBA-16	EDA	0.727	3.06	0.24	15%	60	69
SBA-15	AP	1.54	2.72	0.57	10%	25	70
MS	AP	0.24	1.6	0.15	10%	30	71
MSP	EDA	0.73	0.99	0.73	10%	60	72
SBA-15	Amine-dendrimers	1	1.25	0.40	90%	20	73
SBA-15	Aziridine polymer	4	9.78	0.41	10% (humid)	75	77

stability over hundreds, most likely thousands, of adsorption-desorption cycles.[42]

16.4 Conclusions

Tremendous progress has been achieved in the development of novel chemical CO_2 adsorbents such as amine-modified materials with large surface area. By optimizing the synthesis conditions and using supports with adequate structural properties, it was possible to develop materials with superior CO_2 adsorptive properties, particularly suitable for flue gas treatment. Typically, these materials exhibit large CO_2 adsorption capacity even at low pressure, high rate of adsorption and desorption, and excellent tolerance to moisture in the feed. Furthermore, contrary to physical adsorbents, the selectivity of amine-functionalized materials is not significantly affected by temperature, at least within the range of interest for flue gas treatment. While the stability of this kind of adsorbents has been questioned, it was recently demonstrated that their stability may be dramatically enhanced during extensive adsorption-desorption cycling, provided that the feed and purge gases contain moisture. The role of moisture is to prevent the formation of urea linkages, which are the main source of material deactivation.

This review clearly showed a steady improvement in the CO_2 adsorptive properties of amine-functionalized materials. The course followed so far has resulted in major achievements that may well pave the way for an alternative CO_2 capture technology in the near future.

Acknowledgments

The financial support of the Natural Science and Engineering Research Council of Canada (NSERC) is acknowledged. Abdelhamid Sayari thanks the Federal Government for the Canada Research Chair in Nanostructured Materials for Catalysis and Separation (2001-2015).

References

1. C. Song, *Catal. Today* **115**, 2–32 (2006).
2. D. J. Hofmann, J. H. Butler and P. P. Tans, *Atmos. Environ.* **43**, 2084–2086 (2009).
3. R. Monastersky, *Nature* **458**, 1091–1094 (2009).
4. A. J. Yamakasi, *J. Chem. Eng. Jpn.* **36**, 361–375 (2003).

5. E. F. da Silva and H. F. Svendsen, *Int. J. Greenh. Gas Con.* **1**, 151–157 (2007).

6. H. Audus, *Energy* **22**, 217–221 (1997).

7. Y. Li, *Petroleum Refinery Eng.* **38**, 24–27 (2008).

8. K. Veltman, B. Singh and E. G. Hertwitch, *Environ. Sci. Technol.* **44**, 1496–1502 (2010).

9. M. Hasib-ur-Rahmana, M. Siajb and F. Larachi, *Chem. Eng. Process.* **49**, 313–322 (2010).

10. C. A. Scholes, S. E. Kentish and G. W. Stevens, *Recent Pat. Chem. Eng.* **1**, 52–66 (2008).

11. K. B. Lee, M. G. Beaver, H. S. Caram and S. Sircar, *Ind. Eng. Chem. Res.* **47**, 8048–8062 (2008).

12. J. D. Figueroa, T. Fout, S. Plasynski, H. McIlvried and R. D. Srivastavab, *Int. J. Greenh. Gas Con.* **2**, 9–20 (2008).

13. A. A. Olajire, *Energy* **35**, 2610–2628 (2010).

14. D. Aaron and C. Tsouris, *Separ. Sci. Technol.* **40**, 321–348 (2005).

15. A. Sjostrom and H. Krutka, *Fuel* **89**, 1298–1306 (2010).

16. F. Su, C. Lu, S. C. Kuo and W. Zeng, *Energ. Fuel.* **24**, 1441–1448 (2010).

17. C. Lu, H. Bai, B. Wu, F. Su and J. F. Hwang, *Energ. Fuel.* **22**, 3050–3056 (2008).

18. N. Konduru, P. Lindner and N. M. Assaf-Annid, *AIChE J.* **53**, 3137–3143 (2007).

19. J. Zhang and P. A. Webley, *Environ. Sci. Technol.* **42**, 563–569 (2008).

20. J. Merel, M. Clausse and F. Meunier, *Ind. Eng. Chem. Res.* **47**, 209–215 (2008).

21. Filipe V. S. Lopes, C. A. Grande, A. M. Ribeiro, J. M. Loureiro, E. Oikonomopoulos, V. Nikolakis and A. E. Rodrigues, *Separ. Sci. Technol.* **44**, 104–1073 (2009).

22. M. Cinke, J. Li, C. W. Baushlicher, A. Ricca and M. Meyyappan, *Chem. Phys. Lett.* **376**, 761–766 (2003).

23. F. Su, C. Lu, W. Chen, H. Bai and J. F. Hwang, *Sci. Total Environ.* **407**, 3017–3023 (2009).

24. S. Himeno, T. Komatsu and S. Fujita, *J. Chem. Eng. Data* **50**, 369–376 (2005).

25. L. Huang, L. Zhang, Q. Shao, L. Lu, X. Lu, S. Jiang and W. Shen, *J. Phys. Chem. C* **111**, 11912–11920 (2007).

26. M. T. Ho, G. W. Allinson and D. E. Wiley, *Ind. Eng. Chem. Res.* **47**, 4883–4890 (2008).

27. M. L. Gray, K. J. Champagne, D. Fauth, J. P. Baltrus and H. Pennline, *Int. J. Greenh. Gas Con.* **2**, 3–8 (2008).

28. S. Majumdar, A. Sengupta, J. S. Cha and K. K. Sirkarst, *Ind. Eng. Chem. Res.* **33**, 667–675 (1994).

29. T. C. Drage, A. Arenillas, K. M. Smith and C. E. Snape, *Micropor. Mesopor. Mater.* **116**, 504–512 (2008).

30. X. Xu, C. Song, J. M. Andersen, B. G. Miller and A. W. Scaroni, *Energ. Fuel.* **16**, 1463–1469 (2002).

31. X. Ma, X. Wang and C. Song, *J. Am. Chem. Soc.* **131**, 5777–5783 (2009).
32. X. Xu, C. Song, B. G. Miller and A. W. Scaroni, *Ind. Eng. Chem. Res.* **44**, 8113–8119 (2005).
33. W. J. Son, J. S. Choi and W. S. Ahn, *Micropor. Mesopor. Mater.* **113**, 31–40 (2008).
34. C. Chen, S. T. Yang, W. S. Ahn and R. Ryoo, *Chem. Commun.* **24**, 3627–3629 (2009).
35. C. Chen, W. J. Son, K. S. You, J. W. Ahn and W. S. Ahn, *Chem. Eng. J.* **161**, 46–52 (2010).
36. G. Qi, Y. Wang, L. Estevez, X. Duan, N. Anako, A. H. A. Park, W. Li, C. W. Jones and E. P. Giannelis, *Energy Environ. Sci.* (2010), DOI: 10.1039/c0ee00213e.
37. R. S. Franchi, P. J. E. Harlick and A. Sayari, *Ind. Eng. Chem. Res.* **44**, 8007–8013 (2005).
38. M. B. Yue, Y. Chun, Y. Cao, X. Dong and J. H. Zhu, *Adv. Funct. Mater.* **16**, 1717–1722 (2006).
39. M. B. Yue, L. B. Sun, Y. Cao, Y. Wang, Z. J. Wang and J. H. Zhu, *Chem. Eur. J.* **14**, 3442–3451 (2008).
40. M. B. Yue, L. B. Sun, Y. Cao, Z. J. Wang, Y. Wang, Q. Yu and J. H. Zhu, *Micropor. Mesopor. Mater.* **114**, 74–81 (2008).
41. M. G. Plaza, C. Pevida, B. Arias, J. Fermoso, A. Arenillas, F. Rubiera and J. J. Pis, *J. Therm. Anal. Calorim.* **92**, 601–606 (2008).
42. A. Sayari and Y. Belmabkhout, *J. Am. Chem. Soc.* **132**, 6312–6314 (2010).
43. S. Dasgupta, A. Nanoti, P. Gupta, D. Jena, A. N. Goswami and M. O. Garg, *Separ. Sci. Technol.* **44**, 3973–3983 (2009).
44. P. D. Jadhav, R. V. Chatti, R. B. Biniwale, N. K. Labhsetwar, S. Devotta and S. S. Rayalu, *Energ. Fuel.* **21**, 3555–3559 (2007).
45. J. C. Fisher, J. Tanthana and S. S. C. Chuang, *AIChE J.* **28**, 589–598 (2009).
46. S. Satyapal, T. Filburn, J. Trela and J. Strange, *Energ. Fuel.* **15**, 250–255 (2001).
47. T. Filburn, J. J. Helble and R. A. Weiss, *Ind. Eng. Chem. Res.* **44**, 1542–1546 (2005).
48. S. Lee, T. P. Filburn, M. Gray, J. W. Park and H. J. Song, *Ind. Eng. Chem. Res.* **47**, 7419–7423 (2008).
49. M. L. Gray, Y. Soong, K. J. Champagne, H. Penniline, J. P. Baltrus, R. W. Stevens, R. Khatri, S. S. C. Chuang and T. Filburn, *Fuel Process. Technol.* **86**, 1449–1455 (2005).
50. M. L. Gray, J. S. Hoffman, D. C. Hreha, D. J. Fauth, S. W. Hedges, K. J. Champagne and H. W. Pennline, *Energ. Fuel.* **23**, 4840–4844 (2009).
51. M. G. Plaza, C. Pevida, B. Arias, J. Fermoso, M. D. Casal, C. F. Martin, F. Rubiera and J. J. Pis, *Fuel* **88**, 2442–2447 (2009).
52. M. G. Plaza, C. Pevida, B. Arias, M. D. Casai, C. F. Martin, J. Fermoso, F. Rubiera and J. J. Pis, *J. Env. Eng.* **135**, 426–431 (2009).
53. T. Tsuda and T. Fujiwara, *J. Chem. Soc. Chem. Commun.* **22**, 1659–1661 (1992).

54. O. Leal, C. Bolivar, C. Ovalles, J. J. Garcia and Y. Espidel, *Inorg. Chim. Acta* **240**, 183–189 (1995).

55. H. Y. Huang, R. T. Yang, D. Chinn and C. L. Munson, *Ind. Eng. Chem. Res.* **42**, 2427–2433 (2003).

56. G. P. Knowles, J. V. Graham, S. W. Delaney and A. L. Chaffee, *Fuel Process. Technol.* **86**, 1435–1448 (2005).

57. G. P. Knowles, S. W. Delaney and A. L. Chaffee, *Ind. Eng. Chem. Res.* **45**, 2626–2633 (2006).

58. P. J. E. Harlick and A. Sayari, *Ind. Eng. Chem. Res.* **45**, 3248–3255 (2006).

59. P. J. E. Harlick and A. Sayari, *Ind. Eng. Chem. Res.* **46**, 446–458 (2007).

60. Y. Belmabkhout and A. Sayari, *Adsorption* **15**, 318–328 (2009).

61. A. Sayari, M. Kruk, M. Jaroniec and I. L. Moudrakovski, *Adv. Mater.* **10**, 1376–1379 (1998).

62. A. Sayari, *Angew. Chem. Int. Ed.* **112**, 3042–3044 (2000).

63. R. Serna-Guerrero, Y. Belmabkhout and A. Sayari, *Chem. Eng. J.* **158**, 513–519 (2010).

64. N. Hiyoshi, K. Yogo and T. Yashima, *Micropor. Mesopor. Mater.* **84**, 357–365 (2005).

65. S. N. Kim, W. J. Son, J. S. Choi and W. S. Ahn, *Micropor. Mesopor. Mater.* **115**, 497–503 (2008).

66. R. Serna-Guerrero, E. Da'na and A. Sayari, *Ind. Eng. Chem. Res.* **47**, 4761–4766 (2008).

67. C. Knofel, J. Descarpenteries, A. Benzaouia, V. Zelenak, S. Mornet, P. L. Llewellyn and V. Hornebecq, *Micropor. Mesopor. Mater.* **99**, 79–85 (2007).

68. L. Wang, L. Ma, A. Wang, Q. Liu and T. Zhang, *Chin. J. Catal.* **28**, 805–810 (2007).

69. J. Wei, J. Shi, H. Pan, W. Zhao, Q. Ye and Y. Shi, *Micropor. Mesopor. Mater.* **116**, 394–399 (2008).

70. V. Zelenak, M. Badanicova, D. Halamova, J. Cejka, A. Zukal, N. Murafa and G. Goerigk, *Chem. Eng. J.* **144**, 336–342 (2008).

71. C. Knofel, C. Martin, V. Hornebecq and P. L. Llewellyn, *J. Phys. Chem. C* **113**, 21726–21734 (2009).

72. C. Lu, F. Su, S. C. Hsu, W. Chen, H. Bai, J. F. Hwang and H. H. Lee, *Fuel Process. Technol.* **90**, 1543–1549 (2009).

73. Z. Liang, B. Fadhel, C. J. Schneider and A. L. Chaffee, *Micropor. Mesopor. Mater.* **111**, 536–543 (2008).

74. Z. Liang, B. Fadhel, C. J. Schneider and A. L. Chaffee, *Adsorption* **15**, 429–437 (2009).

75. C. W. Jones, J. C. Hicks, D. J. Fauth and M. Gray, US Patent Application No. US2007/0149398 (2007).

76. J. C. Hicks, J. D. Drese, D. J. Fauth, M. L. Gray, G. Qi and C. W. Jones, *J. Am. Chem. Soc.* **130**, 2902–2903 (2008).

77. J. H. Drese, S. Choi, R. P. Lively, W. J. Koros, D. J. Fauth, M. L. Gray and C.W. Jones, *Adv. Funct. Mater.* **19**, 3821–3832 (2009).

Nanomaterials that Enhance Sensing/Detection of Environmental Contaminants

Chapter 17

Nanostructured ZnO Gas Sensors

Huamei Shang and Guozhong Cao
Department of Materials Science and Engineering
University of Washington, Seattle, WA, USA

17.1 Introduction

A sensor is a form of transducer which converts a physical or chemical quantity into an electrical, optical, or other measurable quantity. For example, a chemical sensor is intended to determine the composition and concentration of the relevant material via an electrical signal.[1] In particular, gas sensors which can detect and monitor, such as oxygen, flammable gases and toxic gases are needed to protect the environment and to monitor production processes. There are three types of solid-state gas sensors currently in large-scale use. They are based on solid electrolytes, on catalytic combustion, and on resistance modulation of semiconducting oxides.[2] For all sensing applications, the selectivity, sensitivity, and response time are critical parameters. In addition, the compactness, low cost, and easiness of integration of these sensors are all important for accommodating multiple sensors. Among these available gas sensors, semiconducting gas sensors are promising candidates for gas sensing development, given their sensitivity to many gases of interest and the ability to fabricate them readily in many forms, such as single crystals, thin and thick films, and nanostructures, including nanoparticles, nanowires (or nanorods), and nanofilms, which are synthesized by various physical and chemical methods. These nanostructured materials used for gas sensors, in particular, have the advantage of fast response times due to short transport distance and high sensitivity as a result of the large surface area, as well as the potential for miniaturization via integration with IC-based technology, leading to low power consumption.[3]

Various semiconducting materials have been investigated for applications in gas sensors, such as ZnO,[4] SnO$_2$,[5,6] WO$_3$,[7] Al$_2$O$_3$,[8] In$_2$O$_3$,[9,10] SiO$_2$,[11] V$_2$O$_5$,[12] Ga$_2$O$_3$,[13] TiO$_2$,[14] CdS,[15] ThO$_2$,[16] γ-Fe$_2$O$_3$,[17] CO$_3$O$_4$,[18] Ag$_2$O,[19] and MoO$_3$.[20] Among these materials, ZnO has been widely studied and is easily fabricated by both chemical and physical methods. The gas-sensitive behavior of ZnO's electrical conductivity was first reported in 1954 by Heiland.[21] Since then, many fundamental investigations on the gas-sensitive nature of single-crystal and polycrystalline ZnO films have been performed.[22–26] The most recent research has been devoted to nanostructured oxides since reactions at grain boundaries and complete depletion of carriers in the grains can strongly modify the material's transport properties.[27–31]

17.1.1 *Structures of ZnO*

ZnO is a II–VI compound semiconductor with a wide direct band gap of 3.37 eV at room temperature. It has a stable hexagonal wurtzite structure in the space group P6$_3$mc with lattice spacing $a = 0.325$ nm and $c = 0.521$ nm. The wurtzite structure of ZnO is composed of a close-packed lattice of oxygen anions with zinc cations filling interstitial positions. All oxygen anions are tetrahedrally coordinated with zinc occupying one half of the tetrahedral sites in the oxygen HCP lattice, achieving maximum cation separation.[32] Zinc d-orbital electrons hybridize with the oxygen p-orbital electrons to create tetrahedral coordination.[33] Comparison of the X-ray diffraction (XRD) patterns of the nanocrystals with the XRD pattern for bulk wurtzite ZnO suggests that the nanocrystals are formed in the wurtzite phase with similar lattice parameters. However, the XRD patterns of the nanocrystals are considerably broadened due to the very small size of these nanocrystals.[34]

The conduction characteristics of ZnO are primarily dominated by electrons generated by the oxygen vacancies and zinc interstitial atoms.[35] The chemical bonding between zinc and oxygen in ZnO is predominately ionic. Zinc gives away its 4s^2 valence electrons which are in turn accepted by oxygen forming a complete 2p^6 valence shell. If the lattice contains a zinc vacancy, then one oxygen atom is missing two electrons, or two oxygen atoms are each missing one electron. The latter may be considered as two individual $h^{\cdot}$ sites; this condition permits electronic conduction as acceptor states. Therefore, a zinc ionic vacancy acts as an acceptor such that unsatisfied bonds may accept electrons moving under the external potential gradient. Conversely, oxygen vacancies will result in donor behavior as a

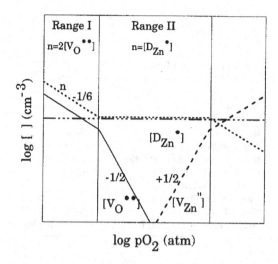

$$\log pO_2 \text{ (atm)}$$

Figure 17.1. The Brouwer diagram of ZnO showing a plateau in electron concentration at intermediate oxygen pressure and an increase in n at reduced pO_2. (Reprinted from G. W. Tomlins, J. L. Routbort and T. O. Mason, *J. Appl. Phys.* **87**, 117 (2000) with permission from AIP.)

result of unbound zinc electrons. The excess valence electrons of a zinc atom may then be thermally excited into the conduction band and considered as donors. This form of intrinsic defect, which leads to n-type conduction, is located approximately 0.01–0.05 eV below the conduction band edge.[36] The Brouwer diagram that best represents the defect structure of ZnO is shown in Figure 17.1.[37] The diagram shows a plateau in electron concentration at intermediate oxygen pressure and an increase in n at reduced pO_2. The plateau is attributed to electroneutrality between electrons and a donor ion, while the upturn in n at low pO_2 is a result of an increase in doubly charged oxygen vacancies, which yields the $-1/6$ slope relationship.

17.1.2 *Properties and applications of ZnO*

ZnO is of great interest for various photonic and electrical applications due to its unique physical and chemical properties, such as a wide band gap (3.37 eV), large exciton binding energy (60 meV) at room temperature, piezoelectricity, and surface chemistry sensitive to the environment.[38] Applications of ZnO include light-emitting diodes,[39] diode lasers,[40] photodiodes,[41] photodetectors,[42] optical modulator waveguides,[43]

Table 17.1.   Physical properties of wurtzite ZnO.

Properties	Value
Lattice constants (T $=$ 300 K)	
a_0	0.32469 nm
c_0	0.52069 nm
Density	5.606 g/cm^3
Melting point	2248 K
Relative dielectric constant	8.66
Gap energy	3.4 eV, direct
Intrinsic carrier concentration	$<10^6$ cm^{-3}
Exciton binding energy	60 meV
Electron effective mass	0.24
Electron mobility (T $=$ 300 K)	200 cm^2/Vs
Hole effective mass	0.59
Hole mobility (T $=$ 300 K)	5–50 cm^2/Vs

photovoltaic cells,[44] phosphor,[45] varistor,[46] data storage,[47] and biochemical sensors.[48] Nanostructured ZnO, nanorods or nanowires in particular, has attracted intensive research, primarily for their large surface area for applications relying on heterogeneous reactions such as sensors and detectors,[47] their light confinement for nanolasers,[40] and their enhanced freedom in lateral dimensions for more sensitive piezoelectric devices.[49]

Table 17.1 lists the basic physical properties of bulk ZnO.[50] It is worth noting that as the dimension of the semiconductor materials continuously shrinks down to nanometer or even smaller scale, some of their physical properties undergo changes known as the "quantum size effects". For example, quantum confinement increases the band gap energy of quasi-one-dimensional (Q1D) ZnO, which has been confirmed by photoluminescence.[51] The band gap of ZnO nanoparticles also demonstrates such size dependence.[52] X-ray absorption spectroscopy and scanning photoelectron microscopy reveals the enhancement of surface states with the downsizing of ZnO nanorods.[53] In addition, the carrier concentration in Q1D systems can be significantly affected by the surface states, as suggested from nanowire chemical sensing studies.[54–57] Although the increase of the band gap in nanostructures is not favorable in most sensor applications, the change of surface states can have significant impacts on the sensing properties. Understanding the fundamental physical properties is crucial to the rational design of functional devices. Investigation of the properties of individual ZnO nanostructures is essential for developing their potential as the building blocks for future nanoscale devices. This chapter will review the up-to-date research progress in the development of ZnO nanostructures and their applications in gas sensing.

17.2 Synthesis of Nanostructured ZnO

ZnO nanostructures have attracted considerable attention for solid-state gas sensors with great potential for overcoming fundamental limitations due to their ultrahigh surface-to-volume ratio. There are a lot of nanostructures, such as nanoparticles, nanorods, nanowires, nanobelts, nanotubes, and nanofilms, that can be used for gas sensors. Various fabrication techniques have been established for the growth of ZnO nanostructures. Aqueous solution growth,[58] electrochemical deposition,[59–61] vapor-liquid-solid (VLS) growth,[40] evaporation-condensation,[62] chemical vapor deposition,[63] carbothermal evaporation,[64] flux growth,[65] and template-based synthesis[66] have all been reported to successfully grow ZnO nanostructures.

The early nanostructure of ZnO used for gas sensors is nanoparticles. The higher activity and fast response of the nanoparticle-based sensors due to the enhancement of surface area at nanodimensions coupled with the possible low fabrication cost, ease of miniaturization, and the compatibility with microelectronic circuitry have provided a renewed interest in the gas sensing properties of metal oxide semiconductors.[67] It has been shown that particle sizes below 20 nm lead to higher values of sensitivity and a more rapid response.[68,69] This has been explained not only by the increase of the specific reactive surface but also a complete depletion of the semiconductors as nanoparticle size approaches the thickness of the space-charge layer.[70] Another possibility is linked to the fact that the density of surface states induced by the chemisorbed oxygen species decreases with decreasing particle size, therefore leading to a lesser degree of Fermi-level pinning. Less Fermi-level pinning means that the surface barrier, and therefore the overall resistance, can undergo larger variations.[71] For fabricating nanoparticle-based gas sensors with superior characteristics and for understanding the size dependence of a gas sensing mechanism, the use of ZnO layers having a well-defined nanoparticle size is a primary requirement.[72] The size and uniformity of ZnO nanoparticles are closely related to sensor characteristics, so in practical sensor production the calcination process should be well-controlled to obtain uniform characteristics. The prevention of excess crystal growth is extremely important for achieving higher sensitivity and quick response of gas sensors. ZnO nanoparticles have been synthesized by various methods, including vapor decomposition, precipitation, and thermal decomposition.[73–78]

ZnO thin and thick films have also been investigated to be used for gas sensors, and they have been grown by many methods both in single crystal and polycrystalline forms. The methods can be chemical vapor deposition,[79] electrochemical deposition,[80] pulse laser deposition (PLD),[81–83] atomic layer epitaxy (ALE) or molecular beam epitaxy

(MBE),[84-86] and metal organic chemical vapor epitaxy (MOVPE)[87-89] and so on. The microstructural and physical properties of ZnO films can be modified by introducing changes into the procedure of its synthesis process. Especially for the application of ZnO as gas sensors, a porous microstructure of the materials with controlled pore size is preferred. The sensitivity and response time of ZnO-based sensors strongly depend on the porosity of the films. The grain size in the films also has a noticeable effect on its gas sensing properties. A study of the influence of thickness on the ppm level has also been done. It is concluded that thinner films of about 100 nm with low carrier concentration have better gas sensitivity than thicker films with higher carrier concentration. The reason is that gas sensitivity is highest when the depletion region generated by chemisorbed oxygen extends through the sensor.[90] It should be noticed that many metal ions, such as Sn, Fe, Cu, In, and Al,[91,92] are doped in ZnO films to modify the microstructure and surface morphology as well as to increase the conductivity and result in enhanced sensitivity. The microstructures are changed from non-oriented growth, for undoped films, to strongly (002) oriented, at intermediate ($\sim$1 at.%) doping levels; and finally again to non-oriented and poor crystallinity, at high (>3 at.%) doping levels. The sensitivity of the films was studied in two steps: first as a function of their temperature (435–675 K) for a fixed concentration and second as a function of concentration (4–100 ppm) for a fixed temperature (675 K). A better sensitivity can be observed for Sn- and Al-doped films.

Furthermore, some mesoporous structures have also been investigated to be used for gas sensing either in the form of mesoporous films or by forming mesopores in ZnO nanorods. Mesoporous ZnO with pore sizes from 2 nm to 50 nm are studied for potential applications in gas sensing.[93,94] Normally, mesoporous structures are composed of amorphous materials and there are few reports of mesoporous structures based on crystalline materials. However, more and more research work on ZnO mesoporous materials has been conducted owing to their huge surface-to-volume ratio for potential applications in solar cells, photocatalysis, and gas sensing.[95,96] Moreover, the synthesis of crystalline ZnO one-dimensional nanomaterials with a mesoporous structure and well-defined size is developed, and it is expected that the mesoporous structure can increase the sensitivity of ZnO gas sensors greatly by increasing the surface-to-volume ratio.[97,98]

The most widely studied nanostructures are one-dimensional ZnO, i.e. nanowires, nanorods, nanotubes, and nanobelts. From the aspect of sensing performance, one-dimensional ZnO is expected to be superior to its thin-film counterpart.[56] Since their diameter is small and comparable to the Debye length, chemisorption-induced surface states effectively affect the electronic structure of the entire channel, thus conferring one-dimensional

ZnO with higher sensitivity than thin-film ZnO. Well-aligned arrays of ZnO nanorods were grown by the vapor-phase process at high temperature on single-crystal substrates such as Si,[99] GaN,[100] and sapphire,[38] which have crystallographic similarity to ZnO. This method places limitations on the process of scaling up because of the expensive single-crystal substrates and high processing temperature. Aligned arrays of [001] ZnO nanorods on glass and silicon substrates have also been readily grown from aqueous solution with nanocrystal seeding;[101] the alignment of nanorods was achieved by evolution selection growth, i.e. the crystal orientation with the higher growth rate and perpendicular to the substrate surface will survive and continue to grow.[102] Well-aligned [001] ZnO nanorod arrays are also synthesized on ITO substrates from aqueous solution by electrochemical deposition or using a two-step growth process with electric-field-assisted nucleation. By using the two-step process, a thin layer of ZnO on the ITO substrates is first grown by electrochemical deposition, and ZnO nanorod arrays are subsequently grown with electrochemical deposited ZnO thin layer as substrate by the spontaneous growth method. A low-temperature hydrothermal method has also been demonstrated to grow high-quality ZnO nanowire arrays.[103] In the following section we will focus our discussion on the synthesis of nanorods and nanowires of ZnO.

17.3 Synthesis of ZnO Nanorods or Nanowires

17.3.1 *Solution-based spontaneous growth*

Two solution-based approaches have been investigated in the literature, one with and the other without the use of a template. Spontaneous growth is a process driven by the reduction of the Gibbs free energy or chemical potential. The reduction of the Gibbs free energy is commonly realized by phase transformation or chemical reaction or the release of stress. For the formation of nanowires or nanorods, anisotropic growth is required, i.e. the crystal grows along a certain orientation faster than in other directions. Uniformly sized nanowires, i.e. with the same diameter along the longitudinal direction of a given nanowire, can be obtained when crystal growth proceeds along one direction, while no growth occurs along other directions. In spontaneous growth, for a given material and growth conditions, defects and impurities on the growth surfaces can play a significant role in determining the morphology of the final products.

According to the classic theory of nucleation and growth,[104] the free energy of forming stable nuclei on a substrate is determined by four factors: the degree of supersaturation S, the interfacial energy between the particle

(c) and the liquid (l) σ_{cl}, the interfacial energy between the particle and the substrate (s) σ_{cs}, and the interfacial energy between the substrate and the liquid σ_{sl}:

$$\Delta G = -RT \ln S + \sigma_{cl} + (\sigma_{cs} - \sigma_{sl})A_{cs} \qquad (17.1)$$

where A_{cs} is the surface area of the particle.

Figure 17.2 is a schematic plot of the number of nuclei (N) as a function of degree of supersaturation (S), which is related to the concentration of the precursor and the solubility in the solution. Figure 17.2 suggests several regions for crystal growth. At a high concentration or a high temperature, homogeneous nucleation is dominant and precipitation in the bulk solution is the main mechanism. This region should be avoided if controlled crystal growth is desired. The region slightly above the solubility line is the heterogeneous nucleation region. In this region heterogeneous

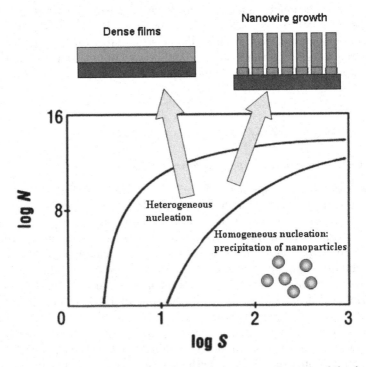

Figure 17.2. Number of nuclei generated in the solution as a function of the degree of supersaturation. The region for growing oriented nanostructured films and 1DNMs is indicated. (Reprinted from B. C. Bunker, P. C. Rieke, B. J. Tarasevich, A. A. Campbell, G. E. Fryxell, G. L. Graff, L. Song, J. Liu and J. W. Virden, *Science* **264**, 48 (1994) with permission from the American Association for the Advancement of Science.)

nucleation on a substrate dominates and therefore it is possible to grow uniform nanostructured films and oriented one-dimensional nanomaterials (1DNMs). Unfortunately, most solution synthesis is carried out at too high concentration so that undesirable precipitation dominates. As a result, oriented nanostructures were difficult to form. From Equation (17.1) and Figure 17.2, we can use the following rules to design a generalized solution synthesis method for oriented nanostructures:

(1) *Controlling the solubility of the precursors and the degree of supersaturation so that massive precipitation is not the dominating reaction.* Similar to VLS growth, the degree of supersaturation has a large effect on the growth behavior. A low degree of supersaturation promotes heterogeneous growth of 1DNMs, while a high degree of supersaturation favors bulk growth in the solution. Experimentally this is accomplished by reducing the reaction temperature, and by reducing the precursor concentrations as much as possible, while at the same time ensuring nucleation and growth can still take place. The formation of the new materials is characterized by the increase in cloudiness of the solution, which can be monitored by light scattering or turbidity measurement. A rapid increase in cloudiness is an indication of rapid precipitation and should be avoided.

(2) *Reducing the interfacial energy between the substrate and the particle.* In most cases, the activation energy for crystal growth on a substrate is lower than that required to create new nuclei from the bulk solution. Still, chemical methods can be used to further lower the interfacial energy. For example, self-assembled monolayers containing surface active groups were used to promote heterogeneous nucleation. The self-assembled monolayers have at least two functions: reducing the surface tension, and stimulating the nucleation of specific crystalline phases. A biomimetic approach was developed to prepare oriented nanostructured ceramic films.[104]

(3) *Controlling crystal growth.* Most crystalline materials, such as ZnO, have anisotropic crystalline structures and specific growth habits. If the growth along certain directions is much faster than in other directions, nanowires or nanorods can be produced. During crystal growth, different facets in a given crystal have different atomic density, and atoms on different facets have a different number of unsatisfied bonds (also referred to as broken or dangling bonds), leading to different surface energy. According to the periodic bond chain (PBC) theory developed by Hartman and Perdok,[105] crystal facets can be categorized into three groups based on the number of broken periodic bond chains on a given facet: flat surface, stepped surface, and kinked surface. The number of broken periodic bond chains can be understood as the number of broken bonds per atom on a given facet.

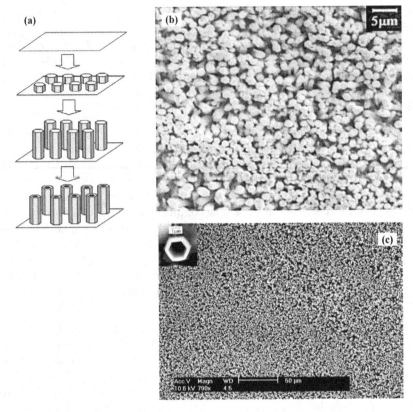

Figure 17.3. ZnO nanostructures from solution synthesis. (a) Schematics of growing oriented rods and tubes on a substrate. (b) Oriented ZnO microrods. (c) Oriented ZnO microtubes after aging. (Reprinted from L. Vayssieres, K. Keis, S.-E. Lindquist and A. Hagfeldt, *J. Phys. Chem.* **B105**, 3350 (2001) and L. Vayssieres, K. Keis, A. Hagfeldt and S.-E. Linquist, *Chem. Mater.* **13**, 4395 (2001) with permission from ACS.)

Vayssieres *et al.* first introduced the concept of purpose-built materials (Figure 17.3(a)) for growing oriented 1DNMs from solutions.[106–108] The same solution approach was extended to ZnO by decomposing the Zn^{2+} amino complex.[107] By heating an equimolar zinc nitrate (0.1 M) and methenamine solution at 95°C, large arrays of oriented ZnO rods, about 1–2 μm in diameter, on various substrates, were produced (Figure 17.3(b)). Longer reaction times led to preferred dissolution of the ZnO rods on the metastable (001) polar surfaces, and produced hollow hexagonal microtubes (Figure 17.3(c)).[108]

Although the unseeded growth is successful in preparing a range of oriented 1DNMs, there is a need to better control the size, to reduce the dimension, to broaden the applicability of the solution-based approaches, and to control the density and the spatial distribution of the 1DNMs. A newly developed seeded growth has the promise to meet these challenges. In the seeded approach, nanoparticles are first placed on the substrates. Such nanoparticles are widely available through commercial sources and even can be readily prepared using techniques reported in the literature. The crystal growth is carried out under mild conditions (low temperature and dilute concentration of the salt). Under these conditions, homogenous nucleation of new nuclei from the bulk solution is not favored and heterogeneous nucleation is prompted. Because the nanoparticles are the same as the materials to be grown, the low activation energy favors the epitaxial growth of the new 1DNMs from the existing seeds. This approach avoids the difficulty in separating the nucleation and growth steps. The size of the seeds and the density will to a large extent determine the size and the population density of the 1DNMs. The seeds can be deposited on the substrate using many mature techniques, such as dip coating, electrophoretic deposition, and stamping, therefore making it possible to micropattern the 1DNMs for device applications. Furthermore, the surface characteristics of the substrate have a large effect on the interfacial energy and nucleation process. These effects are still not well understood despite extensive investigation. The use the seeds bypasses such complications and ensures that the materials produced are reliable and reproducible on different substrates.

Seeded growth was applied to prepare large arrays of ZnO nanowires.[109–111] Uniform films up to wafer size were reported.[111] Three steps for growing oriented ZnO nanorods (Figure 17.4(a)) were revealed[112]: (1) deposition of crystal seeds on the substrate surface, (2) growth of randomly oriented crystals from the seeds, and (3) growth of aligned nanorods after extended reactions. In the early stages of growth, ZnO crystals grew along the fastest growth orientation, the $\langle 001 \rangle$ direction, and these crystals were not aligned. Figure 17.4(b) shows a scanning electron microscopy (SEM) image of the ZnO nanoparticle seeds after 30 minutes of crystal growth, with little sign of crystal growth. These particles mostly consist of rounded rectangular- and rod-shaped crystals with a wide size distribution. After one hour (Figure 17.4(c)) of growth, surface roughness became visible on the ZnO seeds, indicating the initiation of crystal growth. After three hours, short and faceted hexagonal rods were observed (Figure 17.4(d)), although most of these rods were not well aligned yet. However, as the rod-like crystals grew further, randomly oriented crystals began to overlap and their growth became physically limited as the misaligned nanorods began to impinge on neighboring crystals, giving rise to the preferred orientation of the film. Figure 17.4(e) confirms that the

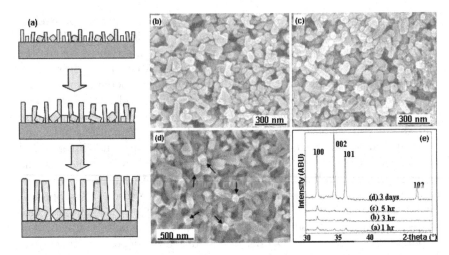

Figure 17.4. Mechanisms of growing oriented 1DNMs with seeded growth. (a) Illustration of growth-controlled alignment of ZnO nanowires. (b) ZnO seed morphology after 30 minutes of growth. (c) ZnO roughening after 1 hour of growth. (d) Initial nanorod growth after 3 hours. (e) XRD patterns at different growth stages, showing alignment at a later stage. (Reprinted from Z. R. Tian, J. A. Voigt, J. Liu, B. McKenzie, M. J. McDermott, R. T. Cygan and L. J. Criscenti, *Nature Mater.* **2**, 821 (2003) with permission from Nature Publishing Group.)

ZnO crystals have a hexagonal wurtzite structure (P6$_3$mc, $a = 3.2495\,\text{Å}$, $c = 5.2069\,\text{Å}$). The results from one to five hours are characteristic of randomly oriented ZnO powders, showing the (100) and (101) reflections as the main peaks. The (002) reflection was only significantly enhanced after extended growth, indicating that the ZnO crystals were mostly randomly oriented at the beginning, but became ⟨001⟩ oriented after a long time of growth.

Figure 17.5(a) shows large arrays of oriented ZnO nanorods formed in a 30 mL aqueous solution of 0.01–0.06 M hexamethylenetetramine (HMT) and 0.01–0.06 M Zn(NO$_3$)$_2$ at 60°C. The nanowires in uniformed arrays have a diameter of ∼250 nm. The SEM images of the tilted samples suggest that these nanowires are about 3 μm long and all stand up on the substrate (Figure 17.5(b)). A control study using no seeds showed that the same synthetic conditions produced randomly oriented rods sporadically scattered on the substrates, about 1 μm in diameter and 10 μm in length (Figure 17.5(c)), or about three to four times that from the seeded growth. The seeded route is a straightforward and reliable method for the synthesis of oriented arrays of nanowires. This method can be easily applied to make patterned nanowires, as shown in Figure 17.5(d). Here, the ZnO seeds were

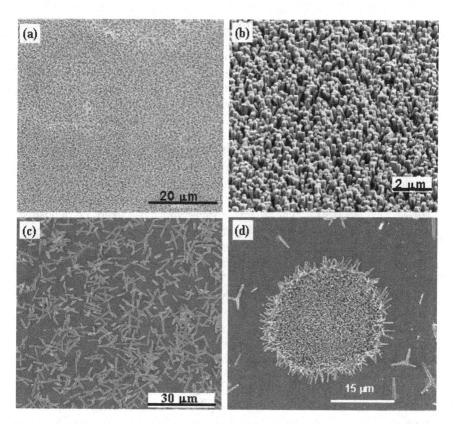

Figure 17.5. Large arrays of oriented and patterned ZnO nanowires using seeded growth. (a) A low-magnification SEM image of large arrays of ZnO. (b) An SEM image of a tilted sample showing good alignment. (c) ZnO microrods without seeds. (d) Preferred ZnO nanorod growth with stamped circular pattern of ZnO nanoseeds. (Reprinted from Z. R. Tian, J. A. Voigt, J. Liu, B. McKenzie, M. J. McDermott, R. T. Cygan and L. J. Criscenti, *Nature Mater.* **2**, 821 (2003) with permission from Nature Publishing Group).

stamped onto a limited area using a ZnO nanocrystal suspension as the ink. The nanowires were mostly formed within this defined area.

17.3.2 *Electrochemical deposition (ECD)*

Electrochemical deposition (ECD), also known as electrodeposition, has also been widely investigated to synthesize ZnO films and nanorod arrays.[113–115] ECD can be understood as a special electrolysis resulting in the deposition of solid material on an electrode. This process involves (1) oriented diffusion of a charged growth species (typically positively charged cations) through a solution when an external electric field is applied, and (2) reduction of the charged growth species at the growth or deposition surface which also serves as an electrode. In general, electrochemical deposition is only applicable to electrical conductive materials such as metals, alloys, semiconductors, and electrical conductive polymers, since after the initial deposition, the electrode is separated from the depositing solution by the deposit and the electrical current must go through the deposit to allow the deposition process to continue. Once little fluctuation yields the formation of small rods, the growth of rods or wires will continue, since the electric field and the density of current lines between the tips of nanowires and the opposing electrode are greater, due to a shorter distance, than that between two electrodes. The growth species will more likely be deposited onto the tips of nanowires, resulting in continued growth.

Epitaxial zinc oxide films were prepared on gallium nitride (0002) substrates by cathodic electrodeposition in an aqueous solution containing zinc salt and dissolved oxygen at 85°C.[114] The films have a hexagonal structure with the c-axis parallel to that of GaN and the [100] direction in ZnO parallel to the [100] direction in GaN in the (0002) basal plane. The crystallographic quality appears to be very good, especially when considering the low temperature deposition and the relatively high growth rate conditions. Such results are related to the growth in the solution environment which, in comparison to the vapor phase, offers additional possibilities for atomic organization during deposition due to easy ionic exchanges between the surface and the solution through near-equilibrium dissolution/precipitation processes.

Cao et $al.$ reported a soft and template-free electrochemical deposition method for preparing wafer-scale ZnO nanoneedle arrays on an oriented gold-film-coated silicon substrate.[115] It has been shown that the ZnO nanoneedles possess a single-crystal wurtzite structure and grow along the c-axis perpendicularly on the substrate. The {111}-oriented Au film/Si substrate results in the formation of {0001}-oriented ZnO nuclei on the film due to the small lattice mismatch between them. The oriented ZnO nuclei serve as seeds and grow preferentially along the c-axis due to the high surface free energy of the {0001} polar plane, leading to the formation of the ZnO nanoneedle array perpendicular to the substrate. This ZnO nanoneedle array is of good crystal quality.

ECD and solution-based spontaneous growth were also combined to grow well-oriented ZnO nanorod arrays.[116] In this method, [001] ZnO nanorod arrays on ITO substrates were fabricated by a two-step process: seeding and subsequent growth. First, ITO substrates were placed in a 0.1 M zinc nitrate aqueous solution for initial growth or deposition. An external

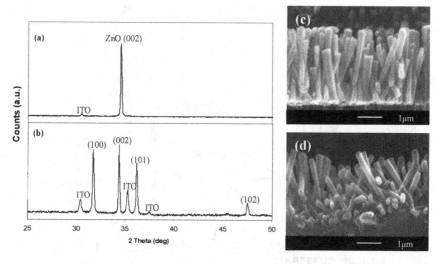

Figure 17.6. XRD and SEM images of ZnO nanorod arrays grown from ECD-assisted nucleation growth. (a) XRD of ZnO nanorods grown with ECD-assisted nucleation. (b) XRD of random nanorods grown without ECD-assisted nucleation. (c) SEM cross-section of ZnO nanorods grown with ECD-assisted nucleation. (d) SEM cross-section of ZnO nanorods grown without ECD-assisted nucleation. (Reprinted from Y. J. Kim, H. M. Shang and G. Z. Cao, *Proceedings of Materials Research Society Symposium* **879E**, Z4.1.1–Z4.1.6 (2005) with permission from MRS.)

electric potential of 1.2 V was applied to the ITOs substrates, as a cathode, with a platinum plate as an anode for 20 minutes. The distance between two electrodes was kept at 1.5 mm. The ITO substrates with ZnO deposits were subsequently heat-treated at 500°C for 30 minutes in air. The ITO substrates with initial ZnO deposits after heat treatment were placed in a mixture solution of 0.015 M zinc nitrate and 0.022 M methenamine at 60°C for 40 hours. $C_6H_{12}N_4$ is a growth-directing agent widely used for the synthesis of ZnO nanorods.[58,61] Figure 17.6 compares XRD spectra and SEM images of ZnO nanorods grown with and without ECD-assisted nucleation. One critical issue of this method is that the nucleation density of ECD deposition will affect the subsequent ZnO rod growth and alignment.[117]

17.3.3 *Vapor-liquid-solid (VLS) growth*

In VLS growth, a second-phase material, commonly referred to as either the impurity or catalyst, is purposely introduced to direct and confine the crystal growth in a specific orientation and within a confined area. The catalyst

H. Shang and G. Cao

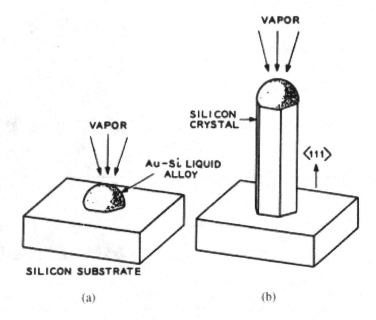

Figure 17.7. Schematic showing the principal steps of the vapor-liquid-solid growth technique: (a) initial nucleation and (b) continued growth.

forms a liquid droplet by itself or by alloying with growth material during the growth, which acts as a trap for the growth species. The enriched growth species in the catalyst droplets subsequently precipitates at the growth surface, resulting in one-directional growth. Wagner *et al.*[118-120] first proposed the VLS theory over 40 years ago to explain the experimental results and observations in the growth of silicon nanowires, and Givargizov[121] further elaborated the experimental observations and models and theories developed regarding the VLS process.

In VLS growth, the process can be simply described as sketched in Figure 17.7. The growth species is evaporated first, and then diffused and dissolved into a liquid droplet. The surface of the liquid has a large accommodation coefficient, and is therefore a preferred site for deposition. Saturated growth species in the liquid droplet will diffuse to and precipitate at the interface between the substrate and the liquid. The precipitation will follow first nucleation and then crystal growth. Continued precipitation or growth will separate the substrate and the liquid droplet, resulting in the growth of nanowires. For a perfect or an imperfect crystal surface, an impinging growth species diffuses along the surface. During the diffusion, the growth species may be irreversibly incorporated into the growth site (ledge, ledge-kink, or kink). If the growth species does not find a preferential site in a given period of time (the residence time), it will escape back to the vapor phase. A liquid surface is distinctly different from a perfect or imperfect crystal surface, and can be considered as a rough surface. A rough surface is composed of only ledge, ledge-kink, or kink sites. As such, every site over the entire surface acts as a trap for the impinging growth species. The accommodation coefficient is unity. Consequently, the growth rate of the nanowires or nanorods by the VLS method is much higher than that without the liquid catalyst. The enhanced growth rate can also be partly due to the fact that the condensation surface area for the growth species in VLS growth is larger than the surface area of the crystal growth. While the growth surface is the interface between the liquid droplet and the solid surface, the condensation surface is the interface between the liquid droplet and the vapor phase. Depending on the contact angle, the liquid surface area can be several times that of the growth surface.

The size of nanowires grown by the VLS method is solely determined by the size of the liquid catalyst droplets. To grow thinner nanowires, one can simply reduce the size of the liquid droplets. A typical method used to form small liquid catalyst droplets is to coat a thin layer of catalyst on the growth substrate and to anneal at elevated temperatures.[122] During annealing, the catalyst reacts with the substrate to form a eutectic liquid and further balls up to reduce the overall surface energy. Au as a catalyst and silicon as a substrate is a typical example. The size of the liquid catalyst droplets can be controlled by the thickness of the catalyst film on the substrate. In general, a thinner film forms smaller droplets, giving smaller diameters of nanowires subsequently grown.

ZnO nanowires have been grown on Au-coated (thickness ranging from 2 to 50 nm) silicon substrates by heating a 1:1 mixture of ZnO and graphite powder to 900–925°C under a constant flow of argon for 5–30 minutes.[123] The grown ZnO nanowires vary with the thickness of the initial Au coatings as shown in Figure 17.8. For a 50 Å Au coating, the diameters of the nanowires are normally 80–120 nm and their lengths are 10–20 μm. Thinner nanowires of 40–70 nm with lengths of 5–10 μm were grown on 30 Å Au-coated substrates. The grown ZnO nanowires are single-crystal with a preferential growth direction of $\langle 001 \rangle$. The growth process of ZnO is believed to be different from that of elementary nanowires. The process involves the reduction of ZnO by graphite to form Zn and CO vapor at high temperatures (above 900°C). The Zn vapor is transported to and reacted with the Au catalyst, which would have already reacted with silicon to form a eutectic Au-Si liquid on silicon substrates, located downstream at a lower temperature to form Zu–Au–Si alloy droplets. As the droplets become supersaturated with Zn, crystalline ZnO nanowires are formed, possibly through the reaction between Zn and CO at a lower temperature. This process can

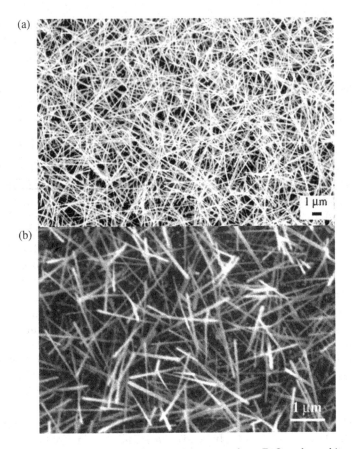

Figure 17.8. (a) SEM image of ZnO nanowires grown from ZnO and graphite powder in an argon flow on the surface of a silicon substrate coated with ∼50 Å thick Au film. (b) SEM image of ZnO nanowires grown under the same conditions as above but using a substrate coated with 30 Å thick Au film. (Reprinted from M. H. Huang, Y. Wu, H. Feick, N. Tran, E. Weber and P. Yang, *Adv. Mater.* **13**, 113 (2001) with permission from John Wiley and Sons.)

be easily understood by the fact that the reaction:

$$ZnO + C \leftrightarrow Zn + CO \qquad (17.2)$$

is reversible at temperatures around 900°C.[124] Although the presence of a small amount of CO is not expected to change the phase diagram significantly, no ZnO nanowires were grown on substrates in the absence of graphite.

ZnO nanowire growth through the VLS mechanism also allows people to pattern these wires into a network.[123] Initial patterning experiments

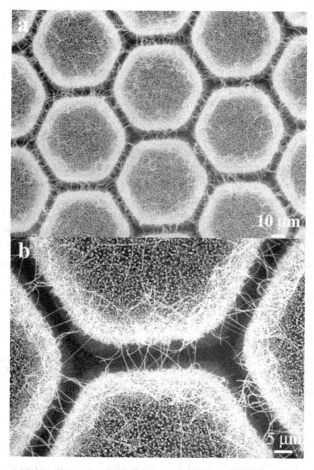

Figure 17.9. (a) SEM image of ZnO nanowire networks grown from the patterned Au islands. (b) SEM image of the same sample at a higher magnification. (Reprinted from M. H. Huang, Y. Wu, H. Feick, N. Tran, E. Weber and P. Yang, *Adv. Mater.* **13**, 113 (2001) with permission from John Wiley and Sons.)

used a 300 mesh copper grid with a hexagonal pattern as a mask during the Au deposition. Most of the wires actually bridge the neighboring metal hexagons and form an intricate network. Figure 17.9 shows SEM images of ZnO nanowire patterns. The nanowires at the edge have a diameter of about 50–200 nm, and can be over 50 μm long. This controlled growth of nanowires to form certain patterns and the eventual control of wire growth at certain locations may have implications for potential applications in nanoscale devices.

17.3.4 *Evaporation-condensation growth*

The evaporation-condensation process is also referred to as a vapor-solid (VS) process. Chemical reactions among various precursors may be involved to produce desired materials. Nanowires and nanorods grown by evaporation-condensation methods are commonly single-crystal with fewer imperfections. The formation of nanowires, nanorods, or nanotubules through evaporation- (or dissolution-) condensation is due to the anisotropic growth. Chemical reactions and the formation of intermediate compounds play important roles in the synthesis of various nanowires by evaporation-condensation methods. Reduction reactions are often used to generate volatile deposition precursors and hydrogen, water, and carbon are commonly used as reduction agents.

Sears[125,126] was the first who demonstrated and explained the growth of nanowires by the evaporation-condensation method. Recently, Wang and co-workers[127] reported the growth of single-crystal nanobelts of various semiconducting oxides simply by evaporating the desired commercially available metal oxides at high temperatures under a vacuum of 300 torr and condensing on an alumina substrate, placed inside the same alumina tube furnace, at relatively lower temperatures. The oxides include zinc oxide of wurtzite hexagonal crystal structure, tin oxide of rutile structure, indium oxide with C-rare-earth crystal structure, and cadmium oxide with NaCl cubic structure. We will just focus on the growth of ZnO nanobelts to illustrate their findings. Figure 17.10 shows the SEM and TEM pictures of ZnO nanobelts.[127] The typical thickness and width-to-thickness ratios of the ZnO nanobelts are in the range of 10–30 nm and 5–10, respectively. Two growth directions were observed: [0001] and [0110]. No screw dislocation was found throughout the entire length of the nanobelt, except a single stacking fault parallel to the growth axis in the nanobelts grown along the [0110] direction. The surfaces of the nanobelts are clean, atomically sharp, and free of any sheathed amorphous phase. A further TEM analysis also revealed the absence of amorphous globules on the tips of nanobelts. These observations imply that the growth of nanobelts cannot be attributed to either screw-dislocation-induced anisotropic growth, or impurity-inhibited growth, and is not due to the VLS mechanism. It seems worthwhile to note that the shape of nanowires and nanobelts may also depend on the growth temperature.

17.3.5 *Chemical vapor deposition (CVD)*

CVD is the process of chemically reacting a volatile compound of a material to be deposited, with other gases, to produce a non-volatile solid

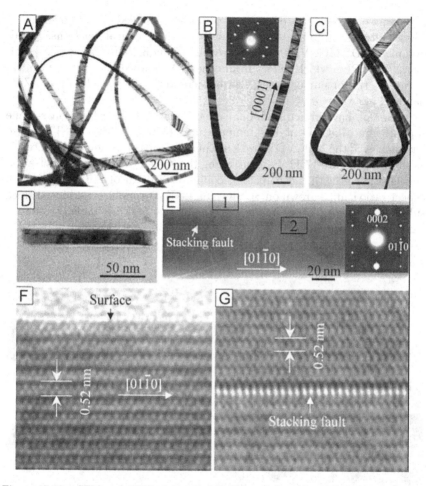

Figure 17.10. SEM and TEM pictures of ZnO nanobelts. (A to C) TEM images of several straight and twisted ZnO nanobelts, displaying the shape characteristics of the belts. (D) Cross-sectional TEM image of a ZnO nanobelt, showing a rectangle-like cross-section. (E) TEM image of a nanobelt growing along [010], showing only one stacking fault present in the nanobelt. (F) HRTEM image from box 1 in (E) showing a clean and structurally perfect surface. (G) HRTEM image from box 2 in (E), showing the stacking fault. (Reprinted from Z. W. Pan, Z. R. Dai and Z. L. Wang, *Science* **291**, 1947 (2001) with permission from the American Association for the Advancement of Science.)

that deposits atomistically on a suitably placed substrate.[128] The CVD process has been very extensively studied and very well documented[129-131] largely due to the close association with solid-state microelectronics. A variety of CVD methods have been developed, depending on the types of precursors used, the deposition conditions applied, and the forms of energy introduced to the system to activate the chemical reactions. Metalorganic CVD (MOCVD), also known as organometallic vapor-phase epitaxy (OMVPE), which differs from other CVD processes by the chemical nature of the precursor gases, has been recently developed to grow ZnO nanorod arrays.

Lu *et al.* have grown high-quality epitaxial ZnO thin films on *r*-plane sapphire by MOCVD at low temperatures (350°C–500°C).[132] They also have achieved the MOCVD growth of ZnO nanotips on various substrates[133] as shown in Figure 17.11. By controlling the ZnO initial growth (nucleation versus other growth mechanisms), columnar growth of ZnO with a high aspect ratio can be grown on Si, SiO_2/Si, GaN *c*-sappphire, and fused silica substrates. Nanotip arrays made with conductive ZnO are single-crystalline, with uniform size and orientation, and show good optical quality. Initially,

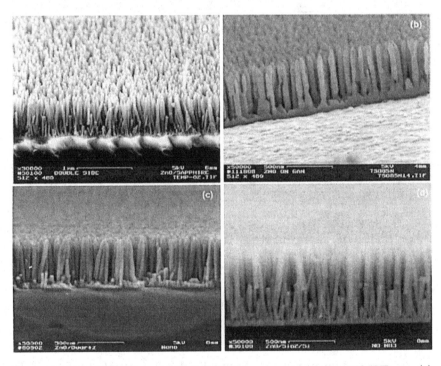

Figure 17.11. SEM images of ZnO nanorod arrays grown by MOCVD on (a) Al_2O_3, (b) epi GaN, (c) fused silica, and (d) SiO_2/Si substrates. (Reprinted from S. Muthukumar, H. Sheng, J. Zhong, Z. Zhang, N. W. Emanaetoglu and Y. Lu, *IEEE Trans. Nanotech.* **2**, 50 (2003) with permission from IEEE.)

they demonstrated the integration of epitaxial ZnO layer and ZnO nanotip arrays on the same SOS (silicon-on-sapphire) chip using the selective growth technology.

Yi and co-workers used catalyst-free MOCVD for growing ZnO nanorod and nanoneedle arrays.[63,134] In this method, no catalyst is employed for ZnO nanorod formation, which leads to preparation of high-purity ZnO nanorods and easy fabrications of nanorod structures. After this research, related research of ZnO nanorod growth via MOCVD was reported by other groups.[135-142]

17.3.6 *Other methods*

ZnO nanostructures have been also synthesized by many other methods, and all of the methods have provided the possibility of synthesizing well-aligned ZnO nanorod arrays with large surface areas to be used for gas sensing and other applications. Heo *et al.* reported on catalyst-driven molecular beam epitaxy (MBE) of ZnO nanorods.[143] Okada *et al.* succeeded in synthesizing ZnO nanorods by pulsed-laser deposition (PLD) without using any catalysts.[144] A novel and simple one-step solid-state reaction in the presence of a suitable surfactant, triethanolamine (TEA), has been developed to synthesize uniform zinc oxide nanorods with an average diameter of 20 nm and length of 200 nm.[145] High-density, well-aligned wurtzite ZnO nanorod arrays with significantly different tip shapes were controllably fabricated by evaporating zinc powders at different temperatures.[146] Well-aligned single-crystalline wurzite ZnO nanowire arrays were successfully fabricated on an Al_2O_3 substrate by a simple physical vapor-deposition method at a low temperature of 450°C.[147] Wang *et al.* used electrophoretic deposition to form nanorods of ZnO from colloidal sols.[148] A template-based method was also combined with other methods, such as, electrodeposition, sol-gel, and polymer-assisted growth, to fabricate well-aligned ZnO nanorod arrays.[149,150] There are many other methods that can be used to fabricate various ZnO nanostructures but are not included here. More details on ZnO fabrication can be found in some good books and review papers published recently.[27,55,96,151-153]

17.4 Nanostructured ZnO for Gas Sensing

The rapid development of modern industries raises a variety of serious environmental problems, for example, the release of various chemical pollutants, such as NO_x, SO_x, CO_2, volatile organic compounds (VOCs), and fluorocarbon, from industry, automobiles, and homes, into the atmosphere, resulting in global environmental issues, such as acid rain, the greenhouse effect, sick house syndrome, and ozone depletion. This atmospheric pollution can cause major disasters within a short period of time, since this type of pollution can diffuse rapidly over large areas. Since the kinds and quantities of pollution sources have also increased dramatically, the development of a method for monitoring and controlling these sources has become very important. Therefore, rapid and reliable detection systems are in great need so as to minimize the damage caused by atmospheric pollution. Solid-state gas sensors which are compact, robust, with versatile applications and a low cost are important devices for monitoring and controlling those pollutants by chemical reactions,[154] As one of the major materials for solid-state gas sensors, bulk and thin films of ZnO have been investigated widely for gas sensing, and the typical detection gases are CO,[155] NH$_3$,[156] alcohol,[157] H$_2$.[158] Furthermore, it can be used not only for detecting the leakage of flammable gases and toxic gases but also for controlling domestic gas boilers.[159]

17.4.1 Gas sensing mechanisms of ZnO semiconductor

N-type semiconducting materials such as SnO$_2$, ZnO, In$_2$O$_3$, and Fe$_2$O$_3$ have been known for the detection of flammable or toxic gases. The detection mechanism is straightforward: oxygen vacancies on metal oxide surfaces are electrically and chemically active. These vacancies function as n-type donors, and often significantly increase the conductivity of the oxide. Upon adsorption of charge-accepting molecules at the vacancy sites, such as NO$_2$ and O$_2$, electrons are effectively depleted from the conduction band, leading to a reduced conductivity of the n-type oxide. On the other hand, molecules, such as CO and H$_2$, would react with surface-adsorbed oxygen and consequently remove it, leading to an increase in conductivity. By measuring these conductance or resistance changes, gas sensors can detect different gases. The effectiveness of gas sensors prepared from semiconducting oxides depends on several factors, including the nature of the reaction taking place at the oxide surface, the temperature, the catalytic properties of the surface, the electronic properties of the bulk oxide, and the microstructure.[2]

Although the detection mechanism of semiconductor sensors generally is based on the resistance changes, their response mechanisms may be a little different depending on detection gases. For the detection of oxygen, the gas sensing element generally responds to changes in oxygen partial pressure at high temperatures (700°C) by exploiting the equilibrium between the composition of the atmosphere and the bulk stoichiometry. The relationship between oxygen partial pressure and the electrical conductivity of the oxide may be represented by[160]:

$$\sigma = A\exp(-E_A/KT)P_{O2}^{1/N} \qquad (17.3)$$

where σ is the electrical conductivity, A is a constant, E_A is the activation energy for conduction, P_{O2} is oxygen partial pressure, and N is a constant determined by the dominant type of bulk defect involved in the equilibrium between oxygen and the sensor.

For the detection of minority gases in air, semiconducting oxides are in an atmosphere of fixed oxygen partial pressure. Bulk changes in oxygen stoichiometry are not relevant to this type of sensing and the materials are normally held at temperatures in the range 300–500°C, where useful surface reactions proceed at a sufficient rate.

The relationship between sensor resistance and the concentration of the deoxidizing gas can be expressed by the following equation over a certain range of gas concentration:

$$R_s = A[C]^{-\alpha} \qquad (17.4)$$

where R_s = electrical resistance of the sensor, A = constant, $[C]$ = gas concentration, and α = slope of the R_s curve. Due to the logarithmic relationship between sensor resistance and gas concentration, semiconductor-type sensors have an advantage of high sensitivity to gas even at low gas concentrations. The excellent stability and performance of the semiconductor-type sensor provides maintenance-free, long-lived, and low-cost gas detection. Various sensors which have different sensitivity characteristics can be manufactured by selecting the most suitable combinations of sensing material, temperature, and activity of sensor materials.

17.4.2 *Grain size and structure effects on gas sensing performance*

The sensitivity of nanostructured ZnO gas elements is comparatively high because of the grain-size effect.[161] For further development of high-sensitivity ZnO gas sensors, understanding the correlation between the sensitivity and the microstructure of nanograin ZnO gas elements is necessary. It is well known that the sensing mechanism of ZnO belongs to the surface-controlled type. Its gas sensitivity is relative to grain size, geometry, surface state, oxygen adsorption quantity, active energy of oxygen adsorption, and lattice defects. Normally, the smaller its grain size, specific surface area, and oxygen adsorption quantity, the higher its gas sensitivity is. In addition, the decrease in grain size of ZnO decreases its working temperature due to the increase in surface activity of ZnO. In general, the working temperature of ZnO is 400–500°C, but that of nanometer ZnO made by emulsion is only 300°C.[161]

For ZnO nanograin-based gas sensors, the sensing properties are influenced not only by the microstructural features, such as the grain size, the geometry, but also by the connectivity between the grains.[162] In the case of neck-grain-boundary-controlled sensitivity, a sensor's sensitivity will decrease with the increasing number of grain boundaries when the width of the space charge layer of the neck in air is comparable with the neck radius. On the ZnO nanograin surface, the adsorbed oxygen extracts the conduction electrons from the near-surface region of the grain leading to the grain-boundary barriers and the neck barriers. Reaction of a reducing gas with the adsorbed oxygen results in its removal, and thereby a reduction in the barrier height. The neck barrier controls the electron-conducting channel through the neck[163] and the electron density in the space charge layer at the neck. Therefore the neck barrier determines the neck resistance. At the same time, the grain-boundary barrier determines the grain-boundary resistance. At a small grain size, the grain-boundary resistance is small and decreases slowly over the entire gas concentration range; therefore its total variation is also small. But the neck resistance decreases steeply and its total variation is large. The grain-boundary resistance may be much smaller than the neck resistance. Corresponding to the resistances, the grain-boundary-controlled sensitivity of nanograin ZnO gas elements is much lower than the neck-controlled sensitivity. It is notable that neck-grain-boundary-controlled sensitivity is lower than neck-controlled sensitivity alone when the width of the space charge layer of the neck in air approaches the radius of the neck. This indicates that decreasing the number of grain boundaries will increase the sensitivity of nanograin ZnO gas elements and implies that the grain-boundary resistance of nanograin ZnO gas elements cannot be ignored though it may be much smaller than the neck resistance. When the gas concentration becomes high enough, the sensitivities begin to saturate. Ma et al. studied the effects both of grain size and grain boundary on gas sensitivity.[162] Figure 17.12 shows that the smaller the grain size of nanograin ZnO gas elements, the higher the neck-controlled sensitivity and the neck-grain-boundary-controlled sensitivity are.[162]

17.4.3 Nanostructured ZnO for gas sensors

ZnO nanostructures have attracted considerable attention for use as solid-state gas sensors with great potential for overcoming fundamental limitations due to their ultrahigh surface-to-volume ratio. Besides ZnO nanoparticles, ZnO nanostructures which have been used for gas sensors

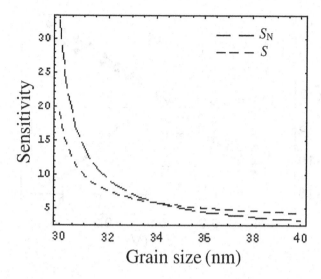

Figure 17.12. The gas sensitivities decrease with increasing grain size for a nanograin ZnO gas sensor. S_N is neck-controlled sensitivity and S is the neck-grain-boundary-controlled sensitivity. (Reprinted from Y. Ma, W. L. Wang, K. J. Liao and C. Y. Kong, *J. Wide Bandgap Mater.* **10**, 113 (2002) with permission from SAGE publications.)

include nanorods, nanowires, nanobelts, and nanotubes. Furthermore, some mesoporous structures have also been investigated to be used for gas sensing either in the form of mesoporous films or by forming mesopores in ZnO nanorods. From the aspect of sensing performance, one-dimensional ZnO, such as nanowires and nanorods, is expected to be superior to its thin-film counterpart.[56] Since their diameter is small and comparable to the Debye length, chemisorption-induced surface states effectively affect the electronic structure of the entire channel, thus conferring one-dimensional ZnO with higher sensitivity than thin-film ZnO.

ZnO nanowires and nanorods can be configured as either two terminal sensing devices or as FETs in which a transverse electric field can be utilized to tune the sensing property. Recently, Wan *et al.*[164] fabricated ZnO nanowire chemical sensors using microelectromechanical system technology. Massive nanowires were placed between Pt interdigitating electrodes. The system exhibited a very high sensitivity to ethanol gas and a fast response time (within 10 seconds) at 300°C, showing a promising application for ZnO nanowire humidity sensors. Electrical transport studies show that ambient O_2 has considerable effect on the ZnO nanowires.[165,166] Fan *et al.* discussed the relationship between oxygen pressure and ZnO nanowire FET performance.[165] It is shown that ZnO nanowires have fairly

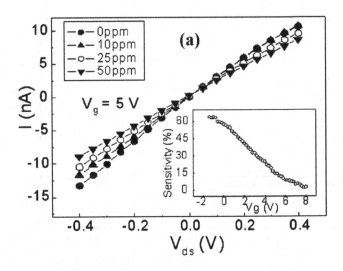

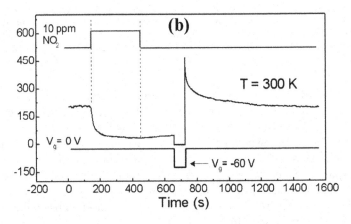

Figure 17.13. (a) I-V curves of a ZnO nanowire under 50 ppm O_2. Inset: gate potential dependence of sensitivity under 10 ppm O_2. (Reprinted from Z. Fan *et al.*, *Appl. Phys. Lett.* **85**, 5923 (2004) with permission from the American Institute of Physics). (b) The nanowire sensing response to 10 ppm NO_2 and the conductance recovery process assisted by a −60 V gate voltage pulse. (Reprinted from Z. Fan *et al.*, *Appl. Phys. Lett.* **86**, 123510 (2005) with permission from the American Institute of Physics.)

good sensitivity to O_2 (Figure 17.13(a)). In addition, it is observed that the sensitivity is a function of back-gate potential, i.e. above the gate threshold voltage of FETs, sensitivity increases with decreasing gate voltage (Figure 17.13(a) inset). This implies that the gate voltage can be used to adjust the sensitivity range. Furthermore, a gate-refresh mechanism was proposed.[55] As demonstrated in Figure 17.13(b), the conductance of nanowires can be recovered by using a negative gate potential larger than the threshold voltage. The gas selectivity for NO_2 and NH_3 using ZnO nanowire FETs was also investigated under the gate-refresh process, exhibiting a gas distinguishability function.

ZnO nanobelts have also attracted researchers' attention in gas sensing under atmospheric conditions. Nanobelts of ZnO with a rectangular cross-section in a ribbon-like morphology are very promising for sensors due to the fact that the surface-to-volume ratio is very high, the oxide is single-crystalline, the faces exposed to the gaseous environment are always the same, and the size is likely to produce a complete depletion of carriers inside the belt. Wen *et al.* measured the NH_3 sensing characteristics.[167] Figure 17.14 shows the current response of the sensor to NH_3 at room temperature at a bias voltage of 5 V. The sensor is very sensitive to ammonia: when the sensor was exposed to 500 ppm NH_3, a current change as high as about 3.5-fold between the on and off states was observed, and this current response to NH_3 could be repeated many times without obvious change in signal intensity. The repeatability and stability of the sensor appear to

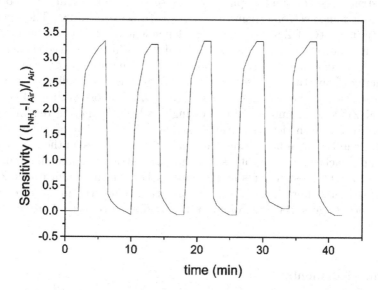

Figure 17.14. Current responses of the ZnO nanobelts sensors when the surrounding gas is switched between air and 500 ppm NH_3. (Reprinted from X. G. Wen, Y. P. Fang, Q. Pang, C. L. Yang, J. N. Wang, W. K. Ge, K. S. Wong and S. H. Yang, *J. Phys. Chem.* **B109**, 15303 (2005) with the permission from ACS.)

be very good. For NH_3 sensors with ZnO, the change in current is mainly caused by the adsorption and desorption of NH_3 and H_2O molecules on the surface of the sensing materials. The high surface-to-volume ratio of our ZnO nanobelt sensor gives rise to high sensitivity to NH_3. In addition, the molecules are easy to adsorb on and desorb from the surface of nanobelts in contrast to the conventional thin-film sensor. Note that the exposed surfaces of ZnO nanobelts are (0001) for the most part, whereas a polycrystalline ZnO film exposes many different crystal planes on the surface, and this may contribute to different sensing properties. For conventional thin-film sensors, an elevated temperature is often used to facilitate molecular desorption. In nanobelt sensors, however, the available interbelt spaces allow the adsorbed molecules to desorb easily in the off state so that they show rapid response and good reversibility.

17.5 Conclusions

ZnO is a unique material and has attracted a lot of research for various applications. ZnO is a wide band gap semiconductor and its nanostructures are being explored for solar cells. ZnO is also piezoelectric and its nanostructures can find applications in sensors and actuators or other micro- or nanoelectromechanical systems (MEMS or NEMS). The electrical conductivity of ZnO is sensitively dependent on its surface chemistry, and thus nanostructured ZnO is an ideal material for surface-chemistry-related sensors and detectors. Making nanostructured ZnO even more attractive from the engineering point of view is the ability to readily fabricate various well-controlled and high-quality nanostructures of single-crystal ZnO under moderate processing conditions, and its readiness to be integrated with device fabrication. There is a myriad of publications available in the open literature. This chapter just focuses on the most common approaches for the synthesis of ZnO nanostructures and the applications of ZnO nanostructures in gas sensing. For sensing applications, ZnO nanowires or nanorods are the most preferred nanostructures, since they offer both large surface-to-volume ratios and easy collection of electrical signals.

Acknowledgments

H. Shang acknowledges the Joint Institute for Nanoscience (JIN) Graduate Fellowship, funded by the Pacific Northwest National Laboratory and the University of Washington, and the Ford Motor Company Fellowship.

References

1. K. Ihokura and J. Watson, *The Stannic Oxide Gas Sensor* (CRC Press, Boca Raton, 1994).
2. P. T. Moseley, *Meas. Sci. Technol.* **8**, 223 (1997).
3. Y. Min, H. L. Tuller, S. Palzer, J. Wöllenstein and H. Böttner, *Sens. Actuators B* **93**, 435 (2003).
4. T. Seiyama, A. Kato, K. Fujishi and M. Nagatami, *Anal. Chem.* **34**, 1502 (1962).
5. N. Taguchi, Japanese Patent S45-38200 (1962).
6. S. C. Chang, Abst. Int. Seminar on Solid-state Gas Sensors (23rd WEH Seminar), 7 (1982).
7. P. J. Shaver, *Appl. Phys. Lett.* **11**, 255 (1967).
8. N. Taguchi, Japanese Patent S50–30480 (1966).
9. J. C. Loh, French Patent 1545292 (1967).
10. H. Steffes, C. Imawan, F. Solzbacher and E. Obermeier, *Sens. Actuators B* **68**, 249 (2000).
11. B. Bott, British Patent 1374575 (1971).
12. C. Imawan, H. Steffes, F. Solzbacher and E. Obermeier, *Sens. Actuators B* **77**, 346 (2001).
13. T. Schwebel, M. Fleischer, H. Meixner and C.-D. Kohl, *Sens. Actuators B* **49**, 46 (1998).
14. T. Tien, H. Stadler, E. Gibbons and P. Zacmanidis, *Am. Ceram. Soc. Bull.* **54**, 280 (1975).
15. M. C. Steele and B. A. Maclver, *Appl. Phys. Lett.* **28**, 687 (1976).
16. M. Nitta, S. Kanefusa, Y. Taketa and M. Haradome, *Appl. Phys. Lett.* **32**, 590 (1978).
17. S. Tao, X. Liu, X. Chu and Y. Shen, *Sens. Actuators B* **61**, 33 (1999).
18. J. R. Stetter, *J. Colloid Interface Sci.* **65**, 432 (1978).
19. N. Yamamoto, *Bull. Chem. Soc. Jpn.* **54**, 696 (1981).
20. C. Imawan, F. Solzbacher, H. Steffes and E. Obermeier, *Sens. Actuators B* **64**, 193 (2000).
21. W. Heywang, *Amorphe und polykristalline Halbleiter* (Springer Verlag, Berlin, 1984), p. 204.
22. F. Lin, Y. Takao, Y. Shimizu and M. Egashira, *Sens. Actuators B* **25**, 843 (1995).
23. W. Göpel, *J. Vac. Sci. Technol.* **15**, 1298 (1978).
24. P. Esser and W. Göpel, *Surf. Sci.* **97**, 309 (1980).
25. W. Göpel, T. A. Jones, M. Kleitz, J. Lundstrom and T. Seiyama, *Sensors, Vol. 2: Chemical and Biochemical Sensors, Part I* (VCH, Weinheim, 1991).
26. W. Göpel, T. A. Jones, M. Kleitz, J. Lundstrom and T. Seiyama, *Sensors, Vol. 3: Chemical and Biochemical Sensors, Part II* (VCH, Weinheim, 1992).
27. Z. L. Wang, *Nanowires and Nanobelts: Materials, Properties and Devices* (Kluwer Academic Publishers, Boston, 2003).

28. G. Sberveglieri, G. Faglia, S. Groppelli, P. Nelli and A. Camanzi, *Semicond. Sci. Technol.* **5**, 1231 (1990).
29. M. Ferroni, V. Guidi, G. Martinelli, E. Comini, G. Sberveglierri, D. Boscarino and G. Della Mea, *J. Appl. Phys.* **88**, 1097 (2000).
30. V. E. Henrich and P. A. Cox, *The Surface Science of Metal Oxides* (Cambridge University Press, New York, 1994).
31. W. Göpel, *Sens. Actuators A* **56**, 83 (1996).
32. W. D. Kingery, H. K. Bowen and D. R. Uhlmann, *Introduction to Ceramics*, 2nd edn. (John Wiley & Sons, Inc., New York, 1976).
33. S. J. Pearson, D. P. Norton, K. Ip, Y. W. Heo and T. Steiner, *J. Vac. Sci. Technol. B* **22**, 932 (2004).
34. R. Viswanatha, S. Sapra, B. Satpati, P. V. Satyam, B. N. Dev and D. D. Sarma, *J. Mater. Chem.* **14**, 661 (2004).
35. F. A. Kröger, *The Chemistry of Imperfect Crystals* (Publishers Inc., Amsterdam, 1964).
36. D. P. Norton, Y. W. Heo, M. P. Ivill, K. Ip, S. J. Pearton, M. F. Chisholm and T. Steiner, *Mater. Today* **7**, 34 (2004).
37. G. W. Tomlins, J. L. Routbort and T. O. Mason, *J. Appl. Phys.* **87**, 117 (2000).
38. M. H. Huang, S. Mao, H. Feick, H. Yan, Y. Wu, H. Kind, E. Weber, R. Russo and P. Yang, *Science* **292**, 1897 (2001).
39. N. Saito, H. Haneda, T. Sekiguchi, N. Ohashi, I. Sakaguchi and K. Koumoto, *Adv. Mater.* **14**, 418 (2002).
40. K. Keis, E. Magnusson, H. Lindstorm, S. E. Lindquist and A. Hagfeldt, *Sol. Energ. Mater. Sol. Cell.* **73**, 51 (2002).
41. J. Y. Lee, Y. S. Choi, J. H. Kim, M. O. Park and S. Im, *Thin Solid Films* **403**, 553 (2002).
42. S. Liang, H. Sheng, Y. Liu, Z. Hio, Y. Lu and H. Shen, *J. Cryst. Growth* **225**, 110 (2001).
43. M. H. Koch, P. Y. Timbrell and R. N. Lamb, *Semicond. Sci. Tech.* **10**, 1523 (1995).
44. K. Keis, E. Magnusson, H. Lindstorm, S. E. Lindquist and A. Hagfeldt, *Sol. Energ. Mater. Sol. Cell.* **73**, 51 (2002).
45. S. J. Pearton, D. P. Norton, K. Ip, Y. W. Heo and T. Steiner, *Superlatt. Microstr.* **34**, 3 (2003).
46. Y. Lin, Z. Hang, Z. Tang, F. Yuan and J. Li, *Adv. Mater. Opt. Electron.* **9**, 205 (1999).
47. Y. C. Kong, D. P. Yu, B. Zhang, W. Fang and S. Q. Feng, *Appl. Phys. Lett.* **78**, 4 (2001).
48. Y. Cui, Q. Wei, H. Park and C. M. Lieber, *Science* **293**, 1289 (2001).
49. P. M. Martin, M. S. Good, J. W. Johnston, G. J. Posakony, L. J. Bond and S. L. Crawford, *Thin Solid Films* **379**, 253 (2000).
50. S. J. Pearton, D. P. Norton, K. Ip, Y. W. Heo and T. Steiner, *Prog. Mater. Sci.* **50**, 293 (2005).
51. X. Wang, Y. Ding, C. J. Summers and Z. L. Wang, *J. Phys. Chem. B* **108**, 8773 (2004).

52. L. M. Kukreja, S. Barik and P. Misra, *J. Cryst. Growth* **268**, 531 (2004).
53. J. W. Chiou, K. P. Krishna Kumar, J. C. Jan, H. M. Tsai, C. W. Bao, W. F. Pong, F. Z. Chien, M.-H. Tsai, I.-H. Hong, R. Klauser, J. F. Lee, J. J. Wu and S. C. Liu, *Appl. Phys. Lett.* **85**, 3220 (2004).
54. H. Chik, J. Liang, S. G. Cloutier, N. Kouklin and J. M. Xu, *Appl. Phys. Lett.* **84**, 3376 (2004).
55. Z. Fan and J. G. Lu, *Appl. Phys. Lett.* **86**, 123510 (2005).
56. A. Kolmakov and M. Moskovits, *Annu. Rev. Mater. Res.* **34**, 151 (2004).
57. Y. Zhang, A. Kolmakov, S. Chretien, H. Metiu and M. Moskovits, *Nano. Lett.* **4**, 403 (2004).
58. L. Vayssieres, K. Keis, S. E. Lindquist and A. Hegfeld, *J. Phys. Chem. B* **105**, 3350 (2001).
59. M. Izaki and T. Omi, *Appl. Phys. Lett.* **68**, 2439 (1996).
60. Th. Pauporte and D. Lincot, *Appl. Phys. Lett.* **75**, 3817 (1999).
61. B. Cao, W. Cai, G. Duan, Y. Li, Q. Zhao and D. Yu, *Nanotechnology* **16**, 2567 (2005).
62. V. A. L. Roy, A. B. Djurisic, W. K. Chan, J. Gao, H. F. Lui and C. Surya, *Appl. Phys. Lett.* **83**, 141 (2003).
63. W. I. Park, D. H. Kim, S. W. Jung and G. C. Yi, *Appl. Phys. Lett.* **80**, 4232 (2002).
64. B. D. Yao, Y. F. Chan and N. Wang, *Appl. Phys. Lett.* **81**, 757 (2002).
65. X. Kong and Y. Li, *Chem. Lett.* **32**, 838 (2003).
66. Y. Li, G. W. Meng and L. D. Zhang, *Appl. Phys. Lett.* **76**, 2011 (2000).
67. W. Y. Chung, J. W. Lim, D. D. Lee, N. Miura and Y. Yamazoe, *Sens. Actuators B* **64**, 118 (2000).
68. C. Xu, J. Tamaki, N. Miura and N. Yamazoe, *Sens. Actuators B* **3**, 147 (1991).
69. M. K. Kennedy, F. E. Kruis, H. Fissan, B. R. Mehta, S. Stappert and G. Dumpich, *J. Appl. Phys.* **93**, 551 (2003).
70. H. Ogawa, M. Nishikawa and A. Abe, *J. Appl. Phys.* **53**, 4448 (1982).
71. C. Malagù, V. Guidi, M. Stefancich, M. C. Carotta and G. Martinelli, *J. Appl. Phys.* **91**, 808 (2002).
72. M. K. Kennedy, F. E. Kruis, H. Fissan and B. R. Mehta, *Rev. Sci. Instrum.* **74**, 4908 (2003).
73. L. Spanhel and M. A. Anderson, *J. Am. Chem. Soc.* **113**, 2826 (1991).
74. E. A. Meulenkamp, *J. Phys. Chem. B* **102**, 5566 (1998).
75. A. Van Dijken, E. A. Meulenkamp, D. Vanmaekelbergh and A. Meijerink, *J. Phys. Chem. B* **104**, 1715 (2000).
76. M. Izaki and T. Omi, *Appl. Phys. Lett.* **68**, 2439 (1996).
77. E. M. Wong and P. C. Searson, *Appl. Phys. Lett.* **74**, 2939 (1999).
78. R. Liu, A. A. Vertegel, E. W. Bohannan, T. A. Sorenson and J. A. Switzer, *Chem. Mater.* **13**, 508 (2001).
79. S. Roy and S. Basu, *Bull. Mater. Sci.* **25**, 513 (2002).
80. F. Hossein-Babaei and F. Taghibakhsh, *Electron. Lett.* **36**, 1815 (2000).
81. S. Hayamizu, H. Tabata, H. Tanaka and T. Kawai, *J. Appl. Phys.* **80**, 787 (1996).

82. M. Okoshi, K. Higashikawa and M. Hanabusa, *Appl. Surf. Sci.* **154–155**, 424 (2000).

83. V. Craciun, J. Elders, J. G. E. Gardeniers and I. W. Boyd, *Appl. Phys. Lett.* **65**, 2963 (1995).

84. P. Yu, Z. K. Tang, G. K. L. Wong, M. Kawasaki, A. Ohotomo, H. Koinuma and Y. Segawa, *Solid State Commun.* **103**, 459 (1997).

85. P. Yu, Z. K.Tang, G. K. L. Wong, M. Kawasaki, A. Ohotomo, H. Koinuma and Y. Segawa, *J. Cryst. Growth* **184/185**, 601 (1998).

86. D. M. Bagnall, Y. F. Chen, Z. Zhu, T. Yao, M. Y. Shen and T. Goto, *Appl. Phys. Lett.* **73**, 1038 (1998).

87. S. Liang, C. R. Gorla and N. Emanetoglu, *J. Electron. Mater.* **27**, L72 (1998).

88. Y. Liu, C. R. Gorla, S. Liang, N. Emanetoglu, Y. Lu, H. Shen and M. Wraback, *J. Electron. Mater.* **29**, 60 (2000).

89. H. Yuan and Y. Zhang, *J. Cryst. Growth* **263**, 119 (2004).

90. N. G. Patel and B. H. Lashkari, *J. Mater. Sci.* **27**, 3026 (1992).

91. F. K. Sahn, Z. F. Liu, G. X. Liu, B. C. Shin and Y. S. Yu, *J. Korean Phys. Soc.* **44**, 1215, (2004).

92. D. F. Paraguay, M. Miki-Yoshida, J. Morales, J. Solis and L. W. Estrada, *Thin Solid Films* **373**, 137 (2000).

93. B. O'Regan, V. Sklover and M. Grätzel, *J. Electrochem. Soc.* **148**, C498 (2001).

94. J. Jiu, K. Kurumada and M. Tanigaki, *Mater. Chem. Phys.* **81**, 93 (2003).

95. T. F. Jaramillo, S.-H. Baeck, A. Kleiman-Shwarsctein and E. W. McFarland, *Macromol. Rapid Commun.* **25**, 297 (2004).

96. Z. L. Wang, *Mater. Today* **7**, 26 (2004).

97. Y. H. Xiao, L. Li, Y. Li, M. Fang and L. D. Zhang, *Nanotechnology* **16**, 671 (2005).

98. X. D. Wang, C. J. Summers and Z. L. Wang, *Adv. Mater.* **16**, 1215 (2004).

99. Y. W. Zhu, H. Z. Zhang, X. C. Sun, S. Q. Feng, J. Xu, Q. Zhao, B. Xiang, R. M. Wang and D. P. Yu, *Appl. Phys. Lett.* **83**, 144 (2003).

100. M. Yan, H. T. Zhang, E. J. Widjaja and R. P. H. Chang, *J. Appl. Phys.* **94**, 5240 (2003).

101. L. Vayssieres, *Adv. Mater.* **15**, 464 (2003).

102. G. Z. Cao, J. J. Schermer, W. J. P. van Enckevort, W. A. L. M. Elst and L. J. Giling, *J. Appl. Phys.* **79**, 1357 (1996).

103. L. E. Greene, M. Law, J. Goldberger, F. Kim, J. C. Johnson, Y. Zhang, R. J. Saykally and P. Yang, *Angew. Chem. Int. Ed.* **42**, 3031 (2003).

104. B. C. Bunker, P. C. Rieke, B. J. Tarasevich, A. A. Campbell, G. E. Fryxell, G. L. Graff, L. Song, J. Liu and J. W. Virden, *Science* **264**, 48 (1994).

105. P. Hartman and W. G. Perdok, *Acta Cryst.* **8**, 49 (1955).

106. L. Vayssieres, N. Beermann, S.-E. Lindquist and A. Hagfeldt, *Chem. Mater.* **13**, 233 (2001).

107. L. Vayssieres, K. Keis, S.-E. Lindquist and A. Hagfeldt, *J. Phys. Chem. B* **105**, 3350 (2001).

108. L. Vayssieres, K. Keis, A. Hagfeldt and S.-E. Lindquist, *Chem. Mater.* **13**, 4395 (2001).
109. Z. Tian, J. A. Voigt, J. Liu, B. McKenzie and M. J. McDermott, *J. Am. Chem. Soc.* **124**, 12954 (2002).
110. J. Liu, Y. Lin, L. Liang, J. A. Voigt, D. L. Huber, Z. R. Tian, E. Coker, B. McKenzie and M. J. McDermott, *Chem. Eur. J.* **97**, 604 (2003).
111. L. E. Greene, M. Law, J. Goldberger, F. Kim, J. C. Johnson, Y. Zhang, R. J. Saykally and P. Yang, *Angew. Chem. Int. Ed.* **42**, 3031 (2003).
112. Z. R. Tian, J. A. Voigt, J. Liu, B. McKenzie, M. J. McDermott, R. T. Cygan and L. J. Criscenti, *Nature Mater.* **2**, 821 (2003).
113. M. Izaki and T. Omi, *Appl. Phys. Lett.* **68**, 2439 (1996).
114. Th. Pauporte and D. Lincot, *Appl. Phys. Lett.* **75**, 3817 (1999).
115. B. Cao, W. Cai, G. Duan, Y. Li, Q. Zhao and D. Yu, *Nanotechnology* **16**, 2567 (2005).
116. Y. J. Kim, H. M. Shang and G. Z. Cao, *J. Sol-Gel Sci. Technol.* **38**, 79–84 (2005).
117. L. Vayssieres, *Int. J. Nanotechnology* **1**, 1 (2004).
118. R. S. Wagner and W. C. Ellis, *Appl. Phys. Lett.* **4**, 89 (1964).
119. R. S. Wagner, W. C. Ellis, K. A. Jackson and S. M. Arnold, *J. Appl. Phys.* **35**, 2993 (1964).
120. R. S. Wagner, in *Whisker Technology*, ed. A. P. Levitt (Wiley, New York, 1970).
121. E. I. Givargizov, *Highly Anisotropic Crystals* (D. Reidel, Dordrecht, 1986).
122. E. I. Givargizov, *J. Vac. Sci. Technol. B* **11**, 449 (1993).
123. M. H. Huang, Y. Wu, H. Feick, N. Tran, E. Weber and P. Yang, *Adv. Mater.* **13**, 113 (2001).
124. D. R. Askkeland, *The Science and Engineering of Materials* (PWS, Boston, 1989).
125. G. W. Sears, *Acta Metall.* **3**, 361 (1955).
126. G. W. Sears, *Acta Metall.* **3**, 367 (1955).
127. Z. W. Pan, Z. R. Dai and Z. L. Wang, *Science* **291**, 1947 (2001).
128. M. Ohring, *The Materials Science of Thin Films* (Academic Press, San Diego, 1992).
129. K. F. Jensen and W. Kern, in *Thin Film Processes II*, eds. J. L. Vossen and W. Kern (Academic Press, San Diego, 1991).
130. K. L. Choy, *Prog. Mater. Sci.* **48**, 57 (2003).
131. P. Ser, P. Kalck and R. Feurer, *Chem. Rev.* **102**, 3085 (2002).
132. S. Muthukumar, N. W. Emanetoglu, G. Patounakis, C. R. Gorla, S. Liang and Y. Lu, *J. Vac. Sci. Technol. A* **19**, 1850 (2001).
133. S. Muthukumar, H. Sheng, J. Zhong, Z. Zhang, N. W. Emanetoglu and Y. Lu, *IEEE Trans. Nanotech.* **2**, 50 (2003).
134. W. I. Park, G. C. Yi, M. Kim and S. J. Pennycook, *Adv. Mater.* **14**, 1841 (2002).
135. W. Lee, H. G. Sohn and J. M. Myoung, *Mater. Sci. Forum* **449**, 1245 (2004).
136. X. Liu, X. H. Wu, H. Cao and R. P. H. Chang, *J. Appl. Phys.* **95**, 3141 (2004).

137. K. Ogata, K. Maejima, S. Fujita and S. Fujita, *J. Cryst. Growth* **248**, 25 (2003).
138. K. Maejima, M. Ueda, S. Fujita and S. Fujita, *Jpn. J. Appl. Phys.* **42**, 2600 (2003).
139. K. S. Kim and H. W. Kim, *Physica B* **328**, 368 (2003).
140. B. P. Zhang, N. T. Binh, Y. Segawa, K. Wakatsuki and N. Usami, *Appl. Phys. Lett.* **83**, 1635 (2003).
141. S. W. Kim, S. Fujita and S. Fujita, *Jpn. J. Appl. Phys.* **41**, L543 (2002).
142. J. J. Wu and S. C. Liu, *Adv. Mater.* **14**, 215 (2002).
143. Y. W. Heo, V. Varadarajan, M. Kaufman, K. Kim, F. Ren, P. H. Fleming and D. P. Norton, *Appl. Phys. Lett.* **81**, 3046 (2002).
144. T. Okada, B. H. Agung and Y. Nakata, *Appl. Phys. A* **79**, 1417 (2004).
145. C. Jin, X. Yuan, W. Ge, J. Hong and X. Xin, *Nanotechnology* **14**, 667 (2003).
146. N. Pan, X. Wang, K. Zhang, H. Hu, B. Xu, F. Li and J. Hou, *Nanotechnology* **16**, 1069 (2005).
147. S. C. Lyu, Y. Zhang, C. J. Lee, H. Ruh and H. J. Lee, *Chem. Mater.* **15**, 3294 (2003).
148. Y. C. Wang, I. C. Leu and M. N. Hon, *J. Mater. Chem.* **12**, 2439 (2002).
149. Y. Li, G. S. Cheng and L. D. Zhang, *J. Mater. Res.* **15**, 2305 (2000).
150. B. B. Lakshmi, P. K. Dorhout and C. R. Martin, *Chem. Mater.* **9**, 857 (1997).
151. Z. R. Tian, J. A. Voigt, J. Liu, B. McKenzie, M. J. McDermott, M. A. Rodriguez, H. Konishi and H. F. Xu, *Nature Mater.* **2**, 821 (2003).
152. G.-C. Yi, C. Wang and W. I. Park, *Semicond. Sci. Technol.* **20**, S22 (2005).
153. G. Z. Cao, *Nanostructures and Nanomaterials: Synthesis, Properties and Applications* (Imperial College Press, London, 2004).
154. D.-D. Lee and D.-S. Lee, *IEEE Sens. J.* **1**, 214 (2001).
155. H.-W. Ryu, B.-S. Park, S. A. Akbar, W.-S. Lee, K.-J. Hong, Y.-Jin Seo, D.-C. Shin, J.-S. Park and G.-P. Choi, *Sens. Actuators B* **96**, 717 (2003).
156. G. Sberveglieri, *Sens. Actuators B* **23**, 103 (1995).
157. G. S. Trivikrama Rao and D. Tarakarama Rao, *Sens. Actuators B* **55**, 166 (1999).
158. X. L. Cheng, H. Zhao, L. H. Huo, S. Gao and J. G. Zhao, *Sens. Actuators B* **102**, 248 (2004).
159. C. H. Kwon, H. K. Hong, D. H. Yun, K. Lee, S. T. Kim, Y. H. Roh and B. H. Lee, *Sens. Actuators B* **24/25**, 610 (1995).
160. P. Kofstad, *Non-stoichiometry, Diffusion and Electrical Conductivity in Binary Metal Oxides* (Wiley-Interscience, New York, 1972).
161. J. Q. Xu, Y. Q. Pan, Y. A. Shun and Z.-Z. Tian, *Sens. Actuators B* **66**, 277 (2000).
162. Y. Ma, W. L. Wang, K. J. Liao and C. Y. Kong, *J. Wide Bandgap Mater.* **10**, 113 (2002).
163. C. Xu, J. Tamaki, N. Miura and N. Yamazoe, *Sens. Actuators B* **3**, 147 (1991).
164. Q. Wan, Q. H. Li, Y. J. Chen, T. H. Wang, X. L. He, J. P. Li and C. L. Lin, *Appl. Phys. Lett.* **84**, 3654 (2004).

165. Z. Fan, D. Wang, P. Chang, W. Tseng and J. G. Lu, *Appl. Phys. Lett.* **85**, 5923 (2004).
166. Q. H. Li, Q. Wan, Y. X. Liang and T. H. Wang, *Appl. Phys. Lett.* **84**, 4556 (2004).
167. X. G. Wen, Y. P. Fang, Q. Pang, C. L. Yang, J. N. Wang, W. K. Ge, K. S. Wong and S. H. Yang, *J. Phys. Chem. B* **109**, 15303 (2005).

Chapter 18

Synthesis and Properties of Mesoporous-based Materials for Environmental Applications

Jianlin Shi, Hangrong Chen, Zile Hua and Lingxia Zhang

Shanghai Institute of Ceramics, Chinese Academy of Sciences
Shanghai, PR China

18.1 Introduction

The synthesis of nanoporous materials has attracted intensive attention in recent years for their wide range of promising applications. Nanoporous materials are a large family of solid-state materials. According to the definition of pore sizes by the International Union of Pure and Applied Chemistry (IUPAC),[1] porous materials can be divided into three classes: microporous materials ($< 2\,nm$), mesoporous materials (2–$50\,nm$), and macroporous materials ($> 50\,nm$). Most or part of these three classes of materials can be included under the so-called nanoporous materials (1–$100\,nm$), which have found more and more important applications in the fields of adsorption, separation, catalysis, supporting materials, optic-electric devices, petroleum and chemical industries, and so on, owing to their well-defined pore structures, large surface areas and high thermal stability, and tunable pore surface characteristics, etc.[2-4]

Ordered mesoporous inorganic materials were first reported in 1992.[5,6] Such novel mesostructured materials templated by surfactant molecular self-assembly have great potential in many fields owing to their special pore characteristics. Several strategies for the fabrication of mesoporous silica materials have been extensively investigated over the past decade, and the approach to the synthesis of mesoporous silica has been extended to other metal oxides with various approaches. Herein we mainly focus on some typical synthesis approaches for mesoporous materials and mesoporous-based composites and their environmental catalysis processes.

473

18.2 Mesoporous-based Nanocomposites for Automotive Exhaust Treatments

Nowadays, automobiles are prevalent around the world as the most popular and necessary vehicles in our daily lives. About 50 million cars are produced every year, and over 700 million cars are used worldwide. Automotive exhaust has become an increasingly serious source of atmospheric pollution. Emission regulations have been applied in European countries since 1988, and more rigorous regulations are being planned in both the USA and Europe. Thus, the use of catalysts, especially the three-way catalysts (TWCs) for purifying exhaust gases, which contain mainly oxides of carbon (CO), oxides of nitrogen (NO_x), hydrocarbons (HC), and particulate matter, is necessary and indispensable in every vehicle. The automotive TWC system generally operates under a certain air-fuel (A/F) ratio range, which is controlled by an oxygen sensor device. Common three-way catalysts consist of noble metals, such as Pt, Rh, and Pd, as active sites; metal oxides as promoter materials; and a transition-alumina support. In order to obtain high conversion rates of CO, HC, and NO_x, the A/F ratio oscillations should be buffered. Therefore, CeO_2-based materials have been pervasively used as an oxygen storage material associated to the redox couple of Ce^{4+}/Ce^{3+}, i.e. ceria can provide oxygen for oxidizing CO and HC under rich A/F conditions and removes it from the exhaust gas phase for reducing NO_x under lean A/F. The addition of zirconia into the ceria lattice can significantly increase the oxygen storage capacity (OSC), as well as the thermal stability. This property is believed to be related to the displacement of the oxygen sub-lattice around zirconium, leading to higher oxygen mobility and easier bulk reduction of the solid solution.[7] Therefore, CeO_2/ZrO_2 composite oxide has been widely used as an automobile exhaust three-way catalyst in the last decade. A great number of approaches have been reported for the synthesis of the high-surface-area CeO_2/ZrO_2 composite, such as co-precipitation,[8] sol-gel,[9] and microemulsion precipitation.[10]

18.2.1 Synthesis of mesoporous CeO_2 catalysts and CO catalytic oxidation property

Cerium oxide is an important rare earth oxide and has been widely investigated for applications in automotive exhaust purification, oxygen storage and release catalysis, and solid oxide fuel cells.[11,12]

Several novel strategies, including template methods, have been developed for the fabrication of mesoporous materials over the past years. The structures and properties of the templates play a very important role with respect to the properties of the replicated porous materials. The principle of

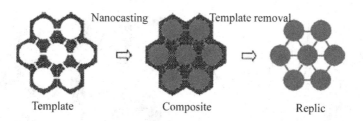

Figure 18.1. Illustration of the nanocasting pathway to produce mesoporous oxides.[13]

the nanocasting pathway with hard templates mainly involves three main steps: (1) formation of the template, (2) the casting step, and (3) removal of the template, as illustrated in Figure 18.1.[13] Generally, a three-dimensional pore network is necessary in the template to create a stable replica. Furthermore, the precursor should be easily converted into the desired composition with as little volume shrinkage as possible. And finally, the template should be easily and completely removed to obtain the true replicas. Here, we give an example of the synthesis of ordered mesoporous CeO_2 using cubic Ia3d mesoporous silica KIT-6 as a hard template.

A typical mesoporous silica with cubic Ia3d symmetry (designated as KIT-6) was prepared according to the literature.[14] To obtain the surfactant-removed hard template (named as KIT-eo) with abundant silanol groups on the interior mesopore surface, the surfactant was removed by solvent extraction using ethanol refluxing under stirring, followed by oxidization using 15 wt.% H_2O_2 solution under stirring at 310 K. The surfactant-removed product was filtered, washed with ethanol, and dried at 333 K in vacuum. A typical synthesis process of cerium oxide mesoporous material with uniform mesopores and fine particle morphology using the as-prepared KIT-eo of Ia3d symmetry as hard template is as follows[15]: First, 4.7 g of $Ce(NO_3)_3 \cdot 6H_2O$ is dissolved in 20 mL of ethanol, then this solution is incorporated into 1 g of KIT-eo by the incipient wetness impregnation technique. After ethanol is evaporated at 333 K, the composite is calcined at 673 K in order to transfer the precursor into cerium oxide. The silica template in the CeO_2/SiO_2 composite is removed by washing with heated 1–2 M NaOH solution for three times. This template-free mesoporous CeO_2 array is collected by filtering, washed with water and ethanol, and dried at 373 K.

The small-angle X-ray diffraction (XRD) pattern of the template-free mesoporous ceria is shown in Figure 18.2(a), which shows the same characteristic Bragg diffraction peaks as the silica template belonging to the same symmetry (cubic Ia3d space group) with a slightly shrunken lattice space of $a = 22.60$ nm. The wide-angle XRD pattern of the mesoporous CeO_2 replica shows slightly broader Bragg diffraction peaks, which can be

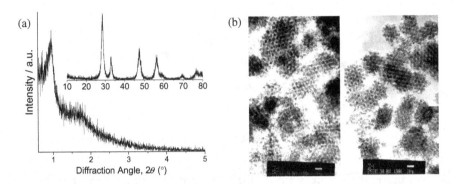

Figure 18.2.   (a) Small- and wide-angle XRD patterns (inset) of ordered mesoporous ceria replica; (b) TEM images of the template-free ordered mesoporous ceria replica.[15]

indexed to a cubic structure (space group Fm3m, JCPD No. 431002) with a lattice space of $a = 0.5411$ nm, indicating a well-crystallized CeO_2 nanorod framework.

The transmission electron microscopy (TEM) images of the replica cerium oxide in Figure 18.2(b) show a well-ordered framework and a good replica of the silica template pore channels.

Compared with the reference sample CeO_2-D, which was obtained by directly calcining $Ce(NO_3)_3 \cdot 6H_2O$ at 823 K for four hours, the mesoporous ceria replica shows higher catalytic activity for the oxidation of CO to CO_2 at a relatively low temperature. After being further loaded with CuO, the composite can greatly enhance the CO oxidation reactivity. The conversions of CO over these samples are shown in Figure 18.3, which shows the strong effect of CuO loading on the CO conversions. The temperatures for 50% conversion of carbon monoxide (T_{50}) for the samples of CeO_2-D, and mesoporous ceria replica and CuO-loaded ceria replica are shown in the inset. The T_{50} of the cerium oxide replica (523 K) is significantly lower than that of CeO_2-D (606 K), meaning that the activity of this replica CeO_2 is higher than that of CeO_2-D. With the increase of the CuO loading amount, the T_{50} is distinctively decreased, and the lowest T_{50} (389 K) of all catalyst samples is achieved at 20% CuO loading, and then the T_{50} (400 K) increases slightly as the CuO loading amount increases to 30%.

Recently, Deshpande et al.[16] demonstrated a novel synthesis of thermally stable, mesoporous ceria using the sols of pure CeO_2 nanoparticles as building blocks in a block-copolymer-assisted assembly process. A hydrogenated polybutadiene-poly(ethylene oxide) block copolymer (PHB-PEO; $H[CH_2CH_2CH_2CH(CH_2\ CH_3)] \times (OCH_2CH_2)yOH$) was used as the structure-directing agent. This synthesis pathway involves the

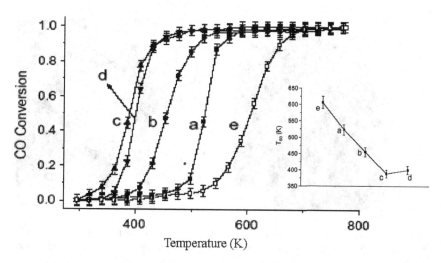

Figure 18.3. Percentage of CO conversion of $CO + O_2$ to CO_2 reaction and T_{50} (inset) value over ordered mesoporous ceria replica loaded with 0% (a), 10% (b), 20% (c), and 30% CuO. (e) is for a reference sample.[15]

addition of the CeO_2 nanoparticle sol in a mixture of ethanol and water to an alcoholic solution of the PHB-PEO block copolymer. Evaporation of the solvent induces the cooperative assembly of the nanoparticles and the PHB-PEO block copolymer micelles. By this approach, they have successfully assembled the nearly monodisperse and highly crystalline ceria nanoparticles (with a diameter of about 3 nm) into highly ordered three-dimensional mesostructures. Furthermore, the mesopores are particularly large and show a high degree of ordering in terms of pore shape and three-dimensional arrangement, as shown in Figure 18.4.[16] The BET surface area obtained for the calcined sample was $87 \, m^2/g$ after calcination at 500°C, and such a relatively low specific surface area value was partly due to the high density of CeO_2 ($7.132 \, cm^3/g$). This mesostructure was stable upon calcination at 500°C and consisted of relatively large mesopores of about 10–12 nm.

18.2.2 Synthesis of mesoporous metal-loaded CeO_2/ZrO_2 composite catalysts and their catalytic properties

Ceria-zirconia mixed oxides are currently targets of intensive research studies for their successful applications in the latest generations of automotive exhaust TWCs due to their unique redox properties, high oxygen storing/releasing capacity, and high thermal stability.[17] Controlling the

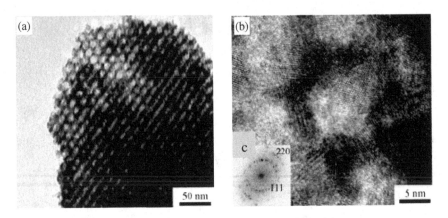

Figure 18.4.   (a) TEM image of the sample calcined at 500° C for two hours; (b) HRTEM image of a 32 × 32 nm^2 selected area; and (c) the corresponding electron diffraction pattern.[16]

porosity of these catalysts is highly desirable, and it was noted that the chemical/physical properties and structural texture are strongly dependent on the preparation procedure. The use of high-surface-area mesoporous oxide supports may give rise to well-dispersed and stable noble-metal nanoparticles on the pore surface and thus can strongly influence their catalytic performances. Here we give an example of highly dispersive pure platinum nanoparticles, which was well confined in the pore channels of ordered mesoporous zirconia. This novel nanocomposite shows good catalytic activity.

The platinum-loaded mesoporous zirconia sample was synthesized by an ion-exchange method.[18] The as-synthesized mesophase zirconia containing surfactants was first post-treated by phosphoric acid to modify the inner pore surface and then sulphuric acid and ethanol were used to extract the surfactants by an ion-exchange route. After the removal of the surfactants, a sucrose solution containing H_2PtCl_6 was filled into the pore channels by ion exchange under vigorous agitation. After being aged and dried at 90°C, the platinum was confined and spontaneously reduced $in\ situ$ during thermal treatment at 773 K. Figure 18.5(a) shows a representative high-resolution transmission electron microscopy (HRTEM) image recorded from a platinum-loaded (0.5 at.%) mesoporous zirconia sample. A large number of homogeneously dispersed metallic platinum nanoparticles in pore channels with uniform size can be directly observed. From the measurement of the Pt nanoparticles in these TEM images, an average particle size of 1.8 ± 0.1 nm was determined. These confined platinum nanoparticles show high catalytic activity for the oxidation reaction of CO. The percent CO

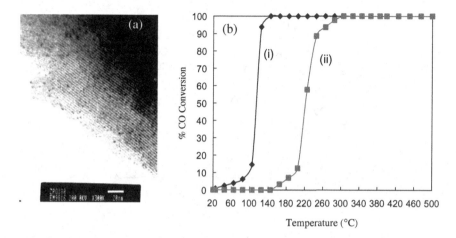

Figure 18.5.   (a) Representative HRTEM image of sample Pt/M-ZrO$_2$ and (b) the percent CO conversion over (i) 0.5 at.% Pt/M-ZrO$_2$ for CO+O$_2$ reaction, and (ii) the reference Pt-loaded mesoporous zirconia sample prepared by the traditional impregnation method.[18]

conversions over the prepared 0.5 at.% Pt/M-ZrO$_2$ sample and the reference Pt-loaded sample prepared by the traditional impregnation method are also shown in Figure 18.5(b) for comparison. The quickly accelerated CO conversion starts below 120°C, and complete CO oxidation is achieved at 145°C over the 0.5 at.% Pt/M-ZrO$_2$ sample, giving CO$_2$ as the product, which can ensure immediate activation of the catalyst on engine start-up, whereas the same reaction under identical conditions requires as high as 300°C for the reference sample.

After further incorporation with ceria, the temperature-programmed reaction profiles for the conversion of CO + NO into CO$_2$ + N$_2$ over such a composite sample CeO$_2$ (5 wt.%)-Pt(1 at.%)/M-ZrO$_2$ are shown in Figure 18.6. The mixed gases start their fast conversion into CO$_2$ + N$_2$ at 120°C, and reach their complete conversion at about 220°C, which indicates that this platinum-loaded and ceria-doped mesoporous zirconia material shows high and sustained catalytic activity for the conversion of NO + CO into CO$_2$ + N$_2$ at a relatively low temperature.

In recent years, hierarchically structured porous materials at multiple length scales have attracted much attention owing to their high surface areas and regular porosity at the nanometer scale, which can provide many novel properties and have important prospects in practical industrial processes such as catalysis, adsorption, separation, and some biological applications.[19-21] In this kind of study, surfactants have been shown to be important in organizing the inorganic species into a variety

J. Shi et al.

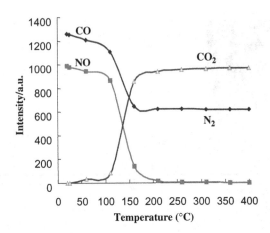

Figure 18.6.   Temperature-programmed reaction profiles for the conversion of $CO + NO$ into $CO_2 + N_2$ over sample CeO_2(5 wt.%)-Pt(1 at.%)/M-ZrO_2 (because the mass number of CO is the same as that of N_2 ($m/z = 28$), their signal positions are at the same curve).[18]

of mesoporous forms via self-assembly or cooperative processes. Thus, the combination of surfactant and the exo-templating methods, such as colloidal polymer latex and emulsion droplets, can allow the fabrication of hierarchical meso/macroporous architectures.[22] Yu and co-workers[23] demonstrated the synthesis of a thermally stable, hierarchically ordered, meso/macrostructured, and palladium-loaded ceria-zirconia catalyst using a combination of surfactant and colloidal crystal templating methods. A typical synthesis process is as follows[1]: A 10 wt.% micellar solution of cetyltrimethyl ammonium bromide (CTAB) is prepared by dissolving the surfactant in water at room temperature under ultrasound irradiation. A meso/macrostructured sample with cerium content $Ce_xZr_{1-x}O_2$ (X = 0.6, denoted as MM60) is prepared by mixing an appropriate ratio of zirconium n-propoxide ($[Zr(OC_3H_7)_4]$) with ammonium cerium(IV) nitrate ($[(NH_4)_2Ce(NO_3)_6]$) in a minimum amount of n-propanol with ultrasound-assisted dissolution. After homogenization, the resulting reddish-brown solution is slowly added to the surfactant solution with gentle stirring to disperse the droplets, and the pH of the solution is kept at 11 using ammonia. The obtained mass is transferred into a teflon-lined autoclave, and heated under static conditions at 80°C for 48 hours. The product is filtered and dried overnight and then calcined at 500–800°C for the removal of the surfactant and crystallization. The reference sample prepared by the co-precipitation method with a cerium content of 60% is denoted as CP60. A 0.8 wt.% Pd in $Ce_xZr_{1-x}O_2$ is prepared by immersing the catalyst support

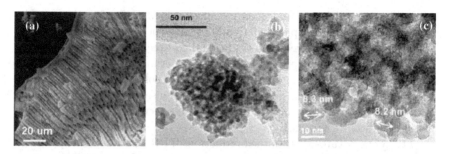

Figure 18.7. SEM (a) and TEM images (b, c) with different magnifications for Pd-MM60 after calcination.[23]

in an aqueous solution of $PdCl_2$ under reduced pressure and ultrasound irradiation. The prepared sample shown in Figure 18.7[23] indicates that both macrochannels (A) and mesoporous structures (B, C) can be observed. The surface area and pore volume are $131.4\,m^2/g$ and $0.14\,cm^3/g$ respectively, which are much higher than those of the reference sample Pd-CP60. The catalytic activity profile of CO conversion over Pd-MM60 and the reference samples show that the T_{50} and T_{90} for Pd-MM60 are 65 and 75°C, respectively, which are much lower than those for Pd-CP60. This may be because Pd-MM60 possesses mesoporous frameworks that provide a high surface-area-to-volume ratio, and hence more active sites are available for CO oxidation. The poor activity of Pd-CP60 can be related to its structural and chemical properties. When calcined at 700°C, the BET surface area of Pd-CP60 decreases sharply to $20.6\,m^2/g$; in addition, Pd-CP60 annealed at 700°C consists of a mixture of cubic and tetragonal phases. Such phase inhomogeneity also strongly influences the oxidation/reduction behavior.

For the sake of energy saving, the ratio of air/fuel fed into internal combustion engines trends toward higher values than the stoichiometric 14.7, which gives rise to decreased NO reduction efficiency of TWCs in such a fuel-lean condition. Therefore, the development of a new generation of TWCs to improve the NO reduction activity under oxygen-rich conditions is becoming urgent. Recently, Wang and co-workers reported[24] a surfactant-controlled synthetic method to obtain a nanophase of mesoporous ceria-zirconia solid solution by using cerium nitrate $(Ce(NO_3)_3 \cdot 6H_2O)$ and zirconium oxychloride octahydrate $(ZrOCl_2 \cdot 8H_2O)$ as precursors and a cationic surfactant, myristyltrimethylammonium bromide $(CH_3(CH_2)_{13}N(CH_3)_3Br)$ as molecular template. The $3\,wt.\%$ $Pd/Ce_{0.6}Zr_{0.4}O_2$ catalyst was prepared by impregnating the mesoporous $Ce_{0.6}Zr_{0.4}O_2$ support, which was calcined at 800°C, with an aqueous solution of $Pd(NO_3)_2 \cdot 2H_2O$. The metal-supported catalyst was dried at

120°C for several hours and then calcined at 600°C for four hours. The surface area of the sample prepared by using such a method is 10–15% higher than those prepared by a traditional co-precipitation method. Compared with pure ceria prepared by the same method with a crystallite size of 31 nm at 800°C, the prepared $Ce_{0.6}Zr_{0.4}O_2$ solid has a much smaller crystallite size of 11.4 nm after calcination at the same temperature, which indicates that zirconium addition to ceria can effectively inhibit the sintering and crystallite growth.

The temperature-programmed reduction (TPR) method was applied to study the reduction properties of oxygen species in the ceria-zirconia support and Pd/ceria-zirconia catalyst. The TPR profile of the ceria-zirconia calcined up to 800°C consisted of two hydrogen consumption peaks: one at 420°C and another at 590°C (Figure 18.8(a)).[24] The peak at the higher temperature (590°C) is ascribed to the reduction of bulk lattice oxygen and the one at 420°C is ascribed to the reduction of the surface oxygen species. On pure CeO_2, two peaks are also observed in the TPR profile; however, the temperature for bulk oxygen reduction in ceria is as high as 900°C.[25] This result shows that addition of zirconium to ceria significantly lowers the oxygen reduction temperature, thus improving the reducibility of bulk oxygen. The TPR profile of PdO/Ce-Zr-O shows a sharp negative peak at about 100°C, which strongly indicates hydrogen generation during the TPR procedure (Figure 18.8(b)), which in turn was related to the formation/decomposition of a palladium hydride phase (b-PdHx). Compared to the TPR profile of the $Ce_{0.6}Zr_{0.4}O_2$ solid, the temperature of the peak maximum in the palladium-supported catalyst markedly shifted to the low temperature range. In addition, a new reduction peak appeared

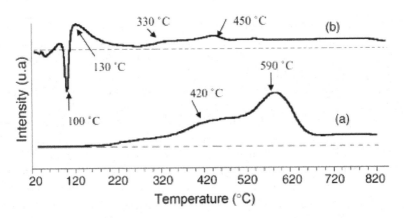

Figure 18.8.   TPR profiles of (a) 3 wt.% PdO/$Ce_{0.6}Zr_{0.4}O_2$ and (b) $Ce_{0.6}Zr_{0.4}O_2$.[24]

Table 18.1. Activity and selectivity of NO reduction by CO over $3\,wt.\%$ $Pd/Ce_{0.6}$ $Zr_{0.4}O_2$ catalyst.[24]

Temperature ($^\circ$C)	CO Conversion (%)	NO Conversion (%)	Selectivity (%)	
			N_2	NO_2
50	32	32	100	0
100	38	33	100	0
150	52	40	100	0
200	67	50	95	5
250	85	60	91	9
300	96	65	93	7
350	100	65	88	12
400	100	70	84	16

at 130°C, which may be associated with the hydrogen spillover from the metal to the support.

Catalytic tests of NO reduction by CO in an oxygen-excess condition showed that the $Pd/Ce_{0.6}Zr_{0.4}O_2$ catalyst had high activity for both CO oxidation and NO reduction, as can be seen in Table 18.1.[24] The light-off temperature (T_{50}) that corresponds to 50% conversion of CO to CO_2 and NO reduction is approximately 130 and 200°C, respectively. At the cool start of the reaction (below 150°C), NO reduction by CO with excess oxygen over the $Pd/Ce_{0.6}Zr_{0.4}O_2$ catalyst showed selectivity of around 100% to N_2. However, a competition between NO reduction by CO and CO oxidation by O_2 was observed: at below 200°C, NO lowered CO oxidation activity; however, at reaction temperatures above 200°C, the high activity of CO oxidation resulted in an inhibited effect on NO reduction. The presence of NO may induce oxidation of the metallic Pd active sites, thus inhibiting the CO oxidation on Pd in the range of low reaction temperatures. On the other hand, NO reduction was also strongly affected by CO oxidation under excess oxygen conditions.

18.3 Electrocatalytic Performance of Mesoporous-based Nanocomposites for Full Cell Reactions

Fuel-cell-based automobiles have gained great attention in recent years due to growing public concern about air pollution and consequent environmental problems.[26] Polymer electrolyte membrane (PEM) fuel cells and direct methanol fuel cells (DMFCs) as well as their accompanying environmental benefits have resulted in intensive studies on the fundamental and applied aspects of their potential as clean and ideal power sources. To increase the catalytic activity of methanol electro-oxidation, much effort has been devoted towards the development of new catalysts. In addition, the development of new carbon materials as supports to help achieve optimum catalytic performance has also attracted great attention, for the common carbon supports in practical use have broad pore size distributions and irregular structures.

18.3.1 *Ordered porous carbons as catalyst supports in direct methanol fuel cell*

DMFCs are now being considered as candidate power sources for both portable power applications and electric vehicles as they provide high energy output and high yield, theoretically without harmful by-products.[27,28] It is now widely accepted that highly dispersed platinum electrodes can provide higher activity and weaken the poisoning effect compared with bulk platinum. Therefore, a high surface area and well-developed porosity are essential for a catalyst support to induce high catalytic activity. Except for carbon black, which has been commercially used as a carbon support, several different carbon materials such as mesostructured carbon and ordered porous carbon were reported as supports for higher dispersion of the catalysts owing to the high surface area and well-defined porosity of these materials. A uniform porous carbon material via colloidal crystal templates has been recently reported by Yu and co-workers[29], as shown in Figure 18.9.

Silica spheres 15–1,000 nm in diameter were used as the colloidal crystal templates for mesoporous carbon spheres. With this approach,[29] uniform

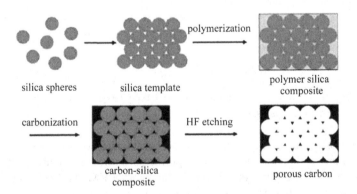

Figure 18.9.   Schematic synthesis procedure for uniform porous carbons via colloidal crystal templates.[29]

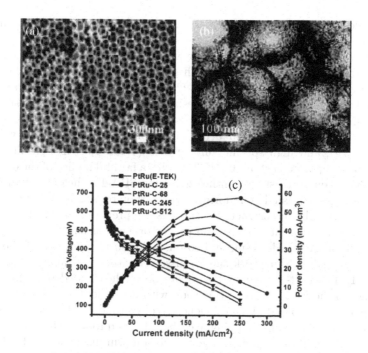

Figure 18.10. SEM (a) and TEM (b) images of the prepared ordered porous carbons. (c) Voltage and power density responses of porous carbon-supported $Pt_{50}Ru_{50}$ alloy catalysts as compared to that of a commercial catalyst (E-TEK) in DMFCs. The DMFCs were operated at $70°C$.[29]

ordered porous carbons of well-defined pore structures with tunable pore sizes and high surface area (from 450 to $1,000\,m^2/g$) can be well obtained by the carbonization of cross-linked phenol and formaldehyde, as can be observed in Figures 18.10(a) and 18.10(b). Figure 18.10(c) shows the unit cell performance of DMFCs at $70°C$ by the porous carbon-supported catalysts as compared to those of the commercial E-TEK catalysts. The catalyst loadings were $3.0\,mg/cm^2$ for each of the supported Pt-Ru anode catalysts, and $5.0\,mg/cm^2$ for the Pt black cathode catalyst, respectively. It can be seen that all the porous carbon-supported Pt-Ru catalysts exhibit much higher specific activity for methanol oxidation than the E-TEK catalyst by about 5–43% in methanol oxidation activity, which was considered to be not only due to the higher surface areas and larger pore volumes, which allow a higher degree of catalyst dispersion, but also to the highly integrated and interconnecting pore systems with periodic order, which allow efficient gas diffusion. The porous carbon with a 25 nm pore diameter (PtRu-C-25) has the highest performance with a power density of $167\,mW/cm^2$ at $70°C$, which corresponds to a 43% increase as compared to that of the commercial catalyst.

As stated earlier, the recent discovery of ordered mesoporous silica provides suitable templates for the synthesis of carbons or other metal oxides with ordered mesoporous structures. Ordered mesoporous carbons such as CMK-1 and CMK-3 have been synthesized by using the three-dimensionally structured mesoporous silica as a hard template. These ordered mesostructured carbons have a regular array of uniform pores, large pore volume (1–2 cm^3/g), and high specific surface area (1,300–2,000 m^2/g). The first ordered mesoporous carbon (CMK-3)[30] with a faithful replica of the ordered silica has been successfully synthesized using SBA-15 silica as a template. The materials consisted of uniformly sized carbon rods arranged in a hexagonal pattern. The mesostructured carbons are prepared by infiltrating the pore system with a suitable carbon precursor, such as sucrose, furfuryl, or a phenol-formaldehyde resin, and so on, followed by pyrolysis of the precursor. The silica framework can be dissolved either by NaOH solution or with HF. Pt or Pt-Ru nanoparticles can be further loaded by impregnating the synthesized mesoporous carbon with a colloidal solution of Pt/Ru nanoparticles prepared by an ethylene glycol method.[31] The energy dispersive X-ray (EDX) image mapping of carbon, Pt, and Ru and the microregion element distribution exhibited a good and uniform dispersion of Pt/Ru in the carbon support with a metal loading of 15.30% by mass.

Very recently, Yi and co-workers[32] proposed a new fabrication method for producing uniformly sized mesoporous Pt-carbon catalysts using mesoporous Pt-alumina as a template with a metal source and using poly(divinylbenzene) as a carbon precursor. Two types of mesoporous Pt-alumina templates were prepared under different calcination conditions (PtAl-A and PtAl-N were produced by calcination in a stream of air or nitrogen, respectively). Compared to the conventional impregnation method, this method does not require an additional Pt deposition step on the support because the template itself serves as a metal source. The Pt-carbon catalysts retained a well-developed mesoporosity with a high surface area (> 590 m^2/g) and a large pore volume (> 0.45 cm^3/g). Furthermore, the mesoporous Pt-carbon (PtC) catalysts showed a higher metal dispersion and better electrocatalytic performance than the impregnated Pt-carbon (Pt/CMK-3) catalyst.[32]

Figure 18.11 shows the catalytic performance of the prepared Pt catalysts. Methanol electro-oxidation occurred at about 0.4 V in all Pt-carbon catalysts, a typical feature for non-alloyed Pt-carbon catalysts.[33] The maximum current density observed at 0.6 V for the PtC catalysts (especially PtC-N) was higher than that for the impregnated Pt-carbon (Pt/CMK-3) catalyst.

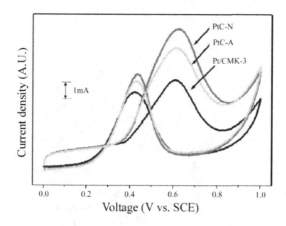

Figure 18.11.   Cyclic voltammograms of PtCs and Pt/CMK-3 catalysts on a glass carbon disk electrode (working electrode) in H_2SO_4 (0.5 M) containing CH_3OH (1 M). The Pt loadings on the working electrode were identical in all the cases (scan rate = 20 mV/s).[33]

18.3.2   *Mesoporous precious-metal catalysts for methanol electro-oxidation*

The development and characterization of poison-tolerant catalysts for the oxidation of methanol has stimulated increased interest in the surface morphology of platinum and platinum-based catalysts. Mesoporous metallic systems appear to be a class of attractive electrocatalysts owing to their high specific surface area and periodic nanostructural texture. For example, platinum produced from the liquid crystal phase was characterized by a specific surface area of approximately $60 \, m^2/g$ (compared to $35 \, m^2/g$ for platinum black) and large particle sizes, as shown in Figure 18.12.[34] The combination of high surface areas, uniform pore diameters, and large particles makes the metallic system considerably interesting for applications in catalysis, batteries, fuel cells, electrochromic windows, and sensors.[35] It was found that the intrinsic activity of platinum for oxygen reduction was more favorable compared to that of commercial platinum. Furthermore, the mesoporous platinum showed potential-dependent tolerance to CO poisoning and high electrocatalytic activity towards formic acid oxidation.[36]

In addition, the production of binary or ternary Pt-base alloys that combine high specific surface areas, controllable pore diameters, and uniform distribution of metal species is expected to affect a wide range of catalytic processes. Recently, a mesoporous Pt-Ru alloy was successfully synthesized by chemical co-reduction of hexachloroplatinic acid and ruthenium

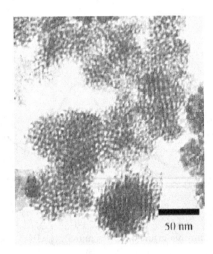

Figure 18.12.   TEM image of mesoporous Pt nanoparticles.[34]

trichloride within the aqueous domains of a normal topology hexagonal
mesophase of oligoethylene oxide surfactants using Zn as the reductant.[36]
The resultant material containing mesoporous structures has a high specific surface area and has been examined as an electrocatalyst towards the
electro-oxidation of methanol using cyclic voltammetry and chronoamper-
ometry. Figure 18.13 shows the results of methanol catalytic oxidation. It
can be seen that the oxidation current increases considerably at increased
potential until a current peak at about 0.75 V. Further increases over the
potential lead to a rapid decay in the oxidation current. Upon reversing
the scan at 0.85 V, the current further decays until 0.80 V at which point a
small peak is seen with further decreases in potential. The current density
of the forward-scan current peak is calculated to be 700 mA/cm^2, within
a factor of two to three of the peak current density value, assuming a
diffusion-controlled irreversible six-electron oxidation reaction under the
same conditions. When the potential scan is reversed at 0.70 V, no peak
or hysteresis loop in the current response is observed in the inset, which
suggests that there is virtually no accumulation of poisons on the electrode
surface when the potential is more negative than 0.70 V over the timescale
of this experiment. It is believed[36] that methanol is oxidized mainly via a
direct pathway to soluble products on the mesoporous Pt-Ru electrode and
steady-state kinetics can be attained even at low potentials. The electrocatalytic activity of the mesoporous Pt-Ru towards methanol oxidation is
more favorable compared to an ultrafine Pt-Ru electrocatalyst with similar
bulk composition.

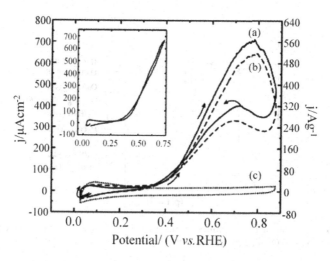

Figure 18.13. Cyclic voltammograms for a mesoporous Pt-Ru electrode in 0.5 mol/dm H_2SO_4 solution containing 0.5 mol/dm methanol at 60°C (dE/dt = 50 mV/s). (a) First scan; (b) second scan; (c) cyclic voltammogram in the absence of any methanol. Inset: cyclic voltammogram under the same conditions over 0 to 0.7 V.[36]

18.4 Synthesis of Mesoporous Titania and its Photocatalytic Applications

Due to their low cost, strong redox power, and high chemical/photocorrosion stability, titanium-dioxide- (TiO_2-) based materials have been extensively used and further studied in the past decades to address the increasing air and water pollution problems through photocatalytic processing, especially for decomposing organic contaminants with low levels. Generally, under UV irradiation from either artificial or solar sources with energy larger than the material's bandgap, electron-hole pairs are generated in the wide-bandgap semiconducting TiO_2 materials because of the electron transition from the valence band to the conduction band. As shown in Figure 18.14, in a pH 7 solution, the redox potentials for photogenerated holes (h^+) and photogenerated electrons (e^-) are +2.53 V and −0.52 V versus the standard hydrogen electrode (SHE), respectively.[38] Then interfacial electron-transfer reactions with adsorbed organic compounds lead to them decomposing to CO_2 and H_2O through radical chain reactions or other mechanisms; and consequently purification of polluted air or wastewater. Since the catalytic reaction is a surface process, a high surface area resulting from smaller particle size or porous structures means better catalytic efficiency. Therefore, with the discovery of ordered mesoporous materials,[5,6] recently the synthesis of TiO_2-based mesoporous materials

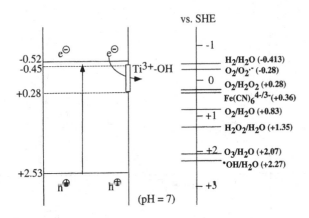

Figure 18.14. Schematic diagram showing the potentials for various redox processes occurring on the TiO_2 surface at pH 7.[38]

has also attracted great attention due to their higher surface areas and easy recovery after reaction compared to that of the ordinary TiO_2-based materials; this probably leads to higher photocatalytic efficiency in the purification of polluted air or wastewater. In this part, preparation of mesoporous TiO_2 with amorphous or crystallized frameworks and TiO_2-containing mesoporous materials and their photocatalytic applications are presented.

18.4.1 *Synthesis of pure TiO_2 mesoporous materials and their photocatalytic applications*

18.4.1.1 *Mesoporous TiO_2 with amorphous framework*

For mesoporous materials synthesis with surfactants as soft templates, the interactions among the inorganic precursors, organic species, and solvent/co-solvent in a bulk solution cooperatively determine the ordering degree and symmetry of the final structures. Thus, since 1992, although ordered mesoporous silica materials with adjustable pore sizes and diverse pore structures have been reported through proper choices of many process parameters[5,6,39−41] (such as surfactant species, reactant molar ratios, pH values, reaction temperature, and reaction time), the successful preparation of transition-metal-oxide-based mesoporous materials, such as TiO_2, is difficult because of the high reactivity of titanium sources towards hydrolysis/condensation. As a result, poorly structured materials with a dense inorganic network were often formed. To lower the precursor hydrolysis/condensation rates and control the formation

Table 18.2.   Characterization and reactivity results for titania materials.[45]

Sample	Surface Area (m^2/g)	Bandgap (eV)	Quantum Yield[a]
Meso-TiO$_2$ (473 K)	712	3.19	0.0026
Meso-TiO$_2$ (973 K)	90	3.25	0.0089
Meso-TiO$_2$, extracted[b]	603	3.15	0.010
ns-TiO$_2$, as-synthesized		3.24	0.0094
ns-TiO$_2$ (873 K), 1 h	39		0.022
ns-TiO$_2$ (873 K), 3 h	36	3.22	0.038
ns-TiO$_2$ (873 K), 12 h	28	3.26	0.27
Anatase (Aldrich)	9	3.28	0.41
Degussa P25	50	3.22	0.45

[a]Quantum yield is defined as the molecules of acetone formed per incident photon.
[b]Surfactant was extracted three times with 8% nitric acid in ethanol.

processes of non-silica-based mesoporous materials, stabilizing agents such as acetylacetone,[42] glycolate,[43] triethanolamine,[44] alkyl phosphate,[45] and primary amine[46] surfactants were used with little water or in the absence of it. Then through calcination or solvent extraction to remove the surfactant templates, mesoporous TiO$_2$ were obtained, though their structural ordering and thermal stability needs to be further improved. It is noted that at this stage most of the synthesized mesoporous TiO$_2$ were of amorphous frameworks or consisting of very small crystallites, which could hardly be recorded in XRD or electron diffraction (ED) patterns. However, it is known that the basic requirement for photoactive TiO$_2$ materials is their high crystallinity. The defects present in the amorphous or small crystallites are the recombination centers for the photogenerated electron-hole pairs, which would cause the decrease of quantum yields. Table 18.2 shows the structural properties and photocatalytic results of acetone for mesoporous TiO$_2$,[45] nanosized TiO$_2$ prepared without surfactant template, and commercial TiO$_2$ powder Degussa P25, a standard titania P25. It is obvious that the quantum yields of the mesoporous samples (0.00xx) are several orders of magnitude lower than that of nanosized TiO$_2$ (0.27) and of P25 (0.45), although the mesoporous one possesses a larger surface area. Thus, high-surface-area mesoporous titania with controllable crystalline framework is required to achieve its photocatalytic applications.

18.4.1.2 *Mesoporous TiO$_2$ with crystallized framework*

Recently, with the development of the so-called evaporation-induced self-assembly (EISA) process,[47] highly organized mesoporous TiO$_2$ materials with nanosized anatase walls have been prepared reproducibly using poly(ethylene oxide)-(PEO-) based templates.[48,49] The resulting materials not only possess high surface area and high thermal stability but also exhibit different and tailored mesostructural symmetry (worm-like, cubic $Im3m$, two-dimensional hexagonal $p6m$). Figure 18.15 is the schematic diagram for this whole process, and various characterization techniques accompanying each step describe the time evolution of the mesostructure and consequently clarify the formation mechanism during EISA.[49] Starting from $TiCl_4$ ethanol solution, the extended X-ray absorption fine structure (EXAFS) and UV-Vis experiments confirmed the appearance of the inter-mediated product of metallic chloroethoxide precursors $[TiCl_{4-x}(OEt)_x]$ with $x \approx 2$. Further, to this highly acidic system, controlled quantities of water were slowly added to induce the hydrolysis of the inorganic moieties. In a certain range of molar ratios of m_{H_2O}/m_{Ti} (> 4), clear sols containing hydrolyzed hydrophilic moieties with proper polymeriza-tion degrees were formed. The dip-coating process was conducted in a cir-cumstance with controlled relative humidity to prepared mesostructured TiO_2 films. In situ characterization with mass spectroscopy (MS), syn-chrotron radiation of small-angle X-ray scattering (SAXS), and interfer-ometry reflected the mesostructure evolution (including the disordered film formation first and then the occurrence of the disorder-to-order transition) with fast evaporation of solvent ethanol and progressive concentration of water, HCl, and other non-volatile components. Post-treatments under con-trolled atmosphere[50] or with gradual heating[49] would continuously enhance the condensation degree of the TiO_2 framework with the slow elimination of water and HCl and eventually result in the formation of more thermally stable structures. After being calcinated at higher than 300°C to remove the surfactant template and thermally treated at elevated temperature (up to 600°C), mesoporous TiO_2 thin films with high surface area and crystallized frameworks were obtained (Figure 18.16). Using a similar process combined with more careful choice of surfactant species or thermal treating programs, crystalline mesoporous TiO_2 materials with a larger pore size ($\sim 10\,nm$)[51] and higher thermal stability (up to 700°C)[52] have been achieved. Instead of a common Ti source of chloride or alkoxide, "pre-synthesized" TiO_2 nanoparticles or other nano-building blocks were also considered to be used to alleviate the system sensitivity to the process conditions.[53-55] However, the structural ordering and the thermal stability of the resulting products needs to be further improved.

Considering the inorganic-inorganic interplay between two or more inor-ganic precursors, a new and general method for the preparation of non-siliceous mesoporous materials has been reported based on the acid-base pair principle.[56,57] For the synthesis of mesoporous TiO_2, according to the relative acidity and alkalinity of inorganic precursors on solvation, an acid precursor (titanium alkoxide) and a base counterpart (titanium chloride)

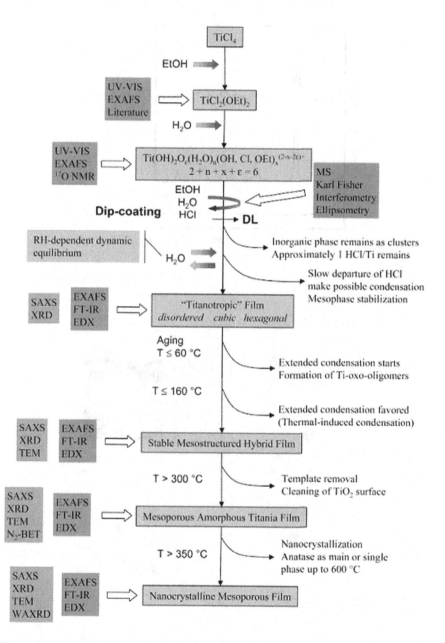

Figure 18.15.   Sequential events in the formation of ordered mesoporous TiO_2 films.[49]

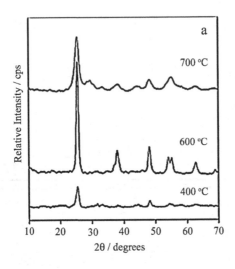

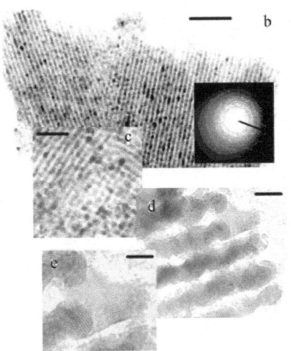

Figure 18.16.   (a) XRD patterns of F127-templated TiO_2 mesoporous films calcined at the temperature indicated. (b–e) HRTEM images of TiO_2 films treated at 600°C at increased magnifications; the inset in part (b) is an ED pattern characteristic of anatase. Scale bars are (b) 100 nm, (c) 50 nm, (d) 10 nm, and (e) 5 nm.[49]

were chosen. In the amphoteric solvents, such as ethanol or tetrahydrofuran, a self-adjusted sol-gel synthesis took place without any extra reagents. Moreover, since metal alkoxide served as an extra oxygen donor, the Ti-O-Ti bridge partially originated from the condensation between Ti-Cl and Ti-OR and the cross-linkage of the inorganic framework in the resulting materials were strengthened. After being calcined at 350°C to remove the surfactant template, polycrystalline diffraction rings were observed in the ED pattern, which confirmed the semicrystallized wall structures in the synthesized TiO_2. However, the material thermal stability at elevated temperatures was not mentioned.

18.4.1.3 *Photocatalytic behavior of mesoporous TiO_2 with crystallized framework*

Under 365 nm UV irradiation, the photocatalytic activity of EISA-prepared mesoporous TiO_2 nanocrystalline films was evaluated by the photodecomposition of acetone,[58] a widely used industrial and domestic chemical. As shown in Table 18.3, the photocatalytic activity increases with the increase in thermal treatment temperature, which could be explained by the gradual crystallization of the titania framework. For sample MT500 calcined at 500°C, specific photocatalytic activity two times higher than that of the conventional TiO_2 film could be obtained, which resulted from the large surface area (85 m^2/g versus 9.1 m^2/g) and its three-dimensionally connected pore structures of the mesoporous catalysts.

Table 18.3. Photocatalytic activity of mesoporous TiO_2 thin films sintered at different temperatures and their comparison with a conventional TiO_2 thin film.[58,a]

Sample	Mass	Degradation[b]	Rate Constant K (min^{-1})	Specific Photoactivity	
	(mg)	Rate (%)		(mol/g/h)[c]	(mol/m^2/h)[d]
MT300	13.2	1.6	$2.6*10^{-4}$	$4.0*10^{-4}$	$3.8*10^{-4}$
MT400	12.5	9.9	$16.4*10^{-4}$	$2.6*10^{-3}$	$2.3*10^{-3}$
MT500	12.2	11.6	$19.4*10^{-4}$	$3.1*10^{-3}$	$2.7*10^{-3}$
MT600	11.8	7.9	$10.8*10^{-4}$	$1.8*10^{-3}$	$1.5*10^{-3}$
TiO_2-500[e]	8.8	4.8	$7.6*10^{-4}$	$1.9*10^{-3}$	$1.2*10^{-3}$

[a]The area of the substrate covered by the TiO_2 monolayer was 140 cm^2.
[b]Average degradation rate of acetone after one hour of photocatalytic reaction.
[c]Acetone degradation amount per unit mass catalyst after one hour of photocatalytic reaction.
[d]Acetone degradation amount per unit film surface after one hour of photocatalytic reaction.
[e]Data from Ref. 59.

Table 18.4. Physical characterizations of NH_3-treated mesoporous titania synthesized with HDA and CTABr in acidic (A) and basic (B) medium. The calculated rate constants are derived from the first-order photocatalytic reaction of rhodamine 6G photodecomposition.[50]

Sample[a]	S_{BET} (m^2/g)	V_p (cm^3/g)	D_{BJH} (Å)	k (min^{-1})
TiO_2–HDA–A–T	258	0.20	26.8	0.016
TiO_2–HDA–B–T	195	0.11	20.0	0.007
TiO_2–CTABr–A–T	340	0.28	34.0	0.029
TiO_2–CTABr–B–T	180	0.25	32.0 (broad)	0.029

[a]T stands for NH_3-treated. HDA and CTABr are the abbreviations for hexadecylamine and cetyltrimethylammoniumbromide, respectively.

Researchers studied the effect of solution conditions on the photocatalytic activity of the synthesized mesoporous titania.[50] As shown in Table 18.4, using the photodecomposition of rhodamine 6G as the model reaction, the photocatalytic activity of mesoporous TiO_2 synthesized under different conditions were compared. It is obvious that with similar or even smaller surface areas, TiO_2-CTABr series samples with a larger pore size (> 3.0 nm) possess better photocatalytic activity than TiO_2-HDA-B-T samples with a smaller pore size (2.0 nm), which reflects the accessibility of the anatase phase for rhodamine 6G and in turn determines the photocatalytic activity of the mesoporous titania. The same reasons could be used to explain the higher photocatalytic activity of TiO_2-HDA-A-T samples than that of TiO_2-HDA-B-T, because under acidic conditions, the slower condensation of the titania precursor leads to larger-pore mesostructures and consequently better accessibility for photocatalysis.

Solvent effects during the EISA process to prepare mesoporous TiO_2 materials were emphasized recently through several experiments.[60,61] Because of the higher hydrophobicity of 1-butanol than the usually used ethanol, the use of 1-butanol could enhance the microphase separation between the template and the inorganic precursor and allow for the formation of a robust, structurally well-ordered mesoporous TiO_2 material with a nanocrystallized framework. It was further found that by just simply varying the solvent and co-solvent (methanol, ethanol, 1-butanol, or 1-octanol), stable mesoporous titania with pure anatase, pure rutile, bicrystalline (anatase and rutile) with controlled phase composition, and tricrystalline (anatase, rutile, and brookite) frameworks could be synthesized.[62] As different phase structures or phase mixtures of TiO_2 showed varied photocatalytic activity for different kinds of organic pollutants, this report opened up new directions of catalyst design for future TiO_2-based photocatalytic materials with higher quantum yields. As an example, mesoporous rutile TiO_2 powder with an average pore diameter of 2.6 nm and surface area of 174.5 m^2/g showed better photocatalytic results than that of nano-anatase titania with a surface area of 313 m^2/g or nano-rutile titania with a surface area of 115.6 m^2/g, when the gas-phase photo-oxidation of a mixture of benzene and methanol was used as the probe reaction.[63]

18.4.1.4 Photocatalytic activity of specially structured mesoporous TiO_2

Hierarchically macro/mesoporous titania is another interesting material for its photocatalytic application. The existence of an additional macroporous structure not only accelerates the mass transport of organic pollutants to the catalytic titania phase and the discharge of photodecomposed products, but also increases the photoabsorption efficiency and consequently the photocatalytic efficiency, because the extinction of light in semiconductors follows the exponential law $[I = I_0 \exp(-\alpha l)]$, where l is the penetration distance of the light and α is the reciprocal absorption length. Combining sol-gel chemistry, multiscale templating approaches, aerosol processing, and specific treatments, spheres of controlled diameter, made of a periodically organized mesoporous titania crystalline network that surrounds spherical macropores, have been prepared.[64] Just using a single surfactant template (e.g. Brij56 or P123) with or without ultrasonic treatment,[65,66] hierarchical macro/mesoporous titania was also obtained (Figure 18.17), in spite of its

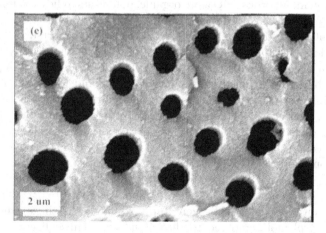

Figure 18.17.   SEM images of hierarchical macro/mesoporous TiO_2 calcined at 500°C.[66]

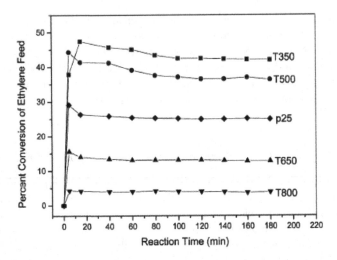

Figure 18.18.   Percent conversions of ethylene on hierarchical macro/mesoporous TiO_2 calcined at 350, 500, 650, and 800°C, and Degussa P25.[66]

decreased mesostructural ordering. Figure 18.18 shows the ethylene conversion on the series hierarchical macro/mesoporous TiO_2 and Degussa P25 with 254 nm UV irradiation.[66] The photocatalytic conversion rates with samples T350 and T500 are 42% and 37%, respectively, which are larger than the ~26% of commercial TiO_2 powder Degussa P25.

In addition to the phase structure, the photodecomposition catalytic activity of mesoporous TiO_2 on organic pollutants relies on its surface area as well. On the other hand, novel photocatalytic activity of mesoporous TiO_2, based on its structural advantages, was reported recently and described as "stick-and-leave transformation".[67] The photocatalytic transformation of various substrates can be found in Table 18.5. D represents the distribution ratio of substrates in the solution and adsorbed on the catalysts while $mTiO_2$ and $nTiO_2$ refer to mesoporous and non-porous TiO_2 materials, respectively. The transformation of benzene into phenol, one of the most difficult synthetic reactions, gives typical results. Because benzene molecules are well adsorbed on $mTiO_2$ ($D = 0.64$) and no phenol adsorption happens, it reduces the by-reaction rates and shows very high selectivity to phenol ($> 80\%$). In contrast, $nTiO_2$ catalysts give very low product yield (1 to approximately 2%) and selectivity ($< 10\%$), resulting from the lower D value of benzene (approximately 0.2 times) and higher D value of phenol compared to that of $mTiO_2$. Through enhancing the reaction selectivity, this process would address the environmental problem of chemical industries as sources of pollution.

Table 18.5. Photocatalytic transformation of various specially structured mesoporous TiO_2 substrates.[67]

Run	Catalyst	Reactant	D (%)	Conv (%)	Product	D (%)	Yield (%)	Select (%)
1	$nTiO_2(100)$		0.17	90		0.01	39	35
2	$nTiO_2(58)$		0	82		0	28	34
3	$mTiO_2(61)$	11	0.11	89	1	0	61	72
4	$nTiO_2(100)$		0.10	91		0.01	16	18
5	$nTiO_2(58)$		0	87		0	11	13
6	$mTiO_2(61)$	14	0.12	90	6	0	65	72
7	$nTiO_2(100)$		0.05	60		0	26	23
8	$nTiO_2(58)$		0	71		0	12	17
9	$mTiO_2(61)$	9	0.09	84	4	0	75	89
10	$nTiO_2(100)$		0.24	26		0.01	2	8
11	$nTiO_2(58)$		0.28	16		0	1	6
12	$mTiO_2(61)$	16	0.64	23	1	0	19	83
13[b]	$mTiO_2(61)$		0.64	42		0	34	81
14[c]	$mTiO_2(61)$		0.64	10		0	8	80

[a]Reaction conditions: reactants, 20 μmol; photoirradiation time, two hours; catalyst, 10 mg; buffered (pH 7) aqueous solution, 10 mL; temperature, 313 K.
[b]Photoirradiation time, six hours.
[c]Reactants, 0.5 mmol (which is not fully dissolved in the aqueous solution).

18.4.2 Synthesis of TiO_2 containing mesoporous materials and their photocatalytic applications

Besides the pure TiO_2 mesoporous materials mentioned, TiO_2-containing mesoporous materials represent a wider range for the choice of new photocatalysts, which include nanosized TiO_2-particle-incorporated silica mesoporous materials,[68-72] mesostructured TiO_2-other semiconductor (or noble metal) composites,[73,74] metal-ion-doped mesoporous TiO_2 materials,[70,75] surface-modified mesoporous materials,[76] and so on. In fact, in view of the complexity during the preparation of pure TiO_2 mesoporous materials and the synergistic effect of structural and functional characteristics of different components in TiO_2-containing mesoporous composites, the composite is more advantageous for practical applications.

18.4.2.1 TiO_2 supported on mesoporous silica

Using ordered mesoporous silica materials as supports, nanosized TiO_2 particles have been incorporated by many post-synthesis methods. Depending on the support structure, the size and amount of incorporated TiO_2 particles could be varied. In comparison with ordinary anatase TiO_2, they showed better photocatalytic activity towards the photodegradation of rhodamine 6G dye,[68] acetophenone,[72] cyanide,[69] and so on. Figure 18.19 compares the kinetics of cyanide photo-oxidation in the presence of series TiO_2-SBA-15 and TiO_2-commercial SiO_2 composite samples.[69] Although at lower TiO_2 content both samples show similar activity, the rate constant increase of the TiO_2-SBA-15 sample is more significant with the content increase of TiO_2. This is caused by the pore limitation of ordered SBA-15 supports, which controls the size and size distribution of incorporated TiO_2 particles.

18.4.2.2 Mesoporous TiO_2 incorporated with metal particles or other metal ions

Since the incorporated metals would enhance electron migration from the semiconductor surface and reduce the recombination probability of photogenerated electrons and holes, mesoporous TiO_2 catalysts incorporating silver nanoparticles were prepared using wet impregnation followed by heat treatment, and their photocatalytic activity towards the decomposition of stearic acid was studied recently.[73] Figure 18.20 shows the photocatalytic activity of three anatase samples: 400°C calcined cubic mesoporous titania with silver (red circles) and without silver (green triangles) and 600°C calcined mesoporous titania which no longer has mesostructured ordering (blue squares). It can be seen that Ag-containing catalysts has an apparent higher initial activity due to the photogenerated electron transfer from titania to silver. However, after 17 minutes, its photocatalytic activity is lower than that of the sample without silver. Possible reasons are suggested to be related to the stearic acid loading amount and deactivation of the silver-containing sample.

La^{3+}-doped mesoporous titania with a highly crystallized framework and long-range order has been prepared by using nano-anatase particles as nano-building units.[75] In this process, La^{3+} plays two important roles. First, it works as the mesostructure stabilizer by restraining the growth of nano-anatase particles, especially in the direction perpendicular to the walls. Second, it increases photocatalytic activity by enhancing the quantum yield. Through the degradation experiment of methyl orange dye under UV light irradiation, 0.25 at.% of La^{3+} was determined as the optimal doping amount, at which the mesoporous titania had the highest photocatalytic activity (shown in Figure 18.21). With higher La^{3+} doping content, the decrease in photocatalytic activity may be caused by the accompanying increase in Ti^{3+} species, which act as photohole traps and enhance the recombination of photogenerated electrons and holes and consequently decrease the quantum yield of the catalytic process.

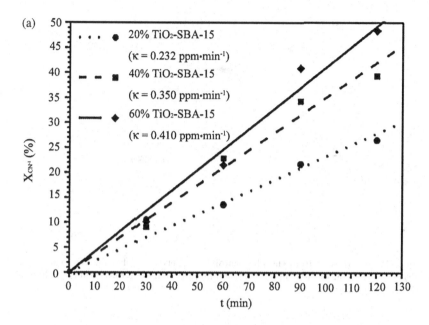

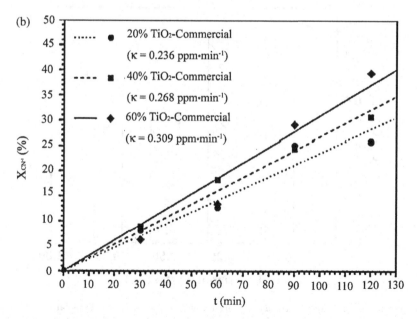

Figure 18.19. Kinetics of cyanide photo-oxidation in the presence of (a) 20–60% TiO$_2$-SBA-15 samples; and (b) 20–60% TiO$_2$-commercial SiO$_2$ materials. κ-values are the respective initial rate constants normalized to the titania content.[69]

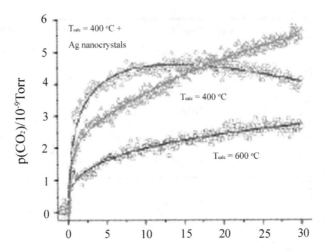

Figure 18.20.   Results from the photocatalytic experiments showing the generation of CO_2 as a function of time upon UV illumination of mesoporous titania with stearic acid.[73]

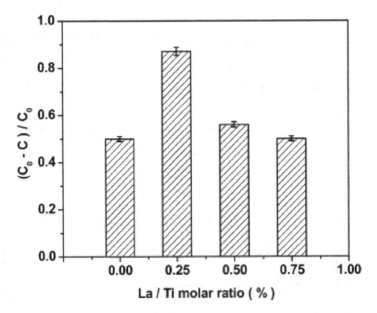

Figure 18.21.   Photocatalytic activity of samples with different La/Ti molar ratios evaluated by the degradation of methyl orange solution illuminated by a UV source for 60 minutes.[75]

18.5 Mesoporous-based Materials as Adsorbents

With the fast development of modern industry, various toxic inorganic and organic chemicals are discharged into the environment as industrial wastes, causing serious water, air, and soil pollution. Adsorption is considered to be a safer and less expensive remedy to handle this problem than precipitation, ion exchange, and membrane filtration. An ideal adsorbent should have uniformly accessible pores, an interlinked pore system, a high surface area, and physical and/or chemical stability.

As stated earlier in this chapter, during the past few years, mesoporous materials have been of great interest as adsorbents for heavy metals, dyes, harmful gases, and so on, owing to their ultrahigh surface area, uniform and tunable pore diameter, and their tunable pore surface and/or pore wall framework characteristics with different functional groups for use as adsorbents. Furthermore, they allow molecular accessibility to large internal surface areas and volumes. Mesoporous adsorbents have shown good adsorption capacity and been proved to have great advantages (higher adsorption capacity or selectivity) over traditional adsorbents, e.g. resins and activated charcoal.

18.5.1 *Heavy and other toxic metals*

The removal of heavy metal ions from wastewater has been a subject of extensive industrial research. Several kinds of materials such as ion-exchange resins, activated charcoal, and ion chelating agents immobilized on inorganic supports have been used as heavy metal adsorbents. However, these materials show many drawbacks because of their wide distribution of pore size, heterogeneous pore structure, low selectivity, and low loading capacities for heavy metal ions. Research on the environmental application of mesoporous silicas as novel heavy metal adsorbents has also been carried out recently. Mesoporous silicas such as MCM-41, HMS, SBA-15, and SBA-1 have been functionalized by several groups to enable the materials to interact strongly with metallic cations, and particularly mercury or metallic anions such as chromate and arsenate. There are two ways, the co-condensation and post-synthesis methods, to introduce functional groups into the mesoporous matrix. In general, the former route leads to a uniform incorporation and dispersion of functional groups in their framework, while the latter gives rise to a functional layer on their pore wall surface. The latter leads to the loss of surface area and reduction of pore diameter and pore volume (as high as 60–84%).[77] The degree of pore volume reduction could be a result of the nature of the ligand, such as the size, and the ligand loading. Moreover, materials prepared by the immobilization method appear to have non-uniformly distributed functional groups, whereas the materials prepared by the one-step sol-gel co-condensation method could obtain mesoporous functional adsorbents with high density and uniformly distributed functional groups and long life cycles.

In 1997, Liu and co-workers functionalized mesoporous silica with mercaptopropylsilane monolayers, which covalently bound to and were closely packed on their pore surface with coverage up to 75%.[78,79] They used mercaptopropylsilyl-functionalized large-pore MCM-41 as adsorbents and obtained a remarkable adsorption capacity of 600 mgHg/g. They washed the mercury-loaded adsorbents with a concentrated HCl (12.1 M) solution to regenerate them, after which 100% mercury was removed and the adsorbents retained a capacity of 210 mgHg/g in the second cycle.[78] The selectivity of the adsorbents was tested in their research work. Mercier and co-workers used 3-mercaptopropyltrialkoxysilane (MPTS) to modify HMS and MSU mesoporous silica samples by either the post-grafting or co-condensation method.[80−82] The authors compared the structure and adsorption properties of these two kinds of composites. Samples prepared by either the one-step co-condensation or grafting method show high selectivity for Hg^{2+} and the Hg^{2+} adsorption capacity is proportional to the thiol content (Hg/S in mol = 1). Lee and co-workers synthesized a new kind of bi-functional porous silicas via the co-condensation of TEOS, MPTS, and 3-aminopropyltrimethoxylsilane (APTS).[83] This adsorbent also has highly selective adsorption for Hg^{2+}, but the existence of amino groups decreases their metal adsorption capacities.

Jaroniec and co-workers reported a propylthiourea-modified mesoporous silica for mercury removal.[84,85] This material was prepared via a two-step modification by attachment of an aminopropyl group and its subsequent conversion into a thiourea ligand. About 1.5 mmol/g of 1-benzoyl-3-propylthiourea groups was attached to the silica surface. The maximum loading of mercury ions from aqueous solution for this material was approximately 1.0 g Hg^{2+}/g, double the capacity of the previously known thiol-containing samples. The authors claimed that the increase in adsorption capacity arose from the multifunctional nature and the deprotonation susceptiblility of the 1-benzoyl-3-propylthiourea ligand. The adsorbent can be regenerated under mild conditions via washing with a slightly acidified aqueous thiourea solution, and over 70% of the initial adsorption capacity is retained.

Thioethers have the same affinity for heavy metal ions as thiols. Shi and co-workers prepared thioether-functionalized organic-inorganic mesoporous composites by one-step co-condensation of tetraethoxysilane (TEOS) and (1,4)-bis(triethoxysilyl)propane tetrasulfide (BTESPTS) with tri-block copolymer poly(ethylene oxide)-poly(propylene oxide)-poly(ethylene oxide)

Table 18.6.   Physicochemical properties and Hg^{2+} adsorption capacity of the thioether-functionalized organic-inorganic mesoporous composite samples.[86]

Sample	A_{BET}^a (m^2/g)	D_{BJH} (nm)	S Content (mmol/g)	Hg Adsorption Capacity (mg/g)	Hg/S
S2	634	5.02	1.50	627	2.08
S6	654	3.74	2.83	1280	2.25
S8	576	3.47	3.48	1860	2.66
S10	532	3.35	3.76	2330	3.09
S12	253	3.26	4.02	2540	3.15
S15	242	3.13	4.24	2710	3.19

$^a A_{BET}$, BET surface area; D_{BJH}, BJH pore diameter calculated from the desorption branch.

Table 18.7.   The concentrations (ppm) of metal contaminants in wastewater solutions before and after treatment with thioether-functionalized organic-inorganic mesoporous composite samples.[86]

Solution	Hg^{2+}	Pb^{2+}	Cd^{2+}	Zn^{2+}	Cu^{2+}
No treatment	10.76	13.07	12.43	6.79	8.78
After treatment with S2	3.80	12.85	12.35	6.76	8.75
After treatment with S6	0.34	12.64	12.32	6.76	8.74
After treatment with S12	0.23	12.47	12.23	6.51	7.16

($P123$, $EO_{20}PO_{70}EO_{20}$) as template.[86] The new materials showed higher selectivity and adsorption capacity for Hg^{2+} (as listed in Table 18.6 and 18.7) than mesoporous adsorbents modified with MPTS or ordinary polythioether chelating resins with an equal amount of S content. Among them, sample S15 (15% BTESPTS in mol) shows an extraordinarily high adsorption capacity of up to $2,700\,mg\ Hg^{2+}/g$, which is about one order higher than that of ordinary polythioether chelating resins. The authors attributed these results to the special stereocoordination chemistry of S with Hg^{2+} when the thioether groups had been incorporated into the mesoporous material framework. This result also confirms the reactivity of bridging organic groups within the frameworks and its difference from that of terminal groups located in the channel space. A similar result was also reported by Mercier and co-workers with the substitution of the ethylenediamine group for the thioether group in the framework.[87] A preferential uptake for Cu^{2+} ions, compared with Ni^{2+} and Zn^{2+} ions, was observed.

Dai and co-workers improved the adsorption selectivity of specific cations by a novel surface imprinting method.[88] Two imprinting complex precursors $[Cu(apts)_xS_{6-x}]$ and $[Cu(aapts)_2S_2]^{2+}$, ($x = 3 - 5$;

S $=$ H$_2$O or methanol; apts $=$ H$_2$NCH$_2$CH$_2$CH$_2$Si(OMe)$_3$; aapts $=$ H$_2$NCH$_2$CH$_2$NHCH$_2$CH$_2$CH$_2$Si(OMe)$_3$) were post-grafted onto the mesoporous silica. The residue was washed with copious amounts of 1 M HNO$_3$ to remove all the Cu^{2+} ions and any excess ligands. The Cu^{2+} adsorption capacity of these imprint-coated mesoporous composite reaches 0.1 mmol/g, which is ten times higher than that of samples prepared by the conventional grafting method as mentioned earlier. The authors attributed this to the stereochemical arrangement of ligands on the surface of mesopores, which optimizes the binding of targeted metal ions. However, this absorbant suffers from high cost and poor regenerability.

Most recently, Sayari and co-workers post-synthesized pore-expanded MCM-41 silica using N,N-dimethyldecylamine and obtained material A with an open pore structure and readily accessible amine groups.[89] However, the surface area of material A was as low as 89 m^2/g due to the large molecules blocking the pore channels, which limited their adsorption capacities. The highest adsorption capacity was only 106 mg Cu^{2+}/g. After the selective removal of the occluded amine they obtained material B, a highly porous, hydrophobic material with a large surface area and pore volume. Experiment found material B to be a fast, sensitive, recyclable adsorbent for organic pollutants. This adsorption is still efficient in the presence of very dilute solutions of adsorbates.

Liu and co-workers reported the synthesis of a composite with Cu(II) ethylenediamine-terminated silane immobilized on mesoporous silica and used it as an efficient and selective anion trap for both arsenate (AsO$_4^{3-}$) and chromate (CrO$_4^{2-}$).[90] These materials can remove arsenate and chromate almost completely from solutions containing more than 100 ppm toxic metal anions. The capacity of these materials is higher than 120 mg (anion)/g.[90]

Nam and Tavlarides synthesized a high-capacity phosphonic-acid-functionalized adsorbent by co-condensation of TEOS and trimethoxysilylpropyl diethylphosphonate (DEPPS).[91] They applied these composites to adsorb chromium(III) from aqueous solutions. Experiment showed an equilibrium chromium(III) adsorption capacity of 82 mg/g at pH 3.6 and rapid adsorption kinetics. Column adsorption at a high space velocity of 156 h^{-1} was achieved with a minimum effluent chromium concentration of 0.01 mg/L. The chromium-loaded adsorbent could be regenerated using 12 M HCl. Kinetics results showed that near equilibrium was reached in as short as two minutes of contact time. Column tests showed dynamic operation at high throughput with sharp breakthrough curves and effluent concentrations (< 0.02 mg/L) far below the USEPA MCL level (0.1 mg/L) of total chromium.

Yi and co-workers reported a post-hydrolysis method to prepare mesoporous alumina (MA) with a wide surface area (307 m^2/g), uniform pore size (3.5 nm), and a sponge-like interlinked pore system.[92] Batch experiment of these materials found that their maximum uptake for As (V) was 121 mg/g and was seven times higher than that of conventional active alumina (AA). The adsorption was also kinetically rapid with complete adsorption in less than five hours as compared to about two days to reach half of the equilibrium value by AA. NaOH of 0.05 M was found to be the most suitable desorption agent. More than 85% of the arsenic adsorbed to the MA was desorbed in less than one hour. They also tested several other activated-alumina adsorbents with different pore properties and concluded that a uniform pore size and an interlinked pore system, rather than the surface area of the adsorbents, were the key factors.

Park and co-workers synthesized lanthanum-impregnated SBA-15 materials and applied them to remove arsenate from water.[93] At an arsenate concentration of 0.667 mmol As/l, the adsorption capacity of 50% lanthanum-impregnated SBA-15 by weight (designated as La50SBA-15) was 1.651 mmol As/g (123.7 mg As/g), which was approximately 10 and 14 times higher than La(III)-impregnated alumina and La(III)-impregnated silica gel, respectively.

18.5.2 *Dyes and other organic pollutants*

There has been increasing scientific concern about the hazardous effects of colored dyes in wastewater from the dyeing process in textile industries. The release of the colored dyes into the ecosystem is a dramatic source of esthetic pollution, causting eutrophication and perturbations in aquatic life. Some azo dyes and their degradation products such as aromatic amines are highly carcinogenic. The traditional treatments (e.g. flocculation, sedimentation, and filtration) have proved to be insufficient to purify a significant quantity of wastewater. Adsorption technology has become one of the most effective technologies for the removal of dyes. Practically, adsorbents for dye removal require the following potential advantages: (1) a large accessible pore volume, (2) hydrophobicity, (3) high thermal and hydrothermal stability, (4) no catalytic activity, and (5) easy regeneration. In the past few years, mesoporous materials have become a promising adsorbent for dye removal.

CTAB-modified mesoporous molecular sieve FSM-16 was tested as an adsorbent for acid dye (acid yellow, AY, and acid blue, AB) removal in comparison with as-prepared FSM-16 and activated carbon (AC).[94] The hydrothermal modification of FSM-16 with CTAB remarkably narrowed the pore openings and therefore improved the acid dye sorption in the low concentration range (>20 ppm) as compared with that of CTAB-free FSM-16 and AC.

Yeung and co-workers prepared ordered mesoporous silica (OMS) adsorbents by grafting amino- and carboxylic-containing functional groups onto MCM-41 for the removal of acid blue 25 and methylene blue dyes from wastewater.[95] The amino-containing OMS-NH$_2$ adsorbent has a large adsorption capacity (256 mg/g) and a strong affinity for acid blue 25. It can selectively remove acid blue 25 from a mixture of dyes (i.e. acid blue 25 and methylene blue). OMS-COOH is a good adsorbent for methylene blue, displaying high adsorption capacity (113 mg/g) and selectivity for the dye. However, the amount of dye adsorbed is much less than that of available functional groups. Regeneration of these adsorbents was also operated by simple washing with alkaline or acid solution. Part of the functional groups and the adsorbed dyes could be recovered.

With regard to organic pollutants of aquatic ecosystems, phenols are harmful to living organisms even at ppb levels. A variety of recyclable inorganic adsorbents, e.g. activated carbons, organically modified clays, and layered double hydroxides (LDHs), have drawn much attention as promising adsorbents particularly for the efficient removal of organic pollutants from aqueous solutions. Mesoporous materials have attracted great interest for the removal of organic pollutants because of their novel structure and chemical characteristics. As-synthesized mesoporous silicas were used for the adsorption of organic substrates.[96] However, these materials with filled channels would exhibit low adsorption capacity, and particularly low rates of adsorption because of mass transfer limitation within the crowded space.

Instead of using as-synthesized MCM-41, Inumaru and co-workers used calcined and surface-modified MCM-41 via grafting alkyltriethoxysilane.[97] This enables the material hydrophobicity to be controlled by the alkyl chain length. The adsorption rate could also be optimized by varying the amount of grafted material. Hanna and co-workers prepared a large-pore MCM-41 using trimethylbenzene (TMB) as swelling agent, then removed the TMB by heating it in air at 115°C.[98] This endowed the hydrophobic material with an open structure, and thus improved its adsorption properties.

In 1998, Albanis and co-workers used mesoporous alumina aluminum phosphates (AAPs) to remove chlorinated phenols from aqueous solution.[99] The AAP material (with a P/Al ratio of 0.6) can adsorb 14.8% of 2,4-dichlorophenol, 27.1% of 2,4,6-trichlorophenol, or 58.3% of pentachlorophenol, respectively. Most recently, Sayari and co-workers used a highly porous, hydrophobic material B with high surface area and pore volume to adsorb organic pollutants.[89] The initial capacities for adsorbing 4-chloroguaiacol and 2-6-dinitrophenol reached 95 and 110 mg/g, respectively, as shown in Table 18.8.

Humic acids are derived from the degradation of plants and microorganisms. They contain aliphatic, aromatic, and hydrophilic functional

Table 18.8.   Adsorption of chloroguaiacol and dinitrophenol on material B.[89]

Pollutant	Concentration (g/L)	% Adsorbed Pollutant	Adsorbent Capacity (mg/g)
Chloroguaiacol	0.11	89.3	96.6
Chloroguaiacol	0.25	89.9	95.6
Chloroguaiacol	0.50	19.9	95.2
Dinitrophenol	0.08	98.7	115.5
Dinitrophenol	0.25	98.5	88.6

groups, such as carboxyl and phenolic groups. Since they cause color in all open water sources and potential health implications, their removal is an important task for improving the water quality. Liu and co-workers synthesized a new mesoporous organosilica material (ß-CD-silica-4%) containing microporous ß-cyclodextrins (ß-CDs) by the co-condensation of a silylated ß-CD monomer with tetraethoxysilane in the presence of CTAB.[100] Adsorption experiments showed that ß-CD-silica-4% material removed up to 99% of humic acid from an aqueous solution containing 50 ppm of humic acid at a solution-to-solid ratio of 100 mL/g. However, the regeneration of these adsorbents is very difficult; after treatment with a concentrated NaCl solution, only 30% of the loaded humic acid could be removed.

18.5.3 *Gases*

Song and co-workers proposed a novel CO_2 "molecular basket" adsorbent using polyethylenimine-(PEI-) modified mesoporous materials.[101] At PEI loading of 50 wt.% in MCM-41-PEI, the highest CO_2 adsorption capacity of 246 mg/g-PEI was obtained, which is 30 times higher than that of MCM-41 and about 2.3 times that of pure PEI. Further, the adsorption/desorption processes of CO_2 for these composites were faster than that for pure PEI. These composites also have good selectivity for CO_2 and recycle stability. Most recently, they used these composites as a low-temperature H_2S adsorbent.[102] PEI, which has a high density of amino groups to interact with H_2S, is therefore expected to show high H_2S adsorption capacity. More importantly, mesoporous MCM-41 acts also as a separation medium for the PEI nanoparticles; such nanoparticles are not likely to aggregate in the application process and therefore retain their original properties. The PEI in MCM-41 is highly dispersed and expected to show a higher adsorption efficiency for H_2S than bulk PEI particles. The H_2S adsorption by a commercial ZnO, which is used in practical applications, was also measured to compare with that of the new adsorbent. The new adsorbent had a higher H_2S adsorption capacity, lower breakthrough time, and much lower operation temperature than the commercial ZnO. Therefore, these composites can be a promising candidate for low-temperature adsorption of sulfur compounds from hydrocarbon fuels in solid oxide fuel cells. With regard to adsorbents and sensors using mesoporous materials for volatile organic compounds and other toxic gases, more references can be found in the following section.

18.5.4 *Others*

Ho and co-workers prepared mesoporous Ti oxohydroxide ($TiO_x(OH)_y$) using dodecylamine as a template, and studied their capacity for fluoride removal.[103] Zirconia and silica had been introduced into the mesoporous Ti oxohydroxide to enhance the ion-exchange capacity. Results showed that zirconia-containing mesoporous Ti oxohydroxide exhibited the highest ion-exchange capacity for fluoride (about $1.2 \, mmol \, F^-/g$), as it has the smallest particle size, with high uniformity among the mesoporous materials prepared.

Shin and co-workers studied aluminum-impregnated mesoporous adsorbents for phosphate removal from water.[104] They found that the Al-impregnated mesoporous materials showed faster adsorption kinetics (reached equilibrium in an hour) as well as higher adsorption capacities (as high as $862 \, \mu mol/g$) as compared with activated alumina. The authors concluded from the study that (a) the uniform mesopores of these adsorbents increase the diffusion rate of the adsorbate in the adsorption process, thus leading to the fast adsorption kinetics; and (b) the high phosphate adsorption capacities are attributed not only to the increase in surface hydroxyl density on Al oxide due to the well-dispersed Al component but also to the decrease in stoichiometric ratio of surface hydroxyl ions to phosphate by the formation of monodentate surface complexes.

18.6 Mesoporous-based Materials as Sensors

18.6.1 *Gas sensors*

Toxic volatile organic compounds emitted from building materials and furniture that stay indoors, especially in new buildings, are suspected to be a cause of health hazards such as sick building or sick house syndrome. With the increasing number of biological and environmental issues, reliable, sensitive, selective, and user-friendly chemical sensors have become more and more attractive. A sensor includes a transducer (electrochemical, piezoelectrical, or optical) whose typical response is significantly altered by the presence of the chemical to be detected. Mesoporous materials, because of their novel structural properties, have become promising candidates not only for adsorbents but also for sensors. Their highly ordered, uniform and three-dimensional cross-linked pore structure, and large specific surface area can undoubtedly enhance the sensitivity of adsorption-type sensors. Sasahara and co-workers employed Pd-loaded mesoporous silica to fabricate a catalytic sensing material of an adsorption/combustion-type sensor for volatile organic compounds (VOCs).[105] Toluene was used as a typical example of VOCs with a pulse heating mode. A large response peak was observed by flash combustion of the toluene molecules adsorbed on the sensing material during the non-heating period and the integral response is nearly proportional to the specific surface area of the sensing materials. The response is stronger than that of a Pd/γ-Al_2O_3 sensor. They also pointed out that thermal conductivity of the sensing material was another important factor to control the sensitivity of this type of sensor. Most recently, Floch and co-workers developed a selective chemical sensor based on the plasmonic response of phosphinine-stabilized gold nanoparticles (NPs) hosted on periodically organized mesoporous silica thin layers (Figure 18.22).[106] Gold NPs stabilized by phosphinine ligands were used. A subsequent substitute of phosphinine ligands with thiol and phosphane resulted in a blueshift of the plasmon peak with a concomitant relative decrease of the peak intensity (Figure 18.23). Such signals show the possibility of the UV-Vis detection of a ligand-exchange reaction at the surface of gold nanoparticles in solution. Thus these nanoparticles can be used as a transducer. They immobilized these Au NPs in mesoporous silica thin films by a one-step EISA approach, which can preserve both good accessibility for the analyte to the detection center and good optical qualities. The properties of the phosphinine ligand enable selective detection of thiols and small phosphanes, so the composite films can act as very sensitive and responsive sensors for small molecules, such as volatile thiols or PH_3. It was claimed that the design of large-scale or more-quantitative detectors is possible by optimizing the gold loading and/or pore size in the film.

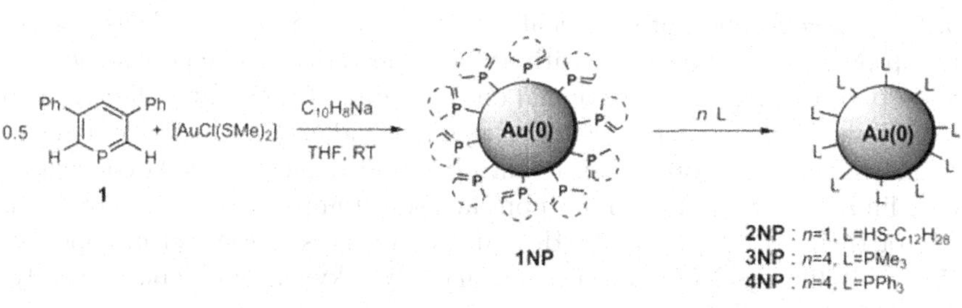

Figure 18.22.   Synthesis diagram of nNP ($n = 1$–4).[106]

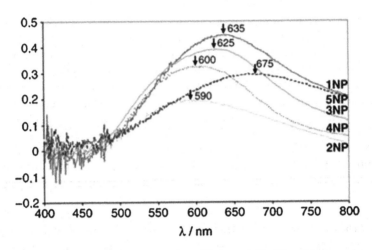

Figure 18.23. UV-Vis spectra recorded for nNP ($n = 1$–5). The diffusing noise of the solution was simulated with an exponential function and subtracted from the recorded spectra in order to determine the peak maxima properly.[106]

Nitrogen oxides generated by combustion are not only dangerous and harmful to health, but also create acid rain that destroys the environment. The development of a highly sensitive, responsive, and portable monitoring technique for these gases is an urgent requirement. Zhou and co-workers applied ordered mesoporous silica films to develop an NO_x gas sensor based on the surface photovoltage (SPV) system, using a metal-insulator-semiconductor (MIS) structure with mesoporous thin film as the insulator layer (Figure 18.24).[107,108] When the rear side of the semiconductor is irradiated by an LED beam, an AC photocurrent can be induced. The adsorption of NO_x on the mesoporous silica layers brings a change of the surface photocurrent on the semiconductor layers. Thus the sensitivity to supplied NO_x gas is estimated by the induced photocurrent through the MIS structure. The large surface area and uniform nanosized pores of ordered mesoporous materials enable SPV devices to show good gas adsorption properties with better sensitivity and selectivity. Cubic pore structures are a better choice than hexagonal ones because of their three-dimensional cross-linked pore channels, which bring much higher gas sensitivity. Tin was added to the mesoporous silica thin films and detectable NO_2 concentration limits were lowered to 300 ppm at room temperature, as reported in one of their recent papers.[108] Hon and co-workers fabricated mesoporous WO_3 thin films as NO_2 gas sensors using poly(alkylene oxide) block copolymer $EO_{100}PO_{64}EO_{100}$, F127, as a template. The film has nanocrystallite domains, a porous structure, with a relative surface area of $143 \, m^2/g$ after

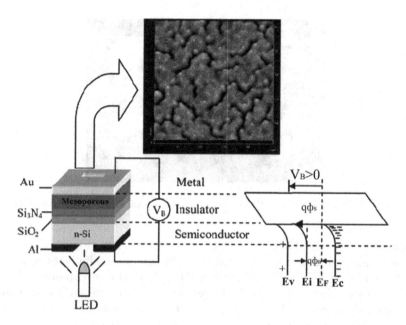

Figure 18.24.   Tin-modified mesoporous silica sensor with MIS structure and its energy band diagram. The AFM image of an electrode shows the porous layer of Au film.[108]

being calcined at 250°C.[109] Upon exposure to NO_2, the electrical resistance of semiconducting mesoporous WO_3 thin films is found to increase dramatically. The mesoporous WO_3 thin-film sensor shows a high sensitivity up to $R_g/R_a = 226$ (R_g and R_a are the electric resistance of the films in test gas and air, respectively) at 100°C for detecting 3 ppm NO_2. The one calcined at 250°C and operated at 35°C shows a high enough sensitivity of $R_g/R_a = 23$; the authors claimed it a unique NO_2 gas sensor for its high sensitivity at such a low temperature.

Semiconductive mesoporous materials are of great interest as gas sensor materials. Hyodo and co-workers synthesized mesoporous SnO_2 and further improved their thermal stability by a phosphoric acid treatment process.[110,111] They found that thick-film sensors prepared from the phosphoric acid- (PA-) treated mesoporous SnO_2 powders of larger pore size showed better H_2 response, in comparison with the sensors prepared from untreated powder and from PA-treated mesoporous powder of smaller pore size. While semiconductor gas sensors exhibit high sensitivity (small change in gas composition causes dramatic change in resistance), their stability and selectivity still remain unsatisfactory for many applications. To solve this problem, Cabot and co-workers used Pd- and Pt-loaded mesoporous silica

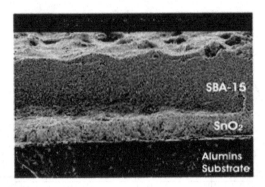

Figure 18.25.   A SEM cross-sectional view of a SnO$_2$ sensor with a mesoporous filter.[112]

as a novel selective filter for a SnO$_2$-based gas sensor (Figure 18.25).[112] Results indicate that SiO$_2$: Pd and SiO$_2$: Pt filters can eliminate response to CO without affecting the response to CH$_4$ when they are exposed to CH$_4$-CO mixtures (Figure 18.26). The large catalytically active surface area and the high resistivity of the mesoporous SiO$_2$ make it suitable as an efficient filter to improve the selectivity and stability of gas sensors.

Dyes are developed as oxygen sensors based upon the principle that oxygen is a powerful quencher of the electronically excited state of dyes. The key factors include the solubility of oxygen and the diffusion coefficient of oxygen in the host medium, and the excited state lifetime of the dye. Mesoporous materials were chosen as a kind of new candidate, which could adsorb sufficient dye to produce a strong photoluminescence signal and at the same time provide rapid access to oxygen molecules from the external atmosphere. Winnik and co-workers prepared mesoporous silica short rods.[113] Sensor films were prepared by adsorbing these particles at sub-monolayer coverage on a thin polymer film with a positively charged surface prepared by a layer-by-layer method. Various dyes were incorporated into the pores of the mesoporous silica particles when these films were dipped into solutions of dyes and the solvent was allowed to evaporate. Dyes incorporated into the mesoporous silica-coated polymer films have a more intense fluorescence than those adsorbed to the layer-by-layer film itself. Therefore, composite films are promising oxygen sensors with a rapid response.

18.6.2 Other sensors

Uranium and its salts, which can cause kidney damage and acute arterial lesions, are both radiologically and chemically toxic. Because concentrations of uranium compounds in contaminated groundwater are typically at very low levels, accurate analyses become more difficult. Lin and co-workers

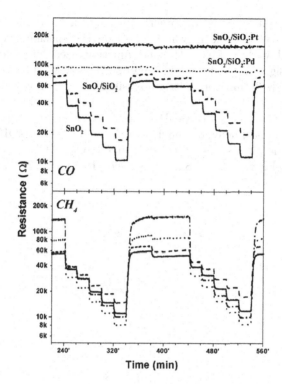

Figure 18.26. Transitory resistances of the SnO_2, SnO_2/SiO_2, SnO_2/SiO_2 : Pd, and SnO_2/SiO_2 : Pt gas sensors when exposed to different concentrations of CO and CH: from 20 to 400 ppm and 200 to 4,000 ppm, respectively. Two different relative humidity conditions are considered: 30 and 50%.[112]

reported a new uranium electrochemical sensor for uranium detection based on an adsorptive square-wave stripping voltammetry technique.[114] This technique of pre-concentrating uranium chelates on an electrode surface prior to the detection step has been shown to be highly sensitive for the detection of trace uranium and its compounds. To prepare the sensor, an acetamide phosphonic acid ligand was immobilized via covalent bonding onto mesoporous silica MCM-41 and the ligand-bearing mesoporous silica was then embedded in a carbon graphite matrix. This sensor is mercury-free, solid-state, and has less ligand depletion than existing sensors. Using this functionalized mesoporous sensors, voltammetric responses for uranium detection are reported as a function of pH value, pre-concentration time, and aqueous-phase uranium concentration. The uranium detection limit is 25 ppb after five minutes of pre-concentration and improves to 1 ppb after 20 minutes of pre-concentration. The relative standard deviations are normally less than 5%.

Martínez-Máñez and co-workers reported a new method for the determination of fluoride in water based on the specific reaction of fluorhydric acid with a MCM-41 solid functionalized with fluorescent or colorimetric signalling units.[115] The mesoporous MCM-41 solid is first functionalized with 3-aminopropyltriethoxysilane in refluxing toluene. Reaction of this amino-functionalized solid with 9-anthraldehyde (in absolute ethanol at 45°C; then with NaBH$_4$ in ethanol at room temperature 4-{2-[4-(dimethylamino)phenyl]diazenyl}benzoic acid (DCC–TsOH, 0°C) and lissamine rhodamine B sulfonyl chloride (in CH$_2$Cl$_2$, 0°C) results in the synthesis of samples S1, S2, and S3, respectively (Figure 18.27). Functionalized

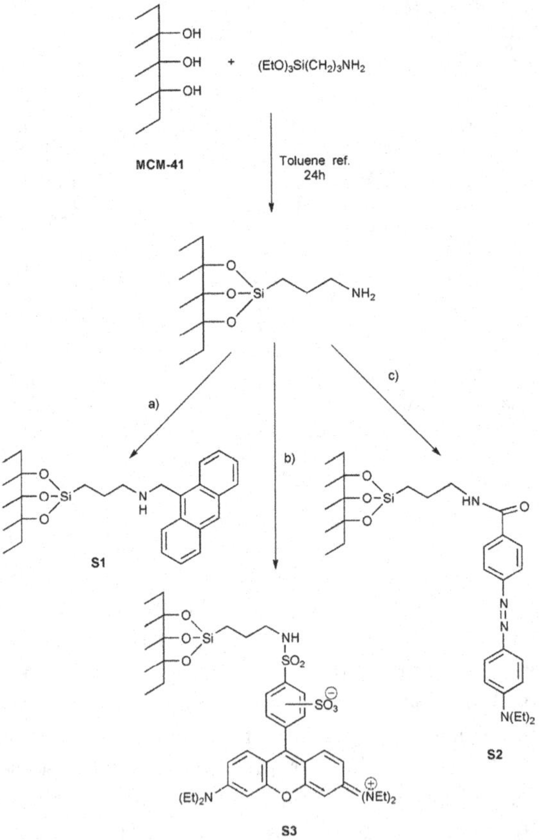

Figure 18.27.   Synthesis of S1, S2, and S3 by the functionalization of an MCM-41 matrix first with 3-aminopropyltriethoxysilane and then with (a) 9-anthraldehyde, (b) lissamine rhodamine B sulfonyl chloride, and (c) 4-{2-[4-(dimethylamino) phenyl]diazenyl}benzoic acid.[115]

monoliths of S3 have been added to buffered solutions containing fluoride. The resulting solution color deepened with the increase in the amount of fluoride. The good results obtained in the determination of fluoride content in toothpaste suggest its potential use as a practical method for the determination of fluoride anions in water samples.

Based on diffusion followed by an immobilizing reaction, Xue and coworkers first reported a new quantitative optical sensor for metal ions using amine-grafted mesoporous silica monoliths with $H_2N(CH_2)_3Si\text{-}(OMe)_3$.[116] The sensor unifies two processes including slow diffusion of the metal ions to the binding sites and fast metal-ligand (ML_n) complexation. When the ligands have been saturated and diffusion of the Cu^{2+} reaches a balance to the metal solution, the adsorbance of the complex ML_n can be observed spectroscopically. They studied its response properties for Cu^{2+} in detail and gave a mathematical model for such a quantitative sensor.

18.7 Summary

As a special kind of nanomaterial, nanoporous materials have attracted a lot of research attention in recent years, among which ordered mesoporous materials are one of the most attractive focal points due to their unique characteristics such as extraordinarily high surface area, uniform pore size distribution, and so on. To meet the ever-increasing needs for environmental protection and environment-friendly processes, mesoporous materials have been expected to play more and more important roles in the fields of environment-related areas due to their high catalytic, adsorbing, and sensing performances. In this chapter, we have reviewed the recent progress in the synthesis and properties of various kinds of mesoporous materials and mesoporous-based nanocomposites by chemical assembly approaches at the nanoscale. Emphasis has been placed on the environment-related catalytic, adsorbing, and sensing performances of these materials.

Section 18.2 focused mainly on the catalytic performances for automobile exhaust treatments of mesoporous zirconia-based three-way catalysts. The metal-loaded composite catalysts show significantly lower ignition temperature for catalysis and higher stability and lower noble metal consumption as compared to the common noble metal-loaded zirconia/ceria powders. The reason is the high dispersion of noble metal particles in the pore channels, which provide abundant active sites for the exhaust conversion and prevent the aggregation of metal particles. The mesoporous materials, especially conductive mesoporous carbons, are promising as electrode reaction catalysts for polymer electrolyte membrane cells. After the catalytically active site loading in mesoporous carbons, the composites show high potential for future applications in such clean energy power sources, as pointed out in Section 18.3. Another active research field is the photocatalysis of titania-based materials for the decomposition of harmful organics. As introduced in Section 18.4, mesoporous titania has attracted much attention due to its high surface area. The section focused on the preparation of phase-structure-controllable mesoporous titania, which leads to attractive photocatalytic performances. The last two parts (Section 18.5 and 18.6) dealt with the adsorbing and sensing performances of mesoporous-based composites, such as mesoporous organic/inorganic hybrids. Again, owing to the very high surface area of the pore channels and the tunable wall and pore surface chemical properties, the materials are very promising for applications in the detection and removal of harmful substances such as heavy metal ions, toxic gases, organic pollutants, and so on.

Further research work is expected on the synthesis of nano/mesoporous materials with controlled structures and/or architectures, and their properties, especially in terms of their environmental applications. The quantity production of these advanced materials with high reproducibility and low cost needs to be explored in the future for practical applications as well.

References

1. K. S. W. Sing, D. H. Evereet, K. H. W. Haul, L. Moscou, R. A. Pierotti and J. Rouquerol, *Pure Appl. Chem.* **57**, 603–619 (1985).
2. J. Y. Ying, C. P. Mehnert and M. S. Wong, *Angew. Chem. Int. Ed.* **38**, 56–77 (1999).
3. F. Schuth and W. Schmidt, *Adv. Mater.* **14**, 629–638 (2002).
4. A. Taguchi and F. Schuth, *Micropor. Mesopor. Mater.* **77**, 1–45 (2005).
5. C. T. Kresge, M. E. Leonowicz, W. J. Roth, J. C. Vartulli and J. S. Beck, *Nature* **359**, 710 (1992).
6. J. S. Beck, J. C. Vartuli, W. J. Roth, M. E. Leonowicz, C. T. Kresge, K. D. Schmitt, C. T.-W. Chu, D. H. Olson, E. W. Sheppard, S. B. McCullen, J. B. Higgins and J. L. Schlenker, *J. Am. Chem. Soc.* **114**, 10834 (1992).
7. P. Fornasiero, E. Fonda, R. Di. Monte, G. Vlaic, J. Kaspar and M. Graziani, *J. Catal.* **187**, 177 (1999).
8. C. Bozo, N. Guilhaume, E. Garbowski and M. Primet, *Catal. Today* **59**, 33 (2000).
9. P. Fornasiero, N. Hickey, J. Kaspar, C. Dossi, D. Gava and M. Graziani, *J. Catal.* **189**, 326 (2000).
10. A. P. Oliveira and M. L. Torem, *Powder Technol.* **119**, 181 (2001).
11. H. Shinjoh, *J. Alloy. Compd.* **408–412**, 1061 (2006).
12. I. M. Hung, H. P. Wang, W. H. Lai, K. Z. Fung and M. H. Hon, *Electrochim. Acta* **50**, 745–748 (2004).
13. A. H. Lu and F. Schuth, *C. R. Chimie* **8**, 609–620 (2005).

14. F. Kleitz, S. H. Choi and R. Ryoo, *Chem. Commun.* 2136 (2003).
15. W. H. Shen, X. P. Dong, Y. F. Zhu, H. R. Chen and J. L. Shi, *Micropor. Mesopor. Mater.* **85**, 157–162 (2005).
16. A. S. Deshpande, N. Pinna, B. Smarsly, M. Antonietti and M. Niederberger, *Small* **1**, 313–316 (2005).
17. M. Ozawa, *J. Alloy. Compd.* **275–277**, 886–890 (1998).
18. H. R. Chen, J. L. Shi, J. N. Yan, Z. L. Hua, H. G. Chen and D. S. Yan, *Adv. Mater.* **15**(13), 1078–1081 (2003).
19. D. B. Kuang, T. Breaesinski and B. Smarsly, *J. Am. Chem. Soc.* **126**, 10534 (2004).
20. A. Corma, P. Atienzer, H. Garcia and J. Y. C. Ching, *Nature* **3**, 394 (2004).
21. H. R. Chen, J. L. Gu, J. L. Shi, Z. C. Liu, J. H. Gao, M. L. Ruan and D. S. Yan, *Adv. Mater.* **17**, 2010–2014 (2005).
22. T. Sen, G. J. T. Tiddy, J. L. Casci and M. W. Anderson, *Angew. Chem. Int. Ed.* **42**, 4649 (2003).
23. C. Ho, J. C. Yu, X. C. Wang, S. Lai and Y. F. Qiu, *J. Mater. Chem.* **15**, 2193–2201 (2005).
24. L. F. Chen, G. Gonzalez, J. A. Wang, L. E. Norena, A. Toledo, S. Castillo and M. M. Pineda, *Appl. Surf. Sci.* **243**, 319–328 (2005).
25. F. Fajardie, J. F. Tempere, J. M. Manoli, G. D. Mariadassou and G. Blanchard, *J. Chem. Soc. Faraday Trans.* **94**, 3727 (1998).
26. Z. Q. Mao, *Chin. J. Power Sources* **27** 179–182 (2003).
27. W. Chrzanowski and A. Wieckowski, *Langmuir* **14**, 1967 (1998).
28. H. D. Dinh, X. Ren, F. H. Garzon, P. Zelenay and S. Gottesfeld, *J. Electroanal. Chem.* **492**, 222 (2000).
29. G. S. Chai, S. B. Yoon, J. S. Yu, J. H. Choi and Y. E. Sung, *J. Phys. Chem. B* **108**, 7074–7079 (2004).
30. R. Ryoo, S. Joo, M. Kruk and M. Jaroniec, *Adv. Mater.* **13**, 677 (2001).
31. Y. Wang, J. Ren, K. Deng, L. Gui and Y. Tang, *Chem. Mater.* **12**, 1622 (2000).
32. H. Kim, P. Kim, J. B. Joo, W. Y. Kim, I. K. Song and J. Yi, *J. Power Sources* **145**, 139 (2005).
33. K. W. Park, J. H. Choi and Y. E. Sung, *J. Phys. Chem. B* **107**, 5951 (2003).
34. J. H. Jiang and A. Kucernak, *J. Electroanal. Chem.* **533**, 153–165 (2002).
35. G. S. Attard, P. N. Bartlett, N. R. B. Coleman, J. M. Elliott, J. R. Owen and J. H. Wang, *Science* **278**, 838 (1997).
36. J. H. Jiang and A. Kucernak, *J. Electroanal. Chem.* **543**, 187–199 (2003).
37. J. M. Orts, E. Louis, L. M. Sander and J. Clavilier, *Electrochim. Acta* **44**, 1221 (1998).
38. A. Fujishima, T. N. Rao and D. A. Tryk, *J. Photochem. Photobiol. C: Photochem. Rev.* **1**, 1 (2000).
39. Q. Huo, D. I. Margolese, U. Ciesla, P. Feng, T. E. Gler, P. Sieger, R. Leon, P. M. Petroff, F. Shüth and G. D. Stucky, *Nature* **368**, 317 (1994).
40. D. Zhao, J. Feng, Q. Huo, N. Melosh, G. H. Fredrickson, B. F. Chmelka and G. D. Stucky, *Science* **279**, 548 (1998).
41. P. T. Tanev and T. J. Pinnavaia, *Science* **267**, 865 (1995).

42. D. M. Antenolli and J. Y. Ying, *Angew. Chem. Int. Ed. Engl.* **34**, 2014 (1995).
43. D. Khushalani, G. A. Ozin and A. Kuperman, *J. Mater. Chem.* **9**, 1491 (1999).
44. S. Cabrera, J. El-Haskouri, A. Beltrán-Portier, D. Beltrán-Portier, M. D. Marcos and P. Amorós, *Solid State Sci.* **2**, 513 (2000).
45. V. F. Stone Jr. and R. J. Davis, *Chem. Mater.* **10**, 1468 (1998).
46. D. M. Antonelli, *Micropor. Mesopor. Mater.* **30**, 315 (1999).
47. Y. Lu, R. Ganguli, C. A. Drewien, M. T. Anderson, C. J. Brinker, W. Gong, Y. Guo, H. Soyez, B. Dunn, M. H. Huang and J. I. Zink, *Nature* **389**, 364 (1997).
48. E. L. Crepaldi, G. J. de A. A. Soler-Illia, D. Grosso and C. Sanchez, *New J. Chem.* **27**, 9 (2003).
49. E. L. Crepaldi, G. J. de A. A. Soler-Illia, D. Grosso, F. Cagnol, F. Ribot and C. Sanchez, *J. Am. Chem. Soc.* **125**, 9770 (2003).
50. E. Beyers, P. Cool and E. F. Vansant, *J. Phys. Chem. B* **109**, 10081 (2005).
51. B. Smarsly, D. Grosso, T. Brezesinski, N. Pinna, C. Boissière, M. Antonietti and C. Sanchez, *Chem. Mater.* **16**, 2948 (2004).
52. D. Grosso, G. J. de A. A. Soler-Illia, E. L. Crepaldi, F. Cagnol, C. Sinturel, A. Bourgeois, A. Brunet-Bruneau, H. Amenitsch, P. A. Albouy and C. Sanchez, *Chem. Mater.* **15**, 4562 (2003).
53. Y. K. Hwang, K. Lee and Y. Kwon, *Chem. Commun.* 1738 (2001).
54. M. S. Wong, E. S. Jeng and J. Y. Ying, *Nano Lett.* **1**, 637 (2001).
55. G. J. de A. A. Soler-Illia, E. Scolan, A. Louis, P. Albouy and C. Sanchez, *New J. Chem.* **25**, 156 (2001).
56. B. Tian, H. Yang, X. Liu, S. Xie, C. Yu, J. Fan, B. Tu and D. Zhao, *Chem. Commun.* 1824 (2002).
57. B. Tian, X. Liu, B. Tu, C. Yu, J. Fan, L. Wang, S. Xie, G. D. Stucky and D. Zhao, *Nature Mater.* **2**, 159 (2003).
58. J. C. Yu, X. Wang and X. Fu, *Chem. Mater.* **16**, 1523 (2004).
59. J. C. Yu, J. G. Yu and J. C. Zhao, *Appl. Catal. B: Environ.* **36**, 31 (2002).
60. S. Haseloh, S. Y. Choi, M. Mamak, N. Coombs, S. Petrov, N. Chopra and G. A. Ozin, *Chem. Commun.* 1460 (2004).
61. S. Y. Choi, M. Mamak, N. Coombs, N. Chopra and G. A. Ozin, *Adv. Funct. Mater.* **14**, 335 (2004).
62. H. Luo, C. Wang and Y. Yan, *Chem. Mater.* **15**, 3841 (2003).
63. Y. Li, N. Lee, E. Lee, J. S. Song and S. Kim, *Chem. Phys. Lett.* **389**, 124 (2004).
64. D. Grosso, G. J. de A. A. Soler-Illia, E. L. Crepaldi, B. Charleux and C. Sanchez, *Adv. Funct. Mater.* **13**, 37 (2003).
65. L. Zhang and J. C. Yu, *Chem. Commun.* 2078 (2003).
66. X. Wang, J. C. Yu, C. Ho, Y. Hou and X. Fu, *Langmuir* **21**, 2552 (2005).
67. Y. Shiraishi, N. Saito and T. Hirai, *J. Am. Chem. Soc.* **127**, 12820 (2005).
68. B. J. Aronson, C. F. Blanford and A. Stein, *Chem. Mater.* **9**, 2842 (1997).
69. R. van Grieken, J. Aguado, M. J. López-muñoz and J. Marugán, *J. Photochem. Photobiol. A: Chem.* **148**, 315 (2002).

70. E. P. Reddy, B. Sun and P. G. Smirniotis, *J. Phys. Chem. B* **108**, 17198 (2004).
71. E. P. Reddy, L. Davydov and P. G. Smirniotis, *Appl. Catal. B: Environ.* **42**, 1 (2003).
72. Y. Xu and C. H. Langford, *J. Phys. Chem. B* **101**, 3115 (1997).
73. M. Andersson, H. Birkedal, N. R. Franklin, T. Ostomel, S. Boettcher, A. E. C. Palmqvist and G. D. Stucky, *Chem. Mater.* **17**, 1409 (2005).
74. M. H. Bartl, S. P. Puls, J. Tang, H. C. Lichtenegger and G. D. Stucky, *Angew. Chem. Int. Ed.* **43**, 3037 (2004).
75. S. Yuan, Q. Sheng, J. Zhang, F. Chen, M. Anpo and Q. Zhang, *Micropor. Mesopor. Mater.* **79**, 93 (2005).
76. Y. Yang, Y. Guo, C. Hu, Y. Wang and E. Wang, *Appl. Catal. A: General* **273**, 201 (2004).
77. A. Walcarius, M. Etienne and B. Lebeau, *Chem. Mater.* **15**, 2161 (2003).
78. X. Feng, G. E. Fryxell, L. Q. Wang, A. Y. Kim, J. Liu and K. M. Kemmer, *Science* **276**, 923–926 (1997).
79. J. Liu, X. D. Feng, G. E. Fryxell, L. Q. Wang, A. Y. Kim and M. L. Gong, *Adv. Mater.* **10**, 161–165 (1998).
80. L. Mercier and T. J. Pinnavaia, *Adv. Mater.* **9**, 500–503 (1997).
81. J. Brown, L. Mercier and T. J. Pinnavaia, *Chem. Commun.* 69–70 (1999).
82. J. Brown, R. Richer and L. Mecier, *Micropor. Mesopor. Mater.* **37**, 41–48 (2000).
83. B. Lee, Y. Kim, H. Lee and J. Yi, *Micropor. Mesopor. Mater.* **50**, 77–90 (2001).
84. V. Antochshuk and M. Jaroniec, *Chem. Commun.* 258–259 (2002).
85. V. Antochshuk, O. Olkhovyk, M. Jaroniec, I. Park and R. Ryoo, *Langmuir* **19**, 3031–3034 (2003).
86. L. X. Zhang, W. H. Zhang, J. L. Shi, Z. L. Hua, Y. S. Li and J. N. Yan, *Chem. Commun.* 210–211 (2003).
87. K. Z. Hossain and L. Mercier, *Adv. Mater.* **14**, 1053–1056 (2002).
88. S. Dai, M. C. Burleigh, Y. Shin, C. C. Morrow, C. E. Barnes and Z. Xue, *Angew. Chem. Int. Ed.* **38**, 1235–1239 (1999).
89. A. Sayari, S. Hamoudi and Y. Yang, *Chem. Mater.* **17**, 212–216 (2005).
90. G. E. Fryxell, J. Liu, T. A. Hauder, Z. Nie, K. F. Ferris, S. Mattigod, M. Gong and R. T. Hallen, *Chem. Mater.* **11**, 2148–2154 (1999).
91. K. H. Nam and L. L. Tavlarides, *Chem. Mater.* **17**, 1597–1604 (2005).
92. Y. Kim, C. Kim, I. Choi, S. Rengaraj and J. Yi, *Environ. Sci. Technol.* **38**, 924–931 (2004).
93. M. Jang, J. K. Park and E. W. Shin, *Micropor. Mesopor. Mater.* **75**, 159–168 (2004).
94. M. M. Mohamed, *J. Colloid Interface Sci.* **272**, 28–34 (2004).
95. K. Y. Ho, G. McKay and K. L. Yeung, *Langmuir* **19**, 3019–3024 (2003).
96. Y. Miyake, T. Yumoto, H. Kitamura and T. Sugimoto, *Phys. Chem. Chem. Phys.* **4**, 2680–2684 (2002).

97. K. Inumaru, Y. Inoue, S. Kakii, T. Nakano and S. Yamanaka, *Phys. Chem. Chem. Phys.* **6**, 3133–3139 (2004).

98. K. Hanna, I. Beurroise, R. Denoyel, D. Desplantier-Giscard, A. Galarneau and F. J. De Renzo, *J. Colloid Interface Sci.* **252**, 276–283 (2002).

99. T. G. Danis, T. A. Albanis, D. E. Petrakis and P. J. Pomonis, *Wat. Res.* **32**(2), 295–302 (1998).

100. C. Liu, N. Naismith and J. Economy, *J. Chromatogr. A* **1036**, 113–118 (2004).

101. X. Xu, C. Song, J. M. Andresen, B. G. Miller and A. W. Scaroni, *Micropor. Mesopor. Mater.* **62**, 29–45 (2003).

102. X. Xu, I. Novochinskii and C. Song, *Energ. Fuel.* **19**(2), 2214–2215 (2005).

103. L. N. Ho, T. Ishihara, S. Ueshima, H. Nishiguchi and Y. Takita, *J. Colloid Interface Sci.* **272**, 399–403 (2004).

104. E. W. Shin, J. S. Han, M. Jang, S. H. Min, J. K. Park and R. M. Rowell, *Environ. Sci. Technol.* **38**, 912–917 (2004).

105. T. Sasahara, A. Kido, H. Ishihara, T. Sunayamaa and M. Egashira, *Sens. Actuators B* **108**, 478–483 (2005).

106. F. Goettmann, A. Moores, C. Boissière, P. L. Floch and C. Sanchez, *Small* **6**, 636–639 (2005).

107. T. Yamada, H. S. Zhou, H. Uchida, M. Tomita, Y. Ueno, T. Ichino, I. Honma, K. Asai and T. Katsube, *Adv. Mater.* **14**, 812–815 (2002).

108. B. Yuliarto, H. S. Zhou, T. Yamada, I. Honma, Y. Katsumura and M. Ichihara, *Anal. Chem.* **76**, 6719–6726 (2004).

109. L. G. Teoh, Y. M. Hon, J. Shieh, W. H. Lai and M. H. Hon, *Sens. Actuators B* **96**, 219–225 (2003).

110. Y. Shimizu, T. Hyodo and M. Egashira, *J. Eur. Ceram. Soc.* **24**, 1389–1398 (2004).

111. Y. Shimizu, A. Jono, T. Hyodo and M. Egashira, *Sens. Actuators B* **108**, 56–61 (2005).

112. A. Cabot, J. Arbiol, A. Cornet, J. R. Morante, F. Chen and M. Liu, *Thin Solid Films* **436**, 64–69 (2003).

113. B. H. Han, I. Manners and M. A. Winnik, *Chem. Mater.* **17**, 3160–3171 (2005).

114. W. Yantasee, Y. Lin, G. E. Fryxell and Z. Wang, *Electroanalysis* **10**, 870–873 (2004).

115. A. B. Descalzo, D. Jiménez, J. E. Haskouri, D. Beltrán, P. Amorós, M. D. Marcos, R. Martínez-Máñez and J. Sotoa, *Chem. Commun.* 562–563 (2002).

116. D. L. Rodman, H. Pan, C. W. Clavier, X. Feng and Z. Xue, *Anal. Chem.* **77**, 3231–3237 (2005).

Chapter 19

Electrochemical Sensors Based on Nanomaterials for Environmental Monitoring

Wassana Yantasee, Yuehe Lin and Glen E. Fryxell

Pacific Northwest National Laboratory
Richland, WA, USA

19.1 Introduction

19.1.1 *Current stage of electrochemical sensors for the analysis of toxic metals*

Environmental monitoring, especially of toxic metal ions in surface and subsurface water sources, as well as occupational monitoring of worker exposure to toxic metals presently rely on the collection of discrete liquid samples for subsequent laboratory analysis using techniques such as ICP-MS and AAS. Sensors that are field-deployable and able to measure part-per-billion (ppb) or nanomolar levels of toxic metal ions will reduce the time and costs associated with environmental and occupational monitoring of hazardous metal species. Electrochemical sensors based on adsorptive stripping voltammetry (AdSV) appear to be a very promising technique that offers desired characteristics such as field deployability, specificity for targeted metal ions, enhanced measurement frequency and precision, robustness, inexpensiveness, and infrequent regeneration of sensor materials.[1-3]

Adsorptive stripping voltammetry usually involves selective preconcentration of metal ions on an electrode surface, followed by quantification of the accumulated species by a voltammetric method that generates a favorable signal-to-background ratio. Preconcentration of metal ion species on an electrode surface prior to the detection step allows sensitive analyses of the metal species that may be present at extremely low levels in contaminated groundwater, wastewater, or body fluids. Conventionally, the preconcentration of metal species has been done on mercury-based electrodes by forming an amalgam of the metal ions, which have issues related to the use and disposal of toxic mercury and the mechanical instability of mercury electrodes (i.e. mercury drop electrodes), making them unfit for routine field applications.[4,5] To develop mercury-free sensors, functional ligands have been immobilized on electrode surfaces for the preconcentration of trace metal ions by employing the specific binding properties of the ligands towards the target metal ions without applying a potential. Immobilization of ligands on the electrode surface can be done by coating monolayer[6-9] or polymeric films[10-12] on the electrode surfaces, or embedding suitable functional ligands in a conductive porous matrix.[13-20] The number of functional groups on the monolayer films is often limited and the stability and durability of the films may be poor. Polymeric films are often affected by shrinking and swelling because of the changes in solution pH or electrolyte concentration.[21,22] Thus a conductive carbon paste mixed with functional ligands is more widely used. However, the ligands in these sensors are in loose association or physical contact with the carbon paste materials, thus degradation of the sensors occurs over time as a result of the depletion of ligand-bearing materials unless certain solvents in which the ligands are insoluble are employed.[19]

19.1.2 Nanostructured silicas and carbon nanotubes (CNTs) in electrochemical sensing of toxic metals

Nanostructured materials, defined as materials that have at least one dimension smaller than 100 nm, are increasingly popular in the development of electrochemical sensors for toxic metal ions because they permit the fabrication of miniature sensing devices that are sensitive, compact, low-cost, low-energy-consuming, and easily integrated into field-deployable units. Two nanomaterials that are the fastest growing in the electrochemical sensing field are ordered mesoporous silicas (OMSs) and carbon nanotubes (CNTs) which will be the subjects of this review. Figure 19.1 shows the structures of (a) ordered mesoporous silica with parallel hexagonal pore structures, and (b) open-end multi-wall CNT in comparison with ordered pyrolytic graphite.

19.1.2.1 Nanostructured silica materials

The uses of silica-based organic-inorganic materials, including silicas coated with inorganic layers supporting catalysts, and silica gels grafted with organic groups or immobilized with enzymes, as electrode modifiers were widely studied in 1990s and have been nicely reviewed by Walcarius.[23]

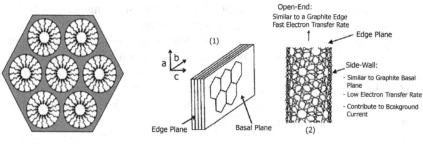

(a) MCM-41 with self-assembled monolayers   (b) ordered pyrolytic graphite (1) and open-end multi-wall CNT (2)

Figure 19.1. Structures of (a) MCM-41 with self-assembled monolayers (SAMMS), (b-1) highly ordered pyrolytic graphite, and (b-2) open-end multi-wall CNT, from Ref. 135, reproduced by permission of The Royal Society of Chemistry.

Also in the past decade, crystalline nanoporous silica, MCM-41, having hexagonal arrays of regular pore structure, has been synthesized by the surfactant templating process.[24] Because of its rigid, open, parallel pore structure, MCM-41 significantly improves the accessibility of analytes in and out of the pores when compared with amorphous silicas.[25] Amorphous materials are not desirable for electrochemical applications since most adsorption of analytes on the functional groups residing inside the pores as well as electrochemical transformations are diffusion-controlled.[26] Unmodified ordered-mesoporous silicas have been dispersed in a carbon paste matrix in order to utilize the reactivity of their silanol groups for the preconcentration of Cu^{2+}, Hg^{2+}, Ag^+, and Pb^{2+}.[27-30] However, only metal hydroxides of mercury(II) are soluble enough to reach and react with silanol groups in the pH range of 4–7.[29] Other metal species that are not hydroxylated can only bind with silanolate groups, which are formed in significant amounts only at above pH 7. Because both the number of immobilized functional groups and the structure of the organically modified silicates are believed to affect the voltammetric responses of metal ions,[27,31] installation of well-designed organic functional groups in the pores of ordered mesoporous silicas opens a new frontier for a new class of electrochemical sensors.

Organically modified OMS materials have been used for a decade as sorbent materials, yet their use in electrochemical sensors is relatively new. Few works have employed electrochemistry methods to study the properties of such materials, including cyclic voltammetry to study the voltammetric responses of electroactive species,[28,29,31-35] potentiometric monitoring of pH to study the protonation kinetics,[36] or the steady-state current method to study the uptake[36] and diffusion of metal ions[37,38] of various mesoporous silicas. At the Pacific Northwest National Laboratory (PNNL), we have exploited the attractive properties of the organically modified OMSs in electrochemical sensors for metal ions that are solid-state and mercury-free.[39–46] Preconcentration of metal ions on the electrode surface by utilizing the binding affinity and selectivity of the functionalized ordered mesoporous silicas at open-circuit potential enables their voltammetric quantitation at ultra-trace levels.

19.1.2.2 *Carbon nanotubes*

There has been enormous interest in exploiting CNTs in electrochemical sensors since they were first introduced in 1991.[47] CNTs are distinguished according to their structural properties[48]: a single-wall CNT (SWCNT) consists of a single graphitic sheet rolled into a cylinder (with 1–2 nm o.d. and several microns in length); and a multi-wall CNT (MWCNT) consists of graphitic sheets rolled into closed concentric tubes (with 50 nm o.d. and several microns in length), each separated by van der Waals forces to have a gap of 3.4 Å. The electron transfer rate of the electrode is dominated by the CNT surface structure. Multi-wall CNTs have a side-wall structure similar to that of the graphite basal plane of highly ordered pyrolytic graphite (HOPG) and the open-end similar to the edge-plane of HOPG (Figure 19.1(b)), thus the electron transfer rate along the open ends of the CNTs is normally faster than along their side-walls.[49]

In the past decade, researchers have emphasized the use of anti-interference layers or artificial electron mediators for improving the selectivity of amperometric biosensors. Because of the electrocatalytic properties of CNTs, the artificial mediators for shuttling electrons between the enzymes and the electrodes are not needed, thereby eliminating the dependence on the mediators and enhancing the reproducibility. CNTs also minimize the surface fouling of biosensors, thus imparting higher stability to these devices. With these advantages, CNTs have been investigated more extensively for electrochemical sensors for biomolecules than for metal ions. With CNTs, the overvoltage for the oxidation of hydrogen peroxide[50,51] and nicotinamide adenine dinucleotide (NADH) is reduced,[50,52] allowing low-potential detections of glucoses,[50,51,53] organophosphorous compounds,[54] and alcohols.[50] Nonetheless, our work and others have shown that CNTs are also very appealing for the development of mercury-free sensors and CNT nanoelectrode arrays for trace metal ion analysis. Given that we have previously reviewed the use of CNTs in electrochemical biosensors,[55,56] this chapter will focus only on its use for the analysis and sensing of metal ions.

19.2 SAMMS-based Electrochemical Sensors: Principles and Configurations

19.2.1 Metal ion preconcentration on SAMMS materials

For electrochemical sensors, successful preconcentration of trace metal ions (in μg/L or ppb) present in complex matrices requires that the sorbent meet a number of important criteria, including (a) high selectivity for target metals, (b) high loading capacity, (c) fast sorption kinetics, (d) excellent stability, and (e) ability to be easily regenerated. The PNNL has been a leader in developing a new class of nanostructured sorbents, the self-assembled monolayer on mesoporous supports (SAMMS), by installation of various organofunctional moieties (Figure 19.2(a) for example) on MCM-41. Initially aimed at facilitating environmental clean-up of complex nuclear/chemical waste, the materials are designed based on the stereochemistry of the ligand, size of the chelation cavity, and hardness/softness of the metal ion and ligand field to selectively sequester a specific target species, including lanthanides,[57,58] actinides,[59-61] heavy and transition metal ions,[21,62-65] radiocesium,[66] radioiodide,[67] and oxometallate anions.[68] SAMMS materials are highly efficient sorbents that meet all the above requirements; their large adsorption capacity and multi-ligand chelation ability enhances their binding affinity and stability, their open, parallel

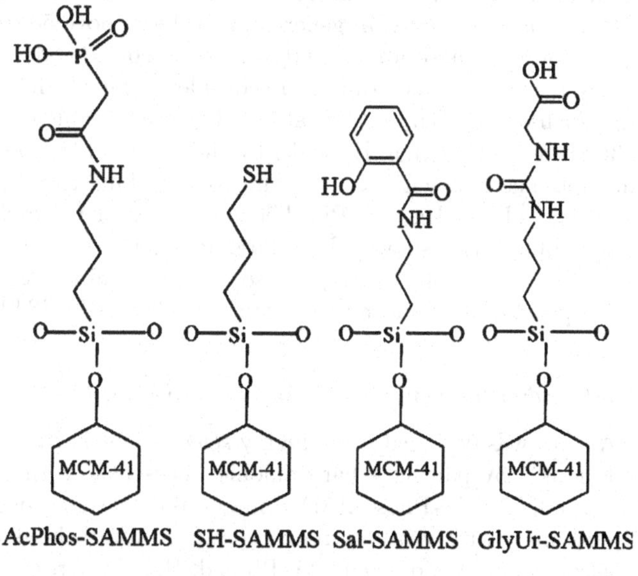

AcPhos-SAMMS   SH-SAMMS   Sal-SAMMS   GlyUr-SAMMS

Figure 19.2.   SAMMS materials with various organosilanes.

pore structure with a pore size of 5.0 nm allows for easy diffusion of analytes into the nanoporous matrix, and their rigid ceramic backbone prevents pore closure due to solvent swelling, resulting in fast sorption kinetics. Equilibrium adsorption on SAMMS is normally reached within minutes. The high surface area of silica substrates ($1,000 \, m^2/g$) and the monolayer self-assembly technique afford a high functional group density up to tenfold higher than that of simple functionalization methods.[41,65] With their superior properties over conventional sorbents, they are highly promising for the preconcentration of metal ions on electrochemical sensors. Figure 19.2 shows the schematics of four SAMMS materials that have been investigated in our laboratory, first as sorbent materials and later as electrode modifiers, including thiol- (SH-) SAMMS, acetamide phosphonic acid- (Ac-Phos-) SAMMS, glycinyl urea- (Gly-Ur-) SAMMS, and salicylamide- (Sal-) SAMMS.

19.2.1.1 *Thiol- (SH-) SAMMS*

SH-SAMMS was designed to be an excellent sorbent for lead and mercury.[21,62] With thiol coverage of up to 82%, SH-SAMMS has a loading capacity for Hg of 0.64 g Hg/g.[21] In the presence of Ag, Cr, Zn, Ba, and Na, SH-SAMMS has demonstrated mass-weighted distribution coefficients (K_d) of Hg and Pb in the order of $10^5 \, mL/g$ at neutral pH.[62] The higher the K_d values, the more effective the sorbent material is at sequestering the target species at trace concentrations. In general, K_d values above 500 mL/g are considered acceptable, those above 5,000 mL/g are considered very good, and K_d values in excess of 50,000 mL/g are considered outstanding.[60] Background ions, such as Na, Ba, and Zn, although present at high concentrations (i.e. 350 times higher than Hg and Pb), did not bind to SH-SAMMS. Other transition metals like Cd, Cu, Ag and Au may bind to SH-SAMMS, but not as effectively as Hg and Pb. The K_d values of Cu and Cd, for instance, are about 25 times lower than those of Hg. The presence of other anions (i.e. CN^-, CO_3^{2-}, SO_4^{2-}, PO_4^{3-}) also did not significantly interfere with the adsorption of lead and mercury ions onto SH-SAMMS.[64]

19.2.1.2 *Acetamide phosphonic acid- (Ac-Phos-) SAMMS*

Ac-Phos-SAMMS was designed to be highly selective for hard Lewis acids like actinide ions[69] by pairing a hard anionic Lewis base with a suitable synergistic ligand. The Ac-Phos-SAMMS material displays excellent selectivity for plutonium with virtually no competition from a large excess of a wide variety of metal cations (e.g. Ni, Pb, Cd, Hg, Cu, Cr, Ca, and Na) and anions (phosphate, sulfate, and citrate).[60] In the absence of actinides,

Ac-Phos-SAMMS containing phosphonic acids is a good sorbent material for heavy and transition metal ions such as Cu, Pb, and Cd; the K_d values are in the order of 10^4 mL/g in excess of 4,000-fold by mole of acetate buffer and at a solution-to-solid ratio of 200 mL/g.[65] It can uptake 99% of 2 mg/L Cd^{2+} within a minute at a solution-to-solid ratio of 200 mL/g.[65]

19.2.1.3 *Glycinyl urea- (Gly-Ur-) SAMMS*

Gly-Ur-SAMMS contains carboxylic acid and amide carbonyl groups that are arranged in a suitable fashion with a large chelating cavity and may bind to metal ions by many different schemes.[41] From pH 4.6 to 6.4 (Table 19.1), Gly-Ur-SAMMS has a very good affinity for Pb^{2+} and Hg^{2+} and moderate affinity for Cu^{2+}: the K_d values are in the order of 10^4 mL/g for Pb and Hg and 10^3 mL/g for Cu even in excess of 50-fold by mole of Ca^{2+} and 10,000-fold by mole of $NaNO_3$ per metal ions.[41] At pH 2, the K_d dropped significantly, suggesting that the regeneration of the sorbent materials is very feasible using an acid wash. After ten cycles of washing in 0.5 M HCl and reuse, Gly-Ur-SAMMS showed no sign of material degradation and the binding affinity of europium remained unchanged.[60]

19.2.1.4 *Salicylamide- (Sal-) SAMMS*

Sal-SAMMS contains a stabilized phenolic group that is in resonance with the adjacent amide carbonyl group, thereby conveniently providing an easy

Table 19.1. Distribution coefficients (K_d) for Gly-Ur-SAMMS at initial metal ion concentration of 0.5 mM in 0.1 M $NaNO_3$ (Ref. 41) and for Sal-SAMMS at initial lanthanide concentration of 2 ppm in 0.1 M $NaNO_3$ (Ref. 57), both with a solution/ solid ratio of 200.

Gly-Ur-SAMMS Metal	pH 2.0	pH 4.6	pH 6.4
Cu	6	3,438	2,111
Hg	2,023	22,274	17,563
Pb	23	39,971	38,071

Sal-SAMMS Cation	pH = 1	pH = 2.5	pH = 4.5	pH = 6.5
La	0	0	17	7,870
Nd	0	0	18	> 100,000
Eu	0	129	17,100	48,000
Lu	6	2	20	> 100,000

anchoring point for metal ions.[57] Because of the size of its ligand cavity, Sal-SAMMS prefers to sequester smaller lanthanide cations such as Lu and Eu over the larger lanthanides like La as demonstrated by the K_d values in Table 19.1.

19.2.2 Sensors based on SAMMS-conductive materials

Carbon is a versatile electrode material that can undergo various chemical and electrochemical modifications to produce suitable surfaces for high electrode responses. Carbon electrodes have a wide useful potential range, especially in the positive direction, because of the slow kinetics of carbon oxidation. Conductive carbon paste and graphite ink have provided easy ways to construct electrodes for many decades, but their use in conjunction with OMSs as electrode materials are still limited. At the PNNL, thiol,[45] acetamide phosphonic acid,[39,42,43,46] glycinyl urea,[41] and salicylamide[40] self-assembled mesoporous silica materials have been incorporated into carbon paste electrodes (CPEs) or screen-printed carbon electrodes (SPCEs) for the electroanalysis of many metal ions important for environmental and occupational monitoring. Figure 19.3 shows various configurations of electrodes and sensors that have been investigated in our laboratory, while Table 19.2 summarizes our work and others which used different types of ordered mesoporous silica materials in electrochemical sensors for metal ions. The lower detection limits of electrochemical sensors employing electrolytic deposition are often estimated from their signal-to-noise ratio (e.g. 3S/N). For SAMMS-based electrodes that preconcentrate metal ions at open circuit, detection limits are obtained experimentally as a function of the preconcentration time.

19.2.2.1 Detection principles

Unlike other classes of electrode modifiers that are intrinsically conductive such as zeolites,[70–72] metal oxides,[73] and carbon nanotubes, SAMMS materials are electronic insulators. To use SAMMS materials in electroanalysis, they must be (a) embedded into the conductive matrix like graphite paste or carbon ink, or (b) in close contact with the conductive surface. On the contrary, since SAMMS particles are not conductive, the high surface of the material does not contribute to the charging current for the sensors, leading to the low background current for the measurements.

    The overall analysis involved a two-step procedure: a preconcentration step at open circuit, followed by medium exchange to a pure electrolyte solution for the voltammetric quantification. During the preconcentration step, metal ions (e.g. M(II)) are accumulated on SAMMS, which are embedded

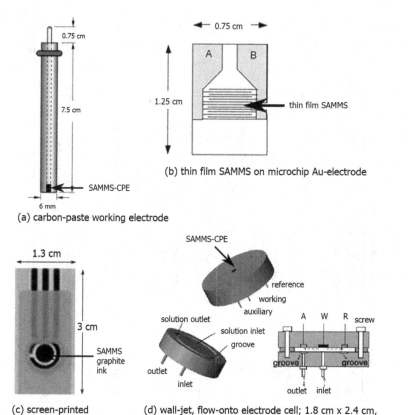

Figure 19.3. Four electrode configurations used at the PNNL, from Ref. 46, reproduced by permission of Elsevier.

on the electrode surface, by complexation with the functional groups (e.g. -SH) on SAMMS:

$$SiO_2\text{----}SH + M(II) \leftrightarrow SiO_2\text{----}S\text{-}M(II). \tag{19.1}$$

The detection step consists of electrolysis and stripping. In the electrolysis step, accumulated M(II) are desorbed in an acidic medium, then cathodically electrolyzed at a negative potential (i.e. $-1.0\,V$) for a certain period (i.e. 60 seconds) to convert metal ion (M(II)) to elemental metal M(0):

$$SiO_2\text{----}S\text{-}M(II) + H^+ \leftrightarrow SiO_2\text{----}SH + M(II), \tag{19.2}$$

$$M(II) + 2\,e^- \leftrightarrow M(0). \tag{19.3}$$

Table 19.2. Ordered mesoporous silica materials in electrochemical sensors for metal ions.

Mesoporous Silicas	Electrode Type	Metal Ions	Detection Limits (Preconcentration Time)	Linear Range (Preconcentration Time)	Reproducibility (% RSD)	Figure	Reference
SH-SAMMS	CPEs	Simultaneous Hg/−Pb	0.5 ppb Pb^{2+}/3 ppb Hg^{2+} (20 min), 10 ppb Pb^{2+}/20 ppb Hg^{2+} (2 min)	10–15 ppb Pb^{2+} (2 min) 2–1600 ppb Hg^{2+} (2 min)	< 5%	3a, 4	45
Gly-Ur-SAMMs	CPEs	Simultaneous Cu/Pb/Hg	1 ppb Pb^{2+} (5 min)	2.5–50 ppb pb^{2+} (2 min)	3.5% (6 samples), 50 ppb	3a, 5	41
Ac-Phos-SAMMS	CPEs	U(VI)	1 ppb U(VI) (20 min) 25 ppb U(VI) (5 min)	25–500 ppb U(VI) (5 min)	< 5%	3a, 6	43
Ac-Phos-SAMMS	CEPs	Simultaneous Cu/Pb/Cd	0.5 ppb Cu^{2+}/Pb^{2+}/Cd^{2+} (20 min) 10 ppb Cu^{2+}/Pb^{2+}/Cd^{2+} (2 min)	10–2000 ppb Cu^{2+}/Pb^{2+}/ Cd^{2+} (2 min)	5% (7 samples), 12% (5 sensors), 50 ppb	3a, 7	42
Monsil	CPEs	Hg^{2+}, Ag^+, Pb^{2+}, Cu^{2+}	2×10^{-4} M Cu^{2+}, 5×10^{-4} M Pb^{2+}, 7×10^{-4} M Ag^+	NA	6% (5 samples)	NA	30
MCM-41	CPEs	Cu^{2+}, Hg^{2+}	3×10^{-8} M Cu^{2+}, 5×10^{-8} M Hg^{2+}, (3S/∧ N)	NA	5% (6 samples), 1×10^{-5} M	NA	27–29

(Continued)

Table 19.2. (*Continued*)

Mesoporous Silicas	Electrode Type	Metal Ions	Detection Limits (Preconcentration Time)	Linear Range (Preconcentration Time)	Reproducibility (% RSD)	Figure	Ref.
Propylammonium-MCM-41	CPEs	Cu^{2+}	1×10^{-5} M Cu(II) (1 min)	NA	NA	NA	80
SO$_3$H-MCM-41	CPEs	Cu^{2+}	1×10^{-5} M Cu(II) (1 min)	NA	NA	NA	81
Thin-film SH-SAMMS	Microchip Au array	Pb^{2+}	25 ppb Pb^{2+} (30 min), 250 ppb Pb^{2+} (5 min)	250–5000 ppb Pb^{2+} (5 min)	< 5%	3b, 8	44
Thin-film mercaptopropyl-MPS	Gold, platinum, GC	Hg^{2+}	1×10^{-5} M Hg^{2+} (1 min)	NA	NA	NA	93
Ac-Phos-SAMMs	CPEs embedded in wall-jet, flow-onto cell	Pb^{2+}	1 ppb Pb^{2+} (3 min)	1–25 ppb Pb^{2+} (3 min)	2.5% (7 samples), 10 ppb	3d, 9	46
Ac-Phos-SAMMS	SPCEs	Pb^{2+}	2.5 ppb Pb^{2+} (5 min)	2.5–500 ppb Pb^{2+} (5 min)	5% (6 samples), 10% (5 sensors), 500 ppb	3c, 10	39
Sal-SAMMS	SPCEs	Eu^{3+}	10 ppb Eu^{2+} (10 min)	75–500 ppb Eu^{3+} (5 min)	10% (5 samples), 10% (5 sensors), 100 ppb	3c, 11	40

In the detection step, the elemental metal M(0) is subsequently detected by an anodic stripping voltammetry technique:

$$M(0) - 2e^- \leftrightarrow M(II). \tag{19.4}$$

From a voltammogram, the peak location is used to identify metal ions (e.g. Hg at 0.3 V, Cd at -0.8 V, and Pb at -0.5 V, approximately), while the response (either peak area in volt-ampere or peak height in ampere) is a function of metal ion concentration in the sample. After the metal preconcentration (step 1), the detection step is performed in a clean medium, thus the interference only affects the preconcentration step and not the detection step unlike other types of electrochemical sensors where the stripping steps are performed in the same solution.

Factors affecting the performance of the SAMMS-modified electrodes were investigated. Table 19.3 summarizes the typical operating conditions. It is worth noting that most of the operating conditions (e.g. electrode compositions, electrolysis/stripping media conditions, electrolysis time) for metal ion detection obtained using SH-SAMMS-CPEs can be used effectively with the CPEs modified with other SAMMS materials with slight modifications. The most sensitive and reliable electrode contains about 10–20 wt.% SAMMS in carbon paste electrodes or 10 wt.% SAMMS in screen-printed carbon electrodes. The protocol is a sequence of a two to five minute preconcentration period, a 60-second electrolysis period of the preconcentrated species at a negative potential in 0.2–0.5 M acid solution, followed by square-wave anodic stripping voltammetry to a positive-potential direction also in the same acid solution. The binding between SAMMS and metal ions is reversible, therefore the SAMMS-CPEs can be easily regenerated without damaging the ligand monolayer by desorption of the preconcentrated species in an acidic solution. By choosing an appropriate acid solution, the electrodes usually are ready to be used again without preconditions. We found that SAMMS-based electrodes can be reliable even after 80 consecutive runs with no need for surface renewal.

19.2.2.2 SH-SAMMS-CPEs

Figure 19.4(a) shows simultaneous detections of Pb^{2+} at -0.5 V and Hg^{2+} at 0.3 V, while Figure 19.4(b) shows their linear calibration curves measured simultaneously at a SH-SAMMS-modified carbon paste electrode.[45] The large differential potential (ΔE) of 0.8 V enables Pb^{2+} and Hg^{2+} to be detected simultaneously without interference with each other. After two minutes of preconcentration, the area of each anodic peak of Pb^{2+} and Hg^{2+} was proportional to their concentrations in the range of 10–1,500 ppb Pb^{2+} and 20-1,600 ppb Hg^{2+}, respectively. Because of the extremely large

Table 19.3. Typical operating parameters for the voltammetric measurements of metal ions at SAMMS-based electrochemical sensors.

Mesoporous Silicas	Electrode Type	% SAMMS	Metal Ions	Preconcentration*	Electrolysis	Stripping	Anodic Peak Location
SH-SAMMS	CPE	20%	Hg/Pb	2 min	-1.0 V, 60s in 0.2 M HNO_3	-1.0 V to 0.6 V	-0.55 V (Pb), 0.25 V (Hg)
Ac-Phos-SAMMS	CPE	20%	Cu/Pb/Cd	2 min	-1.0 V, 60s in 0.2 M HNO_3	-1.0 V to 0.4 V	-0.47 V (Pb), -0.1 V (Cu), 0.23 V (Hg)
Gly-Ur-SAMMs	CPE	12.5%	Cu/Pb/Hg	2 min	-0.8 V, 60s in 0.3 M HNO_3	-0.8 V to 0.6 V	-0.47 V (Pb), -0.1 V (Cu), 0.23 V (Hg)
Ac-Phos-SAMMS	CPE	20%	U	U in acetate buffer (pH 5), 5 min	-0.8 V, 60s in 0.2 M HNO_3	-0.8 V to 0.4 V	-0.37 V
Thin-film-SH-SAMMS	Au-electrode array	100%	Pb	5 min	-1.0 V, 60s in 0.1 M HNO_3	-1.0 V to -0.4 V	-0.48 V
Ac-Phos-SAMMS	CPEs embedded in wall-jet, flow-onto cell	10%	Pb	360 μL flowed at 2μL/s	-1.0 V, 70s in 0.3 M HCl	-0.8 V to -0.2 V	-0.64 V
Ac-Phos-SAMMS	SPCEs	10%	Pb	3 min	-1.0 V, 120s in 0.3 M HCl	-0.7 V to -0.49 V	-0.6 V
Sal-SAMMS	SPCEs	10%	Eu	8 mL, Eu in acetate (pH 4.6), 5 min	-0.9 V, 60s in 20 μL of 0.1M NH_4Cl (pH 3.5)	-0.95 V to -0.4 V	-0.72 V

*Unless specified otherwise, electrodes were immersed in 15 mL of stirred metal ions in DI water (pH 5.5) under open circuit conditions.

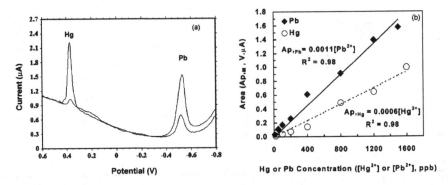

Figure 19.4. Simultaneous voltammetric detection of Hg^{2+} and Pb^{2+} at SH-SAMMS-CPEs: (a) anodic peaks, and (b) linear responses with a function of their solution concentrations, from Ref. 45, reproduced by permission of The Royal Society of Chemistry.

surface area of SAMMS and the self-assembly chemistry, SAMMS contains a high loading density of the functional group, which minimizes the competition for the binding sites among the metal ions and enables simultaneous detection of the metal ions without losing signal intensity.

19.2.2.3 Gly-Ur-SAMMS-CPEs

Figure 19.5 shows the voltammetric response (current) of Pb as a function of (a) preconcentration time and (b) Pb solution concentration at a Gly-Ur-SAMMS-modified carbon paste electrode.[41] A linear response current was obtained from 1 to at least 15 minutes of the preconcentration period (at open circuit) in 25 ppb Pb^{2+} solution and from 2.5 to 50 ppb Pb^{2+} in the solution after two minutes of the preconcentration period. The percent relative standard deviation (%RSD) of six consecutive measurements of 50 ppb Pb^{2+} was found to be 3.5. The same electrode can also be used for simultaneous detection of lead, copper, and mercury as shown in Figure 19.5(c) where the peak is found at $-0.47\,V$, $-0.06\,V$, and $0.23\,V$, respectively.

19.2.2.4 Ac-Phos-SAMMS-CPEs for uranium

Uranium is one of the most important actinides at numerous nuclear sites. The most recent voltammetric detection of uranium has been made successfully at the sub-nanomolar level by employing electrolytic accumulation of uranyl species in complexes with chelating agents like catechol,[74] oxine,[75] chloranilic acid,[76] cufferon,[77] and salicylideneimine[78] at mercury drop electrodes or cufferon[79] at Bi-film electrodes. Because of its excellent affinity and selectivity for actinide ions, Ac-Phos-SAMMS was used to modify a

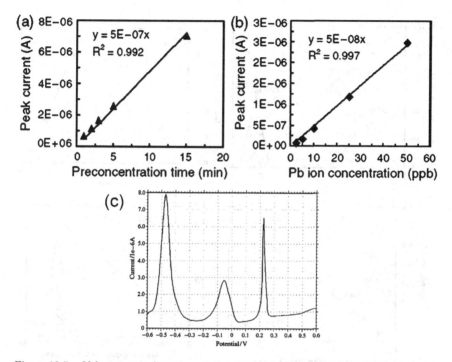

Figure 19.5. Voltammetric detection of metal ions at Gly-Ur-SAMMS-CPEs: (a) linear response of 25 ppb Pb^{2+} as a function of preconcentration time, (b) after two minutes' preconcentration as a function of Pb^{2+} solution concentration, and (c) simultaneous detection of Pb^{2+}, Cd^{2+}, and Hg^{2+}, from Ref. 41, reproduced by permission of American Scientific Publishers.

carbon paste electrode for uranium detection,[43] which will eliminate the potential interferences from heavy and transition metal ions existing at higher concentrations in most wastes.

Uranium in aqueous solution will form complexes with ligands that vary as a function of solution pH. Speciation can greatly affect the voltammetric responses of uranium (matrix effect). The type and concentration of anions, the concentration of uranium, and the pH of the solution are factors that determine which uranium complex will be preferentially formed in the solution and adsorbed on the SAMMS. Sensors that rely on electrodeposition of metal-ligand complexes on electrode surfaces before detecting the metal within the same solution are often affected by the presence of competing ligands in both the deposition and detection steps. Using SAMMS-based electrodes with medium exchange in the detection step after uranium preconcentration can overcome the matrix effect and ligand interference.

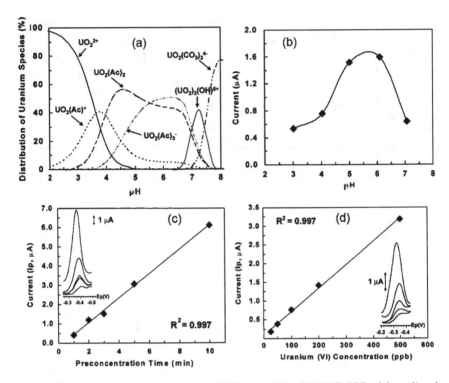

Figure 19.6. Voltammetric detection of U(VI) at Ac-Phos-SAMMS-CPEs: (a) predicted speciation of 500 ppb U(VI) in 0.05 M Na-acetate, (b) voltammetric response of 500 ppb U(VI) after three minutes' preconcentration as a function of pH, (c) linear response of 500 ppb U(VI) at pH 5 as a function of preconcentration time, and (d) linear response after five minutes' preconcentration as a function of U(VI) concentration, from Ref. 43, reproduced by permission of Wiley-VCH.

The pH of the solution was predicted to affect the distribution of uranyl complexes with acetate (used as the buffer), hydroxyl, and carbonate ligands as shown in Figure 19.6(a). In the pH range of 2–5, the dominant uranyl species have positive or neutral charges (e.g. UO_2^{2+}, $UO_2(Ac)^+$), which bind more strongly to the phosphonic acid on Ac-Phos-SAMMS as the pH increases, leading to increased voltammetric response (Figure 19.6(b)). Above pH 6, uranyl species with negative charge appear to cause repulsive forces, resulting in low uranium adsorption and low voltammetric responses. After a five-minute preconcentration period of U(VI) in acetate buffer (pH 5), the current at -0.37 V was proportional to the preconcentration time (Figure 19.6(c)) and U(VI) concentration in the range of 25–500 ppb (Figure 19.6(d)). The uranium detection limits improved significantly with longer preconcentration time (Table 19.2).

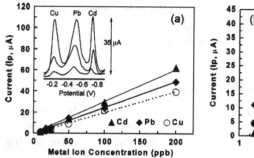 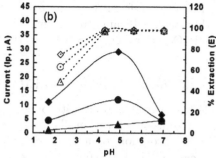

Figure 19.7. Simultaneous detection of metal ions at Ac-Phos-SAMMS-CPEs: (a) voltammetric responses of $Pb^{2+}/Cd^{2+}/Cu^{2+}$ after two minutes' preconcentration as a function of their solution concentrations, and (b) responses of 50 ppb metal ions as a function of pH (two minutes' preconcentration), from Ref. 42, reproduced by permission of Elsevier.

19.2.2.5 *Ac-Phos-SAMMS-CPEs for heavy metal ions*

Ac-Phos-SAMMS has been used to modify a carbon paste electrode for simultaneous detection of cadmium (Cd^{2+}), copper (Cu^{2+}), and lead (Pb^{2+}) in the absence of actinides.[42] Figure 19.7(a) shows the linear voltammetric responses of Cu^{2+}, Pb^{2+}, and Cd^{2+} after two minutes of preconcentration in the metal ion concentration range of 10–200 ppb. The signals of Cu were not affected by the presence of Pb^{2+} and Cd^{2+} in the concentration range studied, owing to the much higher loading capacity of the SAMMS than the total moles of metal ions. Above 200 ppb, the detection of Cu^{2+}, Pb^{2+}, and Cd^{2+} started to interfere with each other because of the close proximity of their signals (i.e. $\Delta E_{Cu-Pb} = 0.3\,V$, $\Delta E_{Pb-Cd} = 0.25\,V$). The sample pHs were found to exert a significant but predictable effect on the preconcentration process. In Figure 19.7(b), the voltammetric responses of Cu, Pb, and Hg followed the adsorption isotherms in the pH range of 2–6. Above pH 6, while the adsorption isotherms still remained at the maximum, the peak currents were negligible due to the formation of metal hydroxide complexes that may be sparingly soluble and not adsorb on the electrode surface.[31] Such information about speciation can be obtained only at electrochemical sensors and not by ICP-MS which yields only the total metal concentrations.

In addition to our work summarized here, the Walcarius group has developed carbon paste electrodes modified with propylammonium-[80] and mercaptopropyl-[81] functionalized mesoporous silica materials for voltammetric detection of Cu^{2+} (Table 19.2). To exploit the amine ligands on the propylammonium moieties, a five-minute electrochemical pretreatment at

−1.3 V was performed in order to deprotonate the N-bearing group.[80] In the other work, mercaptopropyl-functionalized mesoporous silica was oxidized by H_2O_2 to first create SO_3H functional groups.[81]

19.2.3 *Sensors based on SAMMS thin films*

Mesoporous silica films with pore sizes of up to 100 Å have been synthesized from spin-casting of silica sol-gels (normally consisting of silica precursor, acid solution, organic solvent, and water) by a surfactant-templating process in which the pores are formed upon removal of the surfactant by calcinations.[82–86] The precise design and control of the pore structure, pore size/volume, and pore orientation of the silica films can be achieved by using structure-directing agents, such as the non-ionic surfactant Pluronic F-127 ($EO_{106}PO_{70}EO_{106}$). By controlling the film thickness (e.g. changing the composition of the precursor solution, spin rate, and calcination conditions), a continuous, defect-free, mesoporous silica thin film is produced.[44] Spin-casted silica thin films have been found to be stable in pure water and acidic electrolyte solutions, but unstable in basic aqueous solutions due to the dissolution of the mesoporous silica framework.[35] Because electrodes modified with these silica films cannot be used in basic solutions, the silanol groups, inherent to the nanoporous silica, are not useful in binding with positively charged metal ions.[27] Therefore, without functional ligands attached, the spin-cast films of mesoporous silicas are often used as molecular sieves in electrodes by either blocking or improving the voltammetric responses of the analyte species.[35] To remove this limitation, we have used a self-assembly technique to attach organic thiol functional groups onto the mesopore surfaces of the silica thin film that was previously spin-casted and calcined on the surface of a microchip-based gold electrode array, as shown in Figure 19.3(b). This two-step approach has also been used by the Sanchez group.[87,88] Several surface characterization methods were performed on the mesoporous silica thin film before and after thiol functionalization. TEM measurement of the SH-FMS film shows short-range ordered mesopores. The nitrogen adsorption/desorption isotherms of the untreated SiO_2 thin film are of type IV with a clear H_2-type hysteresis loop, which is consistent with the film's large ratio of pore size to pore spacing[89,90] and cubic arrangement of pores.[91] From BET analysis, the calcined film had a large primary pore diameter of 7.7 nm, a porosity of 60%, and a high BET surface area of 613 m^2/g. The Fourier transform infrared (FTIR) spectra indicate that organic thiol monolayers were successfully immobilized inside the nanopores of the mesoporous silica. After the thiol immobilization, the increase in the film's refractive index suggests the decrease in film porosity from 66% to 23%. The similarity of the X-ray diffraction (XRD) patterns before and after thiol functionalization indicates that the structure of the mesoporous film remained unchanged after the thiol attachment.

With the SH-SAMMS thin-film-modified electrode array, the preconcentration of Pb can be performed in slightly acidic electrolyte solutions or neutral non-electrolyte solutions[44] which corresponds to a previous batch sorption study of SH-SAMMS in powder form.[62] In non-basic solutions, the mesoporous silica films are more likely to remain stable for a long period of time. The voltammetric responses increased linearly with Pb^{2+} concentrations ranging from 250 to 5,000 ppb after five minutes of preconcentration and from 25 to 100 ppb after 30 minutes of preconcentration (Figure 19.8 and Table 19.2). The large dynamic range for lead signals is a result of the high loading capacity of the SAMMS thin film in comparison to that of conventional chemically self-assembled thin-film electrodes that often has a smaller linear range. Through the silica-surfactant self-assembly process, conductive additives or binders that are often used to mix with SAMMS to make an electrode can be eliminated. Thus the electrodes can be constructed in a manner that is highly reproducible which is very important for any two-dimensional surface reaction to be successful.[92] The SAMMS thin film can be made to adhere to a wide variety of surfaces. The microchip-based SH-SAMMS electrodes also have low maintenance requirements (no activation and regeneration were required) and can be integrated into microfluidic devices for field applications.

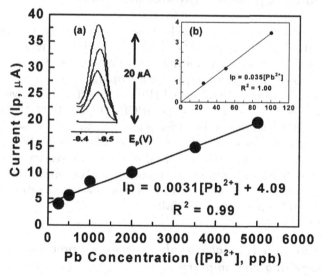

Figure 19.8. Voltammetric response of Pb^{2+} after five minutes of preconcentration on an SH-SAMMS thin film on a microchip-Au-array; insets show (a) the Pb voltammograms, and (b) responses after 30 minutes of preconcentration, from Ref. 44, reproduced by permission of the Royal Society of Chemistry.

In a different approach, the Walcarius group has recently used a one-step technique by spin-casting a sol-gel mixture containing both alkoxysilane and organosilane precursors in order to coat thin films of mercaptopropyl-modified mesoporous silica on gold, platinum, and glassy carbon electrodes.[93] The surfactant was removed by solvent extraction instead of calcination. The electrodes were used to accumulate 10^{-5} M of Hg(II) in 0.1 M HNO$_3$ at open circuit prior to its stripping voltammetry in 3 M HCl (Table 19.2).

Improvement of sensor performances may be accomplished by manipulating the pore structure, size, and orientation as well as the thickness of the SAMMS thin films. Partial pore-closing or long axes of pores being oriented parallel to the plane of the film caused by the exposed side of the film being dried first[94] may often cause slow metal diffusion into the SAMMS thin films. To increase the permeability of SAMMS thin films, PNNL researchers[95] have engineered hierarchical porosity into the films by using cellulose nitrate to generate a system of larger pores (15–30 nm diameter) connected by the smaller pores (2–10 nm diameter) generated using traditional organic surfactants. This will potentially reduce the preconcentration time and increase the detection sensitivity of metal ions at the SAMMS thin-film sensors.

19.3 Field-deployable SAMMS-based Sensors

Once the sensors based on SAMMS were successfully evaluated and optimized in batch experiments using electrode configuration as in Figure 19.3(a), two sensor platforms were investigated for field applications: (1) a portable analyzer employing a stripping voltammetry technique at a wall-jet, flow-onto electrode cell (Figure 19.3(d)), and programmable sequential injection, and (2) reusable SPCEs (Figure 19.3(c)). Lead (Pb) has been used as the model metal for evaluating the two sensor platforms developed from Ac-Phos-SAMMS modification of electrodes because (a) Pb is one of the metals of greatest concern due to its high toxicity and common occurrence of Pb poisoning from environmental and occupational exposures, especially in children,[96,97] and (b) Pb has excellent binding capacity, selectivity, and rate[65] on Ac-Phos-SAMMS.

19.3.1 Automated portable analyzer based on flow-injection/ stripping voltammetry

To develop next-generation analyzers that are portable, fully automated, and remotely controllable, we have integrated a sequential injection analysis

(SIA) method with the nanostructured electrochemical sensors. Since it was first introduced in 1990, SIA has gained popularity because of its economical use of samples and reagents, robustness, inexpensiveness, and simple design of instrumentation.[74,98,99] To detect Pb at low ppb levels, carbon paste modified with Ac-Phos-SAMMS was embedded in a very small wall-jet (flow-onto) electrochemical cell (Figure 19.3(d)) in connection with the SIA system as shown in Figure 19.9.[46] Microliters of samples and reagents were injected into the system by a programmable sequential flow technique. The portable system yielded a linear calibration curve in the useful range of 1 to at least 25 ppb of Pb(II). The automation increased the reproducibility by eliminating human errors and batch-to-batch variation, thus increasing the measurement reproducibility; the %RSD of seven consecutive measurements of Pb^{2+} was reduced from 5% in batch experiments to 2.5% using the automated system.

The effect of interferences on Pb detection was evaluated. An ionic species may be considered as an interference if (a) it can out-compete the target metals for binding sites on SAMMS during the preconcentration step and (b) once preconcentrated on SAMMS its peak response can overlap that of the target metals in the stripping step. Organic molecules and anions can also act as interferences if they can out-compete SAMMS for the target metal ions. From batch competitive sorption experiments,[65] Ac-Phos-SAMMS has an affinity for metal ions in the decreasing order of

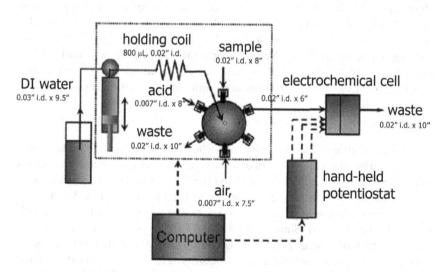

Figure 19.9. Portable metal ion analyzer consisting of programmable sequential injector and wall-jet, flow-onto cell having an Ac-Phos-CPE as the working electrode, from Ref. 46, reproduced by permission of Elsevier.

$Pb^{2+} > Cu^{2+} > Mn^{2+} > Cd^{2+} > Zn^{2+} > Co^{2+} > Ni^{2+} \sim Ca^{2+} \gg Na^{+}$.
This corresponds to the results obtained at an Ac-Phos-SAMMS sensor
where a large excess of 100-fold (by mole) of Ca, 70-fold of Zn, Ni, and Co,
and 10-fold of Mn did not interfere with Pb signals.[46] The electrode was
reliable for at least 90 measurements over 5 days of operation.

19.3.2 Reusable screen-printed sensors

Disposable sensors for the assay of toxic metal ions are gaining popularity
because of their ease of use, simplicity, and low cost.[100–106] Of all the disposable sensors, SPCEs (Figure 19.3(c)) coupled with an AdSV technique
have been increasingly investigated due to their measurement sensitivity,
simplicity during field applications, and ability to be mass produced at
very low costs.[107] Most screen-printed electrodes for the sensitive assay
of metal ions have been based on mercury film,[108–111] or mercury oxide
particles.[107] Disposal of electrodes containing mercury leads to occupational
and environmental heath concerns which may result in future regulation
of mercury-based electrodes. Mercury-free screen-printed electrodes have
been developed by employing gold,[112] silver,[113] or bare carbon electrodes
by applying suitable reduction potential to accumulate the target metals
(e.g. Cu and Pb).[104,114] However, the sensitivity, reliability, and cost com-
petitiveness of such electrodes are yet to reach those of the mercury-based
electrodes. Although chemically modified screen-printed electrodes have
been developed by using ligands such as 1-(2-pyridylazo)-2-naphthol[103]
or calixarene,[102] ligand modification has been done *ex situ* by the drop-
coating technique, followed by the complexation of the ligands and Pb^{2+} in
ammonia buffer. Thus they often have the disadvantages of not being user-
friendly, short electrode life time, and poor inter-electrode reproducibility.

19.3.2.1 Ac-Phos-SAMMS-SPCEs

We have modified SPCEs with Ac-Phos-SAMMS for Pb detection.[39]
A screen-printed sensor consists of a built-in three-electrode system: screen-
printed SAMMS-graphite ink mixture (10 wt.% SAMMS) as a working electrode and screen-printed carbon as a counter electrode, and Ag/AgCl as
the reference electrode, situated on a 1.3 cm × 3 cm × 0.5 mm plastic substrate (Figure 19.3(c)). Similar to SAMMS-CPEs, preconcentration of Pb^{2+}
on SAMMS-based sensors can be accomplished at open-circuit potential
without electrolyte and solution degassing. The linear calibration curve
was found in the range of 0 to at least 100 ppb Pb^{2+} after 5 minutes
of preconcentration, as shown in Figure 19.10(a). Cadmium, lead, and
copper can also be detected simultaneously at the screen-printed sensors

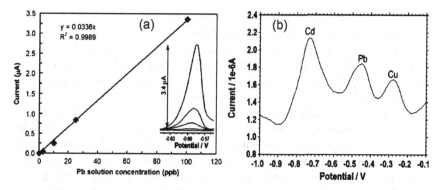

Figure 19.10. Metal ion detection after five minutes of preconcentration on Ac-Phos-SAMMS-SPCEs: (a) linear response of Pb^{2+} as a function of solution concentration, and (b) simultaneous detection of 90 ppb Cd^{2+}/18 ppb Pb^{2+}/18 ppb Cu^{2+}, from Ref. 39, reproduced by permission of Elsevier.

(Figure 19.10(b)). Even when being mixed with graphite ink, SAMMS on the screen-printed sensors still adsorbed Pb well within a short period, which is attributed to the rigid, open, parallel mesopores and the suitable interfacial chemistry of SAMMS that allow easy access to hydrated metal ions.[65] The SAMMS-SPCEs can be reused for tens of measurements with minimal degradation, thereby enabling the establishment of the calibration curve and lowering the costs compared to single-use SPCEs. Reproducibilty (%RSD) was found to be 5% for a single sensor (six measurements) and 10% for five sensors. The inter-electrode reproducibility can be improved through the precision of manufacturing of the sensors, in which SAMMS modification is done *in situ*.

19.3.2.2 *Sal-SAMMS-SPCEs*

Rare earth cations like lanthanides are difficult to electrochemically reduce to elemental forms; the voltammetric detection limits of europium (Eu) in perchlorate solutions after being directly reduced at a boron-doped diamond electrode were only at the millimolar level.[115] In order to detect Eu at the nanomolar level, researchers have electrolytically accumulated Eu in the presence of thenoyltrifluoroacetone[116] or salicylic acid[117] in aqueous solutions at the mercury drop electrodes. The preconcentration of lanthanides on an electrode surface by non-electrolytic methods is more preferable.[118,119] Nafion-coated electrodes have been used to preconcentrate Eu^{3+} at open-circuit potential, but the Eu detection limits by normal voltammetric methods are still at the micromolar level since Eu^{3+} is only weakly incorporated by ion exchange into the Nafion coating[10,11] and is completed by NH_4^+ used as the electrolyte.[120]

Eu is often present in nuclear wastes and used as an Am(III) mimic. It is very similar in size to U(IV), Np(IV), and Pu(IV).[58] Having considered its significance, we have recently developed sensors for Eu based on colloidal-Au-SPCEs modified with salicylamide-SAMMS. In acidic solutions, the electroanalysis of trace Eu is highly challenging because the Eu^{3+}/Eu^{2+} reduction peak can be found at a fairly negative potential (e.g. −1.2) where the hydrogen evolution reaction occurs, resulting in very high background.[115] Ugo et al.[10] and Moretto et al.[120] found that NH_4Cl improved the reversibility of the reduction of rare earth metals at Nafion-modified glassy carbon electrodes. Therefore, after Eu^{3+} preconcentration with Sal-SAMMS-graphite ink on a screen-printed electrode (Figure 19.3(c)), Eu^{3+} was converted to Eu^{2+} by applying −0.9 V for 60 seconds in 20 μL of NH_4Cl solution (pH 3.5), followed by stripping voltammetry in the same solution, yielding an Eu^{3+}/Eu^{2+} oxidation peak at −0.72 V. The peak current was a linear function of Eu^{3+} in the samples as shown in Figure 19.11. After five to ten minutes of preconcentration

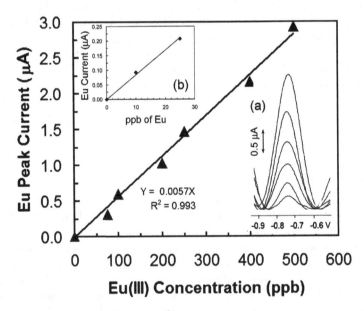

Figure 19.11.   Linear response of Eu^{3+} after five minutes of preconcentration on Sal-SAMMS-SPCEs (with built-in colloidal Au); insets show (a) the corresponding voltammograms, and (b) the linear response after ten minutes of preconcentration, from Ref. 40, reproduced by permission of the Royal Society of Chemistry.

at open-circuit potential, the ppb levels of Eu could be detected. Tenfold higher concentrations of Lu and Nd did not interfere with Eu adsorption, while that of La reduced the Eu peak current by 15%. As the concentration of NH_4Cl increased from $0.05\,M$ to $0.2\,M$, the Eu responses also increased, suggesting NH_4^+ did not compete with Eu for the binding sites of Sal-SAMMS, unlike at the Nafion-film electrodes. Other lanthanides and actinides are less likely to interfere with Eu sorption on Sal-SAMMS at this Eu concentration range because the large capacity of SAMMS will minimize the competitive binding.[42] The Sal-SAMMS sensors had a large working range pH of the Eu solutions, from 2 to 6.5, although the K_d measured from batch sorption was very small at pH below 2.5 (Table 19.1). This is perhaps due to the local pH effect; the ligand field is buried in a graphite ink matrix of the SPCEs which may restrict the transport of H^+ into such a matrix, causing the effective pH at the Sal-ligand interface to be more moderate than it is out in the bulk solution. Because of the close proximity of the Pb peak (found at $-0.6\,V$) to the Eu peak, Pb at a much higher concentration (e.g. fourfold by mole) was found to reduce the Eu peak current by 40%, while the influence of Pb was negligible at a one-to-one molar ratio of Pb and Eu. If needed, separation of Pb from lanthanides prior to the detection may be easily done using materials like Ac-Phos-SAMMS or Gly-Ur-SAMMS.[57] Nevertheless, Sal-SAMMS-SPCEs provide a simple, reproducible, and mercury-free method for Eu detection.

19.4 CNT-based Electrochemical Sensors

19.4.1 *Various methods for preparation of CNT-based sensors*

Although CNTs are relatively new in analytical fields, their unique electronic (electron transfer rate similar to that of edge-plane graphite), chemical (biocompatibility and ability to be covalently functionalized), and mechanical properties (three times stronger than that of steel) make them extremely attractive for chemical and biochemical sensors.[121,122] Various methods have been used to prepare CNT-based sensors: (a) casting of CNT thin films, from the suspension of CNTs in solvents,[51,52,123–129] such as dimethylformamide (DMF),[128] acetone,[129] Nafion,[51,126] dihexadecyl hydrogen phosphate (DHP),[127] and sulfuric acid,[52] prior to being coated on electrode surfaces, (b) using CNTs as paste electrodes or electrode composites,[130–133] and (c) using aligned CNTs as electrode substrates.[53,134–141] In particular for metal ion sensors (summarized in Table 19.4), DHP[127] and Nafion[126] have been used to disperse

Table 19.4.   Carbon nanotubes in electrochemical sensors for metal ions.

Electrode Configuration	Electrode Substrate	Metal Ions	Detection Limits (Preconcentration Time)	Linear Range (Accumulation Time)	Figure	Reference
Hg-film aligned CNT array	Au	Pb	1 ppb Pb^{2+} (3 min)	2–100 ppb Pb^{2+} (3 min)	12	139, 145, 154, 155
Bi-film aligned CNT array	Au	Simultaneous Cd/Pb	0.04 ppb Pb^{2+} (from 3S/N)	0.5–8 ppb Pb^{2+} (2 min)	12, 13	135
MWNT/Nafion film	Glassy carbon	Cd	1×10^{-6} M (4 min)	1×10^{-6} M 4×10^{-6} M (4 min)		126, 127
MWN/DHP film	Glassy carbon	Simultaneous Cd/Pb	6×10^{-9} M Cd, 4×10^{-9} M Pb (5 min)	2.5×10^{-8} M–1×10^{-5} M Cd 2×10^{-8} M to 1×10^{-5} M Pb		127, 128

MWCNTs under ultrasonication prior to being drop-coated on glassy carbon electrodes. The CNT film enables the development of mercury-free electrodes that can detect from 10^{-9} to 10^{-6} M of Cd and Pb.

Most CNT-based sensors take advantage of the bulk properties of CNTs, including increased electrode surface area,[142] fast electron transfer rate,[143] and good electrocatalytivity in promoting electron-transfer reactions of many important species.[55,56] The use of CNTs as nanoelectrodes has been increasingly explored since conventional macroelectrodes (having diameters of millimeters), such as glassy carbon and carbon paste, are known to have slow mass transport. Using nanoelectrodes (having diameters of nanometers) can enhance mass transport[138,139,144]; as electrodes decrease in size, radial (three-dimensional) diffusion becomes dominant and results in fast mass transport and fast electron transfer. The high diffusion rate at nanoelectrodes enables the study of fast electrochemical and chemical reactions.[146] Nanoelectrodes also have higher responsiveness (or higher mass sensitivity) than macroelectrodes, attributed to their lower background (charging) currents.[145] Additionally, they are less influenced by solution resistance due to lower ohmic drop.[139] Despite its advantages, a single nanoelectrode offers an extremely low capacitive current (in pico-amperes), thereby requiring expensive signal amplifiers. To solve this issue, nanoelectrode arrays consisting of millions of nanoelectrodes have been developed at the PNNL in collaboration with Boston College in order to provide magnified signals without the need for a signal amplifier.

High-density aligned CNT electrode arrays (i.e. dense CNT forest) have been reported to show fast electron transfer and electrocatalytic characteristics,[143,147] but they do not maintain the properties of individual nanoelectrodes due to the overlapping of their diffusion layers.[148–152] To make each carbon nanotube on the array work as an individual nanoelectrode, the spacing must be sufficiently larger than the diameter of the nanotubes. Millions of electrodes with nanoscale dimensions will result in an improved signal-to-noise ratio (and hence improved detection limits).

19.4.2 *Fabrication of aligned CNT nanoelectrode array*

The fabrication procedure for a low-site-density aligned CNT nanoelectrode array has been refined.[138,139] Figure 19.12(a) shows the schematic of the fabrication of the CNT nanoelectrode array. Briefly, Ni nanoparticles were first electrodeposited on a Cr-coated silicon (Si) substrate with an area of $1\,cm^2$. The aligned CNT arrays with low site density were subsequently grown from those nickel (Ni) nanoparticles by plasma-enhanced chemical vapor deposition. To take advantage of a faster electron transfer rate at the

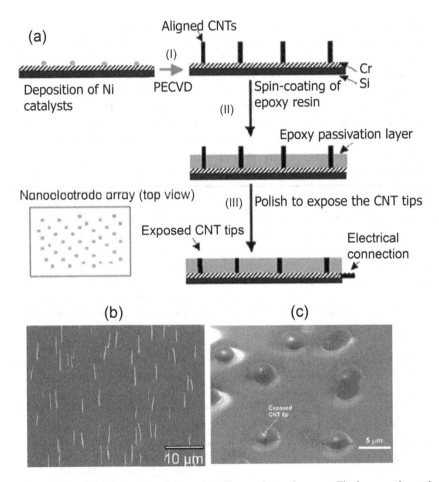

Figure 19.12. (a) Fabrication of aligned CNT nanoelectrode array: (I) plasma-enhanced vapor deposition of CNTs on top of Ni nanoparticles created by electrolytic deposition, (II) spin-coating of epoxy resin, and (III) polishing to expose the tips of CNTs. SEM images (b) after step (I): showing CNT density of $2 \times 10^6 \, cm^2$, and (c) after step III: coating with the epoxy resin and polishing to expose the CNT tips. (a) is from Ref. 56, reproduced with permission of *Frontiers in Bioscience*, and (b) and (c) are from Ref. 139, reproduced with permission of Wiley-VCH.

open ends of CNTs, and to eliminate the background generated from the side-wall of CNTs, an epoxy passive layer was used to block the side-wall of the carbon nanotube.[139] Specifically, an Epon epoxy resin 828 passivation layer (7–$9 \, \mu m$) was coated on the aligned CNT array to preserve the CNTs on the silicon substrate surface using a standard spin-coating technology. The protruding parts of the CNTs were broken and removed by ultrasonication in water. Figures 19.12(b) and 19.12(c) show the SEM images of the aligned CNT array with a site density of 2×10^6 per cm^2 and after coating with the epoxy resin and polishing to expose the CNT tips (white dots), respectively. The CNTs are at least 5 μm from their nearest neighbors, while the diameter of each nanotube is about 50–80 nm. From these low-site-density CNTs, the nanoelectrode array consisting of up to millions of CNTs, each serving as individual nanoelectrodes, was fabricated by adding the electrical connection on the CNT-Si substrate.

19.4.3 *Applications of CNT-based sensors for metal ion monitoring*

With a Hg-coated CNT nanoelectrode array, a linear relationship between the Pb signals and the Pb concentration in the solutions ranging from 1 to 100 ppb (μg/L) has been obtained.[139] Because Hg is highly toxic and not suitable for field-deployable use, the relatively benign bismuth, Bi(III), has been evaluated as a Hg substitute.[79,135,153,154] Bi-based electrodes performed as well as Hg-based electrodes for Cr(VI) quantitation after the accumulation of Cr(VI)-diethylenetriamine pentaacetic acid (DTPA) chelate.[153,155] A highly responsive voltammetric analysis of Cd and Pb at the Bi-based CNT nanoelectrode array was obtained.[135] Figure 19.13(a) shows well-defined peaks of 5 μg/L of Cd and Pb obtained after only two minutes of accumulation, in which Bi was accumulated *in situ* with the target metals at -1.2 V. Figure 19.13(b) shows a linear calibration curve of Cd achieved at a very useful concentration range of 0.5–8 μg/L of Cd.

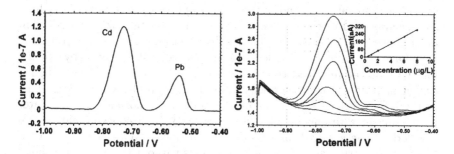

Figure 19.13. Voltammetric responses of metal ions at aligned CNT nanoelectrode array: (a) voltammograms of 5 ppb Cd^{2+} and 5 ppb Pb^{2+} after two minutes of deposition at -1.2 V in the presence of 500 ppb Bi^{3+}, (b) linear response of Cd^{2+} with a function of solution concentration, other conditions the same, from Ref. 135, reproduced by permission of The Royal Society of Chemistry.

19.5 Conclusions and Future Work

This chapter is a review of the work relevant to the two fastest growing nanomaterials in electrochemical sensing of metal ions: organically modified ordered mesoporous silicas and carbon nanotubes. SAMMS materials are highly effective as electrode modifiers; they can be either mixed with conductive materials or spin-casted as a thin film on the electrode surface. The interfacial chemistry of SAMMS can be fine-tuned to selectively preconcentrate the specific metal ions of interest. The functional groups on SAMMS materials enable the preconcentration of target metal ions to be done without mercury, supporting electrolytes, applied potential, and solution degassing, all of which are often required in conventional stripping voltammetric sensors. SAMMS materials grafted with four organic functionalities (thiol, acetamide phosphonic acid, salicylamide, or glycinyl-urea) have been used successfully at the PNNL in electrochemical sensors for heavy and transition metals (cadmium, copper, lead, and mercury), actinide uranium, and lanthanide europium. The high loading capacity and high selectivity of SAMMS materials are desirable for metal ion detection based on the AdSV technique because they minimize the competition for the binding sites of the non-target species, thereby reducing the interference and preserving the signal intensity of the target metal ions. The covalent bonding between the functional groups and the surface of silica prevents sensors from degrading over time due to depletion of ligand-bearing materials. The reversible binding between metal ions and ligands on SAMMS enables many successive uses of SAMMS-based electrodes with virtually no additional work required for surface regeneration.

Since they were first introduced in 1991, CNTs have been widely investigated for electrochemical sensors of many important biomolecules because of their electrocatalytic and antifouling properties, biocompatibility, high surface area, and mechanical strength. For trace metal analysis, CNT thin films created by drop-coating CNT-solvent suspensions on electrode surfaces have been explored in order to develop mercury-free sensors by exploiting the bulk properties of the CNTs. An array of low-site-density aligned carbon nanotubes has been grown on metal substrates by a non-lithographic method. Each CNT serves as a nanoelectrode which normally has a greater mass transfer rate and higher mass sensitivity than conventional macro-electrodes. The array of millions of CNT nanoelectrodes provides magnified voltammetric signals for trace metal ions without the need for a signal amplifier.

Our future work will focus on using other SAMMS materials in electrochemical sensors of other pollutants. The interfacial chemistry and electrochemistry of metal species on the surfaces of SAMMS-based electrodes will be an ongoing study: this fundamental knowledge is needed for predicting how the sensors will perform in real wastes that consist of many interferences and ligands at a broad spectrum of pH levels. Using the SAMMS-based sensors with biological samples for occupational monitoring of toxic metal ions is also desirable. To overcome the electrode fouling by proteins in biological samples, new metal sensors are being developed at the PNNL with the composite of SAMMS as the metal ion preconcentrator and CNT as the conductive matrix. In another approach, to avoid the use of hydrophobic binders and to improve the mass transfer of the metal ions in samples to the SAMMS surface, superparamagnetic SAMMS is being developed to first bind with metal ions in the liquid phase, followed by immobilization of metal-bound magnetic SAMMS on the electrode surface via magnetic or electromagnetic force. The results will be reported in due course.

Acknowledgments

This work was supported by DOE-EMSP and NIH/NIESH/1R01 ES010976-01A2. The Pacific Northwest National Laboratory is operated by Battelle Memorial Institute for the U.S. Department of Energy (DOE). The research was performed in part at the Environmental Molecular Sciences Laboratory (EMSL), a national scientific user facility sponsored by the DOE's Office of Biological and Environmental Research and located at the PNNL. The authors especially thank Dr Charles Timchalk, Prof. Zhifeng Ren, Dean A. Moore, Brad J. Busche, Dr Guodong Liu, Yi Tu, and Xiaohong Li for their contributions.

References

1. Y. Lin, C. A. Timchalk, D. W. Matson, H. Wu and K. D. Thrall, *Biomed. Microdevices* **3**, 331 (2001).
2. Y. Lin, R. Zhao, K. D. Thrall, C. Timchalk, W. D. Bennett and M. D. Matson, *Proc. Soc. Opt. Eng. (SPIE)* **3877**, 248 (1999).
3. J. Wang, Y. Lin and L. Chen, *Analyst* **118**, 277 (1993).
4. J. Wang, J. Lu, D. D. Larson and K. Olsen, *Electroanalysis* **7**, 247–250 (1995).
5. J. Wang, J. Wang, B. Tian and M. Jiang, *Anal. Chem.* **69**, 1657–1661 (1997).
6. N. Muskal and D. Mandler, *Curr. Separations* **19**, 49–54 (2000).
7. I. Turyan, M. Atiya and D. Mandler, *Electroanalysis* **13**, 653 (2001).
8. I. Turyan and D. Mandler, *Anal. Chem.* **66**, 58–63 (1994).
9. I. Turyan and D. Mandler, *Anal. Chem.* **69**, 894–897 (1997).
10. P. Ugo, B. Ballarin, S. Daniele and A. Mazzocchin, *J. Electroanal. Chem.* **291**, 187–199 (1990).

11. P. Ugo, B. Ballarin, S. Daniele and A. Mazzocchin, *J. Electroanal. Chem.* **324**, 145 (1992).

12. N. Akmal, H. Zimmer and H. B. Mark, *Anal. Lett.* **24**, 1431–1443 (1991).

13. R. P. Baldwin, J. K. Christensen and L. Kryger, *Anal. Chem. Commun.* **58**(8), 1790–1798 (1986).

14. T. H. Degefa, B. S. Chandravanshi and H. Alemu, *Electroanalysis* **11**, 1305–1311 (1999).

15. V. S. Ijeri and A. K. Srivastava, *Anal. Sci.* **17**, 605–608 (2001).

16. M. F. Mousavi, A. Rahmani, S. M. Golabi, M. Shamsipur and H. Sharghi, *Talanta.* **55**, 305–312 (2001).

17. Z. Navratilova, *Electroanalysis* **3**(8), 799–802 (1991).

18. B. Ogorevc, X. H. Cai and I. Grabec, *Anal. Chim. Acta* **305**, 176–182 (1995).

19. S. V. Prabhu, R. P. Baldwin and L. Kryger, *Anal. Chem.* **59**(8), 1074–1078 (1987).

20. I. G. Svegl, M. Kolar, B. Ogorevc and B. Pihlar, *Fresen. J. Anal. Chem.* **361**, 358–362 (1998).

21. X. B. Chen, X. D. Feng, J. Liu, G. E. Fryxell and M. Gong, *Separ. Sci. Technol.* **34**, 1121–1132 (1999).

22. H. Ju, Y. Gong and H. Zhu, *Anal. Sci.* **17**, 59 (2001).

23. A. Walcarius, *Chem. Mater.* **13**, 3351–3372 (2001).

24. C. T. Kresge, M. E. Leonowicz, W. J. Roth, J. C. Vartuli and J. S. Beck, *Nature* **359**, 710–712.

25. A. Walcarius, M. Etienne, S. Sayen and B. Lebeau, *Electroanalysis* **15**, 414 (2003).

26. A. Walcarius, *C. R. Chimie* **8**, 693–712 (2005).

27. A. Walcarius, C. Despas and J. Bessière, *Anal. Chim. Acta* **385**, 79–89 (1999).

28. A. Walcarius, C. Despas, P. Trens, M. J. Hudson and J. Bessière, *J. Electroanal. Chem.* **453**, 249–252 (1998).

29. A. Walcarius and J. Bessière, *Chem. Mater.* **11**, 3009–3011 (1999).

30. A. M. Bond, W. Miao, T. D. Smith and J. Jamis, *Anal. Chim. Acta* **396**, 203–213 (1999).

31. A. Walcarius, N. Luthi, J. L. Blin, B. L. Su and L. Lamberts, *Electrochim. Acta* **44**, 4601–4610 (1999).

32. J. F. Diaz, K. J. Balkus, F. Bedioui, V. Kurshev and L. Kevan, *Chem. Mater.* **9**, 61 (1997).

33. J. F. Diaz, F. Bedioui, E. Briot, J. Devynck and K. J. Balkus, *Mater. Res. Soc. Symp. Proc.* **431**, 89 (1996).

34. Y.-X. Jiang, N. Ding and S.-G. Sun, *J. Electroanal. Chem.* **563**, 15 (2004).

35. C. Song and G. Villemure, *Micropor. Mesopor. Mater.* **44**(Sp. Iss.), 679–689 (2001).

36. A. Walcarius, M. Etienne and B. Lebeau, *Chem. Mater.* **15**, 2161 (2003).

37. M. Etienne, B. Lebeau and A. Walcarius, *New J. Chem.* **26**, 384 (2002).

38. A. Walcarius and C. Delacôte, *Chem. Mater.* **15**, 4181 (2003).

39. W. Yantasee, L. A. Deibler, G. E. Fryxell, C. Timchalk and Y. Lin, *Electrochem. Commun.* **7**, 1170–1176 (2005).

40. W. Yantasee, G. Fryxell and Y. Lin, *Analyst* **131**, 1342 (2006).
41. W. Yantasee, G. E. Fryxell, M. M. Conner and Y. Lin, *J. Nanosci. Nanotechnol.* **5**, 1537–1540 (2005).
42. W. Yantasee, Y. Lin, G. E. Fryxell and B. J. Busche, *Anal. Chim. Acta* **502**, 207–212 (2004).
43. W. Yantasee, Y. Lin, G. E. Fryxell and Z. Wang, *Electroanalysis* **16**, 870–873 (2004).
44. W. Yantasee, Y. Lin, X. Li, G. E. Fryxell, T. S. Zemanian and V. V. Viswanathan, *Analyst* **128**, 899–904 (2003).
45. W. Yantasee, Y. Lin, T. S. Zemanian and G. E. Fryxell, *Analyst* **128**, 467–472 (2003).
46. W. Yantasee, C. Timchalk, G. E. Fryxell, B. P. Dockendorff and Y. Lin, *Talanta* **68**, 256–261 (2005).
47. S. Iijima, *Nature* **354**, 56–58 (1991).
48. R. Saito, G. Dresselhaus and M. S. Dresselhaus, *Physical Properties of Carbon Nanotubes* (Imperial College Press, London, 1998).
49. J. M. Nugent, K. S. V. Santhanam, A. Rubio and P. M. Ajayan, *Nano Lett.* **1**, 87 (2001).
50. J. Wang and M. Musameh, *Anal. Chem.* **75**, 2075–2079 (2003).
51. J. Wang, M. Musameh and Y. Lin, *J. Am. Chem. Soc.* **125**, 2408–2409 (2003).
52. M. Musameh, J. Wang, A. Merkoci and Y. Lin, *Electrochem. Commun.* **4**, 743–746 (2002).
53. Y. Lin, F. Lu, Y. Tu and Z. F. Ren, *Nano Lett.* **4**, 191–195 (2004).
54. Y. Lin, F. Lu and J. Wang, *Electroanalysis* **16**, 145–149 (2003).
55. Y. Lin, W. Yantasee, F. Lu, J. Wang, M. Musameh, Y. Tu and Z. F. Ren, *Dekker Encyclopedia of Nanoscience and Nanotechnology* (Marcel Dekker, New York, 2004), pp. 361–374.
56. Y. Lin, W. Yantasee and J. Wang, *Front. Biosci.* **10**, 492–505 (2005).
57. G. E. Fryxell, H. Wu, Y. Lin, W. J. Shaw, J. C. Birnbaum, J. C. Linehan, Z. Nie, K. Kemner and S. Kelly, *J. Mater. Chem.* **14**, 3356–3363 (2004).
58. W. Yantasee, G. E. Fryxell, Y. Lin, H. Wu, K. N. Raymond and J. Xu, *J. Nanosci. Nanotechnol.* **5**, 527–529 (2005).
59. J. C. Birnbaum, B. Busche, Y. Lin, W. Shaw and G. E. Fryxell, *Chem. Commun.* **13**, 1374 (2003).
60. G. E. Fryxell, Y. Lin, S. Fiskum, J. C. Birnbaum, H. Wu, K. Kemner and S. Kelly, *Environ. Sci. Technol.* **39**, 1324–1331 (2005).
61. Y. Lin, S. K. Fiskum, W. Yantasee, H. Wu, S. V. Mattigod, E. Vorpagel and G. E. Fryxell, *Environ. Sci. Technol.* **39**, 1332–1337 (2005).
62. X. D. Feng, G. E. Fryxell, L. Q. Wang, A. Y. Kim, J. Liu and K. Kemner, *Science* **276**, 923–926 (1997).
63. G. E. Fryxell, Y. Lin, H. Wu and K. M. Kemner, in *Studies in Surface Science and Catalysis*, eds., A. Sayari and M. Jaroniec (Elsevier Science, Amsterdam, 2002), pp. 583–590.
64. G. E. Fryxell, J. Liu, S. V. Mattigod, L. Q. Wang, M. Gong, T. A. Hauser, Y. Lin, K. F. Ferris and X. Feng, in *Ceramics Transactions*, Vol. 107, eds.

G. T. Chandler and X. Feng (The American Ceramic Society, Westerville, OH, 2000), pp. 29–37.
65. W. Yantasee, Y. Lin, G. E. Fryxell, B. J. Busche and J. C. Birnbaum, *Separ. Sci. Technol.* **38**, 3809–3825 (2003).
66. Y. Lin, G. E. Fryxell, H. Wu and M. Engelhard, *Environ. Sci. Technol.* **35**, 3962–3966 (2001).
67. S. V. Mattigod, G. E. Fryxell, R. J. Serne, K. E. Parker and F. M. Mann, *Radiochim. Acta* **91**, 539–545 (2003).
68. G. E. Fryxell, J. Liu, T. A. Hauser, Z. Nie, K. F. Ferris, S. Mattigod, M. Gong and R. T. Hallen, *Chem. Mater.* **11**, 2148–2154 (1999).
69. R. G. Pearson, *J. Am. Chem. Soc.* **85**, 3533–3539 (1963).
70. A. Walcarius, T. Barbaise and J. Bessiere, *Anal. Chim. Acta* **340**, 61–76 (1997).
71. S. Kilinc Alpat, U. Yuksel and H. Akcay, *Electrochem. Commun.* **7**, 130–134 (2005).
72. C. Bing, G. Ngoh-Khang and C. Lian-Sai, *Electrochim. Acta* **42**, 595–604 (1997).
73. X. Cui, G. Liu, L. Li, W. Yantasee and Y. Lin, *Sensor Lett.* **3**, 16–21 (2005).
74. J. F. van Staden and R. E. Taljaard, *Talanta* **64**, 1203 (2004).
75. J. Wang, R. Setiadji, L. Chen, J. Lu and S. Morton, *Electroanalysis* **4**, 161 (1992).
76. S. Sander, W. Wafner and G. Henze, *Anal. Chim. Acta* **349**, 93 (1997).
77. J. Wang and R. Setiadji, *Anal. Chim. Acta* **264**, 205 (1992).
78. M. B. R. Bastos, J. C. Moreira and P. A. M. Farias, *Anal. Chim. Acta* **408**, 83–88 (2000).
79. L. Lin, S. Thongngamdee, J. Wang, Y. Lin, O. A. Sadik and S.-Y. Ly, *Anal. Chim. Acta* **535**, 9–13 (2005).
80. S. Sayen and A. Walcarius, *J. Electroanal. Chem.* **581**, 70–78 (2005).
81. V. Ganesan and A. Walcarius, *Langmuir* **20**, 3632–3640 (2004).
82. I. A. Aksay, M. Trau, S. Manne, I. Honma, N. Yao, L. Zhou, P. Fenter, P. M. Eisenberger and S. M. Gruner, *Science* **273**, 892–898 (1996).
83. H. Yang, N. Coombs, I. Sokolov and G. A. Ozin, *Nature* **381**, 589–592 (1996).
84. H. Yang, A. Kuperman, N. Coombs, S. Mamiche-Afara and G. A. Ozin, *Nature* **379**, 703–705 (1996).
85. D. Y. Zhao, J. L. Feng, Q. S. Huo, N. Melosh, G. H. Fredrickson, B. F. Chmelka and G. D. Stucky, *Science* **279**, 548–552 (1998).
86. D. Y. Zhao, Q. S. Huo, J. L. Feng, B. F. Chmelka and G. D. Stucky, *J. Am. Chem. Soc.* **120**, 6024–6036 (1998).
87. F. Cagnol, D. Grosso and C. Sanchez, *Chem. Commun.* 1742 (2004).
88. L. Nicole, C. Boissiere, D. Grosso, A. Quach and C. Sanchez, *J. Mater. Chem. Commun.* **15**, 3598 (2005).
89. Q. Huo, R. Leon, P. M. Petroff and G. D. Stucky, *Science* **268**, 1324–1327 (1995).
90. V. Luzzati, H. Delacroix and A. Gulik, *J. Phys. II* **6**, 405–418 (1996).

91. D. Zhao, P. Yang, N. Melosh, J. Feng, B. F. Chmelka and G. D. Stucky, *Adv. Mater.* **10**, 1380–1385 (1998).
92. J. J. Gooding, V. G. Praig and E. A. H. Hall, *Anal. Chem.* **70**, 2396–2402 (1998).
93. M. Etienne and A. Walcarius, *Electrochem. Commun.* **7**, 1449–1456 (2005).
94. D. Grosso, F. Cagnol, G. J. d. A. A. Soler-Illia, E. L. Crepaldi, H. Amenitsch, A. Brunet-Bruneau, A. Bourgeios and C. Sanchez, *Adv. Funct. Mater.* **14**, 309 (2004).
95. R. E. Williford, G. E. Fryxell, X. S. Li and R.S. Addleman, *Micropor. Mesopor. Mater.* **84**, 201–210 (2005).
96. W. L. Roper, V. Houk, H. Falk and S. Binder, *Preventing Lead Poisoning in Young Children* (Centers for Disease Control and Prevention, Atlanta, GA, 1991)
97. G. A. Wagner, R. M. Maxwell, D. Moor and L. Foster, *J. Am. Med. Assoc.* **270**, 69–71 (2000).
98. J. Ruzicka, G. D. Marshall and G. D. Christian, *Anal. Chem.* **62**, 1861 (1990).
99. J. Ruzicka and G. D. Marshall, *Anal. Chim. Acta* **273**, 329 (1990).
100. I. Palchetti, C. Upjohn, A. P. F. Turner and M. Mascini, *Anal. Lett.* **33**, 1231–1246 (2000).
101. K. C. Honeychurch, D. M. Hawkins, J. P. Hart and D. C. Cowell, *Talanta* **57**, 565–574 (2002).
102. K. C. Honeychurch, J. P. Hart, D. C. Cowell and D. W. M. Arrigan, *Sens. Actuators B* **77**, 642–652 (2001).
103. K. C. Honeychurch, J. P. Hart and D. C. Cowell, *Anal. Chim. Acta* **431**, 89–99 (2001).
104. K. C. Honeychurch, J. P. Hart and D. C. Cowell, *Electroanalysis* **12**, 171–177 (2000).
105. K. C. Honeychurch and J. P. Hart, *Trends Anal. Chem.* **22**, 456–469 (2003).
106. Q. G. Q. Health, *Cadmium* (Queensland Government, Canberra, 2002), pp. 1–4.
107. J.-Y. Choi, K. Seo, S.-R. Cho, J.-R. Oh, S.-H. Kahng and J. Park, *Anal. Chim. Acta* **443**, 241–247 (2001).
108. D. Desmond, B. Lane, J. Alderman, M. Hill, D. W. M. Arrigan and J. D. Glennon, *Sens. Actuators B* **48**, 409–414 (1998).
109. J. Wang and T. Baomin, *Anal. Chem.* **64**, 1706–1709 (1992).
110. J. Wang, J. Lu, B. Tain and C. Yarnitsky, *J. Electroanal. Chem.* **361**, 77–83 (1993).
111. J. Wang, B. Tian, V. B. Nascimento and L. Angnes, *Electrochim. Acta* **43**, 3459–3465 (1998).
112. P. Masawat, S. Liawruangrath and J. M. Slater, *Sens. Actuators B* **91**, 52–59 (2003).
113. J.-M. Zen, C.-C. Yang and A. S. Kumar, *Anal. Chim. Acta* **464**, 229–235 (2002).

114. K. Z. Brianina, N. F. Zakharchuck, D. P. Synkova and I. G. Yudelevich, *J. Electroanal. Chem.* **35**, 165 (1972).

115. S. Ferro and A. D. Battisti, *J. Electroanal. Chem.* **533**, 177–180 (2002).

116. M. Mlakar and M. Branica, *Anal. Chim. Acta* **247**, 89 (1991).

117. M. Mlakar, *Anal. Chim. Acta* **260**, 51–56 (1992).

118. M. Zhang and X. Gao, *Anal. Chem.* **56**, 1917 (1984).

119. X. Gao and M. Zhang, *Anal. Chem.* **56**, 1912 (1984).

120. L. M. Moretto, J. Chevalet, G. A. Mazzocchin and P. Ugo, *J. Electroanal. Chem.* **498**, 117–126 (2001).

121. R. H. Baughman, A. Zakhidov and W. A. De Heer, *Science* **297**, 787 (2002).

122. Q. Zhao, Z. Gan and Q. Zhuang, *Electroanalysis* **14**, 1609 (2002).

123. J. Wang, M. Li, Z. Shi, N. Li and Z. Gu, *Anal. Chem.* **74**, 1993–1997 (2002).

124. J. Wang, M. Li, Z. Shi, N. Li and Z. Gu, *Electroanalysis* **14**, 225–230 (2002).

125. Z. Wang, J. Liu, Q. Liang, Y. Wang and G. Luo, *Analyst* **127**, 653–658 (2002).

126. Y.-C. Tsai, J.-M. Chen, S.-C. Li and F. Marken, *Electrochem. Commun.* **6**, 917–922 (2004).

127. K. Wu, S. Hu, J. Fei and W. Bai, *Anal. Chim. Acta* **489**, 215–221 (2003).

128. H. X. Luo, Z. J. Shi, N. Q. Li, Z. N. Gu and Q. K. Zhuang, *Anal. Chem.* **73**, 915–920 (2001).

129. F. H. Wu, G. C. Zhao and X. W. Wei, *Electrochem. Commun.* **4**, 690 (2002).

130. V. G. Gavalas, S. A. Law, J. C. Ball, R. Andrews and L. G. Bachas, *Anal. Biochem.* **329**, 247–252 (2004).

131. M. D. Rubianes and G. A. Rivas, *Electrochem. Commun.* **5**, 689–694 (2003).

132. F. Valentini, S. Orlanducci, M. L. Terranova, A. Amine and G. Palleschi, *Sens. Actuators B* **100**, 1170–125 (2004).

133. J. Wang and M. Musameh, *Analyst* **128**, 1382–1385 (2003).

134. M. Gao, L. M. Dai and G. G. Wallace, *Electroanalysis* **15**, 1089–1094 (2003).

135. G. Liu, Y. Lin, Y. Tu and Z. Ren, *Analyst* **130**, 1098–1101 (2005).

136. K. P. Loh, S. L. Zhao and W. D. Zhang, *Diam. Relat. Mater.* **13**, 1075–1079 (2004).

137. W. C. Poh, K. P. Loh, W. D. Zhang, T. Sudhiranjan, J. S. Ye and F. S. Sheu, *Langmuir* **20**, 5484–5492 (2004).

138. S. G. Wang, Q. Zhang, R. Wang and S. F. Yoon, *Biochem. Biophys. Res. Commun.* **311**, 572–576 (2003).

139. S. G. Wang, Q. Zhang, R. Wang, S. F. Yoon, J. Ahn, D. J. Yang, J. Z. Tianm, J. Q. Li and Q. Zhou, *Electrochem. Commun.* **5**, 800–803 (2003).

140. J. Wang, G. Liu and R. Jan, *J. Am. Chem. Soc.* **126**, 3010 (2004).

141. J. Gooding, R. Wibowo, J. Liu, W. Yang, D. Losic, S. Orbons, F. Mearns, J. Shapter and D. Hibbert, *J. Am. Chem. Soc.* **125**, 9006 (2003).

142. Y. Tu, Z. P. Huang, D. Z. Wang, J. G. Wen and Z. F. Ren, *Appl. Phys. Lett.* **80**, 4018–4020 (2002).

143. V. Menon and C. Martin, *Anal. Chem.* **67**, 1920 (1995).

144. D. Arrigan, *Analyst* **129**, 1157–1165 (2004).

145. X. Yu, D. Chattopadhyay, I. Galeska, F. Papadimitrakopoulos and J. Rusling, *Electrochem. Commun.* **5**, 408 (2003).
146. R. Feeney and S. P. Kounaves, *Electroanalysis* **12**, 677 (2000).
147. S. Fletcher and M. D. Horne, *Electrochem. Commun.* **1**, 502 (1999).
148. H. J. Lee, C. Beriet, R. Ferrigno and H. H. Girault, *J. Electroanal. Chem. Commun.* **502**, 138 (2001).
149. J. Li, H. T. Ng, A. Cassell, W. Fan, H. Chen, Q. Ye, J. Koehne, J. Han and M. Meyyappan, *Nano Lett.* **3**, 597 (2003).
150. M. E. Sandison, N. Anicet, A. Glidle and J. M. Cooper, *Anal. Chem.* **74**, 5717 (2002).
151. Y. Tu, Y. Lin and Z. F. Ren, *Nano Lett.* **3**, 107–109 (2003).
152. Y. Tu, Y. Lin, W. Yantasee and Z. Ren, *Electroanalysis* **17**, 79–84 (2005).
153. L. Lin, N. S. Lawrence, S. Thongngamdee, J. Wang and Y. Lin, *Talanta* **65**, 144–148 (2005).
154. J. Wang, J. Lu, S. B. Hocevar and B. Ogorevc, *Electroanalysis* **13**, 13 (2001).

Chapter 20

Nanomaterial-based Environmental Sensors

Dosi Dosev*, Mikaela Nichkova† and Ian M. Kennedy*

*Department of Mechanical and Aeronautical Engineering
†Department of Entomology, University of California,
Davis, CA, USA

20.1 Introduction

Environmental science and technology frequently require the measurement and detection of the constituents of soil, water and air. Measurements of these components of the environment are often necessary first steps in the development and validation of models of environmental processes; measurements are also very important during the monitoring of processes in the environment, particularly with regard to monitoring the transport of contaminants, and detection of toxins like pesticides. Air quality modeling requires high-quality data on the concentrations of air contaminants. Typical components of the air of interest are those compounds that are often emitted by combustion sources such as engines. The air pollution that is created from these emissions is well known to have an adverse impact on health.[1,2] Gas-phase species such as carbon monoxide, unburned hydrocarbons, oxides of nitrogen, ozone, and most significantly in recent times, particulate matter, are known to be associated with adverse effects on human health. Over the years, many bench-top analyzers have been developed for detecting these compounds in the ambient air. Those technologies are well established but the instruments are fairly large and expensive. Their size and cost do not permit the deployment of such systems in large-scale arrays of sensors. The thrust of recent sensor research is directed towards the development of smaller, cheaper, portable devices.

The detection of toxins in groundwater and in soil is very important in assessing threats to human health from seepage of waste from contaminated soils, and in monitoring the transport of toxins through the groundwater and eventually consumption by humans. In general, well-established techniques such as liquid and gas chromatography are used to analyze samples that are returned to the laboratory. This method of analysis is slow and expensive. It is certainly not amenable to the deployment of large-scale arrays for detecting the transport of toxins from contaminated sites. The contaminants of soil and water that are of common interest include industrial pollutants such as polyaromatic hydrocarbons (PAH), chlorinated solvents and metal-containing compounds. In some cases, the materials may be widely dispersed, as is the case with dioxins.

The detection of trace compounds in the environment faces problems that are unique to environmental science and technology. First, the detection must take place within a complex matrix of many compounds, all of which may interfere with the detection process. Oxygen, which is generally ubiquitous in environmental settings, can interfere with measurements by quenching fluorescence. Sensors that are based on immunoassay techniques, that typically use antibodies as biorecognition elements, can also suffer from interference from other compounds in the sample due to non-specific binding. In this case, compounds other than the target compounds are bound to the antibody and are detected as the sample analyte. Over many years, methods have been devised to improve the sensitivity and specificity of antibodies for application to immunoassays. However, in environmental samples there may be many closely related chemical compounds that could give rise to this problem. Diminished specificity can be a particular problem if molecularly imprinted polymers (MIPs) are used as substitutes for antibodies; the use of MIPs has been advocated by several research groups.[3-7]

The monitoring of environmental sites generally requires the deployment of many sensors in a network.[8] For example, monitoring the transport of contaminated waste from soils or water requires a fairly large area to be instrumented with sensors. For this reason, environmental applications typically require sensors to be both rugged and relatively cheap. The latter requirement is exacerbated by budgets for environmental clean-up research and technology that are often smaller than the funding that is available for biomedical research. Sensors that are deployed in the field need to be simple enough to be operated by minimally trained technicians, and to involve fairly straightforward technology. They need to be able to withstand a harsh environment, to operate off a self-contained power supply, and ideally to interface to a network with telemetry that can transmit data back to a base station.

Not all sensors that are employed in environmental research and technology need to be used in the field — many important sensing technologies can be used in the laboratory. Research into the exposure of human populations to environmental toxins is typically done using laboratory instrumentation. However, many samples are collected from a large population.[9] In this case, the technology needs to be able to handle a large number of samples in a reasonable time. Consequently, high-throughput technologies become very important in handling large numbers of samples. The requirement for high throughput suggests that it might be desirable to work with small volumes, and to be able to tolerate minimal clean-up or preparation prior to analysis.

Working with very small volumes of sample indicates the need for detection methods with high sensitivity. Recent advances in nanotechnology may be able to provide new analytical approaches that meet the new requirements for ultra-sensitive detection.[10–18] The development of quantum dots offers a good example of nanotechnology applied to biosensors.[19–25] Quantum dots are very small particles made of semiconductor materials such as cadmium sulfide, cadmium selenide, zinc sulfide, as well as other compounds.[26–29] When the semiconductor materials are synthesized into very small particles on the order of 10 nm in diameter, the quantum confinement effect serves to shift their bulk-phase optical emission towards the blue end of the spectrum. Hence, it is possible to synthesize a wide range of particles of different emission wavelengths. This technology has found commercial application with several companies worldwide and is being applied widely in biotechnology. The use of quantum dots as labels in techniques such as immunoassays in environmental samples may provide some advantages: narrow emission characteristics may enable discrimination against broad background fluorescence; the lack of photo-bleaching allows persistent excitation; and strong luminescence offers an excellent signal-to-noise ratio.

Considerable effort has been channeled into the development of new nanostructured surfaces for application to optical biosensors. Nanotechnology can be used to create novel surfaces on which to carry out conventional detection. Surface-enhanced Raman scattering (SERS) has been coupled with nanostructured silver[30] and gold surfaces[31,32] to gain additional signal strength. The characteristics of nanoparticles that are used for SERS may be important,[33] including the refractive index of the materials and the size of the particles. It has been shown that SERS is much more effective on a surface that provides nanoscale features or roughness. In fact, some researchers have found that fractal aggregates of nanoparticles are actually more effective then single nanoparticles by themselves. Other surfaces may contain porous features[34–38] that enable selective capture of analytes within the sample, immobilization of the biological detection elements within the porous matrix, or immobilization of gold nanoparticles for SERS.[39] The matrix serves to protect the detection elements, and screens out many confounding compounds in the sample.

Over the past decade, a great deal of attention has been given to the development of novel materials such as carbon nanotubes, and zinc oxide nanowires and nanobelts. Carbon nanotubes are particularly attractive for sensor applications because their conductivity may change as materials or compounds are adsorbed onto the surface. Zinc oxide is a semiconductor material that also exhibits piezoelectric behavior. When it is synthesized in the form of nanowires and nanobelts, the change in its electrical or piezoelectric properties can be used to detect the presence of adsorbents on the surface. Both of these very interesting materials and morphologies are being applied to the development of sensors, biological and electrochemical.

Novel nanoscale materials are beginning to offer a range of new functionalities and possibilities for environmental sensing. There are many examples of nanoscale materials applied to biosensing in general, with a primary emphasis on applications in biomedical and biological sciences. There are fewer examples of applications to environmental problems. This disparity is being redressed to some extent by research over the past decade. We expect to see many more applications of nanoengineered materials in environmental sensors in the future. This review will cover some of the fundamentals of nanoscale materials applied to sensing. Examples of successful applications of these materials to novel environmental sensors will be discussed.

20.2 Sensor Technologies

We begin our discussion of sensor technologies with a review of the fundamentals of biosensors that make use of a biological molecule for the capture and recognition of specific compounds in a mixture.

20.2.1 *Detecting element for biosensors*

A biosensor is defined as an analytical device that consists of a biological component (enzyme, antibody, receptor, DNA, cell and so on) in intimate contact with a physical transducer that converts the biorecognition process into a measurable signal (electrical, optical and so on) as seen in Figure 20.1. Biosensors are commonly classified as immunosensors, enzymatic (catalytic), non-enzymatic receptor, whole-cell (microbial sensors) or nucleic acid (DNA) biosensors, according to the biological recognition element.

Immunosensors are affinity-based sensors designed to detect the direct binding of an antibody (Ab) to an antigen (Ag) to form an immunocomplex at the transducer surface. Depending on the transducer technology employed, immunosensors can be divided into three principal classes: optical, electrochemical and piezoelectrical. They are based on the principles of

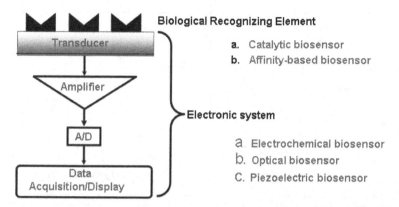

Figure 20.1.   General configuration of biosensors. The biological recognizing element (enzyme, antibody, receptor, DNA, cell, etc.) is in intimate contact with a physical transducer that converts the biorecognition process into a measurable signal (electrical, optical, etc.). Catalytic biosensors have enzymes as the biorecognition element and affinity biosensors are based on antibodies, DNA, cells or other types of receptor.

solid-phase immunoassays, where the immunoreagent (Ab or Ag) is immobilized on a solid support, so that the interaction takes place at the solid-liquid interface.

Antibodies are globular proteins produced by the immune system of mammals as a defense against foreign agents (antigen, Ag). The structure of the antibody molecules is usually typified by the immunoglobulin G (IgG) subclass, which is the most commonly used in immunochemical application. The IgGs (molecular weight of 150 kDa) are composed of four polypeptide chains: two identical heavy (H) and two identical light (L) chains (50 kDa and 23–25 kDa, respectively) interlinked by disulfide bridges (see Figure 20.2).[40] The region that carries the antigen-binding sites is known as the Fab fragment, and the constant (crystallized) region that is involved in immune regulation is termed Fc. Both H and L chains are divided into constant (C) and variable (V) domains based on their amino acid sequence variability. The most important regions of the antibody with regard to the Ab-Ag binding interaction are the variable regions, consisting of the association of the V_H and V_L domains. Within each of these domains there are three distinct areas of even higher sequence variability, known as complementary determining regions (CDRs), bounded by four framework regions (FRs). The CDRs of the V_H and V_L domains form the antigen-binding site. Their amino acid sequence and spatial conformation determine the binding specificity and binding strength (affinity) of the Ab molecule to the Ag molecule. The interaction between Ab and Ag is reversible and it is stabilized by electrostatic forces, hydrogen bonds, hydrophobic and van der

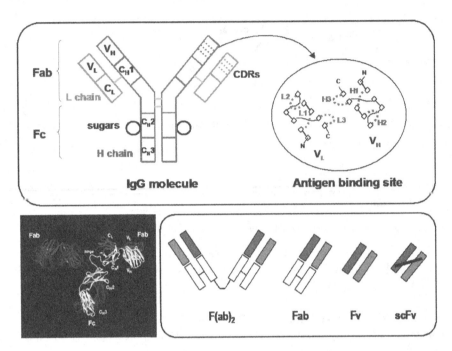

Figure 20.2.   Antibody structure. Constant domains: C_L, C_H1, C_H2 and C_H3; variable domains: V_L and V_H. Antigen-binding site: CDRs (hypervariable regions): L1, L2, L3, H1, H2, H3; constant framework: black line (from Ref. 40 with permission).

Waals interactions.[41] At equilibrium, the affinity constant Ka is defined as:

$$Ka = \frac{[AbAg]}{[Ab] \cdot [Ag]}. \tag{20.1}$$

Antibody fragments can be generated by enzymatic or chemical degradation and by genetic engineering.[42] These include the F(ab)2, Fab, Fd fragments; Fv, the smallest fragment required for complete Ag binding; single-chain scFv fragments, in which Fv is stabilized by a flexible amino acid linker; and others (see Figure 20.2).

Immunosensors can be either direct (where the immunochemical reaction is directly determined by measuring the physical changes induced by the formation of the immune complex) or indirect (where a sensitive detectable label is combined with the Ab or Ag of interest). As environmental contaminants are often small-sized molecules, most of the devices that are described in the literature perform indirect measurements by using competitive immunoassay configurations and/or labels such as enzymes, fluorescent chemicals or electrochemically active substances (see Figure 20.3). The sensitivity and specificity of an immunosensor are determined by the

Figure 20.3. A schematic illustration of competitive immunoassay. The analyte (target compound) competes with the enzyme tracer (competitor with similar molecular structure) for the antibodies (IgG) immobilized on the surface. The amount of tracer is detected by the enzymatic reaction which converts the substrate to a measurable product. The signal is reversibly proportional to the concentration of the analyte in the sample.

same characteristics as in other solid-phase immunoassays, namely, the affinity and specificity of the binding agent, and the background noise of the detection system (transducer). Challenges still encountered in biosensor development include fabrication, immobilization of the binding agent onto a transducer, effective signal generation, miniaturization and integration.

The majority of biosensor research is currently directed towards clinical applications, but a variety of immunosensors for environmental applications have also been developed in the last years.[43–45] Both types of optical and electrochemical transducers have been demonstrated to provide detection limits in the low parts-per-billion to the high parts-per-trillion range. Research in the Ab field is still growing; future developments will make use of recombinant fragments with desired characteristics, better defined chemical structure, stability and so on. As the orientation of the binding molecule (recognition element) after being immobilized onto a transducer is likely to affect the sensitivity, it is desirable to equip the recombinant Ab fragments with suitable tags to facilitate their orientation at the sensor surface.[46]

Even though not based on biological elements, molecularly imprinted polymer (MIP)-based sensors mimic the biological activity of antibodies, receptor molecules, etc.[47] MIPs combine highly selective molecular recognition, comparable to that of biological systems, with the typical properties of polymers such as high thermal, chemical and stress tolerance, and an extremely long shelf life without any need for special storage conditions.[48] However, until now the affinities achieved by MIP biosensors are not comparable to those reached by antibodies. MIP optical biosensors have been developed for detection of pesticides[49] and chemical agents[50] in water.

Immunosensors are limited by problems with matrix effects that arise from the non-specific binding of extraneous compounds from the sample or matrix to the antibody. The regeneration of the antibodies following an assay is also problematic. Implementation and commercialization of immunosensor technology is slow. With further improvement in the fabrication of miniaturized biosensors that are simple, rapid, portable, cost-effective and with the ability to regenerate, it is anticipated that they will provide a powerful tool for environmental and biological monitoring.

20.2.1.1 DNA biosensors

Nucleic acid hybridization is the underlying principle of DNA biosensors. A single-strand DNA (ssDNA) molecule immobilized on a sensory surface is able to seek out, or hybridize to its complementary strand in a sample (Figure 20.4) by binding their complementary bases, i.e. C to G and A to T.[51]

The parallel analysis of large numbers of DNA fragments can be provided by DNA chips or arrays.[52] Currently, most DNA chips use optical detection of fluorescence-labeled oligonucleotides. Despite the dominance of the optical systems, advances in electrochemical detection are quite promising. Besides the voltammetric detection of redox intercalators, new developments have focused on the mediated oxidation of guanidine within the DNA, the amplification of the hybridization event by an enzyme label and impedance analysis, and the electron transport through the DNA

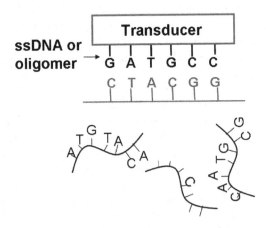

Figure 20.4.   A schematic illustration of ssDNA probes based on the specific DNA hybridization reaction. The ssDNA or oligomer is immobilized on the transducer surface and it hybridizes with the complementary DNA from the sample. (From Ref. 51 with permission.)

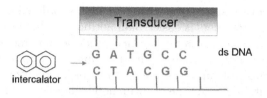

Figure 20.5. A schematic illustration of DNA receptors (dsDNA probes). The double-stranded DNA is immobilized on the surface and small organic compounds intercalate selectively into it.

double helix.[52] The ssDNA probes (see Figure 20.4) have gained considerable interest because of their importance in the early diagnosis of diseases, such as cancer and hypercholesterolemia. They are used for the detection of microbial and viral pathogens (i.e. HIV, *Cryptosporidium*, *Escherichia coli, Giardia, Mycobaterium tuberculosis* and so on) in combination with polymerase chain reaction (PCR) units, detection of mutated genes (gene p53 for tumor suppression is mutated in malign tumors) and analysis of gene sequences. However, for the monitoring of carcinogenic low-molecular-weight compounds — DNA intercalators (endocrine disruptors, polyaromatic hydrocarbons,[53] aromatic amines, drugs,[54] organophosphate pesticides[55] and so on) — a second type of DNA biosensors referred to as DNA receptors (double-strand or dsDNA probes) is applied. In this approach, biosensors monitor the interaction of small pollutants with affinities for DNA with the immobilized dsDNA layer (Figure 20.5). These biosensors may therefore be used as general indicators of pollution when integrated in a panel of tests, since they can give rapid and easy-to-evaluate information on the presence of such compounds.

20.2.1.2 *Aptasensors*

Aptamers are artificial nucleic acid ligands (RNA or DNA) that can be generated *in vitro* from vast populations of random sequences; they are designed to bind to amino acids, drugs, proteins and other molecules by so-called systematic evolution of ligands by exponential enrichment (SELEX).[56,57] Aptamers are ideal candidates for use as biocomponents in biosensors (aptasensors), possessing many advantages over state-of-the-art affinity sensors. Aptamers have a number of advantages compared to antibodies such as smaller size, rationally designed binding affinity and specificity, higher temperature stability, no need for immunization of animal hosts, and a higher degree of purification, thus eliminating batch-to-batch antibody variation. Furthermore, the functional groups of aptamers can be altered for the direct or indirect covalent immobilization on biochips or other support materials, resulting in highly ordered receptor layers. Regeneration of aptamers is easier than antibodies, due to the simple and quite stable structure, which is a key factor that permits repeated use, economy and precision of biosensors. The development of aptasensors is being boosted by using optical and acoustic methods to analyze biological phenomena in solution in real time or by immobilizing the aptamer onto a solid support. Aptasensors for the detection of thrombin, anthrax spores and proteins have been reported.[57] Despite the wealth of literature detailing the use of aptamers in therapeutics, the use of aptamers in *in vitro* diagnostics is a field that is still in its infancy with the main focus on the detection of the aptamer-target binding event using fluorescence detection. However, with the advent of automated platforms for aptamer selection, it is anticipated that aptamers will become increasingly exploited in the biosensor field.

20.2.1.3 *Microbial (whole-cell) biosensors*

Microbial (whole-cell) biosensors can monitor the metabolism of cells by the measurement of pH, O_2 consumption, CO_2 production, redox potential, electric potential on nervous system cells or bioluminescence in bacteria. Viable microbial cells (bacteria and yeasts) have been genetically modified to produce a recombinant organism that exhibits a number of important traits, e.g. expression of cellular degradative enzyme, specific binding proteins, and a reporter enzyme such as bacterial luciferase which is induced in the presence of the target analytes (antibiotics, pesticides, benzene, toluene and xylene).[58] Nerve cells growing on array structures have also been implemented in the development of chips and sensors.[59]

20.2.1.4 *Catalytic (enzymatic) biosensors*

Catalytic biosensors rely on the enzyme-catalyzed conversion of a non-detectable substrate into an optically or electrochemically detectable product or vice versa (see Figure 20.6).[58,60] This process allows the detection of substrates, products, inhibitors and modulators of the catalytic reaction. Several enzyme-catalyzed reactions involve the production or consumption of a detectable product of low-molecular-weight species, such as O_2, CO_2 and ions. The most frequently employed enzymes are oxidoreductases and hydrolases.[61] The activity of cholinesterases is inhibited by a variety of organophosphorous and carbamate insecticides. Tyrosinase is used to detect contamination caused by phenolic compounds. On the other hand, various pollutants such as hydrazines, atrazine, cyanide and diethyldithiocarbamates act as tyrosinase inhibitors. An improvement in sensitivity of the

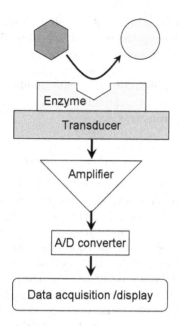

Figure 20.6. A schematic illustration of catalytic biosensors. A non-detectable substrate is converted into an electrochemically or optically detectable product by the enzyme immobilized on the sensor surface.

enzymatic biosensors can be achieved by the use of recombinant enzyme mutants.[62]

20.2.2 *Physical sensors*

It is not necessary to incorporate a biological recognition element in a sensor device; physical adsorption onto a surface can be sufficient to change the conductivity, for example, of a thin film. The change is detected electrically and is related to the concentration of a particular analyte. Physical adsorption lacks the high degree of specificity that is provided by the biological or biomimetic recognition elements that were discussed above. Physical adsorption finds most application in gas sensors.

20.3 Signal Transduction

20.3.1 *Electrochemical transducers*

An electrochemical biosensor, according to the IUPAC definition, is a self-contained integrated device, which is able to provide specific quantitative or semi-quantitative analytical information using a biological recognition element (biochemical receptor); the receptor is held in direct contact with the transduction element.[63] Biosensors based on electrochemical transducers have the advantage of being economic with a fast response; the possibility of automation allows application in a wide number of areas.[64] The electrochemical biosensors can be classified as conductimetric, impedimetric, potentiometric or amperometric.

20.3.1.1 Conductimetric and impedimetric biosensors

These immunosensors are essentially based on the same physical principle and one or another parameter can be measured. Conductimetry describes the measurement of a current as a function of applied voltage while impedance refers to the measurement of a voltage as a function of an applied current. In both cases the conductimetric or impedimetric properties are influenced by a sensing layer placed between a set of two electrodes. For example, in the case of conductimetric immunosensors[60] the measurement is based on the use of an electroconductive polymer doped with analyte, to which specific antibodies have been immobilized. A direct competitive immunoassay produces a change in the conductimetric properties of the polymer.

Impedimetric biosensors have been used for the detection of bacteria and have been applied to sanitation microbiology.[65] These biosensors are based on the principle that microbial metabolism results in an increase in both conductance and capacitance, causing a decrease in the impedance. Impedance is usually measured by a bridge circuit. Often a reference module is included to measure and exclude non-specific changes in the test module. The reference module serves as a control for temperature changes, evaporation, changes in amounts of dissolved gases and degradation of culture medium during incubation.

20.3.1.2 Potentiometric transducers

Potentiometric sensors incorporate a membrane or surface that is sensitive to a desired species and that generates an electrical potential proportional to the logarithm of the concentration of the active species, measured in relation to a reference electrode. The potentiometric devices can measure changes in pH and ion concentration. It is possible to use transistors as electronic signal amplifiers coupled to ion-sensitive electrodes (ISEs), and such transistors are called ion-sensitive field-effect transistors (ISFETs). These biosensors are based on the immobilization of a biologically active material, usually enzymes, antigens or antibodies, on a membrane or on the surface of a transducer such as an ISE that responds to the species formed in the enzymatic reaction or to the formation of an antigen-antibody immunocomplex.

In the case of potentiometric immunosensors, a change in potential occurs after the specific binding of the antibody or the antigen to the immobilized partner. Proteins in aqueous solution are polyelectrolytes and consequently the electrical charge of an antibody can be affected by binding the corresponding antigen. The difference in electrical potential is measured between the working electrode where the specific antibody has been immobilized and a reference electrode. A main disadvantage of this system is that variations in the potential due to the antibody-antigen interaction are small (1–5 mV) and therefore the reliability and sensitivity of the analysis are limited by background effects.[60]

The research in this field has been aimed at getting better limits of detection and selectivity of the ISE, with the aim of supplying the necessary requirements for its application in industry. New developments include sensor arrays, new ionophores, improvement of the detection limit and new electrodes for miniaturization.[66,67]

ISFETs are an important subset of potentiometric sensors. An ISFET is composed of an ion-selective membrane applied directly to the insulated gate of the FET. When such ISFETs are coupled with a biocatalytic or biocomplexing layer, they become biosensors, and are usually called either enzyme (ENFETs) or immunological (IMFETs) field-effect transistors. Operating properties of ENFET- and IMFET-based devices are strongly related to those of the ISE-based biosensors. The importance of the ISFET can be attributed to its capacity for miniaturization and the possibility of exploiting the processes of microelectronics in its manufacture.[68]

20.3.1.3 *Amperometric transducers*

Amperometric sensors measure the electrical current produced by a chemical reaction of an electroactive species with an applied potential, which is related to the concentration of the species in solution. The amperometric sensor is faster, more sensitive, more precise and more accurate than the potentiometric ones. Therefore, it is not necessary to wait until thermodynamic equilibrium is obtained; the response is a linear function of the concentration of the analyte. However, the selectivity of the amperometric devices is only governed by the redox potential of the electroactive species present. Consequently, the electric current measured by the instrument can include the contributions of several chemical species. The first amperometric biosensor[69] for glucose analysis using the glucose oxidase enzyme with the Clark oxygen electrode was based on monitoring the oxygen consumption. The formation of the reaction product or consumption of reagent can be measured to infer the analyte concentration. These biosensors are referred to as the first generation. Amperometric biosensors modified with mediators are referred to as second-generation biosensors. Mediators are redox substances that facilitate the electron transfer between the enzyme and electrode. The direct enzyme-electrode coupling, or mediator-free biosensors based on direct electron transfer mechanisms, are referred to as the third generation. In this case, the electron is directly transferred from the electrode to the enzyme and to the substrate molecule (or vice versa). In this mechanism, the electron acts as a second substrate for the enzymatic reactions and leads to the generation of a catalytic current. The substrate transformation (electrode process) is essentially a catalytic process.[70,71]

20.3.1.4 *Piezoelectric transducers*

Piezoelectric transducers find application as immunosensors. In these devices, an antigen or antibody is immobilized on the surface of a crystal.[72] The interaction of these elements with the analyte is highly specific and can be monitored through the oscillation of the immersed crystal in a liquid which will produce a modification of mass in the crystal, measurable by means of the modification of the frequency of oscillation. Immunosensors based on the wave acoustics principle, among other types, can be used for detection of pathogenic micro-organisms, gases, aromas, pesticides, hormones and other compounds.[73–75]

20.3.2 *Optical transduction devices*

We begin with a general description of the devices that have been developed and implemented as platforms for biosensors, without restricting the discussion to approaches that involve nanoscale materials.

The binding of a target molecule to a receptor can be detected by an interrogating light beam, or by an intrinsic light output from the products of the binding reaction. Optical interrogation can be implemented in several different modes. For example, it is possible to detect subtle differences in the refractive index in a medium at the surface of a sensor on which antibodies have been immobilized. When an antigen-antibody interaction takes place, the refractive index of the medium becomes slightly higher; this change can be measured using a variety of different optical techniques that are sensitive to small changes in refractive indexes. Fluorescence techniques are very commonly used in biosensors. In this case, molecules are labeled with fluorophores that are excited by a light source; the resulting fluorescence is measured by a photodetector. Spectroscopy can be used to detect the presence of an analyte by evaluating the optical spectrum directly; the Raman scattering signal that is observed at a shifted wavelength that is unique to the target compound can be used in a detection method. Spontaneous Raman scattering is normally quite weak. However, nanoscale materials are able to provide a significant enhancement in the Raman signal. In all cases, the optical interrogation will lead ultimately to an electrical signal that provides a calibrated measure of the extent of binding to receptors, and hence of the amount of an analyte that is present in a sample.

20.3.2.1 *Waveguide sensors*

Several variants of biosensors are based on the application of the evanescent wave that appears at the interface between two media that have different refractive indexes. The evanescent wave at the interface can be used in conjunction with waveguides, with surface plasmon waves and with resonant mirrors. In several configurations of biosensors, the evanescent wave is utilized within a microchannel that is filled with sample fluid. The capture or binding of analyte molecules by the biological detection element at the sensor surface is detected through techniques that rely on refractive index changes.

Another family of biosensors uses the fluorescent output from a reporter molecule on a target molecule as the detection mode. Evanescent waves may also be used in this case to excite the labeling fluorophores, as seen in total internal reflection fluorescence (TIRF) devices. Antibodies that are raised against a specific analyte are immobilized on the top of a waveguide. Fluid that contains previously labeled analyte molecules, along with unlabeled sample analyte, is added to the system. Competition for binding to the antibodies takes place. A larger concentration of sample analyte will lead to a lower fluorescence signal as fewer of the labeled molecules are bound to antibodies. The evanescent waveguide field will only excite fluorophores that are in the immediate vicinity of the surface, thereby avoiding the excitation of more distant unbound fluorophores which can contribute to unwanted background signal.

Several configurations have been used for evanescent field detection, including planar waveguides and optical fibers, with signal transduction implemented via fluorescence or interferometry. Sapsford *et al.*[76] have provided a thorough review of fluorescence planar waveguide biosensors; Taitt and Ligler[77] have reviewed developments in evanescent wave optical fiber biosensors; Campbell and McCloskey[78] have reviewed interferometric biosensors.

20.3.2.2 *Surface plasmon resonance (SPR)*

An interface between two optically transparent media can be useful not only in the format of TIRF spectroscopy, but also in application through a phenomenon known as surface plasmon resonance (SPR). These surface waves occur at a metal-dielectric interface. The propagation of the plasmon wave is sensitive to the refractive index of the dielectric medium, and this provides its utility in terms of sensing. The use of SPR for biosensors has been thoroughly reviewed by Homola and colleagues.[79,80]

SPR devices find application in two basic areas: direct sensing of an analyte in samples, and the fundamental study of the kinetics or thermodynamics of biophysical and biochemical reactions. In the first case, binding of an analyte to sensing elements at the surface is detected through the change in the refractive index. The second type of application takes advantage of the fact that SPR biosensors do not require labeling of the analyte or detecting element, as is required in fluorescence-based detection schemes. Therefore, it is possible to measure the kinetics, or time dependence, of reactions between an analyte and sensing element in the biosensor.

20.3.2.3 *Interferometric sensors*

Because the phase of light changes as the refractive index in a medium changes, variations in refractive index caused by binding of analyte molecules can be detected through interference effects. Interferometric biosensors have been fabricated on waveguides where one leg of the interferometer is the sensing part, and a second waveguide channel carries a reference beam. As the refractive index in the detection leg changes, the phase change of light, when recombined with the reference signal, produces interference patterns that can be recorded and interpreted in terms of concentrations of analytes. The use of interferometry for biosensors is reviewed in detail by Campbell and McCloskey.[78]

Interferometric sensors have been used in a number of environmental applications, including competitive immunoassays for the detection of the herbicide, atrazine[81,82]; in this case the limit of detection was about 100 ng/L. In other cases, nanoscale materials have been adopted as part of the sensor. Tinsley-Bown *et al.*[83] used a porous silicon structure with pore sizes greater than 50 nm for an immunoassay. Ryu *et al.*[84] also used a porous silicon interferometer to detect beta-galactosidase that was released from recombinant *E. coli.*

20.3.2.4 *Resonant mirror devices*

The evanescent field above a waveguide can be enhanced when the incident light enters the waveguide at a special angle and is reflected back by a mirror. A resonant condition occurs at this angle, and the device is referred to as a resonant mirror. One of the earliest descriptions of the resonant mirror concept in the literature is due to Cush *et al.*[85] The concept has been commercialized in the IAsys biosensor. The principles and applications of resonant mirror devices have been thoroughly reviewed by Cooper[86] and by Kinning and Edwards.[87]

The device is made of two basic parts: a low refractive index coupling layer, and a high refractive index waveguide layer. An evanescent field is established above the waveguide; it is used to detect changes in the refractive index of the medium that is immediately adjacent to the liquid-waveguide interface. Binding of an analyte to a sensing element such as an antibody will change the refractive index where the evanescent field is propagating. This will affect the angle at which an incident ray of light will resonate within the cavity created by the waveguide. The incident beam is scanned over a range of angles in order to determine when the maximum intensity is observed at the detector. Polarizers are used so that only light that is propagating through the resonant structure is detected, thereby blocking extraneous light from the detector.

20.3.2.5 *Spectroscopy*

Although labeling with a fluorophore is standard procedure in biosensors, in some cases it may be advantageous to carry out label-free detection. This can be achieved by making use of unique spectral signatures that can be obtained from biomolecules through processes such as Raman scattering. A well-known method that uses this technique is surface-enhanced Raman scattering. The normal spontaneous Raman signal is quite weak but its strength can be enhanced if detection is carried out on gold or silver surfaces that support Raman resonances. The surface-enhanced resonant Raman process gives rise to a considerable increase in the Raman cross-section, and therefore in the limit of detection. The nanoscale features of the surfaces are important in determining the extent of the signal enhancement. SERS has been used in basic studies of biochemistry[88] and recently for DNA detection.[89,90]

Raman spectroscopy can also be carried out in small volumes through the use of optical tweezers to control samples.[91,92] With this technique, it has been shown that it is possible to trap cells and other biomolecules within the confocal region of a micro-Raman spectroscopy system. The Raman signature of the biomolecules is unique and allows identification of particular cells and compounds without additional labeling. The equipment to carry this out is rather more complex than simple lab-on-a-chip devices but it offers considerable promise for label-free detection.

20.4 Nanoscale Materials for Sensors

20.4.1 Quantum dots (QDs)

Quantum dots (QDs) represent an early implementation of optically active nanoscale materials. They are made of semiconductor materials with electrons that have energy levels typical of the particular semiconductor material. The exciton Bohr radius represents the average physical separation between excited electron-hole pairs for each material. If the size of a semiconductor crystal becomes as small as the size of the material's exciton Bohr radius, then the electron energy levels become discrete. This phenomenon is called quantum confinement and under such conditions the material is termed a quantum dot. Due to the quantum confinement effect, the absorbance peak of quantum dots and their emission maxima shift to higher energies (shorter wavelengths) with decreasing particle size.[93]

Since "naked" QDs are susceptible to photo-oxidation, they need to be capped by a protective shell of an insulating material or wide-bandgap semiconductor, structurally matched with the core material. Core-shell geometries where the nanocrystal (typically 2–8 nm in diameter) is encapsulated in a shell of a wider-bandgap semiconductor have increased fluorescence quantum efficiencies (> 50%) and greatly improved photochemical stability. The emission quantum yield of these particles increases with increasing thickness of the shell and can reach values of up to 0.85.[31]

In contrast to the organic fluorophores that are conventionally used in biotechnology, quantum dots absorb light over a very broad spectral range. This makes it possible to optically excite quantum dots with different emission spectra (colors) by using a single excitation laser wavelength, which enables simultaneous detection of multiple markers in biosensing and assay applications.[29] Quantum dots are made in CdSe-CdS core-shell structures that emit in the visible region from approximately 550 nm (green) to 630 nm (red). Near-infrared optical emission has been achieved with other material combinations, such as InP and InAs. It is necessary to provide a hydrophobic, chemisorbed surface layer to facilitate bioconjugation via adsorption or covalent binding. Widely used approaches for this purpose include the treatment of the QDs with tri-n-octylphosphine oxide (TOPO) and hexadecylamine; several alternative approaches have also been proposed.[27] Currently, quantum dots with a variety of properties are available commercially.

20.4.2 Carbon nanotubes (CNTs)

Carbon nanotubes (CNTs) are graphitic sheets curled up into seamless cylinders. During the past few years they have attracted a great deal of attention, and have been subject to intense research associated with their possible application as carbon-based electronic devices. Carbon nanotubes are classified into two main types: multi-wall (MWCNTs) and single-wall (SWCNTs). The earliest observations were of micron-long multi-wall nanotubes in 1991.[94] Shortly after that, the synthesis of single-wall tubes was reported.[95] Their diameters could be controllably changed via variations in the growth parameters.

SWCNTs possess a cylindrical nanostructure (with a high aspect ratio), formed by rolling up a single graphite sheet into a tube. The atoms are located using a pair of integers (n, m) and the lattice vector $\vec{C} = n\vec{a}_1 + m\vec{a}_2$ as shown in Figure 20.7.[96]

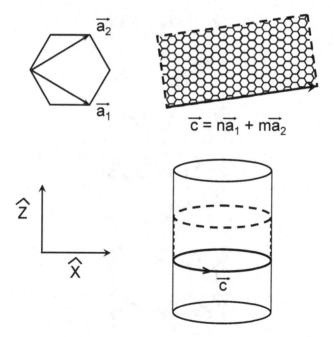

Figure 20.7. Structure of an n, m carbon nanotube. The carbon atoms are at the vertices of the hexagons. (From Ref. 96 with permission.)

A tube can be classified using the pair of integers n and m by viewing the rolling up of the sheet as the "placement" of the atom at 0,0 on the atom at n, m. Hence, different diameter tubes and helical arrangements of hexagons can arise by changing n, m. An example is shown in Figure 20.8.[96]

MWCNTs consist of several layers of graphene cylinders that are concentrically nested like rings of a tree trunk (with an interlayer spacing of 3.4 Å).[97,98] Figure 20.9 compares TEM images of MWCNTs.[94] While SWCNTs typically have diameters ranging from 0.7 nm to 3 nm, they can be several microns in length. They exhibit unusual mechanical properties

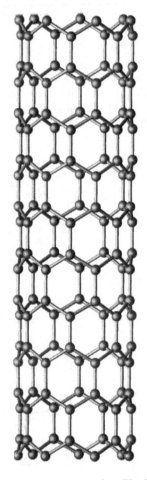

Figure 20.8.   Ball-and-stick model for a nanotube. The balls represent carbon atoms. (From Ref. 96 with permission.)

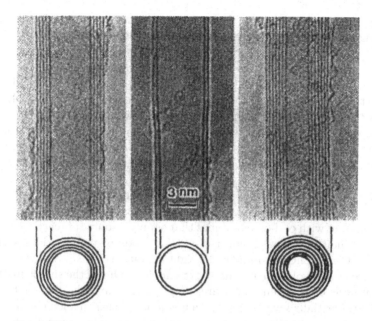

Figure 20.9. TEM images of multi-wall carbon nanotubes. (From Ref. 94 with permission.)

such as high strength and stiffness — the Young's modulus and tensile strength of nanotubes are higher than those of any other material.[99,100]

The unique electrical properties of carbon nanotubes make them extremely attractive for the task of chemical sensors, in general, and electro-chemical detection, in particular.[101] The electronic properties of CNTs are sensitive to the structure of the tube. Carbon nanotubes can be metallic or semiconducting, depending on n, m. For example, if $n - m$ is three times an integer, the carbon nanotube has an extremely small bandgap, and at room temperature, it has metallic behavior. For $n = m$, the tubes are metallic; and for other values of $n - m$, the tubes behave as semiconductors with a bandgap.

20.4.3 *Inorganic nanotubes, wires and belts*

In addition to carbon nanotubes, inorganic nanotubes and nanowires are exciting increasing interest in many areas.[102,103] Like carbon nanotubes, inorganic nanotubes consist of long hollow cylinders that typically have diameters of about 10 nm and wall thicknesses of about 1 nm.[104–107] Nanowires and nanorods are flexible or stiff solid cylinders that are much longer than their diameters. They provide a much greater surface area for sensing applications than the area offered by spherical particles.

These unusual morphologies can be synthesized from a wide variety of materials. Nanotubes have been made from semiconductor metal oxides such as ZnO and TiO_2,[108,109] from sulfides, from phosphides[110] and other materials.[104] Nanorods have also been formed from gold and silver.[111] In general, the synthesis routes follow one of two paths: a gas-phase synthesis in which precursors decompose at high temperature to reform as wires and rods; or a solution-phase process at low temperature. Methods for the synthesis of these materials are described in several recent reviews.[104-106,109,112]

Rao *et al.*[104] describe several of these methods. The so-called carbothermal route makes use of the reaction of metal oxide powders in a reducing environment with carbon nanotubes that supply a source of reactive carbon. This method produces nanowires and nanorods of metal oxides such as SiO_2, InO and ZnO, and under the right conditions of excess nitrogen, can produce nitrides of these metals and oxynitrides. One of the simple methods of synthesis of inorganic nanowires employs the carbothermal route. The method essentially involves heating a metal oxide with an adequate quantity of carbon in an appropriate atmosphere. For example, ammonia provides the atmosphere for the formation of nitrides. An inert atmosphere and a slight excess of carbon yield carbides. In these reactions, carbon helps to form an oxidic species, usually a sub-oxide, in the vapor phase, which then transforms to the final crystalline product.

Rao *et al.*[104] showed that they could react Ga_2O_3 powder with multi-wall carbon nanotubes to yield nanosheets and nanorods. Figure 20.10(a) shows an SEM image of Ga_2O_3 nanowires that were obtained by this reaction at $1,100°C$ in argon gas. The nanowires have a diameter of around 300 nm with lengths extending to tens of microns. The nanowires exhibit good photoluminescence characteristics.[104] Nanowires of other oxides such as SiO_2 (Figure 20.10(b)) have also been prepared by reacting the corresponding metal/metal oxide with carbon in an inert atmosphere.[104] The reaction of NH_3 with a mixture of SiO_2 gel and carbon nanotubes or activated carbon in the presence of a Fe catalyst yielded nanowires of silica.[104] Using silica gel heated with activated carbon at $1,360°C$ in NH_3 yielded SiC nanowires (Figure 20.10(c)). Silicon nanowires have been formed from Si powders or solid substrates with a similar approach (Figure 20.10(d)).[104]

Many synthesis techniques have been reported that use the solution phase or so-called soft chemistry synthesis of nanorods, tubes and wires. Solution-phase chemistry can produce many of the same materials as the gas-phase route, including metal oxides such as ZnO,[113,114] indium tin oxide

Figure 20.10. SEM images of (a) Ga_2O_3 nanowires obtained by the reaction of Ga_2O_3 with MWNTS at $1,100°C$ in 60 sccm Ar; (b) Si_{34} nanowires synthesized by heating silica gel and MWNTs in an NH_3 atmosphere; (c) SiC nanowires obtained by the reaction of silica gel and activated carbon in NH_3 at $1,360°C$ for seven hours; and (d) Si nanowires synthesized using the carbon-assisted method on a Si substrate. (From Ref. 104 with permission.)

(ITO)[115] and others. One of the novel methods for the synthesis of metal oxide nanotubes employs gels derived from low-molecular-weight organic compounds. The gel fibers are coated with oxidic materials followed by the dissolution of the gel in a suitable solvent and by calcination. Figure 20.11 shows TEM images of (a) gel fibers; and nanotubes of (b) titania, (c) ZnO and (d) $ZnSO_4$ obtained by coating the respective precursors on the hydrogel fibers, followed by the removal of the template. The use of hydrogels enabled Rao *et al.*[104] to avoid the use of alkoxides, which are difficult to handle due to their moisture sensitivity. The gel fibers have diameters of 8–10 nm and lengths extending to several hundreds of nanometers. Nanotubes of titania were obtained by the reaction of titanium alkoxide with the gel fibers, followed by the removal of the gel fiber template. The nanotubes generally have inner diameters comparable to the diameter of the gel fibers. ZnO nanotubes were obtained using zinc acetate as the precursor; $ZnSO_4$ nanotubes were obtained by a similar process.

Inorganic nanowires find their greatest application in sensing through the modification of their electrical properties as compounds adsorb onto their surface. Slight changes onto their conductivity can be detected following adsorption. The difficulty lies in attaching electrodes to individual nanowires and reading out the results.

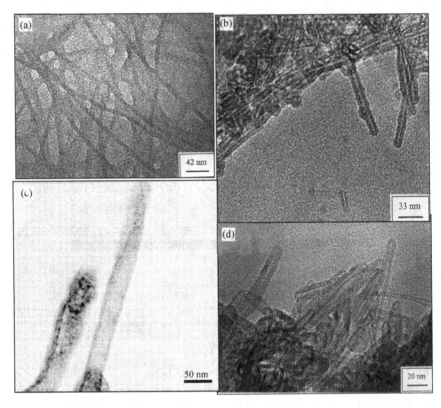

Figure 20.11.   TEM images of (a) gel fibers and nanotubes of (b) titania, (c) ZnO and (d) ZnSO$_4$ obtained by coating the respective precursors on the hydrogel fibers, followed by the removal of the template. (From Ref. 104 with permission.)

20.4.4 Colloidal gold and colloidal silver

Nanoparticles of noble materials such as gold and silver have been used for centuries for coloring glass and silk. The high conductivity of the noble metals supports the presence of a plasmon absorption band. This absorption band results when the incident photon frequency is in resonance with the collective excitation of the conductive electrons of the particle. This effect was termed localized surface plasmon resonance (LSPR). The resonant frequency of the LSPR band is strongly size-dependent. Spherical gold particles of 10–40 nm diameter are red (resonance frequency at about 600 nm). On the other hand, irregularly shaped and bigger particles turn blue (resonance frequency about 450 nm). The full visible spectrum may be covered by gold particles with diameters from 60 to 200 nm. Silver nanoparticles in the same size ranges behave in a similar way. Today, colloidal gold finds application in non-linear optics, supramolecular chemistry, molecular recognition and the biosciences.[31,116]

20.4.5 *Lanthanide-based nanoparticles*

Lanthanides or rare earth metals comprise all elements bearing the atomic numbers 58–71 in the periodic table. Some of the lanthanide ions (e.g. Eu(III), Dy(III), Tb(III) and Sm(III)) have well-determined excitation and emission spectra when they are incorporated into appropriate environments such as chelate molecules or the crystal lattice of oxide. Chelates are organic complexes that serve as antennae which transmit the excitation energy to the lanthanide ion.[117] Oxides like Gd_2O_3 of Y_2O_3 provide ideal hosts for doping with luminescent lanthanide ions;[118] the crystal lattice of the oxide permits efficient excitation and emission of the electronic transitions of their f-electrons (Figure 20.12). The lanthanide emission is narrow, exhibits a large Stokes shift, and yields long luminescence lifetimes (from several μs to ms).

The use of nanoparticles containing luminescent lanthanide ions enables multiple luminescent centers to be incorporated into a single nanoparticle. A single polystyrene nanoparticle of 100 nm may contain several thousand chelate molecules, possibly with different lanthanide ions. The amount of doping into a Gd_2O_3 nanoparticle is even greater. Nevertheless, if the doping exceeds a certain concentration (e.g. Eu^{3+} into Gd_2O_3), the average distance between two Eu^{3+} ions will become too short and they will tend to quench each other instead of emitting light. The direct consequence of this so-called concentration quenching is a decrease in the luminescent lifetime.

The long lifetime of optimally doped nanoparticles offers the opportunity for so-called time-gated measurements: after the sample is excited with a short pulse of UV light the emission signal is accumulated for a suitable time interval with a time delay on the order of several 100 ns. Therefore, interference by short-lived background emission and scattering from the sample can be minimized as seen in Figure 20.13.

20.4.6 *Magnetic nanoparticles*

Magnetic nanoparticles, especially magnetite (Fe_3O_4) nanoparticles, are promising candidates for biomolecule tagging, imaging, sensing and separation.[15,119,120] Magnetite is a common magnetic iron oxide that has a cubic inverse spinel structure with oxygen forming a face-centered cubic (fcc) structure and Fe cations occupying interstitial tetrahedral sites and

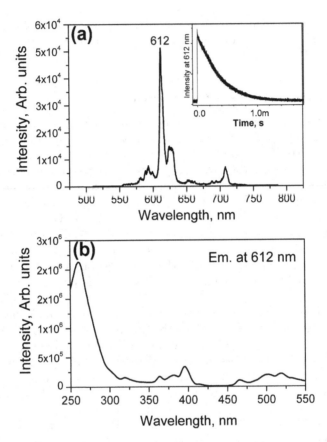

Figure 20.12. (a) Emission spectrum and 1 ms lifetime and (b) excitation spectrum of Eu^{3+} doped into Gd_2O_3 nanoparticles.

octahedral sites. The electrons can hop between Fe^{2+} and Fe^{3+} ions in the octahedral sites at room temperature, thus giving the magnetite properties of half-metallic material. Using chemical precipitation methods, different nanometer-scale MFe_2O_4 materials (M = Co, Ni, Mn, Mg, etc.) can be synthesized and used in studies of nanomagnetism.[121]

For magnetic particles below a certain critical diameter, a single nanoparticle cannot support more than one domain. The critical diameter generally falls in the 10–100 nm range.[122] In a single-domain cluster, the atomic magnetic moments are coupled via exchange interactions to form a large net cluster moment — the cluster is called a superparamagnet.

After surface coating and functionalization, the magnetic nanoparticles can be dispersed into water, forming water-based suspensions. From a

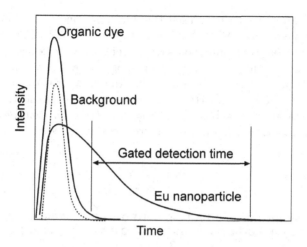

Figure 20.13. Time-gated detection with long-lifetime nanophosphors avoids the short-lifetime background fluorescence from a sample (background shown as dashed line). An organic dye has a short lifetime.

suspension, the particles can interact with an external magnetic field and be positioned to a specific area, facilitating magnetic resonance imaging for medical diagnosis and cancer therapy based on heating with an external AC magnetic field. These applications require that the nanoparticles are super-paramagnetic with sizes of a few nanometers and a narrow size distribution so that the particles have uniform physical and chemical properties.

20.5 The Application of Nanomaterials in Sensors

20.5.1 *Optical transduction using nanoscale materials*

Nanoscale materials offer unique opportunities for optical transduction in biosensors due to the particular interaction between light and matter at the nanoscale. Light can be either absorbed or emitted.

20.5.1.1 *Absorbance-based sensors*

When gold nanoparticles form aggregates, the light absorption spectrum of the aggregate depends strongly on the average distance between the particles within the aggregate. As the interparticle distances in these aggregates decrease to less than approximately the average particle diameter, the color becomes blue. On the other hand, aggregates with interparticle distances substantially greater than the average particle diameter appear red.[123,124]

Several research groups have used this phenomenon to develop biosensor detection schemes for DNA,[125−129] for studying biomolecular interactions in real time[130] and for immunoassays. A refractive index sensor based on colloidal gold-modified optical fiber was reported by Cheng and Chau.[131]

As an example, Figure 20.14 describes the basic principle for detection of DNA using gold nanoparticles. Mercaptoalkyloligonucleotide-modified Au nanoparticles are used as reporter groups. Hybridization results in the binding of an oligonucleotide probe to the target sequence, and in the

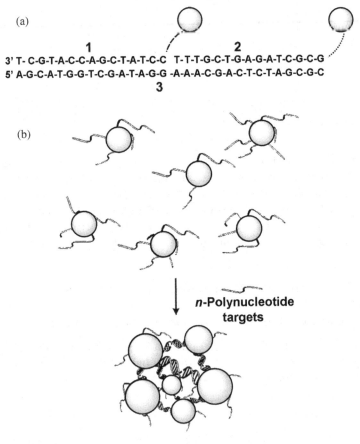

(a)

1   **2**
3' T- C-G-T-A-C-C-A-G-C-T-A-T-C-C T-T-T-G-C-T-G-A-G-A-T-C-G-C-G
5' A-G-C-A-T-G-G-T-C-G-A-T-A-G-G -A-A-A-C-G-A-C-T-C-T-A-G-C-G-C
3

(b)

n-Polynucleotide
targets

Figure 20.14.   (a) Hybridization of two mercaptoalkyloligonucleotide-modified Au nanoparticles and polynucleotide target. (b) Schematic representation of the concept for generating aggregates signaling hybridization of nanoparticle-oligonucleotide conjugates with oligonucleotide target molecules. The nanoparticles and the oligonucleotide interconnects are not drawn to scale, and the number of oligomers per particle is believed to be much larger than depicted. (From Ref. 125 with permission.)

formation of an extended polymeric network in which the reporter units are interlocked.[125]

Localized surface plasmon resonance excitation in silver and gold nanoparticles produces strong extinction and scattering spectra that in recent years have been used for important sensing and spectroscopy applications. Several ultra-sensitive nanoscale optical biosensors based on LSPR have been reported. LSPR nanobiosensors for the biotin-streptavidin system have been developed using gold nanoparticles self-assembled on a functionalized glass surface[130,132] and using triangular silver nanoparticles directly fabricated on the glass by nanosphere lithography.[133] Colloidal gold-enhanced SPR was developed for ultra-sensitive detection of DNA hybridization.[134] Use of the Au nanoparticle tags leads to a greater than tenfold increase in angle shift, corresponding to a more than 1,000-fold improvement in sensitivity for the target oligonucleotide as compared to the unamplified event, and makes it possible to conduct SPR imaging on DNA arrays with a sensitivity similar to that of traditional fluorescence-based methods for DNA hybridization.

20.5.1.2 *Luminescence-based sensors*

Gold nanoparticles are also applied as quenchers for fluorescent labels in a new class of optical biosensors for recognizing and detecting specific DNA sequences.[25] Oligonucleotides labeled with a thiol group at one end and a fluorescent dye at the other end were attached to the surface of gold nanoparticles, spontaneously forming an arch-like conformation (see Figure 20.15). Binding of target molecules resulted in a conformation change and this restored the fluorescence of the quenched fluorophore. The biosensor developed on this basis was able to detect single-base mutations in a homogeneous format.

Recently, the same principle was employed for the development of biosensors for the detection of neurotoxic organophosphates that have found widespread use in the environment for insect control.[55]

A large variety of environmental sensors are based on the photoluminescence quenching in porous silicon and silicon nanocrystals due to adsorption of different molecules on their surface. The advances in this type of sensors were reviewed in detail by Shi *et al.*[135] After different surface treatments of Si nanocrystals, their photoluminescence was found to be sensitive to a variety of organic solvents, amines, aromatic compounds, metal ions and corrosive, toxic and explosive gases.

Quantum dots are very attractive for biolabeling due to properties such as excellent brightness, narrow and precise tunable emission, negligible photo-bleaching, fairly high quantum yields, and photostability.[93,136]

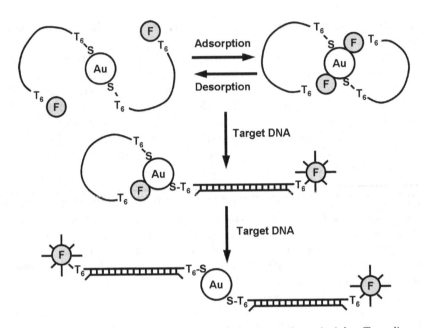

Figure 20.15.   Nanoparticle-based probes and their operating principles. Two oligonucleotide molecules (oligos) are shown to self-assemble into a constrained conformation
on each gold particle (2.5 nm diameter). A T6 spacer (six thymines) is inserted at both
the 3'- and 5'-ends to reduce steric hindrance. Single-stranded DNA is represented by
a single line and double-stranded DNA by a cross-linked double line. In the assembled
(closed) state, the fluorophore is quenched by the nanoparticle. Upon target binding, the
constrained conformation opens, the fluorophore leaves the surface because of the structural rigidity of the hybridized DNA (double-stranded), and fluorescence is restored. In
the open state, the fluorophore is separated from the particle surface by about 10 nm.
See text for detailed explanation. Au, gold particle; F, fluorophore; S, sulfur atom. (From
Ref. 25 with permission.)

Although quantum dots have proven to be suitable labels for biological
imaging, their application in quantitative biosensors is still in an early stage.
A biosensing detection scheme was developed that utilizes donor-acceptor
energy transfer between QDs and receptors for conducting recognition-
based assays.[137] In this configuration, the luminescence of antibody-coated
quantum dots was quenched due to specific binding with quencher-labeled
antigen. Gerion *et al.*[28] have employed quantum dots as fluorescent labels
for DNA microarray analysis while Geho *et al.*[138] reported the use of quantum dots as detection elements for protein microarrays.

Silica nanoparticles doped with organic fluorescent dyes were synthesized by Tapec *et al.*[139] They exhibited very good photostability and minimal dye leakage. The silica matrix of the nanoparticles permits different

methods of surface biomolecular modification for biosensor and bioanalysis applications. Application of dye-doped silica particles for the detection of glutamate was demonstrated.

Nanoparticles of Eu_2O_3 have shown promising application as luminescent tags in immunoassay for detection of atrazine.[140] Recently, a microarray format for an immunoassay using improved Eu-doped Gd_2O_3 nanoparticles was reported.[141] The excellent photostability of these particles allows the samples to be observed and analyzed under UV excitation for prolonged time periods. Their sharp emission spectrum facilitates easy separation of the signal from the background fluorescence. Detection of phenoxybenzoic acid (a generic biomarker of human exposure to pyrethroids) was demonstrated using these nanoparticles.[141] In this configuration, an internal fluorescent standard was introduced by making use of the emission from the nanoparticle as a measure of the number of antibodies that were attached to its surface; the detection of analyte was reported by a second fluorophore of a different wavelength. Miniaturization of a conventional immunoassay offers the possibility for building portable biosensor devices; the lanthanide oxide nanoparticles offer the opportunity to build new detection schemes using their unique luminescent properties such as long lifetime,[142,143] excellent photostability[141] and sharp emission spectra.

20.5.2 *Electrochemical biosensors using nanomaterials*

Nanomaterials can be used in a variety of electrochemical biosensing schemes. New nanoparticle-based signal amplification and coding strategies for bioaffinity assays, along with carbon nanotube molecular wires, have been proposed during the last few years. Gold, silver and metal oxide nanoparticles have been widely used in biosensors because of their large relative surface area and good biocompatibility.

20.5.2.1 *Enzyme-based electrochemical biosensors*

An extremely important challenge in amperometric enzyme electrodes is the establishment of efficient electron transfer between the redox active site of the enzyme (protein) and the electrode surface. The possibility of direct electron transfer between enzymes and electrode surfaces could pave the way for superior reagentless biosensors, as it obviates the need for co-substrates or mediators and allows efficient transduction of the biorecognition event. The conductive properties of some nanomaterials, such as carbon nanotubes and metal nanoparticles, suggest that they could mediate the electrical communication between the redox enzyme and the electrode. The ability of CNTs to promote electron transfer reactions is attributed to the presence of edge-plane defects at their end caps. CNTs associated with carbon paste electrodes or glassy carbon electrodes show excellent electrocatalytic activity toward the oxidation of ascorbic acid, dopamine, phenols, tryptophan, carbohydrates, and the reduction of hydrogen peroxide and nitrite ions.[144,145]

The structural alignment of enzymes on the ends of CNTs organized as an array on a conductive surface was accomplished by Willner's group.[146] An array of perpendicularly oriented carbon nanotubes on a gold electrode was fabricated by covalently attaching carboxylic-acid-functionalized carbon nanotubes to a cystamine-monolayer-functionalized gold electrode. The amino-derivative of the flavin adenine dinucleotide (FAD) cofactor was covalently coupled to the carboxylic groups at the free ends of the aligned CNTs (see Figure 20.16).

Such enzyme reconstitution on the ends of the CNTs represents an extremely efficient approach for plugging an electrode into glucose-oxidase (GOx). Electrons were thus transported across distances greater than

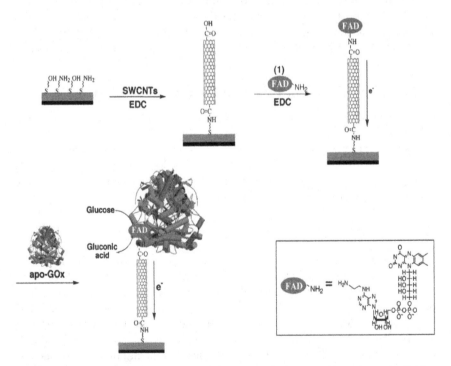

Figure 20.16.   Assembly of CNT electrically contacted glucose oxidase electrode: linking the reconstituted enzyme, on the edge of the FAD-functionalized CNTs, to the electrode surface. (From Ref. 146 with permission.)

150 nm with the length of the CNTs controlling the rate of electron transport. An interfacial electron transfer rate constant of 42 s^{-1} was estimated for a 50 nm long CNT.

The anodic detection of dehydrogenase and oxidase enzymes at ordinary electrodes is often hampered by the large over-voltage encountered during their oxidation. The greatly enhanced redox activity of hydrogen peroxide[147] and NADH[148] at CNT-modified electrodes addresses these over-voltage limitations. Further improvement of the detection of the enzymatically liberated hydrogen peroxide was accomplished by deposition of platinum[145] and palladium[149] nanoparticles onto CNTs. Co-axial nanowires, consisting of a concentric layer of polypyrrole uniformly coated onto aligned CNTs, have provided a template for making glucose sensors with a large amount of electrochemically entrapped GOx in the ultra-thin polypyrrole film.[150] Similarly, amperometric detection of organophosphorus pesticides and nerve agents was performed using a screen-printed biosensor based on co-immobilized acetocholine esterase, choline esterase and CNTs.[151]

The excellent electrocatalytic properties of metal nanoparticles (compared to bulk metal electrodes) can also enhance amperometric enzyme electrodes. Gold nanoparticles were used as "electrical nanoplugs" for the alignment of GOx on conducting supports and wiring its redox center.[152] An improved amperometric biosensing of glutamate was achieved by a transducer built on dispersed iridium nanoparticles ($d = 2$ nm) in graphite-like carbon.[153] Ferrocene-doped silica nanoparticles conjugated with a biopolymer chitosan served as an electron transfer mediator for the determination of glucose in rat brain.[154] ZrO_2 nanoparticles/chitosan composite film was used as an immobilization matrix in an acetylcholinesterase biosensor for the determination of pesticides in real vegetable samples.[36] Glassy carbon electrodes with a film of mercury-doped silver nanoparticles was used for cysteine detection in urine samples.[155] HRP-modified nanoelectrodes formed by arrays of gold nanotubes offered enhanced sensitivity compared to a classical gold macroelectrode.[156]

20.5.2.2 *Bioaffinity (immuno and DNA) electrochemical biosensors*

Nanomaterials can be used as quantitation tags or as amplification carriers in a variety of bioaffinity electrochemical formats. By analogy to fluorescence-based methods, several electrochemical detection methods have been pursued in which DNA sequences and antibodies have been labeled with electroactive nanoparticles. The characteristic electrochemical response of the nanoparticle reporter signals the hybridization event or the immunoreaction, respectively.

Since CNTs exhibit a high conductivity and reveal very high area-to-weight ratio, they have been used as a support for the immobilization of biomolecules and further used in immuno- and DNA sensors. For example, immunosensing systems with electrochemiluminescent detection have been designed using CNTs as supports of the immunorecognition systems.[157] DNA-functionalized carbon nanotubes arrays were constructed, hybridized with the complementary DNA and the redox label daunomycin was intercalated into the double-stranded DNA-carbon nanotube assembly and detected by differential pulse voltammetry.[158] In another configuration, amino-functionalized DNA covalently bound to aligned CNTs on a gold electrode was hybridized with ferrocene-labeled complementary DNA; an enhanced electrochemical signal provided by the high surface area of the CNT-modified electrode was observed.[159] Similarly, the coupling of the CNT nanoelectrode array with the $Ru(bpy)_3^{2+}$-mediated guanine oxidation has facilitated the detection of subattomoles of DNA targets.[160] Using this method, the sensitivity of the DNA detection was improved by several orders of magnitude compared to methods where DNA was immobilized using self-assembled monolayers on conventional electrodes.

Conducting polymer nanowire biosensors have been shown to be attractive for label-free bioaffinity sensing. Real-time monitoring of nanomolar concentrations of biotin at an avidin-embedded polypyrrole nanowire has been demonstrated.[161] Similarly, label-free conductivity measurements of antibodies associated with human autoimmune diseases have been performed using polyethylene-oxide-functionalized CNTs.[162]

Functionalized CNTs have been used as carriers of multiple enzyme labels for electrochemical DNA and immunosensing.[163] For example, CNTs were loaded with about 9,600 alkaline phosphatase molecules per CNT and further modified with an oligonucleotide or an antibody. Such a CNT-derived double-step amplification pathway (of both the recognition and transduction events) allows the detection of DNA down to the 1.3 zmol level and indicates great promise for PCR-free DNA analysis.

Nanoparticle-based amplification schemes have resulted in an improved sensitivity of bioelectronic assays by several orders of magnitude. The use of colloidal gold tags allowed the highly sensitive detection of DNA hybridization by anodic stripping voltammetry.[53,164] The same approach was reported for sandwich immunoassays where immunoglobulins G (IgGs) were detected at concentrations as low as 3×10^{-12} M, which is competitive with colorimetric ELISA or with immunoassays based on fluorescent europium chelate labels.[165] Subsequent signal amplification of the electrochemical method for analyzing sequence-specific DNA by gold nanoparticle DNA probes was achieved by silver enhancement where the sensitivity increased by approximately two orders of magnitude.[166,167] Silver enhancement on the gold nanoparticles was also applied to an electrochemical sandwich-type immunoassay.[168] A new strategy for amplifying particle-based electrical DNA detection based on oligonucleotides functionalized with polymeric beads carrying multiple gold nanoparticle tags allowed the detection of DNA targets down to the 300 amol level, and offered great promise for ultra-sensitive detection of other biorecognition events.[169]

Inorganic nanocrystals offer an electrodiverse population of electrical tags as needed for designing electronic coding. A novel, sensitive DNA hybridization detection protocol, based on DNA-quantum dot nanoconjugates coupled with electrochemical impedance spectroscopy (EIS) detection, has been reported.[170] Due to having more negative charges, space resistance and the semiconductor property, CdS nanoparticle labels on target DNA could improve the sensitivity by two orders of magnitude compared with non-CdS-tagged DNA sequences. Cu-Au alloy core-shell nanoparticles combine the good surface modification properties of Au with the good electrochemical activity of the Cu core and allow their application as oligonucleotide labels for electrochemical stripping detection of DNA hybridization.[171]

An effective method for amplifying the electrical detection of DNA hybridization has been developed using CNT carriers loaded with a large number of CdS QDs (see Figure 20.17).[172] Such use of CNT amplification platforms has been combined with an ultra-sensitive stripping voltammetric detection of the dissolved CdS tags following dual hybridization events of a sandwich assay on a streptavidin-modified 96-well microplate. Anchoring of the monolayer-protected quantum dots to the acetone-activated CNTs was accomplished via hydrophobic interactions. A substantial (approximately 500-fold) lowering of the detection limit has been obtained compared to conventional single-particle stripping hybridization assays, reflecting the CdS QDs loading on the CNT carrier.

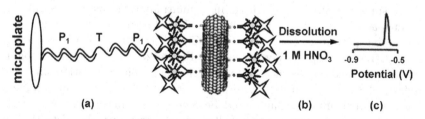

Figure 20.17. Schematic representation of the analytical protocol: (a) Dual hybridization event of the sandwich hybridization assay, leading to capturing of the CdS-loaded CNT tags in the microwell; (b) dissolution of the CdS tracer; (c) stripping voltammetric detection of cadmium at a mercury-coated glassy carbon electrode. P1, DNA probe 1; T, DNA target; P2, DNA probe 2. (From Ref. 172 with permission.)

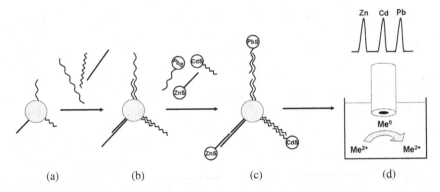

Figure 20.18. Multi-target electrical DNA detection protocol based on different colloid nanocrystal tracers. (a) Introduction of probe-modified magnetic beads. (b) Hybridization with DNA targets. (c) Second hybridization with the QD-labeled probes. (d) Dissolution of QDs and electrochemical detection. (From Ref. 173 with permission.)

Nanoparticles hold particular promise as the next generation of barcodes for multiplexing experiments. The labeling of probes bearing different DNA sequences or antibodies with different nanoparticles enables the simultaneous detection of more than one target in a sample, as shown in Figures 20.18[173] and 20.19.[174] The number of targets that can be readily detected simultaneously (without using high-level multiplexing) is controlled by the number of voltammetrically distinguishable nanoparticle markers. A multi-target sandwich hybridization assay involving a dual hybridization event, with probes linked to three tagged inorganic crystals and to magnetic beads, has been reported (see Figure 20.18).[173] The DNA-connected QDs yielded well-defined and well-resolved stripping peaks at $-1.12\,V$ (Zn), $-0.68\,V$ (Cd) and $-0.53\,V$ (Pb) at the mercury-coated glassy carbon electrode (versus the Ag/AgCl reference electrode).

A similar approach was used in a multi-analyte electrical sandwich immunoassay for beta(2)-microglobulin, IgG, bovine serum albumin and C-reactive protein in connection with ZnS, CdS, PbS and CuS colloidal crystal tracers, respectively, as seen in Figure 20.19.[174] Each biorecognition event yields a distinct voltammetric peak, whose position and size reflects the identity and level, respectively, of the corresponding antigen. Such electrochemical coding could be readily multiplexed and scaled up in multi-well microtiter plates to allow simultaneous parallel detection of numerous proteins or samples. Other attractive nanocrystal tracers for creating a pool of non-overlapping electrical tags for such bioassays are ZnS, PbS, CdS, InAs and GaAs semiconductor particles, in view of the attractive stripping behavior of their metal ions.

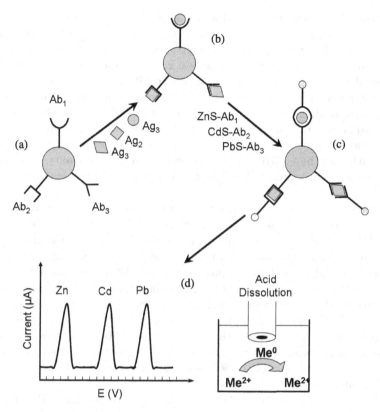

(b)

(a)

Ab₁

Ag₃
Ag₂
Ag₃

ZnS-Ab₁
CdS-Ab₂
PbS-Ab₃

(c)

Ab₂   Ab₃

(d)

Zn   Cd   Pb

Current (μA)

E (V)

Acid
Dissolution

Me⁰

Me²⁺   Me²⁺

Figure 20.19.   Multi-protein electrical detection protocol based on different inorganic colloid nanocrystal tracers. (a) Introduction of antibody-modified magnetic beads; (b) binding of the antigens to the antibodies on the magnetic beads; (c) capture of the nanocrystal-labeled secondary antibodies; (d) dissolution of nanocrystals and electrochemical stripping detection. (Adapted from Ref. 174 with permission.)

20.5.3 *Magnetic sensors*

Magnetic or paramagnetic materials are being used as labels in a new generation of sensors. Detection of the magnetic labels can be accomplished in a variety of methods. Direct detection of magnetic materials can be achieved by a so-called superconducting quantum interference device (SQUID) developed by Chemla *et al.*[175] This technique has been introduced for the rapid detection of biological targets through the use of superparamagnetic nanoparticles. In this technique, target proteins are fixed to a mylar film in the well of a microtiter plate. A suspension of magnetic nanoparticles carrying antibodies is added to the mixture and one-second pulses of magnetic field are applied parallel to the SQUID. This magnetic field magnetizes the nanoparticles; their magnetization relaxes when the field switches off. Nanoparticles that are unbound have the freedom to relax rapidly by Brownian rotation and their magnetization is not detected. Nanoparticles that are specifically bound to the target are restricted to the much slower Néel relaxation and their magnetization (indicative of their presence) is detected by the SQUID. The ability to distinguish between bound and unbound labels makes it possible to run assays which do not require separation and removal of unbound magnetic particles. Although the SQUID is a very sensitive detection modality, it is a large and expensive instrument that does not lend itself to broad deployment as a sensor.

One of the methods for detection of magnetic particles is implemented into the Force Amplified Biological Sensor (FABS) developed by Baselt *et al.*[176] This biosensor measures the forces that bind DNA-DNA, antibody-antigen or ligand-receptor pairs together, at the level of single molecules. Another configuration based on magnetic beads detection is the Bead Array Counter (BARC).[177] It is a multi-analyte biosensor in which DNA probes are patterned onto the solid substrate chip directly above giant magnetoresistive (GMR) sensors. Sample analytes containing complementary DNA hybridize with the probes on the surface (Figure 20.20). Labeled, micronsized magnetic beads are then injected that specifically bind to the sample DNA. A magnetic field is applied, removing any beads that are not specifically bound to the surface.

The beads remaining on the surface are detected by the GMR sensors, and the intensity and location of the signal indicate the concentration and identity of biological targets (such as pathogens) present in the sample. Analyzing millions of transducers offers the possibility to detect or screen thousands of analytes.

A biosensor using magnetic nanoparticles as tags (Figure 20.21) was developed by Kurlyandskaya *et al.*[178] The sensor's working principle is based on a magneto-impedimetric response produced by magnetic nanoparticles to an external magnetic field. This response depends on the value of the applied magnetic field and the amplitude and frequency of the flowing current. This configuration is attractive because it offers the possibility of a sensor with enhanced sensitivity and a two-step sensing procedure with separation of the chemical immobilization part from the sensing process, which may have significant potential in the field of medical and pharmaceutical applications.

Graham *et al.*[179] developed magnetic-field-assisted magnetoresistive-based DNA chips with incorporated spin valve sensors for rapid probe-target hybridization using low target DNA concentrations. A spin

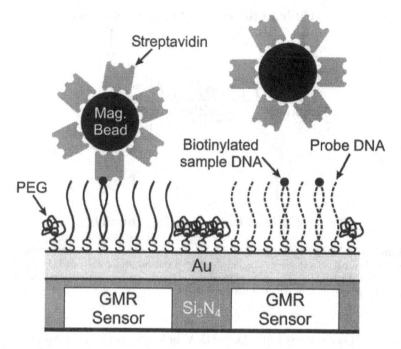

Figure 20.20.   Schematic diagram of the BARC chip surface chemistry and hybridization assay. Thiolated DNA probes are patterned onto a gold layer directly above the GMR sensors on the BARC chip. Biotinylated sample DNA is then added which hybridizes with the DNA probes on the surface when the complementary sequence is present. Unbound sample DNA is washed away. Streptavidin-coated magnetic beads are injected over the chip surface, binding to biotinylated sample DNA hybridized on the BARC chip. Beads that are not specifically bound are removed by applying a magnetic field. Bound beads are detected by the GMR sensors. (From Ref. 177 with permission.)

valve sensor is an electronic component, the conductivity of which is highly sensitive to the external magnetic field. A schematic representation of the operational principles is presented in Figure 20.22. When an electric current of 20 mA is passed through both current lines at either side of the sensor (central), a magnetic field gradient is generated resulting in a magnetic force F_1 that attracts the magnetic labels to the narrow regions of the current lines. When a sufficient number of labels have been concentrated on the current lines, the current is switched off and the magnetic labels are then attracted to the nearby sensor by a magnetic force F_2 resulting from the smaller magnetic field generated by the current through the sensor. Thus, high concentrations of magnetically labeled target DNA are rapidly brought into close contact with sensor-bound probe DNA. The detection

Cu lead for exciting current contacts

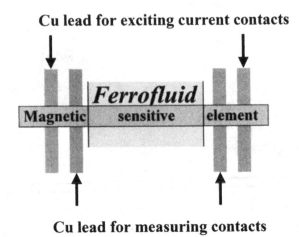

Cu lead for measuring contacts

Figure 20.21. Principal diagram of the MI sensitive element. The magnetic sensitive element is a horizontal sheet. The lightly shaded central area corresponds to the Ferro-fluid® application area. The covered magnetic element is narrower than the Ferro-fluid® application area. (From Ref. 178 with permission.)

range of this chip was 140–14,000 DNA molecules per sensor, equivalent to 2–200 fmole/cm^2.

Grancharov et al.[180] demonstrated the application of a commercially available magnetic tunnel junction (MTJ) field sensor to detect specific binding of biofunctionalized manganese ferrite (MnFe$_2$O$_4$) nanoparticles. The reported detection scheme does not require modulated AC magnetic fields. This fact, combined with the micrometer size of the tunnel junction, suggests that this detection method could potentially be scaled into a multi-analyte "lab-on-a-chip" detector.

Recently, Liu and Lin[181] reported the development of a renewable electrochemical magnetic immunosensor based on the use of magnetic beads and gold nanoparticle labels. Magnetic beads with an anti-IgG antibody-modified surface were attached to a renewable carbon paste transducer surface by a magnet which was incorporated into the sensor. Using a sandwich immunoassay, gold nanoparticles were attached to the surface of the magnetic beads. The captured gold nanoparticles were detected by highly sensitive electrochemical stripping which offers simple and fast quantification, and avoids the use of an enzyme label and substrate. The stripping signal of gold nanoparticles is a measure for the concentration of target IgG in the sample solution. Such particle-based electrochemical magnetic immunosensors could be readily used for simultaneous parallel detection

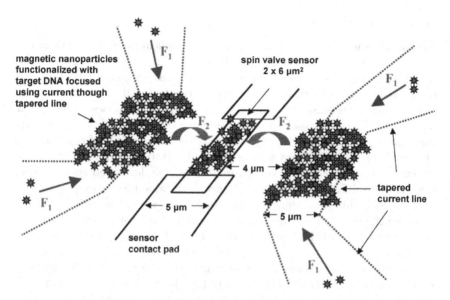

Figure 20.22. A schematic representation of the operational principles of the tapered current line structures with a magnetoresistive spin valve sensor. When 20 mA is passed through both current lines at either side of the sensor (central), a magnetic field gradient is generated resulting in a magnetic force F_1 that attracts the magnetic labels to the narrow regions of the current lines. When a sufficient number of labels has been concentrated on the current lines, the current is switched off and the magnetic labels are then attracted to the nearby sensor by a magnetic force F^2 resulting from the smaller magnetic field generated by the current (8 mA) through the sensor. Thus, high concentrations of magnetically labelled target DNA are rapidly brought into close contact with sensor-bound probe DNA. (From Ref. 179 with permission.)

of multiple proteins by using multiple inorganic metal nanoparticle tracers and are expected to open new opportunities for disease diagnostics and biosecurity.

A piezoelectric immunosensor using magnetic nanoparticles was proposed by Li et al.[182] A goat anti-IgG antibody (IgGAb) as the model analyte was first covalently immobilized to magnetic nanoparticles, which were surface modified with amino groups. The magnetic nanoparticles were attached to the surfaces of quartz crystal with the help of a permanent magnet. The specific binding of the analyte (immunoglobulin G) to the IgGAb-coated magnetic nanoparticles was monitored by measuring the shift of the resonance frequency of the quartz crystal. The piezoelectric immunosensor can determine IgG in the range of 0.6–34.9 μg/ml with a detection limit of 0.36 μg/ml.

20.5.4 Gas sensors

For completeness, we include a review of the application of nanoscale materials to gas detection. Although the detection of gases does not rely on the use of biorecognition elements, nanoscale materials are commonly employed in environmental gas sensors. Semiconductor materials have found wide application as gas sensors, often referred to as electronic noses. Devices that incorporate materials such as tin oxide have been available commercially for some time. Sensors of this type operate on the principle of resistance or electrical conductivity variations as gases adsorb to the surface. Due to the selective nature of this adsorption, sensors can be developed that respond selectively to different gases in a mixture. Typically, the sensing element is made up of a thin film deposited on a substrate. Several investigators have explored the possibility of using nanocrystalline films, or layers of nanoparticles themselves, as the sensing element in the hope of achieving improved sensitivity.

A large variety of nanoscale materials have been examined for application as sensors. Metal oxides constitute a broadly applied family of materials for gas sensing. Tin oxide (SnO_2) has been widely exploited as a suitable semiconductor sensing material, commonly as nanocrystalline films,[35,183] in the form of nanoparticles either by themselves[184] or in conjunction with other nanoparticles such as functionalized copper,[185] or as nanotubes.[186] Tin oxide can be used to detect common air contaminants such as CO, CO_2, NO and NO_2, toxic gases such as H_2S, or ethanol in gasoline. Other nanoscale metal oxides can be used for vapor sensing. Tungsten oxide[187] (WO_3) has been used to detect vapors of H_2S, N_2O and CO; nanostructured thin films of MoO_3 were used by Guidi et al.[188] to detect CO and NO_2; nanocrystalline thin films of n-type $MgFe_2O_4$ were used by Liu et al.[189] to detect vapors of H_2S, CO, liquefied petroleum gas (LPG) and ethanol; α-Fe_2O_3 nanotubes were used by Chen et al.[103] to detect ethanol and H_2.

Non-oxide materials can also be used for vapor detection — metal nanoparticles have been used as vapor sensors by attaching long-chain alkylthiols to the particles. The spacing of the array of nanoparticles that is formed by the chain affects the dielectric constant of the assembly and the electrical conductivity. The network of particles and polymer can accept organic vapor molecules that change the particle spacing. The application of Au nanoparticles to vapor sensing is reviewed by Katz et al.[190]

Baraton and Merhari[184] have explored the use of tin oxide nanoparticles as a sensing element for low-cost, deployable instruments that monitor air quality. The deployment of an array of sensors for air quality monitoring necessitates the development of a technology that is sufficiently cheap and simple to be widely deployed in dispersed arrays. Semiconductor-based gas sensors avoid the difficulties that might be inherent in using optical techniques that require expensive excitation and detection systems. An improvement in the sensitivity and reduction in the cost of production will lead to wider application of sensors of this type.

Baraton and Merhari used laser evaporation of tin oxide starting material to produce nanoparticles of tin oxide with average diameters of about 15 nm; the diameters were measured both directly with TEM and indirectly using BET surface area measurement. The surface reactivity of these nanoparticles is crucial in determining their sensitivity and selectivity to components of a gas mixture that is being analyzed. The authors used FTIR spectroscopy as a screening tool for checking the reactive groups that were attached to the surface of the nanoparticles.

A reduction in the grain size of semiconductor films, or reduction in the size of semiconductor nanoparticles, confers several advantages in the application of these materials as a gas sensor. Ogawa *et al.*[191] showed that as the grain size approaches twice the Debye length of the material, there is a significant improvement in sensitivity towards binding of gas molecules at the surface. For example, tin oxide exhibited a maximum sensitivity with grain sizes of about 6 nm. In addition, the increased curvature of very small particles leads to a greater number of surface defects, providing an increased number of reactive sites for binding of gas molecules. The combination of increased surface area per unit mass with a greater number of reactive sites per surface area to bind yields greater sensitivity for nanoscale materials in gas sensing. The increased sensitivity may also allow the sensor to operate at a lower temperature, thereby reducing power requirements for the instrument.

A Nd : YAG laser was used to evaporate pellets of tin oxide. The powder that was produced was examined with XRD, TEM and FTIR spectroscopy. X-ray diffraction showed that the nanocrystals of tin oxide were formed with average diameters of about 15 nm, in agreement with the TEM images (Figure 20.23).[184] The laser evaporation method provided a production rate that was suitable for the manufacturing of relatively low-cost sensors. Other processes, particularly wet methods that are batch processes, may not be simply scalable to useful levels.

FTIR spectroscopy was used to determine the nature of the reactive groups at the surface of nanoparticles. Absorption bands in the infrared spectrum in the region of 4,000–3,000 cm^{-1} were ascribed to the stretch of OH groups there were bonded to the surface of the particle through exposed tin atoms; bending vibrations of the OH group gave rise to absorption features in the 1,500–1,000 cm^{-1} region of the infrared spectrum. The nature of the tin atom at the surface, particularly its valence state, can affect the

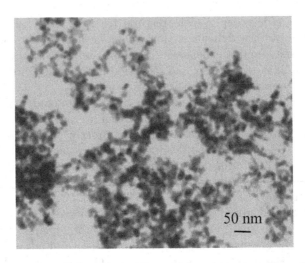

Figure 20.23.　TEM of SnO_2 nanoparticles used for gas sensing. (From Ref. 184 with permission.)

electron distribution in the absorbed group, thereby changing the infrared spectrum, providing a way to ascertain details of the surface state of the nanoparticles. Baraton and Merhari[184] found that infrared spectroscopy was a very useful tool for rapidly screening the sensing capability of different materials as they were synthesized. This offered a relatively high-throughput method for optimizing sensor materials.

The sensitivity of the nanoscale particles to target gases was determined by measuring the infrared absorption spectrum as gases of interest were admitted to a chamber. When carbon monoxide, for example, was admitted to the chamber the absorption spectrum changed as shown in Figure 20.24. The variation in the absorption peaks arises from the release of free carriers at the surface of the nanoparticles due to reactions of the following type:

$$2CO + O_2^- \rightarrow 2CO_2 + e^-$$
$$CO + O^- \rightarrow CO_2 + e^-.$$

(20.2)

Evacuating the chamber and removing the gases regenerated the sensing material. Reversibility of the reactions at the surface is important so that the sensor is reusable.

Screen-printing was used to deposit layers of the nanoparticles on a substrate to fabricate a useful device. The particles were dispersed in a suitable solvent and then applied to the substrate through the screen-printing process. Subsequent treatment was found to be important in providing sufficient sintering of the particles to form an intact nanoscale layer without substantial grain growth; a temperature of about 450°C was found to be suitable. Subsequent layers of nanoparticles were applied in a similar process. The authors optimized the screen-printing process for producing multiple layers of nanoparticles.

The nanoparticle sensors were compared with conventional thin-film semiconductor devices. The nanoparticle sensor response is compared to that of conventional devices in Table 20.1. With the exception of CO, the

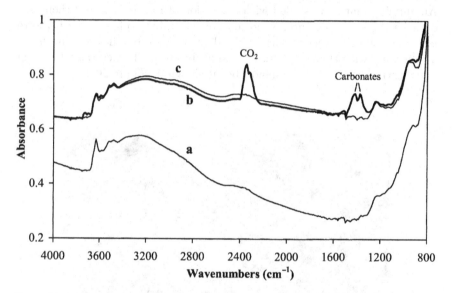

Figure 20.24. Infrared spectrum of the SnO_2 nanoparticles at 300°C: (a) under 50 mbar oxygen; (b, bold line) after addition of 10 mbar CO in the presence of oxygen; (c) after evacuation. (From Ref. 184 with permission.)

Table 20.1. Comparison between the gas detection thresholds of commercial sensors and the gas detection thresholds of nanoparticle SnO_2 sensor prototypes. (From Ref. 184.)

	Typical Detection Limits of Commercial Sensors		
	Semiconductor	Electrochemical	Nanoparticle Sensor
CO	100 ppm	5 ppm	30 ppm
NO_2		600 ppb	500 ppb
No		900 ppb	800 ppb
O_3		200 ppb	200 ppb

nanoparticle sensor was competitive with the electrochemical methods but of course could provide the ease of use that comes with a semiconductor device.

Spheres provide the smallest specific surface area of possible geometries of nanoscale particles, i.e. the smallest area per unit mass. Nanotubes offer much greater surface area per unit mass, and this may provide greater sensitivity when the nanotubes are deployed as gas sensors. Both organic and inorganic nanotubes have been investigated for this application. Chen *et al.*[103] explored the use of inorganic iron oxide nanotubes for gas detection. They used a templating method for the production of the nanotubes. An alumina matrix was filled with a solution of iron nitrate and then processed at high temperature. Nanotubes of the stable α-phase of Fe_2O_3 were produced by this process. They were able to create bundles of nanotubes, 60 μm long, with the template; the nanotubes were characterized by XRD and TEM. An image of the nanotubes is shown in Figure 20.25.

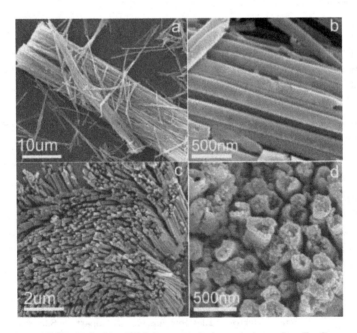

Figure 20.25. Representative SEM images of the as-prepared α-Fe_2O_3 nanotubes: (a) Low-magnification view of the nanotubes; (b) walls of the nanotube bundles; (c) tips of the tube bundles at low magnification; (d) tips of the tube bundles at high magnification. (From Ref. 103 with permission.)

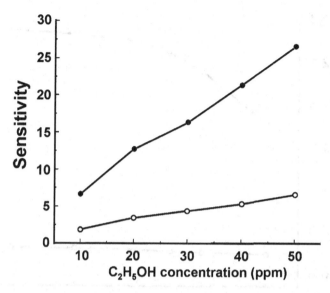

Figure 20.26. Sensor sensitivity to ethanol: nanotubes (filled circles); nanoparticles (empty circles). (From Ref. 103 with permission.)

The performance of the iron oxide nanotube sensor was compared for the detection of C_2H_5OH and H_2 at room temperature with nanoparticles of the same material that were about 200 nm in diameter. The authors measured the change in resistance of the bundle and the nanoparticles when they were exposed to the vapors. The sensitivity is reported as a ratio of resistance in the presence of vapor to the resistance in the absence of the analyte vapor (R_{air}/R_{gas}) where R_{air} is the bundle resistance in the absence of the analyte vapor and R_{gas} is the resistance with the vapor present. The nanotube bundle was about five times more sensitive to ethanol than nanoparticles of the same material (Figure 20.26); the sensitivity to H_2 was much greater than the sensitivity to ethanol (Figure 20.27). Furthermore, they demonstrated that the device could be recycled up to 50 times — reusability and repeatability are necessary properties of a useful gas sensor. The change in geometry from sphere to hollow tube improved the sensitivity and suggested that the tube may be the preferred geometry for gas sensors.

Carbon nanotubes can also be used in gas sensors. Since their first application in field-effect transistors (FETs) in the late 1990s, carbon nanotubes have been found to be sensitive to several gases, including O_2[192] that was able to convert a carbon nanotube from a semiconductor to a metal with

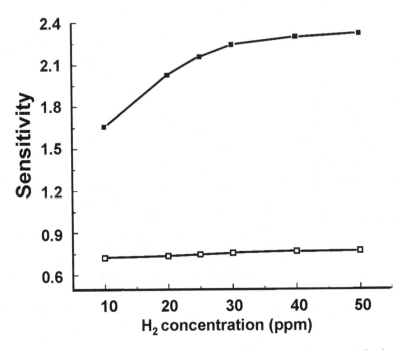

Figure 20.27.   Sensor sensitivity to H_2: nanotubes (filled squares); nanoparticles (empty squares). (From Ref. 103 with permission.)

a concomitant change in resistance. This work highlighted the problems in dealing with carbon nanotubes in air. Kong *et al.*[193] found that other gases such as NO_2 and NH_3 could be detected by carbon nanotubes. Zhang *et al.*[194] found the multi-wall carbon nanotubes were sensitive to the presence of organic solvent vapor.

     The surface of the nanotubes can be functionalized with adducts that provide sensitivity to specific gases. Because a single-wall carbon nanotube behaves as a semiconductor, it can be incorporated into FET devices. Star *et al.*[195] fabricated an FET that incorporated SWCNTs that were fabricated with sequential chemical vapor deposition (CVD) and complementary metal oxide semiconductor (CMOS) processes on an SiO_2 base; the nanotubes served as the conducting source-to-drain channel. The carbon nanotubes were functionalized with a Na^+, K^+ or Ca^{2+} ion-exchanged Nafion polymer that acted as the chemically sensitive layer. Treatment with the Nafion polymer left a thin layer (<10 nm) on the nanotubes as seen by atomic force microscopy (Figure 20.28).

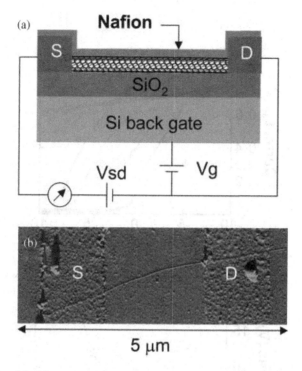

Figure 20.28. (a) Schematic representation of a field-effect transistor device with a carbon nanotube transducer contacted by two Ti/Au electrodes (source and drain) and a silicon back gate. The carbon nanotube conducting channel is exposed to and covered by Nafion. (b) Atomic force microscope (tapping mode) topograph of the actual nanotube device coated with Nafion. (From Ref. 195 with permission.)

Measurements of the electric current between source and drain were made in a flow cell with controllable humidity. The gate voltage was swept from $-10\,\mathrm{V}$ to $+10\,\mathrm{V}$ in the forward direction and from $+10\,\mathrm{V}$ to $-10\,\mathrm{V}$ in the reverse direction. The hysteresis in the source-to-drain current, I_{SD}, that is the difference in the maximum current in the forward gate voltage sweep versus the maximum current in the reverse sweep, is plotted in Figure 20.29 where the sensitivity to humidity is apparent. The effect of gate voltage on the charge-sensitive nanotube structure was found to depend on the relative humidity over the range of $12 \pm 93\%$ RH with millisecond response time. Although Nafion films are used directly in conductance measurements of humidity, the semiconductor device operated through the interaction of the ionic polymer and the underlying semiconductor SWCNT, offering a new mode of measurement using modulation of the gate voltage with

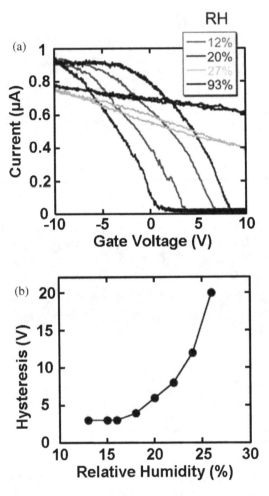

Figure 20.29.   (a) The device characteristics, I_{SD} versus V_G curves (V_{SD} 40 mV), of the NTFET device functionalized with Nafion (Na) membrane measured at different relative humidity (RH) values. (b) Humidity dependence of the reversible hysteresis (forward I_{SD} ± reverse I_{SD}) in the device measured in the range of 20 volts (-10 V to 10 V) at the sweep rate of 4 Hz. (From Ref. 195 with permission.)

the promise of enhanced operating properties. Star *et al.*[195] used a similar FET device for CO_2 detection. In this case, the carbon nanotubes were functionalized with PEI/starch polymers. The response of the sensor to CO_2 concentration was rapid with a wide dynamic range from 500 ppm to 10% in air (Figure 20.30).

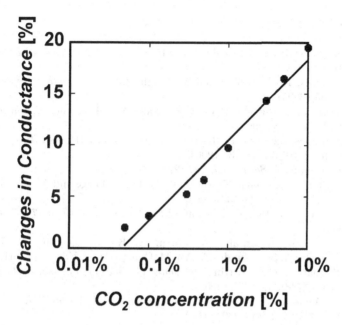

Figure 20.30.   Calibration curve for the CNTFET sensor for CO_2. (From Ref. 195 with permission.)

20.6 Conclusion

Nanoscale materials offer new functionalities for sensing. Although the application of this new class of materials in environmental sensors is in its infancy, significant advances in sensor technology can be expected to flow from the application of carbon nanotubes, inorganic nanotubes, nanoparticles and nanostructured surfaces in sensors. Devices based on these materials are expected to lead to detection devices that are sufficiently cheap and safe to be widely deployed in environmental sensor arrays.

Acknowledgments

This article was made possible in part by the support of NSF Grant DBI-0102662 and the Superfund Basic Research Program with Grant 5P42ES04699 from the National Institute of Environmental Health Sciences, NIH. This material is based partially upon work supported by the Cooperative State Research, Education, and Extension Service, U.S. Department of Agriculture, under Award No. 05-35603-16280.

References

1. J. Seagrave, J. D. McDonald and J. L. Mauderly, *Exp. Toxicol. Pathol.* **57**, 233–238 (2005).
2. C. Sioutas, R. J. Delfino and M. Singh, *Environ. Health Perspect.* **113**, 947–955 (2005).
3. A. Bossi, S. A. Piletsky, P. G. Righetti and A. P. F. Turner, *J. Chromatogr. A* **892**, 143–153 (2000).
4. R. Carabias-Martinez, E. Rodriguez-Gonzalo and P. Revilla-Ruiz, *J. Chromatogr. A* **1056**, 131–138 (2004).
5. S. D. Harvey, *J. Separ. Sci.* **28**, 1221–1230 (2005).
6. S. Piletsky, E. Piletska, K. Karim, G. Foster, C. Legge and A. Turner, *Anal. Chim. Acta* **504**, 123–130 (2004).
7. D. F. Tai, C. Y. Lin, T. Z. Wu and L. K. Chen, *Anal. Chem.* **77**, 5140–5143 (2005).
8. R. Pon, M. A. Batalin, V. Chen, A. Kansal, D. Liu, M. Rahimi, L. Shirachi, A. Somasundra, Y. Yu, M. Hansen, W. J. Kaiser, M. Srivastava, G. Sukhatme and D. Estrin, *Distributed Computing in Sensor Systems, Proceedings* **3560**, 403–405 (2005).
9. G. C. Windham, P. Mitchell, M. Anderson and B. L. Lasley, *Environ. Health Perspect.* **113**, 1285–1290 (2005).
10. K. Andersson, D. Areskoug and E. Hardenborg, *J. Mol. Recogn.* **12**, 310–315 (1999).
11. A. Baeumner, *Food Technol.* **58**, 51–55 (2004).
12. J. R. Chen, Y. Q. Miao, N. Y. He, X. H. Wu and S. J. Li, *Biotechnol. Adv.* **22**, 505–518 (2004).
13. P. Fortina, L. J. Kricka, S. Surrey and P. Grodzinski, *Trends Biotechnol.* **23**, 168–173 (2005).
14. M. J. Friedrich, *JAMA* **293**, 1965–1965 (2005).
15. K. K. Jain, *Expert Rev. Mol. Diagn.* **3**, 153–161 (2003).
16. P. Kohli, M. Wirtz and C. R. Martin, *Electroanalysis* **16**, 9–18 (2004).
17. T. Vo-Dinh, B. M. Cullum and D. L. Stokes, *Sens. Actuators B* **74**, 2–11 (2001).
18. T. Vo-Dinh and P. Kasili, *Anal. Bioanal. Chem.* **382**, 918–925 (2005).
19. L. J. Yang and Y. B. Li, *J. Food Prot.* **68**, 1241–1245 (2005).
20. E. R. Goldman, I. L. Medintz, J. L. Whitley, A. Hayhurst, A. R. Clapp, H. T. Uyeda, J. R. Deschamps, M. E. Lassman and H. Mattoussi, *J. Am. Chem. Soc.* **127**, 6744–6751 (2005).
21. X. J. Ji, J. Y. Zheng, J. M. Xu, V. K. Rastogi, T. C. Cheng, J. J. DeFrank and R. M. Leblanc, *J. Phys. Chem. B* **109**, 3793–3799 (2005).
22. A. R. Clapp, I. L. Medintz, J. M. Mauro, B. R. Fisher, M. G. Bawendi and H. Mattoussi, *J. Am. Chem. Soc.* **126**, 301–310 (2004).
23. C. A. Constantine, K. M. Gattas-Asfura, S. V. Mello, G. Crespo, V. Rastogi, T. C. Cheng, J. J. DeFrank and R. M. Leblanc, *Langmuir* **19**, 9863–9867 (2003).

24. A. D. Sheehan, J. Quinn, S. Daly, P. Dillon and R. O' Kennedy, *Anal. Lett.* **36**, 511–537 (2003).
25. D. J. Maxwell, J. R. Taylor and S. M. Nie, *J. Am. Chem. Soc.* **124**, 9606–9612 (2002).
26. A. P. Alivisatos, W. W. Gu and C. Larabell, *Annu. Rev. Biomed. Eng.* **7**, 55–76 (2005).
27. R. E. Bailey and S. Nie, in *The Chemistry of Nanomaterials: Synthesis, Properties and Applications*, eds. C. N. R. Rao, A. Muller, A. K. Cheetam (Wiley/VCH, Weinheim, 2004), pp. 405–417.
28. D. Gerion, F. Pinaud, S. C. Williams, W. J. Parak, D. Zanchet, S. Weiss and A. P. Alivisatos, *J. Phys. Chem. B* **105**, 8861–8871 (2001).
29. W. C. W. Chan, D. J. Maxwell, X. H. Gao, R. E. Bailey, M. Y. Han and S. M. Nie, *Curr. Opin. Biotech.* **13**, 40–46 (2002).
30. D. L. Stokes, Z. H. Chi and T. Vo-Dinh, *Appl. Spectrosc.* **58**, 292–298 (2004).
31. M. Seydack, *Biosens. Bioelectron.* **20**, 2454–2469 (2005).
32. C. R. Yonzon, C. L. Haynes, X. Y. Zhang, J. T. Walsh and R. P. Van Duyne, *Anal. Chem.* **76**, 78–85 (2004).
33. F. Tam, C. Moran and N. Halas, *J. Phys. Chem. B* **108**, 17290–17294 (2004).
34. L. A. DeLouise, P. M. Kou and B. L. Miller, *Anal. Chem.* **77**, 3222–3230 (2005).
35. J. W. Gong, Q. F. Chen, W. F. Fei and S. Seal, *Sens. Actuators B* **102**, 117–125 (2004).
36. Y. H. Yang, M. M. Guo, M. H. Yang, Z. J. Wang, G. L. Shen and R. Q. Yu, *Int. J. Environ. Anal. Chem.* **85**, 163–175 (2005).
37. L. Rotiroti, L. De Stefano, N. Rendina, L. Moretti, A. M. Rossi and A. Piccolo, *Biosens. Bioelectron.* **20**, 2136–2139 (2005).
38. S. Sotiropoulou and N. A. Chaniotakis, *Anal. Chim. Acta* **530**, 199–204 (2005).
39. D. S. dos Santos, P. J. G. Goulet, N. P. W. Pieczonka, O. N. Oliveira and R. F. Aroca, *Langmuir* **20**, 10273–10277 (2004).
40. E. Harlow and D. Lane, *Antibodies: A Laboratory Manual* (Cold Spring Harbor Laboratory, N.Y., 1988).
41. C. P. Price and D. J. Newman, *Principles and Practice of Immunoassay*, 2nd edn. (Macmillian, Basingstoke, 1997).
42. J. Fotzpatrick, L. Fanning, S. Hearty, P. Leonard, B. M. Manning, J. G. Quinn and R. O'Kennedy, *Anal. Lett.* **33**, 2563–2609 (2000).
43. M. C. Estevez-Alberola and M. P. Marco, *Anal. Bioanal. Chem.* **378**, 563–575 (2004).
44. K. A. Fahnrich, M. Pravda and G. G. Guilbault, *Talanta* **54**, 531–559 (2001).
45. C. R. Suri, M. Raje and G. C. Varshney, *Crit. Rev. Biotechnol.* **22**, 15–32 (2002).
46. B. Hock, M. Seifert and K. Kramer, *Biosens. Bioelectron.* **17**, 239–249 (2002).
47. A. J. Baeumner, *Anal. Bioanal. Chem.* **377**, 434–445 (2003).

48. J. O. Mahony, K. Nolan, M. R. Smyth and B. Mizaikoff, *Anal. Chim. Acta* **534**, 31–39 (2005).

49. A. L. Jenkins, R. Yin and J. L. Jensen, *Analyst* **126**, 798–802 (2001).

50. A. L. Jenkins and S. Y. Bae, *Anal. Chim. Acta* **542**, 32–37 (2005).

51. J. H. Zhai, H. Cui and R. F. Yang, *Biotechnol. Adv.* **15**, 43–58 (1997).

52. M. Campas and I. Katakis, *Trends Anal. Chem.* **23**, 49–62 (2004).

53. F. Lucarelli, L. Authier, G. Bagni, G. Marrazza, T. Baussant, E. Aas and M. Mascini, *Anal. Lett.* **36**, 1887–1901 (2003).

54. Y. Liu and B. Danielsson, *Anal. Chem.* **77**, 2450–2454 (2005).

55. A. L. Simonian, T. A. Good, S. S. Wang and J. R. Wild, *Anal. Chim. Acta* **534**, 69–77 (2005).

56. E. Luzi, M. Minunni, S. Tombelli and M. Mascini, *Trends Anal. Chem.* **22**, 810 818 (2003).

57. V. B. Kandimalla and H. X. Ju, *Anal. Lett.* **37**, 2215–2233 (2004).

58. P. D. Patel, *Trends Anal. Chem.* **21**, 96–115 (2002).

59. C. Ziegler, *Fresen. J. Anal. Chem.* **366**, 552–559 (2000).

60. M. P. Marco and D. Barcelo, *Meas. Sci. Technol.* **7**, 1547–1562 (1996).

61. M. Trojanowicz, *Electroanalysis* **14**, 1311–1328 (2002).

62. H. Schulze, S. Vorlova, F. Villatte, T. T. Bachmann and R. D. Schmid, *Biosens. Bioelectron.* **18**, 201–209 (2003).

63. D. R. Thevenot, K. Toth, R. A. Durst and G. S. Wilson, *Anal. Lett.* **34**, 635–659 (2001).

64. J. H. T. Luong, A. Mulchandani and G. G. Guilbault, *Trends Biotechnol.* **6**, 310–316 (1988).

65. P. Feng, *J. Food Prot.* **55**, 927–934 (1992).

66. P. Buhlmann, E. Pretsch and E. Bakker, *Chem. Rev.* **98**, 1593–1687 (1998).

67. R. Koncki, S. Glab, J. Dziwulska, I. Palchetti and M. Mascini, *Anal. Chim. Acta* **385**, 451–459 (1999).

68. U. J. Krull, *Chemtech* **20**, 372–377 (1990).

69. S. J. Updike and G. P. Hicks, *Nature* **214**, 986 (1967).

70. A. L. Ghindilis, P. Atanasov and E. Wilkins, *Electroanalysis* **9**, 661–674 (1997).

71. L. Habermuller, M. Mosbach and W. Schuhmann, *Fresen. J. Anal. Chem.* **366**, 560–568 (2000).

72. C. K. O'Sullivan, R. Vaughan and G. G. Guilbault, *Anal. Lett.* **32**, 2353–2377 (1999).

73. J. M. Abad, F. Pariente, L. Hernandez and E. Lorenzo, *Anal. Chim. Acta* **368**, 183–189 (1998).

74. K. Bizet, C. Gabrielli and H. Perrot, *Analusis* **27**, 609–616 (1999).

75. D. Ivnitski, I. Abdel-Hamid, P. Atanasov and E. Wilkins, *Biosens. Bioelectron.* **14**, 599–624 (1999).

76. K. Sapsford, C. A. R. Taitt and F. S. Ligler, in *Optical Biosensors*, eds. F. S. Ligler, C. A. R. Taitt (Elsevier, Amsterdam, 2002) pp. 95–121.

77. C. A. R. Taitt and F. S. Ligler, in *Optical Biosensors*, eds. F. S. Ligler, C. A. R. Taitt (Elsevier, Amsterdam, 2002) pp. 57–94.

78. D. P. Campbell and C. J. McCloskey, in *Optical Biosensors*, eds. F. S. Ligler, C. A. R. Taitt (Elsevier, Amsterdam, 2002) pp. 277–304.

79. J. Homola, S. S. Yee and D. Myszka, in *Optical Biosensors*, eds. F. S. Ligler, C. A. R. Taitt (Elsevier, Amsterdam, 2002) pp. 207–251.

80. J. Homola, *Anal. Bioanal. Chem.* **377**, 528–539 (2003).

81. E. F. Schipper, S. Rauchalles, R. P. H. Kooyman, B. Hock and J. Greve, *Anal. Chem.* **70**, 1192–1197 (1998).

82. E. F. Schipper, A. J. H. Bergevoet, R. P. H. Kooyman and J. Greve, *Anal. Chim. Acta* **341**, 171–176 (1997).

83. A. M. Tinsley-Bown, L. T. Canham, M. Hollings, M. H. Anderson, C. L. Reeves, T. I. Cox, S. Nicklin, D. J. Squirrell, E. Perkins, A. Hutchinson, M. J. Sailor and A. Wun, *Phys. Status Solidi A* **182**, 547–553 (2000).

84. C. S. Ryu, S. M. Cho and B. W. Kim, *Biotechnol. Lett.* **23**, 653–659 (2001).

85. R. Cush, J. M. Cronin, W. J. Stewart, C. H. Maule, J. Molloy and N. J. Goddard, *Biosens. Bioelectron.* **8**, 347–353 (1993).

86. M. A. Cooper, *Nat. Rev.* **1**, 515–528 (2002).

87. T. Kinning and E. Edwards, in *Optical Biosensors*, eds. C. A. R. Taitt, F. S. Ligler (Elsevier, Amsterdam, 2002) pp. 253–276.

88. S. Bernad, T. Soulimane and S. Lecomte, *J. Raman Spectrosc.* **35**, 47–54 (2004).

89. F. T. Docherty, P. B. Monaghan, R. Keir, D. Graham, W. E. Smith and J. M. Cooper, *Chem. Commun.* 118–119 (2004).

90. D. Graham, L. Fruk and W. E. Smith, *Analyst* **128**, 692–699 (2003).

91. A. P. Esposito, C. E. Talley, T. Huser, C. W. Hollars, C. M. Schaldach and S. M. Lane, *Appl. Spectrosc.* **57**, 868–871 (2003).

92. J. W. Chan, A. P. Esposito, C. E. Talley, C. W. Hollars, S. M. Lane and T. Huser, *Anal. Chem.* **76**, 599–603 (2004).

93. M. Bruchez, M. Moronne, P. Gin, S. Weiss and A. P. Alivisatos, *Science* **281**, 2013–2016 (1998).

94. S. Iijima, *Nature* **354**, 56–58 (1991).

95. S. Iijima and T. Ichihashi, *Nature* **363**, 603–605 (1993).

96. M. L. Cohen, *Mater. Sci. Eng. C* **15**, 1–11 (2001).

97. J. J. Davis, K. S. Coleman, B. R. Azamian, C. B. Bagshaw and M. L. H. Green, *Chem. Eur. J.* **9**, 3732–3739 (2003).

98. R. H. Baughman, A. A. Zakhidov and W. A. de Heer, *Science* **297**, 787–792 (2002).

99. J. W. Mintmire and C. T. White, *Carbon* **33**, 893–902 (1995).

100. E. W. Wong, P. E. Sheehan and C. M. Lieber, *Science* **277**, 1971–1975 (1997).

101. L. R. Hilliard, X. J. Zhao and W. H. Tan, *Anal. Chim. Acta* **470**, 51–56 (2002).

102. R. Fan, M. Yue, R. Karnik, A. Majumdar and P. D. Yang, *Abstr. Pap. Am. Chem. Soc.* **229**, U141–U142 (2005).

103. J. Chen, L. N. Xu, W. Y. Li and X. L. Gou, *Adv. Mater.* **17**, 582–586 (2005).

104. C. N. R. Rao, A. Govindaraj, G. Gundiah and S. R. C. Vivekchand, *Chem. Eng. Sci.* **59**, 4665–4671 (2004).
105. R. Tenne and C. N. R. Rao, *Phil. Trans. Roy. Soc. Lond. Math. Phys. Sci.* **362**, 2099–2125 (2004).
106. M. Remskar, *Adv. Mater.* **16**, 1497–1504 (2004).
107. R. Tenne, *Chem. Eur. J.* **8**, 5297–5304 (2002).
108. J. Q. Hu, Y. Bando, J. H. Zhan, M. Y. Liao, D. Golberg, X. L. Yuan and T. Sekiguchi, *Appl. Phys. Lett.* **87** (2005).
109. S. M. Liu, L. M. Gan, L. H. Liu, W. D. Zhang and H. C. Zeng, *Chem. Mater.* **14**, 1391–1397 (2002).
110. J. Park, B. Koo, K. Y. Yoon, Y. Hwang, M. Kang, J. G. Park and T. Hyeon, *J. Am. Chem. Soc.* **127**, 8433–8440 (2005).
111. J. Perez-Juste, I. Pastoriza-Santos, L. M. Liz-Marzan and P. Mulvaney, *Coord. Chem. Rev.* **249**, 1870–1901 (2005).
112. R. Tenne, *Angew. Chem. Int. Ed.* **42**, 5124–5132 (2003).
113. F. Zhou, X. M. Zhao, H. G. Zheng, T. Shen and C. M. Tang, *Chem. Lett.* **34**, 1114–1115 (2005).
114. Y. C. Zhang, X. Wu, X. Y. Hu and R. Guo, *J. Cryst. Growth* **280**, 250–254 (2005).
115. K. P. Kalyanikutty, G. Gundiah, C. Edem, A. Govindaraj and C. N. R. Rao, *Chem. Phys. Lett.* **408**, 389–394 (2005).
116. M. C. Daniel and D. Astruc, *Chem. Rev.* **104**, 293–346 (2004).
117. E. P. Diamandis, *Clin. Biochem.* **21**, 139–150 (1988).
118. B. M. Tissue and B. Bihari, *J. Fluoresc.* **8**, 289–294 (1998).
119. M. Safarikova and I. Safarik, *Chem. Listy* **89**, 280–287 (1995).
120. Q. A. Pankhurst, J. Connolly, S. K. Jones and J. Dobson, *J. Phys. D: Appl. Phys.* **36**, R167–R181 (2003).
121. S. H. Sun, H. Zeng, D. B. Robinson, S. Raoux, P. M. Rice, S. X. Wang and G. X. Li, *J. Am. Chem. Soc.* **126**, 273–279 (2004).
122. R. H. Kodama, *J. Magn. Magn. Mater.* **200**, 359–372 (1999).
123. U. Kreibig and L. Genzel, *Surf. Sci.* **156**, 678–700 (1985).
124. W. H. Yang, G. C. Schatz and R. P. Vanduyne, *J. Chem. Phys.* **103**, 869–875 (1995).
125. R. Elghanian, J. J. Storhoff, R. C. Mucic, R. L. Letsinger and C. A. Mirkin, *Science* **277**, 1078–1081 (1997).
126. J. J. Storhoff, R. Elghanian, R. C. Mucic, C. A. Mirkin and R. L. Letsinger, *J. Am. Chem. Soc.* **120**, 1959–1964 (1998).
127. J. J. Storhoff, A. A. Lazarides, R. C. Mucic, C. A. Mirkin, R. L. Letsinger and G. C. Schatz, *J. Am. Chem. Soc.* **122**, 4640–4650 (2000).
128. R. A. Reynolds, C. A. Mirkin and R. L. Letsinger, *J. Am. Chemi. Soc.* **122**, 3795–3796 (2000).
129. R. A. Reynolds, C. A. Mirkin and R. L. Letsinger, *Pure Appl. Chem.* **72**, 229–235 (2000).
130. M. Brasuel, R. Kopelman, M. A. Philbert, J. W. Aylott, H. Clark, J. Sumner, H. Xu, M. Hoyer, T. J. Miller, R. Tjalkens, in *Optical Biosensors*, eds. F. S. Ligler and C. A. R. Taitt (Elsevier, Amsterdam, 2002) pp. 497–536.

131. S. F. Cheng and L. K. Chau, *Anal. Chem.* **75**, 16–21 (2003).
132. N. Nath and A. Chilkoti, *Anal. Chem.* **76**, 5370–5378 (2004).
133. A. J. Haes and R. P. Van Duyne, *J. Am. Chem. Soc.* **124**, 10596–10604 (2002).
134. L. He, M. D. Musick, S. R. Nicewarner, F. G. Salinas, S. J. Benkovic, M. J. Natan and C. D. Keating, *J. Am. Chem. Soc.* **122**, 9071–9077 (2000).
135. J. J. Shi, Y. F. Zhu, X. R. Zhang, W. R. G. Baeyens and A. M. Garcia-Campana, *Trends Anal. Chem.* **23**, 351–360 (2004).
136. W. C. W. Chan and S. M. Nie, *Science* **281**, 2016–2018 (1998).
137. P. T. Tran, E. R. Goldman, G. P. Anderson, J. M. Mauro and H. Mattoussi, *Phys. Status Solidi B* **229**, 427–432 (2002).
138. D. Geho, N. Lahar, P. Gurnani, M. Huebschman, P. Herrmann, V. Espina, A. Shi, J. Wulfkuhle, H. Garner, E. Petricoin, L. A. Liotta and K. P. Rosenblatt, *Bioconjugate Chem.* **16**, 559–566 (2005).
139. R. Tapec, X. J. J. Zhao and W. H. Tan, *J. Nanosci. Nanotechnol.* **2**, 405–409 (2002).
140. J. Feng, G. M. Shan, A. Maquieira, M. E. Koivunen, B. Guo, B. D. Hammock and I. M. Kennedy, *Anal. Chem.* **75**, 5282–5286 (2003).
141. D. Dosev, M. Nichkova, M. Liu, B. Guo, G.-Y. Liu, B. D. Hammock and I. M. Kennedy, *J. Biomed. Opt.* **10**, 064006-064001–064006-064007 (2005).
142. W. O. Gordon, J. A. Carter and B. M. Tissue, *J. Lumin.* **108**, 339–342 (2004).
143. S.-C. Chen, R. Perron, D. Dosev and I. M. Kennedy, *Proceedings of SPIE* **5275**, 186 (2004).
144. M. D. Rubianes and G. A. Rivas, *Electrochem. Commun.* **5**, 689–694 (2003).
145. S. Hrapovic, Y. L. Liu, K. B. Male and J. H. T. Luong, *Anal. Chem.* **76**, 1083–1088 (2004).
146. F. Patolsky, Y. Weizmann and I. Willner, *Angew. Chem. Int. Ed.* **43**, 2113–2117 (2004).
147. J. Wang, *Anal. Chim. Acta* **500**, 247–257 (2003).
148. M. Musameh, J. Wang, A. Merkoci and Y. H. Lin, *Electrochem. Commun.* **4**, 743–746 (2002).
149. S. H. Lim, J. Wei, J. Y. Lin, Q. T. Li and J. KuaYou, *Biosens. Bioelectron.* **20**, 2341–2346 (2005).
150. M. Gao, L. M. Dai and G. G. Wallace, *Electroanalysis* **15**, 1089–1094 (2003).
151. Y. H. Lin, F. Lu and J. Wang, *Electroanalysis* **16**, 145–149 (2004).
152. L. A. Bauer, N. S. Birenbaum and G. J. Meyer, *J. Mater. Chem.* **14**, 517–526 (2004).
153. T. Y. You, O. Niwa, R. Kurita, Y. Iwasaki, K. Hayashi, K. Suzuki and S. Hirono, *Electroanalysis* **16**, 54–59 (2004).
154. F. F. Zhang, Q. Wan, X. L. Wang, Z. D. Sun, Z. Q. Zhu, Y. Z. Xian, L. T. Jin and K. Yamamoto, *J. Electroanal. Chem.* **571**, 133–138 (2004).
155. M. G. Li, Y. J. Shang, Y. C. Gao, G. F. Wang and B. Fang, *Anal. Biochem.* **341**, 52–57 (2005).
156. M. Delvaux, A. Walcarius and S. Demoustier-Champagne, *Anal. Chim. Acta* **525**, 221–230 (2004).

157. J. N. Wohlstadter, J. L. Wilbur, G. B. Sigal, H. A. Biebuyck, M. A. Billadeau, L. W. Dong, A. B. Fischer, S. R. Gudibande, S. H. Jamieson, J. H. Kenten, J. Leginus, J. K. Leland, R. J. Massey and S. J. Wohlstadter, *Adv. Mater.* **15**, 1184-+ (2003).

158. H. Cai, X. N. Cao, Y. Jiang, P. G. He and Y. Z. Fang, *Anal. Bioanal. Chem.* **375**, 287–293 (2003).

159. P. G. He and L. M. Dai, *Chem. Commun.* 348–349 (2004).

160. J. Li, J. E. Koehne, A. M. Cassell, H. Chen, H. T. Ng, Q. Ye, W. Fan, J. Han and M. Meyyappan, *Electroanalysis* **17**, 15–27 (2005).

161. K. Ramanathan, M. A. Bangar, M. Yun, W. Chen, N. V. Myung and A. Mulchandani, *J. Am. Chem. Soc.* **127**, 496–497 (2005).

162. R. J. Chen, S. Bangsaruntip, K. A. Drouvalakis, N. W. S. Kam, M. Shim, Y. M. Li, W. Kim, P. J. Utz and H. J. Dai, *Proc. Nat. Acad. Sci. USA* **100**, 4984–4989 (2003).

163. J. Wang, G. D. Liu and M. R. Jan, *J. Am. Chem. Soc.* **126**, 3010–3011 (2004).

164. J. Wang, D. K. Xu, A. N. Kawde and R. Polsky, *Anal. Chem.* **73**, 5576–5581 (2001).

165. M. Dequaire, C. Degrand and B. Limoges, *Anal. Chem.* **72**, 5521–5528 (2000).

166. J. Wang, R. Polsky and D. K. Xu, *Langmuir* **17**, 5739–5741 (2001).

167. H. Cai, Y. Q. Wang, P. G. He and Y. H. Fang, *Anal. Chim. Acta* **469**, 165–172 (2002).

168. H. S. Guo, J. N. Zhang, P. F. Xiao, L. B. Nie, D. Yang and N. Y. He, *J. Nanosci. Nanotechnol.* **5**, 1240–1244 (2005).

169. J. Wang, A. N. Kawde and M. R. Jan, *Biosens. Bioelectron.* **20**, 995–1000 (2004).

170. Y. Xu, H. Cai, P. G. He and Y. Z. Fang, *Electroanalysis* **16**, 150–155 (2004).

171. H. Cai, N. N. Zhu, Y. Jiang, P. G. He and Y. Z. Fang, *Biosens. Bioelectron.* **18**, 1311–1319 (2003).

172. J. Wang, G. D. Liu, M. R. Jan and Q. Y. Zhu, *Electrochem. Commun.* **5**, 1000–1004 (2003).

173. J. Wang, G. D. Liu and A. Merkoci, *J. Am. Chem. Soc.* **125**, 3214–3215 (2003).

174. G. D. Liu, J. Wang, J. Kim, M. R. Jan and G. E. Collins, *Anal. Chem.* **76**, 7126–7130 (2004).

175. Y. R. Chemla, H. L. Crossman, Y. Poon, R. McDermott, R. Stevens, M. D. Alper and J. Clarke, *Proc. Nat. Acad. Sci. USA* **97**, 14268–14272 (2000).

176. D. R. Baselt, G. U. Lee, K. M. Hansen, L. A. Chrisey and R. J. Colton, *Proc. IEEE* **85**, 672–680 (1997).

177. R. L. Edelstein, C. R. Tamanaha, P. E. Sheehan, M. M. Miller, D. R. Baselt, L. J. Whitman and R. J. Colton, *Biosens. Bioelectronics* **14**, 805–813 (2000).

178. G. V. Kurlyandskaya, M. L. Sanchez, B. Hernando, V. M. Prida, P. Gorria and M. Tejedor, *Appl. Phys. Lett.* **82**, 3053–3055 (2003).

179. D. L. Graham, H. A. Ferreira, N. Feliciano, P. P. Freitas, L. A. Clarke and M. D. Amaral, *Sens. Actuators B* **107**, 936–944 (2005).

180. S. G. Grancharov, H. Zeng, S. H. Sun, S. X. Wang, S. O'Brien, C. B. Murray, J. R. Kirtley and G. A. Held, *J. Phys. Chem. B* **109**, 13030–13035 (2005).
181. G. D. Liu and Y. H. Lin, *J. Nanosci. Nanotechnol.* **5**, 1060–1065 (2005).
182. J. S. Li, X. X. He, Z. Y. Wu, K. M. Wang, G. L. Shen and R. Q. Yu, *Anal. Chim. Acta* **481**, 191–198 (2003).
183. F. Li, L. Y. Chen, Z. Q. Chen, J. Q. Xu, J. M. Zhu and X. Q. Xin, *Mater. Chem. Phys.* **73**, 335–338 (2002).
184. M. I. Baraton and L. Merhari, *J. Nanopart. Res.* **6**, 107–117 (2004).
185. R. S. Niranjan, V. A. Chaudhary, I. S. Mulla and K. Vijayamohanan, *Sens. Actuators B* **85**, 26–32 (2002).
186. Y. Wang, J. Y. Lee and H. C. Zeng, *Chem. Mater.* **17**, 3899–3903 (2005).
187. A. Hoel, L. F. Reyes, P. Heszler, V. Lantto and C. G. Granqvist, *Curr. Appl. Phys.* **4**, 547–553 (2004).
188. V. Guidi, D. Boscarino, L. Casarotto, E. Comini, M. Ferroni, G. Martinelli and G. Sberveglieri, *Sens. Actuators B* **77**, 555–560 (2001).
189. Y. L. Liu, Z. M. Liu, Y. Yang, H. F. Yang, G. L. Shen and R. Q. Yu, *Sens. Actuators B* **107**, 600–604 (2005).
190. E. Katz, I. Willner and J. Wang, *Electroanalysis* **16**, 19–44 (2004).
191. H. Ogawa, M. Nishikawa and A. Abe, *J. Appl. Phys.* **53**, 4448–4455 (1982).
192. P. G. Collins, K. Bradley, M. Ishigami and A. Zettl, *Science* **287**, 1801–1804 (2000).
193. J. Kong, N. R. Franklin, C. W. Zhou, M. G. Chapline, S. Peng, K. J. Cho and H. J. Dai, *Science* **287**, 622–625 (2000).
194. B. Zhang, R. W. Fu, M. Q. Zhang, X. M. Dong, P. L. Lan and J. S. Qiu, *Sens. Actuators B* **109**, 323–328 (2005).
195. A. Star, T. R. Han, V. Joshi, J. C. P. Gabriel and G. Gruner, *Adv. Mater.* **16**, 2049–2052 (2004).

Chapter 21

Carbon Nanotube- and Graphene-based Sensors for Environmental Applications

Dan Du

Key Laboratory of Natural Pesticides and Chemical Biology
Ministry of Education, College of Chemistry, Central China
Normal University, Wuhan, P. R. China

21.1 Introduction

Nanomaterials, particularly carbon nanomaterials, have a significant role to play in new developments in each of the biosensor size domains.[1-3] They have shown great promise in many applications, such as bioscience and biotechnology, energy storage and conversion, and environmental and biomedical applications. There have been many reports of carbon nanomaterial-based electrochemical biosensors for the detection of diverse biological structures such as DNA, viruses, antigens, disease biomarkers, and whole cells.[4] In this chapter, we will focus only on the use of two carbon nanomaterials, carbon nanotubes (CNTs) and graphene, in sensors for environmental applications.

CNTs are well-ordered, hollow graphitic nanomaterials made of cylinders of sp^2-hybridized carbon atoms. They have high aspect ratios, high mechanical strength, high surface areas, excellent chemical and thermal stability, and rich electronic and optical properties.[5] The latter properties make CNTs important transducer materials in biosensors: high conductivity along their length means they are excellent nanoscale electrode materials.[6-8] These materials are classed as single-walled carbon nanotubes (SWCNTs) (Figure 21.1(a)), which are single sheets of graphene rolled into tubes, or multi-walled carbon nanotubes (MWCNTs), each of which contains several concentric tubes that share a common longitudinal axis.[9] As one-dimensional carbon allotropes, CNTs have lengths that can vary from several hundred nanometers to several millimeters, but their diameters

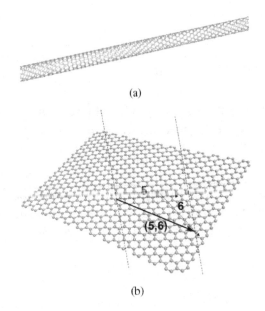

(a)

(b)

Figure 21.1.   The ideal structures of a single-walled carbon nanotube (a) and a graphene sheet (b). (From Ref. 1, reproduced by permission of Wiley-VCH.)

depend on their class: SWCNTs are 0.4–2 nm in diameter and MWCNTs are 2–100 nm in diameter. An important part of the success of CNTs for these applications is their ability to promote electron transfer in electrochemical reactions.

Graphene is a two-dimensional sheet of carbon atoms in a hexagonal configuration with atoms bonded by sp2 bonds. These bonds and this electron configuration are the reasons for the extraordinary properties of graphene, which include a very large surface area (at 2,630 m^2/g, it is double that of SWCNTs), a tunable band gap, room-temperature Hall effect, high mechanical strength (200 times greater than steel), and high elasticity and thermal conductivity.[10] Graphene is the most recent member of the multi-dimensional carbon nanomaterial family, starting with fullerenes as zero-dimensional materials, SWCNTs as one-dimensional nanomaterials, and ending with graphite as a three-dimensional material. Graphene fills the gap for two-dimensional carbon nanomaterials (Figure 21.1(b)). Isolation of individual graphene sheets was long sought, but only in 2004 was it achieved by a surprisingly simple technique.[11] Since then, fundamental research and research on applications have increased rapidly. Graphene is an ideal material for electrochemistry[12–15] because of its very large two-dimensional electrical conductivity, large surface area, and low cost. The use of graphene in electrochemical sensors and biosensors is particularly interesting, with the first articles emerging in 2008. Since then, their number has grown explosively.

21.2 CNT-based Sensors

21.2.1 *CNT solid-phase extraction/electrochemical detection*

Solid-phase extraction (SPE) is the most popular sample preparation method and is very actively used in the field of separation science. Many SPE sorbents have been developed, such as silica-based materials, carbon-based sorbents, ion-pair and ion-exchange sorbents, immunoaffinity extraction sorbents, molecularly imprinted polymers, metal-loaded sorbents, and mixed-mode sorbents.[16-18] The sorptive capacity of conventional carbonaceous sorbents is limited by the density of surface active sites, the activation energy of sorptive bonds, the slow kinetics and non-equilibrium of sorption in heterogeneous systems, and the mass transfer rate to the sorbent surface. The large dimensions of traditional sorbents also limit their transport through low-porosity environments and complicate efforts in subsurface remediation.[19] The CNTs represent a new kind of carbon-based material and is superior to other carbon materials mainly in their special structural feature and unique electronic and mechanical properties.[20] Since their discovery in 1991,[21] extensive interest in CNTs has been shown in the physical, chemical, environmental, and material science fields.[22-26] It is well known that CNTs consist of seamlessly rolled-up graphene sheets of carbon, exhibiting a special sidewall curvature and possessing a π-conjugative structure with a highly hydrophobic surface.[27] These unique properties of the CNTs essentially allow them to interact with some organic compounds, (polynuclear) aromatic compounds, in particular, through π-π electronic and hydrophobic interactions.[28-30] Recently, CNTs are known as ideal SPE sorbents for the separation and detection of organophosphates (OPs) and organic compounds.[31,32] Their porosity and heterogeneity dramatically facilitate the diffusion and solid extraction of analytes. Compared with other different types of mesoporous materials, CNTs are highly conductive and they usually serve as electrocatalytic activity nanomaterials for application in electrochemical detection.[33] Du *et al.* has demonstrated that MWCNTs, serving as selective sorbents, are able to extract nitroaromatic OPs rapidly and effectively.[31] As shown in Figure 21.2, well-defined peaks, proportional to the concentration of methyl parathion (MP), were observed in the range of 0.05–$2.0\,\mu\text{g/mL}$, with a detection limit of $5\,\text{ng/mL}$. The combination of solid-phase extraction to MWCNTs and square-wave

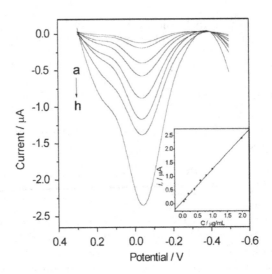

Figure 21.2. Square-wave voltammograms of increasing MP concentrations of 0.05 (a), 0.1 (b), 0.2 (c), 0.4 (d), 0.6 (e), 0.8 (f), 1.0 (g), 2.0 (h) $\mu g/mL$. Inset shows the calibration curve. (From Ref. 31, reproduced by permission of Elsevier.)

voltammetric analysis allows fast, sensitive, and selective determination of MP. Anodic stripping analysis with a very low stripping peak potential avoids the interferences from other electroactive species. MWCNTs are good sorbents for SPE and there are several advantages to MWCNT sorbents. First, the macroporosity and heterogeneity of MWCNTs allow fast extraction of the target analyte. Second, it is possible to control sorption and desorption at MWCNTs by electrochemical methods. Third, both extraction and electrochemical detection at the same MWCNTs is possible. The proposed method is thus expected to open new opportunities for detecting OP pesticides.

Washe *et al.* reported a SWCNT-based potentiometric sensor for the online early detection of aromatic hydrocarbons in coolant water.[34] The potentiometric detection of the uncharged hydrocarbons takes advantage of two features of SWCNTs: (1) high affinity between hydrocarbon and nanotubes; (2) change in the electrical double-layer characteristics of SWCNTs when hydrocarbons are present in an aqueous solution of low constant ionic strength. Compared to existing methods, this sensor is more sensitive, faster, cheaper, and easier to use in online applications. Moreover, it is simple and can be easily miniaturized.

Zhang *et al.* found that the adsorption coverage of anthracene molecules on SWCNTs varied with the aromatic ring substituent.[35] The observed red shifts of emission peaks of the absorptive adduct appear to depend on the energy level of the lowest unoccupied molecular orbital (LUMO) of the adsorbate, consistent with adsorption by a charge-transfer interaction, in which the SWCNTs are the electron donor and anthracene is the acceptor. The anthracene absorptive adducts can be displaced by adsorption of pyrene. CNTs, with their high ratio of surface area to volume, controlled pore size distribution, and heterogeneous surface chemistry, are promising sorbents for various solid-phase extractions.

21.2.2 *Environmental applications of CNT-based sensors*

CNTs with excellent electrical conductivity, high mechanical strength, and good stability have been used extensively in the development of electrochemical biosensors. Incorporation of CNTs onto the electrode surface results in an enhanced electron transfer rate without electrode surface fouling.[36] The electronic properties of CNTs range from metallic to semiconductive, depending on the nanotube's own diameter and chirality.[36-39] In recent years, the promotion or mediation of charge transfer for electroactive species, or electrocatalysis, has received special attention and thus constituted most of the CNT-based chemical sensors or biosensors in environmental applications.[40-43]

21.2.2.1 *Pesticide detection*

OP compounds have been extensively used for pest control due to their high insecticidal activity.[44] The great success in agricultural applications has led to an increase in the production and spread of these pesticides. They exhibit fairly high acute toxicity and can irreversibly inhibit acetylcholinesterase (AChE) that is essential for the functioning of the central nervous system, often causing respiratory paralysis and death.[45] To circumvent these problems the US and European governments have imposed new legislation. Therefore, rapid, sensitive, selective, and reliable quantification of trace levels of OP pesticides is important for health. The highest permissible levels of different pesticides in water for human use go from 0.3 to 400 μg/L. In the last two decades organochlorine insecticides (e.g. DDT, aldrin, and lindane) have been progressively replaced by organophosphorus (e.g. parathion and malathion) and derivatives of carbamic acid (e.g. carbaryl and aldicarb) insecticides that show low persistence in the environment but represent a serious risk due of their high acute toxicity. The electrochemical biosensors used for detecting OPs can be divided into three types depending on the type of enzymes used for constructing biosensors: (1) choline esterase (ChE) and choline oxidase (ChOx) bienzyme-modified biosensors, (2) ChE-modified biosensors, and (3) organophosphorus hydrolase- (OPH-) modified biosensors.

21.2.2.1.1　ChE-ChOx bienzyme biosensors for OPs

A versatile method for pesticide detection is the use of bienzyme biosensors, including AChE and ChOx. The redox potential of acetylcholine (ACh) is too high to be determined directly using electrochemical reaction. For this purpose, AChE and ChOx were immobilized on the surface of the electrode to attain electrochemical analysis according to the following reactions:

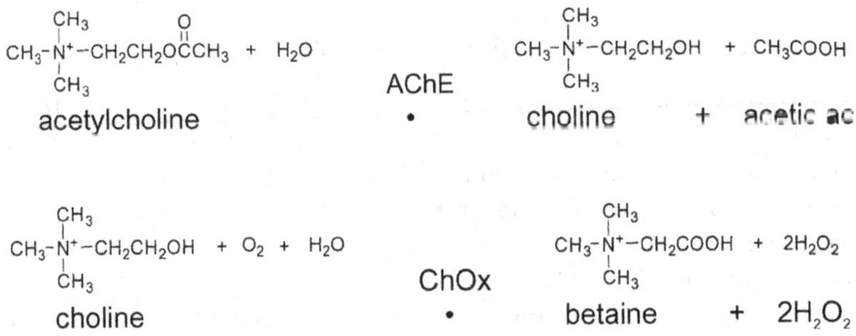

$$\text{acetylcholine} + H_2O \xrightarrow{AChE} \text{choline} + CH_3COOH \; (\text{acetic acid})$$

$$\text{choline} + O_2 + H_2O \xrightarrow{ChOx} \text{betaine} + 2H_2O_2$$

The enzymatic reactions of the AChE-ChOx bienzyme biosensor consist of AChE-catalyzed hydrolysis of ACh into choline and acetic acid and oxidation reaction of choline catalyzed by ChOx. The latter reaction produces H_2O_2, which can be electrochemically oxidized on the electrode to generate an output signal depending on the concentration of acetylcholine. Butyrylcholinesterase (BChE) is sometimes used in place of AChE, in which butyrylcholine is used as substrate. These methods couple an enzyme that is inhibited by the pesticide in conjunction with another enzyme, which uses the product of the first enzymatic reaction as the substrate. Lin *et al.* reported disposable CNT-modified screen-printed biosensors using an AChE/CHO bienzymatic system.[46] It showed a wide dynamic linear range (up to $200\,\mu M$) for detection of methyl parathion, with a detection limit of $50\,nM$. The high sensitivity is attributed to the catalytic activity of the CNTs and the large surface area of the carbon nanomaterials.

　　AChE/ChOx biosensors have also been used for detection of trichlorfon in wastewater.[47] The response of the sensor was evaluated by a successive addition of trichlorfon into a solution of $2\,mM$ ACh. The output current of the sensor decreased with the increasing concentration of trichlorfon over the concentration range of 1×10^{-8} to $2 \times 10^{-5}\,g/ml$. The lower detection limit of the sensor was found to be $1 \times 10^{-9}\,g/ml$ (1 ppb). A key point for constructing high-performance OP sensors in bienzyme systems may be to suitably control the ratio of the catalytic activity of AChE and ChOx in the sensors.

21.2.2.1.2 ChE biosensors for OPs

Bienzyme systems sometimes induce complexity in the design of biosensors originating from different catalytic activity of the enzymes. It is reasonable to assume that, if the reaction products can be oxidized on the electrode directly, one can remove the second enzyme ChOx from the sensor design, and a ChE-modified electrode may be used for detecting OP compounds. This concept uses a ChE-modified electrode as the sensor in the presence of acetylthiocholine in place of acetylcholine as the substrate. Amperometric AChE biosensors based on the inhibition to AChE have shown satisfactory results for pesticide analysis in which the enzymatic activity is employed as an indicator of quantitative measurement of insecticides.[48] When AChE is immobilized on the working electrode surface, its interaction with the substrate of acetylthiocholine generates an electroactive product of thiocholine, which produces an irreversible oxidation peak. The inhibition of OP to AChE is monitored by measuring the oxidation current of thiocholine. The reactions that occurr on the surface of the biosensor are as follows:

$$\text{Acetylthiocholine} + \text{H}_2\text{O} \xrightarrow{\text{AChE}} \text{Thiocholine} + \text{HA (aceticacid)}$$

$$2 \text{ Thiocholine (red)} \rightarrow \text{Thiocholine (ox) (dimeric)} + 2\text{H}^+ + 2\text{e}^-$$

Most reported CNT-based biosensors for detecting OPs are based on immobilization of enzymes on CNT-modified electrodes. These types of sensors take advantage of enzymatic reaction-generated electroactive products, which can be sensitively detected on CNT-modified electrodes. Here, the CNTs on the electrode surface function as both a support for attaching enzymes and a transducer for enhancing electrochemical response. CNTs can promote the electron transfer reaction between the enzyme and the electrode.

Du *et al.* reported a simple and efficient method for immobilization of AChE on an MWCNT-chitosan composite and thus developed a sensitive, fast, and stable amperometric sensor for quantitative determination of triazophos pesticide, a model OP compound.[49] They found that this matrix possessed excellent biocompatibility and good stability, which prevented enzymes from leaving the electrode. MWCNTs promoted electron transfer reactions at a lower potential and catalyzed the electro-oxidation of thiocholine, thus increasing detection sensitivity significantly. As shown in Figure 21.3, upon immersion of AChE-MC/GCE in the standard solution of triazophos at a known concentration for ten minutes, the produced current of ATCl on the AChE-MC/GCE decreased drastically (curve b–d). The response of ATCl at AChEMC/GCE decreased by 34.8% after inhibition with 3.5 μM triazophos. At higher triazophos concentration (28.7 μM), the

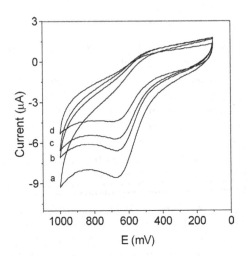

Figure 21.3. Cyclic voltammograms of AChE-MC/GCE in pH 7.0 PBS containing 0.4 mM ATCl after incubation in 0 μM (a), 1.5 μM (b), 3.5 μM (c), and 5.2 μM (d) triazophos solution for 10 minutes. (From Ref. 49, reproduced by permission of Elsevier.)

response decreased by up to 80%. The decrease in peak current increased with the increasing concentration of triazophos. This was because triazophos as one of the OP pesticides exhibited fairly high acute toxicity and was involved in the irreversible inhibition action to AChE, thus reducing the enzymatic activity to its substrate. Under optimal conditions, the inhibition of triazophos was proportional to its concentration in two ranges, from 0.03 to 7.8 μM and from 7.8 to 32 μM, with a detection limit of 0.01 μM.

Based on the change in electrochemical behavior of enzymatic activity induced by pesticide, Du and co-workers[50] further developed a novel electrochemical method for investigating pesticide sensitivity using an AChE biosensor based on an MWCNT-chitosan nanocomposite. Four pesticides (carbaryl, malathion, dimethoate, and monocrotophos) were selected to discuss their inhibition efficiencies to AChE. Figure 21.4(a) illustrates the inhibition curves of these four pesticides. All four pesticides displayed increasing inhibition on AChE with increased immersion time. However, the increase rate of carbaryl (curve a) was much faster than those of malathion, dimethoate, and monocrotophos (curves b, c, and d). For carbaryl, when the immersion time was longer than ten minutes, the peak current trended to a stable value, indicating substrate binding interaction with the active groups of the enzyme reached saturation. However the inhibitions of dimethoate or monocrotophos became stable after 15 minutes. A longer time was observed for monocrotophos to achieve saturated inhibition — after 20 minutes. It is suspected that steric hindrance of different pesticides prevented the

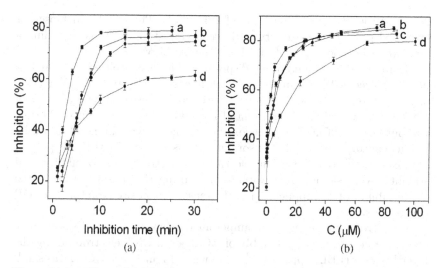

Figure 21.4. Effect of inhibition time on pesticide inhibition after immersion of AChEMC/GCE in 20.0 μM pesticide solutions (a) and inhibition curves (b). Carbaryl (a), malathion (b), dimethoate (c), and monocrotophos (d). (From Ref. 50, reproduced by permission of Elsevier.)

complete inhibition of pesticides to AChE, which resulted in the different inhibition rates. As shown in Figure 21.4(b), their inhibition curves tended towards maximum values at high pesticide concentrations, indicating that interactions with active target groups reached saturation as the pesticide concentration increased, the point at which maximum inhibition rate occurred. The inhibition curves were similar to Michaelis–Menten curves and the Michaelis–Menten constant (K_m) was calculated to be 0.96 μM, 1.78 μM, 1.97 μM, and 4.28 μM, respectively. The inhibition rate was related to the interaction between pesticide and enzyme.

The same group proposed a simple method to immobilize AChE on polypyrrole (PPy) and polyaniline (PANI) copolymer doped with MWCNTs.[51] The synthesized PAn-PPy-MWCNT copolymer presented a porous and homogeneous morphology which provided an ideal size to entrap enzyme molecules. Due to the biocompatible microenvironment provided by the copolymer network, the obtained composite was devised for AChE attachment, resulting in a stable AChE biosensor for screening OP exposure. MWCNTs promoted electron transfer reactions at a lower potential and catalyzed the electro-oxidation of thiocholine, thus increasing detection sensitivity. Based on the inhibition of OPs on the AChE activity, using malathion as a model compound, the inhibition of malathion was proportional to its concentration ranging from 0.01 to 0.5 μg/mL and from 1 to

25 μg/mL, with a detection limit of 1.0 ng/mL. The developed biosensor exhibited good reproducibility and acceptable stability, thus providing a new promising tool for analysis of enzyme inhibitors.

Joshi *et al.* proposed a disposable biosensor based on AChE-functionalized MWCNT-modified thick-film strip electrodes for detection of nM paraoxon.[52] The large surface area and electrocatalytic activity of MWCNTs lowered the overpotential for thiocholine oxidation to 200 mV without the use of mediating redox species and enzyme immobilization by physical adsorption. The biosensor detected as low as 0.5 nM (0.145 ppb) of paraoxon with good precision, electrode-to-electrode reproducibility, and stability. Analysis of a real water sample using the sensor demonstrated the feasibility of the application of the sensor for on-site monitoring of OP compounds.

Lin's group has reported an amperometric biosensor for OPs and nerve agents based on the self-assembly of AChE on a CNT electrode using the layer-by-layer (LBL) approach,[53] as shown in Figure 21.5. AChE was sandwiched with poly(diallyldimethylammonium chloride) (PDDA) layers on the surface of CNTs. The electrocatalytic activity of CNTs leads to a greatly improved electrochemical detection of the enzymatically generated thiocholine product, including a low oxidation overpotential (+150 mV), higher sensitivity, and stability. The biosensor was used to measure as low as 0.4 pM paraoxon with a 6 minute inhibition time under optimal conditions.

21.2.2.1.3 OPH biosensors for OPs

In contrast to the inhibition mode of detection in the biosensors mentioned, an alternative direct biosensing route for the detection of OPs has been proposed by the use of OPH, which can be used for continuous monitoring of OPs in the environment. The foundation of these devices is the hydrolysis of OPs, producing two protons as a result of the cleavage of the P-O, P-F, P-S, or P-CN bonds and an alcohol.[54,55] Electrochemical biosensors in particular have been widely investigated to monitor various pesticides including OP compounds such as paraoxon,[56] sarin,[57] and soman[58] via an enzyme-catalyzed hydrolysis reaction by OPH due to their fast speed, high efficiency, low cost, and small sample size.

Recently, Deo and co-workers[59] described an amperometric biosensor for OP pesticides based on a CNT-modified transducer and OPH biocatalyst. A bilayer approach with the OPH layer at the top of the CNT film was used to prepare the CNT/OPH biosensor. The CNT layer leads to a greatly improved anodic detection of the enzymatically generated *p*-nitrophenol product, including higher sensitivity and stability.

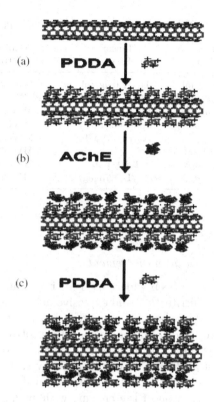

Figure 21.5.   Schematics of layer-by-layer electrostatic self-assembly of AChE on carbon nanotube: (a) assembling positively charged PDDA on negatively charged CNT; (b) assembling negatively charged AChE; (c) assembling the second PDDA layer. (From Ref. 53, reproduced by permission of the American Chemical Society.)

Under optimal conditions, the biosensor was used to measure levels as low as $0.15\,\mu M$ paraoxon and $0.8\,\mu M$ methyl parathion with sensitivities of $25\,nA/\mu M$ and $6\,nA/\mu M$, respectively.

Wang *et al.* prepared OPH/CNT as a submersible biosensor for remote monitoring of OP compounds.[60] The biosensor mounted in a plastic tube was connected to a 50-foot-long, shielding cable via environmentally sealed rubber connectors. The same group recently reported an interesting electrochemical/optical route for detection of OP nerve agents based on OPH-mediated bio-metallization.[61] The widespread interest in OPH-based electrochemical biosensors stems from their simplicity, directness, and speed for determining OPs. The detection limit of OPH-based biosensors could be further improved by either lowering the enzyme Michaelis constant (K_m) or increasing the bimolecular rate constant.

Table 21.1.   CNT-based sensors for environmental applications.

Analytes	Detection Limit	Linear Range	Modified Electrode	Reference
Methyl parathion	5 ng/ml	0.05–2.0 μg/ml	MWCNTs	31
Triazophos	0.01 μM	0.03–32 μM	MWCNT-chitosan	49
Malathion	1 ng/ml	0.01–25 μg/ml	PAn-PPy-MWCNT	51
Paraoxon	0.5 nM	0.5–6.9 nM	MWCNTs	52
Paraoxon	0.4 pM	1.0 pM–10nM	MWCNTs	53
Nitrite	0.1 μM		MWCNTs	64
Arsenite	1.0 ppb	1–500 ppb	MWCNTs	66
Triclosan	16.5 μg/L	50 μg/L–1.75 mg/L	MWCNTs	67

Table 21.1 shows the quantitative detection of environmental pollutants based on CNT sensors.

21.2.2.2 Inorganic toxicant measurement

CNT-based sensors have been widely used for inorganic toxicants, such as Cd(II) and Pb(II),[62] nitrite,[63,64] and explosive detection.[65] Recently, Male et al.[66] developed a biosensor for arsenite using molybdenum-containing arsenite oxidase, prepared from the chemolithoautotroph NT-26 that oxidizes arsenite to arsenate. The enzyme was galvanostatically deposited on the active surface of an MWCNT-modified glassy carbon (GC) electrode. The resulting biosensor enabled direct electron transfer, i.e. effecting reduction and then re-oxidization of the enzyme without an artificial electron-transfer mediator. Arsenite was detected within 10 s at an applied potential of 0.3 V with linearity up to 500 ppb and a detection limit of 1 ppb. Copper, a severe interfering species commonly found in groundwater, did not interfere, and the biosensor was applicable for repeated analysis of spiked arsenite in tap water, river water, and a commercial mineral water.

Yang and co-workers developed an electrochemical sensor based on MWCNTs for rapid detection of triclosan.[67] The electrochemical responses of triclosan were examined, given that its oxidation is irreversible and involves one electron. At the MWCNT electrode, the oxidation signals of triclosan remarkably increased, suggesting that the MWCNTs exhibited a considerable enhancement effect with triclosan. The detection limit was 16.5 μg/L (about 57 nM).

Arribas et al. reported a highly stable and sensitive method for detection of 2,4-dichlorophenol using CNT-modified glassy carbon electrodes in flow injection systems for the stable long-term measurements demanded by industrial activities.[68]

Chromium(VI) is an environmental hazard that exists in nature and in industrial waste. Monitoring and remedy of chromium thus possess environmental significance. Jiang et al.[69] observed an obvious catalytic reduction of Cr(VI) on a PPy-CNT composite in strong acidic media. Dichromate ions exhibited sluggish reduction on the bare GC. With the CNT-based PPy, the reduction potential of dichromate shifted positively by 200–600 mVs, depending on the current maxima of Cr(VI) reduction on the bare GC. The catalytic current demonstrated linearity with Cr(VI). The mechanistic study revealed that the cross-reaction between the reduced polymer composite and Cr(VI) was accompanied by the permeation of Cr(VI) into the microporous composite network.

21.2.2.3 *Gas sensing*

The high surface-to-volume ratio, hollow geometry, and chemical inertness make CNTs potentially attractive for gas molecule adsorption. These may supplement the limitations of conventional metal oxide- or polymer-based gas sensors, in terms of sensitivity and small sizes demanded for miniaturization and construction of massive sensor arrays.[70,73] So far there have been reports regarding CNT-based devices for NH_3, H_2, N_2, He, CO_2, CO, O_2, NO_2, humidity, ethanol, methanol, and HCl vapors.[74–80] By using lithographic or non-lithographic methods, these sensors have been based on isolated SWCNTs,[81,82] SWCNT mats,[83] and most recently, MWCNTs.[80] The latter present complementary features with respect to SWCNTs, especially for their low fabrication cost since they can grow at large scales. The tendency of changed electronic properties of CNTs upon ambient gas adsorption was initially reported by Dai's research group when studying the SWCNT-based field-effect transistor (FET).[81] They found the conductance of a semiconducting SWCNT sample decreased approximately 100 times after exposure to 0.1–1.0% NH_3 for about 10 minutes. Additionally, the conductance of the same device increased 1,000 times upon being exposed to 2–200 ppm NO_2. Later results following this discovery revealed that the electrical resistance and the thermoelectric power of SWCNT bundles are also susceptible to the gas adsorbates. Up to now, this adsorption-changed electronic properties relationship constitutes the major transducing mechanism among the reported gas sensors.

While most of the aforementioned sensors were operated under elevated temperatures to explore adsorption/desorption behavior for different gases, an integration or sensor array might be promising to probe gaseous mixtures. The ultimate goal down this route is to build a prototype of an electronic nanonose. Grimes' group[84] used MWCNTs as sensing components to develop two sensor designs: the first one is a planar interdigital capacitor upon which an MWCNT-SiO_2 composite film is placed, and the second sensor design is an MWCNT serpentine resistor fabricated by photolithographically defining a serpentine SiO_2 path upon silicon with growing MWCNTs. These capacitor- or resistor-based sensors showed very high sensitivity to carbon monoxide, carbon dioxide, and water vapors (reversible) as well as ammonia (irreversible). The chemisorptions of gases on the surface of the semiconducting MWCNTs were found to be responsible for the sensing action. The impedance changes are attributed to p-type conductivity in semiconducting MWCNTs, and the formation of Schottky barriers between the metallic and semiconducting nanotubes. More recently, a similar design by the same group and others for "wireless" or remote sensing has been reported to address corrosive gases or samples in sealed environment.[85,86]

Miniaturization of the gas sensor benefits a variety of applications, such as online process monitoring and gas detection for counter-terrorism. Modi and co-authors recently reported a miniaturized gas ionization nanotube sensor based on aligned MWCNT film.[87] Ionization sensors work by fingerprinting the ionization characteristics of distinct gases, and can therefore overcome some disadvantages of the dominant conductance-based gas sensors, including an inability to identify gases with low adsorption energies, poor diffusion kinetics, or poor charge transfer with nanotubes.[88] Since the conventional ionization gas detectors, such as photo-ionization detectors (PIDs), flame-ionization detectors (FIDs), or electron-capture detectors (ECDs), possess bulky architecture, high power consumption, and risky high-voltage operation, they are not suitable for direct application to gas mixtures and operate in conjunction with a gas chromatographer that separates the mixture into distinct bands that can then be qualitatively and quantitatively analyzed. The proposed miniaturized ionization sensor can be used to monitor gas mixtures without the direct use of a chromatographic setting.[89] This prototype design features the electrical breakdown of a range of gases and gas mixtures at carbon nanotube tips. The sharp tips of nanotubes generate very high electric fields at relatively low voltages, lowering breakdown voltages several-fold in comparison to traditional electrodes, and thereby enabling compact, battery-powered, and safe operation of such sensors. The sensors exhibit good sensitivity and selectivity towards ammonia, oxygen, nitrogen, and helium gases, and are unaffected by extraneous factors such as temperature, humidity, and gas flow. These devices offer several practical advantages over previously reported nanotube sensor systems, which can work independently for environmental monitoring as well as replacing the current bulky FIDs, ECDs, and PIDs in a chromatographic set-up.

21.2.3 Other CNT-based nanocomposites for environmental applications

Recently, functionalized CNTs have been of great interest because of their potential applications in creating next-generation electronic devices and networks. To optimize the potential applications of CNTs, it is essential to modify the CNTs with functional groups to integrate CNTs into the desired structures or attach suitable nanostructures on the nanotubes.[89–93] Metal-nanotube interactions are especially attractive for the development of low-resistance ohmic contacts to these structures. The attachment of gold nanoparticles (NPs) to the sidewalls of CNTs is particularly promising for novel, highly efficient photoelectrochemical cells and sensor devices. The CNTs with gold or platinum NPs attached are primarily based on the self-assembly of NPs on the prefunctioned CNTs by polyelectrolytes.[94] A variety of composites have since been reported for their potential application as photovoltaic device[95] and charge storage.[96] Among these studies, conducting polymers (CPs) received special attention because of their conductive, processible, and compatible properties towards the construction of solid devices. So far, chemically synthesized CNT composites with poly-p-phenylenevinylene,[95] poly-p-phenylenevinylene and polymethylmethacrylate,[96] polypyrrole,[97,98] poly(3,4-ethylenedioxythiophene),[99] and polyaniline[100,101] have been developed in aqueous as well as non-aqueous media.

Mubeen *et al.* demonstrated the fabrication of a hydrogen nanosensor based on electrodeposited Pd NPs on SWCNT networks.[102] SWCNT networks were selected over a single SWCNT sensor, because SWCNT networks have a higher signal-to-noise ratio and ensure reproducible sensor performance. The resulting SWCNT networks showed improved properties for hydrogen sensing.

ZrO_2-CNT composites have been successfully synthesized via decomposition of $Zr(NO_3)_4 \cdot 5H_2O$ in supercritical carbon dioxide-ethanol solution with dispersed CNTs at relatively low temperatures.[103] It was demonstrated that CNTs were fully coated with an amorphous ZrO_2 layer, and the coating layer was nominally complete and uniform. In addition, the thickness of the coating sheath could be readily controlled by tuning the $Zr(NO_3)_4 \cdot 5H_2O$/CNTs ratio used. The chemiluminescent sensor prepared from ZrO_2-carbon nanotube composites exhibited dramatic sensitivity as well as high stability and selectivity to ethanol.

Recently, a highly sensitive and stable $Ru(bpy)_3^{2+}$ ECL sensor was developed for detection of alkylamine compounds using the Pt-CNT-zirconia-Nafion composite film.[104] Incorporation of CNTs and Pt NPs within the zirconia-Nafion composite films increased the ECL response because of the decreased charge transfer resistance in the composite-modified electrode. The present ECL sensor based on the Pt-CNT-zirconia-Nafion films exhibited a detection limit one order of magnitude lower for TPA compared to those based on zirconia-Nafion films without CNTs and Pt NPs. Given the aforementioned benefits, the use of the Pt-CNT-zirconia-Nafion composite provided a useful means for the preparation of not only the present $Ru(bpy)_3^{2+}$ ECL sensor but also the ECL biosensors based on dehydrogenase enzymes because of the excellent electrocatalytic activity of CNTs and Pt NPs towards nicotinamide adenine dinucleotide (NADH) and good biocompatibility of the zirconia-Nafion composite.

21.3 Graphene-based Sensors

Graphene, together with its various derivatives, such as graphene oxide, graphene nanoribbon, chemically reduced graphene oxide, or nitrogen doped graphene,[105-109] has shown fascinating advantages in electrochemistry such as electrochemical devices, capacitors, or transistors due to its remarkable electrochemical properties.[110-113] This area is particularly interesting, with many articles emerging since 2008 focusing on biomolecules. Here we review the use of graphene sensors in environmental analysis, which is a new application of graphene-based sensors. Table 21.2 shows the quantitative detection of environmental pollutants based on graphene sensors.

One example of the use of graphene sensors in environmental analysis is the detection of pesticides. Recently, Lin's group fabricated an electrochemical biosensor based on gold nanoparticles (Au NPs) and chemically reduced graphene oxide nanosheets (cr-Gs) for determination of paraoxon.[114] As shown in Figure 21.6, a nanohybrid of Au NPs and cr-Gs was synthesized by in situ growth of Au NPs on the surface of graphene nanosheets. Then an enzyme nanoassembly (AChE/Au NPs/cr-Gs) was prepared by self-assembling AChE on an Au NP/cr-Gs nanohybrid. By using PDDA as a linker, AChE could be enriched and stabilized on the nanohybrid. Self-assembling was easy and fast to be realized between negatively charged AChE and positively charged PDDA. An electrochemical sensor based on AChE/Au NPs/cr-Gs was further developed for ultrasensitive detection of OP pesticides.

Figure 21.7 is a typical i–t curve acquired at AChE/Au NPs/cr-Gs incubated with different concentrations of paraoxon. Signal (a) is from AChE/Au NPs/cr-Gs for 2 mM ATCh, signals (b) to (h) are generated

Table 21.2.   Graphene-based sensors for environmental applications.

Analytes	Detection Limit	Linear Range	Modified Electrode	Reference
Paraoxon	0.1 pM	0.1 pM–5 nM	Au-graphene	114
Lead	0.02 $\mu g/l$	0.5–50 $\mu g/l$	Nafion-graphene	115
Cadmium	0.005 $\mu g/l$	0.2–15 $\mu g/l$	Nafion-graphene	116
Hydrazine	0.0002 μM	0.001–0.05 μM	Graphene	117

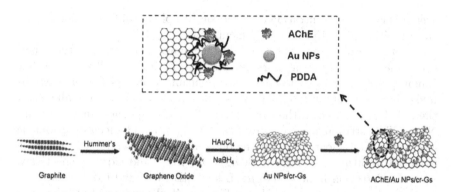

Figure 21.6. Schematic illustration of Au NP/cr-Gs hybrid synthesis and AChE/Au NP/cr-Gs nanoassembly generation by using PDDA. Graphite was used for producing graphene oxide with Hummers' method and Au NPs/cr-Gs was obtained by reducing HAuCl$_4$ on graphene oxide. AChE was stabilized on the surface of the Au NPs/cr-Gs hybrid by self-assembling. (From Ref. 114, reproduced by permission of The Royal Society of Chemistry.)

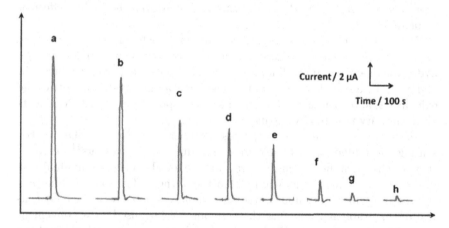

Figure 21.7. Typical amperometric responses of AChE/Au NPs/cr-Gs-based SPEs to paraoxon by using a flow injection analysis system. Signal (a) is from 2 mM ATCh, signals (b) to (h) are from 2 mM ATCh after the AChE/Au NPs/cr-Gs-based SPEs were incubated with paraoxon for 15 minutes with concentrations of 0.1 pM, 0.25 pM, 2.5 pM, 25 pM, 0.25 nM, 2.5 nM, and 5 nM, respectively. The data of amperometric responses are collected and shown together. Working potential: 0.4 V. (From Ref. 114, reproduced by permission of The Royal Society of Chemistry.)

from 2 mM ATCh after the AChE/Au NPs/cr-Gs incubated with different concentrations of paraoxon: 0.1 pM, 0.25 pM, 2.5 pM, 25 pM, 0.25 nM, 2.5 nM, and 5 nM paraoxon for 15 minutes, respectively. Catalysis activity of AChE to ATCh is inhibited apparently after exposure to paraoxon, and even a very low concentration of paraoxon (0.1 pM) could cause the apparent inhibition after 15 minutes of incubation.

Another application of graphene in the environmental field is the detection of heavy metal ions and organic toxicant. Li *et al.*[115,116] reported that Nafion-graphene composite-based electrochemical sensors not only exhibited enhanced sensitivity to metal ions (Pb^{2+} and Cd^{2+}), but also alleviated the interferences due to the synergistic effect of graphene nanosheets and Nafion. The stripping current response was greatly enhanced on graphene electrodes. The linear range for the detection of Pb^{2+} and Cd^{2+} was wide from 0.5 to $50\,\mu g/L$ and 1.5 to $30\,\mu g/L$ for Pb^{2+} and Cd^{2+}, respectively, with the detection limit of $0.02\,\mu g/L$ for both Pb^{2+} and Cd^{2+}, which are more sensitive than those of Nafion-film-modified bismuth electrodes, and ordered mesoporous carbon-coated GCEs, and comparable to Nafion/CNT-coated bismuth film electrodes. The enhanced performance was attributed to the unique properties of the graphene (nanosized graphene sheet, nanoscale thickness of these sheets, and high conductivity), which endowed it with the capability to strongly adsorb target ions, enhanced the surface concentration, improved the sensitivity, and alleviated the fouling effect of surfactants.

Wang and co-workers described the electrochemical properties of reduced graphene sheets (RGSs) for the electrocatalytic properties towards hydrazine oxidation in alkaline media.[117] The RGSs have been produced in high yield by a soft chemistry route involving graphite oxidation, ultrasonic exfoliation, and chemical reduction. The RGSs possess excellent electrocatalytic activity towards hydrazine oxidation (Figure 21.8).

In addition to the above analytes, graphene has been applied in the sensing of individual gas molecules. Schedin and co-workers[118] demonstrated that graphene sensors are capable of detecting individual gas molecules attaching or detaching from the surface of graphene. The process of molecules attaching/detaching changes the local carrier concentration in graphene, allowing the resistance to be used as a convenient means of measurement. Other research work demonstrated that graphene oxide sheets deposited on gold interdigitated electrodes allow the sensing of NO_2, NH_3, and also 2,4-dinitrotoluene via resistance measurements.[119,120] Duan's group[121] studied the adsorption of gas molecules (CO, NO, NO_2, O_2, N_2, CO_2, and NH_3) on graphene nanoribbons (GNRs). The adsorption geometries, adsorption energies, charge transfer, and electronic band structures are obtained. They found that the electronic and transport properties of the GNR with armchair-shaped edges are sensitive to the adsorption of NH_3, and the system exhibits n-type semiconducting behavior after NH_3 adsorption. Other gas molecules have little effect on modifying the conductance of GNRs. Quantum transport calculations further indicate that NH_3

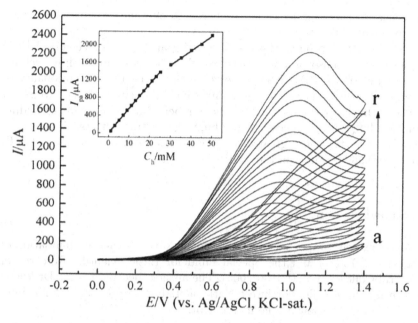

Figure 21.8. Cyclic voltammograms at the RGSs/GC electrode in 0.1 M KOH solution containing an increasing concentration of hydrazine from (a) 1, (b) 3, (c) 5, (d) 7, (e) 9, (f) 11, (g) 13, (h) 15, (i) 17, (j) 19, (k) 21, (l) 23, (m) 25, (n) 30, (o) 35, (p) 40, (q) 45, and (r) 50 mM. Potential scan rate is $0.1\,V\,s^{-1}$. Also shown as an inset is the plot of I_{pa} versus the concentration of hydrazine. (From Ref. 117, reproduced by permission of Elsevier.)

molecules can be detected among these gas molecules by the GNR-based sensor.

21.4 Conclusions and Future Work

This chapter described recent developments of CNT- and graphene-based sensors for environmental applications. In conclusion, it is clear that CNTs and graphene offer many advantages as sensing materials because of their biocompatibility, high conductivity, and abundance of inexpensive source material. Carbon-based sensors are not only relatively cheap to produce, they are also lightweight and compact and hence we expect to see commercial developments in "compact carbon chemistry". We predict a bright future for carbon nanomaterial-based electrochemical sensors in environmental analysis because these sensors can be easily prepared as portable devices for in-field detection. The easy operation of carbon-based biosensors is another merit as compared to chromatographic or spectroscopic measurements.

Despite their promise, both CNTs and graphene still face considerable challenges for sensor applications. Probably the most severe drawback for the application of SWCNTs is heterogeneity. The two-dimensionality of graphene brings other challenges, which include separating the layers and keeping them separated to control the number of layers; minimizing folding and bending during processing; and limiting substrate effects. Clearly, analytical chemists have a great deal to do to address the behavior of CNTs and graphene properly in the future.

Acknowledgments

This work was supported by the National Natural Science Foundation of China (21075047) and the Special Fund for Basic Scientific Research of Central Colleges (CCNU10A02005). The author specially thanks Dr Yuehe Lin from the Pacific Northwest National Laboratory (PNNL) for helpful discussion and reviewing the manuscript.

References

1. W. Yang, K. R. Ratinac, S. P. Ringer, P. Thordarson, J. Justin Gooding and F. Braet, *Angew. Chem. Int. Ed.* **49**, 2114–2138 (2010).
2. M. Pumera, A. Ambrosi, A. Bonanni, E. L. K. Chng and H. Poh, *Trend. Anal. Chem.* **29**, 954–965 (2010).
3. Y. Shao, J. Wang, H. Wu, J. Liu, I. A. Aksay and Y. Lin, *Electroanal.* **22**, 1027–1036 (2010).
4. R. L. McCreery, *Chem. Rev.* **108**, 2646–2687 (2008).
5. P. M. Ajayan, *Chem. Rev.* **99**, 1787–1799 (1999).
6. I. Heller, J. Kong, H. A. Heering, K. A. Williams, S. G. Lemay and C. Dekker, *Nano Lett.* **5**, 137–142 (2005).
7. D. Krapf, B. M. Quinn, M. Y. Wu, H. W. Zandbergen, C. Dekker and S. G. Lemay, *Nano Lett.* **6**, 2531–2535 (2006).
8. J. Justin Gooding, A. Chou, J. Q. Liu, D. Losic, J. G. Shapter and D. B. Hibbert, *Electrochem. Commun.* **9**, 1677–1683 (2007).
9. M. C. Hersam, *Nat. Nanotechnol.* **3**, 387–394 (2008).
10. A. K. Geim and K. S. Novoselov, *Nat. Mater.* **6**, 183–191 (2007).
11. K. S. Novoselov, A. K. Geim, S. V. Morozov, D. Jiang, Y. Zhang, S. V. Dubonos, I. V. Grigorieva and A. A. Firsov, *Science* **306**, 666–669 (2004).
12. M. Pumera, *Chem. Rec.* **9**, 211–223 (2009).
13. M. Liang, L. Zhi, *J. Mater. Chem.* **19**, 5871–5878 (2009).

14. T. Seyller, A. Bostwick, K. V. Emtsev, K. Horn, L. Ley, J. L. McChesney, T. Ohta, J. D. Riley, E. Rotenberg and F. Speck, *Phys. Stat. Sol.* B **245**, 1436–1446 (2008).
15. C. N. R. Rao, A. K. Sood, K. S. Subrahmanyam and A. Govindaraj, *Angew. Chem. Int. Ed.* **48**, 7752–7777 (2009).
16. M. C. Hennion, *J. Chromatogr.* A **856**, 3–54 (1999).
17. C. W. Huck and G. K. Bonn, *J. Chromatogr.* A **855**, 51–72 (2000).
18. M. C. Hennion, *J. Chromatogr.* A **855**, 73–95 (2000).
19. M. S. Mauter and M. Elimelech, *Environ. Sci. Technol.* **42**, 5843–5859 (2008).
20. L. Su, F. Gao and L. Q. Mao, *Anal. Chem.* **78**, 2651–2657 (2006).
21. S. Iijima, *Nature* **354**, 56–58 (1991).
22. Y. Sun, K. Fu, Y. Lin and W. Huang, *Acc. Chem. Res.* **35**, 1096–1104 (2002).
23. P. Avouris, *Acc. Chem. Res.* **35**, 1026–1034 (2002).
24. Q. Zhou, J. Xiao, W. Wang, G. Liu, Q. Shi and J. Wang, *Talanta* **68**, 1309–1315 (2006).
25. J. Riu, A. Maroto and F. X. Rius, *Talanta* **69**, 288–301 (2006).
26. L. Qian and X. R. Yang, *Talanta* **69**, 957–962 (2006).
27. Y. M. Yan, M. N. Zhang, K. P. Gong, L. Su, Z. X. Guo and L. Q. Mao, *Chem. Mater.* **17**, 3457–3463 (2005).
28. W. Chen, L. Duan and D. Q. Zhu, *Environ. Sci. Technol.* **41**, 8295–8300 (2007).
29. R. Q. Long and R. T. Yang, *J. Am. Chem. Soc.* **123**, 2058–2059 (2001).
30. S. Gotovac, Y. Hattori, D. Noguchi, J. Miyamoto, M. Kanamaru, S. Utsumi, H. Kanoh and K. Kaneko, *J. Phys. Chem.* B **110**, 16219–16224 (2006).
31. D. Du, M. H. Wang, J. M. Zhang, J. Cai, H. Y. Tu and A. D. Zhang, *Electrochem. Commun.* **10**, 85–89 (2008).
32. L. H. Shi, X. Q. Liu, H. J. Li, W. X. Niu and G. B. Xu, *Anal. Chem.* **78**, 1345–1348 (2006).
33. M. Valcárcel, S. Cárdenas and B. M. Simonet, *Anal. Chem.* **79**, 4788–4797 (2007).
34. A. P. Washe, S. Macho, G. A. Crespo and F. Xavier Rius, *Anal. Chem.* **82**, 8106–8112 (2010).
35. J. Zhang, J. K. Lee, Y. Wu and R. W. Murray, *Nano Lett.* **3**, 403–407 (2003).
36. M. Melle-Franco, M. Marcaccio, D. Paolucci, F. Paolucci, V. Georgakilas, D. M. Guldi, M. Prato and F. Zerbetto, *J. Am. Chem. Soc.* **126**, 1646–1647 (2004).
37. R. Saito, M. Fujita, G. Dresselhaus and M. S. Dresselhaus, *Appl. Phys. Lett.* **60**, 2204 (1992).
38. J. W. G. Wildöer, L. C. Venema, A. G. Rinzler, R. E. Smalley and C. Dekker, *Nature* **391**, 59 (1998).
39. T. W. Odom, J. L. Huang, P. Kim and C. M. Lieber, *Nature* **391**, 62 (1998).
40. J. H. T. Luong, S. Hrapovic, D. Wang, F. Bensebaa and B. Simard, *Electroanalysis* **16**, 132 (2004).
41. S. Lu, X. Dang, K. Wu and S. Hu, *J. Nanosci. Nanotechnology* **3**, 401 (2003).

42. Z. Xu, X. Chen, X. Qu and S. Dong, *Electroanalysis* **16**, 684 (2004).

43. J. Ye, Y. Wen, W. Zhang, L. M. Gan, G. Xu and F. Sheu, *Electrochem. Commun.* **6**, 66 (2004).

44. J. Kumar, S. K. Jha and S. F. D'Souza, *Biosens. Bioelectron.* **21**, 2100–2105 (2006).

45. A. M. Jiménez and M. J. Navas, *Crit. Rev. Anal. Chem.* **27**, 291–306 (1997).

46. Y. Lin, F. Lu and J. Wang, *Electroanalysis* **16**, 145–149 (2004).

47. H. Shi, Z. Zhao, Z. Song, J. Huang, Y. Yang, J. Anzai and Q. Chen, *Electroanalysis* **17**, 1285–1290 (2005).

48. G. D. Liu and Y. H. Lin, *Anal. Chem.* **77**, 5894–5901 (2005).

49. D. Du, X. Huang, J. Cai and A. D. Zhang. *Sensor. Actuat. B-Chem.* **127**, 531–535 (2007).

50. D. Du, X. Huang, J. Cai and A. D. Zhang. *Biosens. Bioelectron.* **23**, 285–289 (2007).

51. D. Du, X. X. Ye, J. Cai, J. Liu, A. D. Zhang, *Biosens. Bioelectron.* **25**, 2503–2508 (2010).

52. K. A. Joshi, J. Tang, R. Haddon, J. Wang, W. Chen and A. Mulchandani, *Electroanalysis* **17**, 54–58 (2005).

53. G. D. Liu, Y. H. Lin, *Anal. Chem.* **78**, 835–843 (2006).

54. A. Mulchandani, P. Mulchandani, I. Kaneva and W. Chen, *Anal. Chem.* **70**, 4140–4145 (1998).

55. C. A. Constantine, S. V. Mello, A. Dupont, X. Cao, D. Santors, Jr., O. N. Oliveira, Jr., F. T. Strixino, E. C. Pereira, T. Cheng, J. J. Defrank and R. M. Leblanc, *J. Am. Chem. Soc.* **125**, 1805–1809 (2003).

56. S. Kumaran, C. Tranh-Minh, *Anal. Biochem.* **200**, 187–194 (1992).

57. A. Mulchandani, P. Mulchandani, W. Chen, J. Wang and L. Chen, *Anal. Chem.* **71**, 2246–2249 (1999).

58. K. I. Dave, C. E. Miller and J. R. Wild, *Chem. Biol. Interact.* **87**, 55–68 (1993).

59. R. P. Deo, J. Wang, I. Block, A. Mulchandani, K. A. Joshi, M. Trojanowicz, F. Scholz, W. Chen and Y. H. Lin, *Anal. Chim. Acta* **530**, 185–189 (2005).

60. J. Wang, L. Chen, A. Mulchandani, P. Mulchandani and W. Chen, *Electroanalysis* **11**, 866–869 (1999).

61. A. S. Arribas, T. Vazquez, J. Wang, A. Mulchandani and W. Chen, *Electrochem. Commun.* **7**, 1371–1374 (2005).

62. K. B. Wu, S. S. Hu, J. J. Fei and B. Wen, *Anal. Chim. Acta* **489**, 215–221 (2003).

63. Y. D. Zhao, W. D. Zhang, Q. M. Luo and S. F. Y. Li, *Microchem. J.* **75**, 189–198 (2003).

64. P. F. Liu and J. H. Hu, *Sensor. Actuat. B-Chem.* **84**, 194–199 (2002).

65. J. Wang, S. B. Hocevar and B. Ogorevc, *Electrochem. Commun.* **6**, 176–179 (2004).

66. K. B. Male, S. Hrapovic, J. M. Santini and J. H. T. Luong, *Anal. Chem.* **79**, 7831–7837 (2007).

67. J. Q. Yang, P. Wang, X. J. Zhang and K. B. Wu, *J. Agric. Food Chem.* **57**, 9403–9407 (2009).

68. A. S. Arribas, M. Moreno, E. Bermejo, J. A. Perez, V. Roman, A. Zapardiel and M. Chicharro, *Electroanalysis* **23**, 237–244 (2011).
69. M. Jiang, and Y. Lin, in *Encyclopedia of Sensors*, Vol. 2, eds. C. A. Grimes, E. C. Dickey and M. V. Pishko (American Scientific Publishers, Stevenson Ranch, CA, 2006), pp. 25–51.
70. C. Cantalini, W. Wlodarski, Y. Li, M. Passacantando, S. Santucci, E. Comini, G. Faglia and G. Sberveglieri, *Sensor. Actuat. B-Chem.* **64**, 182–188 (2000).
71. C. Cantalini, L. Valentini, I. Armentano, J. M. Kenny, L. Lozzi and S. Santucci, *J. Eur. Ceram. Soc.* **24**, 1405–1408 (2004).
72. B. Y. Wei, M. C. Hsu, P. G. Su, H. M. Lin, R. J. Wu and H. J. Lai, *Sensor. Actuat. B-Chem.* **101**, 81–89 (2004).
73. B. Y. Wei, M. C. Hsu, Y. S. Yang, S. H. Chien and H. M. Lin, *Mater. Chem. Phys.* **81**, 126–133 (2003).
74. C. Cantalini, L. Valentini, I. Armentano, L. Lozzi, J. M. Kenny and S. Santucci, *Sensor. Actuat. B-Chem.* **95**, 195–202 (2003).
75. Y. Lu, J. Li, J. Han, H. T. Ng, C. Binder, C. Partridge and M. Meyyappan, *Chem. Phys. Lett.* **391**, 344–348 (2004).
76. L. Valentini, C. Cantalini, L. Lozzi, S. Picozzi, I. Armentano, J. M. Kenny and S. Santucci, *Sensor. Actuat. B-Chem.* **100**, 33–40 (2004).
77. S. Peng, K. Cho, P. Qi and H. Dai, *Chem. Phys. Lett.* **387**, 271–276 (2004).
78. Y. M. Wong, W. P. Kang, J. L. Davidson, A. Wisitsoraat and K. L. Soh, *Sensor. Actuat. B-Chem.* **93**, 327–332 (2003).
79. C. Cantalini, L.Valentini, L. Lozzi, I. Armentano, J. M. Kenny and S. Santucci, *Sensor. Actuat. B-Chem.* **93**, 333–337 (2003).
80. J. He, C. Chen and J. Liu, *Sensor. Actuat. B-Chem.* **99**, 1–5 (2004).
81. J. Kong, N. R. Franklin, C. W. Zhou, M. G. Chapline, S. Peng, K. Cho and H. Dai, *Science* **287**, 622–625 (2000).
82. J. Zhao, A. Buldum, J. Han and J. P. Lu, *Nanotechnology* **13**, 195–200 (2002).
83. G. U. Sumanasekera, C. K. W. Adu, F. Fang and P. C. Eklund, *Phys. Rev. Lett.* **85**, 1096–1099 (2000).
84. O. K. Varghese, P. D. Kichambre, D. Gong, K. G. Ong, E. C. Dickey and C. A. Grimes, *Sensor. Actuat. B-Chem.* **81**, 32–41 (2001).
85. K. G. Ong, K. Zeng and C. A. Grimes, *IEEE Sensor J.* **2**, 82–88 (2002).
86. S. Chopra, A. Pham, J. Gaillard, A. Parker and A. M. Rao, *Appl. Phys. Lett.* **80**, 4632–4634 (2002).
87. A. Modi, N. Koratkar, E. Lass, B. Wei and P. M. Ajayan, *Nature* **424**, 171–174 (2003).
88. P. Collins, K. Bradley, M. Ishigami and A. Zettl, *Science* **287**, 1801–1804 (2000).
89. J. Liu, A. G. Rinzler, H. Dai, J. H. Hafner, R. K. Bradley, P. J. Boul, A. Lu, T.Iverson, K. Shelimov, C. B. Huffman, F. Rodriguez-Macias, Y. S. Shon, T. R. Lee, D. T. Colbert and R. E. Smalley, *Science* **280**, 1253–1256 (1998).
90. R. J. Chen, Y. Zhang, D. Wang and H. Dai, *J. Am. Chem. Soc.* **123**, 3838–3839 (2001).

91. M. A. Hamon, J. Chen, H. Hu, Y. Chen, M. E. Itkis, A. M. Rao, P. C. Eklund and R. C. Haddon, *Adv. Mater.* **11**, 834–840 (1999).

92. R. Zanella, E. V. Basiuk, P. Santiago, V. A. Basiuk, E. Mireles, I. Puente-Lee and J. M. Saniger, *J. Phys. Chem. B* **109**, 16290–16295 (2005).

93. J. Cao, J. Z. Sun, J. Hong, H. Y. Li, H. Z. Chen and M. Wang, *Adv. Mater.* **16**, 84–87 (2004).

94. B. Kim and W. M. Sigmund, *Langmuir* **20**, 8239–8242 (2004).

95. H. Ago, K. Petritsch, M. S. P. Shaffer, A. H. Windle and R. H. Friend, *Adv. Mater.* **11**, 1281–1285 (1999).

96. B. E. Kilbride, J. N. Coleman, D. F. O'Brien and W. J. Blau, *Syn. Met.* **121**, 1227–1228 (2001).

97. G. Z. Chen, M. S. P. Shaffer, D. Coleby, G. Dixon, W. Zhou, D. J. Fray and A. H. Windle, *Adv. Mater.* **12**, 522–526 (2000).

98. J. H. Chen, Z. P. Huang, D. Z. Wang, S. X. Yang, J. G. Wen and Z. F. Ren, *App. Phys. A* **73**, 129–131 (2001).

99. K. Lota, V. Khomenko and E. Frackowiak, *J. Phys. Chem. Solid.* **65**, 295–301 (2004).

100. Y. Zhou, B. He, W. Zhou, J. Huang, X. Li, B. Wu and H. Li, *Electrochim. Acta.* **49**, 257–262 (2004).

101. V. Barvastrello, E. Stura, S. Carrara, V. Erokhin and C. Nicolini, *Sens. Actuat. B-Chem.* **98**, 247–253 (2004).

102. S. Mubeen, T. Zhang, B. Yoo, M. Deshusses and N. Myung, *J. Phys. Chem. C* **111**, 6321–6327 (2007).

103. Z. Sun, X. Zhang, N. Na, Z. Liu, B. Han and G. An, *J. Phys. Chem. B* **110**, 13410–13414 (2006).

104. S. H. Yoon, J. H. Han, B. K. Kim, H. N. Choi and W. Y. Lee, *Electroanalysis* **22**, 1349–1356 (2010).

105. K. S. Novoselov, A. K. Geim, S. V. Morozov, D. Jiang, M. I. Katsnelson, I. V. Grigorieva, S. V. Dubonos and A. A. Firsov, *Nature* **438**, 197–200 (2005).

106. S. Stankovich, D. A. Dikin, G. H. B. Dommett, K. M. Kohlhaas, E. J. Zimney, E. A. Stach, R. D. Piner, S. T. Nguyen and R. S. Ruoff, *Nature* **442**, 282–286 (2006).

107. A. K. Geim, *Science* **324**, 1530–1534 (2009).

108. X. R. Wang, X. L. Li, L. Zhang, Y. Yoon, P. K. Weber, H. L. Wang, J. Guo and H. J. Dai, *Science* **324**, 768–771 (2009).

109. Y. W. Son, M. L. Cohen and S. G. Louie, *Nature* **444**, 347–349 (2006).

110. W. Y. Kim and K. S. Kim, *Nat. Nanotechnol.* **3**, 408–412 (2008).

111. K. P. Loh, Q. L. Bao, P. K. Ang and J. X. Yang, *J. Mater. Chem.* **20**, 2277–2289 (2010).

112. Y. Shi, W. Fang, K. Zhang, W. Zhang and L. Li, *Small* **5**, 2005–2011 (2009).

113. M. J. Allen, V. C. Tung and R. B. Kaner, *Chem. Rev.* **110**, 132–145 (2010).

114. Y. Wang, S. Zhang, D. Du, Y. Shao, Z. Li, J. Wang, M. H. Engelhard, J. Li and Y. Lin. *J. Mater. Chem.* (2011), DOI: 10.1039/c0jm03441j.

115. J. Li, S. Guo, Y. Zhai and E. K. Wang, *Anal. Chim. Acta* **649**, 196–201 (2009).

116. J. Li, S. Guo, Y. Zhai and E. K. Wang, *Electrochem. Commun.* **11**, 1085–1088 (2009).
117. Y. Wang, Y. Wan and D. Zhang, *Electrochem. Commun.* **2**, 187–190 (2010).
118. F. Schedin, A. K. Geim, S. V. Morozov, E. W. Hill, P. Blake, M. I. Katsnelson and K. S. Novoselov, *Nat. Mater.* **6**, 652–655 (2007).
119. G. H. Lu, L. E. Ocola and J. H. Chen, *Nanotechnology* **20**, 445–502 (2009).
120. J. D. Fowler, M. J. Allen, V. C. Tung, Y. Yang, R. B. Kaner and B. H. Weiller, *ACS Nano.* **3**, 301–306 (2009).
121. B. Huang, Z. Li, Z. Liu, G. Zhou, S. Hao, J. Wu, B. Gu and W. Duan, *J. Phys. Chem. C* **112**, 13442–13446 (2008).

Chapter 22

One-dimensional Hollow Oxide Nanostructures: A Highly Sensitive Gas-sensing Platform

Jong-Heun Lee

Department of Materials Science and Engineering
Korea University, Seoul, Republic of Korea

22.1 Introduction

Hollow oxide structures have versatile applications, such as drug delivery, catalysts, energy storage, dielectrics, and gas sensors.[1-5] Depending on the dimensions of the structures, hollow structures can be divided into one-dimensional hollow nanofibers and three-dimensional hollow spheres. In particular, the hollow morphology of nanostructures is very advantageous for enhancing the performance of oxide semiconductor gas sensors.

Oxide semiconductors exhibit significant conductivity changes due to the adsorption of oxidizing gas or due to the reaction between reducing gases and negatively charged surface oxygen atoms (O^- or O^{2-}).[6] The electron depletion layer in n-type oxide semiconductors at 300–400°C is typically 3–20 nm thick.[7,8] Accordingly, when the nanoparticles are smaller than two times the electron depletion layer thickness, the nanoparticles are fully depleted and the gas response (R_a/R_g; R_a: resistance in air, R_g: resistance in gas) based on the chemoresistive change increases dramatically.[9] However, the dispersion of nanoparticles or nanostructures smaller than ~20 nm is very difficult due to strong van der Waals attractions. In most cases, the nanoparticles form a very large and dense aggregate, which hampers the in-diffusion of analyte gases into the inner parts of secondary particles. This again reduces the gas response because the inner parts of the agglomerates remain insensitive.

The hollow morphology of nanostructures facilitates the rapid and effective diffusion of analyte gas onto the entire sensor surface. Both the inner and outer surfaces of hollow nanostructures can participate in the gas sensing reaction when the shell layer is very thin and nanoporous. In addition, it is relatively convenient to disperse the sub-micrometer-scale or micrometer-scale hollow structures. All of these enhance the gas response significantly. Moreover, the gas selectivity and sensitivity can be tuned further by the addition of metal oxide or metal catalysts. Previously, the present author reported a review on the enhancement of gas sensing characteristics using three-dimensional hollow nanostructures.[10] However, to date, one-dimensional hollow nanostructures for gas sensor applications have not been reviewed.

In this contribution, the gas sensing characteristics of various one-dimensional hollow oxide nanostructures for gas sensor applications have been reviewed. In order to concentrate on the design of gas sensors, only oxide semiconductor gas sensor materials, such as SnO_2, ZnO, TiO_2, Fe_2O_3, In_2O_3, Co_3O_4, Cr_2O_3, Cu_2O/CuO, NiO, and MoO_3, have been investigated. The main focus of this overview is placed on the various synthetic strategies for preparing well-defined one-dimensional hollow nanostructures and the principal parameters for determining the gas sensing characteristics, such as the gas response, selectivity, and response/recovery kinetics.

22.2 Preparation of One-dimensional Hollow Oxide Nanostructures

Recent technological progress in nanoscale fabrication triggered intensive studies on the preparation and functional applications of one-dimensional oxide nanostructures, such as nanowires, nanorods, nanobelts, nanofibers, and nanotubes.[11] Among these, one-dimensional oxide nanostructures with hollow inner structures, such as nanotubes and hollow nanofibers, are promising for accomplishing effective current collection as well as active electrochemical/photochemical reactions without sacrificing high surface area. The representative applications of one-dimensional hollow nanostructures from this viewpoint are electrodes of secondary batteries, catalysts for water photolysis, electrodes of dye-sensitized solar cells, and gas-sensing materials.[12,13]

Table 22.1 summarizes the various synthetic routes for preparing one-dimensional hollow nanostructures for gas sensor applications.[14-69] The representative approaches for preparing one-dimensional hollow nanostructures are template-based physicochemical routes, electrospinning, the Kirkendall effect, anodization, chemical vapor deposition, and solvothermal self-assembly reactions. In the following sections, the preparation of one-dimensional hollow oxide structures and their gas sensor applications will be discussed.

Table 22.1. The synthetic routes of various one-dimensional hollow oxide nanostructures in the literature for gas sensor applications.[14–69]

Materials	Preparation	Procedure	Reference
SnO$_2$	Template method	(c)[a] sol gel coating, (t)[b]cellulose fibers	14
	Template method	(c) sol gel coating, (t) carbonaceous nanofibers	15
	Template method	(c) Electrodeposition, (t) polycarbonate membrane	16
	Template method	(c) Modified LbL[a] assembly, (t) CNTs[d]	17
	Template method	(c) sol gel coating, (t) CNTs	18
	Template method	(c) infiltration casting, (t) AAO[e]	19–22
	Template method	(c) Hydrothermal coating, (t) Si nanotemplates	23
	Electrospinning + Template method	(c) ALD[f], (t) electrospun PAN[g]nanofibers	24
	Electrospinning	Single-capillary electrospinning	25
	Kirkendall effect	Oxidation of Sn nanorods into SnO$_2$ nanotubes	26
	Chemical vapor deposition	Combustion CVD[h]	27
ZnO	Electrospinning + Template method	(c) ALD, (t) electrospun PVA[i]nanofibers	28
	Electrospinning + Template method	(c) sputtering, (t) electrospun PVA nanofibers	29
	Electrospinning	Single-capillary electrospinning	30,31
	Kirkendall effect	Oxidation of Zn NWs[j] into ZnO nanotubes	32,33
	Thermal evaporation	Thermal evaporation of mixed Zn-ZnO powders	34
	Reactive evaporation	Reactive thermal evaporation of Zn powders	35
	Hydrothermal reaction	Hydrothermal reaction of (ZnCl$_2$+methenamine)(aq)	36
	Sonochemical reaction	Sonochemical reaction of (Zn(NO$_3$)$_2$+NaOH+ED[k])(aq)	37
TiO$_2$	Template method	(c) sol gel coating, (t) carbonaceous nanofibers	15
	Template method	(c) sol gel coating, (t) CNTs	38
	Template method	(c) ALD, (t) PC membrane	39
	Template method	(c) LbL coating, (t) Ni nanorods	40
	Template method	(c) sol gel coating, (t) V$_2$O$_5$ nanorods	41
	Template method	(c) sol gel coating, (t) ZnO nanorods/nanowires	42,43
	Template method	(c) sol gel coating, (t) calcite nanorods	44

(Continued)

Table 22.1. (*Continued*)

Materials	Preparation	Procedure	Reference
	Electrospinning + Template method	(c) LbL coating, (t) electrospun PS nanofiber	45
	Electrospinning	Double-capillary (co-axial) electrospinning	46,47
	Hydrothermal method	Hydrothermal treatment of TiO_2 NPs^l + NaOH (aq.)	48
Fe_2O_3	Template method	(c) sol gel coating, (t) carbonaceous nanofibers	15
	Template method	(c) sol gel coating, (t) CNTs	49
	Template method	(c) supercritical fluid-mediated route, (t) CNTs	50
	Template method	(c) infiltration casting, (t) AAO	51
	Electrospinning	Double-capillary (co-axial) electrospinning	52
	Anodization	Sonoelectrochemical anodization of Fe foil	53
	Solvothermal self assembly	Surfactant-mediated solvothermal self assembly reaction	54
In_2O_3	Template method	(c) Modified LbL coating, (t) CNTs	55
	Template method	(c) CVD, (t) AAO	56
	Template method	(c) sol gel coating,(t) AAO	57
	Electrospinning	Single-capillary electrospinning	58
Co_3O_4	Template method	(c) sol gel coating, (t) CNTs	49
	Template method	(c) sonochemical decomposition of $Co_4(CO)_{12}$, (t) CNTs	59
	Kirkendall effect	Oxidation of Co NWs into Co_3O_4 nanotubes	32
Cr_2O_3	Template method	(c) supercritical fluid-mediated route, (t) CNTs	60
CuO	Kirkendall effect	Oxidation of Cu nanowires into CuO nanotubes	61
NiO	Template method	(c) sol gel coating, (t) CNTs	49
	Electrospinning	Double-capillary (co-axial) electrospinning	62
MoO_3	Template method	(c) sol gel coating, (t) CNTs	63
$Pd-SnO_2$	Electrospinning	Single-capillary electrospinning	25
$Pt-TiO_2$, $Pd-TiO_2$	Anodization	Anodization of Ti-Pt and Ti-Pd thin film	64
$RE-In_2O_3$	Electrospinning	Single-capillary electrospinning	65
MoO_3-SnO_2		Hydrothermal reaction between $SnCl_2$ and MoO_3 nanorods	66
$Fe_2O_3-SnO_2$	Template method	(c) sol gel coating, (t) carbonaceous nanofibers	15
$ZnFe_2O_4$	Template method	(c) infiltration casting, (t) AAO	67

(*Continued*)

Table 22.1.   (*Continued*)

Materials	Preparation	Procedure	Reference
LaFeO$_3$	Co-precipitation + molten salt synthesis	Co-precipitation + molten salt synthesis	68
MCo$_2$O$_4$ (M=Ni,Cu,Zn)	Template method	(c) infiltration casting, (t) AAO	69

[a](c): coating, [b](t): template, [c]LbL: layer-by-layer, [d]CNTs: carbon nanotubes, [e]AAO: anodic aluminum oxide, [f]ALD: atomic layer deposition, [g]PAN: polyacrylonitrile, [h]CVD: chemical vapor deposition, [i]PVA: polyvinyl acetate, [j]NWs: nanowires, [k]ED: ethylenediamine, [l]NPs: nanoparticles.

22.2.1   *Template-assisted preparation of one-dimensional hollow nanostructures*

The following five overview papers cover the various template-assisted synthetic routes to prepare one-dimensional hollow oxide nanostructures: (1) the fabrication of one-dimensional inorganic nanostructures via templating against pre-existing one-dimensional nanostructures by Liang et al.[70]; (2) the chemical synthesis of inorganic nanotubes by Xiong et al.[71]; (3) the chemical routes to prepare nanorods, nanowires, and nanotubes by Cao and Liu[72]; (4) the electrochemical routes to prepare one-dimensional nanostructures and porous materials by Lai and Riley[73]; and (5) the preparation of oxide nanotubes via template wetting by Steinhart et al.[74]

A key concept in common among the five comprehensive reviews is the thin coating of nanoscale precursors or oxide layers onto well-defined one-dimensional templates and subsequent removal of the core template by thermal decomposition, dissolution, or by other methods. The precursor/oxide layer is coated using various methods, such as a sol-gel method,[14,15,18,38,41−44,49,57,63] layer-by-layer coating,[40,45,55] infiltration casting,[19,22,51] electrodeposition,[16] sputtering,[29] atomic layer deposition,[39] a supercritical fluid-mediated route,[50,60] chemical vapor deposition,[56] and sonochemical reactions.[59] The materials for the sacrificial cores include the following: (1) polymer (cellulose,[14] carbonaceous,[15] polyacrylonitrile (PAN),[24] polyvinyl acetate (PVA),[28,29] polycarbonate (PC),[39] polystyrene (PS)[45]) fibers; (2) carbon nanotubes (CNTs)[17,38,49,59,60,63]; (3) metal (Ni) nanorods[40]; and (4) metal oxide (anodic aluminum oxide (AAO) membrane[19−22,51,56,57,67,69] and V$_2$O$_5$ nanorod[41]) nanostructures. The overall morphology of one-dimensional hollow structures can be designed by the proper selection of templates and the thickness of the hollow shell layer can be tuned by controlling the thickness of precursors or the oxide coating layer.

J.-H. Lee

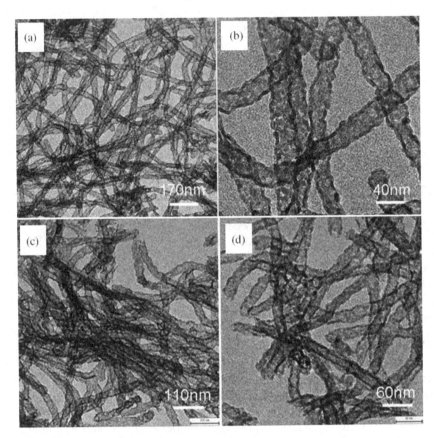

Figure 22.1.   (a, b) TEM images of TiO$_2$ nanotubes and (c, d) SnO$_2$ nanotubes obtained from hydroxide coatings from carbonaceous nanofibers and subsequent removal of the carbonaceous cores (adapted from Ref. 15, used with permission).

Gong *et al.*[15] prepared various oxide nanotubes, such as TiO$_2$, SnO$_2$, Fe$_2$O$_3$, and a SnO$_2$-Fe$_2$O$_3$ composite, by the dispersion of carbonaceous nanofibers in the ethanol solutions containing tetrabutyltitanate, SnCl$_4$, FeCl$_3$, and SnCl$_4$-FeCl$_3$, followed by ultrasonication, the addition of water or ammonia, washing, drying, and heat treatment at 500°C for 1 hour. Polycrystalline TiO$_2$ hollow nanofibers with tube diameters of ∼40 nm and wall thicknesses of ∼5 nm were prepared with this method (Figure 22.1(a, b)). Similar morphologies of SnO$_2$ (Figure 22.1(c, d)), Fe$_2$O$_3$, and SnO$_2$-Fe$_2$O$_3$ composite hollow nanofibers were also prepared. This demonstrates that not only single metal oxides, but also composites of two different oxides can be prepared in the form of one-dimensional hollow nanofibers. The preparation of multi-compositional nanostructures is important in order to design a selective gas sensor because the gas sensing characteristics are closely dependent upon the acid-base properties of sensing materials.[75] This can provide a combinatorial materials library for artificial olfaction to recognize a complex chemical quantity (smell) via a pattern recognition mechanism.[76,77]

Wang *et al.*[22] prepared various one-dimensional SnO_2 nanostructures, such as nanotubes, nanotube-nanorod-hybrids, and nanorods by the infiltration of an $SnCl_4$ aqueous solution in cylinder-shaped AAO membranes and the removal of AAO templates using an NaOH aqueous solution after converting the Sn precursors into SnO_2 by heat treatment at 650°C (Figure 22.2). The diameters of nanotubes ranged from 180 to 250 nm and the shell thickness was in the range of 8–20 nm. The morphologies of the one-dimensional nanostructures could be manipulated by controlling $[SnCl_4]$, that is, the nanotubes and nanorods were prepared by the infiltration of low and high $[SnCl_4]$, respectively (Figure 22.2). This indicates that the thickness of SnO_2 nanotubes can be also controlled by manipulating $[SnCl_4]$.

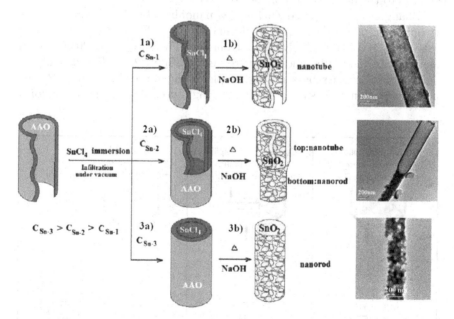

Figure 22.2.   Schematic illustration showing the formation of various one-dimensional SnO_2 nanostructures by the infiltration of an $SnCl_4$ aqueous solution on a cylinder-shaped AAO membrane (reproduced from Ref. 22, used with permission).

22.2.2 *Preparation of one-dimensional hollow nanostructures by electrospinning*

Electrospinning is an economic, versatile, and high-throughput process to produce polymer and metal oxide nanofibers (or nanotubes) using an electric field. The morphology and diameter of nanofibers are closely dependent upon the precursor solution, spinneret configuration, feeding rate of solution, and electric field for electrospinning. The principle and applications of electrospinning have been compiled by Li and Xia[78] and Sigmund *et al.*[79] McCann *et al.*[80] reported an overview of the preparation of core-sheath, hollow, and porous structures via electrospinning. The applications of electrospun nanofibers in acoustic wave sensors, chemoresistive sensors, optical sensors, and biosensors have been compiled by Ding *et al.*[81]

There are three main approaches to prepare hollow nanofibers using electrospinning (Figure 22.3): (1) single-capillary electrospinning of core(polymer)/shell(metal precursor) nanofibers using phase separation of the precursor solution; (2) co-axial two-capillary electrospinning of two immiscible liquids; and (3) the coating or deposition of a metal-containing layer onto the electrospun polymer templates.

Zhang *et al.*[31] prepared ZnO hollow nanofibers by single-capillary electrospinning of an N,N-dimethylformamide (DMF) solution containing PAN ($M_w = 150,000$), polyvinylpyrrolidone (PVP, $M_w = 1,300,000$), and zinc acetate and a subsequent heat treatment (Figure 22.4). Based on observations of the phase separation in a droplet of the mixed PAN and PVP/zinc acetate solution, they suggested that the high-viscosity PAN polymers form

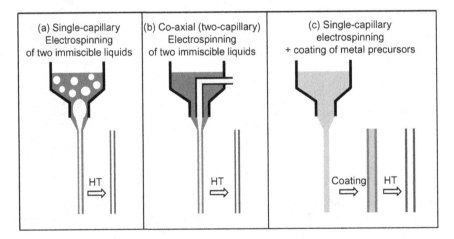

Figure 22.3.   Three main approaches to prepare hollow nanofibers using electrospinning.

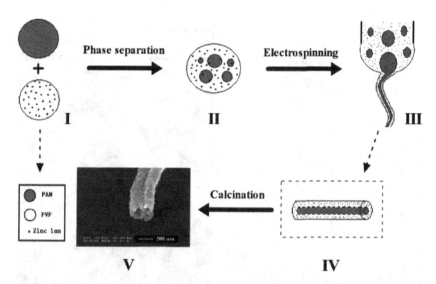

Figure 22.4. An illustrative diagram of the formation mechanism of ZnO hollow nanofibers. ZnO nanofibers were fabricated by electrospinning a precursor solution containing polyacrylonitrile (PAN), polyvinylpyrrolidone (PVP), and zinc acetate and subsequent heat treatment (adapted from Ref. 31 with permission).

discontinuous (dispersed) phases while low-viscosity PVP/zinc acetate composite polymers form a continuous matrix phase. The electrospinning of this precursor solution yielded a core(PAN)/shell(PVP/zinc acetate) configuration, which was finally converted into ZnO hollow nanofibers by a heat treatment. The concentration of the zinc acetate in the PVP/zinc acetate composite polymer determines the shell thickness.

The core-sheath configuration of nanofibers can be formed by co-spinning two immiscible liquids using a co-axial spinneret,[46,80] which can be converted into metal oxide hollow nanofibers by heat treatment. Li and Xia[46] prepared long and uniform TiO_2 (anatase) hollow nanofibers (inner diameter: 200 nm, wall thickness: 50 nm) by co-axial electrospinning of heavy mineral oil (core) with an ethanol solution containing PVP and Ti-isopropoxide (Ti-$(O_iPr)_4$) (sheath), the removal of oily cores by octane, and subsequent heat treatment at 500°C (Figure 22.5). The wall thickness was reduced to 20 nm when the injection rate of the oil phase was increased and the diameters of nanofibers could be tuned by controlling the electric field.

Li *et al.*[82] further suggested various chemical routes to functionalize the inner and outer surfaces of hollow nanofibers. The interiors of TiO_2 hollow nanofibers were functionalized with superparamagnetic iron oxide or SnO_2 nanoparticles by the addition of oil-based ferrofluid or tin isopropoxide to

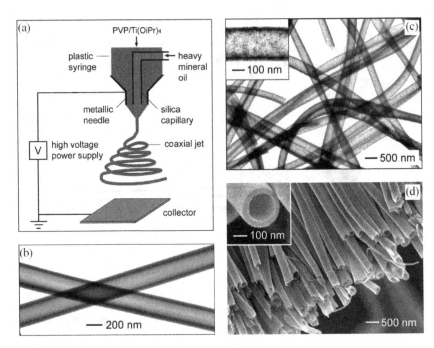

Figure 22.5. (a) Schematic illustration of the setup for co-axial (two-capillary) electro-spinning nanofibers with a core/sheath structure. The spinneret was fabricated from two co-axial capillaries, through which heavy mineral oil and an ethanol solution containing PVP and Ti(OiPr)$_4$ were simultaneously ejected to form a continuous, co-axial jet. (b) TEM image of two as-spun hollow fibers after the oily cores had been extracted with octane. The walls of these tubes were made of a composite containing amorphous TiO$_2$ and PVP. (c) TEM image of TiO$_2$ (anatase) hollow fibers that were obtained by calcining the composite nanotubes in air at 500°C. (d) SEM image of a uni-axially aligned array of anatase hollow fibers that were collected across the gap between a pair of electrodes. These fibers were fractured using a razor blade to expose their cross-sections. In the preparation of all these samples, the feeding rate for heavy mineral oil was 0.1 mL/h and the concentrations of Ti(OiPr)$_4$ and PVP were 0.3 and 0.03 g/mL, respectively. The voltage of electrospinning was 12 kV (reproduced from Ref. 46 with permission from the American Chemical Society).

the mineral oil, the core part of the solution. This can be applied to the control of gas sensing characteristics via compositional control and surface functionalization. Moreover, if p-type semiconductor gas sensor materials, such as Co$_3$O$_4$, NiO, Cr$_2$O$_3$, and CuO, are functionalized on the interior or exterior of n-type semiconductor (SnO$_2$, ZnO, TiO$_2$, In$_2$O$_3$, WO$_3$) hollow nanofibers, the establishment of p-n junctions will open new possibilities for the design of high-performance gas sensors by the extension of the electron depletion layer at the hetero-interface.[83,84] The noble metal catalyst can also be loaded on the surface of hollow nanostructures. Li *et al.*[82] functionalized the inner and outer surfaces of hollow nanofibers with amine-modified functional groups by treating them with NH_2-terminated silanes. They also coated the nanofibers with Au nanoparticles by the electrostatic interaction between the amine-modified surface and citrate-stabilized Au nanoparticles. The Au nanoparticles could be coated only on the outer surface of the TiO_2 hollow fibers when the inner surface was derivatized to be hydrophobic by the addition of octadecyltrichlorosilane to the mineral oil for electrospinning. The facile manipulation of the nanofiber dimensions (diameter and wall thickness) and the functionalization of nanofiber surfaces with other oxides and noble metal nanoparticles, therefore, will provide a valuable tool to control the gas response, gas response/recovery kinetics, and selectivity of gas sensors.

The third approach is based on the combination between a template route and electrospinning (Figure 22.3(c)). In the first step, the long and well-defined polymer templates are fabricated by electrospinning and a thin metal-containing layer is uniformly coated on the surfaces of the electrospun template nanofibers. To date, atomic layer deposition (ALD) of SnO_2^{24} and ZnO,[28] sputtering of ZnO,[29] and layer-by-layer (LbL) coating of TiO_2^{45} on electrospun polymers have been reported. The LbL coating, the successive and alternate coating of oppositely charged polyelectrolytes and inorganic precursors, enables the preparation of well-defined hollow structures with very thin walls and the control of wall thicknesses down to the 5 nm scale. However, one-step coating of multi-compositional and/or catalyst-loaded sensing materials is relatively difficult. In contrast, ALD and sputtering are advantageous for preparing sensing materials with complex compositions.

22.2.3 *Preparation of one-dimensional hollow nanostructures by the Kirkendall effect*

Metal oxide hollow nanostructures can be prepared by a nanoscale Kirkendall effect during the oxidation of metal nanostructures.[85,86] When the out-diffusion of metal atoms is significantly faster than the in-diffusion of oxygen through the initially formed thin metal oxide layer, the inner part of the metal nanostructures becomes vacant as oxidation proceeds. Thus, the overall morphologies and dimensions of hollow oxide nanostructures prepared by the Kirkendall effect are primarily determined by those of metal nanostructures. Metal oxide hollow spheres are formed by the oxidation of metal spheres and metal oxide nanotubes are formed by the oxidation of metal nanowires or nanorods. Raidongia *et al.*[32] prepared ZnO nanotubes (outer diameter: 90–400 nm, wall thickness: ~20 nm, Figure 22.6(a, b)) and

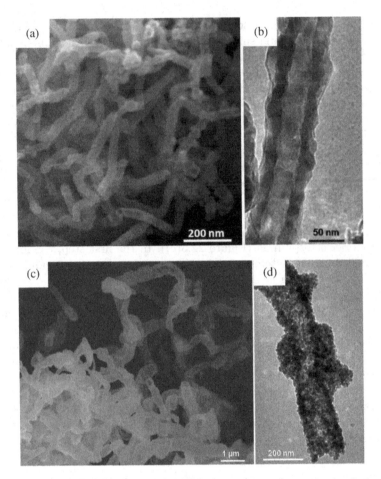

Figure 22.6. (a, b) FESEM and TEM images of ZnO nanotubes and (c, d) FE-SEM and TEM images of $ZnCr_2O_4$ nanotubes (adapted from Ref. 32, used with permission).

Co_3O_4 nanotubes (outer diameter: ~40nm, wall thickness: ~10 nm) by the oxidation of Zn and Co nanowires at 703 and 773 K, respectively. Because the wall thickness is primarily determined by the amount of source metal materials for oxidation, metal oxide hollow spheres with relatively thick walls (50–100 nm) are often prepared by the oxidation of sub-micrometer-scale or microscale metal spheres. In contrast, the wall thickness can be reduced to several nanometers when nanofibers or nanowires with diameters of <20 nm are used for oxidation. This is undoubtedly related to the smaller dimension of metal nanowires and nanorods compared to those of the sub-micrometer-scale metal spheres. Raidongia et al.[32] also prepared

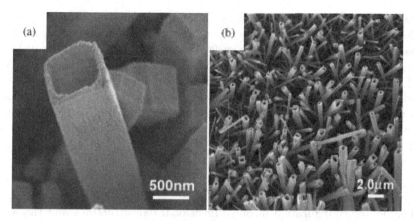

Figure 22.7. (a) SnO_2 nanotube with a fully open top end and (b) SnO_2 nanotubes with fully open tips (adapted from Ref. 27, used with permission).

$ZnCr_2O_4$ nanotubes (wall thickness: 20–30 nm) by reacting Zn nanowires with CrO_2Cl_2 (Figure 22.6(c, d)). This demonstrates that single- and multi-component nanotubes with nanoscale wall thicknesses can be also prepared by the Kirkendall effect.

22.2.4 *Preparation of one-dimensional hollow nanostructures by chemical vapor deposition*

Liu and Liu[27] have prepared SnO_2 nanotubes with square or rectangular cross-sections by a combustion chemical vapor deposition method. For this, sub-microscale mists of an ethanol solution containing tin(II)-ethylhexanoate was combusted in an open atmosphere at a high temperature. Figure 22.7 shows the morphology of the SnO_2 nanotubes grown at 1,150°C. The polycrystalline SnO_2 thin-film layer formed at the beginning stage, which subsequently provided the energetically favorable planes for the nucleation of SnO_2 tubes as well as the pathways for surface diffusion.

22.3 Gas Sensing Characteristics of One-dimensional Hollow Nanostructures

Gas sensing characteristics are determined by many factors, such as dimension, morphology, and donor density of materials, contact configuration between sensing materials, sensing temperature, porosity, surface chemistry, the presence of a catalyst, the heterojunction between different sensing materials, and so on. The distinctive features of one-dimensional hollow oxide nanostructures should be derived from the nanoscale wall thicknesses, high surface areas, high porosities, non- or less-agglomerated network structures, and uniform loadings of catalysts. All of these can enhance the gas response, responding/recovering speed, and selectivity of gas sensors. In the following sections, the effects of the hollow configuration, wall thickness, porosity, composition, and catalyst loading on the gas sensing characteristics of one-dimensional hollow oxide nanostructures will be discussed.

22.3.1 Hollow configuration and wall thickness

The hollow configuration in one-dimensional nanostructures and the wall thickness in one-dimensional hollow oxide structures are very important for controlling electron depletion and chemoresistive changes. In single-crystalline nanotubes, both the outer and inner surfaces of n-type semiconductors form negatively charged surface oxygen by the effective diffusion of oxygen via the channel structures. Thus, full depletion can be accomplished when the wall thickness (t_{wall}) is smaller than two times the electron depletion layer ($2\lambda_D$) (Figure 22.8(a)). In this configuration, the gas response, i.e. the increase of electron density due to the reaction between a reducing gas and the negatively charged surface oxygen, can be maximized. At $t_{wall} > 2\lambda_D$, the wall is partially depleted (Figure 22.8(b)). The resistance of the core part of the wall remains unchanged. Accordingly, the gas response in the configuration of Figure 22.8(b) is lower than that of configuration of Figure 22.8(a). Finally, most of the thick nanowires with solid inner structures remain semiconducting and only a very small portion near the surface changes into the electron depletion layer (Figure 22.8(c)), which leads to a very low gas response. Thus, the gas response can be best maximized by thin-wall nanotubes.

Varghese et al.[87] prepared a TiO$_2$ nanotube array. They controlled the pore size and wall thickness by changing the anodization voltage during anodic oxidation of Ti foil and made a variety of nanotubes: large-pore nanotubes (pore size: 53 ± 10 nm, wall thickness: 17 ± 5 nm); intermediate-pore nanotubes (pore size: 76 ± 15 nm, wall thickness: 27 ± 6 nm); and small-pore nanotubes (pore size: 22 ± 5 nm, wall thickness: 13 ± 2 nm). As the pore size decreased from 76 nm to 22 nm, the conductance variation ($[G_g - G_0]/G_0$, G_g: conductance in H$_2$-N$_2$, G_0: conductance in N$_2$) at 1,000 ppm H$_2$ at 290°C increased 200 times (Figure 22.9). The dramatic enhancement of the H$_2$ response in the small-pore (pore size: 22 nm) nanotubes was attributed to the increased surface area, the enhanced contribution of the space charge layer, the increased contact between nanotubes

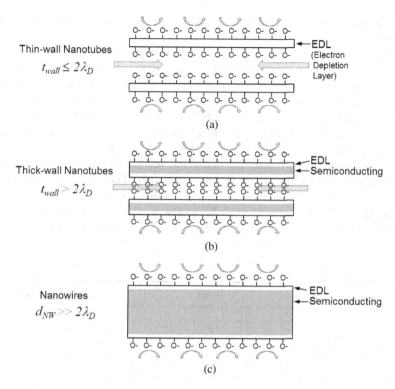

Figure 22.8. The configurations of electron depletion layers in one-dimensional n-type oxide semiconductors; (a) thin-wall nanotubes, (b) thick-wall nanotubes, and (c) nanowires (t_{wall}: wall thickness of nanotubes, λ_D: the thickness of electron depletion layer, d_{NW}: diameter of nanowires).

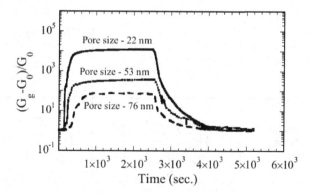

Figure 22.9. Normalized change in electrical resistance of TiO nanotube arrays of 76 nm, 53 nm, 22 nm diameters to 1,000 ppm hydrogen at 290°C (G_g: conductance in H_2-N_2, G_0: conductance in N_2) (reproduced from Ref. 87 with permission from Wiley-VCH).

due to the smaller pores, the thinner wall thickness, and the higher number density of nanotubes.

In polycrystalline nanofibers, principally speaking, the full depletion layer is determined not by wall thickness but by the size of primary nanoparticles within the nanotubes. Nevertheless, the hollow morphology with thin-wall configuration is still very important. If the wall of a nanofiber is very thin, regardless of the location, all of the primary nanoparticles located at the outer and inner parts of the hollow nanofiber can react with the analyte gases, which enhances the gas response (Figure 22.10(a)). If the wall is thick and dense, the primary particles that are not exposed to the ambient atmosphere remain insensitive (Figure 22.10(b)). The gas-sensing sites can be increased when the wall becomes more porous (Figure 22.10(c)). In contrast, it is difficult to achieve a high gas response when the nanofibers with solid inner structures are very dense and thick because only the

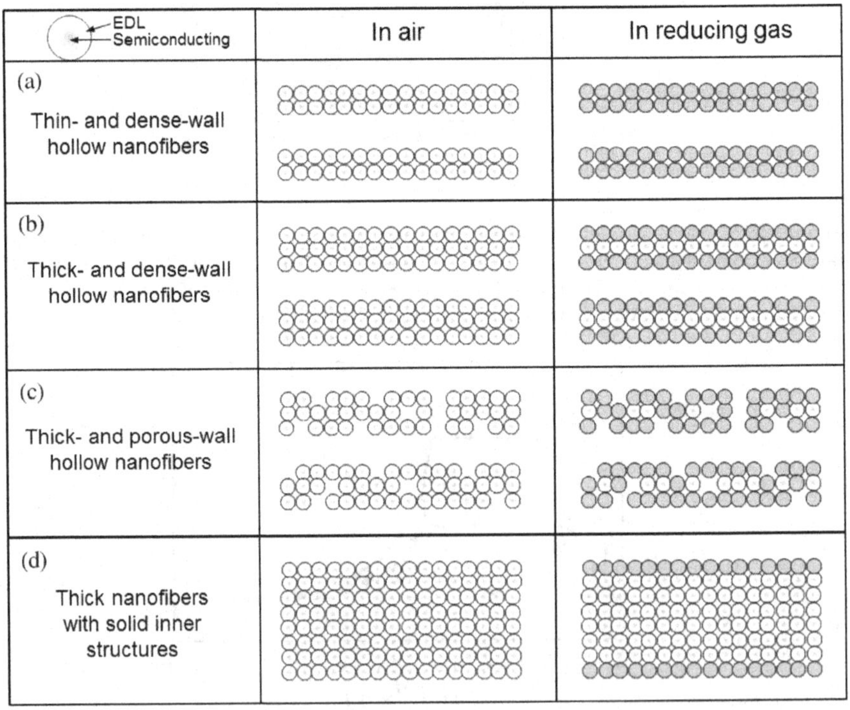

EDL Semiconducting	In air	In reducing gas
(a) Thin- and dense-wall hollow nanofibers		
(b) Thick- and dense-wall hollow nanofibers		
(c) Thick- and porous-wall hollow nanofibers		
(d) Thick nanofibers with solid inner structures		

Figure 22.10.   The gas response behaviors of hollow and solid nanofibers: (a) thin- and dense-wall hollow nanofibers; (b) thick- and dense-wall hollow nanofibers; (c) thick- and porous-wall hollow nanofibers; and (d) thick nanofibers with solid inner structures.

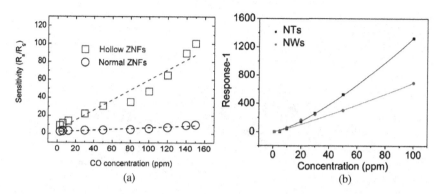

Figure 22.11.   (a) Comparison of CO sensitivity (R_a/R_g) for hollow ZnO nanofibers (hollow ZNFs) and normal ZnO nanofibers with solid inner structures (normal ZNFs) at 300°C (adapted from Ref. 28, used with permission); (b) [gas response (R_a/R_g)] to 1–100 ppm H_2S for In_2O_3 hollow nanofibers (NTs) and In_2O_3 nanofibers with solid inner structures (NWs) at room temperature (adapted from Ref. 58 with permission).

primary particles located near the atmosphere can have resistance changes (Figure 22.10(d)).

Park *et al.*[28] prepared hollow ZnO nanofibers (hollow ZNFs in Figure 22.11(a)) by the ALD of ZnO on sacrificial PVA nanofibers and subsequent core removal by heat treatment. The gas responses (R_a/R_g) to CO of the hollow ZnO nanofibers at 300°C were significantly higher than those of ZnO nanofibers with solid inner structures (normal ZNFs in Figure 22.11(a)) prepared from normal electrospinning of a Zn-precursor solution (Figure 22.8(a)). The responses (R_g/R_a) to 2–20 ppm NO_2 at 300°C were also enhanced significantly by employing nanotubular morphology (Figure 22.11(a)). Xu *et al.*[58] prepared one-dimensional In_2O_3 nanostructures by electrospinning a DMF solution containing In-nitrate and PVP (Mw = 90,000). The nanotubes with hollow inner structures and nanofibers with solid inner structures were prepared from solutions of In/PVP = 0.36 and 0.6, respectively. They reported that the responses (R_a/R_g) to 1–100 ppm H_2S at room temperature of In_2O_3 hollow nanofibers were higher than those of In_2O_3 nanofibers with solid morphologies (Figure 22.11(b)).

Choi *et al.*[29] prepared non-aligned and quasi-aligned hollow ZnO nanofibers by sputtering ZnO on sacrificial PVA nanofibers and subsequent core removal by heat treatment. For the alignment of nanofibers, they used the technique suggested by Li *et al.*,[88] that is, the connection of two strips of aluminum wires set up along the opposite edges on the substrates to the ground terminal (Figure 22.12(a)). The non-aligned nanofibers exhibited

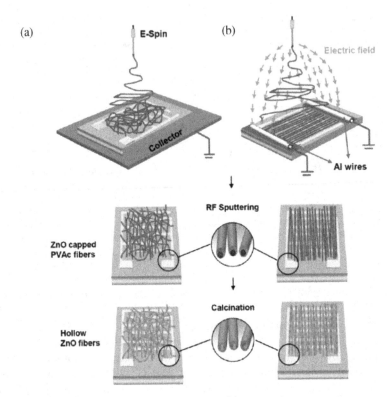

Figure 22.12.   Schematic diagram illustrating the fabrication procedure of an array of
(a) non-aligned and (b) quasi-aligned hollow ZnO fibers via template- (PVA-) assisted
electrospinning and (c) the resistance response R/R_0 (I_0/I where I_0 is the baseline
current in dry air at the beginning of the measurement and I is the DC current at time
t during the measurement) during cyclic exposure to increasing NO_2 concentrations at
$350°$C of sensors comprising a network of non-aligned or quasi-aligned hollow ZnO fibers
and a reference ZnO thin-film sensor. Note that the responses of the three specimens
are shifted vertically by one unit in order to separate them from each other. The NO_2
gas concentration profile is shown on top. The baseline current (I_0) levels were 6.0,
0.9, and 1.6 nA for the quasi-aligned fibers, non-aligned fibers, and thin-film specimens,
respectively. The inset shows a SEM micrograph of quasi-aligned fibers (adapted from
Ref. 29 with permission).

notable gas responses (R_g/R_a) even for 2 ppm NO_2 at 300°C, while ZnO
thin films exhibited weak gas responses at >8 ppm (Figure 22.12(b)). The
enhanced gas response in hollow nanofibers was attributed to the high surface area. It is also interesting that the NO_2 responses of quasi-aligned
nanofibers are ∼2 times higher than those of non-aligned nanofibers. However, the detailed reason for this remains unclear and might be related to
the diffusion of analyte gases and interconnections between nanofibers.

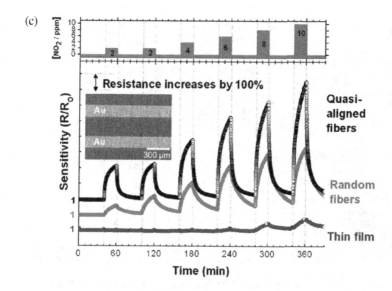

Figure 22.12. (*Continued*)

These results clearly demonstrate that the hollow configuration is very important to achieve a high gas response in one-dimensional nanostructures. The enhanced gas response can be attributed to the effective diffusion of analyte gas to the entire sensing surface through the hollow channels as well as the full depletion of sensing materials with thin-wall configurations.

Kim *et al.*[24] prepared SnO_2 hollow nanotubular fibers by ALD of thin SnO_2 layers onto electrospun PAN nanofiber templates and subsequent core removal by heat treatment at 700°C. They controlled the thickness of SnO_2 shell layers from 8 to 37 nm by increasing the number of ALD cycles from 100 to 500 (Figure 22.13(a–d)).

Overall, the gas responses to 100 ppm C_2H_5OH of SnO_2 nanotubes were significantly higher than the commercial SnO_2 nanopowders at 400–500°C (Figure 22.13(e)). The gas response and response speed tend to increase as the ALD cycles decrease, that is, as the thickness of the SnO_2 shell layers decreases (Figure 22.13(f)). This is due to the full electron depletion of the nanofibers in the thinner configuration and rapid and effective diffusion of C_2H_5OH to the inner and outer surfaces of the hollow nanostructures. This clearly shows that both a high gas response and a rapid response speed can be accomplished simultaneously using hollow nanostructures with very thin walls.

For identical diameters and wall thicknesses, the diffusion of analyte gases will be dependent upon the porosity of the shell layers. For example,

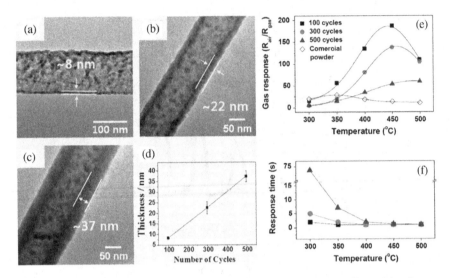

Figure 22.13.   Low-magnification TEM images of SnO$_2$ nanotubes after calcination at 700°C; (a) 100, (b) 300, and (c) 500 ALD cycles, and (d) wall thickness of SnO$_2$ nanotubes as a function of ALD cycles measured by TEM. (e) Gas response and (f) response time of SnO$_2$ nanotube network sensors to 100 ppm ethanol as a function of sensing temperature (adapted from Ref. 24 with permission).

the diffusion of analyte gases will occur more rapidly and effectively for nanoporous hollow structures and the porosity effect will become more significant when the wall is very thick. However, for extremely thin walls (for example, <20 nm) and substantial channel sizes (>20 nm), the nanometer-scale porosity of the nanofibers seems to be less important because the gas can diffuse to the inner surface via the hollow channels. This is an important strength of one-dimensional hollow structures for gas sensor applications.

22.3.2 Porosity

Seo et al.[48] prepared TiO$_2$ nanotubes with different pore sizes by hydrothermal treatment of TiO$_2$ powders (Degussa P25 powders) in an NaOH solution. The peak pore sizes of the TiO$_2$ nanotubes prepared by hydrothermal treatment temperatures of 160, 200, and 230°C were 20.4, 138.6, and 201.3 nm, respectively (Figure 22.14(a–d)). At the sensor temperature of 500°C, the sensors using commercial TiO$_2$ powders exhibited the highest responses to 500 ppm CO, 500 ppm H$_2$, 47 ppm C$_2$H$_5$OH, and 50 ppm toluene (Figure 22.14(e)). However, the relative gas responses were significantly dependent upon the pore size of the nanotubes. That is, as the

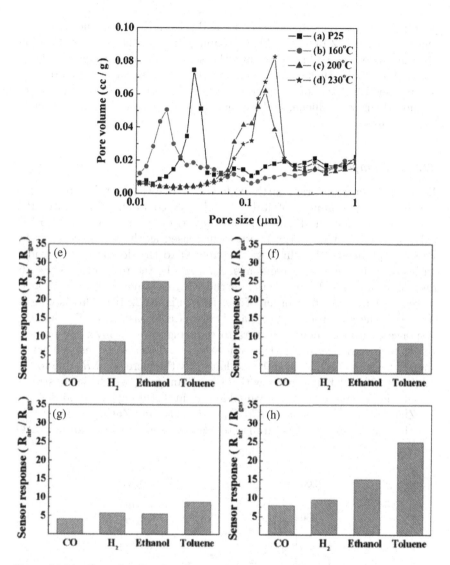

Figure 22.14. Pore size distribution of the sensing films composed of (a) commercial TiO$_2$ nanoparticles (P-25) and those hydrothermally treated at (b) 160, (c) 200, and (d) 230°C. Sensor responses to CO (500 ppm), H$_2$ (500 ppm), ethanol (47 ppm), and toluene (50 ppm) gases at 500°C for the devices using (e) commercial TiO$_2$ nanoparticles (P-25) and those that were hydrothermally treated at (f) 160, (g) 200, and (h) 230°C. The sensors were heat-treated at 600°C (adapted from Ref. 48 with permission).

peak pore size increased, the response and selectivity to toluene tended to increase (Figure 22.14(e–h)). They attributed this to the effective diffusion of large molecules (toluene) toward the sensor surface through the large pores. This shows that pore size engineering has a significant role to play in the selective detection of large molecules, such as volatile organic compounds (benzene, toluene, and xylene), that are among small-sized interference gases.

22.3.3 *Composition*

Xu *et al.*[65] prepared rare-earth-oxide-doped In_2O_3 nanotubes by single-capillary electrospinning of a DMF solution containing In-nitrate, RE (rare earth)-nitrate (RE = Gd, Tb, Dy, Ho, Er, Tm, Yb), and PVP (Mw = 90,000). The energy bandgap increased as the atom order of the REs except In_2O_3:Tb, which was attributed to the doping of RE oxides with gradually increasing bandgaps. As a result, the resistance in the air decreased and the H_2S responses of the RE:In_2O_3 nanotubes at <100°C increased with the atom order of the REs (Figure 22.15). This is consistent with the receptor-function-based theoretical consideration[89] that the response of n-type semiconductors to reducing gases is inversely proportional to the square root of the donor density. Thus, the enhancement of the gas response (R_a/R_g) can be explained by the increase of R_a due to the addition of the RE materials with higher bandgaps, which demonstrates that bandgap engineering is a valuable tool in designing gas responses.

Zhang *et al.*[69] prepared $CuCo_2O_4$, $NiCo_2O_4$, and $ZnCo_2O_4$ nanotubes via the infiltration of a metal-nitrate aqueous solution into porous AAO

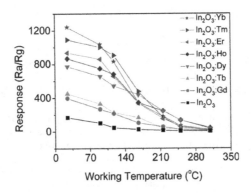

Figure 22.15.   Responses of the In_2O_3:RE and In_2O_3 NTs to 20 ppm H_2S as a function of the working temperature (reproduced from Ref. 65 with permission from the American Chemical Society).

Table 22.2.   Response results of sensors made by as-synthesized $CuCo_2O_4$, $NiCo_2O_4$, and $ZnCo_2O_4$ nanoparticles/nanotubes to various gases at 400 ppm at 300°C (reproduced from Ref. 69 with permission).

Detected Gas	$CuCo_2O_4$		$NiCo_2O_4$		$ZnCu_2O_4$	
	Nanoparticle	Nanotube	Nanoparticle	Nanotube	Nanoparticle	Nanotube
CH_3COOH	1.27	1.40	1.12	35.28	0.944	1.455
SO_2	1.61	41.65	—	3.78	1.26	2.83
C_2H_5OH	1.38	1.11	1.12	7.63	1.00	6.17
CO	1.08	8.37	1.00	1.61	1.35	1.69
Cl_2	1.39	2.72	1.49	1.50	1.27	2.48
NO_2	1.07	1.37	1.01	1.51	1.50	1.72

templates and subsequent removal of the templates (see Table 22.2). The gas responses of all the nanotube sensors were significantly higher than those of their nanoparticle counterparts, which again showed that one-dimensional hollow structures are very promising gas-sensing platforms. The $CuCo_2O_4$, $NiCo_2O_4$, and $ZnCo_2O_4$ nanotube sensors exhibited selective detection of SO_2, CH_3COOH, and C_2H_5OH, respectively. These clearly show that the selective detection to a specific gas can be manipulated by the compositional control of nanotubes.

22.3.4   *Catalyst loading*

The loading of noble metal or metal oxide catalysts are very effective to promote a gas sensing reaction on the surface.[90-94] The role of a catalyst in a gas sensing reaction can be divided into two categories: chemical sensitization and electronic sensitization. The chemical sensitization promotes the gas sensing reaction via the spillover of the sample gas. Thus, in n-type semiconductors, the enhancement of the gas response emanates from the promotion of the reaction between the reducing analyte gas and the adsorbed oxygen. The typical example of a chemical sensitizer is Pt. The electronic sensitization is due to the electronic interaction between metal or metal oxide and semiconductor gas sensors. In general, electronic sensitization accompanies a significant change of the sensor resistance via the shift of its work function. Typical examples of electronic sensitizers are Ag_2O and PdO. The loading of noble metals can also change the gas response or recovery kinetics or the sensing temperature for the maximum gas response. For example, in n-type semiconductors, it can promote the surface reactions between the analyte and the adsorbed oxygen during the response reaction and/or the adsorption of negatively charged oxygen during the recovery reaction. The selective detection of a specific gas can be also accomplished by the promotion of the sensing reaction for a specific gas.[95-97] Thus, catalyst loading is a very effective approach to control gas response, gas selectivity, and the gas sensing temperature.

Joo et al.[64] prepared Pt- and Pd-added TiO_2 nanotube films by the anodization of Pt-Ti and Pd-Ti alloy thin films. Although undoped TiO_2 nanotubes exhibited resistance variations upon exposure to 0.1, 1, and 10% H_2, resistance changes that occurred very slowly and over a very long time (600 s) were insufficient to reach the saturation of signals (Figure 22.16(a)). In contrast, the resistances of 2 wt% Pt-doped and 2 wt% Pd-doped TiO_2 nanotube sensors changed rapidly with changing H_2 concentrations (Figure 22.16(b,c)). The time required to obtain 90% of the resistance variation upon exposure to 1,000 ppm H_2 was very short (10–20 s). Moreover, the gas responses of 2 wt% Pd-doped and 2 wt% Pt-doped TiO_2 nanotube sensors were markedly higher than undoped TiO_2 nanotube sensors. This shows that the loading of catalytic materials can improve not only the gas response, but also the gas response kinetics.

Lee et al.[25] prepared undoped, 0.08 wt% Pd-doped, and 0.4 wt% Pd-doped SnO_2 hollow nanofibers by single-capillary electrospinning and measured the gas responses to 100 ppm H_2, 100 ppm CO, 500 ppm CH_4, and 100 ppm C_2H_5OH at 330–440°C (Figure 22.17). At 385 and 440°C, the undoped and 0.08 wt% Pd-doped sensors exhibited the selective detection of C_2H_5OH. The response of the 0.4 wt% Pd-doped sensor to 100 ppm C_2H_5OH at 330°C ($R_a/R_g = 1,020.6$) was also significantly higher than its responses to 100 ppm H_2, 100 ppm CO, and 500 ppm CH_4 (4.3–41.8). However, at 440°C, the 0.4 wt% Pd-doped sensor exhibited selective detection to H_2 with the minimum cross-sensitivity to C_2H_5OH. Thus, the gas selectivity can be designed by concurrent control of catalyst loading and sensor temperature.

The loading of catalysts to n-type oxide semiconductors often makes the recovery reaction faster, probably due to the promotion of oxygen adsorption during recovery.[98] However, still, it takes a very long time for the complete recovery of such systems, which makes the real-time monitoring of harmful or toxic gases very difficult. Lee et al.[99] prepared SnO_2 hollow spheres whose interiors are functionalized with NiO by coating Sn precursors onto spherical Ni templates and the partial dissolution of core Ni components after heat treatment. The sensor exhibited ultra-fast response and recovery (90% response and recovery times ≤5 s). The fast response was explained by rapid gas diffusion through the thin and porous shells. It is known that a full monolayer of oxygen ions can be adsorbed on the surface of NiO because the Ni^{2+} ions can be oxidized into Ni^{3+}.[100] Accordingly, the ultra-fast recovery was attributed to the promotion of a surface reaction that involved the adsorption of oxygen by the assistance of the NiO inner

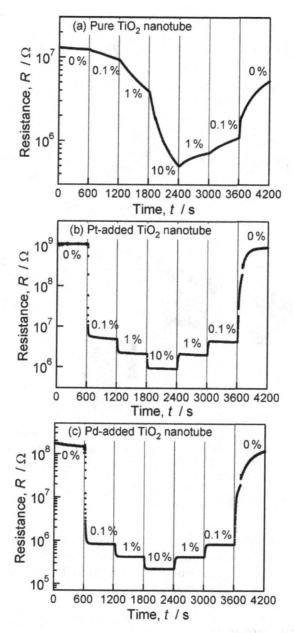

Figure 22.16. Typical response of TiO_2 nanotube sensors prepared from thin films of (a) pure Ti, (b) Ti–2% Pt, and (c) Ti–2% Pd upon exposure to H_2 with concentrations of 0.1, 1.0, and 10 vol% at 290°C (reproduced from Ref. 64 with permission).

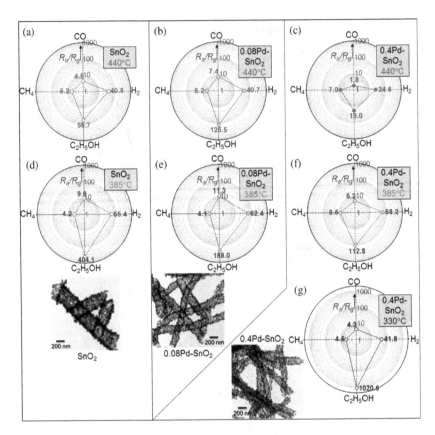

Figure 22.17. TEM images and gas responses to 100 ppm H$_2$, 100 ppm CO, 500 ppm CH$_4$, and 100 ppm C$_2$H$_5$OH of the hollow SnO$_2$ and Pd-SnO$_2$ nanofiber sensors: (a) SnO$_2$, T $=$ 440°C; (b) 0.08Pd–SnO$_2$ (0.08 wt% Pd-doped SnO$_2$), T $=$ 440°C; (c) 0.4Pd–SnO$_2$ (0.4 wt% Pd-doped SnO$_2$), T $=$ 440°C; (d) SnO$_2$, T $=$ 385°C; (e) 0.08Pd–SnO$_2$, T $=$ 385°C; (f) 0.4Pd–SnO$_2$, T $=$ 385°C; (g) 0.4Pd–SnO$_2$, T $=$ 330°C (adapted from Ref. 25 with permission).

layer with abundant negatively charged surface oxygen atoms. This shows that the partial removal of Ni templates can be very useful for achieving ultra-fast recoveries for gas sensing reactions as well as for the synthesis of well-defined hollow nanostructures.

22.4 Future Prospects

The main strengths of one-dimensional hollow oxide nanostructures for gas sensor applications are the high and rapid responses, which emanate from the high surface areas, nanoscale wall thicknesses, less-agglomerated configurations, and the rapid and effective diffusion of analyte gases via hollow channels. The remaining challenges include selective gas detection, rapid recovery, facile sensor fabrication, and sensor stability.

In gas sensors using thin and thick films, the gas selectivity is designed or accomplished by the following: (1) the addition of catalytic materials; (2) the compositional control of sensing materials; (3) the employment of selective filtering layers; and (4) the control of the sensor temperature. The catalytic materials can be loaded in the one-dimensional hollow structures either during the fabrication or after the fabrication of hollow nanostructures. Although the gas sensing characteristics of undoped one-dimensional hollow nanostructures were studied intensively, catalyst-loaded one-dimensional hollow structures have been minimally investigated as gas sensors. In the future, catalyst-loaded one-dimensional hollow structures could be valuable material platforms for selective and sensitive gas sensors. Concurrent engineering of pore size and catalysts in one-dimensional hollow structures is also very promising. As suggested by Seo *et al.*,[48] the selective detection of large-molecule gases can be accomplished by the control of pores or hollow diameters. Therefore, the combination of catalyst loading and pore size engineering will be undoubtedly very effective for detecting a specific gas in a selective manner.

The coating of different metal oxide layers on one-dimensional hollow oxide nanostructures can be used as another factor for designing high-performance gas sensors. At first, the surface functionalization will endow new gas sensing properties. The outer oxide layer can also play the role of a selective filtering layer to remove specific interference gases. To achieve this, a catalyst-loaded porous layer can be coated and its thickness should be tuned precisely. Moreover, the junction between two different sensing materials with different work functions will form a new electron depletion layer at their interface, which can be used to enhance the gas response. The preparation of multi-compositional hollow structures is another important research area. The acid-base property of sensing materials is very important for controlling their reactivity to the different analyte gases,[75] which is a key factor to be controlled in designing highly selective gas sensors. For this, high-throughput processes such as electrospinning will be advantageous.

Simple, economic, and large-scale fabrication of gas sensors is essential for real applications. For mass production, the screen printing of slurries containing hollow nanofibers is the most practical method.[75] Near-field electrospinning of nanofibers directly onto electrodes can also be used for sensor fabrication.[101] When two different grounded electrodes are used, the direct electrospinning of well-aligned one-dimensional nanostructures can also be attained.[28,88]

The performance of oxide semiconductors can be affected by siliceous contamination,[102] oil contamination,[103] and sintering between nanostructures.[104] The contamination corresponds to the chemical poisoning of the sensor and catalyst surfaces[102,105,106] and the sintering is related to the change in the physical configuration of the nanostructures. Degradation of a sensor by the poisoning of the sensor surface in the one-dimensional hollow structures will be similar to that in the conventional oxide semiconductor gas sensors. The sensor stability related to the configuration of one-dimensional hollow nanostructures is, therefore, related to the maintenance of the hollow and porous nanostructure during the sensor operation at high temperatures. Heat treatment of a sensor at a temperature higher than its sensing temperature and long-term aging of the sensor at its sensing temperature are effective methods for avoiding thermal degradation of the sensor. The addition of an insulating second phase to suppress the coarsening of the nanoparticles can also be considered as an effective approach.[7]

22.5 Conclusion

One-dimensional hollow oxide nanostructures provide high surface areas, effective diffusion paths of analyte gases, and less agglomerated configurations, all of which enhance the gas response and its speed significantly. Various physicochemical routes using the template-based method, electrospinning, the Kirkendall effect, anodization, and chemical vapor deposition can be employed to prepare well-defined one-dimensional hollow oxide nanostructures. For gas sensor applications, the thickness and porosity of walls and the diameters of hollow channels are key parameters for tuning the gas response, response kinetics, and selectivity. The gas sensing characteristics can be improved further when the inner and/or outer surface of one-dimensional hollow nanostructures are functionalized or coated by noble metal catalysts or metal oxide materials. One-dimensional hollow oxide nanostructures, therefore, are very promising materials for high-performance gas sensors.

Acknowledgments

This work was supported by the Korea Science and Engineering Foundation (KOSEF) NRL program grant funded by the Korean government (no. R0A-2008-000-20032-0).

References

1. F. Caruso, *Adv. Mater.* **13**, 11–22 (2001).
2. F. Caruso, *Chem. Eur. J.* **6**, 413–419 (2000).
3. F. Caruso, R. A. Caruso and Möhwald, *Science* **282**, 1111–1114 (1998).
4. R. Meyer Jr., H. Weitzing, Q. Xu, Q. Zhang and R. E. Newnham, *J. Am. Ceram. Soc.* **77**, 1669–1772 (1994).
5. S. Han, B. Jang, T. Kim, S. M. Oh, and T. Hyeon, *Adv. Funct. Mater.* **15**, 1845–1850 (2005).
6. N. Yamazoe, *Sens. Actuat. B-Chem.* **108**, 2–14 (2005).
7. N. Yamazoe, *Sens. Actuat. B-Chem.* **5**, 7–19 (1991).
8. N. Barsan and U. Weimer, *J. Electroceram.* **7**, 143–167 (2001).
9. C. N. Xu, J. Tamaki, N. Miura and N. Yamazoe, *Sens. Actuat. B-Chem.* **3**, 147–155 (1991).
10. J.-H. Lee, *Sens. Actuat. B-Chem.* **140**, 319–336 (2009).
11. Y. Xia, P. Yang, Y. Sun, Y. Wu, B. Mayers, B. Gates, Y. Yin, F. Kim and H. Yan, *Adv. Mater.* **15**, 353–389 (2003).
12. J. Li, Z. Tang and Z. Zhang, *Electrochem. Commun.* **7**, 62–67 (2005).
13. G. K. More, O. K. Varghese, M. Paulose, K. Shankar and C. A. Grimes, *Sol. Energ. Mat. Sol. Cells* **90**, 2011–2075 (2006).
14. J. Huang, N. Matsunaga, K. Shimanoe, N. Yamaozoe and T. Kunitake, *Chem. Mater.* **17**, 3513–3518 (2005).
15. J.-Y. Gong, S.-R. Guo, H.-S. Qian, W.-H. Xu and S.-H. Yu, *J. Mater. Chem.* **19**, 1037–1042 (2009).
16. M. Lai, J.-H. Lim, S. Mubeen, Y. Rheem, A. Mulchandani, M. A. Deshusses and N. V. Myung, *Nanotechnology* **20**, 185602 (2009).
17. N. Du, H. Zhang, B. Chen, X. Ma, X. Huang, J. Tu and D. Yang, *Mater. Res. Bull.* **44**, 211–215 (2009).
18. Y. Jia, L. He, Z. Guo, X. Chen, F. Meng, T. Luo, M. Li and J. Liu, *J. Phys. Chem. C* **113**, 9681–9587 (2009).
19. Y. Wang, J. Y. Lee and H. C. Zeng, *Chem. Mater.* **17**, 3899–3903 (2005).
20. W. Zhu, W. Wang, H. Xu and J. Shi, *Mater. Chem. Phys.* **99**, 127–130 (2006).
21. G. X. Wang, J. S. Park, M. S. Park and X. L. Gou, *Sens. Actuat. B-Chem.* **131**, 313–317 (2008).
22. Y. Wang, M. Wu, Z. Jiao and J. Y. Lee, *Nanotechnology* **20**, 345704 (2009).
23. J. Ye, H. Zhang, R. Yang, X. Li and L. Qi, *Small* **6**, 296–306 (2010).
24. W.-S. Kim, B.-S. Lee, D.-H. Kim, H.-C. Kim, W.-R. Yu and S.-H. Hong, *Nanotechnology* **21**, 245605 (2010).
25. J.-K. Choi, I.-S. Hwang, S.-J. Kim, J.-S. Park, S.-S. Park, U. Jeong, Y. C. Kang and J.-H. Lee, *Sens. Actuat. B-Chem.* **150**, 191–199 (2010).
26. N. Du, H. Zhang, B. Chen, X. Ma and D. Yang, *Chem. Commun.* 3028–3030 (2008).

27. Y. Liu and M. Liu, *Adv. Funct. Mater.* **15**, 57–62 (2005).

28. J. Y. Park, S.-W. Choi and S.-S. Kim, *Nanotechnology* **21**, 475601 (2010).

29. S.-H. Choi, G. Ankonina, D.-Y. Youn, S.-G. Oh, J.-M. Hong, A. Rothschild I.-D. Kim, *ACS Nano* **3**, 2623–2631 (2009).

30. W.-S. Chen, D.-A. Huang, H.-C. Chen, T.-Y. Shie, C.-H. Hsieh, J.-D. Liao and C. Kuo, *Cryst. Growth Des.* **9**, 4070–4077 (2009).

31. Z. Zhang, X. Li, C. Wang, L. Wei, Y. Liu and C. Shao, *J. Phys. Chem. C* **113**, 19397–19403 (2009).

32. K. Raidongia and C. N. R. Rao, *J. Phys. Chem. C* **112**, 13366–13371 (2008).

33. Y. Qiu and S. Yang, *Nanotechnology* **19**, 265606 (2008).

34. Y. J. Xing, Z. H. Xi, Z. Q. Xue, X. D. Zhang, J. H. Song, R. M. Wang, J. Xu, Y. Song, S. L. Zhang and D. P. Yu, *Appl. Phys. Lett.* **83**, 1689 1691 (2003).

35. T.-J. Hsueh, S.-J. Chang, C.-L. Hsu, Y.-R. Lin and I.-C. Chen, *J. Electrochem. Soc.* **159**, K152–K155 (2008).

36. J. X. Wang, X. W. Sun, Y. Yang and C. M. L. Wu, *Nanotechnology* **20**, 465501 (2009).

37. Y.-J. Chen, C.-L. Zhu and G. Xiao, *Sens. Actuat. B-Chem.* **129**, 639–642 (2008).

38. D. Eder, I. A. Kinloch and A. H. Windle, *Chem. Commun.* 1448–1450 (2006).

39. H. Shin, D.-K. Jeong, J. Lee, M. M. Sung and J. Kim, *Adv. Mater.* **16**, 1197–1200 (2004).

40. K. S. Mayya, D. I. Gittins, A. M. Dibaj and F. Caruso, *Nano Lett.* **12**, 727–730 (2001).

41. J. Yu, J. Yu, B. Cheng and S. Liu, *Nanotechnology* **18**, 065604 (2007).

42. J. H. Lee, I. C. Leu, M. C. Hsu, Y. W. Chung and M. H. Hon, *J. Phys. Chem. B* **109**, 13056–13059 (2005).

43. J. Qiu, W. Yu, X. Gao and X. Li, *Nanotechnology* **17**, 4695–4698 (2006).

44. D. Liu and M. Z. Yates, *Langmuir* **23**, 10333–10341 (2007).

45. T. Zhang, L. Ge, X. Wang and Z. Gu, *Polymer* **49**, 2898–2902 (2008).

46. D. Li and Y. Xia, *Nano Lett.* **4**, 933–938 (2004).

47. S. Zhan, D. Chen, X. Jiao and C. Tao, *J. Phys. Chem. B* **110**, 11199–11204 (2006).

48. M.-H. Seo, M. Yuasa, T. Kida, J.-S. Huh, K. Shimanoe and N. Yamazoe, *Sens. Actuat. B-Chem.* **137**, 513–520 (2009).

49. H. Ogihara, S. Masahiro, Y. Nodasaka and W. Ueda, *J. Solid State Chem.* **182**, 1587–1592 (2009).

50. Z. Sun, H. Yuan, Z. Liu, B. Han and X. Zhang, *Adv. Mater.* **17**, 2993–2997 (2005).

51. J. Chen, X. Lina, W. Li and X. Gou, *Adv. Mater.* **17**, 582–586 (2005).

52. S. Zhan, D. Chen, X. Jiao and S. Liu, *J. Colloid Interface Sci.* **308**, 265–270 (2007).

53. S. K. Mohapatra, S. E. John, S. Banerjee and M. Misra, *Chem. Mater.* **21**, 3048–3055 (2009).

54. L. Liu, H.-Z. Kou, W. Mo, H. Liu and Y. Wang, *J. Phys. Chem. B* **110**, 15218–15223 (2006).
55. N. Du, H. Zhang, B. Chen, X. Ma , Z. Liu, J. Wu and D. Yang, *Adv. Mater.* **19**, 1641–1645 (2007).
56. X.-P. Shen, H.-J. Liu, X. Fan, Y. Jiang, J.-M. Hong, and Z. Xu, *J. Cryst. Growth* **276**, 471–477 (2005).
57. B. Cheng and E. T. Samulski, *J. Mater. Chem.* **11**, 2901–2902 (2001).
58. L. Xu, B. Dong, Y. Wang, X. Bai, Q. Liu and H. Song, *Sens. Actuat. B-Chem.* **147**, 531–538 (2010).
59. N. Du, H. Zhang, B. Chen, J. Wu, X. Ma, Z. Liu, Y. Zhang, D. Yang, X. Huang and J. Tu, *Adv. Mater.* **19**, 4505–4509 (2007).
60. G. An, Y. Zhang, Z. Liu, Z. Miao, B. Han, S. Miao and J. Li, *Nanotechnology* **19**, 035504 (2008).
61. Y. Chang, M. L. Lye and H. C. Zeng, *Langmuir* **21**, 3746–3748 (2005).
62. Y. Li and S. Zhan, *J. Disper. Sci. Technol.* **30**, 246–249 (2009).
63. B. C. Satishkumar, A. Govindaraj, M. Natha and C. N. R. Rao, *J. Mater. Chem.* **10**, 2115–2119 (2000).
64. S. Joo, I. Muto and N. Hara, *J. Electrochem. Soc.* **157**, J221–J226 (2010).
65. L. Xu, B. Dong, Y. Wang, X. Bai, J. Chen, Q. Liu and H. Song, *J. Phys. Chem. C* **114**, 9089–9095 (2010).
66. A. A. Firooz, A. R. Mahjoub and A. A. Khodadadi, *J. Nanosci. Nanotechnol.* **10**, 6155–6160 (2010).
67. G. Zhang, C. Li, F. Cheng and J. Chen, *Sens. Actuat. B-Chem.* **120**, 403–410 (2007).
68. D. Wang, X. Chu and M. Gong, *Nanotechnology* **17**, 5501–5505 (2006).
69. G.-Y. Zhang, B. Guo and J. Chen, *Sens. Actuat. B-Chem.* **114**, 402–409 (2006).
70. H.-W. Liang, S. Liu and S. H. Yu, *Adv. Mater.* **22**, 3925–3937 (2010).
71. Y. Xiong, B. T. Mayers and Y. Xia, *Chem. Commun.* 5013–5022 (2005).
72. G. Cao and D. Liu, *Adv. Colloid Interface Sci.* **136**, 45–64 (2008).
73. M. Lai and D. J. Riley, *J. Colloid Interface Sci.* **323**, 203–212 (2008).
74. M. Steinhart, R. B. Wehrspohn, U. Gösele and J. H. Wendorff, *Angew. Chem. Int. Ed.* **43**, 1334–1344 (2004).
75. T. Jinkawa, G. Sakai, J. Tamaki, N. Miura and N. Yamazoe, *J. Mol. Catal. A Chem.* **155**, 193–200 (2000).
76. R. A. Potyrailo and V. M. Mirsky (eds.), *Combinatorial Methods for Chemical and Biological Sensors* (Springer, New York, 2009), p. 295.
77. F. Rock, N. Barsan and U. Weimar, *Chem. Rev.* **108**, 705–725 (2008).
78. D. Li and Y. Xia, *Adv. Mater.* **16**, 1151–1170 (2004).
79. W. Sigmund, J. Yuh, H. Park, V. Maneeratana, G. Pyrgiotakis, A. Daga, J. Taylor and J. C. Nino, *J. Am. Ceram. Soc.* **89**, 395–407 (2006).
80. J. T. McCann, D. Li and Y. Xia, *J. Mater. Chem.* **15**, 735–738 (2005).
81. B. Ding, M. Wang, X. Wang, J. Yu and G. Sun, *Materials Today* **13**, 16–27 (2010).
82. D. Li, J. T. McCann and Y. Xia, *Small* **1**, 83–86 (2005).

83. I.-S. Hwang, J.-K. Choi, S.-J. Kim, K.-Y. Dong, J.-H. Kwon, B.-K. Ju, J.-H. Lee, *Sens. Actuat. B-Chem.* **142**, 105–110 (2009).

84. X. Xue, L. Xing, Y. Chen, S. Shi, Y. Wang and T. Wang, *J. Phys. Chem. C* **112**, 12157–12160 (2008).

85. Y. Yin, R. M. Rioux, C. K. Erdonmez, S. Hughes, G. A. Somorjai and A. P. Alivisatos, *Science* **30**, 711–714 (2004).

86. H. J. Fan, Y. Gosele and M. Zacharias, *Small* **3**, 1660–1671 (2007).

87. O. K. Varghese, D. Gong, M. Paulose, K. G. Ong, E. C. Dickey and C. A. Grimes, *Adv. Mater.* **15**, 624–627 (2003).

88. D. Li, Y. Wang and Y. Xia, *Nano Lett.* **3**, 1167–1171 (2003).

89. N. Yamazoe and K. Shimanoe, *Sens. Actuat. B-Chem.* **138**, 100–107 (2009).

90. A. Kolmakov, D. O. Klenov, Y. Lilach, S. Stemmer and M. Moskovits, *Nano Lett.* **5**, 667–673 (2005).

91. A. Cabot, A. Dieguez, A. Romano-Rodriguez, J. R. Morante and N. Barsan, *Sens. Actuat. B-Chem.* **79**, 98–106 (2001).

92. Y. Shimizu, N. Matsunaga, T. Hyodo and M. Egashira, *Sens. Actuat. B-Chem.* **77**, 35–40 (2001).

93. M. Yuasa, T. Masaki, T. Kida, K. Shimanoe and N. Yamazoe, *Sens. Actuat. B-Chem.* **136**, 99–104 (2009).

94. Y. C. Lee, H. Huang, O. K. Tan and M. S. Te, *Sens. Actuat. B-Chem.* **132**, 239–242 (2008).

95. J. Tamaki, K. Shimanoe, Y. Yamada, Y. Yamamoto, N. Miura and N. Yamazoe, *Sens. Actuat. B-Chem.* **49**, 121–125 (1998).

96. N .V. Hieu, H.-R. Kim, B.-K. Ju and J.-H. Lee, *Sens. Actuat. B-Chem.* **133**, 228–234 (2008).

97. U.-S. Choi, G. Sakai and N. Yamazoe, *Sens. Actuat. B-Chem.* **107**, 397–401 (2005).

98. J.-H. Park and J.-H. Lee, *Sens. Actuat. B-Chem.* **136**, 151–157 (2009).

99. H.-R. Kim, K.-I. Choi, K.-M. Kim, I.-D. Kim, G. Cao and J.-H. Lee, *Chem. Commun.* **46**, 5061–5063 (2010).

100. D. Kohl, *J. Phys. D: Appl. Phys.* **34**, R125–R149 (2001).

101. W.-Y. Wu, J.-M. Ting, P.-J. Huang, *Nanoscale Res. Lett.* **4**, 513–517 (2009).

102. D. E. Williams,G. S. Henshaw and F. E. Pratt, *J. Chem. Soc. Faraday Trans.* **91**, 3307–3308 (1995).

103. G. K. Mor, M. A. Carvalho, O. K. Varghese, M. V. Pishko and C. A. Grimes, *J. Mater. Res.* **19**, 628–634 (2004).

104. V. V. Sysoev, T. Schneider, J. Goschnick, I. Kiselev, W. Habicht, H. Hahn, E. Strelcov and A. Kolmakov, *Sens. Actuat. B-Chem.* **139**, 699–703 (2009).

105. J.-J. Ehrhardt, L. Colin and D. Jamois, *Sens. Actuat. B-Chem.* **40**, 117–124 (1997).

106. M. Matsumiya, W. Shin, F. Qiu, N. Izu, I. Mastubara and N. Murayama, *Sens. Actuat. B-Chem.* **96**, 516–522 (2003).

Chapter 23

Preparation and Electrochemical Application of Titania Nanotube Arrays

Peng Xiao[*], Guozhong Cao[†] and Yunhuai Zhang[‡]

[*]College of Physics, Chongqing University, Chongqing, P. R. China

[†]Department of Materials Science and Engineering, University of Washington, Seattle, WA, USA

[‡]College of Chemistry and Chemical Engineering, Chongqing University, Chongqing, P. R. China

23.1 Introduction

Titanium dioxide (titania, TiO_2) is one of the most important transition metal oxides, with potential applications in UV-ray shielding, paint, paper industries, self-cleaning devices, electrochromic devices, and solar cells, etc. Moreover, its ability to decompose pollutants and protect the environment has increased the interest in novel photocatalytic applications. Nanostructured TiO_2 materials, with a typical dimension of less than 100 nm, include spheroidal nanocrystallites and nanoparticles together with elongated nanotubes, nanosheets, and nanofibers. A major feature that distinguishes these various types of nanostructured TiO_2 is their dimensionality. Among these various nanostructured forms of TiO_2, titania nanotubes have attracted significant attention due to their high surface-to-volume ratios and size-dependent properties. Several studies have indicated that titania nanotubes have improved properties compared to any other form of titania for application in photocatalysis,[1,2] sensing,[3,4] photoelectrolysis,[5] and photovoltaics.[6,7] Titania nanotubes have been produced by a variety of methods including the template method,[8–10] hydrothermal processes,[11–13] and the anodization method.[14,15] A timeline showing the development and application of titanate nanotubes is shown in Figure 23.1.[16] However, of

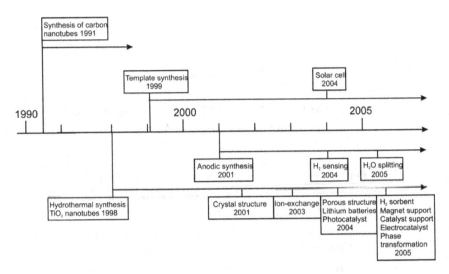

Figure 23.1.   Simplified timeline describing the development of TiO₂-related nanotubular structures.[16]

these nanotube fabrication routes, the architecture demonstrating by far
the most remarkable properties is a highly ordered nanotube array made
by anodization of titanium in a fluoride-based bath, the dimensions of which
can be precisely controlled. In 2001, Grimes and co-workers[14] first reported
the formation of a uniform titania nanotube array via anodic oxidation
of titanium in a hydrofluoric (HF) electrolyte. After that, titania nanotube arrays of various pore sizes (22–200 nm), lengths (0.2–1,00 μm), and
wall thicknesses (7–34 nm) were successfully grown by tailoring the electrochemical conditions. Herein we mainly focus on the anodic fabrication,
material properties, and environmental applications of titania nanotube
arrays.

23.2   Fabrication of TiO₂ Nanotube Arrays

Titania nanotube arrays can be fabricated via anodic oxidation of titanium foil in a fluoride-based solution. Electrolyte composition plays a critical and essentially unexplored role in determining the resultant nanotube
nanoarchitecture and its chemical composition. The effect of other synthesis parameters such as current density, pH value, applied voltage, and
time of anodic oxidation has been critically studied. In all cases, a fluoride
ion-containing electrolyte is needed for TiO₂ nanotube array formation.

23.2.1 Fabrication of TiO$_2$ nanotubes based on aqueous-solution electrolytes

Grimes' group[14] first observed the formation of TiO$_2$ nanotube arrays on a thin titanium foil after anodization treatment in HF-containing aqueous solutions of different concentrations. Arrays of nanotubes with constant length and variable diameter (25–65 nm) could be produced under variable anodizing voltages. As the voltage was increased, a porous or sponge-like structure, a particulate or nodular structure, and finally discrete, hollow, cylindrical, tube-like features were observed. The structural evolution at different voltages and the cross-sectional view of the nanostructures in 0.5 wt% HF using FE-SEM are shown in Figure 23.2.

The uniform titania nanotube arrays were regularly obtained under anodizing voltages ranging from 10 to 40 V, depending on the HF concentration, with relatively higher voltages needed to achieve the tube-like structures in more dilute HF solutions. In all cases, the final length of the nanotubes was found to be independent of the anodizing time. Nanotube samples have, respectively, inner diameters from 25 to 65 nm and lengths from 200 to 400 nm after anodizing for 45 minutes. However, the addition of

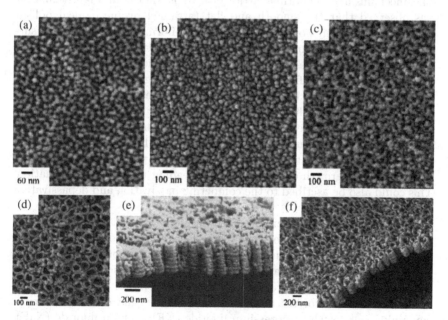

Figure 23.2. FE-SEM top-view images of porous titanium oxide films anodized in 0.5 wt% HF solution for 20 minutes under different voltages: (a) 3 V, (b) 5 V, (c) 10 V, and (d) 20 V; (e), (f) cross-sectional images of titania nanotubes from the sample was anodized in 0.5 wt% HF solution at 20 V for 20 minutes.[14]

chromium trioxide was not found to influence the formation of the titania nanotubes.

The overall reactions for anodic oxidation of titanium can be represented as

$$2H_2O \rightarrow O_2 + 4e + 4H^+ \tag{23.1}$$

$$Ti + O_2 \rightarrow TiO_2 \tag{23.2}$$

$$TiO_2 + 6F^- + 4H^+ \rightarrow TiF_6^{2-} + 2H_2O. \tag{23.3}$$

At the start of the process, an oxide layer forms on the surface of titanium as a result of reactions (23.2) and (23.3). In the presence of F^-, the chemical dissolution (reaction 23.3) reduces the thickness of the oxide layer (barrier layer) keeping the electrochemical etching (field-assisted oxidation and dissolution) process active. Furthermore, a detailed study of titania nanotube array fabrication using anodic oxidation in HF electrolyte was reported by Zhao et al.[17] They systematically studied the formation mechanism, morphology, and the dimensions of the nanotubes as a function of applied voltage, electrolyte concentration, and oxidation time. The mechanism of nanotube array growth is shown in Figure 23.3. It was observed that the smooth and flat (attained by mechanical polishing) foils of titanium would have some non-uniform oxide layer on the surface. Selective etching of the oxide by HF was attributed to non-uniform stresses existing in the thin oxide film. The oxide layer color changed during growth swiftly from an initial purple to green via blue and yellow colors in between, the reason being the increase in thickness of the oxide layer leading to different interference phenomena. The current density was found to decrease in the initial stage, then again increase, and a similar periodic fluctuation could be observed throughout the anodization process. This variation in the current density was consistent with the dissolution and growth of the oxide layer leading to nanotube formation. This could also be ascribed to the competing passivation and depassivation reactions.

The nanotube length increases until the electrochemical etch rate equals the chemical dissolution rate of the top surface of the nanotubes. The chemical dissolution rate is determined by the F^- concentration and solution pH (reaction 23.3). With increasing F^- and H^+ concentrations, chemical dissolution increases. Therefore, controlling the dissolution rate by adjusting the pH of the electrolyte through additives and using different fluorine-containing salts, such as potassium fluoride (KF), sodium fluoride (NaF), and NH_4F, instead of HF are effective ways to increase the nanotube length. A number of papers describing the fluorine-containing salt electrolytes for the preparation of TiO_2 nanotubes by anodization of titanium

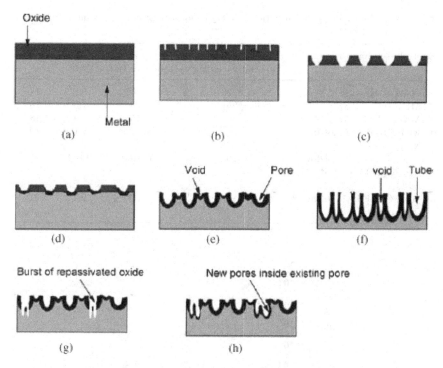

Figure 23.3. Schematic diagram of formation of titania nanotube arrays: (a) oxide layer formation, (b) burst of oxide by the formation of crystallites (pore formation), (c) growth of the pores due to field-assisted dissolution of titania, (d) immediate repassivation of pore tips, (e) void formation in the metallic part between the pores, (f) formation of nanotubes of titania, (g) burst of repassivated oxide, (h) formation of new pores inside existing pores.[17]

have been published. The morphological properties of the resultant nanotubes are summarized in Table 23.1. The authors determined the dissolution rate, R_{diss}, of the anodic TiO_2 depending on the pH value as shown in Figure 23.4.

It was also reported that the nanotube formation was very sensitive to the sweep rate.[15] When using a low sweep rate, only short pores (nanotubes) were obtained. A fast sweep rate led to the growth of a thick porous layer. However, the surface in this thick porous layer was very rough and irregular and the structure itself was loosely cross-linked and not tubular.

Synthesis of ordered titania nanotubular structures by dealloying and subsequent anodization of a Ti–8 at% Al alloy[22] was also achieved. Al was selectively dissolved in 1 M NaOH at a critical potential creating a nanoporous structure, which on further anodization in 1 M H_3PO_4+1–1.5%

Table 23.1. Morphological properties of TiO$_2$ nanotubes produced by anodizing Ti foil in various electrolytes.

Electrolyte Composition	Electrode Potential vs. SCE (V)	Inner Diameter (nm)	Length (μm)	Reference
0.5–3.5%wt HF (pH < 1)	3–20	25–65	0.2	14
0.1 MKF (pH = 2.8)	25	115	1.5	18
0.1 MKF (pH = 4.5)	25	115	4.4	18
0.5 wt% NH$_4$F in 1 mol dm^{-3} (NH$_4$)$_2$SO$_4$	20	90–110	0.5–0.8	19
0.5 wt% NH$_4$F in 1 mol dm^{-3} (NH$_4$)H$_2$PO$_4$, 1 mol dm^{-3} H$_3$PO$_4$	20	40–100	0.1–4	20
0.1–1 wt% NaF in 1 mol dm^{-3} Na$_2$SO$_4$	20	100	2.4	21

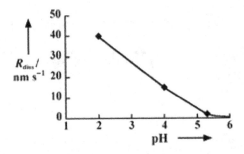

Figure 23.4. Experimental determination of the dissolution rate (R_{diss}) of the anodic TiO$_2$ depending on the pH value.[15]

HF resulted in titania nanotubes with 40–140 nm diameters and lengths of almost 180 nm.

23.2.2 *Fabrication of TiO$_2$ nanotubes based on organic electrolytes*

In aqueous electrolytes, it was further established[20,23] that the thickness of the nanotubes was a result of equilibrium between the electrochemical formation of TiO$_2$ at the bottom and the chemical dissolution of TiO$_2$. For a given rate of nanotube formation, the chemical dissolution of the oxide at the tube mouth by F$^-$ determines the tube length. To control this dissolution reaction, the key point is to minimize water content in the anodization bath to less than 5%. Macak *et al.* and Paulose *et al.* reported long TiO$_2$ nanotube arrays fabricated in organic electrolytes.[24,25] With organic electrolytes, the donation of oxygen is more difficult in comparison to water, thus reducing the tendency to form oxide. At the same time, the reduction in water content reduces the chemical dissolution of the oxide in the fluorine-containing electrolytes and hence aids the formation of longer nanotubes. For example,[25] in a 0.25% NH_4F and ethylene glycol solution at an anodization potential of 60 V, a 134 μm long TiO_2 nanotube array was fabricated after anodizing for 17 hours, with an inner diameter of about 140 nm. After anodizing in DMSO containing 2% HF at 60 V for 69 hours, the nanotubes had a length of approximately 93 μm and a pore diameter of 200 nm. In formamide and/or N-methylformamide solutions containing 1–5 wt% of deionized water and 0.3–0.6 wt% NH_4F, the nanotubes were nearly 70 μm long. In addition to electrolyte composition, anodization duration, and applied field, the length and the quality of the resulting nanotubes are dependent upon the previous usage of the electrolyte solution (the used solution effect),[26] that is, the number of times and duration it had been previously used for anodizing samples under similar conditions.

By employing double-sided electrochemical oxidation of titanium in a once previously used electrolyte comprised of water, NH_4F, and ethylene glycol, Grimes' group[26] obtained two highly ordered, hexagonal, close-packed titania nanotube arrays 360 μm in length that were separated by a thin compact oxide layer; the individual nanotubes in each array had an aspect ratio of ~2,200. A maximum individual nanotube array length of over 1,000 μm was obtained upon anodizing 1.0 mm thick Ti foil at 60 V for 216 hours (9 days) in 0.6 wt% NH_4F and 3.5% water in ethylene glycol. The nanotube array film was separated from the underlying Ti substrate and preserved its flatness when dried in a critical point dryer. They opened the closed end of the tubes after the barrier layer side of the membrane was immersed in a dilute hydrofluoric acid/sulfuric acid solution to etch the oxide layer. Reduced hydroxyl ion injection from the electrolyte, which enabled faster high-field ionic conduction through the barrier layer, was responsible for the high nanotube growth rates achieved.

23.2.3 *Fabrication of transparent TiO_2 nanotubes array*

Titania nanotube arrays have proven useful for a wide range of applications: photocatalysis, solar cell, gas sensing, and biomaterials. However, most of these nanotubes are fabricated from a Ti foil, which ultimately limits their application in functional microdevices. For example, at elevated temperatures, the metal electrodes deposited atop the nanotube array for electrical contact diffuse into the Ti foil underneath the nanotubes, resulting in an electrical short-circuit. Another problem with Ti-foil-based devices is their susceptibility to mechanical shock/vibrations, with the stress at the ceramic-metal interface causing the device to fail. Therefore, it is imperative to develop methods for fabricating highly ordered titania nanotube arrays from Ti thin films atop robust substrates. One of the most straightforward solutions is to deposit Ti as a thin film on an adequate substrate by different depositing methods before electrochemical anodization.

Djenizian et al.[27] deposited 1.92 μm thick Ti thin films by cathodic sputtering using a DC triode system on Si substrates engraved from p-type Si(100) wafers with a resistivity of 1–10 Ωcm. Electrochemical anodization of the Ti films was carried out by potentiostatic experiments in 1 M H_3PO_4 + 1 M NaOH + 0.5 wt% HF electrolyte at room temperature. The TiO_2 nanotubes on a semiconductor substrate have an average tube length of approximately 560 nm, diameter on the order of 80 nm, and wall thickness of approximately 20 nm. However, the Ti thin film was not completely oxidized under these electrochemical conditions. The thicknesses of the TiO_2 nanotube layer and metallic Ti thin film on the Si substrate were estimated to be 0.56 and 1.36 μm, respectively.

An effective way to further improve the photoconversion efficiency is to fabricate transparent TiO_2 nanotube arrays on transparent substrate enabling front-side illumination, so the glass and fluorine-doped tin oxide (FTO)-coated glass was used to deposit Ti-metal films by the RF sputtering technique.[28,29] The quality of a deposited film and its adhesion to the substrate appeared to be interrelated functions of film thickness and substrate temperature during deposition. Films deposited at room temperature using RF sputtering were found to have poor substrate adhesion, peeling off when immersed in the electrolyte. Adhesion was not a problem for films deposited at either 250 or 500°C. Films deposited by RF sputtering at a temperature of about 250°C were granular in nature and did not yield well-defined tubular structures. However, in the case of films sputtered at 500°C, the Ti film structure was highly dense with tight packing of particles; ordered nanotube arrays were formed on these films upon anodization. Transparent anatase titania nanotube arrays with an index of refraction $n = 1.66$ were obtained after annealing at 500°C in oxygen ambient.

By anodizing titanium films in aqueous solutions mixed with hydrofluoric acid, the maximum obtainable thickness of the titania nanotubes was around 200–500 nm, because of the corrosion of the resultant anodic titania films by the HF solution. Fabricating titania nanotube films with no limit on thickness is not easy to realize because of the difficulty in establishing the formation-dissolution equilibrium of anodic titania in aqueous electrolytes. Anodizing Ti-metal films in organic electrolytes could decrease the chemical dissolution at the tube mouth and increase the nanotube length. We deposited 2 μm thick titanium films on FTO substrate by electron beam evaporation (EBEAM). Then ultra-long titania nanotube

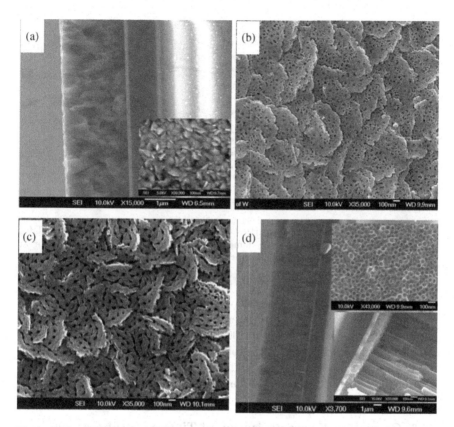

Figure 23.5.   SEM images of TiO$_2$ nanotube arrays/FTO electrode anodizing for different times. (a) Side view and (inset) top view before anodizing; after anodizing for (b) 3 minutes (top view); (c) 15 minutes (top view); (d) 60 minutes (side view, inset is the top view and enlarged side view).

arrays/FTO electrode was fabricated by potentiostatic anodization of the titanium film in C$_2$H$_6$O$_2$ + 25% NH$_4$F + 3% water electrolyte at room temperature. Figure 23.5 shows the typical anodization behavior of the Ti-metal films anodized at 60 V in the electrolyte. Film deposited by EBEAM has a dense, flaky structure before anodization. Within a few minutes of application of the voltage, fine pits or cracks formed on the oxide surface and acted as pore nucleation sites. These pits and cracks arose due to the chemical and field-assisted dissolution of the oxide at local points of high energy. After 15 minutes, due to a corresponding increase in porous structure depth, a porous structure was clearly seen with pore diameters of ~30 nm.

The nanotubular structure was formed after anodizing for 60 minutes with an inner diameter of ~43 nm, wall thickness of ~25 nm, tube length

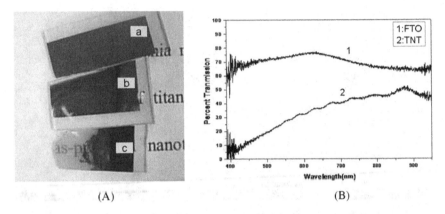

(A)   (B)

Figure 23.6. (a) Transparent properties of (a) Ti film/FTO, (b) as-prepared TiO_2 nanotube arrays/FTO, and (c) TiO_2 nanotube arrays/FTO after annealing in air at 400°C for 3 hours. (b) Transmission spectra and absorbance spectra of TiO_2 nanotube arrays/FTO.

of ∼5.4 μm, and oxide barrier layer of ∼200 nm. The anatase phase of titania nanotube arrays/FTO transparent electrode which was 5 μm long was obtained after annealing in air at 400°C for 3 hours. The transmittance of the titania nanotubes array in visible wavelength range was 45% according to the transmission spectra, and an absorbance peak was observed at 400 nm wavelength from the absorbance spectra (Figure 23.6).

23.3 Crystalline Characterization of TiO_2 Nanotube Arrays

One of the most interesting aspects of titania nanotubes is the different phases with different crystal structures. Titanium dioxide crystallizes in three major different structures: rutile (tetragonal, D_{4h}^{14}-P4$_2$/mnm, $a = b = 4.584$ Å, $c = 2.953$ Å),[30] anatase (tetragonal, D_{4h}^{19}-I4$_1$/amd, $a = b = 3.782$ Å, $c = 9.502$ Å), and brookite (rhombohedral, D_{2h}^{15}-Pbca/mnm, $a = 5.436$ Å, $b = 9.166$ Å, $c = 5.135$ Å).[31] However, only rutile and anatase play any role in the applications of TiO_2, where the anatase phase is preferred in charge-separation devices such as DSSCs and photocatalysts, while rutile is used predominately in gas sensors and as dielectric layers. Rutile has minimum free energy in comparison to other titania polymorphs. Hence, given the necessary activation energy, all other polymorphs including anatase transform into rutile through first-order phase transformation. However, the temperature at which the transformation of metastable anatase to stable rutile takes place depends upon several factors, including impurities present in the anatase, primary particle size, texture, and strain in the structure. Hence, porosity and/or surface area reduction occurs due to the sintering effects associated with the nucleation-growth type of phase transformation.[32–34]

As-anodized titania nanotubes are amorphous, and crystallized by a high-temperature anneal. A detailed study of the crystal structure, high temperature stability, and phase transformation of the titania nanotubes is presented in Figure 23.7.[35]

The as-synthesized nanotubes were annealed at various temperatures (up to 950°C) in oxygen ambient. The sample was crystallized in anatase phase at a temperature between 250 and 280°C. At a temperature near

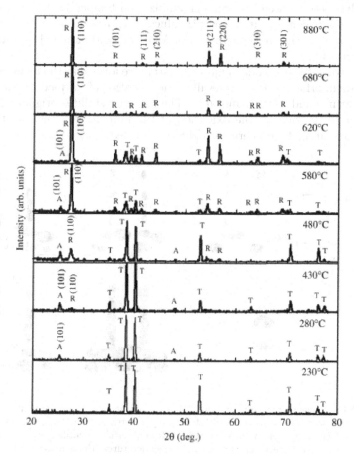

Figure 23.7.   GXRD patterns of the nanotube samples annealed at temperatures ranging from 230 to 880°C in dry oxygen ambient for 3 hours. (A — anatase; R — rutile; T — titanium).[35]

430°C a rutile phase appears in the XRD pattern. Complete transformation to rutile occurred in the temperature range of 620–680°C. The evolution of the surface morphology as a result of high-temperature annealing has also been studied. For nanotube arrays atop Ti foil, the structure of the 20 V sample was found to be stable till around 580°C. No discernible change in the pore diameter or wall thickness was observed even after annealing for 3 hours at this temperature. It was observed that at temperatures in the range of 550–580°C, depending on the sample, small protrusions came out through the porous structure. Above this temperature the tubular structure completely collapsed, leaving dense rutile crystallites. It may be noted that crystallization of the samples prepared using KF- (or NaF-) based electrolyte also showed the same crystallization temperature. Apparently, the electrolyte concentration and pH value have no influence on the crystallization temperature of the nanotubes. A crystallization model is described in Figure 23.8.

According to this model, anatase crystals are formed at the nanotube-Ti substrate interface region as a result of the oxidation of the metal at elevated temperatures and in the nanotubes. The rutile crystallites originate in the oxide layer (formed by the oxidation of titanium metal) underneath the nanotubes at high temperatures through nucleation and growth as well

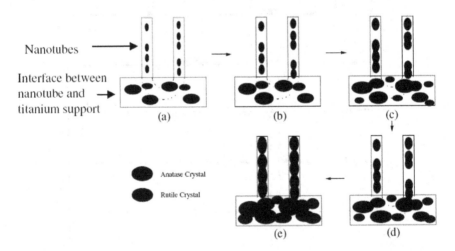

Figure 23.8. Schematic representation of the crystallization of titanium oxide nanotube arrays: (a) nucleation of anatase crystals at temperatures around 280°C; (b) growth of the anatase crystals at elevated temperatures; (c) nucleation of rutile crystals at around 430°C; (d) growth of rutile crystals at higher temperatures (the anatase crystals in the walls in contact with rutile crystals in the interface region are transformed to rutile at temperatures above approximately 480°C); (e) complete transformation of crystallites in the walls to rutile at temperatures above approximately 620°C.[35]

as phase transformation of anatase crystallites existing in the region. The constraints imposed by the nanotube walls, however, make the critical radii needed for rutile nucleation very large.[18,34,36] This prevents the anatase phase at the nanotube walls from undergoing transition to the rutile phase. Therefore, nanotubes annealed at temperatures between 530 and 580°C can be considered as anatase crystallites stacked in a cylindrical shape on a rutile foundation.

23.4 Doped TiO$_2$ Nanotube Arrays

Titania nanotube arrays have been of much interest in a broad range of photoelectrochemical applications because of the high orientation, uniformly stable structure, large internal surface area, and excellent electron percolation pathways for vectorial charge transfer between interfaces. However, only 2–3% of solar light can be utilized because of the wide bandgap of TiO$_2$ electrodes of 3.0–3.2 eV. To enhance the efficiency in the visible range, considerable efforts have been taken involving dye sensitization, doping with suitable species, and decorating with metal or compound materials.

23.4.1 *Doping with suitable species*

Recently, bandgap engineering of TiO$_2$ by anionic doping has been receiving attention. Asahi and co-workers[37] calculated the densities of states (DOSs) of the substitutional doping of C, N, F, P, or S for O in the anatase TiO$_2$ crystal, and identified nitrogen as the most effective dopant due to its comparable ionic radius and because its p-states contributed to bandgap narrowing by mixing with the p-states of oxygen. The common approaches to N-doping TiO$_2$ nanotube arrays include sputtering in a gas mixture of N$_2$ with Ar,[38] annealing in pure ammonia gas,[39] and electrochemical incorporation.[40] Schmuki's group introduced TiO$_2$ nanotube arrays into the ammonia environment in the range of 300–700°C and investigated the feasibility of doping them with nitrogen.[39] The nanotubular layer kept its structural integrity after exposure to hot ammonia gas with no significant morphological change. The N-doped samples were slightly yellow, whereas untreated samples had a grey tone. The structure of doped sample evidently consisted in all cases of anatase with various amounts of rutile. Two different peaks could clearly be observed in the N1s XPS spectrum of the nanotubes. The peak at 400 ± 0.2 eV could be ascribed to the γ-N state, which was molecularly chemisorbed N$_2$, and the other peak at 396 ± 0.2 eV corresponded to the β-N state, which was essentially atomic N in the form of mixed titanium oxide-nitride (TiO$_{2-x}$N$_x$). Another research work implanted anatase titania nanotubes with nitrogen using ion bombardment at 60 keV and a nominal dose of 1×10^{16} ions/cm^2. They found that ion implantation led to amorphization, where a very high number of defects were introduced into the structure that acted as traps for the photogenerated electron-hole pairs, while an adequate subsequent thermal treatment could be used to anneal out the defects. By the wet chemical method,[40] nitrogen dopant can also be incorporated into titania nanotubes in the anodization process. It was reported that N-doped titania nanotubes were fabricated in electrolytic solutions containing 0.07 M HF and varying concentrations of NH$_4$NO$_3$ from 0.2 to 2.5 M, with ammonium hydroxide added to adjust the pH to 3.5. The nitrogen peak at 396.8 eV was observed in the XPS spectra of the nanotubes and assigned to atomic β-N, indicating a chemically bound N$^-$ state, and the nitrogen was substitutional on the oxygen site. The doping of nitrogen was found to be inhomogeneous with the maximum nitrogen being incorporated close to the surface, then linearly decreasing with increasing depth inside the film.

Annealing titania nanotube arrays in a carbon monoxide flame forms carbon-doped titania nanotubes with significantly enhanced conductivity and trivalent titanium ions which modify the surface chemistry, favoring the adsorption of biomolecules. We annealed as-prepared titania nanotube arrays at 500°C in a tube furnace, under a flow of dry O$_2$ and CO for 3 hours.[41] The XPS spectra is shown in Figure 23.9.

The C1s spectra revealed three peaks positioned at 289.0 eV, 285.1 eV, and 281.7 eV for titania nanotube arrays annealed in CO (Figure 23.9(a)(i)),

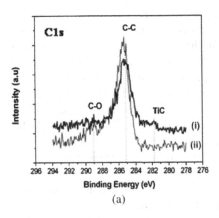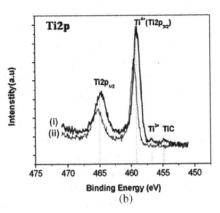

Figure 23.9.   C1s and Ti2p XPS spectra of titania nanotubes annealed in CO and O$_2$ at different temperatures. (a) C1s XPS spectra of samples annealed in CO at 500°C (i) and in O$_2$ at 500°C (ii). (b) T2p XPS spectra of samples annealed in CO at 500°C (i) and in O$_2$ at 500°C (ii).

but two peaks (289.0 eV and 285.1 eV) for nanotubes annealed in O_2 (Figure 23.9(a)(ii)). The peak position of 285.1 eV was close to the position of graphitic sp2-hybridized carbon (284.97 eV), as reported by Gu *et al.*[42] and the peak at 289.0 eV could be assigned to C/O groups resulting the absorption of CO_2 and CO onto the surface of the nanotubes. The peak at 281.7 eV in the nanotubes annealed in CO could possibly be ascribed to TiC, suggesting possible carbon doping. In the Ti2p spectra, in addition to two characteristic peaks of Ti2p1/2 at ∼465 eV, and Ti2p3/2 at ∼459 eV, there were two extra peaks at 456.8 eV and 454.9 eV found in the nanotubes annealed in CO (spectrum a), but not in the nanotubes annealed in O_2 (spectrum b). These two peaks were assigned to Ti^{3+} at ∼456.8 eV and Ti-C at 454.9 eV, though there were energy shifts in the peak positions for Ti^{3+} (2.0 eV) and TiC (3.9 eV). Both C1s and Ti2p XPS spectra strongly suggested that CO annealing likely resulted in the introduction of carbon to the crystal structure of TiO_2 and the formation of trivalent titanium cations, at least on the surface of TNT arrays. When the annealing temperature increased from 400 to 700°C, the doping concentration of carbon was increased accordingly from ∼1.3% to ∼6.8% and the Ti^{3+} defect concentration increased from 0% to 2.2%. The electrochemical properties of titania nanotube arrays annealed under different conditions were examined by cyclic voltammetry (CV) with 10 mM $K_3[Fe(CN)_6]$ as probe at a scan rate 0.1 Vs^{-1}. For O_2-annealed nanotubes, there was a reduction peak at 0.1 V and no oxidation peak was observed in this scan range; this indicated that the electrochemical reaction of the electrode was irreversible. However, for the nanotubes annealed in N_2 and CO, both the oxidation peak and reduction peak were observed during the scan process, and the peak separation was 0.21 V and 0.136 V respectively; this showed that the two electrode reactions were quasi-reversible. According to the peak separation, the average apparent heterogeneous electron transfer rate constant (k) was calculated to be 2.18×10^{-3} cm/s for N_2-annealed titania nanotubes and 1.34×10^{-3} cm/s for CO-annealed titania nanotubes. These results can be matched with the electron transfer rate constants of the carbon nanotube electrode (7.53×10^{-4} cm/s) and boron-doped diamond electrode (1.06×10^{-5} cm/s).

23.4.2 *Decorating with metal or chemical compounds*

Compared with TiO_2 powders, TiO_2 nanotube semiconductors have shown unique chemical and physical properties because the nanotubes have much more free space in their interior as well as outer space that can be filled with active materials such as chemical compounds and noble metals, giving them a fundamental advantage over powders.

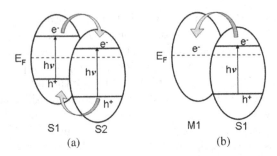

Figure 23.10. The working principles of a semiconductor-semiconductor system and a semiconductor-metal system.

When a small-bandgap semiconductor (S1) is coupled with a large-bandgap semiconductor (S2) with a more positive conduction band (CB) level, their valence bands (VBs) and conduction bands are suitably disposed, which align as shown in Figure 23.10(a). If both of them can be illuminated and activated simultaneously, the electron transfer and hole transfer can occur from one semiconductor to the other. Because the CB level of S1 is more negative than that of S2, and the VB level of S2 is more positive than that of S1, the photogenerated electrons will transfer to the CB of S2, while photogenerated holes transfer to the VB of S1. Thereby, charge carrier recombination can be hindered. Metal with a high work function coupled with a semiconductor (S1) can make the best use of the oxidation ability of VB holes in S1. The interface of the metal and semiconductor forms a Schottky barrier. When S1 is activated, photogenerated electrons will immediately transfer to the metal, while the holes will move to the surface of S1, and thus recombination is effectively suppressed.

Although chemical compounds and noble metals can be coated onto titania nanotubes by some physical methods, such as spray pyrolysis deposition, the sputtering method, and the filtered cathodic vacuum arc technique,[43-45] these methods require costly equipment and a complex fabrication process. Therefore, electrodepositing synthesis is an efficient way to decorate the titania nanotube array with active materials with good distribution because of its significant technical and economical advantages, such as simplicity, low cost, and, most importantly the ability to be performed at low temperature and in different media.

Grimes' group deposited a uniform layer of CdS nanoparticle clumps 40–100 nm in diameter on the crystallized TiO_2 nanotube arrays by cathodic reduction,[46] using a conventional three-electrode system comprising an Ag/AgCl reference electrode and Cd counter electrode. While they had not been able to detect CdS nanoparticles residing within the nanotube

arrays, Macak *et al.* developed an electrodeposition method for filling titania nanotubes with copper by a two-step process involving the reduction
of Ti^{4+} to Ti^{3+} at the bottom of the tubes and then electrodeposition
into the nanotubes.[47] Cai *et al.* fabricated a gold-platinum nanoparticle-
decorated titania nanotubular electrode and CdSe nanoparticles decorated
on the inner and outer surfaces of $4\,\mu m$ long TiO_2 nanotubes through
a simple direct current (DC) electrotechnique.[48] Obviously, the metal or
chemical compound nanoparticle size was not uniform, and ranged from
20 to 50 nm.

All of the research utilized the DC deposition method where there is
only one parameter, namely current density (I), which can be varied. So
it is difficult to control the deposited film composition and morphology.
Compared with DC deposition, pulse electrodeposition (PED)[49–52] favors
the initiation of grain nuclei and greatly increases the number of grains per
unit area, resulting in finer-grained deposits with better properties than
conventionally plated coatings. In PED, the potential or current is alter-
nated swiftly between two different values. This results in a series of pulses
of equal amplitude, duration, and polarity, separated by zero current. Each
pulse consists of an ON-time (T_{ON}) during which potential and current is
applied, and an OFF-time (T_{OFF}) during which zero current is applied. In
electroplating, a negatively charged layer is formed around the cathode as
the process continues. When using PED, the output is periodically turned
off, causing this layer to discharge somewhat. This allows easier passage
of the ions through the layer and onto the part. So it is possible to control
the deposited film composition and thickness in atomic order by regulating
the pulse amplitude and width. Using the PED technique, we successfully
prepared a Ni/TiO_2 nanotube nanocomposite by a two-step process.[53] The
first step is the anodic growth of self-organized TiO_2 nanotubes from a Ti
foil, while the second step is the pulse electrodeposition of Ni nanoparticles
on the nanotubular TiO_2 layer. The deposition experiments were performed
in a three-electrode system (the TiO_2/Ti as a working electrode, nickel plate
as a counter electrode, and Ag/AgCl electrode as a reference electrode). The
electrolyte of electrodeposited Ni was a mixture of $300\,g/L$ $NiSO_4 \cdot 6H_2O$,
$45\,g/L$ $NiCl_2 \cdot 6H_2O$, $37\,g/L$ H_3BO_3, pH 4.4. The pulse-current waveform
during the deposition is shown in Figure 23.11.

First, a current-limited negative pulse was applied to deposit Ni. Then
a short positive pulse was applied to discharge the capacitance of the
barrier layer. The cycle was repeated after T_{OFF} in order to refresh the
concentration of Ni ions at the deposition interface. By varying the amplitude of the negative and positive currents, current off-time, and deposition
time, different morphologies of Ni/TNT nanocomposites were formed. For
example, a thin layer of Ni was deposited on the nanotubular surface at a

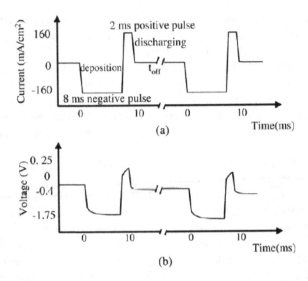

Figure 23.11. (a) Current-time and (b) voltage-time curves for PED with negative pulse (-160 mA/cm^2, 8 ms), positive pulse ($+160$ mA/cm^2, 2 ms) and current off-time (1,000 ms).

current off-time of 100 ms (Figure 23.12(a)). Increasing the current off-time to 1,500 ms resulted in a progressive decrease in the uniform particle size of the deposits down to 25 nm which were evidently embedded in the tube (Figure 23.12(d)). Mohapatra et al. also developed a hematite (α-Fe$_2$O$_3$) nanostructure on a titania nanotube template by the PED technique.[54] Changing the deposition profile led to the surface deposition of Fe in various morphologies. Decreasing the current density from 70 mA/cm^2 led to the surface deposition of Fe in the form of a thin film, nanoparticles, and nanoleaves. As the free energy of formation is different for different morphologies of nanomaterials, it is believed that there will be a particular deposition profile for a definite nanostructure of metal.

23.5 Application of TiO$_2$ Nanotube Arrays

23.5.1 Photocatalytic activity of TiO$_2$ nanotube arrays

The photoactivity of TiO$_2$ is one of its most attractive properties. In 1972, Fujishima and Honda reported light-induced water splitting on TiO$_2$ surfaces.[55] Since then, as a kind of typical semiconductor material, TiO$_2$ has been intensively investigated for applications in the photocatalytic degradation of organic pollutants in wastewaters owing to the stability of its unique

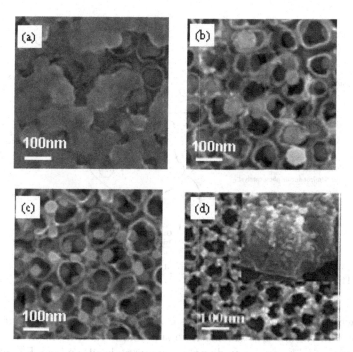

Figure 23.12.   FE-SEM images of Ni/TiO$_2$ nanocomposite obtained at constant current amplitude (70 mA/cm^2), pulse time of both negative and positive currents (8 ms, 2 ms), and different current off-times: (a) 100 ms, (b) 500 ms, (c) 1,000 ms, and (d) 1,500 ms.

chemical structure, optical and electrical properties, and non-toxicity.[56–59] During the past ten years, there were more than 4,000 articles concerning TiO$_2$ published in the environmental research field.

The principle of TiO$_2$ photocatalytic decomposition is outlined in Figure 23.13.[60] The reactions accomplished through photocatalysis can involve oxidations and oxidative cleavages, reductions, geometric and valence isomerizations, substitutions, condensations, and polymerizations. When under light irradiation, the TiO$_2$ photocatalytic reactions will be activated by absorption of a photon with sufficient energy (equal or higher than the bandgap energy of TiO$_2$). The absorption of photons leads to a charge separation due to promotion of an electron (e$^-$) from the valence band of the TiO$_2$ to the conduction band, thus generating a hole (h$^+$) in the valence band, and charge carrier pairs are formed. These charge carriers migrate to the TiO$_2$ surface, where they can react with adsorbed molecules. In the process of migration, the recombination of the electron and the hole should be prevented as much as possible, because it is the main reason leading to the reduction of catalytic efficiency.[61] The photogenerated electrons

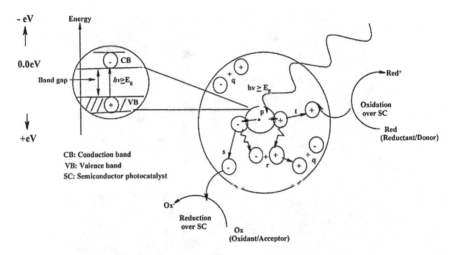

Figure 23.13. Schematic diagram of photocatalytic process initiated by photon acting on the TiO$_2$.[60]

react with electron acceptors such as O$_2$ adsorbed on the Ti(III) surface or dissolved in the water, reducing it to superoxide radical anion O$_2$·.[62] The photogenerated holes can oxidize the organic molecule to form R$^+$, or react with OH$^-$ or H$_2$O, oxidizing them into OH· radicals. Together with other highly oxidant species (peroxide radicals), they are reported to be responsible for the heterogeneous TiO$_2$ photodecomposition of organic substrates. The resulting OH· radical, being a very strong oxidizing agent, can oxidize essentially all organic molecules present in the solution into carbon dioxide and water.

According to this, the relevant reactions at the TiO$_2$ surface causing the degradation of organic pollutants can be expressed as follows:

$$TiO_2 + h\nu(\text{UV}) \rightarrow TiO_2(e_{CB}^- + h_{VB}^+) \qquad (23.4)$$

$$TiO_2(h_{VB}^+) + H_2O \rightarrow TiO_2 + H^+ + OH\cdot \qquad (23.5)$$

$$TiO_2(h_{VB}^+) + OH^- \rightarrow TiO_2 + OH\cdot \qquad (23.6)$$

$$TiO_2(e_{CB}^-) + O_2 \rightarrow TiO_2 + O_2^-\cdot \qquad (23.7)$$

$$O_2^-\cdot + H^+ \rightarrow HO_2\cdot \qquad (23.8)$$

$$\text{organic molecules} + OH\cdot \rightarrow \text{degradation products} \qquad (23.9)$$

$$\text{organic molecules} + h_{VB}^+ \rightarrow \text{oxidation products} \qquad (23.10)$$

$$\text{organic molecules} + e_{CB}^- \rightarrow \text{reduction products} \qquad (23.11)$$

where hv is the photon energy required to excite the semiconductor electrons from the VB region to the CB region. It has been observed that the photocatalytic activity of TiO_2 is completely suppressed in the absence of an electron scavenger such as molecular oxygen. Because the CB of TiO_2 is almost completely isoenergetic with the reduction potential of oxygen in inert solvents, adsorbed oxygen serves as an efficient trap for photogenerated electrons.

In recent years, many groups have focused on the photocatalytic activity of nanostructured TiO_2.[63-66] It has been proven that the annealed TiO_2 nanotube layer shows considerably higher reaction efficiency than a compacted Degussa P25 layer or TiO_2 nanofilm under similar conditions of illumination, as shown in Figure 23.14. What's more, TiO_2 nanotube arrays with high mechanical stability and integrity would be of great practical benefit because they allow repeated use and avoid the sedimentation or fixation process in photoreaction when using photocatalysts in powder form.

At the surface of TiO_2 nanotube arrays, photocatalysts can effectively degrade and mineralize most of the organic pollutants in wastewater, such as dyes,[68] halogenated compounds,[69] and acids,[44] to H_2O, CO_2, PO_4^{3-}, SO_4^{2-}, NO_3^-, and halide ions. In the photocatalytic reaction of the degradation of organic pollutants, the crystallinity, length, tube wall thickness, and calcination temperature of TiO_2 nanotube arrays have an impact on the photodegradation.

Lai et al.[71] found that the UV-Vis absorption edges shifted toward shorter wavelengths with the decrease in the nanotube inner diameter, which might result from the quantum effect due to the thickness of the nanotube wall being only several nanometers. The kinetic behavior of methylene blue photodegradation by titania nanotube arrays fabricated at 20 V

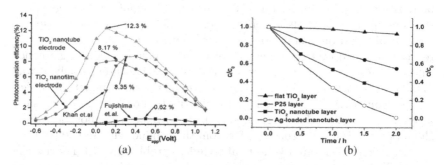

Figure 23.14. (a) Photoconversion efficiency as a function of applied potential. Calculation is shown for study of TiO_2 nanotubes and TiO_2 nanofilm electrode.[68,55] (b) Decomposition rates of acid orange 7 on different UV-illuminated TiO_2 surfaces.[64]

and calcined at 450°C shows the highest photocatalytic activity due to an increase in the anatase crystal phase and an increase in surface area.

Liang and Li[70] investigated the effect of tube wall thickness and calcination temperature and found that the photocatalytic activity of TiO_2 nanotubes decreased with increased nanotube wall thickness but was not proportional to their length. The optimal annealing temperature was 500°C owing to the high degree of crystallinity at this temperature. In addition, the pH, dopant content,[71-73] and dissolved oxygen of reaction solution[74] all can influence the photocatalytic processes.

Liang et al.[75] found that 2,3-DCP in alkaline solution was degraded and dechlorinated faster than that in acidic solution whereas dissolved organic carbon removal presented opposite dependence on pH. It was concluded that the pH influences both the surface state of titania and the ionization state of ionizable organic dyes. For pH values higher than the point of zero charge (PZC) of titania, the surface becomes negatively charged. In contrast, when the pH is lower than the PZC of titania, the surface is positively charged. The cationic dyes, such as methylene blue and methylene green, have a greater affinity for the surface of the TiO_2 nanotube catalyst, resulting in a faster degradation rate compared to P25. The negative charge on the surface of TiO_2 nanotube-based catalysts at pH 6 results in repulsion of the anionic dye (such as indigo carmine and rose bengal), causing the lower activity. To further study the photodegradation mechanism of titania nanotubes, the performance of reagent molecules should be directly determined by their accessibility to the active sites for adsorption onto and reaction with the surfaces of the nanotubes.

Naito et al. investigated the photocatalytic activity of individual TiO_2 nanotubes by the single-molecule counting of OH· using a specific fluorescent probe.[76] By determining the threshold of the fluorescence intensity, the apparent quantum yield of the generation of OH· in the TiO_2 nanotube can be roughly estimated to be 7.4×10^{-6}, which was significantly larger than that obtained by the bulk experiment (1.1×10^{-8}). The time- and space-resolved observation of emissive fluorescein generated by the photocatalytic reaction revealed that the transport of reagents inherent in a single nanotube was closely related to the photocatalytic activity. Interestingly, it was found that there were three highly active sites on the single nanotube, which was possibly attributed to the intrinsic distribution of surface defects such as oxygen vacancies that were present in the respective nanotube, which should mediate the electron transfer from the conduction band to O_2.

At present, there are two critical technical problems which restrict the use of TiO_2 nanotube arrays in large-scale industrial applications. The first one is that TiO_2 is inactive under visible light irradiation due to its large bandgap, which impedes the use of TiO_2 nanotube arrays as a solar-energy-harvesting photocatalyst. The other is that the photon efficiency is too low due to rapid recombination of photogenerated electrons and holes. In order to solve these problems, the work on TiO_2 nanotube array photocatalysts has focused on the modification of TiO_2, such as doping TiO_2 with non-metallic ions (e.g. N, C, C_{60}, S, B) and metal or metallic ions (e.g. Ag, Au, Fe, Sn, Ni, Fe^{3+}) or coupling TiO_2 with other catalysts (e.g. CdS, Fe_2O_3, ZnO, WO_3, WO_x). At the same time, seeking novel synthesis methods yielding TiO_2 with excellent properties is also an effective way.[76] These efforts have achieved good results.

Sun *et al.* fabricated highly ordered Fe^{3+}-doped TiO_2 nanotube array films by the electrochemical anodic oxidation of pure titanium in an HF electrolyte solution containing iron ions.[73] Compared with the undoped TiO_2 nanotube array films, the photocurrent of Fe^{3+}-doped TiO_2 nanotube array films increased obviously and the absorption edge of Fe^{3+}-doped TiO_2 nanotube array films appeared to be red-shifted, which was attributed to the effective separation of photogenerated electron-hole pairs upon the substitutional introduction of an appropriate Fe^{3+} amount into the anatase TiO_2 structure.

Schmuki *et al.* investigated the photocatalytic activities of metallic nanoparticles (Ag and Au) with a diameter of $\sim$10 nm (Ag) and $\sim$28 nm (Au) loaded TiO_2 nanotube layers.[78] Acid orange 7 was chosen as the decomposition model. It was clearly seen that the Ag-loaded nanotube TiO_2 layer showed more than a doubling in the rate constant in comparison with the unloaded tubes. This can be explained by considering the noble metal particles to form locally Schottky junctions with a higher potential gradient established by the Schottky barrier than at the TiO_2/electrolyte interface. Therefore with a sufficient amount of surface decoration, an efficient charge separation of the light-generated electron-hole pairs can be achieved.

With the development of modified work, some groups have made some progress in extending the photoactive region of TiO_2 nanotube arrays to visible light. Based on the principle of sensitizing large-bandgap with smaller-bandgap semiconductors to extend the photoresponse of TiO_2 to visible light, many semiconductors have been employed to sensitize TiO_2, including CdS,[43] CdSe,[78] and PbS.[79] CdSe is one of the most important II–VI group semiconductors; photocatalytic degradation of anthracene-9-carboxylic acid (ACA), one of the derivants of persistent organic pollutants (POPs), was successfully achieved on CdSe/TiO_2 nanotubes when exposed to the 550 nm green monochromatic light. Only 4% of ACA could be removed on the TiO_2 nanotubes, while it could be totally degraded in the presence of CdSe/TNTs; the photodegradation rate was 6.3 μg cm^{-2} h^{-1} on 50 nm diameter CdSe/TiO_2 NTs, suggesting that CdSe played a key role in the photocatalysis under visible light.[78] Lead sulfide (PbS) is also an attractive semiconductor because of its small bandgap (0.41 eV) and large exciton Bohr radius of 20 nm, which leads to extensive quantum size effects. When PbS quantum dots (QDs) were decorated on titania nanotubes, the size of the QDs played an important role in electron injection from PbS QDs to TiO_2 nanotubes. The most favorable case to have charge transfer from the PbS QDs to TiO_2 nanotubes was when the LUMO level of QDs was above the electron affinity of TiO_2 nanotubes. Because the electron affinity of the nanotubes was approximately -4.2 eV, the electron transfer from QDs to TNTs could be observed with a maximum quantum dot size of up to $\sim$6 nm.[79]

Nanocomposite metal oxides such as ZnO/TiO_2 nanocomposite materials can also enhance the quantum efficiency of photocatalysts for applications in water purification.[80] But the total surface exposure area for both metal oxides affects their properties as high-efficiency photocatalysts. For example, the corresponding quantum efficiency of a double-layered ZnO/TiO_2 system could be degraded because the photogenerated electrons accumulated in the TiO_2 underlayer may be unavailable to participate in the photocatalytic reactions.[81,82] Lau et al. fabricated ZnO nanocrystals outside the TiO_2 nanotubes by room-temperature deposition of ZnO plasma using the filtered cathodic vacuum arc technique.[45] This composite nanotube structure allowed both metal oxides to have a large surface exposure area to the surroundings and achieve high quantum efficiency. The photocatalytic degradation results of humic acid (HA) by this ZnO/TiO_2 nanocomposite showed that the reaction rate constant k increased by two times compared with bare TiO_2 nanotubes and was only slightly reduced by 10% after being reused for 20 cycles. It was believed that the combined effects of a large surface area and heterojunctions led to the improvement of photoactivity.

It is known that carbon has different electronic properties related to its graphitic nature. In combination with graphitic carbon, TiO_2 photocatalysts would possess high carrier mobility, and therefore high photoactivity. Lin et al.[84] prepared fullerene- (C_{60}-) modified TiO_2 nanotube arrays by the electrophoresis deposition technique. The as-prepared samples showed high efficiency for the photoelectric catalytic degradation of methylene blue; under some conditions the highest degradation activity of C_{60}-modified TiO_2 nanotubes was 2.3 times the highest activity of unmodified nanotubes. Cai et al.[85] synthesized a tube-in-tube C-TiO_2 composite nanotube array through a simple carbonizing process. The Raman spectra showed a predominate Raman band at $1,589\,cm^{-1}$ corresponding to the ordered graphitic structure. Under UV light irradiation, the degradation rate of methyl orange on C-TiO_2 nanotubes was 2.18 times that on TiO_2 nanotubes, which was $0.409\,g\,cm^{-2}\,h^{-1}$ on C-TiO_2 nanotube arrays and $0.195 \, \mathrm{g \, cm^{-2} \, h^{-1}}$ on unmodified TiO_2 nanotube arrays. The graphitized carbon nanotubes formed in TiO_2 nanotubes not only significantly enhanced the adsorption capacity for pollutants, but also provided a large network to collect and rapidly transfer the photoexcited electrons from the conduction band of TiO_2 during the photocatalytic process, which reduced the recombination between photoexcited electrons and holes and left more holes for the oxidization reaction of the adsorbed pollutants. Finally, we can say that developing TiO_2 nanotube array photocatalysts with high activity under visible light irradiation is the direction for future research.

23.5.2 Application as biosensors and biomaterials

The trend of using novel materials in biosensing systems is continuing, with their success largely due to the continuous design and development that meet the needs of modern electrochemical (bio)sensor technology. TiO_2 nanotubes have gradually received attention due to their one-dimensional nanostructures and uniform nanochannels which may provide a unique reaction vessel for analytes, electronic conductivity, and larger specific surface area which may greatly enhance their activity and sensitivity as sensors. Additionally, TiO_2 nanotubes are chemically inert, rigid, and thermally stable, and therefore attractive for the development of nanotube biosensors.[85–87]

Durrant *et al.* reported that a range of proteins may be readily immobilized on preformed nanoporous TiO_2 electrodes. The binding was mainly electrostatic and controlled by the pH, the protein charge, and the solution ionic strength.[89–91] The nanoporous structure of the films greatly enhanced the active surface available for protein binding over the geometrical area. Liu and Chen developed a novel H_2O_2 electrochemical biosensor involving the co-adsorption of horseradish peroxidase (HRP) and thionine on short TiO_2 nanotube arrays.[92] The linear response range of the sensor to the H_2O_2 concentration was from 1.1×10^{-5} to $1.1 \times 10^{-3} \, \mathrm{M}$, and the detection limit was estimated as $1.2 \times 10^{-6} \, \mathrm{M}$. Because the surface area of the nanomaterials affected the adsorption of proteins or cells, we investigated the adsorption properties and biosensing abilities of different lengths of TiO_2 nanotube arrays. The lengths were $500 \, \mathrm{nm}$, $1.8 \, \mu\mathrm{m}$, and $12 \, \mu\mathrm{m}$. The $1.8 \, \mu\mathrm{m}$ long nanotubes had the maximum adsorption amount of HRP of $0.5 \, \mu\mathrm{g/mm^2}$, while the $12 \, \mu\mathrm{m}$ long TNT adsorbed the maximum amount of thionine of $3.8 \, \mu\mathrm{g/mm^2}$ and the minimum amount of HRP of $0.2 \, \mu\mathrm{g/mm^2}$. The sublinear response range to the H_2O_2 concentration ranged from 1×10^{-5} to $3 \times 10^{-3} \, \mathrm{M}$ for $1.8 \, \mu\mathrm{m}$ long TiO_2 nanotubes; in contrast, there was no response when using $12 \, \mu\mathrm{m}$ long TiO_2 nanotubes due to the poor conductivity.[92] The conductivity of the tubes, as a consequence of their electronic properties, is all-important for the role of these unique nanomaterials in electrochemistry. Annealing in different gases and doping are considered to be one viable approach to narrow the bandgap and enhance the electrical conductivity.[93] The annealing treatment may help TiO_2 nanotubes to form different point defects, such as oxygen vacancies, titanium interstitials, and more complex structures, which will increase the conductivity and have a great effect on the properties of the electrode. Houlihan *et al.* reported on the photoresponse of ceramic TiO_2 photoanodes reduced in CO/CO_2 atmospheres and attributed the observed changes in the spectral dependence of the photocurrent to the different defect states.[95] We studied the electrochemical properties, conductivity, and photocurrent response of TiO_2 nanotube arrays annealed in CO, N_2, and O_2, respectively. Figure 23.15 gives the cyclic voltammetry results of TiO_2 nanotubes calcined in O_2, N_2, and CO at 500°C for 3 hours with 10 mM $K_3[Fe(CN)_6]$ as electroactive probe molecules in a potential range of -0.5 V to 0.8 V versus Ag/AgCl.

For the N_2-annealed TiO_2 nanotubes and CO-annealed TiO_2 nanotubes, both the oxidation peak and reduction peak were observed during the scan process; the peak separation was 0.21 V and 0.136 V respectively, which showed that the two electrode reactions were quasi-reversible. According to the peak separation, the average apparent heterogeneous electron transfer rate constant (k) was calculated to be 2.18×10^{-3} cm/s for N_2-annealed nanotubes and 1.34×10^{-3} cm/s for CO-annealed nanotubes. These results can be matched with the electron transfer rate constants of the carbon nanotube electrode (7.53×10^{-4} cm/s) and boron-doped diamond electrode

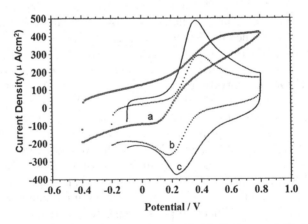

Figure 23.15.   CV curves for TNT electrodes annealed in O_2 (curve a), N_2 (curve b), and CO (curve c) at 500°C for 3 hours. The measurements were taken in 1 M KCl solution at 0.1 Vs^{-1} in the presence of 10 mM $K_3[Fe(CN)_6]$.

$(1.06 \times 10^{-5}\,\mathrm{cm/s})$. However, for O_2-annealed TiO_2 nanotubes there was only a reduction peak at $0.1\,\mathrm{V}$ and no oxidation peak was observed in this scan range, which indicated that the electrochemical reaction of the electrode was irreversible. These results showed that the TNT arrays calcinated in CO and N_2 possessed higher electrical conductivity than that calcined in O_2. Furthermore, CO-annealed nanotubes adsorbed the highest surface concentration of HRP, almost three times than that adsorbed on as-grown and O_2-annealed nanotube arrays. Nevertheless, the amount of thionine adsorbed on CO-annealed nanotubes was approximately 50% higher than that on as-grown nanotubes, and 30% higher than that on O_2-annealed nanotubes.

The amperometric response of CO-annealed TiO_2 nanotubes was approximately an order of magnitude larger for both as-grown and O_2-annealed nanotubes, as shown in Figure 23.16. Such significant enhancement in sensitivity in CO-annealed TiO_2 nanotubes biosensor could be attributed to both the enhanced electrical conductivity and efficient adsorption of both HRP and Th enzymes. Not only the doping and surface defects, but also the surface functional end group will affect the protein and cell adsorption on titanium.

Cai *et al.* studied the influence of surface chemistry and surface electric charge properties on protein adsorption and cell proliferation for titanium

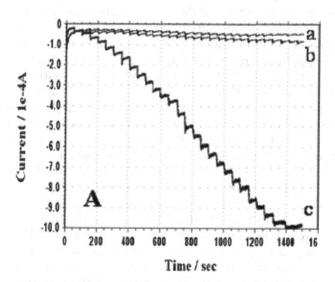

Figure 23.16. The amperometric response of TNT biosensors to successive additions of $50\,\mathrm{mM}$ H_2O_2 solution for (a) as-grown TNT/HRP/Th, (b) TNT/O2/ HRP/Th, and (c) TNT/CO/HRP/Th.

films. The titanium thin films were surface-functionalized with different functional end groups, such as -CH=CH$_2$, -NH$_2$, and -COOH groups. The results showed that Ti-COOH samples displayed a lower protein adsorption than all other groups, such as -NH$_2$, -CH=CH$_2$-terminated titanium thin films. However, a lower cell proliferation on COOH-terminated titanium films was observed compared with NH$_2$-terminated titanium films.[96] It was also reported that modifying the topography, chemistry, and surface energies of titanium surfaces had a strong impact on different cellular responses and modulated the *in vivo* success rate of implantations.[96,97] Some groups showed enhanced cell behavior using the anatase phase of nanotubes within 70–100 nm diameters compared to flat Ti surfaces.[98,99]

Schmuki *et al.* showed significantly enhanced cellular activity of mesenchymal stem cells (MSCs) on 15 nm amorphous nanotubes in comparison to 70–100 nm nanotubes.[101,102] Yet in another study, proliferation of smooth muscle cells was higher on flat TiO$_2$ surfaces than on 30 nm nanotubes, while endothelial cells showed the opposite reaction.[102] Part of the conflict may be due to the different methods used to generate TiO$_2$ surfaces, resulting in different surface topology, crystalline status, and chemical composition. Thus, Schmuki *et al.*[104] investigated the behavior of three different cell types relevant for blood vessel formation (endothelial cell line (mlEND), bone marrow mesenchymal cells (MSCs), and human cord blood endothelial progenitor cells) on TiO$_2$ nanotube surfaces with respect to (i) nanoscale tube diameters, (ii) the crystalline structure of TiO$_2$ as anatase versus amorphous material, and (iii) the content of the remaining fluoride in the nanotubes resulting from the preparation procedure. They concluded that TiO$_2$ nanotube surfaces in the range of 15–30 nm in diameter provide specific topographical cues to support adhesion, proliferation, migration, and differentiation of endothelial cells, and proposed the utilization of such nanostructured surfaces as suitable biomimetic materials for the use of vascular grafts. The responses of endothelial cells and MSCs to TiO$_2$ nanotube surfaces were largely independent of the fluoride content and the crystalline structure of TiO$_2$ but underlying was stringent control by the nanoscale topography. Furthermore, the authors fabricated amphiphilic TiO$_2$ nanotubes by a double anodization procedure combined with organic monolayer grafting. These tubes can be used as "capped" biomolecule carriers with $\sim$4.4 $\times$ 10^{-11} nmol per tube. By utilizing the excellent photocatalytic ability of TiO$_2$, the controllable release for enzyme molecules and a large range of proteins can be achieved. Once the pore surface is sufficiently hydrophilic, capillary forces will allow the entry of the environment and the release of the payload. Additionally, the amphiphilic characteristics with the hydrophobic outside provide a surface condition counteracting non-specific protein adsorption.[104]

23.5.3 *Application as anode for lithium batteries*

Rechargeable lithium batteries have become the dominant power source for portable consumer electronic devices because of their superior energy density. They are also the technology of choice for future hybrid electric vehicles, which are central to the reduction of CO_2 emissions arising from transportation. The worldwide market for rechargeable lithium batteries is now valued at 10 billion dollars per annum and growing.[105] The rechargeable lithium battery is a lithium-ion device, comprising a graphite negative electrode (anode), a non-aqueous liquid electrolyte, and a positive electrode (cathode) formed from layered $LiCoO_2$. Replacing the graphite powder negative electrode with nanomaterials would increase the power of the battery due to their reduced dimensions and high surface area. For example, carbon nanotubes exhibited twice the lithium storage compared with graphite, but similar problems of surface-layer formation and safety are present.[106,107] In the search for alternatives to graphite, titanium oxides aroused attention due to their inherent protection against lithium deposition, low cost, and low toxicity. In this regard, spinel-type $Li_4Ti_5O_{12}$ and anatase TiO_2 were well studied in the literature.[108-111] TiO_2 (B) nanowires were also found to be promising negative electrode materials for Li-ion batteries.[110]

In $Li_4Ti_5O_{12}$ three Li per formula unit can be inserted reversibly at 1.5 V, corresponding to a maximum theoretical specific capacity of 175 mA·h/g whereas in the case of TiO_2 polymorphs (anatase, rutile, TiO_2 (B)) a theoretical specific capacity of 335 mA·h/g is possible, corresponding to the insertion of one Li per formula unit. The voltage corresponding to the Ti^{4+}/Ti^{3+} redox couple vs. Li^+/Li in these TiO_2 polymorphs varies from 1 V to 1.8 V. The low voltage and high theoretical specific capacity make TiO_2 an attractive negative electrode material. TiO_2 exists in three crystalline modifications, viz. rutile, anatase, and brookite. At room temperature, anatase can accommodate 0.5–1 Li per formula unit without undergoing any major structural changes. The amount of Li that can be inserted into anatase TiO_2 varies with the crystallite size of the host.[111,112] To the best of our knowledge, there are no detailed reports of Li insertion into the brookite phase. Lithium insertion into rutile TiO_2 is negligible at room temperature.[113,114] However, Li can be inserted into rutile TiO_2 at 120°C in lithium polymer electrolyte cells.[115] The low reactivity of rutile TiO_2 towards Li under ordinary electrochemical conditions can be attributed to the dense close packing of the structure. TiO_2 nanotube arrays fabricated by anodization of pristine titanium have the advantage of ordered structure with uniform tube orientation, which not only facilitates the close contact between nanotubes and electrolyte but also favors both electron and lithium-ion transport.

We annealed TiO_2 nanotube arrays at 300, 400, and 500°C in N_2. Lithium-ion intercalation measurements revealed that annealing in nitrogen resulted in much enhanced lithium-ion insertion capacity and improved cyclic stability.[116] TiO_2 nanotube arrays annealed at 300°C exhibited the best lithium-ion intercalation property with an initial high discharge capacity of up to 240 mA·h/g at a high current density of 320 mA/g. The excellent discharge capacity at a high charge/discharge rate could be attributed to the large surface area of the nanotube arrays and a short facile diffusion path for lithium-ion intercalation as well as improved electrical conductivity. As the annealing temperature increased, the discharge capacity decreased, but the cyclic stability improved; 400°C annealed TiO_2 nanotube arrays possessed an initial discharge capacity of 163 mA·h/g and retained a capacity of 145 mA·h/g at the 50th cycle. Such temperature dependence is most likely due to the change in crystallinity and micro- or nanostructures.

To enhance the intercalation capacities of TiO_2 nanotube arrays at high charge/discharge rates, efforts have been devoted to improve the electrical conductivity of anatase TiO_2.[117] Doping and creating point defects like Ti^{3+} ions accompanied with O^{2-} vacancies on TiO_2 surfaces are effective ways to improve the conductivity of TiO_2 electrodes, so Ti^{3+} defects could possibly facilitate the lithium-ion intercalation and result in a much higher reversible intercalation capacity. We investigated the influences of point defects on the lithium-ion intercalation properties of CO-annealed TiO_2 nanotube arrays as compared to those of N_2-annealed TiO_2 nanotube arrays.[118] The electrochemical impedance spectra showed that N_2-annealed arrays possessed an electrode resistance of 66 Ω and a charge-transfer resistance of 38 Ω while CO-annealed arrays possessed an electrode resistance of 60 Ω and a reduced charge-transfer resistance of around 26 Ω, indicating a higher charge-transfer rate of Li^+ in the electrode. The CO-annealed TiO_2 nanotube array possessed much higher intercalation capacities, approximately double that of the N_2-annealed TiO_2 nanotube array at high current densities, i.e. possessing a capacity of 101 mA·h/g at 10 A/g. Also it demonstrated a less sensitive intercalation capacity; for example, an intercalation capacity of 261 mA·h/g decreased to 223 mA·h/g, a less than 20% reduction, when the current density increased from 100 mA/g to 320 mA/g. Comparatively, the N_2-annealed nanotube array was found to be more sensitively dependent on the current density: the capacity decreased from 245 mA·h/g to 164 mA·h/g when the current density increased from 100 mA·h/g to 320 mA·h/g. Both arrays exhibited excellent cycle stability over long-term cycles. The N_2-annealed nanotubes started with a capacity of 164 mA·h/g, retaining a capacity of 160 mA·h/g at the tenth cycle and 145 mA·h/g at the 50th cycle, with a capacity loss rate of ~0.2% per cycle. The CO-annealed nanotubes had an initial capacity of 223 mA·h/g, retaining a capacity of

210 mA·h/g after ten cycles. After 50 cycles, the remaining capacity was as high as 179 mA·h/g, with a capacity loss rate of ~0.9% per cycle.

Electrochemical intercalation comprises of three simultaneous and sequential processes: (1) redox reaction at the interface between the intercalation host and the electrolyte, (2) nucleation and growth of the new phase starting at the interface, and (3) charge and mass transfer. When the reaction rate exceeds the charge and/or mass transport properties of the solid electrode, charge accumulation and local (or surface) polarization occurs, which in turn hinders further intercalation reaction and leads to a low intercalation capacity. Improved charge and mass transfer properties would significantly enhance the intercalation capacity by eliminating or reducing the local polarization. So the improved electrical conductivity of the CO-annealed TiO_2 nanotube array could facilitate charge transport accompanying rapid intercalation reactions at the interface, allowing large intercalation capacities under high current densities. The presence of defects may also contribute to the improved intercalation capacity of the CO-annealed nanotubes. In the titania system, both intercalation and deintercalation processes involve phase transition between tetragonal TiO_2 and orthorhombic $Li_x TiO_2$ through the following reaction:

$$x Li^+ + x e^- + TiO_2 \leftrightarrow Li_x TiO_2. \qquad (23.12)$$

Phase transition occurs through nucleation at the interface and subsequent growth from the interface towards the interior. The presence of defects on the surface of TiO_2 nanotubes could serve as nucleation sites so as to promote the phase transition between TiO_2 and $Li_x TiO_2$, resulting in excellent intercalation capacities.

23.6 Conclusions and Future Work

This chapter is a review of the work relevant to the fabrication of titania nanotube arrays by anodization and its applications in photoelectrochemical and electrochemical fields, such as photocatalysts, biomaterials, and negative electrode materials of lithium batteries.

In principle, titania nanotube arrays can be fabricated via anodic oxidation of titanium foil in a fluoride-based solution. By adjusting the pH of the electrolyte or using organic electrolytes, the nanotube length can be increased from 500 nm to 134 μm. Titanium films deposited on an FTO substrate can also serve as a foundation for TiO_2 nanotube arrays, giving rise (after anodizing and annealing) to anatase nanotube arrays that were 5 μm long.

As-anodized titania nanotubes are amorphous, and are crystallized in anatase phase at a temperature between 250 and 280°C. At a temperature near 430°C the rutile phase appears; complete transformation to rutile occurs in the temperature range of 620–680°C. Above 580°C, the tubular structure completely collapses. To enhance the properties of titania nanotubes, some efforts have been made involving doping with suitable species by annealing in different ambient conditions and decorating with metal or compound particles by electrochemical deposition.

Besides their high orientation, uniformly stable structure, large internal surface area, excellent electron percolation pathways for vectorial charge transfer between interfaces, titania nanotube arrays are environmentally friendly. Thus, titania nanotube arrays have attracted broad attention in electrochemical and photoelectrochemical applications. Here we reviewed its applications in photocatalysts, biomaterials, and lithium ion intercalation. Future work will focus on the following aspects: First, how to enlarge the tube diameter during the anodization process. The tube length can be controlled exactly from several nanometers to 1,000 μm by adjusting electrolytes, whereas the tube diameter is only from 20 nm to 200 nm. Until now there has been no effective way to increase the tube width. Enlarging the tube diameter is helpful to trap more incident rays and enhance the optical property in its application. The second aspect is fabricating the special structure of titania nanotube-based composite material by different methods and exploring its photoelectrochemical properties and environmental applications. We believe that this material warrants extended and in-depth study and hope this review can serve as a useful resource for researchers in their future work.

References

1. M. Adachi, Y. Murata, M. Harada and S. Yoshikawa, *Chem. Lett.* **8**, 942–943 (2000).
2. S. Z. Chu, S. Inoue, K. Wada, D. Li, H. Haneda and S. Awatsu, *J. Phys. Chem. B* **107**, 6586–6589 (2003).
3. O. K. Varghese, D. Gong, M. Paulose, K. G. Ong, E. C. Dickey and C. A. Grimes, *Adv. Mater.* **15**, 624–627 (2003).
4. G. K. Mor, M. A. Carvalho, O. K. Varghese, M. V. Pishko and C. A. Grimes, *J. Mater. Res.* **19**, 628–634 (2004).
5. G. K. Mor, K. Shankar, M. Paulose, O. K. Varghese and C. A. Grimes, *Nano Lett.* **5**, 191–195 (2005).
6. S. Uchida, R. Chiba, M. Tomiha, N. Masaki and M. Shirai, *Electrochemistry* **70**, 418–420 (2002).

7. M. Adachi, Y. Murata, I. Okada and S. Yoshikawa, *J. Electrochem. Soc.* **150**, G488–G493 (2003).
8. P. Hoyer, *Langmuir* **12**, 1411–1413 (1996).
9. B. B. Lakshmi, P. K. Dorhout and C. R. Martin, *Chem. Mater.* **9**, 857–862 (1997).
10. S. Kobayashi, N. Hamasaki, M. Kimura, H. Shirai and K. Hanabusa, *J. Am. Chem. Soc.* **124**, 6550–6551 (2002).
11. T. Kasuga, M. Hiramatsu, A. Hoson, T. Sekino and K. Niihara, *Langmuir* **14**, 3160–3163 (1998).
12. Q. Chen, W. Z. Zhou, G. H. Du and L. H. Peng, *Adv. Mater.* **14**, 1208–1211 (2002).
13. B. D. Yao, Y. F. Chan, X. Y. Zhang, W. F. Zhang, Z. Y. Yang and N. Wang, *Appl. Phys. Lett.* **82**, 281–283 (2003).
14. D. Gong, C. A. Grimes, O. K. Varghese, W. Hu, R. S. Singh, Z. Chen and E. C. Dickey, *J. Mater. Res.* **16**, 3331–3334 (2001).
15. J. M. Macak, H. Tsuchiya and P. Schmuki, *Angew. Chem. Int. Ed.* **44**, 2100–2102 (2005).
16. D. V. Bavykin, J. M. Friedrich and F. C. Walsh, *Adv. Mater.* **18**, 2807–2824 (2006).
17. J. L. Zhao, X. H. Wang, R. Z. Chen and L. T. Li, *Solid State Commun.* **134**, 705–710 (2005).
18. Q. Y. Cai, M. Paulose, O. K. Varghese and C. A. Grimes, *J. Mater. Res.* **20**, 230–236 (2005).
19. L. V. Taveira, J. M. Macak, H. Tsuchiya, K. F. P. Dick and P. Schmuki, *J. Electrochem. Soc.* **152**, B405–B410 (2005).
20. A. Ghicov, H. Tsuchiya, J. M. Macak and P. Schmuki, *Electrochem. Commun.* **7**, 505–509 (2005).
21. J. M. Macak, K. Sirotna and P. Schmuki, *Electrochim. Acta* **50**, 3679–3684 (2005).
22. F. M. Bayoumi and B. G. Ateya, *Electrochem. Commun.* **8**, 38–44 (2006).
23. J. M. Macak, H. Tsuchiya, L. Taveira, A. Ghicov and P. Schmuki, *J. Biomed. Mater. Res. A* **75A**, 928–933 (2005).
24. J. M. Macak, H. Tsuchiya, L. Taveira, S. Aldabergerova and P. Schmuki, *Angew. Chem. Int. Ed.* **44**, 7463–7465 (2005).
25. M. Paulose, K. Shankar, S. Yoriya, H. E. Prakasam, O. K. Varghese, G. K. Mor, T. A. Latempa, A. Fitzgerald and C. A. Grimes, *J. Phys. Chem. B* **110**, 16179–16184 (2006).
26. M. Paulose, H. E. Prakasam, O. K. Varghese, L. Peng, K. C. Popat, G. K. Mor, T. A. Desai and C. A. Grimes, *J. Phys. Chem. C* **111**, 14992–14997 (2007).
27. Y. D. Premchand, T. Djenizian, F. Vacandio and P. Knauth, *Electrochem. Commun.* **8**, 1840–1844 (2006).
28. G. K. Mor, O. K. Varghese, M. Paulose and C. A. Grimes, *Adv. Funct. Mater.* **15**, 1291–1296 (2005).
29. G. K. Mor, K. Shankar, M. Paulose, O. K. Varghese and C. A. Grimes, *Nano Lett.* **6**, 215–218 (2006).

30. F. A. Grant, *Rev. Mod. Phys.* **31**, 646–674 (1959).
31. G. V. Samsonov, *The Oxide Handbook*, 2nd edn. (IFI/Plenum Press, USA, 1982).
32. O. J. Whittemore and J. J. Sipe, *Powder Technol.* **9**, 159–164 (1974).
33. K. N. P. Kumar, K. Keizer and A. J. Burggraaf, *J. Mater. Chem.* **3**, 1141–1449 (1993).
34. K. N. P. Kumar, K. Keizer, A. J. Burggraaf, T. Okubo and H. Nagamoto, *J. Mater. Chem.* **3**, 1151–1159 (1993).
35. O. K. Varghese, D. Gong, M. Paulose, C. A. Grimes and E. Dickey, *J. Mater. Res.* **18**, 156–165 (2003).
36. O. K. Varghese, M. Paulose, K. Shankar, G. K. Mor and C. A. Grimes, *J. Nanosci. Nanotechnol.* **5**, 1158–1165 (2005).
37. R. Asahi, T. Morikawa, T. Ohwaki, K. Aoki and Y. Taga, *Science* **293**, 269–271 (2001).
38. A. Ghicov, J. M. Macak, H. Tsuchiya, J. Kunze, V. Haeublein, L. Frey and P. Schmuki, *Nano Lett.* **6**, 1080–1082 (2006).
39. R. P. Vitiello, J. M. Macak, A. Ghicov, H. Tsuchiya, L. F. P. Dick and P. Schmuki, *Electrochem. Commun.* **8**, 544–548 (2006).
40. K. Shankar, K. C. Tep, G. K. Mor and C. A. Grimes, *J. Phys. D: Appl. Phys.* **39**, 2361–2366 (2006).
41. Y. H. Zhang, P. Xiao, X. Y. Zhou, D. W. Liu, B. B. Garcia and G. Z. Cao, *J. Mater. Chem.* **19**, 948–953 (2009).
42. Y. L. Gu, L. Y. Chen, Y. T. Qian, W. Q. Zhang and J. H. Ma, *J. Amer. Ceram. Soc.* **88**, 225–227 (2005).
43. K. Shin, S. Seok, S. H. Im and J. H. Park, *Chem. Commun.* **46**, 2385–2387 (2010).
44. A. K. M. Kafi, G. Wu and A. Chen, *Biosen. Bioelectron.* **24**, 566–571 (2008).
45. H. Y. Yang, S. F. Yu, S. P. Lau, X. Zhang, D. D. Sun and G. Jun, *Small* **5**, 2260–2264 (2009).
46. S. Chen, M. Paulose, C. Ruan, G. K. Mor, O. K. Varghese, D. Kouzoudis and C. A. Grimes, *J. Photoch. Photobio. A* **177**, 177–184 (2006).
47. J. M. Macak, B. G. Gong, M. Hueppe and P. Schmuki, *Adv. Mater.* **19**, 3027–3031 (2007).
48. Q. Kang, L. Yang and Q. Cai, *Bioelectrochemistry* **74**, 62–65 (2008).
49. J. Y. Fei and G. D. Wilcox, *Electrochim. Acta* **5**, 2693–2698 (2005).
50. T. Houga, A. Yamada and Y. Ueda, *J. Jpn. I. Met.* **64**, 739–742 (2000).
51. Y. Ueda, N. Hataya and H. Zaman, *J. Magn. Magn. Mater.* **156**, 350–352 (1996).
52. M. S. Chandrasekar and M. Pushpavanam, *Electrochim. Acta* **53**, 3313–3322 (2008).
53. Y. Zhang, Y. Yang, P. Xiao, X. Zhang, L. Lua and L. Li, *Mater. Lett.* **63**, 2429–2431 (2009).
54. S. K. Mohapatra, S. Banerjee and M. Misra, *Nanotechnology* **19**, 315601 (2008).
55. A. K. Fujishima and K. Honda, *Nature* **238**, 37–38 (1972).
56. A. K. Linsebigler, G. Lu and J. T. Yates, *Chem. Rev.* **95**, 735–758 (1995).

57. M. R. Hofmann, S. T. Martin, W. Choi and D. W. Bahnemann, *Chem. Rev.* **95**, 69–96 (1995).
58. F. Zhang, J. Zhao, T. Shen, H. Hidaka, E. Pelizzetti and N. Serpone, *Appl. Catal. B* **15**, 147–156 (1998).
59. J. Krysa, M. Keppert, G. Waldner and J. Jirkovsky, *Electrochim. Acta* **50**, 5255–5260 (2005).
60. U. I. Gaya and A. H. Abdullah, *J. Photochem. Photobiol. C* **9**, 1–12 (2008).
61. U. G. Akpan and B. H. Hameed, *J. Hazard. Mater.* **170**, 520–529 (2009).
62. I. K. Konstantinou and T. A. Albanis, *Appl. Catal. B: Environ.* **49**, 1–14 (2004).
63. Y. Xie, *Adv. Funct. Mater.* **16**, 1823–1831 (2006).
64. J. M. Macak, M. Zlamal, J. Krysa and P. Schmuki, *Small* **3**, 300–304 (2007).
65. S. Anandan, P. S. Kumar, N. Pugazhenthiran, J. Madhavan and P. Maruthamuthu, *Sol. Energ. Mater. Sol. Cell.* **92**, 929–937 (2010).
66. Y. Lai, L. Sun, Y. Chen, H. Zhuang, C. H. Lin and J. W. Chin, *J. Electrochem. Soc.* **153**, D123–D127 (2006).
67. H. Zhuang, C. H. Lin, Y. Lai, L. Sun and J. Li, *Environ. Sci. Technol.* **41**, 4735–4740 (2007).
68. S. U. M. Khan, M. Al-Shahry and W. B. Ingler, *Science* **297**, 2243–2245 (2002).
69. S. Liu, L. Yang, S. Xu, S. Luo and Q. Cai, *Electrochem. Commun.* **11**, 1748–1751 (2009).
70. H. Liang and X. Li, *J. Hazard. Mater.* **162**, 1415–1422 (2009).
71. Y. Lai, L. Sun, Y. Chen, H. Zhuang, C. H. Lin and J. W. Chin, *J. Electrochem. Soc.* **153**, D123–D127 (2006).
72. F. Dong, W. R. Zhao and Z. B. Wu, *Nanotechnology* **19**, 365607 (2008).
73. L. Sun, J. Li, C. L. Wang, S. F. Li, H. B. Chen and C. J. Lin, *Sol. Energ. Mater. Sol. Cell.* **93**, 1875–1880 (2009).
74. N. K. Shrestha, M. Yang, Y. C. Nah, I. Paramasivam and P. Schmuki, *Electrochem. Commun.* **12**, 254–257 (2010).
75. H. Liang, X. Li, Y. Yang and K. Sze, *Chemosphere* **73**, 805–812 (2008).
76. K. Naito, T. Tachikawa, M. Fujitsuka and T. Majima, *J. Am. Chem. Soc.* **131**, 934–936 (2009).
77. N. K. Allam, K. Shankar and C. A. Grimes, *Adv. Mater.* **20**, 3942–3946 (2008).
78. I. Paramasivam, J. M. Macak and P. Schmuki, *Electrochem. Commun.* **10**, 71–75 (2008).
79. L. Yang, S. Luo, R. Liu, Q. Cai, Y. Xiao, S. Liu, F. Su and L. Wen, *J. Phys. Chem. C* **114**, 4783–4789 (2010).
80. C. Ratanatawanate, Y. Tao and K. J. Balkus, *J. Phys. Chem. C* **113**, 10755–10760 (2009).
81. D. Barreca, E. Comini, A. P. Ferrucci, A. Gasparotto, C. Maccato, C. Maragno, G. Sberveglieri and E. Tondello, *Chem. Mater.* **19**, 5642–5649 (2007).
82. Z. Liu, D. Sun, P. Guo and J. O. Leckie, *Nano Lett.* **7**, 1081–1085 (2007).

83. H. Y. Zhu, X. P. Gao, Y. Lan, S. X. Song, Y. X. Xi and J. C. Zhao, *J. Am. Chem. Soc.* **126**, 8380–8381 (2004).

84. J. Lin, R. Zong, M. Zhou and Y. Zhu, *Appl. Catal. B* **89**, 425–431 (2009).

85. L. Yang, S. Luo, S. Liu and Q. Cai, *J. Phys. Chem. C* **112**, 8939–8943 (2008).

86. Q. Heng, B. Zhou, J. Bai, L. Li, Z. Jin, J. Zhang, J. Li, Y. Liu, W. Cai and X. Zhu, *Adv. Mater.* **20**, 1044–1049 (2008).

87. D. V. Avykin, E. V. Milsom, F. Marken, D. H. Kim, D. H. Marsh, D. J. Riley, F. C. Walsh, K. H. El-Abiary and A. A. Lapkin, *Electrochem. Commun.* **7**, 1050–1058 (2007).

88. S. Oh, R. R. Finones, C. Daraio, L. Chen and S. Jin, *Biomaterials* **26**, 4938–4943 (2005).

89. E. Topoglidis, A. E. G. Cass, G. Gilardi, S. Sadeghi, N. Beaumont and J. R. Durrant, *Anal. Chem.* **70**, 5111–5113 (1998).

90. E. Topoglidis, T. Lutz, R. L. Willis, C. J. Barnett, A. E. G. Cass and J. R. Durrant, *Faraday Discuss.* **116**, 35–46 (2000).

91. E. Topoglidis, A. E. G. Cass, B. O'Regan and J. R. Durrant, *J. Electroanal. Chem.* **517**, 20–27 (2001).

92. S. Liu and A. Chen, *Langmuir* **21**, 8409–8413 (2005).

93. P. Xiao, B. B. Garcia, Q. Guo, D. Liu and G. Cao, *Electrochem. Commun.* **9**, 2441–2447 (2007).

94. P. Xiao, D. Liu, B. B. Garcia, S. Sepehrib, Y. Zhang and G. Cao, *Sens. Actuat. B-Chem.* **134**, 367–372 (2008).

95. J. F. Houlihan, R. F. Bonaquist, R. T. Dirstine and D. P. Madacsi, *Mater. Res. Bull.* **16**, 659–667 (1981).

96. K. Cai, M. Frant, J. Bossert, G. Hildebrand, K. Liefeith and K. D. Jandt, *Colloids Surf. B* **50**, 1–8 (2006).

97. K. Cai, M. Müller, J. Bossert, A. Rechtenbach and K. D. Jandt, *Appl. Surf. Sci.* **250**, 252–267 (2005).

98. K. Cai, J. Bossert and K. D. Jandt, *Colloids Surf. B* **49**, 136–144 (2006).

99. K. C. Popat, L. Leoni, C. A. Grimes and T. A. Desai, *Biomaterials* **28**, 3188–3197 (2007).

100. K. S. Brammer, S. Oh, J. O. Gallagher and S. Jin, *Nano Lett.* **8**, 786–793 (2008).

101. J. Park, S. Bauer, K. von der Mark and P. Schmuki, *Nano Lett.* **7**, 1686–1691 (2007).

102. J. Park, S. Bauer, K. A. Schlegel, F. W. Neukam, K. V. Mark and P. Schmuki, *Small* **6**, 666–671 (2009).

103. L. Peng, M. L. Eltgroth, T. J. LaTempa, C. A. Grimes and T. A. Desai, *Biomaterials* **30**, 1268–1272 (2009).

104. J. Park, S. Bauer, P. Schmuki and K. Mark, *Nano Lett.* **9**, 3157–3164 (2009).

105. Y. Song, F. Schmidt-Stein, S. Bauer and P. Schmuki, *J. Am. Chem. Soc.* **131**, 4230–4232 (2009).

106. P. G. Bruce, B. Scrosati and J. Tarascon, *Angew. Chem., Int. Ed.* **47**, 2930–2946 (2008).

107. R. S. Morris, B. G. Dixon, T. Gennett, R. Raffaelle and M. J. Heben, *J. Power Sources* **138**, 277–280 (2004).
108. Z. Zhou, J. J. Zhao, X. P. Gao, Z. F. Chen, J. Yan, P. V. Schiever and M. Morinaga, *Chem. Mater.* **17**, 992–1000 (2005).
109. T. Ohzuku, A. Ueda and N. Yamamoto, *J. Electrochem. Soc.* **142**, 1431–1435 (1995).
110. K. Nakahara, R. Nakajima, T. Matsushima and H. Majima, *J. Power Sources* **117**, 131–136 (2003).
111. A. R. Armstrong, G. Armstrong, J. Canales, R. Garcia and P. G. Bruce, *Adv. Mater.* **17**, 862–865 (2005).
112. S. Y. Huang, L. Kavan, I. Exnar and M. Gratzel, *J. Electrochem. Soc.* **142**, L142–L144 (1995).
113. M. A. Reddy, M. S. Kishore, V. Pralong, V. Caignaert, U. V. Varadaraju and B. Raveau, *Electrochem. Commun.* **8**, 1299–1303 (2006).
114. T. Ohzuku, Z. Takehara and S. Yoshizawa, *Electrochim. Acta* **24**, 219–222 (1979).
115. L. Kavan, D. Fattakhova and P. Krtil, *J. Electrochem. Soc.* **146**, 1375–1379 (1999).
116. W. J. Macklin and R. J. Neat, *Solid State Ionics* **694**, 53–56 (1992).
117. D. Liu, P. Xiao, Y. Zhang, B. B. Garcia, Q. Zhang, Q. Guo, R. Champion and G. Cao, *J. Phys. Chem. C* **112**, 11175–11180 (2008).
118. D. Liu, Y. Zhang, P. Xiao, B. B. Garcia, Q. Zhang, X. Zhou, Y. H. Jeong and G. Cao, *Electrochim. Acta* **54**, 6816–6820 (2009).

Index